Lecture Notes in Computer Science 8638

Commenced Publication in 1973
Founding and Former Series Editors:
Gerhard Goos, Juris Hartmanis, and Jan van Leeuwen

T0210960

Sergey Balandin   Sergey Andreev
Yevgeni Koucheryavy (Eds.)

Internet of Things, Smart Spaces, and Next Generation Networks and Systems

14th International Conference, NEW2AN 2014
and 7th Conference, ruSMART 2014
St. Petersburg, Russia, August 27-29, 2014
Proceedings

 Springer

Volume Editors

Sergey Balandin
FRUCT Oy
Kissankellontie 20B, 00930 Helsinki, Finland
E-mail: sergey.balandin@fruct.org

Sergey Andreev
Tampere University of Technology
Department of Electronics and Communications Engineering
Korkeakoulunkatu 1, 33720 Tampere, Finland
E-mail: sergey.andreev@tut.fi

Yevgeni Koucheryavy
Tampere University of Technology
Department of Electronics and Communications Engineering
Korkeakoulunkatu 1, 33720 Tampere, Finland
E-mail: yk@cs.tut.fi

ISSN 0302-9743   e-ISSN 1611-3349
ISBN 978-3-319-10352-5   e-ISBN 978-3-319-10353-2
DOI 10.1007/978-3-319-10353-2
Springer Cham Heidelberg New York Dordrecht London

Library of Congress Control Number: 2014945955

LNCS Sublibrary: SL 5 – Computer Communication Networks
and Telecommunications

Typesetting: Camera-ready by author, data conversion by Scientific Publishing Services, Chennai, India

Printed on acid-free paper

Springer is part of Springer Science+Business Media (www.springer.com)

Preface

We welcome you to the joint proceedings of the 14th NEW2AN (Next Generation Teletraffic and Wired/Wireless Advanced Networks and Systems) and 7th Internet of Things and Smart Spaces ruSMART (Are You Smart) conferences held in St. Petersburg, Russia, during August 27–29, 2014.

Originally, the NEW2AN conference was launched by ITC (International Teletraffic Congress) in St. Petersburg in June 1993 as an ITC-sponsored Regional International Teletraffic Seminar. The first edition was entitled "Traffic Management and Routing in SDH Networks" and held by R&D LONIIS. In 2002, the event received its current name, NEW2AN. In 2008, NEW2AN acquired a new companion in Smart Spaces, ruSMART, hence boosting interaction between researchers, practitioners, and engineers across different areas of ICT. From 2012, the scope of the ruSMART conference has been extended to cover the Internet of Things and related aspects.

Presently, NEW2AN and ruSMART are well-established conferences with a unique cross-disciplinary mixture of telecommunications-related research and science. NEW2AN/ruSMART is accompanied by outstanding keynotes from universities and companies across Europe, USA, and Russia.

The 14th NEW2AN technical program addressed various aspects of next-generation data networks. This year, special attention was given to advanced wireless networking and applications as well as to lower-layer communication enablers. In particular, the authors demonstrated novel and innovative approaches to performance and efficiency analysis of ad hoc and machine-type systems, employed game-theoretical formulations, Markov chain models, and advanced queuing theory. It is also worth mentioning the rich coverage of graphene and other emerging materials, photonics and optics, generation and processing of signals, as well as business aspects.

The 7th Conference on Internet of Things and Smart Spaces ruSMART 2014 provided a forum for academic and industrial researchers to discuss new ideas and trends in the emerging areas of Internet of Things and Smart Spaces that create new opportunities for fully customized applications and services. The conference brought together leading experts from top affiliations around the world. This year, there was active participation by industrial world-leader companies and particularly strong interest from attendees representing Russian R&D centers, which have a good reputation for high-quality research and business in innovative service creation and applications development.

This year, the first day of NEW2AN/ruSMART Technical Program starts with the keynote talk on "Advanced Features of Enhanced Multimedia Broadcast Multicast Services" by Dr. Alexey Anisimov of Motorola Solutions. The second day is featured by the talk "Smart Processing of Disruptively Big Data in the Internet of Things" by Prof. Arcady Zaslavsky, from the Commonwealth

Scientific and Industrial Research Organization (CSIRO), Australia's national science agency.

We would like to thank the Technical Program Committee members of both conferences, as well as the associated reviewers, for their hard work and important contribution to the conference. This year, the conference program met the highest quality criteria with an acceptance ratio of around 30%.

The current edition of the conferences was organized in cooperation with Open Innovations Association FRUCT, ITC (International Teletraffic Congress), IEEE, Tampere University of Technology, St. Petersburg State University of Telecommunications, St. Petersburg State Polytechnical University, and Popov Society. The support of these organizations is gratefully acknowledged.

We also wish to thank all those who contributed to the organization of the conferences. In particular, we are grateful to Roman Florea for his substantial work on supporting the conference website and his excellent job on the compilation of camera-ready papers and interaction with Springer.

We believe that the 14th NEW2AN and 7th ruSMART conferences delivered an informative, high-quality, and up-to-date scientific program. We also hope that participants enjoyed both the technical and social conference components, the Russian hospitality, and the beautiful city of St. Petersburg.

August 2014

Sergey Balandin
Sergey Andreev
Yevgeni Koucheryavy

Organization

NEW2AN International Advisory Committee

Igor Faynberg	Alcatel Lucent, USA
Jarmo Harju	Tampere University of Technology, Finland
Villy B. Iversen	Technical University of Denmark, Denmark
Andrey Koucheryavy	Giprosviaz, Russia
Kyu Ouk Lee	ETRI, Republic of Korea
Sergey Makarov	St. Petersburg State Polytechnical University, Russia
Mohammad S. Obaidat	Monmouth University, USA
Andrey I. Rudskoy	St. Petersburg State Polytechnical University, Russia
Konstantin Samouylov	People's Friendship University of Russia, Russia
Manfred Sneps-Sneppe	Ventspils University College, Latvia
Michael Smirnov	Fraunhofer FOKUS, Germany
Sergey Stepanov	Sistema Telecom, Russia

NEW2AN Technical Program Committee

Hassen Mohammed Alsafi	IIUM, Malaysia
Sergey Andreev	Tampere University of Technology, Finland
Francisco Barcelo-Arroyo	Universitat Politecnica de Catalunya (UPC), Spain
Torsten Ingo Braun	University of Bern, Switzerland
Paulo Carvalho	University of Minho, Portugal
Wei Koong Chai	University College London, UK
Jiajian Chen	Qualcomm Research Silicon Valley, USA
Chrysostomos Chrysostomou	Frederick University, Cyprus
Roman Dunaytsev	The Bonch-Bruevich St. Petersburg State University of Telecommunications, Russia
Markus H. Fidler	Leibniz Universität Hannover, Germany
Dieter Fiems	TELIN, Ghent University, Belgium
Roman Florea	Tampere University of Technology, Finland
Ivan Ganchev	Telecommunications Research Centre (TRC), University of Limerick, Ireland
Norton González	DeVry University, Brazil
Jarmo Harju	Institute of Communications Engineering, Tampere University of Technology, Finland

Yevgeni Koucheryavy	Tampere University of Technology, Finland
Tatiana Madsen	NetSec, Electronic Systems, Aalborg University, Denmark
Maja Matijasevic	University of Zagreb, Croatia
Paulo Mendes	SITI, COPELABS, Lusofona University, Portugal
Edmundo Monteiro	University of Coimbra, Portugal
Dr. Nitin	Jaypee University of Information Technology, India
Antonino Orsino	DIIES, University Mediterranea of Reggio Calabria, Italy
Stoyan Atanasov Poryazov	Institute of Mathematics and Informatics, Bulgarian Academy of Sciences, Bulgaria
Simon Pietro Romano	University of Naples Federico II, Italy
Zhefu Shi	University of Missouri - Kansas City, USA
Sai Ganesh Sitharaman	Zscaler, Inc., USA
Ozgur B. Akan	Koc University, Turkey

ruSMART Executive Technical Program Committee

Sergey Boldyrev	Nokia, Helsinki, Finland
Nikolai Nefedov	Nokia Research Center, Zurich, Switzerland
Ian Oliver	Nokia, Helsinki, Finland
Alexander Smirnov	SPIIRAS, St. Petersburg, Russia
Vladimir Gorodetsky	SPIIRAS, St. Petersburg, Russia
Michael Lawo	Center for Computing Technologies (TZI), University of Bremen, Germany
Michael Smirnov	Fraunhofer FOKUS, Germany
Dieter Uckelmann	University of Applied Sciences in Stuttgart, Germany
Cornel Klein	Siemens Corporate Technology, Germany

ruSMART Technical Program Committee

Sergey Balandin	FRUCT, Finland
Michel Banatre	IRISA, France
Mohamed Baqer	University of Bahrain, Bahrain
Sergei Bogomolov	LGERP R&D Lab, Russia
Gianpaolo Cugola	Politecnico di Milano, Italy
Alexey Dudkov	NRPL Group, Finland
Kim Geun-Hyung	Dong Eui University, Republic of Korea
Didem Gozupek	Bogazici University, Turkey
Victor Govindaswamy	Texas A&M University, USA

Table of Contents

Video Solutions for Smart Spaces

NEW2AN

Advances in Wireless Networking

Ad Hoc Networks and Enhanced Services

Sensor- and Machine-Type Communication

Networking Architectures and Their Modeling

Traffic Analysis and Prediction

Analytical Methods for Performance Evaluation

Materials for Future Communications

Generation and Analysis of Signals

Business Aspects of Networking

Progress on Upper Layers and Implementations

Modeling Methods and Tools

Techniques, Algorithms, and Control Problems

Photonics and Optics

Signals and Their Processing

ruSMART

Linked Data-Driven Smart Spaces

Oscar Rodríguez Rocha[1], Cristhian Figueroa[1], Iacopo Vagliano[1],
and Boris Moltchanov[2]

[1] Politecnico di Torino, corso Duca degli Abruzzi 24, 10129. Turin, Italy
{oscar.rodriguezrocha,cristhian.figueroa,iacopo.vagliano}@polito.it
[2] Telecom Italia, via G. Reiss Romoli, 274, 10148. Turin, Italy
boris.moltchanov@telecomitalia.it

Abstract. In this paper, we present an approach to exploit Linked Data
in Smart Spaces, doing more than just using RDF to represent information. In particular, we rely on knowledge stored in DBpedia[1], a dataset
in the Web of Data. We also provide a platform to implement such an
approach and a eTourism use case, both developed in collaboration with
a mobile operator. Finally, we provide also a performance evaluation of
the main component of the platform.

Keywords: Smart Spaces, Linked Data, Recommender Systems,
Semantic Annotation, eTourism.

1 Introduction

The idea of extending the capabilities of the Web in order to publish structured
data on it exists from its own creation. While the Semantic Web is the goal,
Linked Data (LD) provides the means to make it reality [1]. It refers to a set of
best practices for publishing and connecting structured data on the Web in order
to increase the number of data providers and consequently accomplish the goals
of the Web of Data. In this way, Linked Data makes possible to semantically
interlink and connect different resources at data level regardless the structure,
location etc. Data published on the Web using Linked Data has resulted into a
global data space called the Web of Data.

On the other hand and according to [2], a Smart Space (SS) can be defined
as any real or virtual location equipped with passive and active artifacts. These
artifacts can be any kind of sensors and actuators, mobile devices, software
agents, autonomous vehicles, management systems, and also human beings. By
interacting with each other, these artifacts can provide a benefit.

Due to the large amount of data that exist in the Web of Data, it is possible
to find related structured information about many of the components (artifacts) that are part of a specific Smart Space. At the same time, it is possible
that some components of the Smart Space (e.g. users' devices) enrich the existing information about other components by generating semantic annotated

---

[1] http://dbpedia.org

S. Balandin et al. (Eds.): NEW2AN/ruSMART 2014, LNCS 8638, pp. 3–15, 2014.

User-generated Content (UGC). To achieve this, we integrate an enabler of the FI-WARE EU project[2].

Given this scenario, the main goal of this paper is to demonstrate how information available in the Web of Data can be exploited in order to improve the interaction between these components, offering more benefits to the end user. We present a platform to test this approach and an eTourism use case, which was modeled and developed on top of it, in conjunction with a mobile operator.

This paper is structured as follows: section 3 presents our approach to exploit LD in order to improve interactions in SS, in section 4 the platform is introduced and each component briefly described, while in section 5 the eTourism use case is provided. Finally, an evaluation of the recommendation algorithm is available in section 6, before the relevant related works (section 2) and the conclusions.

2   Related Work

The focus on this paper was to exploit the knowledge present in the Web of Data into SS as according to the best of our knowledge there are no works trying to do this. Usually in the literature, RDF and ontologies are used to represent information and to share it but not to provide services as our recommender does, thus we compare our recommender with others after briefly discussing related works addressing SS.

Smart-M3 [3] is the reference platform for SS applications, it exploits Semantic Web technologies but it does not rely on external LD datasets. The work of Smirnov et al. [4] was a source of inspiration for this work. They provide an application on top of Smart-M3 with an eTourism use case, focusing on providing recommendation for trips and integrating other tourist services, while we exploit LD for recommendation and the main service is providing information to users and allow them to add content. Kiljander et al. [5] proposed a method to identifies real word objects, while we deal with the problem to manages objects already represented in the Web of Data.

Focusing on recommendation, the principal works related to LDR are reported. Passant [6] presents *"DBRec"* a RS for the music domain to recommend artists and bands based solely on transversal links between them within a preselected subset of the DBpedia dataset. To do this, a Linked Data Semantic Distance (*LDSD*) is defined, which considers the number of input and output links for each artist/band. Kaminskas et al. [7] suggest musicians regarding to a point of interest (POI). They build a knowledge base from DBpedia by linking musicians with POIs, and then using a spreading activation algorithm rank the musicians based on their similarity with a given POI.

In [8], they recommend experts for solving problems in an Open Innovation (OI) scenario based on concept recommendation techniques. In this case a textual-based problem description is analyzed to extract key concepts, which are then related with entities on the DBpedia dataset. Then, the SR is able

---

[2] FI-WARE is a project funded by the European Commission. More information in http://www.fi-ware.org.

to identify concepts that can be seen as topics related with the experts to be recommended. To do this, Damljanovic et al. studies two recommendation techniques supported on the links of DBpedia: the *hyProximity* measure, that takes into account hierarchical links (SKOS categories); and the transversal distance measure, that relies on the number of links between concepts of the data set constituted of transversal properties relevant to the OI scenario.

The main problem with LDRs is that they require a curated sub-set of a LD dataset reduced to only concepts for a specialized domain of interest, which normally is performed manually by domain experts. Additionally, although Damljanovic et al. studies both the transversal and hierarchical measures, they do not studied their combination to generate recommendations.

3 Relating Smart Spaces and Linked Data

As mentioned earlier in the introduction, our proposal is based on improving the interaction between the components of a SS, making available the most useful information about each of them, which translates into more benefits for the end user. Due to the increase in the amount of LD that are present in the Web, it is increasingly possible to find information related to Smart Space components and increase the information associated to them (in form of UGC) by means of Semantic Annotation techniques.

For example, by considering the case of a tourist SS, where components are the tourist attractions of a city and other POIs (like monuments, hotels, etc.). More detailed information about each of them can be extracted from linked datasets. Thus, when a mobile user is in a tourism situation like visiting a city, the processing of the information associated tourist attractions of a city or other POIs could enable more reliable recommendations for more interesting places. This use case is detailed in section 5.

This is possible because any resource present in Web of Data has a URI, i.e. it is uniquely identified. Hence, it is possible to access it in order to get the information about the object it represents and nowadays it is also simpler to publish information into the web. Any user can easily insert new content in textual or multimedia format (with the use of contemporary mobile devices, it is becoming more and more real time, e.g. probably it is possible that some kind of news appear on Twitter before than on any news portal). Consider a car accident in a city: witnesses can post information or content about it even before the local press agency arrives. Last but not least, it is also easier to link information with already existing resources. In fact, much effort has been done by the LD community to provide more ways to increase LD publication. There are plenty of tools to annotate raw data and link them, e.g. DBpedia spotlight[3] or the FI-WARE Semantic Annotator GE (for which details are provided in section 4.4).

Another element to consider is that the structure of LD can be exploited in a given SS. In the use case of a SS made up of tourist attractions, it is possible

---

[3] http://dbpedia-spotlight.github.io/demo

to identify different inner SSs, e.g. it may be useful consider only hotels, or only parks within a city SS. This can be done since LD is structured: for example considering DBpedia, any resource belongs to one or more categories and categories of hotels and monuments exist. In addition, categories are organized in a tree hierarchy, thus it is possible consider bigger or smaller SS by referring to more general or more specific categories. E.g. a SS can include only monuments, another only hotels, or both can be included considering building category. Hence, it is possible to associate categories with SS, if we refer to Saint Petersburg to exemplify a SS, then we can also consider another SS within it: the hotels SS (associated to `Category:Hotels in Saint Petersburg`), monuments SS (`Category:Monuments and memorials in Saint Petersburg`) and so on. The whole SS may be associated to `Category:Buildings and structures in Saint Petersburg`.

Finally, information in Linked Data and its structure can be used to increase the user experience. Possible ways of exploiting it are decision making and SS coordination, and the eTourism use case given in section 5 suggests how take advantage of this.

4   Reference Platform

The overall system is shown in figure 1. Our platform communicates with mobile devices and with the Web of Data by means of web interactions and it is made up of three main modules: a recommender, a UGC manager and a semantic annotator. The first can provide information from Web of Data and UGC stored, while the second stores and retrieves UGC, which is also linked to resources belonging to the Web of Data by means of the annotator. In the following, we provide a description of each components of the platform.

4.1   Linked Data-Driven Recommender

Recommender systems (hereafter RSs) are software tools and techniques that suggest items to users [9]. Items can be of different kinds such as songs, news, posts on social networks, persons, services, etc. In this paper, we refer to only one type of RS known as Linked Data-driven Recommenders (LDRs). LDR are recommenders based on knowledge, but unlike traditional knowledge-based recommenders, they are based on datasets modeled, built, and maintained by different organizations and communities around the world. Those datasets contain knowledge from different domains and sources; and can be published on the Web of Data according the LD principles [10].

4.2   LD-Driven Recommendations

LDR systems suggest items by measuring semantic distances based on the relationships of concepts in a dataset. Normally, items are associated to concepts of a dataset by using natural language processing techniques, which analyze textual information of items, and extract keywords that can be matched with concepts to identify different inner SSs, e.g. it may be useful consider only hotels, or only parks within a city SS. This can be done since LD is structured: for example considering DBpedia, any resource belongs to one or more categories and categories of hotels and monuments exist. In addition, categories are organized in a tree hierarchy, thus it is possible consider bigger or smaller SS by referring to more general or more specific categories. E.g. a SS can include only monuments, another only hotels, or both can be included considering building category. Hence, it is possible to associate categories with SS, if we refer to Saint Petersburg to exemplify a SS, then we can also consider another SS within it: the hotels SS (associated to `Category:Hotels in Saint Petersburg`), monuments SS (`Category:Monuments and memorials in Saint Petersburg`) and so on. The whole SS may be associated to `Category:Buildings and structures in Saint Petersburg`.

Finally, information in Linked Data and its structure can be used to increase the user experience. Possible ways of exploiting it are decision making and SS coordination, and the eTourism use case given in section 5 suggests how take advantage of this.

4   Reference Platform

The overall system is shown in figure 1. Our platform communicates with mobile devices and with the Web of Data by means of web interactions and it is made up of three main modules: a recommender, a UGC manager and a semantic annotator. The first can provide information from Web of Data and UGC stored, while the second stores and retrieves UGC, which is also linked to resources belonging to the Web of Data by means of the annotator. In the following, we provide a description of each components of the platform.

4.1   Linked Data-Driven Recommender

Recommender systems (hereafter RSs) are software tools and techniques that suggest items to users [9]. Items can be of different kinds such as songs, news, posts on social networks, persons, services, etc. In this paper, we refer to only one type of RS known as Linked Data-driven Recommenders (LDRs). LDR are recommenders based on knowledge, but unlike traditional knowledge-based recommenders, they are based on datasets modeled, built, and maintained by different organizations and communities around the world. Those datasets contain knowledge from different domains and sources; and can be published on the Web of Data according the LD principles [10].

4.2   LD-Driven Recommendations

LDR systems suggest items by measuring semantic distances based on the relationships of concepts in a dataset. Normally, items are associated to concepts of a dataset by using natural language processing techniques, which analyze textual information of items, and extract keywords that can be matched with concepts to identify concepts that can be seen as topics related with the experts to be recommended. To do this, Damljanovic et al. studies two recommendation techniques supported on the links of DBpedia: the *hyProximity* measure, that takes into account hierarchical links (SKOS categories); and the transversal distance measure, that relies on the number of links between concepts of the data set constituted of transversal properties relevant to the OI scenario.

The main problem with LDRs is that they require a curated sub-set of a LD dataset reduced to only concepts for a specialized domain of interest, which normally is performed manually by domain experts. Additionally, although Damljanovic et al. studies both the transversal and hierarchical measures, they do not studied their combination to generate recommendations.

3   Relating Smart Spaces and Linked Data

As mentioned earlier in the introduction, our proposal is based on improving the interaction between the components of a SS, making available the most useful information about each of them, which translates into more benefits for the end user. Due to the increase in the amount of LD that are present in the Web, it is increasingly possible to find information related to Smart Space components and increase the information associated to them (in form of UGC) by means of Semantic Annotation techniques.

For example, by considering the case of a tourist SS, where components are the tourist attractions of a city and other POIs (like monuments, hotels, etc.). More detailed information about each of them can be extracted from linked datasets. Thus, when a mobile user is in a tourism situation like visiting a city, the processing of the information associated tourist attractions of a city or other POIs could enable more reliable recommendations for more interesting places. This use case is detailed in section 5.

This is possible because any resource present in Web of Data has a URI, i.e. it is uniquely identified. Hence, it is possible to access it in order to get the information about the object it represents and nowadays it is also simpler to publish information into the web. Any user can easily insert new content in textual or multimedia format (with the use of contemporary mobile devices, it is becoming more and more real time, e.g. probably it is possible that some kind of news appear on Twitter before than on any news portal). Consider a car accident in a city: witnesses can post information or content about it even before the local press agency arrives. Last but not least, it is also easier to link information with already existing resources. In fact, much effort has been done by the LD community to provide more ways to increase LD publication. There are plenty of tools to annotate raw data and link them, e.g. DBpedia spotlight[3] or the FI-WARE Semantic Annotator GE (for which details are provided in section 4.4).

Another element to consider is that the structure of LD can be exploited in a given SS. In the use case of a SS made up of tourist attractions, it is possible

---

[3] http://dbpedia-spotlight.github.io/demo

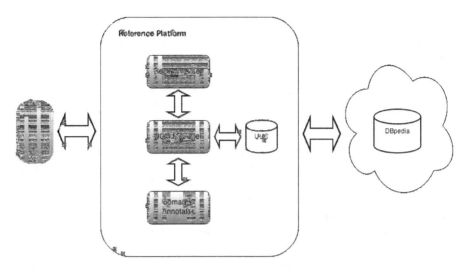

Fig. 1. Overall system architecture

(or entities) of the dataset. In this way, an item is more likely to be suggested if its related concepts have lower semantic distance with the initial item than other concepts in the dataset.

Semantic distances are used by LDRs to exploit hierarchical or transversal relationships between concepts. Hierarchical relationships are links established over hierarchical properties that organize concepts according the categories they belong to. For example, DBpedia dataset supports three kinds of hierarchical relationships: SKOS (Simple Knowledge Organization System)[4], YAGO (Yet Another Great Ontology)[11], or WordNet[5] categories. Transversal relationships are links connecting concepts without the aim of establishing a classification or hierarchy [8].

4.3 Recommendation Algorithm

Our LD-driven recommender combines both hierarchical and transversal relationships to generate more accurate recommendations. Additionally, it uses a category-based organization in order to detect a set of contexts in which concepts are arranged according to the application domain they belong to. In this way, the user can access easier to recommended items organized by broader categories, which can be seen also as an explanation for the recommendation. The hierarchical and transversal relationships were obtained from DBpedia because it is being established as the central interlinking hub for the Web of Data, enabling also access to many other datasets in the Linked Open Data Cloud[12] [13].

---

[4] http://www.w3.org/2004/02/skos
[5] http://en.wikipedia.org/wiki/WordNet

The algorithm 1 represents our LD-driven recommendation approach. It starts by creating a category graph ($categoryGraph$) based on hierarchical information extracted from an initial concept (URI_{in}) until reach a maximum level ($maxLevel$) of categories in the category tree of DBpedia. The $maxLevel$ value is used to limit the levels of super categories that the algorithm extract when navigating the category tree (Lines 1 - 4). Those categories are extracted using the hierarchical relationship `skos:broader` from the SKOS model of DBpedia.

Next, the algorithm extracts subcategories for all the super categories found in the last step for the $categoryGraph$, in order to go one level down to increase the possibility for finding more concept candidates (Lines 5 - 8). Then, it obtains the concepts for each category in the $categoryGraph$ (including sub-categories), and arranges them in a map $resultMap$ that relates each concept with the set of categories it belongs to (Lines 9 - 13). The $resultMap$ is the set of candidate concepts found by the recommender, which still needs to be organized before being displayed to the user. For each candidate concept in the $resultMap$ the algorithm calculates a transversal distance (td) counting the input/output properties that the candidate concept shares with the initial concept. Those distances are added to a map in order to make them available for generating the final ranking ($distancesMap$) (Lines 14 - 17).

Finally, by using the hierarchical information of the $categoryGraph$, the algorithm organizes the results by context categories and ranks the concepts for each category in descendent order of td (Line 18).

Algorithm 1. Recommendation Algorithm

Require: An input URI: URI_{in}, $maxLevel$
Ensure: A ranked set of recommended concepts RC classified by categories
1: $C_{in} = getCategories(URI_{in})$
2: **for all** $c \in C_{in}$ **do**
3:    $SC = getSuperCategoriesUntilLevel(c, maxLevel)$
4:    $categoryGraph.add(SC)$
5:    **for all** $sc \in SC$ **do**
6:       $subC = getSubCategories(sc)$
7:       $categoryGraph.add(subC)$
8:    **end for**
9:    **for all** $cg \in CategoryGraph$ **do**
10:       $subC = getConcepts(cg)$
11:       $resultMap.add(subC, cg)$
12:    **end for**
13: **end for**
14: **for all** $candidateConcept \in result$ **do**
15:    $td = transversalDistance(candidateConcept, URI_{in})$
16:    $distancesMap.add(candidateConcept, td)$
17: **end for**
18: $classifyResultsByContextCategories(results)$
19: **return** $ranking$

4.4   Semantic Annotator

Thanks to FI-WARE project, it is possible to reuse existing components for creating modern applications and services. In this case, we have reused the Semantic Annotator GE, which is publicly available from the FI-WARE catalogue[6]. The decision of reusing an existing component or creating a new one was taken after a careful analysis, in which it was concluded that the component that provides the service offers greater advantages and reduces the time of integration on the platform. In addition, it encourages the standardization of the present modules.

In order to extract semantic information, each plain text information associated to received content is analyzed by the Semantic Annotator GE (which architecture is shown in Figure 2). First, the Text Processor module identifies the source language, then a morphological analysis is performed using FreeLing[7] configured with the identified language. From this analysis, proper nouns lemmas are extracted while other part-of-speech are discarded. At this time, non-numeric proper nouns lemmas with a score of at least 0.2 are preserved and merged with plain tags to compute a well-defined list of unique (multi) words. At this stage, the module uses term frequency to further process the title and to extract other potential relevant words.

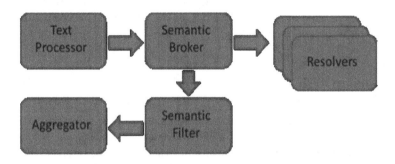

Fig. 2. Semantic Annotator GE at a glance

The next step involves the Semantic Broker, which is assisted by a set of resolvers that perform full-text or term-based analysis based on the previous output. Such resolvers are aimed at providing candidate semantic concepts referring to Linked Data as well as additional related information if available. Resolvers may be domain or language-specific, or general purpose. For term-based analysis, each word of the previously computed list is individually processed to identify a list of candidate Linked Data resources to match with. A set of predefined services, such as DBpedia, Sindice[8] and Evri[9] are invoked in parallel.

---

[6] http://catalogue.fi-ware.org/enablers/semantic-annotation
[7] http://nlp.lsi.upc.edu/freeling/
[8] http://sindice.com
[9] http://www.evri.com

The Semantic Filtering module processes candidate Linked Data resources received by the broker and performs a disambiguation based on the DBpedia score and the string similarity between each surface form and its corresponding list of candidates, which is based on the Jaro Winkler distance. This function aims at maximizing both values to identify the preferred candidate. In this process, after several empirical tests, candidates with distance lower than 0.8 are discarded at this stage, unless their DBpedia score is the maximum. Automatic annotation is performed using the preferred candidate identified during this step.

4.5   UGC Management

This layer of the platform is responsible for saving User-generated Content and its semantically enriched version. Thus, for each given UGC entry, we have:

- UGC itself (any given multimedia file);
- Original associated plain information (stored in a SQL database);
- Associated semantic information, represented in RDF (stored in a triple store).

Additionally, this layer offers the means to retrieve specific content, either through the invocation of a REST API (to get the multimedia contents attached) or performing more complex SPARQL queries in the public endpoint.

5   eTourism Use Case

The previously described platform can be applied in a tourism scenario and can be exploited by an application that assists tourists by providing them suggestion about places to visit, accommodations, and other points of interest (POIs). In addition, it allows users to share their tourist experience by providing content such as pictures, videos, reviews and comments about places they have been. Then, this content is available to other users and can be exploited it for enriching information about tourist attractions and facilities.

The SS is the entire city that is being visited and it is made up of users' mobile devices, monuments, museums, theatres, parks, hotels and other POIs. For example, if a user is in Saint Petersburg, the SS includes (but is not limited to) the Peter and Paul Fortress, the Saint Isaac's Cathedral and the Hermitage museum. All of these tourist attractions have a corresponding representation in DBpedia, thus they support user interactions, since users can decide if visit them or not by consulting the information provided by DBpedia. Basically, there are two possible interactions:

1. user device receives information about nearby tourist attractions;
2. user publishes information regarding the tourist attractions from his device (generates content).

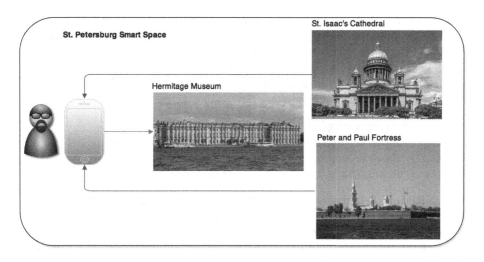

Fig. 3. Possible interactions in a eTourism use case

The first enables users to be provided with recommendations about tourist attractions, while the latter allows users to share their experiences and increase the amount of available information to other tourists.

Figure 3 illustrates the two different possible interactions for a user present in the Saint Petersburg SS. He is at the Hermitage museum and his device receives recommendations about the Peter and Paul Fortress and the Saint Isaac's Cathedral, which are nearby. He can access the DBpedia information related to each of them and also the related UGC shared by other users (incoming arrows). Then he decides to take a picture with his friends and make it available to other tourists, thus he adds information to the Hermitage museum by using the semantic annotator and UGC manager (outgoing arrows).

This use case has been developed together with Telecom Italia, the major network provider in Italy, and it shows how information of real word objects stored in Web of Data can be exploited into Smart Spaces. In this case information about tourist attractions and POIs allows to users to decide how to behave, e.g. if visit or not a particular monument.

6   Experimental Results

Usually, there are two main parameters used to evaluate recommender systems: performance and relevance. Performance measures the execution time needed to generate recommendations, while relevance the degree of correspondence between recommendations and results expected by users.

Results presented in this paper are focused on the performance evaluation of our algorithm, as at the moment of writing, we are still conducting a relevance evaluation which requires a more extensive study with final users in order to know the items that are considered as relevant by them regarding to a initial item. Those relevant items can be used to calculate the precision and recall measures, which are widely used in information retrieval.

The performance evaluation starts by selecting four URIs, acting as different initial concepts (in this case points of interest in St. Petersburg) to execute the recommendations, which are:

1. http://dbpedia.org/resource/Winter_Palace
2. http://dbpedia.org/resource/Saint_Petersburg_Metro
3. http://dbpedia.org/resource/Hermitage_Museum
4. http://dbpedia.org/resource/Palace_Embankment

These URIs are then given one by one to the LD-driven algorithm, which generate recommendations for four different values D of the maximum level $maxLevel$ (1 - 4). In this way, we were able to evaluate the execution time for different values of the maximum depth that the algorithm can explore in the category graph to get super categories.

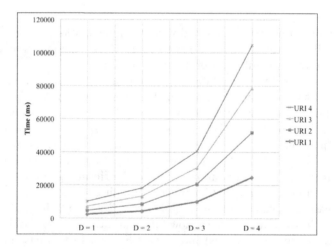

Fig. 4. Execution time (milliseconds) for different values of D

Figure 4 shows the execution time for four values of D. As can bee seen, by increasing the value of D, the execution time rises exponentially from 2,5 seconds ($D = 1$) to 26 seconds ($D = 4$). This behavior was expected, because as D increases the recommender has to get more super categories in order to produce an increment in the execution time. Additionally, we evaluated the number of candidate concepts generated by our algorithm for the four URIs, and also considering the different values of D (Figure 5).

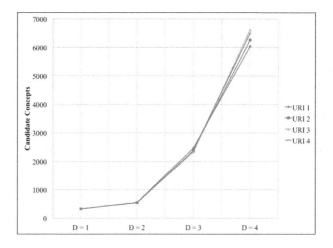

Fig. 5. Candidate Concepts generated for different values of D

We have evaluated our algorithm with four values of D, because of the exponential behavior of the algorithm when D increases. The selection of D depends on the number of candidate concepts that are needed for the specific application domain. According to our study we found that $D = 2$ is the best value because it can obtain an enough number of results (approximately 500) while consuming a reasonable execution time (approximately 4 seconds). The time and number of candidates as well as it significance for users was assessed by some experts of Telecom Italia who told us that the number of results was reasonable and results were relevant for them. However, they expressed the execution time was still long in order to produce a good experience for the final user. The problem in this case was not our algorithm but the time to execute SPARQL queries in the DBpedia endpoint, for this reason we are working to make our approach as independent as possible in order to reduce the execution time. Finally, it is important to notice also that the input URI impacts on computational time, while the number of items found it is quite the same for any of the tested URIs.

7   Conclusions and Future Work

We presented a new approach to integrate LD into SS based on the association of resources on the Web of Data with SS components. This association allows obtaining information about each component, which can be used to determine the mode of interaction between components, i.e. a component can make decisions about how to act by basing on the information it receives about the other components. Such interactions can be influenced through the use of a LD-driven recommender that was also introduced. Some SS components such as mobile devices can generate content and enrich the information of another SS components by means of a semantic annotation process. We provided the description of our reference platform that implements this approach and was used for an eTourism use case. The present work was done in collaboration with a mobile operator.

The relevance of the information obtained from LD and recommended by the LD-driven Recommender is still a work in progress. The results of this evaluation by means of precision and recall techniques will be reported in future publications regarding this project. In order to extend the context-awareness level of SS components, we're planning to introduce a Context Broker. The idea is to provide a more detailed analysis of the context and improve the recommendation of information. For doing this, we plan to reuse the Context Management GE of the FI-WARE project, as it provides an already available and mature solution to manage context information. In addition, we are investigating on how to allow users to add new real world objects into the Web of Data, since at now they can add information about resources that already exist but they cannot create new resources.

References

1. Bizer, C., Heath, T., Berners-Lee, T.: Linked data - the story so far. Int. J. Semantic Web Inf. Syst. 5(3), 1–22 (2009)
2. Moltchanov, B., Mannweiler, C., Simoes, J.: Context-awareness enabling new business models in smart spaces. In: Balandin, S., Dunaytsev, R., Koucheryavy, Y. (eds.) ruSMART/NEW2AN 2010. LNCS, vol. 6294, pp. 13–25. Springer, Heidelberg (2010)
3. Honkola, J., Laine, H., Brown, R., Tyrkko, O.: Smart-m3 information sharing platform. In: Proceedings of the The IEEE Symposium on Computers and Communications, ISCC 2010, pp. 1041–1046. IEEE Computer Society, Washington, DC (2010)
4. Smirnov, A., Kashevnik, A., Balandin, S.I., Laizane, S.: Intelligent mobile tourist guide - context-based approach and implementation. In: Balandin, S., Andreev, S., Koucheryavy, Y. (eds.) NEW2AN/ruSMART 2013. LNCS, vol. 8121, pp. 94–106. Springer, Heidelberg (2013)
5. Kiljander, J., Ylisaukko-oja, A., Takalo-Mattila, J., Etelper, M., Soininen, J.P.: Enabling semantic technology empowered smart spaces. Journal Comp. Netw. and Communic. 2012 (2012)
6. Passant, A.: dbrec — Music Recommendations Using DBpedia. In: Patel-Schneider, P.F., Pan, Y., Hitzler, P., Mika, P., Zhang, L., Pan, J.Z., Horrocks, I., Glimm, B. (eds.) ISWC 2010, Part II. LNCS, vol. 6497, pp. 209–224. Springer, Heidelberg (2010)
7. Kaminskas, M., Fernández-Tobías, I., Ricci, F., Cantador, I.: Knowledge-based music retrieval for places of interest. In: Second International ACM Workshop on Music Information Retrieval with User-Centered and Multimodal Strategies, MIRUM 2012, p. 19. ACM Press, New York (2012)
8. Damljanovic, D., Stankovic, M., Laublet, P.: Linked Data-Based Concept Recommendation: Comparison of Different Methods in Open Innovation Scenario. In: Simperl, E., Cimiano, P., Polleres, A., Corcho, O., Presutti, V. (eds.) ESWC 2012. LNCS, vol. 7295, pp. 24–38. Springer, Heidelberg (2012)
9. Manning, C.D., Raghavan, P., Schutze, H.: Introduction to Information Retrieval. Computational Linguistics 35, 307–309 (2009)

10. Berners-Lee, T.: Linked Data - Design Issues (2006),
    http://www.w3.org/DesignIssues/LinkedData.html
11. Suchanek, F.M., Kasneci, G., Weikum, G.: Yago: A core of semantic knowledge.
    In: Proceedings of the 16th International Conference on World Wide Web, WWW
    2007, pp. 697–706. ACM, New York (2007)
12. Bizer, C., Lehmann, J., Kobilarov, G., Auer, S., Becker, C., Cyganiak, R., Hell-
    mann, S.: DBpedia - A crystallization point for the Web of Data. Journal of Web
    Semantics 7, 154–165 (2009)
13. Chiarcos, C., Hellmann, S., Nordhoff, S.: Towards a linguistic linked open data
    cloud: The open linguistics working group. TAL 52(3), 245–275 (2011)

Resource Allocation Policies
for Smart Energy Efficiency in Data Centers

Marina Khludova

St. Petersburg State Polytechnic University, Russia,
Institute of Applied Mathematics and Mechanics,
Department of Telematics
mvkhludova@rambler.ru

Abstract. A data center must serve arriving requests in such way that the energy usage, reliability and quality of service performance should be balanced. This work is devoted to on-line resource allocation policies in data centers. We study a Markovian queueing system with controllable number of servers in order to minimize energy consumption and thus maximize the average revenue earned per unit time. An analytical model approach based on continuous-time Markov chain is proposed. The model is tested by simulation. Combining on-line measurement, prediction and adaptation, our techniques can dynamically determine the number of servers to handle the predicted workload. The policies comply with energy efficiency and service level agreements even under extreme workload fluctuations.

Keywords: Allocation Policy; Energy Efficiency; Data Center.

1 Introduction

This work was primarily motivated by the growing energy consumption by data centers (DC) and their inefficiency. According [1] DC consumes several megawatts for powering the equipment and cooling systems (that protect facilities in order to prevent hardware damages). A global IT electricity consumption is about 0.5% of would production. Thus, reducing energy consumption is a high priority task in DC industry. In this paper we will present technique based on controlled queueing system and prediction algorithm to improve energy efficiency in DC. Such technique is one of the directions of *smart energy efficiency* study.

Our problem of controlling a queue of tasks serviced by a fixed size set of homogeneous servers that can be turned on and/or be off in order to minimize the operational cost per unit time has been motivated by applications in computer systems. This problem has been studied in the literature during several decades and it is still active today. The existing queueing literature contains several articles on two special cases of the system we study here. Namely, several authors performed research work on finding efficient strategies to vary the number of servers as found in the *call center* problem literature [9, 10] that has obvious applications in practice. In

S. Balandin et al. (Eds.): NEW2AN/ruSMART 2014, LNCS 8638, pp. 16–28, 2014.
© Springer International Publishing Switzerland 2014

addition, a significant amount of research work have provided results on the single server with variable rate [8].

In this paper we have presented a model with the aim to balance performance and *smart energy efficiency* in DC. The approach described in [6] uses an economic model similar to our utility-based approach. The work here is in a preliminary stage, but the performance of several heuristics, in particular the Threshold-based Heuristic, is very encouraging. More work will be needed in examining this performance. We have presented Stochastic Markov model for performance evaluation of resources allocation policies. Currently we have received following results:

(i)   a technique to uniformly construct and compare allocation policies is chosen,

(ii)   an effects of variations in workload is quantified,

(iii)   dynamic resource allocation techniques that can handle changing workloads in DC is presented.

2   Model Description and Problem Formulation

In this section we present the controlled queuing system used in our study, and provide some technical preliminaries.

2.1   System Model

We consider a model where there are K homogeneous servers. These can be in one of two states: turned on or turned off. When a server is in turned on, it can service incoming requests. When a server is in turned off, it can't process any requests, but will consume less (or no) power. The service time requirements of tasks are exponentially distributed with parameter b. The QoS constraints are homogeneous among all the applications. We denote the number of powered on servers at time t by $k(t)$, $0<k(t) \leq K$, each running at rate $\mu(t)$. At first we consider an arrival process with fixed rate λ (so called homogeneous Poisson process). Then we consider non-homogeneous Poisson process, which counts events that occur at a variable rate. In this case, the generalized rate function is given as $\lambda(t)$. The tasks are independent, non-preemptable. Tasks are treated as statistically identical and scheduled FCFS. Each task requires for its processing only one server. We assign a negative 'holding cost', h_c, to keep a task in the system for one unit of time. Conversely we assign a profit e to keep a server switched off for a unit of time. This should reflect the relative energy savings. We assign a profit g to complete a task in accordance with its QoS need and penalty z otherwise. A server can be switched on or off. This will take an amount of time, assumed to be exponentially distributed. Of course, a "real" system will also not have an infinite queue length. In this case, we use a finite queue length model and study the system's performance under a given queue size limitation. Our abstract model is applicable to many hardware and software resources found on a server.

2.2   Problem Formulation

The resource allocation policies considered in this work are greedy policies that assign all active servers as soon as possible. Let $n(t)$ be the number of tasks in the system at time t. The objective of this paper is to present and evaluate resource allocation policies for smart energy efficiency in DC.

These policies are based on the state of the system. The system state may be described by $\{\lambda(t); n(t); \mu(t); k(t)\}$. We assume that an *allocation* module (see Fig. 5) has *control* the number of active servers $k(t)$. Since inter-arrival and service times are assumed to be exponential, we can assume that the decisions about the *control* variable are going to be made only whenever a task arrives or leaves the system.

3   Resource Allocation Policies

In this section we consider some policies we can implement to balance the performance of the system with its power consumption. The policies are modeled using continuous time Markov chains and performance results are obtained by solving the global balance equations. For small Markov process the simplest way to represent the process is often in terms of its state transition diagram [2]. In diagram we represent each state of the process as a node in a graph. The arcs in the graph represent possible transitions between states of process. The arcs are labeled by the transition rates between states. Since every transition is assumed to be governed by an exponential distribution, the rate of the transition is the parameter of the distribution.

3.1   Static Allocation Policy

We will do static allocation by examining a queue with $k(t)$ servers. It is possible to calculate the mean number of tasks in the system n_k and mean throughput r_k. The performance of the system is measured by the average revenue earned per unit time:

$$\Phi(n_{k*}, k^*, \ h_c, e, g, z) = h_c\, n_{k*} + e\,(K - k^*) + g\, r_{k*}\, P_{success} - z\, r_{k*}(1 - P_{success}), \qquad (1)$$

where $P_{success}$ is the portion of tasks which are successfully completed in accordance with their QoS needs. This is then the optimal static allocation of servers. Since we can repeat this process for any number of servers' k, we can easily determine the optimal number k^* that maximizes average revenue.

3.2   Queues with Hysteresis Behavior

At first we consider DC as 2-server queue system with master and reserve servers. There are additional parameters p and q: p means the probability of event "turn reserve server on after arrival of new task in the system", q means the probability of event "turn reserve server off after leaving the system by completed task". To calculate and compare the resource allocation process we chose such technique as continuous time first order Markov chains. The Markov diagram for resource allocation policy with hysteresis behavior is depicted in Fig.1.

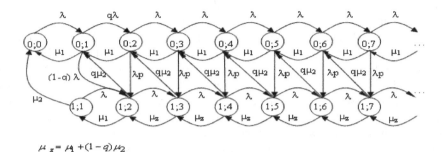

$\mu_z = \mu_1 + (1-q)\mu_2$

Fig. 1. State transition diagram for policy with hysteresis behavior

The tasks arrive at the system with the rate λ. States in the model are labeled as (i,j), where i denotes the number of reserve servers, j denotes the number of tasks in the system. The continuous-time version of the Chapman-Kolmogorov equations is:

$$
\begin{cases}
-\lambda P_{00}(t) + \mu_1 P_{01}(t) + \mu_2 P_{11}(t) = \dfrac{dP_{00}(t)}{dt} \\[2mm]
-(\lambda+\mu_1)P_{01}(t) + \lambda P_{00}(t) + \mu_1 P_{02}(t) + q\mu_2 P_{12}(t) = \dfrac{dP_{01}(t)}{dt} \\[2mm]
-(\lambda+\mu_1+\lambda p)P_{02}(t) + q\lambda P_{01}(t) + \mu_1 P_{03}(t) + q\mu_2 P_{13}(t) = \dfrac{dP_{02}(t)}{dt} \\[2mm]
-(\lambda+\mu_1+\lambda p)P_{0i}(t) + \lambda P_{0i-1}(t) + \mu_1 P_{0i+1}(t) + q\mu_2 P_{1i+1}(t) = \dfrac{dP_{0i}(t)}{dt}, \forall i \in [3;9] \\[2mm]
-(\mu_1+\lambda p)P_{010}(t) + \lambda P_{09}(t) = \dfrac{dP_{010}(t)}{dt} \\[2mm]
-(\lambda+\mu_2)P_{11}(t) + \mu_1 P_{12}(t) = \dfrac{dP_{11}(t)}{dt} \\[2mm]
-(\lambda+\mu_1+q\mu_2)P_{12}(t) + (1-q)\lambda P_{01}(t) + \lambda p P_{02}(t) + \lambda P_{11}(t) + (\mu_1+(1-q)\mu_2)P_{13}(t) = \dfrac{dP_{12}(t)}{dt} \\[2mm]
-(\lambda+(\mu_1+(1-q)\mu_2)+q\mu_2)P_{1i}(t) + \lambda p P_{0i}(t) + \lambda P_{1i-1}(t) + (\mu_1+(1-q)\mu_2)P_{1i+1}(t) = \dfrac{dP_{1i}(t)}{dt}, \forall i \in [3;9] \\[2mm]
-((\mu_1+(1-q)\mu_2)+q\mu_2)P_{110}(t) + \lambda p P_{010}(t) + \lambda P_{19}(t) = \dfrac{dP_{110}(t)}{dt}
\end{cases}
$$

The first active server is running at rate μ_1 and reserve server is running at rate μ_2. The allocation policy is considered as policy with hysteresis behavior. A reserve server can be switched on or off. It takes an amount of time, assumed to be exponentially distributed.

3.3   Threshold-Based Heuristic

The Idle heuristic policy follows the policy of powering down any server that is idle and powering up a server, if possible, when there are tasks in the queue that are not currently being served by any server. It does not take account of switching times. That is, we power up a server, if possible, when the number of tasks in the system is bigger than the number of servers currently servicing a task.

The Threshold-based Heuristic is a generalization of the Idle Heuristic policy. In common case queues are controlled by a sequence of forward thresholds and a sequence of backward thresholds such that when the current queue length exceeds a forward threshold, the service parameters are changed [2, 3]. In the same way, when a backward threshold is reached, the parameters are changed again. In this case services are speeded up by turning on reserve server. For this heuristic we choose some threshold, i_0 . Additional servers are then switched on when there are more or equal to i_0 tasks waiting to be served and switched off otherwise. Choosing threshold value should depend on both the differential between the holding cost and the power savings, and the switching times. The Idle Heuristic policy is equivalent to setting $i_0=0$.

States in the model in Fig. 2 are labeled as *(i,j)*, where i denotes the number of tasks in the system, j denotes the number of active servers. There is additional parameter i_0. The first active server is running at rate μ_1 and reserve server is running at rate μ_2. The allocation policy is considered as threshold-based policy with hysteresis behavior. We study the system's performance under a given queue size limitation.

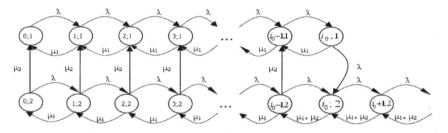

Fig. 2. State transition diagram for threshold-based policy

3.4 Global Balance Equations

Our primary objective with respect to a Markovian model is to calculate the probability distribution of random variable $X(t)$ over the space S, as the system settles into a regular of behavior. This is termed the steady state probability distribution.

The performance results may be derived from solving the global balance equations of the Markovian models [4]. When the model is in steady state, we must assume that the total probability flux out of a state is equal to the total probability flux into the state. Roughly speaking there three ways in which performance can be derived from the steady state distribution of a Markov process: state-based measures, e.g. utilization of reserve server, rate-based measures, e.g. mean throughput, other measures, e.g. mean response time. For other category of measures we may use one of the operational laws to derive the information we need. Thus, applying Little's Law we can get the average sojourn time experienced by each task in the system.

4   Performance Evaluation Result

The performance figures derived from solving the global balance equations of Markovian models and from simulation are reported in this section.

The heuristics with hysteresis and Threshold-based policy, described in the previous section are compared in performance under different scenarios (Fig. 3, 4).

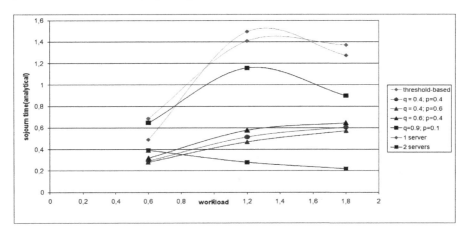

(a)

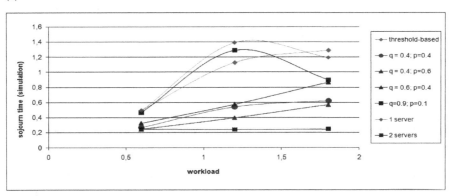

(b)

Fig. 3. Average sojourn time (in seconds) derived (a) from solving global balance equations (b) from simulation model

The goal of the analysis is to investigate the performance trade-off of the proposed policies with respect to commonly used policy with 2 servers. Therefore, the ratio of the average sojourn time under proposed policies to the average sojourn time under commonly used policy is may be considered. In this system model we show the effect of changing offered workload. We assume that $\mu_1 = 4$ (1/second), $\mu_2 = 5$ (1/second), $i_0 = 10$. Queue size is limited. The algorithm was implemented in Matlab 2012.

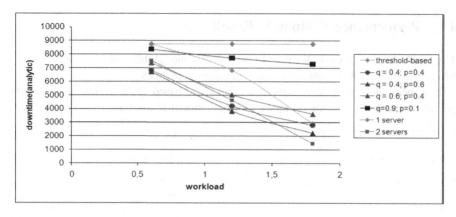

(a)

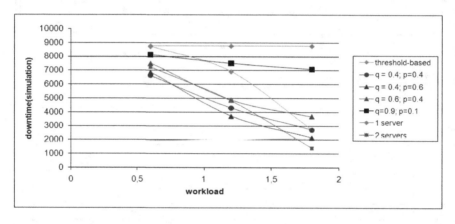

(b)

Fig. 4. Downtime (in hours) for reserve server per year for proposed and commonly used policies (a) analytical model (b) simulation

The simulation results (derived by GPSS model) corroborate analytical outcomes. In order to compute the average revenue earned per unit time, one needs first to perform the steady state analysis of the continuous-time Markov chain and then calculate the value of utility-based function proposed by Eq. 1.

5   Dynamic Resource Allocation Techniques

Dynamic resource allocation techniques that can handle changing application workloads in DC are the focus of this section. To perform dynamic resource allocation we need to employ (Fig. 5) three components: (i) a *monitoring* module that measures the workload and the performance metrics of each application (such as parameters of its request arrival rate, average response time, etc.), (ii) a *prediction* module that uses the measurements from the monitoring module to estimate the workload characteristics in the near future, and (iii) an *allocation* module that uses these workload estimates to determine average revenue and the number of active servers.

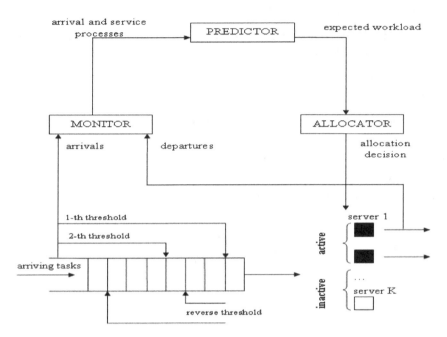

Fig. 5. System architecture combining online measurements with prediction and allocation

5.1   On-Line Monitoring and Measurement

The on-line monitoring module is responsible for measuring various system and application metrics. These metrics are used to estimate the system model parameters and workload characteristics. These measurements are based on the following time intervals (Fig. 6). Suppose tasks arrive during the busy hour as a non-stationary Poisson process. In particular, some authors assume that the busy hour can be partitioned into sub-periods where within each sub-period λ is constant but between sub-periods λ jumps in value. We contend that trend estimation is more fruitful. By using trend estimation it is possible to construct a model which is independent of anything known about the nature of the process. This model can be used to describe the behavior of the observed data.

Measurement interval: $[0,T]$ is the interval over which various parameters of interest are sampled. For instance, the monitoring module tracks the number of request arrivals Nk in k-th interval and records this value.

Let $0 \le t_1 < t_2 < \ldots < t_N \le T$ be the random variables representing consecutive arrival times. The simplest statistic function for trend estimation is $y = \dfrac{1}{T} \sum\limits_{i=1}^{N} \dfrac{t_i}{T}$.

Let $\lambda(t) = c + d \dfrac{t}{T}, 0 \le t \le T$ (2)

is rate function for non-stationary Poisson process. We will determine coefficients c and d with help of statistics:

$$v = \frac{N}{T}, \quad y = \frac{1}{T} \sum_{i=1}^{N} \frac{t_i}{T}.$$ (3)

Mathematical expectations of these functions are:

$$M\{v\} = \frac{1}{T} \int_{0}^{T} \left(c + d \frac{t}{T} \right) dt = c + \frac{d}{2} \text{ and } M\{y\} = \frac{1}{T} \int_{0}^{T} \frac{t}{T} \left(c + d \frac{t}{T} \right) dt = \frac{c}{2} + \frac{d}{3}.$$

Let us denote the estimation of the unknown coefficients c, d as $\hat{c}$ and $\hat{d}$. After some calculations we obtain:

$$\hat{c} = 4 \frac{N}{T} - 6 \sum_{i=1}^{N} \frac{t_i}{T}, \qquad \hat{d} = -6 \frac{N}{T} + \frac{12}{T} \sum_{i=1}^{N} \frac{t_i}{T}.$$ (4)

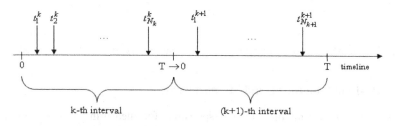

Fig. 6. Time intervals used for monitoring and measurement

Let us extend our observation for some non-overlapping adjoining time intervals. For k-th interval: $\hat{\lambda}_k(t) = \hat{c}_k + \hat{d}_k \left(\dfrac{t}{T} \right); 0 \le t \le T; k = 0,1,\ldots$,

$\hat{\lambda}_{k+1}(0) = \hat{\lambda}_k(T); k = 0,1,\ldots$ and therefore $\hat{c}_{k+1} = \hat{c}_k + \hat{d}_k; k = 0,1,\ldots$

Workload in a DC often fluctuates significantly on the timescale of hours or days, expressing a large "peak-to-mean" ratio. To demonstrate the property of our monitoring and measurement module, we conducted a controlled experiment using synthetic workloads (Fig. 7) with arrival rate measured in arrivals per second. Also presented two versions of our estimations for this arrival process with $T=300$ seconds and $T=100$ seconds.

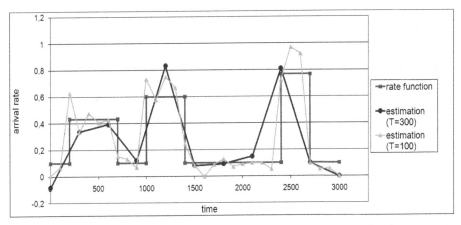

Fig. 7. Rate function for non-stationary Poisson process and its estimation

The choice of value T for a particular measurement interval $[0,T]$ depends on the desired responsiveness from the system. If the system needs to react to workload changes on a fine time-scale, then a small value of T (e.g., $T = 10$ seconds) should be chosen. On the other hand, if the system needs to adapt to long term variations in the workload over time scales of hours, then a coarse-grain measurement interval of some minutes may be chosen.

History (H): The history represents a sequence of recorded values for each parameter of interest. Our monitoring module maintains a finite history consisting of the most recent H values for each such parameter (Eq. 3, 4); these measurements form the basis for predicting the future values of these parameters.

Sliding Window (SW): The sliding window is the time interval between two successive invocations of the adaptation algorithm. Thus the past measurements are used to predict the workload for the next SW time units, and the system adapts over this time interval. As we would see in the next sub-section, our queuing model description considers a time period equal to the sliding window to estimate the average revenue (Eq. 1), and this model is updated every SW time units.

5.2   Workload Prediction Techniques

The workload seen by an application can be characterized by two complementary distributions: the *request arrival process* and the *service demand distribution*. Together these distributions enable us to capture the workload intensity and its variability. Our technique measures the various parameters governing these distributions over a certain time period and uses these measurements to predict the workload for the next window.

To estimate $\hat{\lambda}_k$, the monitoring module measures statistics (Eq. 3) in each measurement interval $[0,T]$. The sequences of these values form a time series. Using these time series to represent a stochastic process (Eq. 2), our prediction module attempts to predict the value of statistics (Eq. 3) for the next *SW*. To predict *v, y*, we model the process as an AR(1) process (*autoregressive* of order 1). This is a simple linear regression model in which a sample value is predicted based on the previous sample value.

To estimate the service demand for an application, the prediction module computes the probability distribution of the per-request service demands. This distribution is represented by a histogram of the per-request service demands. Upon the completion of each request, this histogram is updated with the service demand of that request. The distribution is used to determine the expected request service demand for requests in the next *SW*. Value $\hat{b}$ could be computed as the mean, the median, or a percentile of the distribution obtained from the histogram.

5.3   Allocating Resources

The *allocation* module has *control* the number of active servers. Since there are **K** servers in the system, we can consider any distribution of *k(t)* powered up servers, serving the queue, and **(K−k(t))** powered down servers. The allocation module is invoked periodically (every *SW*) to make decision dynamically. We considered some approaches to find the number of active servers [5].

Let $\hat{\lambda}_k$ denote the estimated request arrival rate and $1/\hat{b}$ denote the estimated service rate in the next *SW* (i.e., over the next **W** time units). The formula for the mean waiting time for *delayed* tasks for this **M/M/n** queue is quite well known (see [4]). It can be used to define QoS needs for tasks:

$$E[\hat{T}_W] = \frac{E[\hat{b}]}{k(t) - \hat{\lambda}_k E[\hat{b}]} .$$

Also we can employ the formula for a percentile of mean waiting time distribution:

$$m_{T_W}(y) = \frac{E[\hat{b}]}{k(t)(1-\rho)} \ln\left(\frac{100C(k(t),\rho)}{100-y}\right),$$

where $C(k(t),\rho)$ is Erlang C formula for the probability that all active servers in the system are busy and workload of each server is $\rho = \hat{\lambda}_k E[\hat{b}] / k(t)$.

So that we can easily find the number k^* that maximizes average revenue (Eq. 1). We assume that $\hat{\lambda}_k E[\hat{b}] \le k^*$ and system is stable.

6   Conclusion and Further Work

In this work we considered techniques for dynamic resource allocation in DC. We model queue of tasks serviced by a fixed size set of homogeneous servers that can be turned on and/or be off in order to minimize the operational cost per unit time.

The main goal of our techniques is to react to transient system overloading and underloading in such way that the energy usage, reliability and QoS performance should be balanced.

We make three specific contributions in this paper. First, have for an object to capture the transient behavior of application workloads, we model the set of homogeneous servers using a *controlled queuing system*. In order to comply with energy efficiency we use *threshold-based resource allocation policies* with hysteresis and present an efficient technique for solving the corresponding analytical models and computing various performance measures of interest.

A second contribution of our work is a *prediction algorithm* that estimates the workload parameters of applications in the near future using on-line measurements. Our prediction algorithm uses time series analysis techniques for workload estimation.

Third, this model dynamically relates the total resource requirements of all applications to their workload characteristics. The advantage of this model is that it does not make steady-state assumptions about the system and adapts to changing application behavior. Combining on-line measurement, prediction and adaptation, our techniques can dynamically determine the number of servers to handle the predicted workload based on (i) its Quality of Service (sojourn time) needs, (ii) the observed workload and (iii) average revenue earned per unit time.

The main directions for further investigation are:

    i.   hardware reliability is affected by state changes in DC, as components degrade faster when they are switched on/off than when they operate continuously, therefore, it incurs the following amortized cost;

    ii.   wavelet analysis [7,11] can be used to estimate the system model parameters and workload characteristics.

References

1. Kumar, R.: Data Center Power and Cooling Scenario through 2015. Gartner, Inc. (March 2007)
2. Luh, H., Viniotis, I.: Threshold control policies for heterogeneous server systems. Mathematical Methods of Operations Research 55, 121–142 (2002)
3. Khludova, M.: On-line Parallelizable Task Scheduling on Parallel Processors. In: Malyshkin, V. (ed.) PaCT 2013. LNCS, vol. 7979, pp. 229–233. Springer, Heidelberg (2013)
4. Kleinrock, L.: Queuing Systems, vol. 1. Wiley Interscience (1975)
5. Grassmann, W.K.: Finding the Right Number of Servers in Real-World Queuing Systems. Interfaces 18(2), 94–104 (1988)
6. Chase, J., Anderson, D., Thakar, P., Vahdat, A., Doyle, R.: Managing energy and server resources in hosting centers. In: Proceedings of the Eighteenth ACM Symposium on Operating Systems Principles (SOSP), pp. 103–116 (2001)
7. Bozhokin, S.V.: Wavelet analysis of learning and forgetting of photostimulation rhythms for a nonstationary electroencephalogram. Technical Physics 55(9), 1248–1256 (2010)
8. George, J.M., Harrison, J.M.: Dynamic control of a queue with adjustable service rate. Operations Research 49, 720–731 (2001)

9. Mazalov, V.V., Gurtov, A.: Queuing System with On-Demand Number of Servers. Mathematica Applicanda 40(2), 1–12 (2012)
10. Koole, G., Mandelbaum, A.: Queueing models of call centers: An introduction. Annals of Operations Research 113(1), 41–59 (2002)
11. Kwon, D., Ko, K., Vannucci, M., Reddy, A., Kim, S.: Wavelet methods for the detection of anomalies and their application to network traffic analysis. Quality and Reliability Engineering International 22(8), 953–969 (2006)

Design of a Framework
for Controlling Smart Environments

Stefan Nosović, Sebastian Peters, and Bernd Bruegge

Technical University Munich,
Boltzmannstr. 3, 85748 Garching, Germany
http://www1.in.tum.de

Abstract. Controlling a smart environment consisting of fixtures that communicate using different protocols is challenging. In this paper, a design of a highly extensible framework for controlling smart environments is proposed. The design is based on low-coupled Web Services communicating over an Enterprise Service Bus (ESB). The framework is independent from the fixture (adressable sensor, addressable actuator) communication protocols. It provides communication protocol abstractions that enable the framework to be extended with arbitrary fixture communication protocol implementation. The framework enables users to control fixtures using a uniform REST-based protocol. In addition, the proposed framework contains components that enable smart environment automation, collect user activity and context data and manage user rights.

Keywords: Smart environments, uniform communication protocols, pattern-based architectures.

1 Introduction

The technology advancements in the last 20 years have pushed Mark Weiser's[1] vision of non-intrusive availability of computers throughout the physical environment to become a reality. One of the implementations of the ubiquitous computing concept is the smart environment. Lupiana et al.[2] defined the smart environment as a *highly integrated* computing and sensory environment that *effectively reasons* about the physical and *user context* of the space to *transparently act* on human desires. In particular, a *highly integrated* environment is an environment that is saturated with devices and sensors that are fully integrated with wireless networks. The *effective reasoning* of the environment is based on a reasoning mechanism for the environment as a whole, not just for an individual device or component. It takes into the account also the *user context* that is created using the information on an individual's profiles, policies, current location and mobility status. Based on the result of the effective reasoning, the environment *transparently acts* to adapt to the inhabitant's needs without requiring their intervention.

The goal of creating a smart environment is to increase the comfort of the inhabitants. Inhabitant's comfort is increased by automatically adopting the state

S. Balandin et al. (Eds.): NEW2AN/ruSMART 2014, LNCS 8638, pp. 29–39, 2014.

of the environment (temperature, humidity or brightness) to his/her preferences. Also, an inhabitant's comfort is increased by enabling him/her to control the environment in an intuitive way. Today, this can be done by using mobile devices (phones, tablets, laptops etc.) or by issuing voice or even gesture-based commands. Another goal of the smart environment is to help the inhabitants to optimize their energy consumption. A scenario in which the heating is turned on or off based on the inhabitants' presence in the environment and increased or decreased based on the inhabitants' preferred temperature of the environment. In addition, smart environment also enhances the abilities of its inhabitants in executing tasks. An example for that can be enabling the voice triggered answering the phone if the inhabitant's hands are busy.

However, when setting up a smart environment, numerous different fixture (addressable sensor, addressable actuator) manufacturers are using different communication protocols. This causes incompatibility between fixtures because they are not able to exchange environment information among themselves. This means that no device in the smart environment knows the states of all the fixtures. Without the knowledge about the state of the fixtures in the environment, the creation of the environment context is not possible. The context of the environment is created by analysis and extraction of useful knowledge from user behaviour data and environment states data. Another challenge when having fixtures produced by different manufacturers using different protocols is that the data representation is different. For example, one manufacturer can use a light bulb brightness scale from 0 to 255 while another one can use a scale from 0 to 10000. In order to extract useful data, different values reported by the same types of fixtures (e.g. light bulbs) have to be normalized. This means that the gathered data may require time-consuming preprocessing before it can be used for context creation. The use of different communication protocols may increases the initial cost for the smart environment by forcing the use of proprietary fixture controllers. Furthermore, the fixture manufacturers have done little to increase the usability of their products[3].

In this paper, a framework that provides the user with a uniform protocol for communicating with fixtures that use different communication protocols is proposed. The framework gathers information about the state of all the fixtures while pre-processing the received values before storing them into the database. In addition, the framework supports automatic reacting to the changes in the environment. The proposed solution will increase the usability of the smart environment by enabling the user to control the smart environment using a single controller for all fixtures. Enabling the smart environment to be controlled by a single controller, the need for proprietary controllers is mitigated, which may lower the initial cost of setting up a smart environment.

The remainder of this paper is organized as follows: a short overview of the related work is given in Section 2. In Section 3, the Smart Environment Integration Framework along with the description of the requirements that it has to fulfill are presented. Section 4 describes the application of the presented framework in a smart office environment. In Section 5, the advantages and disadvantages of the presented design are discussed. Conclusion and description of the future work is given in Section 5.

2 Related Work

Wu et al.[4] introduce a model based peer-to-peer (P2P) architecture based on multiple Open Services Gateway Initiative (OSGi) platforms in which service-oriented mechanisms are used for system components to interact, and mobile agent technology is applied to augment the interaction mechanisms. The solution introduced by Wu et al. requires the installation of OSGi framework on the fixtures. The advantage of the solution presented in this paper is that it is designed to work with off-the-shelf fixtures, which means no changes to the fixtures have to be made in order to integrate them into a smart environment.

Li et al.[5] propose a client/server architecture for controlling a smart environment equipped with devices using ZigBee protocol. The solution proposed in this paper also supports dynamic web pages, but its main goal is to provide support for mobile applications for interaction with smart environments. In contrast to the solution proposed by Li et al. which focuses on controlling devices using only the ZigBee protocol, the solution proposed in this paper supports controlling of devices that use a variety of communication protocols.

Koß et al.[6] present an architecture for a smart environment control system that was to achieve the goals of working in a heterogeneous hardware landscape and provide a generalized interface for usage in higher-level domains such as data-analysis and event processing. The architecture of the system presented in this paper is based on the architecture proposed by Koß et al. Unlike the requirements that are presented in the paper by Koß et al., which are related to the optimization of energy consumption, the requirements defined in this paper focus on providing a uniform protocol for controlling fixtures that use different communication protocol. Unlike wrapping different communication protocol implementations into separate web services, the solution proposed in this paper is based on one web service that handles all the communication with fixtures, no matter which communication protocol the fixtures use.

Wille et al.[7] introduce a Tiny Smart Environment Platform (TinySEP). TinySEP combines the concepts of having different device drivers for communication with the devices using different protocols. The drivers produce signals (data) that are combined to create services (e.g. send a sms if the environment is not occupied and the windows are open). The devices, based on the examples presented in the paper by Wille et al., refer to different type of sensors. Unlike the solution presented in this paper, the TinySEP does not support actuators in the environment and thus does not enable the user user to remotely control or influence the state of the smart environment, but rather only get information about it.

OpenHab[8] is a vendor and technology agnostic open source automation software based on service bus architecture. The event bus is the base service of open-HAB end enables adding and removing of different protocol implementations as

OSGi bundles. The advantage of the approach presented in this paper is that it provides a mechanism for user authentication and request based access control. This mechanism enables the assignment of priorities to different users so that their requests get serviced in case of a conflicting request coming from another user (e.g. one user wants to open a window while another wants to close it). In contrast to OpenHab, the solution presented in this paper provides better security of the system by checking access rights for every request and a conflict resolution mechanism.

3   Smart Environment Integration Framework (SEIF)

The solution proposed in this paper is the creation of Smart Environment Integration Framework (SEIF), a framework for controlling smart environments. SEIF is a framework that acts as a broker between the user and the fixtures. The broker architectural pattern was formalized by Buschmann et al. in 1996 as follows:

"The broker architectural pattern can be used to structure distributed software systems with decoupled components that interact by remote service invocations. A broker component is responsible for coordinating communication, such as forwarding requests, as well as for transmitting results and exceptions."[9]

SEIF will provide the user with a uniform protocol for controlling the smart environment while hiding the fixture communication details from the user.

3.1   Requirements

One of the requirements for the SEIF is to enable the controlling of all fixtures (no matter which communication protocol they use) in the smart environment using a unified communication protocol. The goal of creating a unified communication protocol is to facilitate development of multiple types of client applications (iOS, Android, web browser based applications etc.) that would be able to control the fixtures.

The goal of providing support for multiple types of client applications is to enable the users to chose which platform they want to use for controlling the smart environment. SEIF is intended to be used by multiple users. Having multiple users that have access to the smart environment, SEIF has to enforce user authentication and user access control to protect the smart environment, or a part of it, from unauthorised use or misuse.

SEIF should support adding, removing or changing fixtures in a smart environment. If the added fixture uses a communication protocol that is supported by the SEIF, information about the fixture properties have to be provided to the framework. If a new fixture uses some other communication protocol than the ones that are supported by the SEIF, the framework should be easy to extend with the implementation of the newly required communication protocol. In addition, the design of the SEIF has to support extensibility with respect to new components of the framework which could provide information or perform actions which were not foreseen in this paper.

What makes an environment smart is the ability to adapt to the inhabitants' needs to increase their comfort or to minimize energy consumption without causing discomfort. In order to enable the smart environment to adapt to its inhabitants' needs, SEIF has to support automated fixture control. In order to gain information about the inhabitants' perception of comfort (e.g. what temperature they prefer, how bright they like the room to be, how loud the music should be etc.), the smart environment has to gather and analyze the state of the environment data, the behavior of its inhabitants and the changes in the environment that they cause. The gathering and analysis of such user and environment data has to be supported by the SEIF.

An additional requirement is to minimize coupling between the components of the SEIF, so that all of the components can be developed independently.

3.2   Smart Environment Design

A smart environment is made up of three distinct types of components which can be seen in Figure 1:

1. **Client components** - consists of a variety of applications enabling the user to control the smart environment.
2. **Fixture components** - consists of a number of fixtures that make up the smart environment.
3. **SEIF component** - makes communication possible by translating user requests into commands understandable to the fixtures, enables environment and user data collection and smart environment automation.

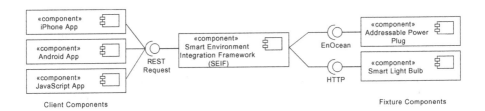

Fig. 1. Smart environment component diagram

Fixture components can be of many different types, but for the simplicity reasons were omitted from the component diagram in Figure 1. The focus of this paper is on the SEIF component of the smart environment and thus the other components will not be discussed any further.

3.3   Smart Environment Integration Framework Components

The purpose of the SEIF component is to address the problem of the heterogeneity of the fixtures and the heterogeneity of their protocols. SEIF introduces a REST-based protocol to hide the device-specific properties from the users. It receives the REST-based requests and translates them to device-specific protocol commands. The component diagram of the SEIF can be seen in Figure 2.

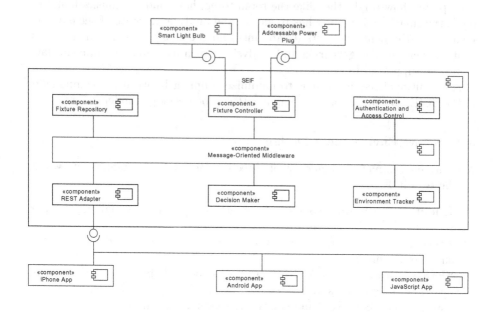

Fig. 2. SEIF component diagram

As shown in Figure 2, SEIF is made up of seven components. Each component is implemented as a web service and the communication between the components is performed over a service bus using SOAP. The architecture of the SEIF is based on the Enterprise Service Bus (ESB) architectural pattern [10].

REST Adapter component is responsible for providing a unified REST-based protocol for communication with the smart environment. Upon receiving a request from the client component, the REST Adapter processes the requests and forwards it to appropriate components such as Fixture Controller or Environment Tracker for servicing. After getting a response from the aforementioned component, the REST Adapter forwards the response to the client component. Two types of REST requests are supported:

- GET - used for requesting for the status of a fixture, a group of fixtures or the location of fixtures
- POST - used for changing the state of a fixture or a group of fixtures as well as for user authentication

The structure of REST-based protocol request is as follows:

$$\text{https://}SEIF\_\ address : port\_\ number/\text{REST}/\text{api}/\text{rest}/user\_id/fixture\_$$
$$name/command\_name$$

As shown in the example above, every request has to contain the *user_id* which is obtained after a user authenticates to the SEIF and is session specific. The fixtures are uniquely identified by their names. Every request has to contain the name of the command which is to be executed.

Fixture Repository component is responsible for storing fixture properties that are required for a successful communication. The fixture properties encompass the information about the ID of the fixture, the type of protocol that it is using and the location of the fixture. To provide fixture location information, the Fixture Repository stores the information about the building layout and the location of every fixture in the building. The facility manager is responsible for providing the SEIF with the fixture information. These information are provided to the SEIF during runtime, meaning that it does not have to be restarted before using the newly provided information.

Fixture Controller component is responsible for communication with the fixtures. It contains the implementations of all the protocols needed for a successful communication with fixtures. The functioning of the Fixture Controller is independent from the concrete implementations of the protocols. Fixture Controller provides the abstractions of the fixture communication protocols that facilitate the implementation of new protocols and make it's functioning independent from the concrete implementations. When the Fixture Controller receives a request for changing the state of a fixture, either coming from the REST Adapter (user request) or Decision Maker (environment automation), the appropriate communication protocol is chosen based on the fixture properties fetched from the Fixture Repository.

Decision Maker component is responsible for the smart environment automation. It contains the event-response based rules that are triggered by the events in the environment (e.g. a change in temperature). The response is the action that can be performed by the actuators in the environment (e.g. turning the heating on). The Decision Maker is informed about the events in the environment by the Fixture Controller. The event-response based rules can be provided manually by the inhabitants or can be suggested or generated automatically by the machine-learning module. This module contains machine-learning algorithms that analyse the environment and user data gathered by the Environment Tracker and try to recognize a pattern of behaviour of the inhabitants. If a pattern of behaviour is recognized, an environment automation rule will be suggested to the user or imported to the rule base with the goal of increasing his comfort.

Authentication and Access Control component is responsible for storing user credential, access rights and enforcing them. Before a user is allowed to issue a request to the SEIF, he/she first has to provide his/her credentials to the Authentication and Access Control component of the framework. If a user provides valid credentials and issues a request to the SEIF, the component that is responsible for servicing the request checks with the Authentication and Access Control if the user is granted rights to perform the requested action. Authentication and Access Control also contains a simple priority-based mechanism for resolving conflicting requests. In case of a conflicting request, the requests from a user with a higher priority is permitted. Based on the response from the Authentication and Access Control, the request is either serviced or rejected.

Environment Tracker component is responsible for tracking user activities and changes in the smart environment. Changes in the smart environment include the changes of fixture states caused by the user (e.g. a light turned on or off) and changes caused by external factors (e.g. change of temperature or brightness of the room caused by sunset). Every time a fixture detects a change in the environment, it reports the change to the Fixture Controller, which in turn reports the change to the Environment Tracker. Every reported change in the environment is pre-processed (e.g. data coming from window sensors that use different protocols are transformed into boolean values stating if the window is open or closed) before being stored into the database. If a user issues a request to the SEIF, the request is reported to the Environment Tracker by the component which is servicing the request.

Message-oriented Middleware component is responsible for providing connectivity and message handling between SEIF components. The components can send either peer-to-peer messages or broadcast them to all components. This component enables the implementation of the ESB architectural pattern. The ESB lowers coupling between the components and enables the framework to be easily extensible with new components when necessary.

SEIF architecture can be represented as an open four layered architecture which is shown in Figure 3. The layers of this architecture can be identified as:

1. **Interface Layer** manages client connections. Clients establish HTTP connections and send REST-based requests. Interface layer processes the incoming requests, and forwards them to the appropriate components for servicing. Interface layer forwards the system responses to the clients.
2. **Application Logic Layer** communicates with the fixtures. This layer is responsible for translating user's requests to appropriate commands using a protocol that the fixture can understand. This layer also contains a component for the smart environment automation.
3. **Data Layer** provides the upper layers with the information about user access rights, fixture details and the state of the smart environment.
4. **Storage Layer** persists the data that the data layer provides to the upper layers.

The advantage of using a layered architecture lays in the fact that the implementation of an entire layer can be changed without effecting the other layers, as long as the interface of the layer does not change. In case of the architecture show in Figure 3, if a layer interface is changed, the change influences only the layers above.

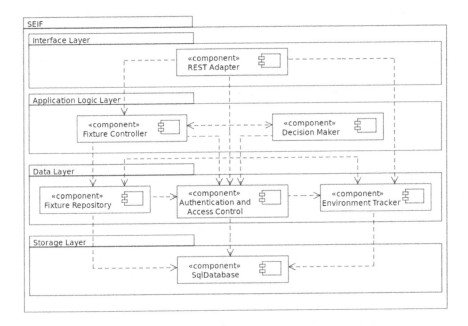

Fig. 3. Layered architecture of SEIF

4   SEIF Application

The SEIF has been implemented and deployed in the two offices at the Chair for Applied Software Engineering of the Technical University Munich. These offices are equipped with actuators for controlling blinds, light fixtures and heating, as well as sensors for measuring temperature, movements, CO_2 and brightness. A prototype web site was developed for controlling the fixtures. The development of mobile iOS and Android applications for controlling these devices is in progress. The development of such applications is made easier by creating a unified REST-based protocol for controlling heterogeneous fixtures that are using different communication protocols.

The SEIF is being used on a daily basis. The employees use the web site to control the light, blinds and the heating in their offices. They create rules in the Decision Maker based on their preferences and needs. In the evening, the lights at the offices are automatically turned off unless an employee is present in the office. In case the CO_2 level reaches a certain threshold in the office, the light in the room starts turning on and off.

So far, no complaints by about the performance of the SEIF have been reported by the users. Further performance and usability testing and evaluation of the framework will be performed after the development of the applications for controlling the smart spaces is finished.

By using SEIF, the initial cost for the smart environment was lowered by not acquiring the proprietary controllers. At the same time, the usability of the smart environment was improved by enabling the user to set the lights or the heating in their office without leaving their seat. In addition, the usage of a single controller for controlling all the lights was enabled, which mitigates the need of having a set of switches on the wall that can cause confusion.

5   Discussion

The benefits of using the presented design are related to the development of the components and their deployment. Due to the decoupling of components, after defining the interfaces of the SEIF components they can be developed independently, as long as the interfaces do not change. This is important because the Fixture Controller has to be extended every time when support for a new communication protocol is required. The proposed design also facilitates the integration of new components into the SEIF by just connecting the component to the bus and using the services of the existing components.

Not only that components can be developed independently, they can also be deployed to the different machines without changing the implementation of the components. This property of the system enables every component to be migrated to a separate machine if the performance of the system decreases due to the inability of a single machine to handle the workload in a reasonable amount of time.

The disadvantage of the design presented in this paper is that the response time of the SEIF may vary depending on the performance of the network connecting the components, due to the intensive communication between the components. This disadvantage represents an acceptable tradeoff between the performance and the low coupling of the SEIF components that makes the system easily extensible.

6   Conclusion and Future Work

In this paper, a framework that enables the user to control different fixtures (that use different protocols) using a unified, REST-based protocol is proposed. The proposed framework was designed to be extensible with respect to fixtures that it communicates with, framework components and communication protocols. In addition, it is shown that such a framework can decrease the cost of the initial investment for creating a smart environment while increasing its usability at the same time.

The emphasis of future work will be on the implementation of use cases to reduce the energy consumption in a smart environment and to increase the comfort of the inhabitants. The focus will be on enhancing the Decision Maker component so that it can generate new automation rules based on the analysis of environment and user data gathered by the Environment Tracker. The goal of the automation rules will be to minimize the energy consumption while maximizing the comfort of the smart environment inhabitants. The future work will also focus on creating more complex conflict resolution mechanisms than the priority-based currently used.

Acknowledgements. We would like to thank Emitzá Guzmán Ortega and Juan Haladjian for providing useful feedback on this paper. Special thanks go to Yang Li who did multiple rounds of proofreading and providing useful feedback.

References

1. Weiser, M.: Hot topics-ubiquitous computing. Computer 26 10, 71–72 (1993)
2. Lupiana, D., Omaray, Z., Mtenzi, F., O'Driscoll, C.: Smart Spaces in Ubiquitous Computing. In: 4th International Conference on Information Technology (2009)
3. Gann, D., Barlow, J., Venables, T.: Digital Futures: Making Homes Smarter (1999)
4. Wu, C.-L., Liao, C.-F., Fu, L.-C.: Service-Oriented Smart-Home Architecture Based on OSGi and Mobile-Agent. IEEE Transactions on Systems, Man and Cybernetics, Part C: Applications and Reviews, 193–205 (2007)
5. Li, L., Xie, W., He, Z., Xu, X., Chen, C., Cui, X.: Design of Smart Home Control System Based on ZigBee and Embedded Web Technology. In: Lei, J., Wang, F.L., Deng, H., Miao, D. (eds.) AICI 2012. LNCS (LNAI), vol. 7530, pp. 67–74. Springer, Heidelberg (2012)
6. Koß, D., Bytschkow, D., Gupta, P.K., Schatz, B., Sellmayr, F., Bauereiß, S.: Establishing a smart grid node architecture and demonstrator in an office environment using the soa approach. In: 2012 International Workshop on Software Engineering for the Smart Grid (SE4SG), pp. 8–14 (2012)
7. Wille, S., Shcherbakov, I., de Souza, L., Wehn, N.: TinySEP–A Tiny Platform for Ambient Assisted Living. In: Ambient Assisted Living, pp. 229–243. Springer (2012)
8. OpenHab Project, http://www.openhab.org/index.html (accessed May 2014)
9. Buschmann, F., Henney, K., Schimdt, D.: Pattern-oriented Software Architecture: On Patterns and Pattern Language, vol. 5. John Wiley & Sons (2007)
10. Erl, T.: SOA design patterns. Pearson Education (2008)

Smart Space-Based Tourist Recommendation System
Application for Mobile Devices

Alexander Smirnov[1,2], Alexey Kashevnik[1], Andrew Ponomarev[1], Nikolay Teslya[1], Maksim Shchekotov[1], and Sergey I. Balandin[3]

[1] SPIIRAS, St. Petersburg, Russia
{smir,alexey,ponomarev,teslya,shekotov}@iias.spb.su
[2] ITMO University, St. Petersburg, Russia
[3] FRUCT Oy, Helsinki, Finland
Sergey.Balandin@fruct.org

Abstract. The paper presents smart space-based tourist recommendation system that allows acquiring information about places of interests around the tourist from different internet sources (like wikipedia, wikivoyage, wikitravel, panoramio, flickr). The system implements ranking acquired attractions according to the tourist preferences and current situation in the tourist location. Tourists can rate attractions that they like or dislike. Based on these ratings a recommendation service clusters tourists into groups with similar interests and uses evaluations of tourists belonging to the same group for ranking attractions around the tourist. The paper presents a prototype service for these purposes that is based on smart space technology. The prototype has been developed for Android devices and is available for free downloaded from Google Play market.

Keywords: e-tourism, recommendation system, tourist service, mobile devices.

1 Introduction

The modern world provides tourists with huge possibilities for searching information about places of interests through the Internet. Modern mobile technologies allow tourists to get interesting information via the Internet during their trips. Smartphones are mainstream with active iOS and Android devices surpassing 700 million globally by now. Global Mobile data traffic is growing rapidly to an impressive share of 13% of Internet traffic in 2012 [1]. There are a lot of services and applications that allow to simplify this search, proactively provide information about interesting attractions during the trip, provide user feedback about attending places of interests, etc. In accordance with [1] at the moment German Apple Store accounted around 780.000 apps and 36.000 travel apps (category Travel) representing a market share of 4,62% of all available apps.

Therefore, development of mobile applications that can provide access to the interesting for the tourist attractions nearby taking into account current situation in the location region is an actual task with good business potential. For these purposes it is

S. Balandin et al. (Eds.): NEW2AN/ruSMART 2014, LNCS 8638, pp. 40–51, 2014.

needed to develop a set of methods and algorithms for generating recommendations for the tourist.

The system is service-based and uses the smart space technology, which allows providing information for sharing between different services of the system. This technology [2], [3] aims in the seamless integration of different device by developing of ubiquitous computing environments, where different services can share information with each other, make different computations and interact for joint tasks solving.

The rest of the paper is structured as follows. Section 2 reviews related work in the area of mobile application for tourists. Section 3 briefly describes tourist recommendation system architecture. In Section 4 the recommendation service is overviewed. Section 5 summarizing tourist recommendation system implementation is followed by conclusion section that summarizes main results and findings of the study.

2   Related Work

A mobile application for tourists COMPASS is described in [4]. Authors propose the application that makes context-aware recommendation based on tourist's interests and context. The application is built upon the WASP platform that provides generic supporting services combined with semantic web technology.

Kramer at el. [5] present context-driven mobile tourist guide that has been developed for Windows Mobile operation system. The study presents methodology, implementation and evaluation of mobile tourist guide.

GoTour [6] is an Android-based mobile application for providing tourism and geographic services in Istanbul city. Application has internal attraction database and provides possibilities of searching places of interests around using the Variable Neighborhood-based algorithm.

World Around Me [7] is a Windows Phone 7 application that shows the user photos around the user location. Photos are automatically downloaded from Flickr and Panoramio and presented to the user.

Raptis at el. [8] present a context based design for guiding visitors in museums using mobile devices.

A set of existing systems have been considered compared and classified according to the presented context-based approach. P. Jeon proposes a context-aware mobile platform for use of various surrounding local services with minimal user intervention in the forthcoming ubiquitous environment [9]. Context management technology is used for modeling current situation and proposing the user appropriate services.

Luyten and Coninx [10] describe implementation of a context-aware mobile guide for outdoor as well as indoor locations. It uses GPS to identify user's location in outdoor environments, communicates with other objects in the environment through Bluetooth. The information that is shown in the user interface can be obtained in two different ways: stored on the mobile guide, or queried from the artifacts that are in the direct surroundings of the mobile guide through wireless communication.

The travel guide Triposo [11] is a free mobile guide service available for Apple and Android devices. A user can download the application and appropriate database (which is updated ones each two months) to the mobile device beforehand and use it during the trip without Internet connection. The application supports logging of travelling. It includes databases from the following sources World66, Wikitravel, Wikipedia, Open Street Maps, TouristEye, Dmoz, Chefmoz and Flickr [12]. Each guide contains information on sightseeing, nightlife, restaurants and more.

Millions of traveler reviews, photos, and maps can be accessible in TripAdvisor [13]. Tourists can plan their trips taking into account over 100 million reviews and opinions by travelers. TripAdvisor makes it easy to find the lowest airfare, best hotels, great restaurants, and fun things to do, wherever you go. The mobile application is free, it supports all mobile platforms.

Smart Travelling [14] is an online travel guide that supports about 30 cities worldwide including the most interesting destinations in European countries and USA. The guide includes a database of restaurants, cafes, hotels, shopping-tips and other places of interests. The mobile application for iPhone is accessible in AppStore. Integration with Google maps allows user to see the current location in the map and helps to navigate to each and every tip in destination cities. Application allows the user to download the content and use guide without Internet connection.

ARTIZT [15], an innovative museum guide system, where a ZigBee [16] protocol is used for determine user's position information. Visitors use tablets to receive personalized information and interact with the rest of the elements in the environment. The system achieves a location precision of less than one meter. The context is used to provide needed at the moment personalized information to the user.

The carried out analysis of the mentioned above systems shows that they can be divided into three main groups:

- applications that implement search for information around the tourist;
- applications that have own databases with information about attractions and provide this information to the user;
- applications that collect estimation information about attraction estimations and suggest tourists if this attraction is good or not.

Developed tourist recommendation system as opposed to another implements search of information about attraction in different Internet sources. It ranks attractions based on tourist preferences and context information. The system also allows making routes (pedestrian, car and public transport) from the tourist location to an interesting attraction. Decentralized smart space in the proposed approach allows mobile device of every tourist acquire up-to-date information from different services (e.g., attraction, region, or worldwide services) and based on it make own decision about the suggestions to the tourist taking into account his/her preferences and current situation in the region.

3   Tourist Recommendation System Architecture

Tourist recommendation system architecture (Fig. 1) consists of the following main components:

- client application installed to the tourist mobile device that shares tourist context with the smart space and provides the tourist results of guide application operation;

- attraction information service that implements retrieving and caching the information about attractions;

- recommendation service that evaluates attraction/image/description scores based on ratings that have been saved to internal database earlier [17];

- region context service that acquires and provides information about current situation in the considered region (e.g., weather, traffic jams, closed attractions);

- ridesharing service that finds matching routes of the tourist and preferred attraction locations with accessible in the region drivers routes; the service provides the tourist possibilities of comfort transportation to the preferred attraction [18];

- public transport service that finds information about public transport applicable to reach the preferred attraction [19].

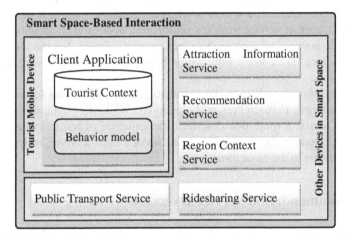

Fig. 1. General architecture of smart space-based tourist recommendation system

The client application is located in the tourist mobile device while other services use powerful backend computer systems. So, the main engineering task in the tourist recommendation system development is implementation of resource-intensive operations using other system, which is located in computational servers. Thereby, the main tasks of client application are: collect and provide information about tourist's context, profile, and actions; communication with smart space; provide results to the tourist; and share tourist ratings of attended attractions, browsed descriptions and images with the smart space in according with its behavior model.

The attraction information service is responsible for six main tasks. The first task is extraction of attractions titles around the tourist from external sources and sharing this information with the smart space. The second task is caching acquired information for quick access by tourists in the same location without additional requests to external sources at nearest future (live time of cached attraction is set by guide application administrator). The third task is providing internal identifiers for attractions. The fourth task is extraction of default images for attractions that are stored in the internal database. Default images are defined by the recommendation service as the best images for an attraction based on tourists ratings. The fifth task is extraction of attraction details (lists of images and text descriptions) from external sources and sharing them with the smart space. And the last task is to set and refresh default images in the internal database for the attractions based on the information available in the smart space from the recommendation service.

The recommendation service is responsible for ranking attractions, images, and descriptions for providing the tourist with the best attractions to see and the best images and description of chosen attraction for acquaintance. It stores the following mappings in the internal database: (internal attraction identifier, tourist context, region context -> attraction rating), (image URL -> image rating), (description URL, description text -> description rating). The internal database also keeps similarities between tourists, which are calculated as background process. The recommendation service as the core of the tourist recommendation system is described in details in the next section.

The region context service acquires information about tourist location area from region specific services as soon as the client application publishes the tourist context to the smart space. The context information is published to the smart space and used by other services.

The public transport and ridesharing services provide alternatives for transportation means for the tourist to reach preferred attractions. While the public transport service provides the tourist a route that describes the sequence of transportation means to reach the preferred attraction, the ridesharing service finds a driver, which goes from the tourist location to the same direction.

4   Recommendation Service Overview

To improve the quality of information presented to the tourist by the tourist recommendation system the list of attractions found by the attraction information service should be ordered with respect to a predicted degree of interestingness for the tourist as well as reachability (taking into account the current situation in the area).

The attraction's degree of interestingness is estimated by the recommendation service. The service takes tourist ratings associated with each attraction by all tourists as an input. According to the conventional classification (e.g. in [20]), it performs user-based collaborative filtering. One of the promising directions to improve the predictive quality of recommendation systems in general (and collaborative filtering systems among them) is context-awareness [21]. The context describes conditions in which the tourist rates an object or asks for recommendations.

The main scheme of recommendation process for tourist recommendation system in presented in Fig. 2.

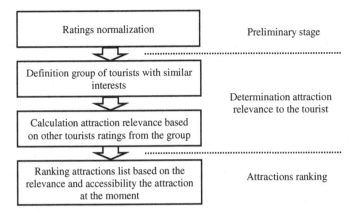

Fig. 2. The main scheme of recommendation process for tourist recommendation system

In the proposed tourist recommendation system the following context attributes are distinguished:

a) time;

b) company in which the tourist visited the attraction (alone, with a friend or with the family);

c) weather (sunny, rainy, etc).

Values are assigned to these attributes in mostly automated fashion. For example, the tourist opens the attraction evaluation screen being near to that particular attraction (according to the mobile device's GPS sensor). In this case the time attribute is filled in with the current time and current weather is queried from the context service. However, there is also a possibility to set the values of context attributes manually in the evaluation screen of the mobile application. It is convenient, for example, if a tourist wants to rate the attractions seen during the day upon returning to the hotel in the evening. To facilitate deferred evaluation the proposed system tracks attractions the tourist visits and shows unrated visited attractions in a special screen. The tourist does not have to assign values to each context attribute. If a context attribute is not given a value, it is assumed to have value "any".

There are three general approaches to take context into account in recommendation systems [21]: (a) contextual pre-filtering; (b) contextual post-filtering; (c) contextual modelling.

The advantage of the contextual pre-filtering and post-filtering approaches is that they are compatible with classical (not context-aware) recommendation algorithms. The context awareness in these approaches comes true by transformation of either input or output of the classical recommendation algorithm. In the contextual pre-filtering approach, the rating data that is not related to the context is filtered out before applying the recommendation algorithm. On the other hand, in the contextual post-filtering approach the resulting list of recommendations is ordered or filtered taking into account context values.

In the contextual pre-filtering approach all the ratings that are irrelevant to the context discarded from the rating matrix before the recommendation algorithm is applied.

For example, if in some attraction recommendation service the context includes weather conditions, then making recommendations in a rainy day should not use ratings assigned in sunny days. This approach aggravates the important problem inherent to collaborative filtering systems – rating matrix sparcity. The main goal pursued by contextual pre-filtering methods is to take into account the context, but not let rating matrix to become too sparse.

In the proposed system the context generalization method [22] (one of the contextual pre-filtering methods) is used for taking context into account. In this method, the rating matrix is filtered not only by exact values of context attributes, but also by its possible generalizations. To use this method the context model has to support context generalization. In most general form, it means that at least one context attribute must be defined on a set with a strict partial order relation of generalization ($\rightarrow$). Let A be a set of attribute values and $a_i, a_j \in A$. Then notation $a_i \rightarrow a_j$ means that value a_j is a generalization of a_i. A context is usually represented by m attributes. Let $c = (c_1, \ldots, c_m)$ and $c' = (c'_1, \ldots, c'_m)$ are two contexts. We define c' as a generalization of c ($c \rightarrow c'$) if there exists at least one $i \in \{1,..,m\}$, such that $c_i \rightarrow c'_i$. We call context c *incompatible* with c' if neither $c \rightarrow c'$ nor $c = c'$. In most cases, the generalization relation forms some kind of a hierarchy (or multiple hierarchies).

In the proposed system the context generalization is enabled by following:

a) The set of Time attribute (see Fig. 3) values includes not only exact date and time values but also "any" value and aggregate values for each season, day type (working day or weekend) and time of day (morning, afternoon, evening). The generalization relation is defined naturally.

b) The set of Company attribute values includes values "alone", "with friends", "with family" and "any". "Any" value is defined to be a generalization of any other value.

c) The set of Weather attribute values includes values "sunny", "rainy", "cloudy", "snowy" and "any". "Any" value like in (b) is defined to be a generalization of any other value.

Fig. 3. Context generalization example

For example, the exact context could be (Time: "July 31, 2013 17:30"; Company: "with family"; Weather: "sunny"). This context can be generalized to (Time: "summer"; Company: "with family"; Weather: "sunny") or even to (Time: "summer"; Company: "any"; Weather: "any").

It is obvious that a context can be generalized in several ways and directions. In systems with many attributes and many levels of granularity of attributes, enumerating all possible context generalizations is a problem and various heuristics are used for picking appropriate generalizations [22]. In the proposed system, there are not so many possible generalizations, so all of them are enumerated through implicit directed graph traversal procedure. The nodes of this graph are attribute values and the arcs are generalization relations.

A tourist rates attractions on a five-point scale (1 – bad, 5 – excellent). The rating obtained from the tourist (*raw* rating) is normalized to reduce individual bias in assessment: some tourists tend to put relatively high ratings to all attractions, others in contrary tend to put relatively low ratings. Normalized rating $\tilde{r}_{uj}$ given by tourist u to attraction j is defined by formula:

$$\tilde{r}_{uj} = r_{uj} - \frac{1}{|K_u|+1}\left(3 + \sum_{k \in K_u} r_{uk}\right),$$

here, r_{uj} is raw rating of the attraction j given by tourist u, and K_u is a set of all attractions rated by tourist u. The idea of normalization is to shift from user-oriented five-point scale to calculations-oriented zero-centered scale. The sign of the normalized rating corresponds to general attitude of the tourist (whether it is positive or negative) and the absolute value of the rating corresponds to the strength of that attitude. The straightforward way to normalize ratings is to subtract scale average (i.e. "3") from each rating. It would work nice if tourists normally used all the range of five-point scale. However, most tourists in fact rate items using some subset of the scale, e.g., only "3", "4" and "5". In this case subtracting scale average would result in non-negative normalized ratings missing the fact that the tourist definitely likes items he/her rated "5" and probably doesn't like items rated "3". Hence, the normalization procedure should capture not only the scale characteristics but also the observed usage of this scale. Therefore, a popular method of normalization is subtracting average tourist rating from all his/her ratings. This method works well in most cases but have some subtle drawback which turns out when there are only a few ratings. For example, when the tourist rated only two items – both with "5" – then normalization over the average tourist rating would turn these ratings into zeroes. I.e. *a priori* notion of five-point scale with "5" as the best mark is lost in favor of adaptation to the observed usage of this scale. To alleviate this drawback in the proposed system we use slightly modified version of the normalization over the average tourist rating. During the normalization we add one fake rating of "3" (scale average) to the set of tourist ratings having a purpose to stick other ratings to the original notion of the scale. This modification is significant when there are a few ratings (in the example above two "5" ratings become positive) but its contribution to the normalized ratings vanishes as the number of tourists' ratings grows.

Attraction rating estimation for a given tourist is performed in two steps:

1) a group of tourists with ratings similar to the given tourist's is determined;

2) rating of attraction is estimated based on ratings of this attraction assigned by tourists of the group.

While building the list of recommendations, several possible generalizations of the context is used. For each context generalization ratings received in contexts incompatible with this generalization are not taken in to account.

Tourist group is determined by k-Nearest Neighbors method (kNN). The similarity between tourists u and v is calculated as a cosine measure between normalized ratings vectors of tourists according to the following formula:

$$s_{uv} = \frac{\sum\limits_{o \in O} \tilde{r}_{uo} \tilde{r}_{vo}}{\sqrt{\sum\limits_{o \in O} \tilde{r}_{uo}^2} \sqrt{\sum\limits_{o \in O} \tilde{r}_{vo}^2}}.$$

Here O is a set of attractions rated by both tourists u and v.

Attraction rating estimation for the tourist is based on ratings of that attraction assigned by other tourists of the group with respect to their similarity to the tourist. It is calculated as a weighted average of normalized ratings among group members:

$$r_{uj}^* = \frac{\sum\limits_{v \in G} \tilde{r}_{vj} s_{uv}}{\sum\limits_{v \in G} |s_{uv}|},$$

here G is the group of the tourist.

The resulting list of attractions L presented to the tourist u is sorted in descending order of:

$$s_j = k r_{uj}^* + (1-k)\left(1 - \frac{d_j^w}{\max\limits_{i \in L} d_i^w}\right),$$

here $k \in [0,1]$ is a model parameter correlated to the importance of the attraction rating estimation in favor of its reachability; d_j^w is the estimation of time to reach the attraction j.

5   Implementation

Implementation of the proposed services has been developed based on Smart-M3 information platform [23, 24] which is open source and accessible at Sourceforge [25]. Service implementations have been developed using Java KPI library [26]. Mobile clients have been implemented using Android Java Development Kit [27].

Fig. 5, left screenshot shows the main application screen, where the tourist can see images extracted from accessible internet sources around, clickable map with his/her location, context situation (weather), and the best attractions around ranked by the recommendation service described in Section 4. By pressing "menu" button guide application allows to search information for worldwide attractions by choosing another area (country, region, and city) and access the settings page of the mobile tourist guide application. In the status bar the tourist can search for attractions worldwide. The middle screenshot (Fig. 5) shows description of an attraction that the tourist is interested in. It consists of the list of images and text description. The right screenshot in Fig. 5 shows pedestrian path from the tourist location to the interesting attraction.

Fig. 4. Prototype Screenshots: main screen, attraction details, and route to the attraction

Tourist can browse information about the best attractions around presented by the mobile tourist guide in the main screen and click button "More" to see more attractions (see two screenshots in Fig. 5). They show attractions recommended by the system for the tourist (left one without recommendation service sorted by distance and right one has been generated using the recommendation service).

Fig. 5. Prototype Screenshots: without and with recommendation service

6   Conclusion

The paper presents the smart space-based tourist recommendation system and application for Android smartphones, which acquires information about places of interests around the tourist from different Internet sources and implements their ranking according to the tourist interests and current situation in the tourist location region. The paper presents a set of methods and algorithms for generating recommendations for a user that allow significant increase of the system usability. Proposed study has been successfully implemented as distributed service-based application. For the tourist recommendation system a smart space infrastructure has been created that includes: attraction information service, context service, recommendation service, ridesharing service, public transport service and Smart-M3 information sharing platform for integration these services. Mobile client that interacts with created smart space has been developed and published in Google Play Market[1].

Acknowledgements. This research is a part of Karelia ENPI CBC programme grant KA322 "Development of cross-border e-tourism framework for the programme region (Smart e-Tourism)", co-funded by the European Union, the Russian Federation and the Republic of Finland. The presented results are also a part of the research carried out within the project funded by grant #13-07-12095 and 13-07-00336 of the Russian Foundation for Basic Research. This work was partially financially supported by Government of Russian Federation, Grant 074-U01. Authors are grateful for DIGILE IoT SHOK program that provided required support of this research.

References

1. Wagner, S., Franke-Opitz, T., Schwartze, C., Bach, F.: Mobile Travel App Guide: Edition 2013 powered by ITB., Pixell Online Marketing GMBH (2013),
   http://www.itb-berlin.de/me-dia/itb/itb_media/itb_pdf/publikationen/MTAG_2013.pdf
2. Cook, D.J., Das, S.K.: How smart are our environments? an updated look at the state of the art. Pervasive and Mobile Computing 3(2), 53–73 (2007)
3. Balandin, S., Waris, H.: Key properties in the development of smart spaces. In: Stephanidis, C. (ed.) UAHCI 2009, Part II. LNCS, vol. 5615, pp. 3–12. Springer, Heidelberg (2009)
4. van Setten, M., Pokraev, S., Koolwaaij, J.: Context-Aware Recommendations in the Mobile Tourist Application COMPASS. In: De Bra, P.M.E., Nejdl, W. (eds.) AH 2004. LNCS, vol. 3137, pp. 235–244. Springer, Heidelberg (2004)
5. Kramer, R., Modsching, M., Hagen, K.: Development and evaluation of a context-driven, mobile tourist guide. International Journal of Pervasive Computing and Communications 3(4), 378–399 (2005)
6. Al-Rayes, K., Sevkli, A., Al-Moaiqel, H., Al-Ajlan, H., Al-Salem, K., Al- Fantoukh, N.: A Mobile Tourist Guide for Trip Planning. IEEE Multidisciplinary Engineering Education Magazine 6(4), 1–6 (2011)
7. Vdovenko, A., Lukovnikova, A., Marchenkov, S., Sidorcheva, N., Polyakov, S., Korzun, D.: World Around Me Client for Windows Phone Devices. In: Proc. 11th FRUCT Conf., pp. 206–208 (2012)

---

[1] https://play.google.com/store/apps/details?id=ru.nw.spiiras.tais

8. Raptis, D., Tselios, N., Avouris, N.: Context-based design of mobile applications for museums: a survey of existing practices. In: Proceedings of the 7th International Conference on Human Computer Interaction with Mobile Devices & Services, Salzburg, Austria (September 2005)

9. Jeon, P.B.: Context Aware Intelligent Mobile Platform for Local Service Utilization. In: The 2012 IEEE/WIC/ACM International Conference on Intelligent Agent Technology, Macau, pp. 635–638 (December 2012)

10. Luyten, K., Coninx, K.: ImogI: Take Control over a Context Aware ElectronicMobile Guide for Museums. HCI in Mobile Guides. University of Strathclyde, Glasgow (September 2004)

11. New travel guide Triposo brings algorithms to apps, "The Independent (September 2011), http://www.independent.co.uk/travel/news-and-advice/new-travel-guide-triposo-brings-algorithms-to-apps-2353352.html (last access date May 08, 2013)

12. Triposo, http://www.triposo.com (last access date May 08, 2013)

13. Tripadvisor, http://www.tripadvisor.ru (last access date May 08, 2013)

14. Smart Travelling web page, http://www.smart-travelling.net/en/ (last access date May 08, 2013)

15. García, O., Alonso, R.S., Guevara, F., Sancho, D., Sánchez, M., Bajo, J.: ARTIZT: Applying Ambient Intelligence to a Museum Guide Scenario. In: Novais, P., Preuveneers, D., Corchado, J.M. (eds.) Ambient Intelligence - Software and Applications. AISC, vol. 92, pp. 173–180. Springer, Heidelberg (2011)

16. ZigBee Alliance, http://www.zigbee.org/ (last access date May 08, 2013)

17. Smirnov, A., Kashevnik, A., Ponomarev, A., Shilov, N., Shchekotov, M., Teslya, N.: Recommendation System for Tourist Attraction Information Service. In: Proc. FRUCT Conf., pp. 140–147 (November 2013)

18. Smirnov, A., Shilov, N., Kashevnik, A., Teslya, N., Laizane, S.: Smart Space-based Ridesharing Service in e-Tourism Application for Karelia Region Accessibility. Ontology-based Approach and Implementation. In: Proc. 8th Int. Joint Conference on Software Technologies, Reykjavik, Iceland, July 29-31, pp. 591–598 (2013)

19. Smirnov, A., Kashevnik, A., Balandin, S.I., Laizane, S.: Intelligent Mobile Tourist Guide: Context-Based Approach and Implementation. In: Balandin, S., Andreev, S., Koucheryavy, Y. (eds.) NEW2AN/ruSMART 2013. LNCS, vol. 8121, pp. 94–106. Springer, Heidelberg (2013)

20. Balabanovic, M., Shoham, Y.: Fab: Content-based, Collaborative Recommendation. Communications of the ACM 40(3), 66–72 (1997)

21. Adomavicius, G., Mobasher, B., Ricci, F., Tuzhilin, A.: Context-aware recommender systems. AI Magazine 32(3), 67–80 (2011)

22. Adomavicius, G., Sankaranarayanan, R., Sen, S., Tuzhilin, A.: Incorporating contextual information in recommender systems using a multidimensional approach. ACM Trans. Inf. Syst. 23(1), 103–145 (2005)

23. Honkola, J., Laine, H., Brown, R., Tyrkko, O.: Smart-M3 Information Sharing Platform. In: Proc. IEEE Symp. Computers and Communications (ISCC 2010), pp. 1041–1046. IEEE Comp. Soc. (June 2010)

24. Morandi, F., Roffia, L., D'Elia, A., Vergari, F., Cinotti, T.: RedSib: a Smart-M3 semantic information broker implementation. In: Proc. 12th Conf. of Open Innovations Association FRUCT and Seminar on e-Tourism, SUAI, pp. 86–98 (2012)

25. Smart-M3 at Sourceforge, http://sourceforge.net/projects/smart-m3 (last access date May 08, 2013)

26. Smart-M3 Java KPI library at Sourceforge, http://sourceforge.net/projects/smartm3-javakpi/ (last access date May 08, 2013)

27. Android Java Development Kit, http://developer.android.com/sdk/index.html (last access date May 08, 2013)

Smart Space-Based in-Vehicle Application for e-Tourism

Technological Framework and Implementation for Ford SYNC

Alexander Smirnov[1,2], Alexey Kashevnik[1], Nikolay Shilov[1], and Andrew Ponomarev[1]

[1] SPIIRAS, St. Petersburg, Russia
{smir,alexey,nick,ponomarev}@iias.spb.su
[2] ITMO University, St. Petersburg, Russia

Abstract. Recently, the development of in-vehicle services has become more and more popular. Many people spend a lot of time in vehicles and producers propose new in-vehicles systems to provide information for drivers and entertain them. Drivers travel from one location to another and usually would like to know information about places of interests around. Integration of e-tourism services with vehicle multimedia system opens new possibilities for delivering useful information about places of interests around to the driver. The paper presents a technological framework and implementation of mobile tourist guide and in-vehicle system integration (on the example of Ford SYNC system). The system consists of several services that find, extract and process potentially useful for the tourist information about places of interests around and provide it through user-friendly vehicle interface. The smart space technology is used for providing interaction possibilities and interoperability of these service.

Keywords: smart space, e-tourism, in-vehicle systems, mobile devices.

1 Introduction

Recently, the development in-vehicle systems become more and more popular. They have transformed from simple audio players to complex solutions that allow to communicate with popular smartphones, share information from different vehicle sensors and provide possibilities to deliver information through in-vehicle screen or stereo system (like Ford SYNC[1], GM OnStar MyLink[TM2], Chrysler UConnect[®3], Honda HomeLink[4], Kia UVO[5], Hyundai Blue Link[6], MINI Connected[7], Totyota Entune[8], and BMW ConnectedDrive[9] systems).

---

[1] http://www.ford.com/technology/sync/
[2] https://www.onstar.com
[3] http://www.chryslergroupllc.com/innovation/pages/uconnect.aspx
[4] http://www.homelink.com/
[5] https://www.myuvo.com/
[6] https://www.hyundaiusa.com/technology/bluelink/
[7] http://www.mini.com/connectivity/
[8] http://www.toyota.com/entune/
[9] http://www.bmw.com/com/en/insights/technology/connecteddrive/
   2013/index.html

S. Balandin et al. (Eds.): NEW2AN/ruSMART 2014, LNCS 8638, pp. 52–61, 2014.

Integration of different mobile applications with in-vehicle system is an interesting and promising task. Presented in the paper e-tourism application is a mobile tourist guide that allows to acquire interesting for the driver information related to his/her location from different internet sources (like Wikipedia, Wikivoyage, Wikitravel, Panoramio, Flickr). Integration with in-vehicle system allows to provide this information in a convenient for the driver form.

In the paper, the authors propose an application that allows the driver getting the following information using personal mobile device connected to the in-vehicle system:

- generate list of interesting places around, based on the driver, vehicle, and region contexts; driver context includes information about the driver (e.g., his/her preferences, company); vehicle context contains such information as vehicle location, hardware and etc.; region context includes such information as weather, traffic jams, etc.
- provide information about attractions from different sources and recommend the tourist the best for him/her attraction descriptions;
- allow the driver to rate attractions, their images, and descriptions.

Due to the mobile device restrictions (limited computational capacities and power consumption) it is not reasonable to perform a complex computations in a mobile device. In this case, an infrastructure is needed that allows different devices to interact with each other for distribution of computations during solving their tasks.

The smart spaces technology [1, 2, 3, 4, 5] aims at the seamless integration of different devices by developing ubiquitous computing environments, where different services can share information with each other, perform computations, and interact with each other for joint task solving.

The open source Smart-M3 platform [6, 7] has been used as a core for in-vehicle e-tourism application development. Usage of this platform makes it possible to significantly simplify further development of the system, include new information sources and services, and to make the system highly scalable. The key idea of this platform is that the formed smart space is device-, domain-, and vendor-independent. Smart-M3 assumes that devices and software entities can publish their embedded information for other devices and software entities through simple, shared information brokers. The Smart-M3 platform consists of two main parts: information agents and kernel. The kernel consists of two elements: Semantic Information Broker (SIB) and information storage. Information agents are software entities installed on mobile devices of the smart space users and other devices that hosted smart space services. These agents interact with SIB through the Smart Space Access Protocol. SIB is the access point for receiving the information to be stored, or retrieving the stored information. All this information is stored in the information storage as a graph that conforms with the rules of the Resource Description Framework [8]. In accordance with these rules all information is described by triples "Subject - Predicate - Object".

The rest of the paper is structured as follows. Section 2 presents related work in the considered area. Section 3 considers the technological framework for smart space-based in-vehicle application for e-tourism. Section 4 describes the system service interactions in the smart space. The implementation is presented in section 5. Main results and findings are summarized in the conclusion.

2   Related Work

There are exist some research efforts in the area of intelligent driver support integrated with in-vehicle information systems. The Intelligent Refueling Advisor is one of such applications [9]. The Refueling advisor integrates the expected routes over time, estimated fuel consumption, current fuel level, and current and forecasted gas prices at individual gas stations along the given route, and generates corresponding recommendations (when, where, and how much fuel to get). The system can provide recommendations both upon request, as well as proactively.

The approach proposed in [10] extends such systems as ERA-GLONASS and eCall via service network composition enabling not only transmitting additional information but also information fusion for defining required emergency means as well as planning for a whole emergency response operation. The main idea of the approach is to model the cyber physical human system components by sets of services representing them. The services are provided with the capability of self-contextualisation to autonomously adapt their behaviors to the context of the car-driver system.

Some research in the area of theoretical aspects of using proactive recommending systems in BMW cars is carried on in the Technical University of Munich [11]. However, this work is aimed at development of proactive support systems with underlying algorithms, and does not deal with issues of interoperability, information integration and smart spaces.

For e-Tourism applications there are a lot of related works have been considered in previous authors publication related to the considered topic [12], the authors provided the description of related work in the area of tourist guides [13–16]. There are several state-of the-art papers that evaluate different mobile tourist guides. The following applications accessible for downloading have been considered: GoTour [17], Tourist Attractions [18], Foursquare [19], World Explorer [20], Smart Museum [21], My Tourist Guide [22], World Around Me [23].

3   Technological Framework

A smart space enables devices to share information independently of their locations. It is used for interaction of mobile devices with each other and implemented via an open source software platform (Smart-M3) [7]. The platform aims at providing a Semantic Web information sharing infrastructure between software entities and devices. In this platform, the ontology is represented via RDF.

Every e-tourism application user installs a special client to his/her mobile device. The client consists of smart space module, vehicle module, and behaviour model (see Fig. 1). Behaviour model determines the client functionality in terms of ontology, whereas smart space and vehicle modules responsible for communication with smart space services and in-vehicle system correspondingly. Vehicle module communicate with the in-vehicle system for getting context information (such as location, speed) from vehicle sensors and visualise location-based service results in convinient for the driver form (using vehicle screen or text to speech function of in-vehicle system). The smart space module shares this context information with the smart space for other services which process it and share results back.

In-vehicle system provides information about vehicle state (sensors), allow to display on the in-vehicle system screen location-based system results, and provides possibilities to convers text information to speech for providing it to the driver in convinient form.

The client application is located in the user mobile device while other services use powerful computer systems. So, the main engineering task in the in-vehicle e-tourism application development is implementation of resource-intensive operations using powerful computers as much as possible. For these purposes smart space services are used:

- attraction information service that implements retrieving and caching the information about attractions;
- recommendation service that evaluates attraction/image/description scores based on ratings that have been saved to service database earlier [24];
- region context service that acquires and provides information about current situation in the considered region (e.g., weather, traffic jams, closed attractions);

The attraction information service is responsible for the following tasks. The first one is extraction attractions titles around the driver from external sources and shares it in smart space. The second task is caching acquired information for quick access by drivers in the same location without additional requests to external sources at nearest future (live time of cached attraction is set by administrator). The third task is extraction of default images for attractions that are stored in the service database. Default images are defined by the recommendation service as the best images for an attraction based on drivers ratings. The fourth task is extraction of attraction details (lists of images and text descriptions) from external sources and sharing them with the smart space. And the last task is to set and refresh default images in the internal database for the attractions based on the information available in the smart space from the recommendation service.

Communication between in-vehicle system and client application is implemented via Bluetooth network that allows to exchange information with the speed up to 2 Mb/sec. Communication between client application and smart space services is implemented via cellular network. The quality and speed of this kind of information transfer depends on the vehicle location.

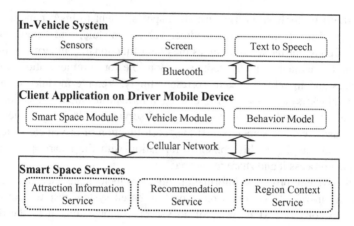

Fig. 1. Technological framework of in-vehicle e-tourism application

The recommendation service is responsible for ranking attractions, images, and descriptions for providing the driver the best attractions to see and the best images and description of chosen attraction for acquaintance. It stores the following mappings in the internal database: (internal attraction identifier, driver & vehicle context, region context -> attraction rating), (image URL -> image rating), (description URL, description text -> description rating). The internal database also keeps similarities between users, which are calculated as background process. The attraction's degree of interestingness is estimated by the following way. The service takes user ratings associated with each attraction by all users as an input. According to the conventional classification (e.g. [25]), it performs user-based collaborative filtering. One of the promising directions to improve the predictive quality of recommendation systems in general (and collaborative filtering systems among them) is context-awareness [26]. The context describes conditions in which the user rates an object or asks for recommendations.

The region context service acquires information about vehicle location area from region specific services as soon as the client application shares the vehicle context with the smart space. The ontology slice that describes a service at a certain point of time is its abstract context (Fig. 2). It is formed automatically (or reused) applying ontology slicing and merging techniques [27]. The purpose of the abstract context formation is to collect and integrate knowledge relevant to the current task (situation). The information sources defined in the abstract context provide the information that instantiates the context and forms the operational context.

A concrete description of the current situation is formed, and the problem at hand is augmented with additional data. On the knowledge representation level, the operational context is a set of RDF triples to be added to the smart space by an appropriate service. Therefore, other services can discover these RDF triples and understand the current problem.

In Fig. 2, service *i* queries up-to-date information from the operational context through smart space in accordance with the task specified in the service's ontology. Services *j* and *n* are involved in solving a particular task. They form the operational context related to this task and based on the abstract contexts of the services. This operational context is described by the smart space service ontology, which also corresponds to the current task and integrates abstract contexts of the involved services.

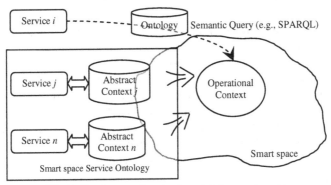

Fig. 2. Ontology and context models in smart space

4   Application Services Interaction

The smart space-based in-vehicle application service interaction sequence for getting a list of recommended attractions for the vehicle driver is shown in Fig. 3. Every service publishes and subscribes for triples. Subscribe for triples determine notifications when a required information for the service appears in the smart space.

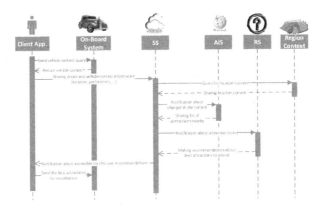

Fig. 3. Application services interaction

Smart space services contributing to the creation of the operational context and involved in the scenario include: attraction information system (AIS) providing textual and multimedia information about the attractions, attraction recommendation service (RS) analysing appropriateness of attractions and points of interests to user preferences, region context service shares with the smart space information about current situation in the location region (e.g., weather, traffic jams). Any other services can also be connected to the smart space to enrich the decision support possibilities.

The smart space module of client application queries context information from in-vehicle system append it by the user context and shares it with the smart space. The context service subscribes to changes in a driver & vehicle context and generates the region-specific context, which is published with the smart space. The attraction information service also subscribes for changes in a driver and vehicle context for searching the list of attractions around, when these changes appears in the smart space. The recommendation service subscribes for the list of attractions shared with the smart space by the attraction information service.

The recommendation service also queries an information about the driver, car and region contexts, evaluates the attraction scores, and shares the ordered list of attractions with the smart space [24]. Smart space module of the client application receives notification for its subscription and presents the results to the driver using graphical interface (see case study section). Vehicle module of the client application sends to the in-vehicle system the best attraction for visualization.

5   Implementation

Android OS is used for implementation of the in-vehicle e-tourism application. At the moment, this mobile OS is one of the most popular operating systems for mobile devices today [28]. For the interaction of client application with in-vehicle system the AppLink system for Ford SYNC is used. The AppLink provides the current vehicle location and other car information to the client application.

The main application screen of the client application is shown in Fig. 4 (left). In this screen the driver can see images extracted from accessible internet sources around, clickable map with his/her location, context situation, and the best attractions around ranked by the recommendation service. The same attractions driver can see in the in-vehicle system display (Fig. 5). Driver can browse information about the best attractions around presented by the application in the main screen and press button "More" to see more attractions (see Fig. 4 middle).

In Fig. 4 right, images and description for attraction are shown. The driver can leaf over images to browse them. Images are extracted from different Internet sources ranked by the recommendation service according to their estimations. The developed context service searches for weather conditions in a user location at the moment. It uses World Weather Online [29] resource for getting this information. The resource API allows to implement not more than 500 requests per hour for free and has premium API for an increased number of requests. Weather information is updated every 3-4 hours.

For development vehicle module the AppLink emulator[10] v2.3 and SyncProxy SDK[11] v.1.6.1 for Android have been used. Fig. 5 shows example of attraction visualisation on in-vehicle system (Ford SYNC). Ford SYNC allows to show only two lines of text at the moment in the screen. In this case it is proposed to show the driver only three main attractions (header and first one in the first screen and second & third

---

[10] https://developer.ford.com/content/restricted/ALE_Win7_2.3_20140414.zip

[11] https://developer.ford.com/content/restricted/SyncProxy_1.6.1_Android.zip

one in the second screen). For the attractions visualisation the following interfaces have been used: show information in the car screen (function *proxy.show()*) and text to speech function for providing audio information of recommended attractions (function *proxy.speak()*). Driver can choose interesting attarction and lissen information about it (e.g. information about St. Andrew's Cathedral is presented in Fig. 4 right).

Fig. 4. Main application screen and settings page

Implemented experiments show that speed of developed application for in-vehicle system the same as usual Android-based application (shown in Fig. 3). Application for in-vehicle system shows the driver only the main interesting infromation due to the AppLink system restrictions. Using of in-vehicle system through Bluetooth network increases the mobile device power consumption but it does not matter because car drivers usually have mobile device chargers inside the cars.

Fig. 5. Example of in-vehicle e-tourism application result

6   Conclusion

The paper proposes a smart space-based technological framework for in-vehicle e-tourism application and its implementation for Ford SYNC. The application consists of several services that extract and process useful for the driver information about places of interests around and provide it in a user-friendly form using in-vehicle multimedia system. These services that allow to implement resource-intensive operations on powerful computers and implement communications using the smart space technology. Implementation shows that results generated by the developed location-based information system can be provided to the driver in a convenient form.

Acknowledgements. The presented results are part of the research carried out within the project funded by grants # 13-07-00336, 13-07-12095, 13-07-00271, 12-07-00298, 14-07-00345, 13-01-00286 of the Russian Foundation for Basic Research. This work was partially financially supported by Government of Russian Federation, Grant 074-U01.

References

1. Cook, D.J., Das, S.K.: How smart are our environments? an updated look at the state of the art. Pervasive and Mobile Computing 3(2), 53–73 (2007)
2. Balandin, S., Waris, H.: Key properties in the development of smart spaces. In: Stephanidis, C. (ed.) UAHCI 2009, Part II. LNCS, vol. 5615, pp. 3–12. Springer, Heidelberg (2009)
3. Kiljander, J., Ylisaukko-oja, A., Takalo-Mattila, J., Eteläperä, M., Soininen, J.-P.: Enabling semantic technology empowered smart spaces. Computer Networks and Communications 2012 (2012)
4. Ovaska, E., Cinotti, T.S., Toninelli, A.: The design principles and practices of interoperable smart spaces. In: Advanced Design Approaches to Emerging Software Systems: Principles, Methodologies, and Tools, pp. 18–47. IGI Global (2012)
5. Korzun, D.G., Balandin, S.I., Gurtov, A.V.: Deployment of Smart Spaces in Internet of Things: Overview of the Design Challenges. In: Balandin, S., Andreev, S., Koucheryavy, Y. (eds.) NEW2AN/ruSMART 2013. LNCS, vol. 8121, pp. 48–59. Springer, Heidelberg (2013)
6. Smart-M3 at Sourceforge, http://sourceforge.net/projects/smart-m3
7. Honkola, J., Laine, H., Brown, R., Tyrkko, O.: Smart-M3 Information Sharing Platform. In: Proc. ISCC 2010, pp. 1041–1046. IEEE Comp. Soc. (2010)
8. Resource Description Framework (RDF). W3C standard, http://www.w3.org/RDF/
9. Klampfl, E., Gusikhin, O., Theisen, K., Liu, Y., Giuli, T.J.: Intelligent Refueling Advisory System. In: Proceedings of 2nd Workshop on Intelligent Vehicle Control Systems, pp. 60–72 (2008)
10. Smirnov, A., Kashevnik, A., Shilov, N., Makklya, A., Gusikhin, O.: Context-Aware Service Composition in Cyber Physical Human System for Transportation Safety. In: Proceedings of the 13th International Conference on ITS Telecommunications (ITST 2013), Tampere, Finland, pp. 139–144 (2013)
11. Bader, R., Neufeld, E., Worndl, W., Prinz, V.: Context-aware POI recommendations in an automotive scenario using multi-criteria decision making methods. In: Workshop on Context-awareness in Retrieval and Recommendation, Palo Alto, CA, pp. 23–30 (2011)
12. Smirnov, A., Kashevnik, A., Balandin, S.I., Laizane, S.: Intelligent Mobile Tourist Guide: Context-Based Approach and Implementation. In: Balandin, S., Andreev, S., Koucheryavy, Y. (eds.) NEW2AN/ruSMART 2013. LNCS, vol. 8121, pp. 94–106. Springer, Heidelberg (2013)

13. Kenteris, M., Gavalas, D., Economou, D.: Evaluation of Mobile Tourist Guides, In: Lytras, M.D., Carroll, J.M., Damiani, E., Tennyson, R.D., Avison, D., Vossen, G., De Pablos, P.O. (eds.) WSKS 2008. CCIS, vol. 19, pp. 603–610. Springer, Heidelberg (2008)
14. Baus, J., Cheverst, K., Kray, C.: A Survey of Map-based Mobile Guides. In: Map-based Mobile Services, pp. 193-209 (2005)
15. Schwinger, W., Grun, C., Proll, B., Retschitzegger, W., Schauerhuber, A.: Context-Awareness in Mobile Tourist Guides. In: Handbook of Research on Mobile Multimedia, 2nd edn., p. 19 (2009)
16. van Setten, M., Pokraev, S., Koolwaaij, J.: Context-Aware Recommendations in the Mobile Tourist Application COMPASS. In: De Bra, P.M.E., Nejdl, W. (eds.) AH 2004. LNCS, vol. 3137, pp. 235–244. Springer, Heidelberg (2004)
17. Al-Rayes, K., Sevkli, A., Al-Moaiqel, H., Al-Ajlan, H., Al-Salem, K., Al- Fantoukh, N.: A Mobile Tourist Guide for Trip Planning. IEEE Multidisciplinary Engineering Education Magazine 6(4), 1–6 (2011)
18. Google Play Market, Tourist Attractions application, `https://play.google.com/store/apps/details?id=com.uknowapps.android.attractions&hl=ru`
19. Foursquare application, `http://foursquare.com`
20. Apple iTunes, World Explorer, `https://itunes.apple.com/ru/app/world-explorer-russkij-putevoditel/id381581095?mt=8`
21. Smart Museum, `http://smartmuseum.ru/`
22. Google Play Market, My Tourist Guide application, `https://play.google.com/store/apps/details?id=com.mytourguideanmytou&hl=ru`
23. Vdovenko, A., Lukovnikova, A., Marchenkov, S., Sidorcheva, N., Polyakov, S., Korzun, D.: World Around Me Client for Windows Phone Devices. In: Proc. 11th FRUCT Conf., pp. 206–208 (2012)
24. Smirnov, A., Kashevnik, A., Ponomarev, A., Shilov, N., Shchekotov, M., Teslya, N.: Recommendation System for Tourist Attraction Information Service. In: Proc. FRUCT Conf., pp. 140–147 (2013)
25. Balabanovic, M., Shoham, Y.: Fab: Content-based, Collaborative Recommendation. Communications of the ACM 40(3), 66–72 (1997)
26. Adomavicius, G., Mobasher, B., Ricci, F., Tuzhilin, A.: Context-aware recommender systems. AI Magazine 32(3), 67–80 (2011)
27. Smirnov, A., Pashkin, M., Chilov, N., Levashova, T.: Constraint-driven methodology for context-based decision support. Design, Building and Evaluation of Intelligent DMSS 14(3), 279–301 (2005)
28. International Data Corporation (IDC), Press Release, Android and iOS Combine for 91.1% of the Worldwide Smartphone OS Market in 4Q12 and 87.6% for the Year, `http://www.idc.com/getdoc.jsp?containerId=prUS23946013`
29. World Weather Online, `http://worldweatheronline.com/`

Geo-Coding and Smart Space Platforms Integration Agent Performance Testing and Analysis

Kirill Yudenok

Saint-Petersburg State Electrotechnical University "LETI",
Saint-Petersburg, Russia
kirill.yudenok@gmail.com

Abstract. Internet of Things, Smart Spaces and Geo-coding technologies are fastest growing directions in modern mobile market and urban environments [1, 2]. This is due to the advent of various services that using common technologies, as well as to develop common requirements and architectures for using geo-contextual services in semantic data processing. Location is a mandatory requirement for the Internet of Things and Smart Spaces directions products, because geo-context is a one of the factor to determine the location of subjects in various environments. As a result, it was decided to integrate geo-coding and smart spaces platforms, for the possibility of using geo-context in the semantic space. As an implementation used two open source software platforms – Geo2Tag and Smart-M3. The article discusses an integration agent performance testing and its analysis, provided recommendations for integration mechanisms optimization.

Keywords: Geo-coding, Smart Spaces, Internet of Things, Smart-M3, Geo2Tag, LBS.

1   Introduction

Geo-coding[1] and Smart Space [3] directions are gaining momentum every day. The appearance of various services that adapt to constantly changing conditions, services, that allowing to markup various objects of the world (virtual or real objects). This is the actual day services, that enjoyed worldwide.

Geo-coding systems markup real or virtual objects by adding the geographical coordinates and time. In turn, smart space provides access to distributed semantic information and communication field for software services, which is being run on various type of devices (personal, autonomous computers, robots etc.). These two directions are mutually different from each other, but together complement each other and adds a new features to the overall system such as, pro-activeness, context awareness, machine-to-machine interactions, platforms possibilities [4]. By combining these platforms will be reached the ability to define and discover objects in a described semantic space. Geo-coding capabilities will not only markup the existing objects, but

---

[1] Geo-coding – http://en.wikipedia.org/wiki/Geocoding

S. Balandin et al. (Eds.): NEW2AN/ruSMART 2014, LNCS 8638, pp. 62–69, 2014.

also expand the space with new data, handle the situations with the new objects, as well as track their location in space and time.

Geo-context is not replaceable component for the Internet of Things and Smart Spaces directions. In our case, access to the geo-context is obtained by the Geo2Tag [5] platform, this is a geographical context marked up on a virtual world map. Access to the geo-context may also be obtained by using positioning technologies or special sensors.

Geo-context use cases in the Internet of Things and Smart Spaces directions, for example, assistance to find a parking space, monitoring energy in the city, assistance in finding electric car charger stations, notification of bus arrival, assistance after a car crash, notification of car traffic jam, location based dating, location based marketing etc [2, 3].

In this article we will talk about testing and performance analysis of the basic mechanisms for the geo-coding and smart space platforms integration agent. The paper is structured as follows: Chapter 2 briefly discusses integration platforms and their main features, in Chapter 3 describes the performance testing methodology of the platforms integration agent, Chapter 4 discusses the performance testing details, in Chapter 5 shows the performance testing results, its brief analysis and provides the optimization recommendations for the basic integration mechanisms, Chapter 6 summarizes the work.

2   Integration Platforms – Geo2Tag and Smart-M3

To support the geo-coding possibilities in the smart space as a geo-coding platform serves a Geo2Tag[2] system, which implement the basic functionality for working with geo-data. Its main features – users and data channels management, basic operations with geo-data (load tags, write tags etc.), multiple geo-data filtering mechanisms (spatial and temporal filtration). Geo2Tag platform includes a full server that handles and stores all geo-data. Platform is implemented using REST API technology, all logic is written on high-level programming language – C++/Qt, Java. Granted API allow to develop services for a variety desktop and mobile platforms (Windows, Linux, Android, J2ME, Web).

Interaction with the smart space provides Smart-M3[3] platform [6, 7, 8]. Its main task is to provide the infrastructure for the exchange of semantic information between different entities (software or hardware). The platform provides a distributed data storage in a special semantic information broker (SIB) and it is processing by means of developed agents – knowledge processors (KP). Programming interfaces allow to develop KP in the following languages – C, C++/Qt, C#, Python, Java, PHP, Javascript.

Common requirements, integration agent architecture and its detailed description has been presented in [9]. The main functional requirements are mechanisms for

---

[2] Geo2Tag – http://geo2tag.org
[3] Smart-M3 – http://en.wikipedia.org/wiki/Smart-M3

converting data from one platform format to another, spatial and temporal filtration. The main non-functional requirements are high-performance solution for handling large amounts of data (cloud based massive offline processing and local context indexing/caching) and compatibility with Geo2Tag and Smart-M3 platforms interfaces (i.e SSAP or REST).

As a result of the platforms integration have been developed a special agent which main tasks are:

1. providing an interface to the semantic and geo-data;
2. distributed storage for semantic and geographical information;
3. interface for the association of semantic objects with geo-data;
4. spatial and temporal filtration (Smart-M3 and Geo2Tag).

Geo-space conceptual model with additional knowledge processors (Cloud-backend, Big Data, context management, sensors) is shown at Fig. 1.

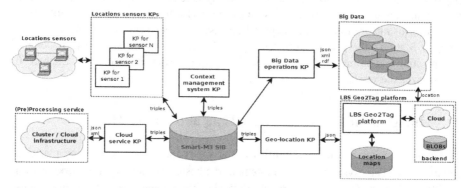

Fig. 1. Geo-space conceptual model

3   Performance Testing Methodology of the Platforms Integration Agent

Geo2Tag and Smart-M3 platforms integration agent was tested on a dedicated virtual machine with installed Smart-M3 platform and below listed characteristics, access to the Geo2Tag platform is performed over the Internet using a HTTP/REST protocol:

- CPU – Intel i7, 3.4 Mhz, 4 cores;
- RAM – 8 Gb;
- OS – Ubuntu 14.04 LTS;
- Geo2Tag – 0.31 version (Qt API 4.8);
- Smart-M3 – 0.9.01 (redland-1.0.16-unibo (Virtuoso[4]), redsibd-0.9.01_time, sib-tcp 0.81, Libwhiteboard Qt API[5]).

---

[4] Virtuoso – http://virtuoso.openlinksw.com/
[5] Libwhiteboard Qt API – http://sourceforge.net/projects/smart-m3/files/Smart-M3_v0.9.5-beta/

The testing object is a geo-coding and smart space platforms integration agent and its basic data processing mechanisms. Testing was conducted of two types – functional testing of an integration agent mechanisms and integration agent load and stress performance testing.

For functional testing were developed unit tests that verify the basic mechanisms of the platforms integration such as platforms data conversion mechanisms, data filtering techniques and the basic mechanisms of the geo-coding and smart space platforms, such as login and load data from the Geo2Tag system, query data from the Smart-M3 platform and others.

For performance testing have been prepared tests (scripts) that test system under a certain load. As a system load acts the different scenarios of the system, as a permanent data conversion from one format to another and vice versa, repetitive data filtering methods, query or insert triples. Stress testing was performed for the main agent repetitive mechanisms - conversion, filtering and data querying.

The main integration agent performance metrics are:

1. query (operation) execution time;
2. the number of operations performed in 1 second;
3. the amount of consumed CPU and memory.

4 Platforms Integration Agent Performance Testing

Performance testing was carried out using a specially designed tool for the Geo2Tag platform called *Profiler* [10], whose tasks are:

1. definition and implementation of load tests for any program operation;
2. creation a separate thread for each test with the counting system performance metrics:

   ○ query (operation) execution time;
   ○ the number of operations performed in one second.

Listing 1 shows a performance test script example of the filtration tags throw the Smart-M3 platform interface:

```
#!/bin/bash
result_dir="./results_r_`date +'%d_%om_%Y_%H_%M_%S'`/";
if [[ "$#" == "2" ]]
then
  steps_count=$1;
  read_requests_count=$2;
else
  steps_count=100;       # number of iterations
  read_requests_count=500;# number of queries (for Geo2Tag operations)
fi
mkdir "$result_dir"  # results directory
```

```
for (( i=0; i<=steps_count; i++ )) ;
do
echo "$i iteration"
cool_num=`printf "%08d" $i`;
./profiler $read_requests_count g2tFilter 2>"$result_dir/$cool_num"
done
```

As a result for each integration mechanism was create a separate test, test selection is performed by a passing name of the operation as a *Profiler* testing tool parameter. For each test was created a separate script that deals with an environment setup and required parameters for the testing tool.

Each test evaluates two system performance metrics – the execution time of the operation (query) and the number of operations performed in one second. Operation execution time metric measures the operation runtime since the beginning of the function execution using the Qt API *msecsTo()* function, the number of operations performed in one second metric counts the number of operations performed during one second of time, based on its runtime according to the formula:

$$number\_of\_operations\_in\_1\_second = 1000 \, / \, operation\_time\_in\_milliseconds \quad (1)$$

Each test can be performed a number of times, depending on the *steps_count* parameter of the script passed to the test main loop. In our case, tests are performed hundred times to collect the necessary statistics. After the test, all characteristics recorded to a file with the number of iteration.

As a result of testing were obtained the necessary statistical data that allowing to say how much time was spent on the operation and the approximate number of operations performed in a one second. In order to ensure that the statistics is true, for each operation statistical data calculated the necessary characteristics – the mathematical expectation, variance and standard deviation, which allow to understand the spread of statistical data and their deviation. Also, all the statistical data verified by the *"three sigma"*[6] rule, confirming that all the random variables are normally distributed.

5   Platforms Integration Agent Performance Testing Analysis

The main platforms integration agent performance tests are:

1. loading geo-data from the Geo2Tag platform (basic filter by radius);
2. triples filtering through the Geo2Tag platform;
3. geo-data filtering through the Smart-M3 platform;
4. convertion tags to triples and vice versa;
5. insert data to the Smart-M3 platform;
6. query data from the Smart-M3 platform.

---

[6] Three sigma rule – http://en.wikipedia.org/wiki/Standard_deviation

Table 1. Integration agent performance testing summary results

Test case	Mean value (ms)	Standard deviation (ms)
Load tags from the Geo2Tag platform by radius	538	79.59
Triples filtering through the Geo2Tag platform	133	31.17
Geo-data filtering through the Smart-M3 platform	1621	424.97
Conversion 1000 triples to the tags	31.01	32.72
Conversion 1000 tags to the triples	27.24	15.73
Insert triples to the Smart-M3 platform	3015.4	173.16
Query triples from the Smart-M3 platform	1302.19	90.16

Summary results of the obtained agent integration performance testing characteristics are presented in Table 1.

From the the summary table of obtained characteristics seen that some integration mechanisms takes a long processing time intervals, among them:

1. geo-data filtering through the Smart-M3 platform – 1-2 seconds;
2. insert triples to the Smart-M3 platform – 3-4 seconds;
3. query triples from the Smart-M3 platform – 1-1.5 seconds.

As a result, we analyzed the execution time of the main data integration mechanisms. As an function calls analysis used tool – *callgrind[7]*, which is part of the profiler – *valgrind[8]* tool and *kcachegrind[9]* tool.

As a result, it was revealed two types of major problems:

1. Multithreaded data processing in the Smart-M3 platform components.
2. Processing and parsing of obtained results for the basic Smart-M3 operations – insert, update, query (Libwhiteboard Qt API).

In the first case the profiling showed that the Smart-M3 insert and query tests incorrectly handle threads. As a result, the platform integration agent and Smart-M3 platform were subjected to analysis using the Intel Threads Profile, which is a part of Intel Inspector XE 2013[10] tool. Where it was found that the *redsib daemon* and *sib-tcp* Smart-M3 platform components have errors while working in multi-threaded mode when performing basic operations. Smart-M3 *whiteboard daemon* component and platforms integration agent (GCSS) are single-threaded.

The second type of the problem associated with the processing of the basic Smart-M3 operations using libwhiteboard Qt API. At first, each Smart-M3 operation (insert, update, query) generates string results output which is converted into XML-tree

---

[7] Callgrind – http://valgrind.org/docs/manual/cl-manual.html
[8] Valgrind – http://valgrind.org/
[9] Kcachegrind – http://kcachegrind.sourceforge.net
[10] Intel Inspector XE 2013 toolset – https://software.intel.com/en-us/intel-inspector-xe

results of the operation. Thus, the XML-tree composition for a large number of triples and their periodical query lead to loss of performance while performing basic operations. The solution to this problem is to use special data structures or switch to a binary data transfer protocol, such as KSP [11].

Also have been fixed error in the tags–triples conversion mechanism while removing duplicates triples, operation time reached ~ 50 ms. It should be noted that the usage of the Smart-M3 platform together with Virtuoso significantly increase performance in the basic triples processing operations. Performance boost when performing basic operations reached approximately 50%, but the libwhiteboard Qt API insert and query operations are fairly slower than in Python API.

According to the above analysis we may suggest the following recommendations for the optimization integration mechanisms:

1. multithreading errors correction for the Smart-M3 platform components – *redsibd*, *sib-tcp*;
2. replacing the current Smart-M3 platform SSAP protocol to the binary protocol, for example, KSP;
3. use Smart-M3 platform with Virtuoso (increase productivity of Smart-M3 platform operations ~ 50%);
4. use Smart-M3 Python API or optimization of the basic platform operations (insert, update, query) for the Libwhiteboard Qt API;
5. use SparQL[11] queries instead of the usual query (Libwhiteboard Qt API does not support SparQL queries);

6   Conclusion

This article presented the performance testing results and its analysis of the basic geo-coding and smart spaces platforms integration mechanisms. The testing revealed that some of the agent integration mechanisms are need to be improved, and some one depends on the platform, protocol and API. Profiling showed that the integration agent has the following problems – Smart-M3 platform miltithreading problem, unsuitable protocol for data exchange, outdated Qt API. The first two problems are mandatory, because they determine the overall platform performance. Recommendations for the mechanisms optimization will help to increase the agent performance.

Still open questions for further development – performance of the whole system (multithreading errors correction, binary data exchange protocol for the Smart-M3 platform, optimization of basic operations in Qt API), cloud computing.

References

[1] Vermesan, O., Friess, P.: Internet of Things: Converging Technologies for Smart Environments and Integrated Ecosystems, Aalborg. River publishers series in communications (2013) ISBN: 978-87-92982-73-5

---

[11] SparQL – http://en.wikipedia.org/wiki/SPARQL

[2] van der Zee, E., Scholten, H.: Application of geographical concepts and spatial technology to the Internet of Things. Vrije University, Faculty of Economics and Business Administration, Amsterdam, Research Memorandum 33 (2013)

[3] D2.2 Requirements, Specifications and Localization and Context-acquisition for IoT Context-Aware Networks. uBiquitous, secUre inTernet-of-things with Location and contExt-awaReness (BUTLER), Jacobs University Bremen gGmbH (JUB), Project number 287901 (October 2012)

[4] Balandin, S., Waris, H.: Key properties in the development of smart spaces. In: Stephanidis, C. (ed.) Universal Access in HCI 2009, Part II. LNCS, vol. 5615, pp. 3–12. Springer, Heidelberg (2009)

[5] Perera, C., Zaslavsky, A., Christen, P., Georgakopoulos, D.: Context Aware Computing for The Internet of Things: A Survey. IEEE Communications Surveys and Tutorials PP(99), 1–44 (2013)

[6] Bezyazychnyy, I., Krinkin, K., Zaslavskiy, M., Balandin, S., Koucheravy, Y.: Geo2Tag Implementation for MAEMO. In: 7th Conference of Open Innovations Framework Program FRUCT, Saint-Petersburg, Russia (2010)

[7] Honkola, J., Laine, H., Brown, R., Tyrkkö, O.: Smart-M3 Information Sharing Platform. In: 1st Workshop on Semantic Interoperability in Smart Spaces (2010)

[8] Honkola, J., Laine, H., Brown, R., Oliver, I.: Cross-Domain Interoperability: A Case Study. Nokia Research Center, Helsinki (2009)

[9] Korzun, D., Balandin, S., Luukkala, V., Liuha, P., Gurtov, A.: Overview of Smart-M3 Principles for Application Development, AIS (2011)

[10] Krinkin, K., Yudenok, K.: Geo-coding in Smart Environments: Integration Principles of Smart-M3 and Geo2Tag. In: Balandin, S., Andreev, S., Koucheryavy, Y. (eds.) NEW2AN 2013 and ruSMART 2013. LNCS, vol. 8121, pp. 107–116. Springer, Heidelberg (2013)

[11] Zaslavsky, M., Krinkin, K.: Geo2tag Performance Evaluation. In: Proceedings of the 12th Conference of Open Innovations Association FRUCT and Seminar on e-Travel, Oulu, Finland (2012)

[12] Kiljander, J., Morandi, F., Soinen, J.-P.: Knowledge Sharing Protocol for Smart Spaces. International Journal of Advanced Computer Science and Applications 3(9) (2012)

Predicting Human Location Based on Human Personality

Seung Yeon Kim and Ha Yoon Song

Department of Computer Engineering,
Hongik University,
Seoul, Korea
dosa0107@gmail.com, hayoon@hongik.ac.kr

Abstract. It is generally believed that human personality affects human mobility patterns. Human personality factors, especially the Big Five factors, allow for the future location of a person to be probabilistically predicted in combination with personal mobility model. For this purpose, we collected the Big Five factors and positioning data for five volunteer participants. Human positioning data can be modeled under an individual human mobility model. With these personality factors and the human mobility model, a person's near future location can actually be predicted using a back propagation network.

Keywords: Human Mobility Model, Personal Mobility Model, Personality Factors, Big Five Factors, Space-Time Analysis, Location Prediction.

1 Introduction

It is usual conjecture that a person's location in everyday life can be affected by his or her personality. Of course, location can vary from time to time as people moves. However the variation in location can be modeled with an individual mobility model for that person. In this paper, we will demonstrate the prediction of human location based on a pre-built personal mobility model and pre-acquired human personality factors. A back propagation network (BPN) is the primary method used to calculate the probability of a person being at a specific location at a given time.

According to previous research [1], human mobility patterns depend on human personality. For example, a person with extroversion may prefer outdoor activities more than an introvert does [2]. Among various methods used to represent human personality with psychological tools, the Big Five factors (BFF) are the most distinguished and the most general and BFF can be acquired by using the Big Five Inventory (BFI) [3,4,5]. Mobile devices with positioning functionality carried by volunteers can be used to acquire another type of positioning data, The set of positioning data can be transformed into a personal mobility model as described by Kim and Song [6]. According to previous research human mobility can be predicted with considerable accuracy [7,8], and these two sets of

S. Balandin et al. (Eds.): NEW2AN/ruSMART 2014, LNCS 8638, pp. 70–81, 2014.

human dependent data, which is collected by five volunteers, will be processed to predict personal locations.

Previous research performed by Luger and Burbey [9,10] utilized Markov chains as the basis for analysis. However, we introduce BPN as a methodology for probabilistic prediction. This indicates that the change in personal location is considered to be a continuous flow rather than a sequence. However the major difference of our research compared to previous research, is that we introduced personality data in the form of BFF. The research conducted by Schmitt, Allik, McCrae and Benet-Mart shows the relationship between BFF and locations [3].

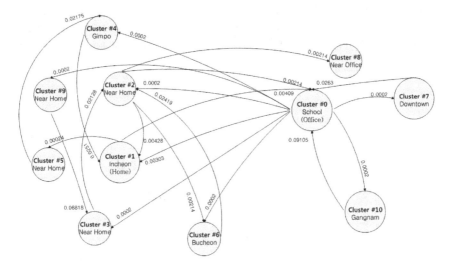

Fig. 1. An Example of Markov Chain Representation of Personal Mobility Model

Combining our past knowledge and methods, we introduce a methodology that combines human mobility patterns and personality data. In Section 2, we will discuss our two bases: BFF and individual position identification. In Section 3, we will discuss the method to predict a person's location at a given time using BFF and the human mobility model. Section 4 will discuss the results of the experiments from Section 3 and the results will be analyzed as well. We will conclude this paper in Section 5.

2 Background

2.1 Big Five Personality Trails

In the field of Psychology, BFF is a method that is used to analyze and present an individual's personality [11]. In the first stage, BFF shows sixteen personalities, factors to present an individual's personality. Fiske showed in a later model that five factors are sufficient to present personality [12]. Tupes and Christal then presented those five factors of human personality [13].

The Big Five factors are:

○ Openness (inventive/curious vs. consistent/cautious)
○ Conscientiousness (efficient/organized vs. easy-going/careless)
○ Etraversion (outgoing/energetic vs. solitary/reserved)
○ Agreeableness (friendly/compassionate vs. cold/unkind)
○ Neuroticism (sensitive/nervous vs. secure/confident)

BFF personality data can be acquired using the Big Five inventory, which is composed of 44 questions and was first presented by John et. al [14]. The most outstanding characteristics of BFF are that each of the five personality factors can be presented as values within a certain range. Due to these characteristics, BFF values can be easily adapted for computerized algorithms and thus are used in this paper.

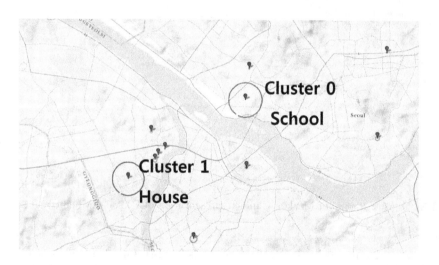

Fig. 2. Mapping of Clusters and Raw Data Using KML viewer

2.2   Human Mobility Model Construction

Each volunteer participant carried his or her own mobile device with positioning functionality turned on. The positioning data acquired by the mobile device was in the usual form of latitude, longitude, time with extra attributes, including the identifier of the device. This sequence of positioning data can be divided into two categories: transient and stationary. The transient state occurs when the device is not staying, but is passing through a position. In the stationary state, the device stays within in a certain area or moves around within this certain area. By using clustering techniques the characteristics of the stationary area, called the location cluster, can be identified and the transition probability between location clusters can be identified.

The set of location clusters and the staying and transition probabilities can be represented as a Markov chain as shown in Figure 1. For example, two location clusters are shown in Figure 2. In this paper we will use the human mobility model constructed through this process as a basis for the location data. The human mobility model is constructed from raw positioning data and it actually shows the position of a participant at a given time. Kim and Song had previously detailed the process for constructing the human mobility model [6].

Table 1. Participant1: Raw Position Data

Time	School	House	Etc.
0	0.4079	0.5750	1.6E-06
1	0.5154	0.4748	4.7E-07
2	0.5378	0.4542	3.6E-07
3	0.5420	0.4501	3.4E-07
4	0.5433	0.4461	3.7E-07
5	0.5640	0.3119	1.6E-06
6	0.7297	0.0027	0.1258
7	0.5229	0.1346	0.2587
8	0.4405	0.4342	1.3E-01
9	0.4365	0.4512	1.4E-01
10	0.4354	0.4527	1.5E-01
11	0.4319	0.4558	1.5E-01
12	0.4142	0.4721	1.4E-01
13	0.3317	0.5528	1.0E-01
14	0.1404	0.7726	4.0E-03
15	0.0377	0.9262	0.0009
16	0.0188	0.9603	0.0037
17	0.0153	0.9651	0.0075
18	0.0204	0.8301	0.0846
19	0.0473	0.0661	0.9612
20	0.0224	0.9302	0.0965
21	0.0234	0.9675	0.0215
22	0.0434	0.9439	0.0034
23	0.1598	0.8169	6.0E-05

3   Positioning and Personality Data Collection

One of the basic goals is to figure out the relationship between the personality factors and location.

Five volunteers participated in this research, and we collected their positioning data for more than six months. The positioning data set has been transformed into a human mobility model that is composed of a set of location clusters

Table 2. BFF values for Each Participant

Participant	1	2	3	4	5
Openness	3.75	3.60	3.70	4.20	3.30
Conscientiousness	3.25	3.30	3.30	4.33	3.80
Extraversion	3.20	2.75	3.75	3.50	3.25
Agreeableness	4.20	3.20	2.60	3.50	3.66
Neuroticism	1.70	2.75	4.00	2.60	2.60

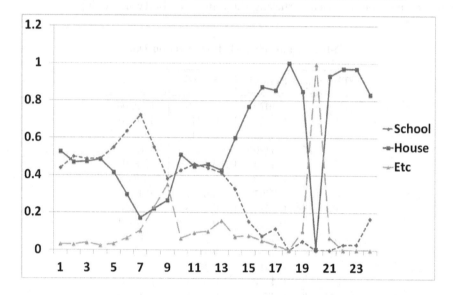

Fig. 3. participant1: Actual Position Data

and transition probability between these location clusters. Figure 2 shows an example of location clusters from actual positioning data set. Each location cluster corresponds to the specific locations of the volunteers, and the more frequently they visit a location cluster, the smaller the cluster number is.

In Figure 2, Cluster 0 stands for the home of the participant, and Cluster 1 stands for the school of the participant. These two location clusters are the most frequent locations for the participant. Since these two clusters are the most common for all participants and other clusters are sparser, we choose these clusters as the basis of our location clusters.

Table 1 shows probability that a participant is at a specific location at a given time. For example, the probability that a participant is home at 21 hrs. is 0.967586, or more than 95%. This means that the participant is mostly at home at 21 hrs. Table 2 shows the BFF of the five participants. The larger the value is, the higher the participant's corresponding personality. For example, Participant1 shows conscientiousness of 4.22 and neuroticism of 1.7, i.e., Participant1 is less sensitive than usual person.

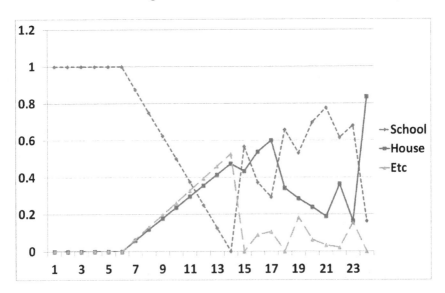

Fig. 4. participant2: Actual Location Data

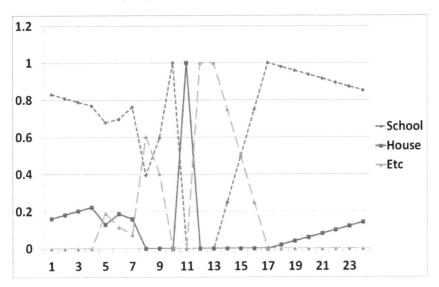

Fig. 5. participant3: Actual Location Data

Figure 3 shows the graphical representation of the data shown in Table 1. Interpretation of the data shows that Participant1 is equally likely to be at home or at school at dawn, and is highly likely to be at home in the afternoon. Figures 4 to 7 show the corresponding graphical representations for Participant 2, 3, 4, and 5. Figure 7 shows an anomaly. It indicates the frequent commuting

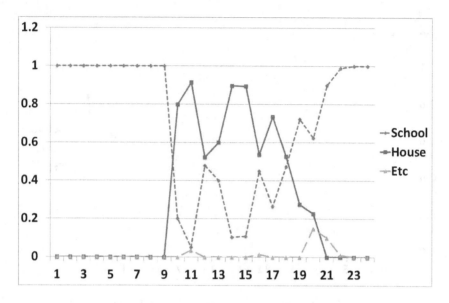

Fig. 6. participant4: Actual Location Data

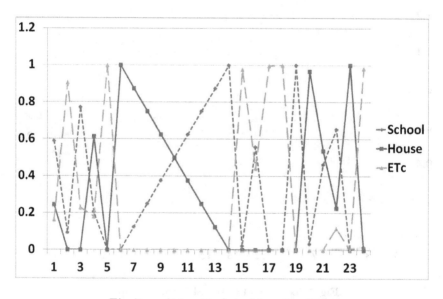

Fig. 7. participant5: Actual Location Data

between home and school within a single day, which is contradictory to common sense. We guess it directly reflects an anomaly in the collected raw data. Methods to filter or correct abnormal raw data must will be discussed in section 5.

4 Experimental Procedure and Results

With the data for five person collected as shown in section 3, BPN is used to process the data. The procedure of learning by BPN is a procedure to create a function that maps input data to spcific result data. For example, such function can be represented as:

$$BPN(T, B_i) = P_j \tag{1}$$

which contains two input data and one output data. The first input data T is time data in unit of hours, i.e. 0 to 24 hours. The second input data B_i is personality data represented by BFF for each participant i. The output data P_j represents positioning data in a range between 0 and 1. The higher P_j is, the higher the probability to be a position is. In our experiment positioning data has one of the values of School, House, and Etc. and represented by j. For example, P_1 stands for school. Then the BPN stands that a person i with personality B_i can be located at position j with probability P_j at time T. With this definition of BPN, the pattern of positioning data can be obtained by the BPN. In order to obtain the meaningful result by BPN, every possible combination of T and B_i must be fully applied to the BPN.

Table 3. Participant1: Prediction of Location

Time || School	House	Etc.
0	0.4171	0.5561	0.0070
1	0.4305	0.5439	0.0069
2	0.4384	0.5392	0.0067
3	0.4668	0.5227	0.0060
4	0.5576	0.4701	0.0042
5	0.7215	0.3685	0.0020
6	0.7969	0.2787	0.0020
7	0.6270	0.2313	0.0250
8	0.3498	0.3162	0.2509
9	0.3549	0.4989	0.0715
10	0.4869	0.5045	0.0215
11	0.4514	0.5262	0.0244
12	0.3606	0.5801	0.0350
13	0.2185	0.6746	0.0671
14	0.1177	0.7609	0.1292
15	0.0828	0.7995	0.1786
16	0.0736	0.8110	0.1976
17	0.0713	0.8141	0.2029
18	0.0707	0.8150	0.2044
19	0.0705	0.8157	0.2040
20	0.0681	0.8277	0.1765
21	0.0449	0.9296	0.0222
22	0.0392	0.9515	0.0083
23	0.1742	0.8027	0.0074

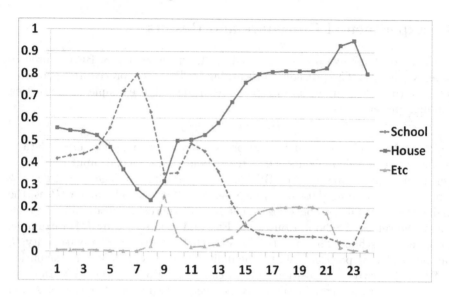

Fig. 8. participant1: Location Predicted

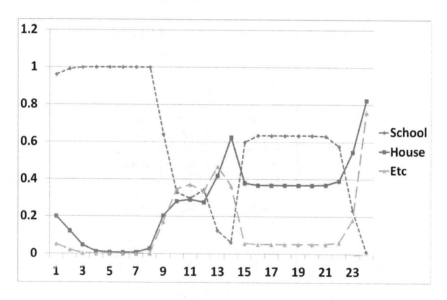

Fig. 9. participant2: Location Predicted

Table 3 shows the probabilistic prediction of the position at a given time after the BPN training process with the BFF inputs of each participants, time of day, and the probability for the position data presented in Table 1. The 120 inputs from 24 hours are combined, producing the corresponding position data (home, school, etc.) for the five participants that are used for training in BPN. In total, 120 training procedures were executed. For each participant, time and position

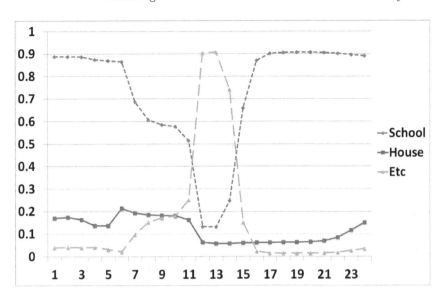

Fig. 10. participant3: Location Predicted

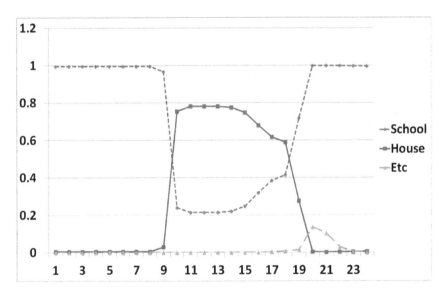

Fig. 11. participant4: Location Predicted

data are required in order to predict the position data. The output is a prediction of the position at each time combining personality data. For example, Table 3 has a probability value of 0.92 for Participant1 to be at home at 21 hrs.

Figure 8 shows the graphical representation of the predicted position data from Table 3. The horizontal axis stands for time in a day and the vertical axis stands for the probability values. Relative to Figure 3, Figure 8 has small

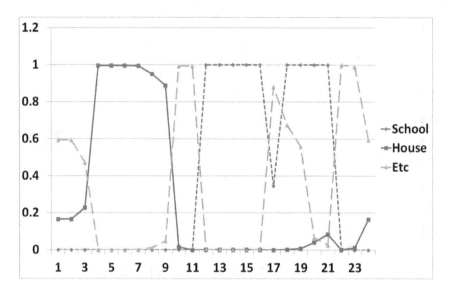

Fig. 12. participant5: Location Predicted

difference. This is due to the normalization process of the BPN training. Every participant's BFF is used as input for the BPN training in order to normalize the personality factors of each participant. Then, the positioning data is affected by specific, distinguished BFF values. Therefore, the shape of the predicted graph is actually smoothened, eliminating raw data outliers. Table 3 shows minute differences to Table 1 as a result of this normalization process. In summary, Figures 9, 10, 11, and 12 show the predicted position of Participants 2, 3, 4 and 5, respectively.

5   Conclusion

We introduced method to probabilistically predict a person's location at a given time by combining personality data and location data. The combination of different knowledge successfully leads to a prediction method of human location. In particular, the results of the prediction are presented in a manner of space-time probability. It is expected we would be able to provide better precision in the results with more volunteer participants since five raw data sets are somewhat deficient. The next step is to gather more participants and more raw data.

Another study will be focus to cover anomalous results shown in Fig 7, which could be deduced, from the participant's personality or due to an insufficient amount of positioning data. More detailed psychological models other than BFF can also be useful to enable more detailed prediction. It is expected that data analysis can have a high computational cost if there is a larger number of participants. In such case, we would consider utilizing a more efficient methodology related to BPN [15].

Acknowledgements. This work was supported by a grant from the National Research Foundation of Korea funded by Korean government (MEST). (NRF-2012R1A2A2A03046473)

References

1. Becker, R., Cáceres, R., Hanson, K., Isaacman, S., Loh, J.M., Martonosi, M., Rowland, J., Urbanek, S., Varshavsky, A., Volinsky, C.: Human mobility characterization from cellular network data. Communications of the ACM 56(1), 74–82 (2013)
2. Carrus, G., Passafaro, P., Bonnes, M.: Emotions, habits and rational choices in ecological behaviours: The case of recycling and use of public transportation. Journal of Environmental Psychology 28(1), 51–62 (2008)
3. Schmitt, D.P., Allik, J., McCrae, R.R., Benet-Martínez, V.: The geographic distribution of big five personality traits patterns and profiles of human self-description across 56 nations. Journal of Cross-Cultural Psychology 38(2), 173–212 (2007)
4. Pervin, L.A., John, O.P.: Handbook of personality: Theory and research (1999)
5. Salleh, A., Al-kalbani, M.S.A., Mastor, K.A.: Testing the five factor personality model in oman
6. Kim, H., Song, H.Y.: Formulating human mobility model in a form of continuous time markov chain. Procedia Computer Science 10, 389–396 (2012)
7. Song, C., Qu, Z., Blumm, N., Barabási, A.-L.: Limits of predictability in human mobility. Science 327(5968), 1018–1021 (2010)
8. Hidalgo, C.A., Gonzalez, M.C., Barabasi, A.-L.: Understanding individual human mobility patterns. Nature 453(7196), 779–782 (2008)
9. Luger, G.F.: Artificial intelligence: Structures and strategies for complex problem solving (2008)
10. Burbey, I.E.: Predicting future locations and arrival times of individuals. PhD thesis, Virginia Polytechnic Institute and State University (2011)
11. Poropat, A.E.: A meta-analysis of the five-factor model of personality and academic performance. Psychological Bulletin 135(2), 322 (2009)
12. Fiske, D.W.: Consistency of the factorial structures of personality ratings from different sources. The Journal of Abnormal and Social Psychology 44(3), 329 (1949)
13. Tupes, E.C., Christal, R.E.: Recurrent personality factors based on trait ratings (1961)
14. John, O.P., Donahue, E.M., Kentle, R.L.: The big five inventory–versions 4a and 54. University of California, Institute of Personality and Social Research, Berkeley (1991)
15. Riedmiller, M., Braun, H.: A direct adaptive method for faster backpropagation learning: The rprop algorithm. In: IEEE International Conference on Neural Networks 1993, pp. 586–591. IEEE (1993)

On Open Source Mobile Sensing

Dmitry Namiot[1] and Manfred Sneps-Sneppe[2]

[1] Lomonosov Moscow State University,
Faculty of Computational Mathematics and Cybernetics,
Moscow, Russia
dnamiot@gmail.com
[2] ZNIIS, M2M Competence Center,
Moscow, Russia
manfreds.sneps@gmail.com

Abstract. The paper discusses phone as a sensor model. For many applied tasks smart phones are an ideal platform for collecting and processing context-related data. The most popular example is, probably, computational social science. Phones can collect data for conducting various social researches about people's social behavior. This paper presents an attempt to describe and categorize existing open source libraries for mobile sensing, describe architecture and design patterns as well as discover directions for the future development.

Keywords: Smartphone, sensing, open source, data mining.

1   Introduction

The ubiquity of sensor-rich and computationally powerful smart phones makes them an ideal platform for collecting and processing context-related data. For example, the modern proliferation of smart phones with sensing capabilities creates a lot of opportunities for collecting data about people's social behavior (computational social science) [1]. We can use the collected data for conducting various social researches. There are many examples of experimental and real business applications, based on phone sensing. For example, beyond data collection, smart phones can implement and test behavior change theories [2] as well as play a mediating role between people and their therapists [3]. Examples may include understanding traffic conditions in a city, understanding environmental pollution levels, or measuring obesity trends. Sensors in the possession of large numbers of individuals enable exploiting the crowd for massively distributed data collection and processing [4]. We should mention in this context the original development in Reality Mining [5]. Actually, it is based on the recording data from mobile devices about context and activities. In general, this application is designed to track important trends in the behavior of people in the city.

What are the typical tasks for sensors collecting tools? Actually, it is a non-trivial task to design applications that appropriately balance between energy efficiency, data collection, storage, and transmission continues [6]. Technically, any sensing library

S. Balandin et al. (Eds.): NEW2AN/ruSMART 2014, LNCS 8638, pp. 82–94, 2014.
© Springer International Publishing Switzerland 2014

should collect sensor data (of course), let users configure this collection process (define what should be collected and how), format data, store data, transmit data to remote servers (asynchronously and synchronously), configure notifications, trigger events based on sensor data. Some of the applications may follow to the broad definition of "sensor". They can include records for social interactions too (internet browsing, making calls, sending messages, etc.).

We think that such a big interest in this area should certainly lead to some standardization of the design process of data collecting tools. For technical and economic reasons, it will be impossible to start each program development, accumulating data sensors from scratch. Building sensor data collection tools requires an understanding of the varying sensor programming interfaces as well as a mobile operational system interfaces related to collecting data from the system. As an example of such standardization in a similar area, we can mention the various attempts to develop (create) software standards for M2M and IoT.

This paper presents an attempt to describe and categorize existing open source libraries for sensing, describe architecture and design patterns as well as discover directions for the future development.

The rest of the paper is organized as follows. Section 2 describes the main problems in mobile sensing, Section 3 describes existing libraries. We close by discussing related services, applications, and future directions for research.

2   On Challenges for Mobile Phone Sensing

On the first hand, we should mention here phones' batteries as a major challenge in achieving social sensing. The challenge is the significant impact that continuous sensing has on the phone's battery life. For example, in [7] authors provide the power overhead for every sensor is expressed as a percentage of the power consumed by the HTC phone:

Accelerometer: 7.4%
Temperature: 2.78%
Barometer: 22.2%
Compass: 29.63%

There are many papers devoted to energy effective data collection (mostly, they cover location info) [8][9].

The common conclusion is actually very simple - user context collection in the simplest form (turn on and reads the sensors periodically) on a smart phone could easily lead to a rapid drain of the smart phone battery. The next conclusion is almost obvious too. Some of the sensors could be cheaper in terms of energy consumption. As it follows from the above-mentioned data, accelerometer is cheaper than Wi-Fi scanning in terms of energy consumption. In the most cases, it is some balancing: data precision versus energy consumption. Actually, these simple conclusions are very important, because they are forming the basic requirements for software development tools (libraries, frameworks) in mobile phone sensing. So, for the each framework (library) in a mobile phone sensing area we should see at least two options: scheduling and sensors reconfiguration. In general, the framework should activate sensors only when they are required.

Authors in [8] enumerate several methods to reduce battery consumption, such as the minimum sensor selection, adjustments for sensor sampling rate and duty cycle, sharing common information with others.

The classical example is user staying on the office desk. In this case GPS and accelerometer provide almost no information about the user's status. The classical example for energy saving in location-based system is collaborative positioning [10]. Location information sharing approaches are popular, because they based on the simple conclusion that users spend most of time with others.

In the same time, we should note that most of research only focused on a single sensor optimization, rather than on entire energy consumption of mobile device.

For selecting services, authors in [8] propose so called context templates. Template describes conditions for the context as well as enumerates sensors for data pulling (Figure 1).

```
user state:
    name : "meeting",
    condition context : "stay"
    condition context : "speech"
    condition context : "located at the office"
sensor reconfiguration :
    user state : "meeting"
    sensor : "accelerometer",
    sampling rate : low
```

Fig. 1. Context-template [8]

The classical instrument for top level description is SensorML [11]. SensorML provides a modular way of describing data processing chains. Chains are constructed from reusable processing components, called Process Models. Process Models provide all metadata useful for components discovery. Each model defines inputs, outputs and parameters, as well as a method for own executing. These methods need to be defined inline or in a library before the model can be used in a Process Chain [12]. It is illustrated in Figure 2.

But usually, such a low level is not enough and we need some ontology and semantic-based tools. For example, OntoSensor [13] ontology presents a general knowledge-base of sensors for query and inference.

The ontology is built around the central notion of a sensor. There are three important clusters of concepts referenced by the sensor: domain concepts, abstract sensor properties and concrete properties [14].

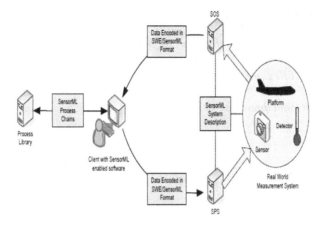

Fig. 2. SensorML environment [12]

Separating concrete and abstract aspects of sensors lets us share and reuse descriptions among specifications, provide multiple concrete descriptions for a single sensor type.

Ontology plays a significant role in FI-WARE projects (EU Future Internet) [15]. The main problem we can highlight here is the parallel existence of two worlds: academic research and practical development. The formal methods for sensor definitely require the automation tools. It is the main conclusion that follows from our investigation. For example, none of below described open source libraries support automatic configuration.

The next big problem for mobile phone sensing is the high level of diversification in mobile sensors. Different sensors cause different usage patterns. We mean here different software APIs and different ways of processing handling. Some data we can handle directly on mobile phones. Other data may require interaction with the server. The same sensors depending on the measurement method may be handled differently. The typical example is wireless sensor. We may collect information about access points directly on the phone. It leads to some services like SpotEx [16]. Alternatively, we can use the information from mobile devices (from the same wireless module) – so called Wi-Fi probe request right on the server for server side processing. So, the whole processing will be performed out of the phone [17].

The typical example of different approaches to data processing is iBeacon. It is programming interface to low energy sensors from Apple (iBeacons). As per original approach for data processing, developers should obtain a list of visible iBeacons right on the phone and request information, associated with obtained devices. It could be performed directly (e.g., on Android 4 mobile OS) or via some API, provided by third party vendors. E.g., here is a code snippet from Estimote [18]. There is BeaconManager object that plays a role of proxy and hides details of connection:

```
(void)beaconManager:(ESTBeaconManager *)manager
  didDetermineState:(CLRegionState)state
      forRegion:(ESTBeaconRegion *)region
{
  if(state == CLRegionStateInside)
  {
    [self setProductImage];
  }
  else
  {
    [self setDiscountImage];
  }
}
```

In the same time, EU project FI-WARE follows to the own common model and own standard Interface to Devices [19]. Figure 3 illustrates the common model:

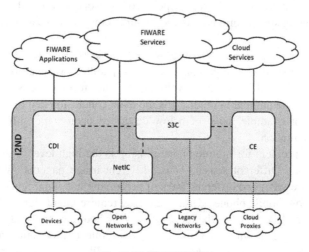

Fig. 3. FI-WARE I2ND

The whole description of FI-WARE is out of the scope of this paper, but the main difference could be explained very shortly. For the same sensor any application should deal with the whole chain of intermediate services: Connected Device Interfacing (CDI), Service, Capability, Connectivity and Control (S3C) Generic Enabler, Network Information & Control (NetIC) Generic Enabler and Cloud Edge [20].

The questions about the place for accumulating data and place for data processing are important for sensing libraries. Yes, it is simpler to implement data processing right on the device (e.g., [21]). But server-side processing could be more flexible, more main table (it is easier to update code, for example) [22]. But any cloud based solution (like the above-mentioned FI-WARE) complicates the start of data collection.

So, as soon as we discuss the libraries (tools), the above mentioned features become requirements for them. It is a key question for any mobile phone sensing tool: can we work directly with sensors or our tool requires some cloud-based mediator. Practically, it dictates how collected data could be processed: can we do that right on the mobile device or selected tools require some cloud-based installation.

A good survey of mobile phone sensing could be found in [23] (Figure 4).

Fig. 4. Mobile phone sensors [23]

From the whole set of available sensors we will highlight the following elements. On the first hand, it is Wi-Fi sensor. It is presented in every smart phone and Wi-Fi access points are available almost everywhere. The next key moment is Bluetooth. Again, because of the wide spread. Nowadays, it is getting the second breath due to Bluetooth Low Energy standard [24] and Apple's iBeacons [25].

The accelerometer is a very important tool for navigation and audio is a much underrated channel.

3   Open Source Libraries for Mobile Phone Sensing

In this section we would like to list some Open Source tools for mobile phone sensing. Frameworks are listed in alphabetical order.

AWARE Framework [26]. AWARE is an Android framework dedicated to instrument, infer, log and share mobile context information, for application developers, researchers and mobile phone users. AWARE is open-sourced under GPLv3.

There are AWARE Client and AWARE server. AWARE Client does not require programming skills. It provides context dashboard that allows users to enable or disable context information, for each context source individually. The data is saved locally on user's mobile phone. Privacy is enforced by design, so AWARE does not log personal information, such as phone numbers or contacts information.

The client collects the wide list of data, such as accelerometer, barometer, battery level, visible neighbor Bluetooth and Wi-Fi devices, gravity and hyroscope profiles, magnetometer profile, location info, temperature and traffic data.

AWARE server lets collect data in a centralized database. There are two versions AWARE Server MQTT and AWARE Server with web services. MQTT client lets share information in real-time with other AWARE Clients and AWARE Servers. It allows servers and clients to push information using a publish-subscribe approach. Web services let synchronize the local data with a remote database with AWARE Server web services. The synchronization can be scheduled, triggered locally with an event or requested remotely. Figure 5 illustrates AWARE server:

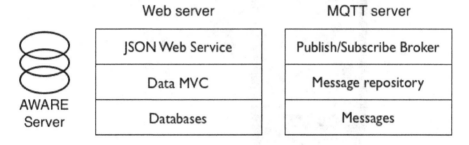

Fig. 5. AWARE server

There is an interesting conception of plugins. For example, Mode of Transportation Context Plugin detects 5 modes of transportation: standing/still, walking, running, biking and riding/driving a car (or bus) with an accuracy of 92%, Noise Level Context Sensor detects how noisy an environment is (measured in dB) [26].

Other plugins are: Pedometer - detects the users steps when walking, Polar HRMontext – detects the users' current heart rate,   Screen Orientation - detects the screen orientation (face up or down), Social – detects contacts' online status, Audio State – detects audio settings as they are changed by the user (volume, vibration), Notification Content – detects notifications and which application sent them, Weather – location-aware weather conditions (via external OpenWeather API's), Google Activity Recognition: recognizes users' mode of transportation using Google Play Services API (still, on foot, in vehicle, on a bicycle, unknown or tilting - moving the phone around randomly, Google Fused Location – provides 24/7 location tracking with less than 1% battery impact per hour, using the default settings,   Browser History – records browser history for standard browser, Settings Sensor – records all the phone settings on the devices and tracks changes.

The architecture is open, so it is possible to create own context sensors. The whole system design is illustrated on Figure 6.

EmotionSense presents a Mobile Phones based Adaptive Platform for Experimental Social Psychology Research [27]. The system is based on several sensor

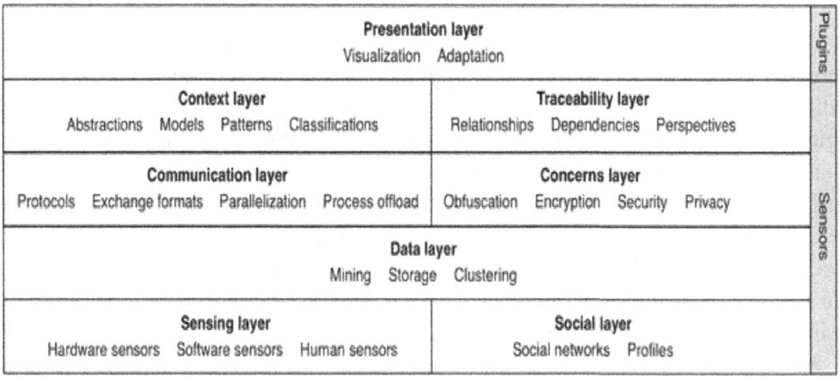

Presentation layer Visualization Adaptation		Plugins
Context layer Abstractions Models Patterns Classifications	**Traceability layer** Relationships Dependencies Perspectives	Sensors
Communication layer Protocols Exchange formats Parallelization Process offload	**Concerns layer** Obfuscation Encryption Security Privacy	
Data layer Mining Storage Clustering		
Sensing layer Hardware sensors Software sensors Human sensors	**Social layer** Social networks Profiles	

Fig. 6. Sensors in AWARE [26]

monitors, such as accelerometer monitor, Bluetooth monitor, Location monitor. In general, it looks like an applied system (detect emotions), rather than a standalone tool for data collection.

Funf [28] is, probably, the most elaborated mobile phone sensing tool. The Funf Open Sensing Framework is an extensible sensing and data processing framework for mobile devices. The core concept is to provide an open source, reusable set of functionalities, enabling the collection, uploading, and configuration of a wide range of data signals accessible via mobile phones [29].

There are several components. Funf Journal is a personal data collection application. It saves data locally. Funf in a Box lets users create own mobile sensing Android app. It does not require programming skills. This application saves data in Drop box account. And third level is developers API.

Funf collects rich sets of data (it is configured). The whole list includes location and wireless info, accelerometer data, Cell Tower IDs, call and SMS logs, browser history and batter status.

Open Data Kit (ODK) [30] is a free and open-source set of tools which help organizations author, field, and manage mobile data collection solutions. ODK provides an out-of-the-box solution for users to:

-   Build a data collection form;
-   Collect the data on a mobile device and send it to a server;
-   Aggregate the collected data on a server and extract it in useful formats.

ODK contains so called ODK Build. It is a form designer with a drag-and-drop user interface. The Build is an HTML5 web application and works best for designing simple forms. It is a standard form for configuring data collection process. ODK Collect renders forms into a sequence of input prompts that apply form logic, entry constraints, and repeating sub-structures. ODK ODK Aggregate provides a ready-to-deploy server and data repository. ODK Aggregate can be deployed on Google's App Engine or on a Tomcat server (or any servlet-compatible web container) backed with a MySQL or PostgreSQL database server.

In terms of data collection ODK supports all built-in Android sensors.

From own experiments we can recommend AWARE and FUNF as the most elaborated frameworks. The practical applications include redefined check-ins [31]. Traditionally, place in social networks is a pair from geo coordinates (latitude, longitude) and some description. Social networks introduce a special kind of records (statuses) – check-ins. Check-in record in social networks is some message (post, status) linked to some location (place) [32]. We introduce a new conception of places, where location is replaced by sensing information (e.g., proximity info) [33]. This concept was developed for indoor applications and IoT projects [34].

In another experiment, we redesigned Geo Messages conception [35], and share sensing info with standard messages.

4   The Model and Patterns

In this section we would like to discuss the possible model and design patterns for mobile phone sensing.

Firstly, we believe that in the current state the software tools available for mobile phone sensing are more focused on laboratory work. Processes of data retrieval and subsequent processing are not fully automated.

We think, that mobile phone sensing application should follow the same design principles as a modern Internet of Things (IoT) capable device. Mobile phone sensing libraries should not create own communication stack. It should follow the same practice as FI-WARE, for example. So, the mobile phone as a device could provide, for example, 6LoWPAN and CoAP, etc. In other words, we think that the conception should follow to the M2M standards, rather than inventing own approaches. For IoT the whole communication stack (e.g. standard OSI model) should be redesigned. On the lowest level we have sensors. There is no need to divide this layer. These devices are embedded devices that directly interact with the physical world. For mobile phones such "devices" are embedded sensors. For pure IoT systems they could have some networking functionality, for the phone as a sensor we can use networking functionality from the phone itself.

The next entity is Network Transport layer. In this model, the network transport contains a set of technologies. It is actually a combination of the OSI data link, network and transport layers. The typical example is TCP/IP over Wi-Fi. Note, that the typical application layer stuff like REST in this model is also a part of Network Transport.

The next entity in this new model is the Data layer. Data layer may present individual data from sensors as well as aggregate data from several devices.

And on the top of Data layer we should have some API (Application Program Interfaces). But here is another interesting issue. In the modern world APIs could be interesting in case of ability to mash up various data sources only. Another important question is the ability for API to control data flow in the above-mentioned stack. The classic example from IoT world is some temperature measurement. It could be a complex state that depends on data from several sensors. So, it may require acting on several aspects of a climate control system to change, etc.

The next important moment is the structure of operations with sensors. Traditionally, Application Program Interfaces for various sensors are mentioned in this context. But on practice, most of the applications just collect data. There is simply no space for full range API. It means, that the main goal for mobile phone sensing could be described as Data Program Interfaces (DPI), rather than as API. We can describe DPI as an interface on the device that exposes and consumes data. There are two moments. Many devices (sensors) very often do not support commands (instructions). Many of sensors just provide some data and nothing more. And second question is the security. Obviously, that data reading without any ability to update something is more safety in many cases. In this connection we can mention several projects. On the first hand, it is a Webinos project [37]. It is a European Union project, which aims at developing software components for the future Internet, in the form of Web Runtime Extensions. We can mention also OpenRemote project [38] in this connection. The above-mentioned MQTT protocol from IBM is a good example of data centric communications. In the publish/subscribe communication model components which are interested in obtaining data register their interest. In publish/subscribe systems it is called subscription. Components which want to share some information, do so by publishing their data. There is also a broker - the entity which ensures that the data gets from the publishers to the broker. This actor coordinates subscriptions. In our case this broker is personalized. It is an in-application proxy. There are three principal types of pub/sub systems: topic-based, type-based and content-based [39]. By topic-based systems, the list of topics is usually known in advance. E.g., in our case we know the tags and their data. In type-based systems, a subscriber states the type of data it is interested in (e.g., temperature data). It means that we need to maintain some directory of tags with metadata. In content-based systems the subscriber describes the content of messages it wants to receive. Subscriber describes types of data for receiving and condition for the delivery (e.g., the temperature is below a certain threshold). Publish/subscribe it is a good example for DPI too.

There is a wide set of papers and implementations devoted to the Web of Sensors with linked data and HTTP based REST protocol [40]. By our opinion, it is the direction for mobile sensing. The models assume that the HTTP server is deployed on each sensor node, making it an independent and autonomous Web device. It could be a limitation for the independent M2M (IoT) devices with limited capabilities, but for mobile sensing the computational power of any smart phone is enough for running web server.

So, in our view, the core of mobile sensing should be a web server on the mobile phone. DPI in this case is a set of CGI (Common Gateway Interfaces) scripts for obtaining data from build-in sensors. There is no API for setting data. And there should be no API for getting context information (history of calls, current activities, etc.). All this is just by the security reasons. Smart phone here is just a computer engine for hosting data accusation software. Context-aware data accusation, like Reality Mining [5] should be a separate issue.

As per API for the above-mentioned stack of protocols, we see the REST-based requests and publish subscribe model. MQTT is, probably, the best candidate for the distribution. As the foundation for data formatting, we see JSON only. It is the most developer-friendly format; it allows flexible integration and mashing up. Also, it could be self-descriptive and semantic data could be provided on demand.

Semantics for sensors could be based on SSN ontology that describes everything about sensors and sensing [41]. This enables broad interoperability between different sensor systems and platforms, based on common data models and descriptions. For abstracting the complexity of sensors, web server on the smart phone will add a layer of semantic discovery and linkage. This decision enables the sensors (phones) and actuators on our sensor nets to be easily combined to build integrated applications. The common data model for sensor nets could be based on IETF CoRE application protocol and sensor descriptions [42]. CoRE provides standard ways for common types of sensors to be discovered by their attributes, and standard ways for the data to be linked into applications. It could be done exactly via providing descriptions of the JSON data structure the sensor provides as their output. And W3C Linked Data standard [43] is a good candidate to provide web representations of data models for sensor data. Linked data provide capabilities for graph-based access control, database-like queries, and big data analysis. In general, it should create a foundation for an event-action programming model and distributed software components. It allows the data sources and software components to easily be assembled and make data flow connections. We see the future in this area as an event-driven architecture. This architecture will provide self-describing software objects, representing sensors, actuators, as well as user interaction endpoints. At the same time, components-based model lets create a community of developers around it.

We are going to create the service enablement layer as open source as well, so that any one of the companies can help

And another conclusion that we can make from our analysis is the role of mobile OS vendors. We believe that mobile sensing will be a part of mobile OS finally. Actually, the above mentioned iBeacon is a first example. It is a special native support on the OS kernel level for the separate sensor. And, probably, the first system that can make the serious progress in this area is Firefox OS.

5   Conclusion

This paper discusses phone as a sensor model. For many applied tasks smart phones are an ideal platform for collecting and processing context-related data. The most known example is, probably, computational social science. Smart phones can collect data for conducting various social researches about people's social behavior. This paper presents an attempt to describe and categorize existing open source libraries for mobile sensing. We describe architecture and design patterns as well as discuss directions for the future development.

References

[1] Lazer, D., et al.: Life in the network: The coming age of computational social science. Science 323(5915), 721 (2009)
[2] Hekler, E., Klasnja, P., Froehlich, J., Buman, M.: Mind the Theoretical Gap: Interpreting, Using, and Developing Behavioral Theory in HCI Research. In: ACM CHI, Paris, France (2013)
[3] Lathia, N., Pejovic, V., Rachuri, K., Mascolo, C., Musolesi, M., Rentfrow, P.: Smartphones for Large-Scale Behaviour Change Interventions. IEEE Pervasive Computing (May 2013)
[4] Aggarwal, C., Abdelzaher, T.: Integrating sensors and social networks. In: Social Network Data Analytics, ch. 14, Springer (2011)
[5] Eagle, N., Pentland, A.: Reality Mining: Sensing Complex Social Systems. J. Personal and Ubiquitous Computing (2005)

[6] Rachuri, K., Efstatiou, C., Leontiadis, I., Mascolo, C., Rentfrow, P.: METIS: Exploring Mobile Phone Sensing Offloading for Efficiently Supporting Social Sensing Applications. In: IEEE PerCom, San Diego, USA (2013)

[7] Priyantha, B., Lymberopoulos, D., Liu, J.: Littlerock: Enabling energy-efficient continuous sensing on mobile phones. IEEE Pervasive Computing 10(2), 12–15 (2011)

[8] Han, Y., Kang, J.M., Seo, S.S., Mehaoua, A., Hong, J.W.K.: An energy efficient user context collection method for smartphones. In: 2013 15th Asia-Pacific on Network Operations and Management Symposium (APNOMS), pp. 1–6. IEEE (September 2013)

[9] Nawaz, S., Efstratiou, C., Mascolo, C.: ParkSense: A smartphone based sensing system for on-street parking. In: Proceedings of the 19th Annual International Conference on Mobile Computing & Networking, pp. 75–86. ACM (September 2013)

[10] Rao, H., Fu, W.T.: A General Framework for a Collaborative Mobile Indoor Navigation Assistance System. In: Proceedings of the 3rd International Workshop on Location Awareness for Mixed and Dual Reality, pp. 21–24 (March 2013)

[11] Botts, M., Robin, A.: OpenGIS Sensor Model Language (SensorML) implementation specication. OpenGIS Implementation Specication OGC 07-000, The Open Geospatical Consortium (July 2007)

[12] Robin, A., Botts, M.E.: Creation of Specific SensorML Process Models. Earth System Science Center-NSSTC, University of Alabama in Huntsville (UAH), HUNTSVILLE, AL 35899 (2006)

[13] Russomanno, D., Kothari, C., Thomas, O.: Sensor ontologies: From shallow to deep models. In: 37th Southeastern Symposium on System Theory (2005)

[14] Compton, M., Neuhaus, H., Taylor, K., Tran, K.N.: Reasoning about Sensors and Compositions. In: SSN, pp. 33–48 (2009)

[15] Usländer, T., Berre, A.J., Granell, C., Havlik, D., Lorenzo, J., Sabeur, Z., Modafferi, S.: The future internet enablement of the environment information space. In: Hřebíček, J., Schimak, G., Kubásek, M., Rizzoli, A.E. (eds.) ISESS 2013. IFIP AICT, vol. 413, pp. 109–120. Springer, Heidelberg (2013)

[16] Namiot, D., Sneps-Sneppe, M.: Proximity as a service. In: 2012 2nd Baltic Congress on Future Internet Communications (BCFIC), pp. 199–205. IEEE (April 2012)

[17] Namiot, D., Sneps-Sneppe, M.: Geofence and Network Proximity. In: Balandin, S., Andreev, S., Koucheryavy, Y. (eds.) NEW2AN 2013 and ruSMART 2013. LNCS, vol. 8121, pp. 117–127. Springer, Heidelberg (2013)

[18] Estimote API, http://estimote.com/api/ (retrived June 2014)

[19] FI-WARE Interface to Network and Devices, http://forge.fi-ware.eu/plugins/mediawiki/wiki/fiware/index.php/FI-WARE_Interface_to_Networks_and_Devices_%28I2ND%29 (retrieved June 2014)

[20] Usländer, T., Berre, A.J., Granell, C., Havlik, D., Lorenzo, J., Sabeur, Z., Modafferi, S.: The future internet enablement of the environment information space. In: Hřebíček, J., Schimak, G., Kubásek, M., Rizzoli, A.E. (eds.) ISESS 2013. IFIP AICT, vol. 413, pp. 109–120. Springer, Heidelberg (2013)

[21] Namiot, D.: Context-Aware Browsing – A Practical Approach. Next Generation Mobile Applications. In: 2012 6th International Conference on Services and Technologies (NGMAST), pp. 18–23 (2012), doi:10.1109/NGMAST.2012.13

[22] Sneps-Sneppe, M., Namiot, D.: About M2M standards and their possible extensions. In: 2012 2nd Baltic Congress on Future Internet Communications (BCFIC), pp. 187–193. IEEE (April 2012), doi:10.1109/BCFIC.2012.6218001

[23] Lane, N.D., et al.: A survey of mobile phone sensing. IEEE Communications Magazine 48(9), 140–150 (2010)

[24] Gupta, N.: Inside Bluetooth Low Energy. Artech House (2013)

[25] Dilger, D.E.: Inside iOS 7: iBeacons enhance apps' location awareness via Bluetooth LE. AppleInsider (2013)

[26] Aware Framework, http://www.awareframework.com/home/ (retrieved: June 2014)

[27] Rachuri, K.K., Musolesi, M., Mascolo, C., Rentfrow, P.J., Longworth, C., Aucinas, A.: EmotionSense: A mobile phones based adaptive platform for experimental social psychology research. In: Proceedings of the 12th ACM International Conference on Ubiquitous Computing, pp. 281–290. ACM (September 2010)

[28] Aharony, N., Pan, W., Ip, C., Khayal, I., Pentland, A.: Social fmri: Investigating and shaping social mechanisms in the real world. Pervasive and Mobile Computing (2011)

[29] Funf, http://funf.org (retrieved: February 2014)

[30] OpenDataKit, http://opendatakit.org/ (retrieved June 2014)

[31] Namiot, D., Sneps-Sneppe, M.: Wireless Networks Sensors and Social Streams. In: 2013 27th International Conference on Advanced Information Networking and Applications Workshops (WAINA), pp. 413–418. IEEE (March 2013)

[32] Namiot, D., Sneps-Sneppe, M.: Customized check-in Procedures. In: Balandin, S., Koucheryavy, Y., Hu, H. (eds.) NEW2AN 2011 and ruSMART 2011. LNCS, vol. 6869, pp. 160–164. Springer, Heidelberg (2011)

[33] Namiot, D., Sneps–Sneppe, M.: Social streams based on network proximity. International Journal of Space-Based and Situated Computing 3(4), 234–242 (2013)

[34] Volkov, A.A., Namiot, D.E., Schneps-Schneppe, M.A.: Building an Effective Infrastructure for Environment. International Journal of Open Information Technologies 1(7), 1–10 (2013) (in Russian)

[35] Namiot, D.: Geo messages. In: 2010 International Congress on Ultra Modern Telecommunications and Control Systems and Workshops (ICUMT), pp. 14–19 (2010), doi:10.1109/ICUMT.2010.567666

[36] Ra, M.R., Liu, B., La Porta, T.F., Govindan, R.: Medusa: A programming framework for crowd-sensing applications. In: Proceedings of the 10th International Conference on Mobile Systems, Applications, and Services, pp. 337–350. ACM (June 2012)

[37] Fuhrhop, C., Lyle, J., Faily, S.: The webinos project. In: Proceedings of the 21st International Conference Companion on World Wide Web, pp. 259–262. ACM (April 2012)

[38] Bjelica, M.Z., Teslic, N.: A concept and implementation of the Embeddable Home Controller. In: 2010 Proceedings of the 33rd International Convention MIPRO, pp. 686–690. IEEE (May 2010)

[39] Eugster, P.T., Felber, P.A., Guerraoui, R., Kernmarrec, A.-M.: The many faces of publish/subscribe. ACM Computing Surveys 35(2), 114–131 (2003)

[40] Guinard, D., Trifa, V.: Towards the web of things: Web mashups for embedded devices. Workshop on Mashups, Enterprise Mashups and Lightweight Composition on the Web (MEM 2009). In: Proceedings of WWW (International World Wide Web Conferences), Madrid, Spain (2009)

[41] Michael, C., et al.: The SSN ontology of the W3C semantic sensor network incubator group. Web Semantics: Science, Services and Agents on the World Wide Web 17, 25–32 (2012)

[42] Shelby, Z.: Embedded web services. IEEE Wireless Communications 17(6), 52–57 (2010)

[43] Heath, T., Bizer, C.: Linked data: Evolving the web into a global data space. Synthesis lectures on the Semantic Web: Theory and Technology 1(1), 1–13 (2011)

Effective Waste Collection with Shortest Path Semi-Static and Dynamic Routing

Theodoros Vasileios Anagnostopoulos[1] and Arkady Zaslavsky[1,2]

[1] Department of Infocommunication Technologies, ITMO University, St. Petersburg, Russia
thanag@di.uoa.gr
[2] CSIRO Computational Informatics, CSIRO, Box 312, Clayton South, VIC, 3169, Australia
Arkady.Zaslavsky@csiro.au

Abstract. Smart cities are the next step in human habitation. In this context the proliferation of sensors and actuators within the Internet of Things (IoT) concept creates a real opportunity for increasing information awareness and subsequent efficient resource utilization. IoT-enabled smart cities will generate new services. One such service is the waste collection from the streets of smart cities. In the past, waste collection was treated with static routing models. These models were not flexible in case of segment collapse. In this paper we introduce a semi-static and dynamic shortest path routing model enhanced with sensing capabilities through the Internet connected objects in order to achieve effective waste collection.

Keywords: Smart Cities, Internet of Things, Waste Collection, Dynamic Routing.

1 Introduction

Smart cities are the next step in human habitation aiming at efficient integration with sustainable environment. Future internet and IoT enable tiny devices to be implanted within the backbone of the human society in real life objects like the waste bins [1], [2]. These objects (i.e., smart things) are applied with distributed artificial intelligence which make them capable of communicating with each other thus forming sensor networks that could be wireless. The new era of these internet-connected things enables them not only to passively sense the environment around them but acting as well, thus having the capability to be sensors and actuators at the same time [3], [4], [5]. In this context the proliferation of sensors and actuators within the IoT concept has become a real challenge. This implies new services that arise and could be provided through cloud service layers to stakeholders (i.e., municipality, citizens) [6], [7]. One such service is the solid waste collection from the smart cities neighborhoods [8], [9]. The past years waste collection was treated with static routing models. These models were not flexible in case of segment collapse in the waste network infrastructure. Recent research that uses dynamic methods for routing the waste collection using sensor-equipped garbage bins and trucks thus outperforms the static methods.

S. Balandin et al. (Eds.): NEW2AN/ruSMART 2014, LNCS 8638, pp. 95–105, 2014.

In this paper we intend to extend the dynamic routing problem in the context of IoT and especially in the concept of smart cities, where the dynamic routing response execution time and the overall waste collection procedure is of high environmental significance. We introduce both an upper layer semi-static shortest path routing model and a lower layer dynamic shortest path routing model in order to achieve effective waste collection. Specifically we threat the waste collection problem as a dual layer problem where the upper layer constructs a waste collection terrain for each city sector, while the lower layer handle the dynamic requirements of real time routing in case of emergency (i.e., a road under construction, unexpected traffic congestion). The network of the waste collection consists of a set of garbage bins distributed in the city terrain located in several sectors of the smart city. Both semi-static and dynamic routing should be used in order the fleet of refuse vehicles to collect the solid waste from every place of the city and empty them to the depot at every network traversal when necessary. The routing models are highly related by means that when an incident occurs the semi-static shortest path routing model invokes the dynamic shortest path routing model to solve the problem. In absence of the dynamic shortest path routing model the overall process would collapse because of routing inconsistency.

This paper is structured as follows. Section 2 reports related work. Section 3 discusses the semi-static and dynamic routing models. Section 4 presents evaluation performance, while Section 5 concludes the paper.

2   Related Work

We report on methods which adopt dynamic routing waste collection. The author in [10] develops a system for dynamic routing for waste collection in the city area. It is used a mobile measuring system on the vehicle in order to perform a stochastic dynamic routing which in turn can make corrections during or after the execution of the existing routes. In [11] the author incorporates discrete event simulation in order to achieve dynamic waste collection from underground containers. The model uses a dynamic planning which exploit information transmitted through motion sensors that are embedded in the underground containers. The authors in [12] propose a heuristic method for waste dynamic routing taking into consideration several tunable parameters. The use of sensors enables reverse inventory routing in more dense waste networks. Such a heuristic method deals with uncertainty of daily and seasonal effects. In [13] the author evaluates dynamic planning methods that used for refuse collection of underground containers. Furthermore the model aims at reducing the amounts of carbon dioxide released in the city environment from the refuse vehicles by making the dynamic routing more effective.

The model in [14] is specialized in the refuse collection of plastic waste which is differentiated from the other municipal solid waste. In this approach is achieved heuristic redesign of the collection routes using an eco-efficiency metric which balances the trade-off between the costs and environmental issues. The dynamic routing model in [15] uses level sensors and wireless communication infrastructure to be aware of the status of the waste containers. It is incorporated analytical modeling and discrete-event simulation in order to achieve real-time dynamic routing. The authors in [16] use improved dynamic route planning for waste collection. It is enhanced a guided variable neighborhood threshold meta heuristic which adapts to the problem of refuse collection.

Routing with time windows is achieved in [17], where the authors analyze the logistics activity within a city. The model finds the cost optimal routes in order the waste vehicles to empty the garbage bins with an adaptive large neighborhood search algorithm. In [18] a rollon-rolloff waste collection routing is achieved serving multiple disposal facilities with huge amounts of garbage at construction sites and shopping districts. It is used large neighborhood search with iterative heuristics algorithms in order to model the problem. The authors in [19] propose a routing model which incorporates Ant Colony System (ACS) algorithm in order to achieve waste routing. They treat the location of the waste bins as a spatial network where they apply k-means in order to cluster the waste bins distribution into a set of partial clusters. In [20] the authors propose a dynamic waste collection routing model based on fuzzy demands by assuming the demands of the customers as fuzzy variables. The model incorporates a heuristic approach based on fuzzy credibility theory.

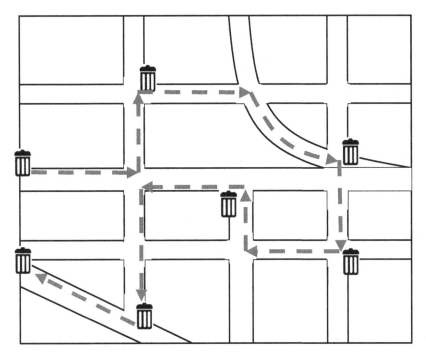

Fig. 1. Upper Layer Semi-Static route traversal for all the waste bins for a certain smart city sector S_i

A heuristic solution is proposed in [21] where the authors state the waste collection problem as a periodic vehicle routing problem with intermediate facilities. The model incorporates variable neighborhood search and dynamic programming in order to achieve optimal solution. The authors in [22] propose a memetic algorithm in order to perform waste collection routing enforced with time windows and conflicts context. The model incorporates a combination of flow and set partitioning formulation in order to achieve multi-objective optimization. In [23] the authors propose a genetic algorithm in order to solve the waste collection dynamic routing problem. Specifically the model assumes that the waste collection problem could be treated as a Traveling Salesman Problem (TSP). Then the genetic algorithm solves the TSP optimally.

We designed a dual layer shortest path model that uses sensor and actuator information of the refuse bins in order to perform real time semi-static and dynamic shortest path routing. The model aggregates data generated by the sensors and actuators located in the waste bins and then updates the refuse vehicles infrastructure. The refuse vehicles in turn decide on real time which route they should follow in case of (i) ordinary schedule, or (ii) network collapse schedule.

3   Semi-Static and Dynamic Routing Models

Consider a smart city with a number of separate bins for solid waste, i.e., organic, plastic, paper, metal and uncategorized. Every waste bin has a level sensor and an actuator. The level sensor is responsible to measure the level of the solid waste in the bin. The actuator is responsible to send a message to the dedicated refuse vehicle when the level of the solid waste reaches some threshold. The refuse vehicle has to do a route from the depot to the smart city in order to take the garbage of several waste bins and then come back to the depot and get empty and ready for the next route. The refuse vehicles are scheduled to route at any time a certain threshold of actuator messages occur. The smart city refuse collection terrain is divided to a set of discrete city sectors. Each city sector is served with a corresponding waste vehicle, which is responsible to collect the solid waste of a specific city sector. For each city sector our model consists of a dual layer model. The upper layer model is a semi-static model where the lower layer model is a dynamic model. The upper layer model constructs a global route traversal for all the waste bins for a certain smart city sector. The lower layer model dynamically constructs a route between two specific waste bins of a certain smart city sector whenever the connection of the road segment between them collapses due to emergency reasons (i.e., a road under construction, unexpected traffic congestion).

3.1   Upper Layer Semi-Static Model

In order to describe the semi-static upper layer routing traversal model we consider a smart city waste collection terrain with a set of city sectors. See Fig. 1. Let $S = \{s_i\}, i = 1, ..., |S|$, be a set of smart city sectors. Let $N^{(i)} = \left\{n_j^{(i)}\right\}, j = 1, ... |N^{(i)}|$, be

Table 1. Semi-Static Upper Layer Algorithm

Input: $N^{(i)}; E^{(i)}; W^{(i)}$
Output: $R^{(i)}$
$R^{(i)} = ()$ /*empty route*/
$j \leftarrow 1; n^f \leftarrow n_j^{(i)}; R^{(i)} \leftarrow R^{(i)} \cup n^f$ /*initialization with the first waste bin*/
While $(N^{(i)} \neq \emptyset)$ **Do**
   $j \leftarrow j + 1; n^s \leftarrow n_j^{(i)}$ /*get the next waste bin*/
   $m \leftarrow 1; w^{(f,s)} \leftarrow \{w_{f,s}^{(i,m)}\}$/*get the first distance weight between n^f and n^s*/
   While $(w_{f,s}^{(i,m)} \neq \emptyset)$ **Do**
      If $(w^{(f,s)} == min)$ **Then** /*get the minimum distance weight between n^f and n^s*/
         $R^{(i)} \leftarrow R^{(i)} \cup n^s$/*update route with the n^s waste bin*/
         $R^{(i)} \leftarrow R^{(i)} \cup e^{(f,s)}$/*update weight list with the $e^{(f,s)}$ road segment*/
         $R^{(i)} \leftarrow R^{(i)} \cup w^{(f,s)}$/*update weight list with the $w^{(f,s)}$ distance weight*/
      Else
         $m \leftarrow m + 1; w^{(f,s)} \leftarrow \{w_{f,s}^{(i,m)}\}$/*get the next distance weight between n^f and n^s*/
      End If
   End While
   $n^f \leftarrow n^s$/*set n^s as the next n^f*/
End While

---

the number of waste bins in the smart city sector s_i. Let $d(n_j^{(i)}, n_{j+1}^{(i)})$ be the distance between the waste bins $n_j^{(i)}$ and $n_{j+1}^{(i)}$. Let $E^{(i)} = \{e_{k,l}^{(i,m)}\}, k, l, m = 1, \ldots |N^{(i)}| - 1$, be the number of road segments in the smart city sector s_i. Where $e_{k,l}^{(i,m)}$ is the number m of road segments between the waste bins $n_k^{(i)}$ and $n_l^{(i)}$ in the smart city sector s_i, for which $d(n_k^{(i)}, n_l^{(i)}) \leq \theta$. Where $\theta \geq 0.5$ is a spatial distance threshold between the waste bins $n_k^{(i)}$ and $n_l^{(i)}$. Let $W^{(i)} = \{w_{k,l}^{(i,m)}\}$ be the distance weights in the smart city sector s_i. Where $w_{k,l}^{(i,m)}$ is the number m of weights of the road segments $e_{k,l}^{(i,m)}$ between the waste bins $n_k^{(i)}$ and $n_l^{(i)}$ in the smart city sector s_i, for which $w_{k,l}^{(i,m)} \leq d(n_k^{(i)}, n_l^{(i)})$. Let,

$$R^{(i)} = \left\{ \left(n_j^{(i)}, e_{k,l}^{(i,m)}, w_{k,l}^{(i,m)} \right) : min \left(\sum_{m=1}^{|N^{(i)}|-1} w_{k,l}^{(i,m)} \right) \right\} \tag{1}$$

be the shortest path in the smart city sector s_i, such that it contains the set of distance weights having the minimum distance weight average with respect to the shortest path waste bins and road segments, respectively.

It holds that $R^{(i)}$ traverses all the waste bins through the road segments with the minimum sum of distance weights. The shortest path semi-static upper layer algorithm is has the following parameters. Let the parameters be $N^{(i)}$, $W^{(i)}$. Then the algorithm is presented in Table 1. The upper layer model is semi-static by means that when a collapse occurs it gives the control to the lower layer model and then incorporates the dynamic routes returned from the lower level model.

3.2   Lower Layer Dynamic Model

In case that a road segment collapses, i.e., is under construction or there is unexpected traffic congestion, then the lower layer dynamic model takes control of the situation. See Fig 2. The model degenerates to the routing between two specific waste bins. Let s_i be the smart city sector where the refuse vehicle is located at the waste bin $n_k^{(i)}$. Consider that the refuse vehicle has collected the garbage from waste bin $n_k^{(i)}$ and on the route to the waste bin $n_l^{(i)}$ the c road segment $e_{k,l}^{(i,c)}$ collapses because it is unexpectedly under construction. It holds that $c \in \{1, ... |N| - 1\}$. Note that before the collapse the distance weight of the c road segment $w_{k,l}^{(i,c)}$ was the minimum distance weight between the waste bins $n_k^{(i)}$ and $n_l^{(i)}$, i.e., due to the semi-static upper layer shortest path algorithm $R^{(i)}$ route traversal. Then the waste vehicle cannot pass through this c road segment in order to collect the garbage from the waste bin $n_l^{(i)}$. In this case a dynamic routing should be done in order to reach the waste bin $n_l^{(i)}$ from another road segment, than c, which do not suffer from collapse. In this case we set the new value of the distance weight of the c road segment $w_{k,l}^{(i,c)} = +\infty$ since it cannot be further used and it must have a high value in order the shortest path dynamic lower layer algorithm not to route from waste bin $n_k^{(i)}$ to waste bin $n_l^{(i)}$ through the c road segment $e_{k,l}^{(i,c)}$. The algorithm returns a shortest path $R'^{(i)} \subset R^{(i)}$ between the waste bins $n_k^{(i)}$ and $n_l^{(i)}$ through another p road segment $e_{k,l}^{(i,p)}$, such that $p \neq c, p \in \{1, ... |N| - 1\}$, with $w_{k,l}^{(i,p)} < w_{k,l}^{(i,m)} \ll w_{k,l}^{(i,c)}$. Then it holds that,

$$R'^{(i)} = \left\{ \left(n_k^{(i)}, n_l^{(i)}, e_{k,l}^{(i,p)}, w_{k,l}^{(i,p)} \right) : \min \left(\sum_{p=1}^{m} w_{k,l}^{(i,p)} \right) \right\} \qquad (2)$$

is the shortest path between waste bins $n_k^{(i)}$ and $n_l^{(i)}$ of the smart city sector s_i, such that it contains the minimum distance weight $w_{k,l}^{(i,p)}$ and the road segment $e_{k,l}^{(i,p)}$, respectively.

It holds that $R'^{(i)}$ routes between two specific waste bins $n_k^{(i)}$ and $n_l^{(i)}$ through the road segment $e_{k,l}^{(i,p)}$ with the minimum distance weight $w_{k,l}^{(i,p)}$, given that the prior minimum road segment $e_{k,l}^{(i,c)}$ is collapsed with $w_{k,l}^{(i,c)} = +\infty$. The shortest path dynamic lower layer algorithm is has the following parameters. Let the parameters be $n_k^{(i)}, n_l^{(i)}, w_{k,l}^{(i,m)}$. Then the algorithm is presented in Table 2. The lower layer model is dynamic by means that when a collapse occurs it handles dynamically the collapse in real-time and then it parses this information to upper layer model in order to redefine its route and direct the refuse vehicle.

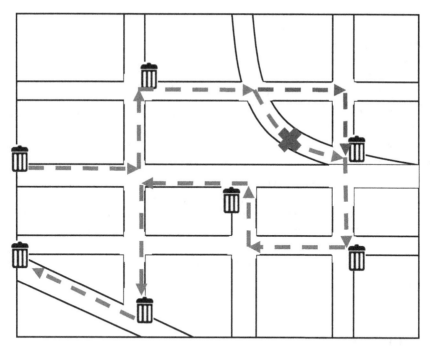

Fig. 2. Lower Layer Dynamic route between two specific waste bins $n_k^{(i)}$ and $n_l^{(i)}$ for a certain smart city sector s_i

3.3 Computational Complexity

In the general case the upper layer semi-static shortest path algorithm has computation complexity $O\left(N^{(i)^2}\right)$ for a set of waste bins $N^{(i)}$ of a smart city sector s_i. However we constrain this complexity considering that from a certain waste bin $n_k^{(i)}$ the algorithm can visit its neighbor waste bin $n_l^{(i)}$, iff,

$$w_{k,l}^{(i,m)} \leq d(n_k^{(i)}, n_l^{(i)}) \leq \theta \tag{3}$$

Let the set of the waste bins visited from $n_k^{(i)}$ under the constraint of the Eq. 3 be $N'^{(i)} = \left\{n_l^{(i)}\right\}$. Then it holds that $N'^{(i)} \subset N^{(i)}$ and subsequently we can infer that the computation complexity of the shortest path upper layer semi-static algorithm is $O\left(N^{(i)} N'^{(i)}\right) \ll O\left(N^{(i)^2}\right)$. For the case of the lower layer dynamic shortest path algorithm the computational complexity is degenerated to $O\left(N^{(i)}\right)$, since there is only the waste bin $n_l^{(i)}$ to be visited from $n_k^{(i)}$.

Table 2. Dynamic Lower Layer Algorithm

Input: $n_k^{(i)}$; $n_l^{(i)}$; $e_{k,l}^{(i,m)}$; $w_{k,l}^{(i,m)}$

Output: $R'^{(i)}$

$R'^{(i)} = ()$ /*empty route*/

$n^f \leftarrow n_k^{(i)}$/*get the first waste bin*/

$R'^{(i)} \leftarrow R'^{(i)} \cup n^f$ /*initialization with the first waste bin*/

$n^s \leftarrow n_l^{(i)}$/*get the second waste bin*/

$m \leftarrow 1$; $w^{(f,s)} \leftarrow \{w_{f,s}^{(i,m)}\}$/*get the first weight between n^f and n^s*/

While $(w_{f,s}^{(i,m)} \neq \emptyset)$ **Do**

   If $(w^{(f,s)} ==$ min) **Then** /*get the minimum distance weight between n^f and n^s*/

      $R^{(i)} \leftarrow R^{(i)} \cup n^s$/*update route with the n^s waste bin*/

      $R^{(i)} \leftarrow R^{(i)} \cup e^{(f,s)}$/*update weight list with the $e^{(f,s)}$ road segment*/

      $R^{(i)} \leftarrow R^{(i)} \cup w^{(f,s)}$/*update weight list with the $w^{(f,s)}$ distance weight*/

   Else

      $m \leftarrow m + 1$; $w^{(f,s)} \leftarrow \{w_{f,s}^{(i,m)}\}$/*get the next distance weight between n^f and n^s*/

   End If

End While

---

4   Evaluation Performance

We evaluated our semi-static upper layer model and dynamic lower layer model in order to assess their performance. In our experiments, the overall number of waste bins in the smart city is 500 waste bins. The smart city terrain is divided to 10 sectors, thus $S = \{s_i\}, i = 1, ...,10$. We assume that each city sector has the same number of 50 waste bins, thus $N^{(i)} = \{n_j^{(i)}\}, j = 1, ...50$. The distance $d(n_j^{(i)}, n_{j+1}^{(i)})$ between the waste bins $n_j^{(i)}$ and $n_{j+1}^{(i)}$ is measured in km. We assume that the number of the final route traversal road segments in the city sector is 49, thus $E^{(i)} = \{e_{k,l}^{(i,m)}\}, k, l = 1, ... 49$. The number of road segments $e_{k,l}^{(i,m)}$ between the waste bins $n_k^{(i)}$ and $n_l^{(i)}$ in each city sector is $m = 3$, assuming that there are three alternative routes between each pair of waste bins. We set the spatial distance threshold between two waste bins $n_k^{(i)}$ and $n_l^{(i)}$ to be 1 km, thus $\theta = 0.5$, then it holds that $d(n_k^{(i)}, n_l^{(i)}) \leq 0.5$ km.

In order the simulation to be realistic we set the distance weights (i.e., in km) $W^{(i)} = \{w_{k,l}^{(i,m)}\}$ of the smart city sector to be randomly selected. We also incorporated time weights (i.e., in min) and fuel weights (i.e., in lit) to measure time spend and fuel consumption in the transition between two waste bins. These weights are also selected randomly by means of generalization. In order to measure the capacity of solid waste collected from the waste bins we used randomly selected capacity weights that measures the current capacity (i.e., in kg) of each waste bin. See Table 3.

Table 3. Routing Simulation

City Sector	Distance (km)	Time (min)	Fuel (lit)	Capacity (kg)
s_1	15.00	56.55	4.830	3310
s_2	19.90	50.40	5.505	3340
s_3	19.80	51.60	5.925	4470
s_4	16.95	54.60	5.400	3570
s_5	17.70	51.30	5.145	3270
s_6	15.80	49.65	5.490	3620
s_7	16.65	52.35	5.340	3340
s_8	20.30	48.75	5.190	3500
s_9	17.55	52.20	5.550	3310
s_{10}	15.00	44.85	4.575	3270

Finally, in order to compare the effectiveness between the semi-static upper layer model and the dynamic lower layer model we test the case where a number of road segments collapse due to a network problem (i.e., a road under construction, unexpected traffic congestion). We compare the CPU elapse time (i.e., in sec) needed to run the semi-static upper layer model algorithm versus the local run of the dynamic lower layer model algorithm for a set of collapsed network segments (i.e., 10) for a set of specific city sectors (i.e., $s_1, ..., s_5$). See Table 4. CPU elapse time is a metric of how fast the routing model algorithms are. This is important in time critical IoT routing because the response time of a waste collection decision affects the overall environmental revenue process of the smart city.

Table 4. CPU Elapse Time Comparison

CPU (sec) s_i	Semi-Static Model	Dynamic Model Collapsed Network Segments									
		1	2	3	4	5	6	7	8	9	10
s_1	0.4056	≈ 0	0.0039	0.0044	0.0052	0.0062	0.0078	0.0104	0.0117	0.0312	0.0390
s_2	0.4680	≈ 0	0.0031	0.0046	0.0053	0.0061	0.0079	0.0108	0.0118	0.0314	0.0395
s_3	0.4680	≈ 0	0.0033	0.0048	0.0055	0.0064	0.0073	0.0107	0.0114	0.0318	0.0392
s_4	0.5772	≈ 0	0.0038	0.0043	0.0057	0.0065	0.0072	0.0102	0.0116	0.0317	0.0391
s_5	0.4836	≈ 0	0.0034	0.0045	0.0056	0.0063	0.0077	0.0103	0.0115	0.0316	0.0394

5   Conclusions

We proposed two routing models in order to achieve solid waste routing in a smart city. The algorithms that introduced are the semi-static upper layer model and the dynamic lower layer model. The upper layer model is semi-static by means that it is built once and runs every time, assuming that there is no network segment collapse. In addition the semi-static model incorporates the changes that handled by the dynamic lower layer model. The dynamic lower layer model is used only in case of a problematic situation (i.e., a road under construction, unexpected traffic congestion) and can handles the routing problem in real time. We evaluated our model by means of effective measures, i.e., distance covered, time spent, fuel consumption and solid waste capacity. We also experimented on certain city sectors with a disastrous scenario of some incrementally collapsed road segments, thus measuring the CPU elapse time of the two models. Future research will be done in the areas of time critical scheduling when waste bins are full and need to empty fast by available waste vehicles. The Internet of Things offers a great promise for various Smart Cities scenarios, including waste collection, which we have just demonstrated.

References

1. Vasseur, J.P., Dunkels, A.: Interconnecting Smart Objects with IP. Morgan Kaufmann, San Francisco (2010)
2. Hernández-Muñoz, J.M., Vercher, J.B., Muñoz, L., Galache, J.A., Presser, M., Gómez, L.A.H., Pettersson, J.: Smart Cities at the Forefront of the Future Internet. In: Domingue, J., et al. (eds.) Future Internet Assembly. LNCS, vol. 6656, pp. 447–462. Springer, Heidelberg (2011)
3. Gubbia, J., et al.: Internet of Things (IoT): A vision, architectural elements, and future directions. Future Generation Computer Systems 29, 1645–1660 (2013)
4. Fadlullah, Z.M., et al.: Toward intelligent machine-to-machine communications in smart grid. Communications Magazine 49(4), 60–65 (2011)
5. Kwang-Cheng, C., Shao-Yu, L.: Machine-to-machine communications: Technologiesand challenges. Ad Hoc Networks 18, 3–23 (2014)
6. Vermesan, O., Friess, P.: Internet of Things: Converging Technologies for Smart Environments and Integrated Ecosystems. River Publishers, Aalborg (2013)
7. Xiaotie, Q., Yuesheng, G.: Data Fusion in the Internet of Thinks. Procedia Engineering 15, 3023–3026 (2011)
8. Ghiani, G., et al.: Capacitated location of collection sites in an urban waste management system. Waste Management 32, 1291–1296 (2012)
9. Felice, P.D.: Integration of Spatial and Descriptive Information to Solve the Urban Waste Accumulation Problem. Procedia Social and Behavioral Sciences (2013) (in Press) (available Online)
10. Milić, P., Jovanović, M.: The Advanced System for Dynamic Vehicle Routing in the Process of Waste Collection. FactaUniversitatis, Series: Mechanical Engineering 9(1), 127–136 (2011)
11. Mes, M.: Using Simulation to Assess the Opportunities of Dynamic Waste Collection. In: Use Cases of Discrete Event Simulation, pp. 277–307. Springer (2012)
12. Mes, M., et al.: Inventory routing for dynamic waste collection. In: Beta Conference, WP No. 431, Eindhoven, Netherlands (2013)
13. Stellingwerff, A.: Dynamic Waste Collection: Assessing the Usage of Dynamic Routing Methodologies, Master Thesis, Industrial Engineering & Management, University of Twente, Twente Milieu (2011)
14. Bing, X.: Vehicle routing for the eco-efficient collection of household plastic waste. Waste Management 34(4), 719–729 (2014)
15. Johansson, O.M.: The effect of dynamic scheduling and routing in a solid waste management system. Waste Management 26, 875–885 (2006)

16. Nuortio, T., et al.: Improved route planning and scheduling of waste collection and transport. Expert Systems with Applications 30, 223–232 (2006)
17. Buhrkal, K., Larsen, A., Ropke, S.: The Waste Collection Vehicle Routing Problem with Time Windows in a City Logistics Context. Procedia Social and Behavioral Sciences 39, 241–254 (2012)
18. Juyoung, W., Byung-In, K., Seongbae, K.: The rollon–rolloff waste collection vehicle routing problem with time windows. European Journal of Operational Research 224(3), 466–476 (2013)
19. Reed, M., Yiannakou, A., Evering, R.: An ant colony algorithm for the multi-compartment vehicle routing problem. Applied Soft Computing 15, 169–176 (2014)
20. Nadizadeha, A., Nasaba, H.: H.: Solving the Dynamic Capacitated Location-Routing Problem with Fuzzy Demands by Hybrid Heuristic Algorithm. European Journal of Operational Research (in Press 2014) (available Online)
21. Hemmelmayr, V., et al.: A heuristic solution method for node routing based solid waste collection problems. Journal of Heuristics 19(2), 129–156 (2013)
22. Minh, T.T., Van Hoai, T., Nguyet, T.T.N.: A Memetic Algorithm for Waste Collection Vehicle Routing Problem with Time Windows and Conflicts. In: Murgante, B., Misra, S., Carlini, M., Torre, C.M., Nguyen, H.-Q., Taniar, D., Apduhan, B.O., Gervasi, O. (eds.) ICCSA 2013, Part I. LNCS, vol. 7971, pp. 485–499. Springer, Heidelberg (2013)
23. Von Poser, I., Awad, A.R.: Optimal Routing for Solid Waste Collection in Cities by using Real Genetic Algorithm. In: Information and Communication Technologies, ICTTA, vol. 1, pp. 221–226. IEEE (2006)

Trust and Reputation Mechanisms
for Multi-agent Robotic Systems

Igor A. Zikratov[1], Ilya S. Lebedev[1], and Andrei V. Gurtov[2]

[1] ITMO University, Russia
[2] Helsinki Institute for Information Technology HIIT and
Department of Computer Science and Engineering, Aalto University, Finland

Abstract. In this paper we analyze the functioning of multi-agent robotic systems with decentralized control in conditions of destructive information influences from robots-saboteurs. We considered a type of hidden attacks using interception of messages, formation and transmission of misinformation to a group of robots, and also realizing other actions which have no visible signs of invasion into a group of robots. We analyze existing models of information security of the multi-agent information system based on a measure of trust, calculated in the course of interaction of agents. We suggest a mechanism of information security in which robots-agents produce levels of trust to each other on the basis of the situation analysis developing on a certain step of an iterative algorithm with the use of onboard sensor devices. For improving the metric of likeness of objects relating to one category ("saboteur" or "legitimate agent") we suggest an algorithm to calculate reputation of agents as a measure of the public opinion created in time about qualities of robots of the category "saboteur" in a group of legitimate robots-agents. It is shown that inter-cluster distance can serve as a metric of quality of trust models in multi-agent systems. We give an example showing the use of the developed mechanism for detection of saboteurs in different situations in using the basic algorithm of distribution of targets in a group of robots.

Keywords: Information security, groups of robots, multi-agent robotic systems, attack, vulnerability, modeling.

1   Introduction

Groups of robots implementing a complex system which consists of many simple devices is a new and actively developing direction of group robotic technology. We assume that desirable group behavior arises from interaction of robots-agents among themselves and their interaction with the environment. The interaction of agents happens in the environment out of a controlled territory that is in conditions where there is a possibility of physical access to robots by the attacker. In such system agents possess several important properties [1]:

S. Balandin et al. (Eds.): NEW2AN/ruSMART 2014, LNCS 8638, pp. 106–120, 2014.

- autonomy: agents are at least partially independent;
- limited view: none of agents have a view of whole system, or the system is so complicated that the knowledge of it has no practical application for the agent;
- decentralization: there are no agents who control all group.

Unique features of a multi-agent robotic system (MRS) complicate the use of existing mechanisms of information security (IS) and give opportunity to attackers to impact on group algorithms (adaptive behavior). The need for research in information security (IS), and also the qualitative description of main threats and features of their implementation in relation to MRS led to appearance of several publications [2, 3]. One of unique threats inherent for MRS as a multi-agent system is the use by attackers of robots-saboteurs who realize harmful actions. We understand robot-saboteur activities as a harmful information influence (attack) directed on implementation of a threat to information security concerning robots-agents R_j ($j = \overline{1, N}$) and realized with the use of information tools and technologies as a result of which the new action selected by agents won't promote an increase of system functionality in available conditions.

In this article we consider mechanisms of soft security directed on detection and neutralization of hidden attacks which do not have identified signs unlike attacks which are carried out by jamming of communication links, DDoS-attacks, cracking and compromising of ciphers, etc. In case of the hidden attacks, robots, their systems and communication links function in a standard mode. Realizing a hidden attack, robots-saboteurs of a warring party can provide false or misleading information, and traditional mechanisms of security can't protect users from this type of threats.

For protection against such hidden attacks we can use a method of the protected agent states, methods of mobile cryptography, a method of Ksyudong [6], Buddy Security Model [7, 8], which matches well with the principles of creation of decentralized systems. Besides, for providing of protection of the user from such threats, we use mechanisms of social monitoring, namely trust and reputation systems. These mechanisms are based on calculation of trust of agents to each other, realized in the course of monitoring of actions of an agent in the system [8, 9, 10, 11, 12]. Distinction in ways of computation of the trust level is caused by features of the domain where interaction of participants takes place. It can be the electronic markets, peer-to-peer networks, on-line social networks, etc. As a result, in existing models of trust there are different treatments of the concept of trust and reputation, different subjects and objects of trust are considered.

The goal of this paper is development of a method of protection of MRS from hidden attacks of robots-saboteurs, based on computation of a measure of trust and reputation to robots-agents in a group of robots in case of decentralized control.

The rest of the paper is organized as follows. In Section 2, we provide a brief survey on multi-agent robotic systems. In Section 3, we develop a model of multi-agent decision making using trust and reputation. In Section 4, we describe implementation of the model as well as its simulated functionality. Finally, Section 5 concludes the paper.

2   Functioning of Systems with Decentralized Planning of Actions

Robots-agents of MRS, unlike agents of MS, are equipped with the onboard sensor and measuring device from which the robot receives information about environment, and also a radio channel intended for information exchange in the course of execution of the target. We consider MRS actions when using the most widespread iterative procedure of optimization of a group decision, distribution of targets in a group of robots [13]. MRS functioning for this goal in a general form looks as follows.

Assume there are M targets and a group of robots which consists of N robots R_j $(j = \overline{1, N})$. A squad of forces (a number of robots for target execution) shall be selected for each target. The target is *provided* when it is selected by the necessary number of robots. The remained robots will form a reserve cluster. Each robot-agent knows coordinates of a target, its own coordinates, and a required squad of forces for each target. The robot "R" estimates efficiency of its actions for each target and tells an array of the estimates $D_j = [d_{j1}, \quad d_{j2}, \quad \dots, \quad d_{jM}]$ to remaining members of group. Matrix "D" with dimensionality (N, M), which elements are estimates of efficiency of the robot "j" for target "l", is created in a processor device (CPU) of each robot. Iterative procedures of formation of the group plan as a result of which for each target $T_l \in \mathbf{T}_c$ the equation maximum is provided, begins after matrix formation

$$\mathbf{Y}_c = \sum_{j,l=1}^{N} d_{jl} n_{jl} \rightarrow max, \tag{1}$$

in case of restrictions

$$\sum_{l=1}^{N} n_{jl} = 1,$$

$$\sum_{j=1}^{N} n_{jl} = n_l^{max},$$

$$d_{jl} \geq 0,$$

where

$$n_{jl} = \begin{cases} 1, & \text{if "}j\text{" robot selects "}l\text{" target,} \\ 0, & \text{otherwise.} \end{cases}$$

Here $j = \overline{1, N}$, $l = \overline{1, N}$, a n_l^{max} is a necessary number of robots which must select "l" target.

The basis for iterative procedures is the analysis by every robot-agent of an array of estimates of efficiency and a selection of a target for which the value of an assessment of the efficiency is maximal. Then there is an information exchange about the selected decisions, the analysis and "discussion" of the decisions made by other robots. The agents with the value d_l, select the suitable target "*l*", "eliminate" from a matrix **D** the provided targets and the robots which have selected the target according to an equation (1). As in the memory of all robots there are identical matrices **D**, and results of computation will match. The procedure repeats until all targets of a set *M* are provided. There is a modification of this algorithm which allows to consider not only estimates d_{jl}, but also a possibly of changes of a goal function if a robot R_j refuses the target selected from the current iterative loop and will select other target. A minor modification of the algorithm allows to resolve a situation when there are some agents with identical estimates of efficiency on one target.

Let's review a trivial example. Assume a group of seven robots (*N*=7) needs to distribute two targets (*M*=2) A and B. It is known that each target should be provided with two agents. We will consider the distance from a robot to a target as a metric of efficiency of the target. That is, the closer the robot is located to the target, the higher is its efficiency. Assume the matrix **D** with estimates of efficiency looks as follows:

	A	B
D_1	3.2	1.0
D_2	1.9	2.5
D_3	0.7	5.4
D_4	3.6	3.5
D_5	5.8	3.4
D_6	4.2	5.6
D_7	5.8	1.4

Then as a result of algorithm operation, the target A will be provided with agents R_2 and R_3, and the target B will be provided with agents R_1 and R_7.

It is obvious that destructive information influences of robots-saboteurs can include transmission to members of a group of a vector of the estimates containing false information, violations of the rules, made in discussion of decisions (unreasonable announcements about a selection of the targets), etc. As a part of a squad of forces intended for the target, there could be saboteurs who would not execute actions required from the legitimate agents concerning the target. Carrying out such attacks can result to not reaching of the maximum by (1), and/or appearance of actually not provided targets. For example, if robot R_5 is the saboteur, it can realize "soft" influence to the target A:

$$D_5 = [0.8, \quad 3.4].$$

As a result of their attack robots R_5 and R_3 will be assigned to the target A, and this target won't be provided with a required number of legitimate agents.

3   Model of Information Security for Multi-agent Robotic Systems on the Basis of Reputation and Trust Computation

Definition 1. *The trust in this case is a measure which characterizes readiness of the subject to interact with an object in this situation.* According to the trust relationships policy, an agent can be blocked when trust to this agent is below some preset value. Then the low level of trust won't allow the saboteur to make a destructive impact on decision-making by agents. It follows from this that actions of the saboteur on increase of trust are assumed by involvement of the robot in achievement of the target of MRS, and it contradicts the logic of its use from the point of view of the attacker.

Trust level computation process in a general form is the following [14]. After the start of an iterative loop the robot "j" (robot-object of the trust) ($j = \overline{1, N}$), receives in the active phase of the current iteration in the disposal communication link and access to processor devices (CPU) of robots-members of the group. Based on available information about states and the current actions of members of group, the object decides an action in case of which the value of the goal function is maximal, and provides access on writing information about the made decision in robots-subjects. Remaining robots-agents (subjects of trust), after having received this information, check:

a) the acquired information regarding compliance;
b) "usefulness" of the action selected by the robot-object from the point of view of an increment the goal function.

If the robot "i" (robot-subject) ($i \neq j$) as a result of a check received the positive decision, it gives the positive vote for the robot-object "j", and reports about it to remaining subjects. Each subject, having received data on results about the check of an object by other subjects, counts the number of the positive and negative votes given for it, calculating trust of object "j".

However in MRS there can be an implementation of groups of saboteurs which highly appreciate each other, and lowly appreciate other members of a group. Discrediting of legitimate agents can be a consequence of such actions [14]. For the solution of this problem it is necessary to introduce a concept of reputation in the mechanism of IS.

Definition 2. *Reputation is a public opinion created in time about qualities of this or that agent-subject.* Then in case of a count of positive and negative votes given for object, the reputation of voting subjects by summing of their estimates will be considered. In this case influence of agents with low reputation on trust computation process to an object will be smaller, than subjects with high reputation. We note that the reputation value depends on history of interaction of the agent in a group, and on time of stay in it.

Thus, the concepts of trust and reputation of multi-agent systems are actually used for recognition in a group of robots of robots-saboteurs implemented by an attacker. Then for the solution of the task of recognition of entered signs of trust and reputation should provide the greatest similarity of objects within a group (cluster) and the greatest distance between groups (clusters). In the simple case we will speak about two clusters: "legitimate agents" and "robots-saboteurs".

4   Implementation of Model of Information Security on the Basis of Trust Level Computation

We will show model implementation using the already considered example of distribution of targets in a group of robots.

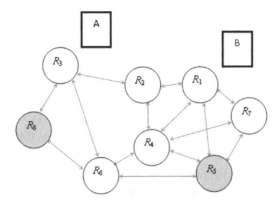

Fig. 1. Distribution of targets in the presence of saboteurs

Assume, that in a group of robots in Figure 1 there are two robots-saboteurs (robots 5 and 8), their task is prevention of selection of targets by a squad of forces.

In Figure 1 the relative positioning of robots and targets is shown, and also arrows between agents designate communications by means of onboard sensor and measuring devices, for example visual communication. We assume that all agents of the considered group have a radio communication channel for information exchange.

Step 1. Each robot-agent creates a vector of efficiency estimates, and tells the estimates to all members of a group. Assume robots-saboteurs carry out an attack which includes   misinformation of agents concerning their distance to the target: $D_5 = [0.8, \quad 3.4]$, $D_8 = [3.1, \quad 0.2]$. As a result in the CPU of each robot the matrix **D** of efficiency estimates is created, which looks like follows

D_1	3.2	1.0
D_2	1.9	2.5
D_3	0.7	5.4
D_4	3.6	3.5
D_5	**0.8**	3.4
D_6	4.2	5.6
D_7	5.8	1.4
D_8	3.1	**0.2**

From the second step, the actions of information security directed on detection of destructive influences are executed.

Step 2. Agents by means of their OSMD execute verification of data in an array **D**. The robot "*j*" writes results of verification into an array of estimates $V_j = [v_{j1}, \quad v_{j2}, \quad \dots, \quad v_{jM}]$, and tells it to members of the group. Here: $v_{ji} = -1$, if the information transferred by the robot "*i*", is not confirmed by data of OSMD of robot "*j*": $v_{ji} = 1$ otherwise. If the robot "*i*" doesn't watch robot "*j*" by means of its OSMD then $v_{ji} = 0$. For example, for the situation in Figure 1, the robot R_1 will make the following array $V_1 = [1,1,0,1,-1,0,1,0]$. As the robot-saboteur R_5 is in an area of coverage of the onboard sensor and measuring unit, the robot R_1 found out that robot R_5 is from the target A at the distance exceeding the value specified in an array D_5. Agents R_3, R_6 and R_8 are out of coverage of the OSMD of the robot R_1, that caused appearance of zeros on the appropriate positions of an array. It is necessary to note that saboteurs R_5 and R_8 can act in coordination. In this case they can realize the following actions:

1. To give each other marks "1" confirming reliability of transferred data even in a case when they are not in the area of coverage of their OSMD.
2. To give marks "-1" to remaining members of the group for their discrediting in case of detecting by their OSMD.

Thus, as a result of execution of step 2 in the CPU of each robot the array **V** is created. This array for a reviewed example looks like follows:

Table 1. Array of action estimates of members of a group

	1	2	3	4	5	6	7	8
1	1	1	0	1	-1	0	1	0
2	1	1	1	1	0	0	0	0
3	0	1	1	0	0	1	0	-1
4	1	1	0	1	-1	1	1	0
5	-1	0	0	-1	1	-1	-1	1
6	0	0	1	1	-1	1	0	-1
7	1	0	0	1	-1	0	1	0
8	0	0	-1	0	1	-1	0	1

Apparently from the table, a column represents a set of estimates from all members of a group of the certain agent. Value of trust of this agent in a simple case can be equal to division of quantity of the positive voices γ^+ on total quantity of voices $\gamma = \gamma^+ + \gamma^-$ [9]:

$$w_i = \frac{\gamma^+}{\gamma^+ + \gamma^-} . \tag{2}$$

For a reviewed example (Table 1) the trust levels of agents the group are calculated with formula (2), will have values: **T** = [0.8, 1.0, 0.75, 0.83, 0.33, 0.6, 0.75, 0.33].

Step 3.Computation of agents' reputation.

If on step 2 agents estimated actions of those objects which were in the area of coverage of their OSMD (that is direct interactions of agents), the actions on step 3 can be regarded as the analysis of interaction of agents with the remaining members of the group who have expressed opinions about observed objects.

We will consider an array of estimates **V** in Table 1. The analysis of this array shows that there are objects of an assessment which are watched by the OSMD of several robots; for example, robots 1 and 2 watched actions of robot 4 and expressed the estimates. Then if the robot's "i" assessment concerning action of object "k", matches the mark which has been stated by robot "j" concerning the same action of object "k", it will be the base of an increase of the reputation level. Otherwise there is a reduction. In the reviewed example, the analysis of the table shows that interaction of robots 1 and 2 can be considered as follows:

1. The reputation increases by 1, if robots 1 and 2 are in the area of coverage of their OSMD, and gave to each other the positive marks.
2. The reputation increases by 1, if robots 1 and 2 watched robot 4 actions by means of their OSMD, and their estimates of its actions matched.
3. The total reputation of the robot 2 received in case of interaction with robot 1, and total reputation of the robot 1 received in case of interaction with robot 2 is equal 2.

The reputation is calculated in the analysis of interaction of robots 3 and 1, will be equal 1, as without watching each other, these agents watched actions of robot 2 and their estimates of its actions matched. Having carried out the similar analysis of the array **V** each robot creates an array **S** of reputation level estimates.

Table 2. Array **S** of reputation level estimates of agents

	1	2	3	4	5	6	7	8
1		2	1	4	-3	2	3	-1
2	2		1	2	-2	2	2	-2
3	1	1		2	-1	2	0	-2
4	4	2	2		-4	2	3	-1
5	-3	-2	-1	-4		-2	-3	1
6	2	2	2	2	-2		2	-2
7	3	2	0	3	-3	2		0
8	-1	-2	-2	-1	2	-2	0	

From here it is possible to calculate the level of reputation of each agent q_j, as a result of the relations to it of all members of a group in the course of their direct interaction, and interaction with neighbors. Using formula (2), we obtain the following values of a vector **Q** = [0.75, 0.69, 0.66, 0.72, 0.12, 0.77, 0.11].

Step 4. Accounting of change of reputation level.

We note that values of vector **Q** do not match the reputation from Definition 2, because components of a vector consider "opinion" of a group about objects, created based on analysis results of only one situation. For the accounting of a factor of time in operations [15, 16] it is suggested to use strictly increasing functions. It is known that function and frequency curve of the random value which characterize duration of functioning of a complex system, an enterprise or a living being, etc. can be described by Veybulla-Gnedenko's function as follows:

$$F(t) = 1 - e^{-at^k}, \tag{3}$$

where "a" determines the scale, and "k" the type of a frequency curve. In case of constant intensity of iterative procedures in the algorithm of distribution of the targets it is possible to assume k=1. If assumed iteration number as the parameter of time, the type of function of time will looks like as in Figure 2.

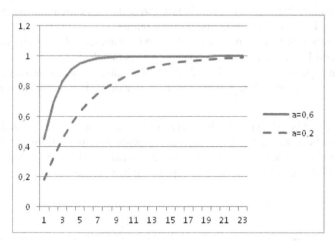

Fig. 2. Influence of parameter "a" on the reputation level with an increase in the number of iterations "l"

From Figure 2 it is visible that setting the parameter according to the trust relationships policy accepted in system, it's possible to control the growth of speed of object reputation.

Thus, a scalar multiplication of vector **Q** on value F(l), where l is the number of the current iteration of target distribution algorithm, will allow to control influence of beginners with a small level of reputation on the process of estimation of agents' trust level in the current situation.

Step 5. Taking into account aforesaid the formula for calculation of trust level (2) will looks as follows:

$$w_i = \frac{p_i}{p_i + n_i}, \tag{4}$$

where

$$p_i = \sum_{j=0}^{N} h_{ij} \cdot q_j \cdot F(l),$$

$$n_i = \sum_{j=0}^{N} g_{ij} \cdot q_j \cdot F(l).$$

Values h_{ij} and g_{ij} are defined by the analysis of estimates v_{ij} of array **V**:

$$h_{ij} = \begin{cases} 1, & \text{if robot "}j\text{" positively estimated actions of robot "}i\text{",} \\ 0, & \text{otherwise} \end{cases}$$

$$g_{ij} = \begin{cases} 1, & \text{if robot "}j\text{" negatively estimated actions of robot "}i\text{",} \\ 0, & \text{otherwise.} \end{cases}$$

Then for a reviewed example we will finally receive component values of a vector of the trust level, calculated with $\mathbf{W} = [0.96, 1.0, 0.94, 0.97, 0.071, 0.9, 0.95, 0.08]$ (Fig.3).

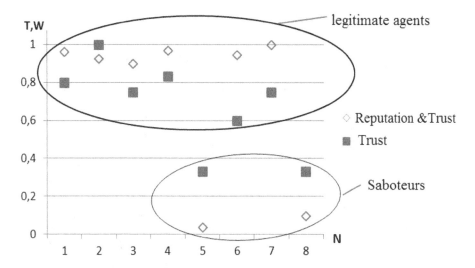

Fig. 3. Characteristics of agents on the trust level **T** and reputation & trust level **W**

From Figure 3 and calculations it is visible that when using measures of trust and reputation (formulas 3 and 4) cluster X_{la} "legitimate agents", to which robots 1-4, 6 and 7 belong, is at bigger distance from cluster X_d "saboteurs" (robots 5 and 8), than when using only the trust measure.

$$\left| X_{\psi la}^1 - X_{\psi d}^1 \right| = 0{,}45 < \left| X_{\psi la}^2 - X_{\psi d}^2 \right| = 0.88,$$

where $X_{\psi la}^i$ and $X_{\psi d}^i$ are the centers of clusters, which calculated as $X_\psi = \Sigma w_i / n$ with use of a formula (2) or formulas (3) and (4). As a result of execution of step 5 there is detection of saboteurs by the criterion of recognition accepted in the system, and further steps which directed on execution of the basic algorithm of the targets distribution without the information transferred by robots-saboteurs.

It is possible to show that further actions of saboteurs as a part of a group lead to an increase of intercluster distance between objects of cluster X_{la} "legitimate agents" and objects of cluster X_d "saboteurs". So for a situation in Figure 4, when robots changed their positions in space, and the following iteration of a target distribution is carried out, we obtain the following results:

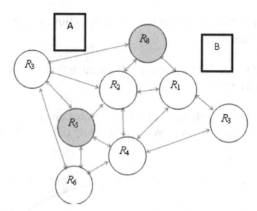

Fig. 4. Situation development on the second step of iterative process of target distribution

Table 3. Array **V** of action estimates of members of a group in the second step of iteration

	1	2	3	4	5	6	7	8
1	1	1	0	1	0	0	1	-1
2	1	1	1	1	-1	0	0	-1
3	0	1	1	0	-1	1	0	-1
4	1	1	0	1	-1	1	1	0
5	0	-1	-1	-1		-1	0	1
6	0	0	1	1	-1	1	0	0
7	1	0	0	1	0	0	1	0
8	-1	-1	-1	0	1	0	0	1

Table 4. Array **S** of reputation estimates of agents in the second step of iteration

	1	2	3	4	5	6	7	8
1		3	2	3	-2	1	2	-2
2	3		3	3	-3	3	2	-3
3	2	3		3	-3	2	0	-2
4	3	3	3		-3	2	2	-2
5	-2	-3	-3	-3		-3	-1	2
6	1	3	2	2	-3		1	-1
7	2	2	0	2	-1	1		-1
8	-2	-3	-2	-2	1	-1	-1	

Then the vector of the trust level will be equal **W** = [0.95, 0.93, 0.904, 0.98, 0.052, 0.97, 1.00, 0.093], and intercluster distance, calculated on formulas (3) and (4), increases.

$$\left|X_{\text{ц}la}^3 - X_{\text{ц}d}^3\right| = 1.075 > \left|X_{\text{ц}la}^2 - X_{\text{ц}d}^2\right| = 0.88.$$

We will consider an algorithm for operation in case of appearance of new objects.

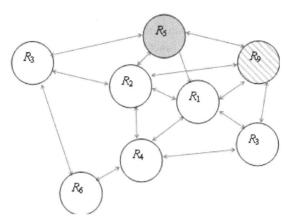

Fig. 5. Operation in case of appearance of a new agent

We assume that as a result of development of a situation there was a situation which is shown in Figure 5. Here R_9 is a new object. As for subjects of a group the interaction history with R_9 absent, its reputation at the moment from the point of view of a group is equal to zero.

We will assume that as a result of execution of steps 1-3 we receive the following arrays **V** and **S** of estimates:

Table 5. Array **V** of action estimates of members of a group on the third step of iteration

	1	2	3	4	5	6	7	9
1	1	1	0	1	-1	0	1	1
2	1	1	1	1	-1	0	0	1
3	0	1	1	0	-1	1	0	0
4	1	1	0	1	0	1	1	0
5	-1	-1	-1	0	1	0	0	-1
6	0	0	1	1	0	1	0	0
7	1	0	0	1	0	0	1	1
9	1	1	0	0	-1	0	1	1

In Table 6 we have a vector $\mathbf{Q} = [0.83, 0.80, 0.82, 0.87, 0.067, 0.85, 0.83, 0.80]$. Considering that for objects $R_1 - R_7$ current iteration is the first, and for R_9 the first, the coefficients calculated on formula 3, when $k=1$ and $a=0.6$ will be equal $F(3) = 0.835$ and $F(1) = 0.451$. Then the reputation of agents will be equal:

$$\mathbf{Q} = [0.695, \ 0.667, \ 0.682, \ 0.73, \ 0.056, \ 0.71, \ 0.69, \ 0.36].$$

Apparently, robots with the smallest reputation are: robot-saboteur R_5 ($q_5 = 0.056$) and agent R_9 ($q_9 = 0.36$). If in the first case the low level of reputation is caused by the fact of detection by members of group of destructive actions from the robot R_5, in the second case the reason is the factor of time which is entered by function 3. It is obvious that changing a function parameter "a" makes possible to settle influence of a time factor.

Table 6. Array S of reputation estimates of agents on the third step of iteration

	1	2	3	4	5	6	7	9
1		3	2	3	-3	1	3	3
2	3		2	4	-4	2	2	3
3	2	2		2	-2	1	0	2
4	3	4	2		-2	1	2	2
5	-3	-4	-2	-2		-1	-2	-3
6	1	2	1	1	-1		1	0
7	3	2	0	2	-2	1		2
8	3	3	2	2	1	0	2	

Using a formula (4) we will receive component values of a vector of the trust level $\mathbf{W} = [0.98, \ 0.983, \ 0.97, \ 0.1, \ 0.03, \ 1.0, \ 1.0, \ 0.98]$. From here it is visible that the trust to agent-beginner is high, and the measure of closeness of this subject to object of a cluster "legitimate agents" is less than to an object of a cluster "saboteurs".

$$\left| X_{\text{цла}} - X_9 \right| = 0.013 < \left| X_{\text{цd}} - X_9 \right| = 0.95.$$

It is obvious that higher quality of recognition of the agents which make destructive information influences, with use of measures of trust and reputation is accompanied by the increasing volume of computing resources. So, in case of operation of the standard algorithm in the CPU of the agent it is necessary to create a matrix $\mathbf{D}$ of efficiency estimates with dimensionality (N, M). When using algorithm "1" it is necessary to create an array $\mathbf{V}$ of actions estimates of members of group with dimensionality (N, N), and when using algorithm "2" it is necessary to create an array $\mathbf{S}$ of reputation level estimates with same dimensionality. However from the point of view of the recognition quality, when categories of object closeness of one cluster can serve as a measure of this recognition quality, and remoteness between clusters, the scoring in use of such character space is obvious.

5   Conclusion

The developed model represents an approach to information security of MRS in which access control of a robot-agent to a group of robots is carried out on the basis of a measure of the trust level. It is produced by members of a group by the analysis of the situation which has developed on the certain step of an iterative process, taking into account the previous history of their interaction. Thus, the members of a group who for the first time appear in the coverage zone of the onboard sensor device of a robot-agent possess the minimum reputation. For increasing the trust level, an agent needs not only to execute functions for serving a target but also to give a correct feedback on the actions of other robots.

References

1. Wooldridge, M.: An Introduction to MultiAgent Systems, 366 p. John Wiley & Sons Ltd., paperback (2002) ISBN 0-471-49691-X
2. Higgins, F., Tomlinson, A., Martin, K.M.: Threats to the Swarm: Security Considerations for Swarm Robotics. International Journal on Advances in Security 2(2&3), 288–297 (2009)
3. Zikratov, I.A., Kozlova, E.V., Zikratova, T.V.: Analysis of vulnerability of robotic complexes with swarm intellect. Scientific and Technical Journal of Information Technologies, Mechanics and Optics 5(87), 149–154 (2013)
4. Neeran, K.M., Tripathi, A.R.: Security in the Ajanta MobileAgent system. Technical Report. Department of Computer Science, University of Minnesota (May 1999)
5. Sander, T., Tschudin, C.F.: Protecting Mobile Agents against malicious hosts. In: Vigna, G. (ed.) Mobile Agents and Security. LNCS, vol. 1419, pp. 44–60. Springer, Heidelberg (1998)
6. Xudong, G., Yiling, Y., Yinyuan, Y.: POM-a mobile agent security model against malicious hosts. In: Proceedings of High Performance Computing in the Asia-Pacific Region, vol. 2, pp. 1165–1166 (2000)
7. Page, J., Zaslavsky, A., Indrawan, M.: A Buddy model of security for mobile agent communities operating in pervasive scenarios. In: Proceeding of the 2nd ACM Intl. Workshop on Australian Information Security & Data Mining, vol. 54 (2004)
8. Page, Zaslavsky, A., Indrawan, M.: Ountering security vulnerabilities using a shared security buddy model schema in mobile agent communities. In: Proc. of the First International Workshop on Safety and Security in Multi-Agent Systems (SASEMAS 2004), pp. 85–101 (2004)
9. Schillo, M., Funk, P., Rovatsos, M.: Using trust for detecting deceitful agents in artificial societies. Applied Artificial Intelligence 14, 825–848 (2000)
10. Golbeck, J., Parsia, B., Hendler, J.: Trust Networks on the Semantic Web. In: Klusch, M., Omicini, A., Ossowski, S., Laamanen, H. (eds.) CIA 2003. LNCS (LNAI), vol. 2782, pp. 238–249. Springer, Heidelberg (2003)
11. Garcia-Morchon, O., Kuptsov, D., Gurtov, A., Wehrle, K.: Cooperative security in distributed networks. Computer Communications (COMCOM) 36(12), 1284–1297 (2013)
12. Ramchurn, S.D., Huynh, D., Jennings, N.R.: Trust in multi-agent systems. The Knowledge Engineering Review 19(1), 1–25 (2004)

13. Kalyaev, I.A., Gayduk, A.R., Kapustyan, S.G.: Models and algorithms of collective control in groups of robots, 280 p. Physmathlit, Moscow (2009)
14. Zikratov, I.A., Zikratova, T.V., Lebedev, I.S.: Confidential model of information security of multi-agent robotic systems with decentralized control. Scientific and Technical Journal of Information Technologies, Mechanics and Optics 2(90), 47–52 (2014)
15. Carter, J.: Reputation Formalization for an Information-Sharing Multi-Agent System. Computational Intelligence 18(2), 515–534
16. Beshta, A., Kipto, M.: Creation of model of trust to objects of an automated information system for preventing of destructive impact on system. News of Tomsk Polytechnic University 322(5), 104–108 (2013)

A Notification Model for Smart-M3 Applications

Ivan V. Galov[1] and Dmitry G. Korzun[1,2]

[1] Department of Computer Science, Petrozavodsk State University (PetrSU)
33, Lenin Ave., Petrozavodsk, 185910, Russia
{galov,dkorzun}@cs.karelia.ru
[2] Helsinki Institute for Information Technology (HIIT) and
Department of Computer Science and Engineering (CSE), Aalto University
P.O. Box 15600, 00076 Aalto, Finland

Abstract. Smart-M3 platform supports development of applications consisting of autonomous knowledge processors that interact by sharing information in a smart space. In this paper, we introduce a notification model for ontology-based design and programming of interactions in such applications. Our model is based on the two Smart-M3 fundamentals: subscription operation and RDF representation. The applicability is demonstrated on the case study of SmartScribo system for multi-blogging and on simulation experiments for performance evaluation.

Keywords: Smart spaces, Smart-M3, Publish/Subscribe, RDF.

1 Introduction

The smart spaces paradigm aims at development of ubiquitous computing environment that acquires and applies knowledge to adapt services to the inhabitants [1]. Smart-M3 interoperability platform [2] provides means for creating and deploying smart spaces. (M3 stands for Multidevice, Multidomain, and Multivendor.) Examples of Smart-M3 applications are SmartScribo system [3] for personalized semantic multi-blogging and SmartRoom system [4] for collaborative work in a spatially localized digital environment. In Smart-M3, a smart space realizes a shared knowledge base for use by applications [2,5]. Each application consists of knowledge processors (KP) that interact in the smart space using blackboard [6] and publish/subscribe [7] interaction models. The information representation is RDF-based, employing Semantic Web technologies [8].

Application developers are faced with problems of design and programming of interacting KPs. In addition to blackboard-based read/write primitives, advanced interaction is based on publish/subscribe: whenever one KP publishes data in the smart space some other KPs are notified about the changes due to subscription. When many KPs participate and much data are shared such interaction becomes complicated for implementation. In this paper we analyze this design and programming problem. We introduce a notification model that systematizes the interaction part on the application level and provides properties to implement on each individual KP. Notification model allows construction of

S. Balandin et al. (Eds.): NEW2AN/ruSMART 2014, LNCS 8638, pp. 121–132, 2014.

information flows coupling a publisher KP with its subscriber KPs. Notification model is ontology-based, enhancing the ontology of the whole application.

The rest of the paper is organized as follows. Section 2 states the problem of design and programming of knowledge processors interaction in a Smart-M3 application. Section 3 describes the notification model and its design properties. Section 4 analyses the applicability of the model using SmartScribo system as a case study. Section 5 evaluates the performance using simulation experiments. Section 6 concludes the paper.

2   Smart-M3 Application: Interaction in Smart Space

Each smart space provides a shared information store for its participants. Blackboard model [6] is used for data exchange: participants write/read data to/from the shared information store. The model is extended with publish/subscribe [7]: participants subscribe on specific content and receive updates made by others.

Smart-M3 platform [2] is open-source platform for implementing smart spaces. The key architectural component of Smart-M3 is Semantic Information Broker (SIB) that provides access to the shared content. Participants are knowledge processors (KPs); they are software agents running on devices of the environment. Publish/subscribe operation supports advanced interaction of KPs when subscriber KPs make persistent queries and react on asynchronously incoming updates [9]. The idea of such interaction is shown in Fig. 1.

Shared content is represented using Resource Description Framework (RDF) of the Semantic Web [8]. An RDF model consists of a set of RDF triples, each has the form of "subject–predicate–object". Such a representation forms a directed graph with subjects and objects as nodes and predicates as links.

Every KP has its own description of content the KP shares participating in the smart space. This content forms a partial RDF graph stored in the smart space (maintained by SIB). An RDF model is machine-oriented and does not provide human-oriented mechanisms to describe semantics of the content nor determine problem-aware content representation structure. For this purpose, KP developers apply ontologies at the design phase. Application ontology declaratively describes the application domain. Overlapping of individual KP ontologies

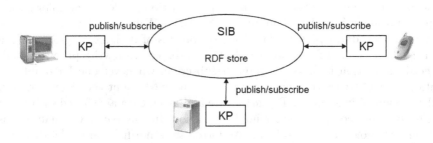

Fig. 1. Smart-M3 concept

makes interaction between KPs possible: each KP tracks changes in shared content if the latter is described within the KP's ontology.

Ontologies can be described with web ontology language (OWL). On one hand, an OWL ontology is serializable to RDF triples. On the other hand it allows describing application domain data and processes in terms of classes and their properties, which provides a high abstraction level for Smart-M3 application developers [10, 11].

From the OWL point of view smart space content consists of linked objects, called individuals. Any individual is an instance of particular ontological class of the application ontology. From the RDF point of view, an individual is a set of RDF triples with the same value of triple subject (i.e., an RDF subgraph). This triple subject is called the identifier of individual. We shall use descriptions in RDF as it allows to see what is actually stored in the smart space. (Note that SIB always operates on the RDF level an RDF triple store is used on the bottom.) Nevertheless we still exploit the term "individual" for simpler intuition.

One KP (or interaction of several KPs) constructs a service. There are KPs involved into service provision and KPs acting as service consumers. An example with two KPs—a client KP and a service KP—is shown in Fig. 2. Each user runs her/his client KP, which changes some properties of different individuals. The service KP has the constraint: to process modified individuals correctly the KP needs to receive the whole updated individual, not just updated properties separately. With subscription on separate properties of individuals, however, service KP subscription would run several times (each time with one edited property). It is not obvious how to estimate a moment when individual modification is finished and it is ready for processing on the service KP. The situation becomes more complicated when several individuals are modified and needed to be processed at the same time. In this case it is more convenient when one KP notifies other KPs about the need of processing specific individuals when they are already modified in the smart space. Thus we prevent reading information that is not ready for processing. It can be considered as some kind of access control mechanism. Another approach of access control is presented in [12] where undesirable changing of information is prohibited.

Let KP sender and KP receiver need to interact. The KP sender needs to pass a portion of information to the KP receiver to attain a required action.

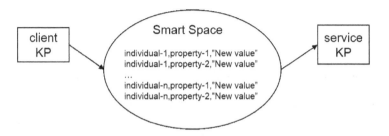

Fig. 2. Schematic example of subscription on several properties

This information can be represented either text data or individuals from the application ontology. The result of interaction is information transfer or service delivery. We consider the following problems where interaction of KPs appears: 1) service provision, 2) service composition, 3) service information dissemination.

In service provision, one KP consumes a service from another KP, and the former usually runs on a personal user device. Service composition implies that a service provides input for another service, forming a new composed service. Besides, services can disseminate some information between each other.

In general, several KPs are involved into interactive activity. To solve this design problem of interacting KPs we introduce a notification model. It focuses on implementation of basic pair-wise interaction between the involved KPs.

3   Notification Model

Given a Smart-M3 application, its notification model describes possible situations for KPs to interact with each other. We consider the following classes of situations where one KP (receiver, denote KP_{rcv}) or more ones are involved into interaction by another KP (sender, denote KP_{snd}).

Request: KP_{rcv} performs a given operation (service) based on data provided by KP_{snd}.

Event: KP_{rcv} reacts on a particular event (informational fact) that KP_{snd} is disseminating.

These situations defines two functional viewpoints on KP_{rcv} in a Smart-M3 application: a data processor or a reacting unit. The former is closer to procedure call programming and the latter follows event-driven programming.

Consider the following pairwise interaction where asynchronous communication applies the publish/subscribe model

$$KP_{snd} \leftrightarrow KP_{rcv}. \tag{1}$$

The following properties are achieved for interacting KPs.

One-to-many: A single KP_{snd} can affect many KP_{rcv} at once.
Decoupling: Sending and receiving do not block KP_{snd} and KP_{rcv}.
Anonymity: KP_{snd} and KP_{rcv} do not need to know each other.

We define a *notification* an abstract informational message to be sent by KP_{snd} and to be received by KP_{rcv} if these KPs need to attain required interaction (performing operation, returning the result, informing on event). In our model, a notification is represented as an RDF triple or a linked set of them. Note that the basic subscription mechanism of Smart-M3 operates on the RDF level, though OWL-aware extensions are available for application developers [9].

Notification process in interaction (1) is depicted in Fig. 3. The steps are implemented in application logic of both KP_{snd} and KP_{rcv}. Before the process, the KPs subscribe for the notifications defined in the model.

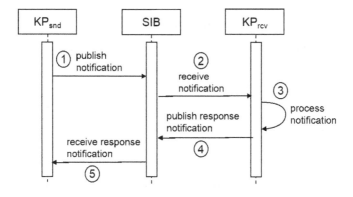

Fig. 3. Notification process for interacting KP_{snd} and KP_{rcv}

1. KP_{snd} publishes the notification in the smart space to start the required interaction with all appropriate KP_{rcv}.
2. KP_{rcv} recognizes the act of publishing due to the subscription and receives the notification data.
3. KP_{rcv} performs a corresponding operation based on the notification data.
4. KP_{rcv} publishes a response notification with the operation result.
5. KP_{snd} receives a response notification due to subscription mechanism.

Steps 1–3 are mandatory for interaction (1) since they implement forward actions in the interaction loop. Steps 4–5 implement feedback actions for the response notification, which are omitted if not needed by the application logic.

We distinguish the following design properties of a notification: representation, activation, function, response, clearing, performance. The KP developer implements these properties when programming the logic of a given KP.

Representation: A simple notification is represented with a single RDF triple. The general form is

$$\langle \text{notification\_id} \rangle, \langle \text{notification\_name} \rangle, \langle \text{value} \rangle$$

where $\langle \text{notification\_id} \rangle$ is the identifier of notification individual, $\langle \text{notification\_name} \rangle$ is the name of operation or event, $\langle \text{value} \rangle$ is the value of parameter (string data or identifier of some individual from the application ontology). The parameter holds data needed to pass to the notification receiver.

A compound notification consists of several RDF triples. It allows passing several parameters within the notification. The general form is

$$\langle \text{notification\_id} \rangle, \langle \text{notification\_name} \rangle, \langle \text{individual\_id} \rangle$$
$$\langle \text{individual\_id} \rangle, \langle \text{parameter1} \rangle, \langle \text{value1} \rangle$$
$$\ldots$$
$$\langle \text{individual\_id} \rangle, \langle \text{parameterN} \rangle, \langle \text{valueN} \rangle$$

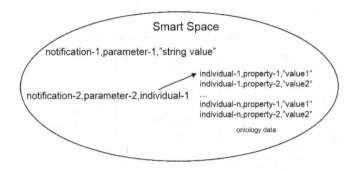

Fig. 4. Relation between a notification and the application ontology

where ⟨notification_id⟩ is the identifier of notification individual, ⟨notification_name⟩ is the name of operation or event, ⟨individual_id⟩ is identifier of the additional notification individual which stores other notification parameters, ⟨parameterN⟩ is the notification parameter name, ⟨valueN⟩ is the notification parameter value.

For a given application, its notification model implements an extension to the application ontology. Recall that every notification has parameter(s) and the value is stored as an object of notification RDF triple(s). The value can be either string data or ontology data. Ontology data are represented as a set of individuals in the smart space. Therefore, a notification can be linked with individuals in the smart space as schematically shown in Fig. 4. Thereby, interaction between KPs is performed due to changes in application ontology extension and not ontology data themselves. Otherwise, passing information from one KP to another requires to change individuals properties directly, which is not appropriate for data transferring. Moreover notifications can act as information exchange protocol between KPs operating on completely different ontologies.

Activation: Notifications are divided into two types depending on the reactive and proactive styles of activation at the KP sender side.

Reactive notification is used when a particular command (e.g., user action) directs KP_{snd} to call KP_{rcv} to perform operation (service). In this case, KP_{snd} typically awaits for a response notification with the operation result.

In proactive notification, KP_{snd} sends the notification with no explicit command from the user. Typically, KP_{snd} does not wait for response. The user context forms implicit dependence on user activity: KP_{snd} analyzes the context in the background (in parallel with primary user activity) and then activates events for other KPs.

Consider an example of reactive notification. A user consumes a service from her/his client KP. The service is constructed by another KP (service KP). When the user makes a control action (e.g., pushing a button) then the client KP sends reactive notification to the service KP and waits for the result. In this case, user is aware that a specific operation is going to be executed and it starts after the particular control command.

An example of proactive notification uses two KPs: client KP and recommender KP. The latter analyzes the data the user is working with and suggests related data from available sources. On the client KP the user browses service information and in parallel the KP sends a notification to the recommender KP (no explicit user action). The recommender KP reacts and finds a list of recommendations. In this case, user can be unaware that something is happening during her/his primary activity. A recommender KP can be even turned off: the system continues with a reduced service set. Proactive notification are usually used in the event-based interaction between KPs.

Function: A notification for interaction (1) can be a request notification or event notification. Request notification is used for another KP to perform certain operation. Event notification is used for informing another KP about an event. The previous examples applied request notifications: the client KP requests the service KP to perform some operations. Event notification is used to notify service KPs about the current user activity, e.g., a notification is sent when the user is reading. Then the service KP provides additional information (as a recommender KP). Also, the fact of reading can be used to block the service KP to deliver its service (e.g., reading is non-interrupted activity).

Response: Depending on the notification destination its KP sender is waiting or not for a response notification (contains operation result). While each client KP notifies a service KP to perform an operation, waiting for a response notification can be needed to show user whether operation was succeed or not. On the contrary, a KP client sending notification about user reading event does not wait for a response as it is possible that no KP will receive and process it. Request interaction between KPs most likely includes response notification while event interaction does not.

Clearing: A request notification is removed by its receiver after completion of required operation. An event notification is removed by its sender, which determines itself the time the notification is kept published.

Performance: Subscription is resource-consuming, thus every KP needs to minimize the number of subscriptions. Notification processing can be implemented within one subscription using RDF triple template with a fixed subject:

$$\langle \text{notification\_id} \rangle, *, * \tag{2}$$

where $*$-mask represents any value. Each KP has its own unique $\langle \text{notification\_id} \rangle$.

In summary, our notification model is a universal solution to make interaction implementation between KPs easier. Note that 100% delivery of subscription updates to subscribers is not guaranteed. For instance, some packets are lost in the communication network. Each KP may assume the best-effort delivery only. Additional resilience mechanisms should be built into application KP logic.

4   Case Study: SmartScribo System

Let us consider a particular Smart-M3 application to show how our notifica-tion model can be applied. SmartScribo system [3] aims at semantic mobile multi-blogging: mobile users interact with multiple blogs at many blog services simultaneously. The smart space stores data related to user blogs and personal information. The architecture is shown in Fig. 5. There are three types of KPs: KP client, KP blog processor, and KP mediator. Each KP client is installed on a personal mobile device. KP blog processor implements interaction with a particular blog service. For each blog service the system has a separate KP blog processor. KP mediator is responsible for additional processing of smart space content (e.g., personalized blog recommendation).

The user workflow is organized as follows. On the KP client its user specifies information about her/his blogs. The KP blog processors receive information about posts from that blogs. Then the user can read/edit existing posts or cre-ate new ones. When a new post is created or existed one is updated the blog processors reflect these changes at the blog service.

The smart space can keep a lot of blogs and posts of many users. For a blog processor it becomes difficult to detect which posts were updated using subscription on selected properties of individuals. For example, a post individual has such properties as title, text, tags. Blog processor can subscribe on triples: ⟨post_id⟩–title–* and similarly for other properties. It can even subscribe on triple ⟨post_id⟩–*–*. Both cases have the problem: the KP receives subscription updates separately on each property.

When can updated post individual be transferred to blog service? Moreover, posts are created and deleted dynamically in the smart space, and it is not easy to set/unset subscription on each individual every time. Although a blog

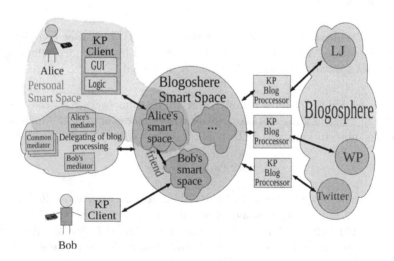

Fig. 5. SmartScribo system architecture, from [3]

Table 1. The set of notifications in SmartScribo system

Notification	Parameters	Description
refreshAccount	account individual identifier	receive blog account information
refreshPosts	account individual identifier	receive posts for given account
sendPost	account individual identifier, post individual identifier	send post to account
editPost	old post individual identifier, new post individual identifier	update old post to a new one
delPost	account individual identifier, post individual identifier	delete post
refreshComments	account individual identifier	receive comments for all posts of given account
sendComment	account individual identifier, comment individual identifier, parent individual identifier	send comment to parent message (post or comment) of given account
delComment	account individual identifier, comment individual identifier, parent individual identifier	delete comment for given parent message

processor can subscribe only on properties related to a post individual, i.e. *–title–*. However it does not solve the problem: subscription will inform about property changes separately and such updates will be mixed among several posts. Therefore a solution is to inform a blog processor about the need of performing particular action with a particular post.

SmartScribo applies the notification model in interaction between KP client and KP blog processor. The list of all notifications is presented in Table 1. The notifications are request and reactive as they are sent after user actions and require blog processor to perform operations. Every notification requires a response notification which informs the user about operation result. As parameters every notification has individual identifier, i.e., subject of triple to represent a particular account, post, or comment.

Notification "refreshPosts" has the form

$$\text{Notification-}\langle service \rangle, \text{refreshPosts}, \langle account\_id \rangle$$

where $\langle service \rangle$ is a blog service type (e.g., "LJ" for LiveJournal), and $\langle account\_id \rangle$ is an account individual identifier. Different service identifiers are used to distinguish notification for different blog processors. According to (2) each blog processor subscribes on a triple template where the subject is the service identifier while predicate and object are of any value. For LiveJornal blog processor the template is Notification-LJ–*–*.

A SmartScribo user specifies her/his blog account credentials using KP client and presses button to refresh posts list on that account. KP client publishes the account individual with all necessary properties for accessing it on the blog service and publishes "refreshPosts" notification. A KP blog processor receives notification with subscription, extracts account properties from the smart space, receives a posts list from the blog service and removes the notification. Then the blog processor publishes the posts list in the smart space and sends a response notification to KP client to inform about the operation result.

Notification "sendPost" has the form

$$\text{Notification-}\langle\text{service}\rangle, \text{sendPost}, \langle\text{notif\_ind}\rangle$$
$$\langle\text{notif\_ind}\rangle, \text{postAcc}, \langle\text{account\_id}\rangle$$
$$\langle\text{notif\_ind}\rangle, \text{postId}, \langle\text{post\_id}\rangle$$

where $\langle\text{service}\rangle$ represents a blog service type, $\langle\text{notif\_ind}\rangle$ is an identifier of additional notification individual storing all notification parameters, $\langle\text{account\_id}\rangle$ is an account individual identifier and $\langle\text{post\_id}\rangle$ — post individual identifier. A user (from KP client) creates a new post and then publishes this post and its notification to the smart space. A KP blog processor receives notification with subscription, and extracts account and post identifiers from the individual. Then KP extracts post properties using post identifier and sends this post to the blog service. Then the blog processor removes the processed notification and publishes a response notification for the KP client.

5   Performance Evaluation

Let $t(n)$ be the time elapsed since sending a notification from KP_{snd} until receiving the result at KP_{rcv}. We consider two types of experiments: 1) all n parameters are retrieving in one subscription query based on (2), 2) each parameter $i = 1, 2, \ldots, n$ is received with a separate query (n iterations query) with the triple template

$$\langle\text{individual\_id}\rangle, \langle\text{parameter}_i\rangle, *$$

The first type represents the best way for KP to subscribe to n-parameter notification. The second type shows the worst case: KP_{rcv} executes a loop to query value of each of n parameters. These two types represent lower and upper bounds on the performance. Let $t_{\text{bst}}(n)$ and $t_{\text{wst}}(n)$ be the time metrics to estimate experimentally.

The experimental setup includes SIB running on a server machine. Another computer hosts KP_{snd} and KP_{rcv}. Both KPs are implemented in Python. The use of the same host computer for the KPs simplifies the time measurement since no synchronization is needed between different computers. The server machine and KP computer are located in different LANs with RTT $= 3$ ms on average.

Measurement cycle consist of 1) KP_{snd} sends an n-parameter notification to SIB, 2) KP_{rcv} receives the notification from SIB, retrieves values of all n parameters, and removes the notification. There are 100 samples for each n.

The plot in Fig. 6 (a) shows the average values for $t_{\text{bst}}(n)$ and $t_{\text{wst}}(n)$. The values fit well to linear regression, which is constructed for n up to 100. The shown plot is a truncated version for $n \leq 50$, since $t_{\text{wst}}(n)$ for $n > 50$ behaves similarly and the growing difference $t_{\text{wst}}(n) - t_{\text{bst}}(n)$ hides the details of $t_{\text{bst}}(n)$.

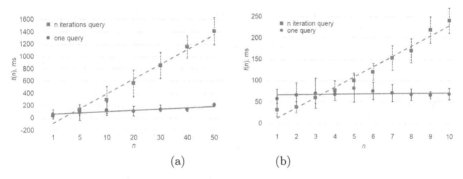

Fig. 6. Experimental behavior of $t_{\mathrm{wst}}(n)$ and $t_{\mathrm{bst}}(n)$

The linear grows $t_{\mathrm{wst}}(n)$ is essentially faster compared to $t_{\mathrm{bst}}(n)$. It is a clear consequence of n iterations query implemented in a loop. Notably that it also leads to higher variability (for each average value its standard deviation is shown as a vertical bar). The observed linear grows of $t_{\mathrm{bst}}(n)$, although low, is due to more processing at the SIB side and more data to transfer when n increases. Therefore, we conclude that our notification model preserves reasonable performance even for large n.

In Smart-M3 applications, the typical case is relatively small n, see Section 4. The plot in Fig. 6 (b) shows the finer-grained measurements for $n = 1, 2, \ldots, 10$. The behavior is again linear, though with less slopes in linear regression. We conclude that $t_{\mathrm{bst}}(n)$ is low and almost constant for this typical case.

The observed grows of slopes of linear regression indicates that there is some non-linear effect in the performance when n increases. We expect that the effect is due to 1) search algorithm complexity at the SIB side and 2) data transfer resources the network provides to KP.

6   Conclusion

The paper presented our notification model for use in Smart-M3 applications. The model supports coordination of interacting KPs, which are autonomous distributed entities communicating via information sharing in the smart space. The RDF nature of the model makes it applicable for almost any Smart-M3 application. The model extends the application ontology with possible requests and events that KPs use in interaction. We classified the key design properties of notification, and a KP developer can consider them as programming patterns in interaction logic of KPs. Our case study—SmartScribo system—shows the feasibility of the model. Further steps of this research are development of resilience and performance optimization mechanisms related to the underlying subscription operation of Smart-M3.

Acknowledgment. This research is financially supported by project # 1481 (basic part of state research assignment # 2014/154) of the Ministry of Education and Science of the Russian Federation. The reported study was partially supported by RFBR, research project # 14-07-00252. The work is a part of project 14.574.21.0060 of Federal Target Program "Research and development on priority directions of scientific-technological complex of Russia for 2014-2020".

References

1. Cook, D.J., Das, S.K.: How smart are our environments? an updated look at the state of the art. Pervasive and Mobile Computing 3(2), 53–73 (2007)
2. Honkola, J., Laine, H., Brown, R., Tyrkkö, O.: Smart-M3 information sharing platform. In: Proc. IEEE Symp. Computers and Communications (ISCC 2010), pp. 1041–1046. IEEE Computer Society (June 2010)
3. Korzun, D.G., Galov, I.V., Balandin, S.I.: Proactive personalized mobile mutliblogging service on SmartM3. Journal of Computing and Information Technology 20(3), 175–182 (2012)
4. Korzun, D., Galov, I., Balandin, S.: Smart room services on top of M3 spaces. In: Balandin, S., Trifonova, U. (eds.) Proc. 14th Conf. of Open Innovations Association FRUCT, SUAI, pp. 37–44 (November 2013)
5. Korzun, D.G., Balandin, S.I., Gurtov, A.V.: Deployment of Smart Spaces in Internet of Things: Overview of the design challenges. In: Balandin, S., Andreev, S., Koucheryavy, Y. (eds.) NEW2AN 2013 and ruSMART 2013. LNCS, vol. 8121, pp. 48–59. Springer, Heidelberg (2013)
6. Corkill, D.D.: Collaborating Software: Blackboard and Multi-Agent Systems & the Future. In: Proc. the Int'l Lisp Conference (October 2003) (invited paper)
7. Eugster, P.T., Felber, P.A., Guerraoui, R., Kermarrec, A.-M.: The many faces of publish/subscribe. ACM Comput. Surv. 35, 114–131 (2003)
8. Chen, H., Finin, T., Joshi, A., Kagal, L., Perich, F., Chakraborty, D.: Intelligent agents meet the semantic web in smart spaces. IEEE Internet Computing 8, 69–79 (2004)
9. Lomov, A.A., Korzun, D.G.: Subscription operation in Smart-M3. In: Balandin, S., Ovchinnikov, A. (eds.) Proc. 10th Conf. of Open Innovations Association FRUCT and 2nd Finnish–Russian Mobile Linux Summit, SUAI, pp. 83–94 (November 2011)
10. Palviainen, M., Katasonov, A.: Model and ontology-based development of smart space applications. In: Pervasive Computing and Communications Design and Deployment: Technologies, Trends, and Applications, pp. 126–148 (May 2011)
11. Korzun, D., Lomov, A., Vanag, P., Honkola, J., Balandin, S.: Generating modest high-level ontology libraries for Smart-M3. In: Proc. 4th Int'l Conf. Mobile Ubiquitous Computing, Systems, Services and Technologies (UBICOMM 2010), pp. 103–109 (October 2010)
12. D'Elia, A., Honkola, J., Manzaroli, D., Cinotti, T.S.: Access control at triple level: Specification and enforcement of a simple RDF model to support concurrent applications in smart environments. In: Balandin, S., Koucheryavy, Y., Hu, H. (eds.) NEW2AN 2011 and ruSMART 2011. LNCS, vol. 6869, pp. 63–74. Springer, Heidelberg (2011)

Elimination of Distorted Images
Using the Blur Estimation at the Automatic Registration
of Meeting Participants

Irina V. Vatamaniuk[1], Alexander L. Ronzhin[2], Anton I. Saveliev[2],
and Andrey L. Ronzhin[2,3]

[1] SUAI, 67, Bolshaya Morskaia, St. Petersburg, 199000, Russia
[2] SPIIRAS, 39, 14th line, St. Petersburg, 199178, Russia
[3] ITMO University, 49 Kronverkskiy av., St. Petersburg, 197101, Russia
{vatamaniuk,ronzhinal,saveliev,ronzhin}@iias.spb.su

Abstract. The methods for estimation of blur and other quality metrics of digital images are discussed. The classification of modern methods of blur estimation used for real-time systems is presented. Several methods of image patch segmentation were applied for enhancement of the processing speed and reliability of image quality assessment. The proposed method of preliminary extraction of face area on the image and estimation of its blur was successfully used for elimination of distorted images before the face recognition stage in a system of automatic registration of participants in a smart meeting room.

Keywords: Blur estimation, face detection, face recognition, smart meeting room, multimodal interfaces.

1   Introduction

The image quality estimation algorithms are demanded in almost all application spheres of computer vision technology, such as: systems of photo and video surveillance and registration, automatic control systems (industrial robots, autonomous vehicle), systems modeling of objects and environment, human-computer interaction systems, quality control systems, etc.

The human's eye perceives any image comprehensively, but can evaluate its quality proceeding from: brightness, contrast, tint, sharpness and noisiness. During digital image processing these parameters can be estimated either separately or in conjunction. In [1] three components of an image are reviewed: contour, gradient and texture, which are semantically significant for perception of the human visual system. Also types of image defects are discussed, for example, light and color edges blurriness, moirй pattern appeared owing to space sampling noise, false contours caused by insufficient amount of quantization levels, etc. The image restoration method [1] is based on sharpness enhancement on edges of gradient intensity changes and noise filtering on gradient regions.

During the processing of an array of same type images, for example at photo registration, not all the quality parameters can be varied. When taking a photo of some

S. Balandin et al. (Eds.): NEW2AN/ruSMART 2014, LNCS 8638, pp. 133–143, 2014.

moving objects by fixed camera brightness, contrast, noisiness can be varied insignif-icantly. In this case the estimation of image sharpness comes to the forefront. The sharpness can be degraded owing to video signal recording and transmission hardware inaccuracy or software errors at digital image processing. The image sharpness can be increased by objective lens focusing on photographing object or using additional im-age processing methods.

The image blur appears at diffusive merging of two contrast colors. The human eye percepts and evaluates image sharpness by the presence of contrast bright or tint tran-sition between adjacent image regions. If camera lens is defocused or if shooting object is moving during capturing photo, the obtained image will be partially or com-pletely blurred. This problem is very urgent for many applications and actively inves-tigated now [2, 3].

Further the problem of image blurriness and methods capable for determination of the low quality of unfocused images will be discussed in Section 2. In Section 3 three methods of selection of the area of interest in image will be presented. Section 4 pre-sents the method for face recognition based on histogram calculation for local binary patterns. The results of experiments of blur estimation and face recognition applied for the photographs of participants in intelligent meeting room are shown in Section 5.

2   Image Blur Estimation Methods

The main causes of image distortions that lead to the degradation of sharpness are: 1) limited resolution of recording equipment; 2) defocusing; 3) the movement of the camera relative to the capturing object [4]. Image sharpness is characterized by reproduction of small parts and determined by resolution of forming system. At the defocusing a point is reproduced in the shape of a spot (blur circle), and two closely spaced points on the original image are merged into one observable point. The size of the blur circle depends on the focal length of the lens as well as the distances from the lens to the object and to the plane of the formed image. Discrete image will be sharp (focused), if the blur circle diameter does not exceed the sampling of the ob-served image. Otherwise, the linear distortion becomes visible [4].

Thus, image blur can be estimated by focus estimation. In addition, a comprehen-sive quality assessment also helps to reveal a blurred image. For blur estimation is advisable to convert color images into achromatic ones and explore a rectangular matrix of brightness values of image pixels. The size of the matrix can have same size as the original image in pixels, or correspond to the size of the selected area.

In [5] the focus evaluation methods are divided into the six groups, where analysis of gradient, Laplacian, wavelet transform, statistical characteristics, discrete cosine transform and miscellaneous methods that does not belong to any of these are applied. These types of methods have different complexity and time processing, so the optimal method of image processing is selected depending on the subject area and dedicated computing resources and time. The method is considered optimal when it allows to select the greater part of images of good quality during the lesser processing time.

Let us consider some of the most perspective methods. One of them, the method based on measuring local contrast is described in [6] (we denote it MIS1, because it belongs to the miscellaneous category). The local contrast is calculated as the ratio of the intensity of each pixel of the grayscale image and the mean gray level in the neighborhood of this pixel.

$$R(x,y) = \begin{cases} \dfrac{\mu(x,y)}{I(x,y)}, I(x,y) \le \mu(x,y) \\ \dfrac{I(x,y)}{\mu(x,y)}, I(x,y) \ge \mu(x,y) \end{cases}, \tag{1}$$

where $I(x, y)$ is the intensity of the current pixel; $\mu(x, y)$ is its neighborhood. The neighborhood size $\mu(x, y)$ centered at the point (x, y) is determined heuristically. The image blur is the sum of values $R(x, y)$ throughout the image or the selected area.

Next, consider another miscellaneous method, which measures the blurriness by the image curvature [6, 7], we denote it MIS2. The grayscale image matrix of pixel brightness is represented as a three-dimensional surface. The coordinates of each point of this surface are the two coordinates of each pixel and the value of its brightness. The surface curvature corresponds to transitions between pixels and is approximated by the following function:

$$f(x,y) = ax + by + cx^2 + dy^2. \tag{2}$$

The image is focused if the curvature value is high. The coefficients a, b, c, d are approximately calculated by the method of least squares by convolution of the original image and the matrices $M1, M2$:

$$M_1 = \frac{1}{6}\begin{pmatrix} -1 & 0 & 1 \\ -1 & 0 & 1 \\ -1 & 0 & 1 \end{pmatrix}, \quad M_2 = \frac{1}{5}\begin{pmatrix} 1 & 0 & 1 \\ 1 & 0 & 1 \\ 1 & 0 & 1 \end{pmatrix}, \tag{3}$$

$$a = M_1 * I, \qquad b = M_1' * I,$$

$$c = \frac{3}{2}M_2 * I - M_2' * I, \qquad d = \frac{3}{2}M_2' * I - M_2' * I, \tag{4}$$

where I is the original image, $M1', M2'$ are the transposed matrices $M1, M2$. The sum of absolute values of the coefficients a, b, c, d reveals the image blurriness:

$$G_C = |a| + |b| + |c| + |d|. \tag{5}$$

Another perspective method is NIQE (No-Reference Image Quality Assessment), which uses 'blind' image analysis [8]. The NIQE method is based on a statistical study of natural images. The natural images are images that are not artificially generated and not distorted by artificial noises, e.g. images obtained using photography, screen capture of video and so on. In natural monochrome images the matrix of normalized coefficients of pixel brightness tends to a normal distribution. Any noise, including blurring, leads to a deviation from the normal distribution. The idea of the method consists in comparing two multivariate Gaussian models of features: calculated for the test image and constructed on the basis of a pre-arranged set of images.

To calculate the features for building the model, the brightness coefficients of image pixels are normalized by subtracting the local mean of the original coefficient matrix of grayscale image brightness and then dividing by the standard deviation:

$$\hat{I}(i,j) = \frac{I(i,j) - \mu(i,j)}{\sigma(i,j) + 1}, \tag{6}$$

where $i \in \{1,2...M\}$, $j \in \{1,2...N\}$ are spatial indexes, M, N are the image height and width, $\mu(i,j)$ is the expected value and $\sigma(i,j)$ is the variance:

$$\mu(i,j) = \sum_{k=-K}^{K} \sum_{l=-L}^{L} w_{k,l} I(i+k, j+l), \tag{7}$$

$$\sigma(i,j) = \sqrt{\sum_{k=-K}^{K} \sum_{l=-L}^{L} w_{k,l} (I(i+k, j+l) - \mu(i,j))^2}, \tag{8}$$

where $w = \{w_{k,l} / k = -K...K, l = -L...L\}$ is two-dimensional circularly symmetric Gaussian weighting function.

This normalization can significantly reduce the dependence between the brightness coefficients of the neighboring pixels, leading them to a form, suitable for the construction of a multivariate Gaussian model. Since the entire image sharpness is often limited by the depth of field of the camera, so the image is split into patches of PxP pixels and then the local sharpness of each patch is estimated. The sharpest patches are selected for further analysis. Local sharpness is calculated by the variance $\sigma(i,j)$:

$$\delta(b) = \sum \sum_{(i,j)} \sigma(i,j). \tag{9}$$

The local area sharpness of the patches is estimated relatively to a threshold, which is selected heuristically. The deviation from the generalized Gaussian model of the image can be revealed by analyzing the products of pairs of normalized coefficients of adjacent pixel brightness in four directions: horizontal, vertical, main and secondary diagonals:

$$\hat{I}(i,j)\hat{I}(i,j+1),$$
$$\hat{I}(i,j)\hat{I}(i+1,j),$$
$$\hat{I}(i,j)\hat{I}(i+1,j+1),$$
$$\hat{I}(i,j)\hat{I}(i+1,j-1), \tag{10}$$

where $i \in \{1,2...M\}$, $j \in \{1,2...N\}$. These parameters follow the asymmetric generalized normal distribution:

$$f(x; \gamma, \beta_l, \beta_r) = \begin{cases} \dfrac{\gamma}{(\beta_l + \beta_r)\Gamma\left(\frac{1}{\gamma}\right)} exp\left(-\left(\dfrac{-x}{\beta_l}\right)^{\Gamma}\right), \forall x < 0 \\ \dfrac{\Gamma}{(B_l + B_r)\Gamma\left(\frac{1}{\Gamma}\right)} exp\left(-\left(\dfrac{x}{B_r}\right)^{\Gamma}\right), \forall x \geq 0 \end{cases}, \qquad (11)$$

where γ is a parameter that controls the shape of the distribution curve, β_l, β_r are parameters that control the spread on the left and the right side respectively, $\Gamma(\cdot)$ is the gamma function:

$$\Gamma(a) = \int_0^\infty t^{a-1}e^{-t}dt, a > 0. \qquad (12)$$

The coefficients γ, β_l, β_r can be effectively evaluated using the method of moments [9]. We have a set of features as a result of the calculations given above. These features are used for construction of the multivariate Gaussian model, which is compared with another multivariate Gaussian model trained on a set of various images with known quality:

$$f_X(x_1, ..., x_k) = \frac{1}{(2p)^{\frac{k}{2}}|\Sigma|^{\frac{1}{2}}} exp\left(-\frac{1}{2}(x - H)^T \Sigma^{-1}(x - H)\right), \qquad (13)$$

where $(x_1,...,x_k)$ is a set of calculated features, v and Σ are mean and covariance matrix of multivariate Gaussian model respectively, which are calculated by the method of maximum likelihood. The image quality coefficient is calculated by the following formula:

$$D(H_1, H_2, \Sigma_1, \Sigma_2) = \sqrt{\left((H_1 - H_2)^T \left(\frac{\Sigma_1 + \Sigma_2}{2}\right)^{-1}(H_1 - H_2)\right)}, \qquad (14)$$

where v_1, v_2 are mean vectors of template multivariate Gaussian model and of the model constructed for the test images respectively, Σ_1, Σ_2 - covariance matrices of these models. The coefficient D indicates difference between the models. When its value is smaller, the distribution of the test image is closer to the normal distribution; otherwise the image contains a noise which can be a blur too.

The method of image blur estimation described above was taken into the base of the developed image processing means at the preliminary processing stage of photos of participants in smart meeting room.

3   Image Patch Selection Approaches

Selection of an area for blur assessment depends on subject domain, in which image processing is implemented. This area may be either a whole image or the contoured object or the area in which the object is located.

Also image resizing and selection of the important content perceived by user has gained significant importance concerning with the diversity and versatility of display devices [10]. New effective mobile image retrieval method based on a weighted combination of color and texture utilizing spatial-color and second order statistics was applied for region-based queries on mobile devices [11].

In our case, during solving the problem of automatic registration of participants in the intelligent meeting room the areas which contain person faces are most important in the image [12, 13]. The examples of the segmentation of areas of interest in the image are shown in Table 1. The captured face database was received by video monitoring system based on detection of occupied sits by analysis of zone of chairs and AdaBoosted cascade classifier for face detection [14, 15].

Table 1. The examples of selection of interest areas during the automatic registration of participants in the intelligent meeting room

Description of the selected area	Example of segmentation
The whole image is analyzed without preliminary face detection	
The selected face area is analyzed the size of which is not less than 200x200 pixels	
The selected face area bounded by the size of 200x200 pixels is analyzed	

These three types of regions in the image were used to compare the performance of the four methods described above. The photos, in which blur face region is detected, are eliminated from the further image processing and only quality image goes into the face recognition process. The process of face recognition based on analysis of histograms of local binary patterns of image is presented below. Selection of area of interest can increase the robustness of quality assessment, because distorted areas, which are located out of the region of interest, are not analyzed. Also the speed of image processing significantly increases by reducing the amount of data being processed.

4   Face Recognition Based on Local Binary Patterns

The main part of the automatic system for participant registration in smart meeting room is the face recognition procedure. Among the all considered methods for face recognition, the principal component analysis (PCA), Linear Discriminant Analysis (LDA), local binary patterns (LBP) analysis should be mentioned as more popular in face processing task. The description of approach based on PCA or eigenfaces are described in [16, 17]. Using LDA method for face recognition is presented in [18]. Also the methods of face database training based on video processing have gaining attention [19]. The processing of extra information available in a video makes the face recognition methods more robust to changes in illumination and pose. In our experiments the LBP method gave the best results, and so its short description is provided below.

One of the first researches dedicated to texture description based on LBP is [20]. The LBP operator describes a pixel 3x3-neighborhood in the binary form. When neighbor's value is greater than the center pixel's one, it's labeled "1". Otherwise, it's labeled "0". This gives an 8-digit binary number. The LBP is considered uniform pattern, if it contains two or less bitwise transitions from 0 to 1 or vice versa when the bit pattern is traversed circularly. The LBP operator can be extended to use circular neighborhoods of any radius and number of pixels by bilinear interpolation of pixel values.

A histogram of the labeled image $f_l(x, y)$ contains information about the distribution of the local micro patterns, such as edges, spots and flat areas, over the whole image. It can be defined as:

$$H_i = \sum_{x,y} I\{f_l(x, y) = i\}, i = 0, \dots, n - 1, \tag{15}$$

in which n is the number of different labels produced by the LBP operator, and $I\{A\}$ is 1 if A is true and 0 if A is false. For efficient face representation, three different levels of locality are used: a pixel-level (labels for the histogram), a regional level (the labels sum over a small region), and a global description of the face (the regional histograms). For this purpose the image is divided into regions $R_0, R_1, \dots, R_{m-1}$ and the spatially enhanced histogram is defined as:

$$H_{i,j} = \sum_{x,y} I\{f_i(x,y) = i\} I\{(x,y) \in R_j\}, i = 0, \ldots, n-1, j = 0, \ldots, m-1. \qquad (16)$$

Obviously, some of the regions contain more information, useful for face recognition than others. Taking this into account, each region is assigned a weight depending on the importance of the information it contains.

The described methods of face recognition used in conjunction with blur estimation methods in the system of preparation and support of meetings carried out in the smart room speed up the registration process.

5 Experiments

In the experiments the photo database of participants of events held in the intelligent meeting room was analyzed. The participant images were captured by the camera model AXIS 215 of the hardware-software complex of audiovisual monitoring of the intelligent meeting room [21, 22]. The resolution of obtained photographs is 640x480 pixels.

At image blur estimation four methods were compared: NIQE [8], the local contrast method (MIS1) [6], the image curvature method (MIS2) [7] described above, and method Tenengrad, which estimates the image gradient [23].

During the experiments three segmentation types of the image part were applied: the whole image, the face area and the face area of 200x200 pixels. The annotated database consisted of 50 sharp and 50 blurred photos. The blurriness coefficients of blurred photos were used to calculate the threshold value for each evaluation method. Then, each blur estimation method was applied for the set of sharp images and the percentage of photos with blur coefficient exceeded the threshold was estimated. The accuracy of every method is determined as a number of the sharp photos passed the threshold. The experimental results with accuracy and processing time estimates are presented in Table 2.

Table 2. The experiment results of image blur estimation

Blur estimation method	Accuracy, %			Processing time, ms		
	Whole image	Face area	200x200 pixels	Whole image	Face area	200x200 pixels
Tenengrad	4	22	26	13.81	5.23	4.81
MIS1	0	0	30	12.80	5.81	4.86
MIS2	2	34	30	10.42	7.70	7.30
NIQE	10	54	94	31.65	38.08	35.08

As seen from Table 2, the best results for majority of the methods are revealed by the small segment found on the selected face region. Naturally, the processing speed is improved by decreasing of size of analyzed area. Furthermore, the accuracy of the method also increases while the analyzed region is decreasing. For example, the blur evaluation method based on the local contrast assessment does not work on large areas, however, shows an acceptable result for the area of 200x200 pixels. The most accurate method of the above is NIQE, but it also needs the longest time.

The proposed procedure of prior search of faces area in the frame and evaluating its blurring based on the NIQE method of statistical analysis of the coefficients of pixel brightness allowed us to determine 94% of undistorted frames received during the automatic registration of the participants in the intelligent meeting room.

The developed automatic participant registration system has two levels of face capture with different quality. At the first level the rapid procedure of face recognition is used. It based on capturing one photo, which includes view of all participants with low resolution and following face recognition. At this stage the image patches with participant faces has resolution around of 30x30 pixels. The faces unrecognized during the first level of processing further are separately captured by pan-tilt-zoom camera with high resolution at the second level of registration system work. Recurrently captured face region has resolution higher than 200x200 pixels.

For the experimental evaluation of the automatic participant registration system working at the first processing level during the events in the smart meeting room around 40,000 photos with resolution around of 30x30 pixels for 30 participants was accumulated. The face recognition was carried out by the following three methods: 1) PCA; 2) LDA; 3) LBP. The training database contains 20 photos for each participant. Table 3 presents average values of face recognition accuracy, as well as first (False Alarm (FA) rate) and second (Miss Rate (MR)) errors type for each used method.

Table 3. Face recognition experiment results

Method	FA, %	MR, %	Accuracy, %
LBP	10,5	10,1	79,3
PCA	0,8	25	74,1
LDA	7,6	15,3	77,1

Next experiment was performed to test the hypothesis about necessity to use images with different head participant orientations relative to the camera. For this aim the training face database was extended and for the testing the database with 55,920 images was used. Average results of the second experiment are shown in Table 4.

Table 4. Face recognition experiment results with extented database

Method	FA, %	MR, %	Accuracy, %
LBP	12	8,5	79,5
PCA	1,3	23,5	75,2
LDA	19,2	7,8	73

The high value of false alarm and miss rate errors were obtained owing to the fact that the photos were captured in the course of actual of events in smart meeting room. During the capturing participants can freely move, and so their faces in the photos could be blurred or partially hidden. The additional factors, which degrade the face recognition performance, are different distances from the camera to a participant, lighting, mobility of participants. As a result method LBP showed the best results with face recognition accuracy of 79,5%.

Introduction of the blur estimation procedure as preliminary stage of photo processing allows the registration system to exclude 22% photos with high resolution but insufficient quality from face recognition stage, as a result the speed of the whole system were significantly increased. Implementation of blur estimation procedure on the first level of processing of the photos with resolution around 30x30 pixels did not give positive results, because such low resolution is insufficient to make decision about image blurriness.

6   Conclusion

The blur evaluation is a necessary step in the processing systems working with large input stream of visual information. Preliminary assessment allows a system to exclude images of poor quality carrying no useful information from the further analysis, thus saving the process memory and speed of automatic vision systems. The image blurriness caused by unsatisfactory shooting conditions occurs owing to the wrong focusing or unexpected movement of the subject. Blur value can be estimated by various methods, particularly by evaluation of the brightness gradient of image, the ratio of the brightness values of pixels on a certain region, the statistical analysis of the brightness coefficients of pixels.

The discussed methods quantify the magnitude of blurring, which is one of the criteria of image quality. These methods are convenient to use in the modeling and processing of different visual information. Preliminary extraction of face area in the image and its blurriness estimation were successfully applied for elimination of distorted photos in the system of automatic registration of participants in the intelligent meeting room that significantly increases the speed of photo processing.

Acknowledgments. This work was partially financially supported by Government of the Russian Federation, Grant 074-U01 and RFBR (grant 13-08-0741-a).

References

1. Krasil'nikov, N.N.: Principles of Image Processing Based on Taking into Account Their Semantic Structure. Information Control Systems 1(32), 2–6 (2008) (in Russ.)
2. Serir, A., Beghdadi, A., Kerouh, F.: No-reference blur image quality measure based on multiplicative multiresolution decomposition. Journal of Visual Communication and Image Representation 24, 911–925 (2013)
3. Soleimani, S., Rooms, F., Philips, W.: Efficient blur estimation using multi-scale quadrature filters. Signal Processing 93, 1988–2002 (2013)
4. Gruzman, I.S., Kirichuk, V.S., Kosykh, V.P., Peretiagin, G.I., Spektor, A.A.: Digital Image Processing in Information Systems. NGTU, Novosibirsk (2002) (in Russ.)

5. Pertuz, S., Puig, D., Garcia, M.A.: Analysis of Focus Measure Operators for Shape-from-focus. Pattern Recognition 46(5), 1415–1432 (2013)
6. Helmli, F., Scherer, S.: Adaptive Shape from Focus with an Error Estimation in Light Microscopy. In: Proceedings of International Symposium on Image and Signal Processing and Analysis, pp. 188–193 (2001)
7. Mendapara, P.: Depth Map Estimation Using Multi-focus Imaging. Electronic Theses and Dissertations (2010)
8. Mittal, A., Soundarajan, R., Bovik, A.C.: Making a 'Completely Blind' Image Quality Analyzer. IEEE Signal Processing Letters 20(3), 209–212 (2013)
9. Sharifi, K., Leon-Garcia, A.: Estimation of Shape Parameter for Generalized Gaussian Distributions in Subband Decompositions of Video. IEEE Transactions on Circuits and Systems for Video Technology 5(1), 52–56 (1995)
10. Zi, L., Du, J., Hou, L., Sun, X., Lee, J.: Perception-Driven Resizing for Dynamic Image Sequences. ComSIS 10(3), 1343–1357 (2013)
11. Lee, Y.-H., Kim, B., Rhee, S.-B.: Content-based Image Retrieval using Spatial-color and Gabor Texture on a Mobile Device. ComSIS 10(2), 807–823 (2013)
12. Ronzhin, A.L., Budkov, V.Y., Karpov, A.A.: Multichannel System of Audio-Visual Support of Remote Mobile Participant at E-Meeting. In: Balandin, S., Dunaytsev, R., Koucheryavy, Y. (eds.) ruSMART/NEW2AN 2010. LNCS, vol. 6294, pp. 62–71. Springer, Heidelberg (2010)
13. Yusupov, R.M., Ronzhin, A.L.: From Smart Devices to Smart Space. Herald of the Russian Academy of Sciences, MAIK Nauka 80(1), 63–68 (2010)
14. Schiele, B., Schiele, J.L.: European Conference on Computer Vision: Object recognition using multidimensional receptive field histograms, vol. 1, pp. 610–619 (April 1996)
15. Viola, P., Jones, M., Snow, D.: Detecting pedestrians using patterns of motion and appearance. In: Proceedings of IEEE ICCV, pp. 734–741 (2003)
16. Turk, M.A., Pentland, A.P.: Face recognition using eigenfaces. In: Proceedings of IEEE Conference on Computer Vision and Pattern Recognition - CVPR, pp. 586–591 (1991)
17. Georgescu, D.: A Real-Time Face Recognition System Using Eigenfaces. Journal of Mobile, Embedded and Distributed Systems 3(4), 193–204 (2011)
18. Taheri, S., Patel, V.M., Chellappa, R.: Component-based recognition of faces and facial expressions. IEEE Transactions on Affective Computing 4(4), 360–371 (2013)
19. Patel, V.M., Chen, Y.-C., Chellappa, R., Phillips, P.J.: Dictionaries for image and video-based face recognition. Journal of the Optical Society of America A 31(5), 1090–1103 (2014)
20. Ojala, T., Pietikainen, M., Harwood, D.: A comparative study of texture measures with classification based on feature distributions. Pattern Recognition 29, 51–59 (1996)
21. Ronzhin, A.L., Budkov, V.Y.: Multimodal Interaction with Intelligent Meeting Room Facilities from Inside and Outside. In: Balandin, S., Moltchanov, D., Koucheryavy, Y. (eds.) NEW2AN/ruSMART 2009. LNCS, vol. 5764, pp. 77–88. Springer, Heidelberg (2009)
22. Ronzhin, A.L., Saveliev, A.I., Budkov, V.Y.: Context-Aware Mobile Applications for Communication in Intelligent Environment. In: Andreev, S., Balandin, S., Koucheryavy, Y. (eds.) NEW2AN/ruSMART 2012. LNCS, vol. 7469, pp. 307–315. Springer, Heidelberg (2012)
23. Lorenzo-Navarro, J., Děniz, O., Santana, M.C., Guerra, C.: Comparison of Focus Measures in Face Detection Environments. In: ICINCO-RA vol. 2. INSTICC Press, pp. 418–423 (2007)

Citywatcher: Annotating and Searching Video Data Streams for Smart Cities Applications

Alexey Medvedev[1], Arkady Zaslavsky[2], Vladimir Grudinin[1],
and Sergey Khoruzhnikov[1]

[1] ITMO University, Kronverkskiy pr, 49., St. Petersburg, Russia
{alexey.medvedev,grudinin}@niuitmo.ru, xse@vuztc.ru
[2] CSIRO, Melbourne, Australia
arkady.zaslavsky@csiro.au

Abstract. Digital pervasive video cameras can be abundantly found everywhere these days and their numbers grow continuously. Modern cities have large networks of surveillance cameras including CCTV, street crossings and the like. Sometimes authorities need a video-recording of some road accident (or of some other event) to understand what happened and identify a driver who may have been at fault. In this paper we discuss the challenges of annotating and retrieving video data streams from vehicle-mounted surveillance cameras. We also propose and evaluate the CityWatcher application – an Android application for recording video streams, annotating them with location, timestamp and additional context in order to make them discoverable and available to authorized Internet of Things applications

Keywords: Smart City, video streaming, Internet of Things (IoT), crowdsensing, Intelligent Transportation Systems, Vehicle-Mounted Surveillance Camera (VMSC).

1 Introduction

Digital pervasive video cameras can be abundantly found everywhere these days and their numbers grow continuously. Modern cities have large networks of surveillance cameras including CCTV, street crossings and the like. Sometimes we need a video-recording of some road accident (or of some other event) to understand what happened and identify a driver who may have been at fault. Many car owners use Vehicle-Mounted Surveillance Cameras (VMSC, or in other words, car black boxes, video registers or smartphones) for recording their driving and events of interest. These video-recordings are potentially important sources of data. The drivers can use video recording from their devices as evidence in case of a road accident. However, video recordings from their devices can't be retrieved by others and shared. Therefore we highlight the challenge of retrieving such video fragments for evidence collection and the methods of annotating, discovering, retrieving and processing video data streams. In particular, we argue that smartphones can be used for evidence collection in case of a road accident. Besides, smartphone owners can use the proposed

S. Balandin et al. (Eds.): NEW2AN/ruSMART 2014, LNCS 8638, pp. 144–155, 2014.

CityWatcher application for sending their automatically or manually annotated reports about road accidents, or any other city problems, e.g., road potholes and cracks.

Functioning of a modern city strongly depends not only on the city infrastructure, but also on the availability and quality of information for citizens and authorities about different aspects of city life. Various applications enhancing and making easier the city life lead to the concept of a 'smart city', which has received a lot of attention in last few years. The "Smart City" concept essentially means efficiency, effectiveness and resource optimization. Efficiency is largely based on the intelligent management and integrated ICT (Information and Communication Technology) infrastructure, and active citizen participation. This implies a new form of governance, including genuine citizen involvement in public policy in the form of crowdsourcing and crowdsensing [1], [2], [3].

Applications for smart cities are usually divided into six main areas: smart living, smart governance, smart economy, smart environment, smart people and smart mobility [4]. A very important feature for all these types of applications is the data collection, which is primarily based on sensors and sensor networks. Smart city applications strongly need access to this public data through the web for visualizing, transforming and making use of it. Active participation of all stakeholders is also very important for the smart city.

As the number of cars continues to grow, road problems become one of the main issues for city management. In smart cities, Intelligent Transportation Systems (ITS)[5] are used for solving transportation and congestion problems using Information and Communication Technology (ICT). ITSs help in fast problem detection, violations of regulations, traffic analysis, evidence collection, safety provisioning, reducing costs and delays and much more. Therefore, ITSs appear to be one of the most important parts of a smart city.

Rapid problem solving is really important for the smart city. If a citizen encounters a problem, he/she should be able to draw city management attention to it without taking any complicated actions. On the other side, all city services must be ready to receive such requests and take measures to solve the problem. Besides, they must have instruments to ask for some help from citizens. This help is called crowdsourcing. If this help takes a form of getting any information from computers, smartphones or other devices that incorporate sensors, than we can speak about Urban Crowdsensing. Working in close contact with citizens is very important part of moving towards smart cities.

Every year road accidents cause loss of lives, loss of money and serious delays on city streets and freeways. Approximately 50%–60% of the delays on urban freeways are caused by road accidents [6]. Reducing delays by faster accident analysis is a challenging task. This task includes building an evidence collection system that can help the police and city authorities to make more accurate and faster decisions. The best way to achieve it is to provide video recordings from different angles and sources to the decision maker. If such videos are available, cars that participated in an accident can be faster moved out of the road and unblock traffic.

During last decade vehicle-mounted surveillance cameras (VMSC) gained significant popularity[7]. VMSC can also be referenced as cars video registers or black boxes or dash cams. They can help to prove innocence in case of a road accident or just make a video of something interesting and uncommon, like falling of a meteor or a tree, or a tornado. *"In the USA, 15% of 200 million cars carry a black box, while 80% of the compact cars built since 2004 mount black boxes. In Japan, about 6 million of the black boxes mounted on vehicles include 40,000 business vehicles and 20,000 regular passenger cars" [7]*.

By now, VMSCs already have a feature of annotating video with street name, as they have a GPS module onboard. The most modern VMSCs have Internet access. It can be predicted, that in future most of such smart cameras will have Internet access. VMSCs are produced by various manufacturers, for example, Garmin[1], Prestigio [2], HP[3]. A typical VMSC is shown in Fig.1(left). Millions of surveillance cameras are deployed, many of them are private. Some people have such cameras at their windows. Also, many people just make recordings with their smartphones. As a smartphone has a camera, it can easily act as a VMSC if it is fixed on a front window.

Many smartphone apps are available in Google Play and Apple AppStore. Examples of such applications include: DailyRoads Voyager[4], Axel Voyager[5], AVR[6], AutoBoy BlackBox[7], CaroO Pro[8] etc. their characteristics will be briefly discussed in the 'related work' section. A smartphone acting as a VMSC is shown in Fig.1(right).

Eventually, the cloud storage would keep videos of many events that can be of interest for city management, police, road services and other governmental organizations. The challenge is in retrieving relevant video streams when they are really needed. Owners of the surveillance camera may not be informed, that an accident happened just in his/her camera view. Owner of a VMSC could pass by without stopping, when an accident happens in front of their car. In such a situation, police has to understand what happened just by listening to stories of accident participants, that not always lead to correct analysis and respective decisions.

Fig. 1. VMSC (left) and smartphone acting as a VMSC (right)

[1] https://buy.garmin.com/en-US/US/shop-by-accessories/other/garmin-dash-cam-20/prod146282.html

[2] http://www.prestigio.com/catalogue/DVRs/Roadrunner_300

[3] http://www.shopping.hp.com/en_US/home-office/-/products/Accessories/Camera_photo_video/H5R80AA?HP-f210-Car-Camcorder

[4] http://www.dailyroads.com/voyager.php

[5] https://play.google.com/store/apps/details?id=net.powerandroid.axel

[6] https://play.google.com/store/apps/details?id=com.at.autovideosregistrator&hl=uk

[7] https://play.google.com/store/apps/details?id=com.jeon.blackbox&hl=ru

[8] https://play.google.com/store/apps/details?id=com.pokevian.prime

According to [8], surveillance video is now the biggest source of Big Data. It is predicted, that the percentage will increase by 65 percent by 2015. As we are adding video recordings from users Internet Connected Objects (ICO), using the Internet of Things terminology, the amount of data increases significantly. One of the most critical challenges is how to transmit and store all this video. First of all, it seems to be very hard to transmit and store all the video recordings from ICOs to some central storage, whether it's cloud or a video repository. Secondly, not many people might want to share their video recordings in some way, as there's no certainty what it will be used for. It is a serious privacy concern. The solution can be found in storing data in the device, where it was recorded. If we would be able to make requests to all possible ICOs, the need to transfer and store the video data will be reduced or eliminated. So, as data stays in ICOs, we are to look at massively distributed Internet of Things system with challenging indexing, search and processing requirements. The challenge is also to communicate with millions of ICOs, make requests and get videos of events that we are interested in. As these ICOs are heterogeneous, opportunistically available and spread all over the region (however big), building such a system becomes a global IoT challenge. One of the key concepts of the system is crowdsourcing [9].

2   Related Work

The process of getting video from surveillance cameras is already well investigated [10]. We focus not on existing road or traffic intersection cameras, as is the case with most current applications, but mainly, on opportunistically available mobile data sources, which belong to citizens. The problem of crowdsourcing with smartphones is discussed in [11]. A prototype of a crowdsourced evidence collection system is presented in [7]. Though, some principles are similar with the proposed CityWatcher system, their prototype is developed for laptops, which is not as ubiquitous as using smartphones and besides, as a major distinction, does not incorporate annotating and discovering relevant video recordings. The number of possible use cases is limited only to requesting the video and the research challenge of how to deal with the heterogeneous world of IoT is not discussed. The problem of crowdsourced video annotation is widely discussed in [12].

As it was mentioned above, there are a number of applications, which allow using an Android smartphone as a VMSC. We will discuss their characteristics and main features, typical for applications of this class. Main characteristics of a typical VMSC emulating application include but aren't limited to: (a) Free or not?; (b) Video format, possibility to choose the format; (c) Cycle video recording; (d) Possibility of making photos; (e) Annotation with timestamp; (f) Annotation with GPS coordinates; (g) Annotation with location information (street name); (h) Map, navigation; (i) Acting as a car's computer (measuring maximum and average speed, etc); (j) Possibility of loading video to the Internet via wireless networks; (k) Indexing significant moments manually or by a sensor; (l) Resource consumption; (m) Possibility of working in background. Some of the mentioned apps are briefly discussed below:

- DailyRoads Voyager – free, easy to use application. Axel Voyager – complicated full featured application, supports navigation functionality and can make indexes of significant moments

- AVR – usual video recording features, image stabilization function, navigation, can send video to the Internet
- AutoBoy BlackBox – full featured video player, displays additional information (speed, compass), very big choice of settings
- CaroO Pro - additional functionality includes features, usual for cars computer

Any of these apps can be used as a VMSC, but none of them support answering requests to share video recordings and make their video recordings discoverable. In the proposed CityWatcher application we focus on request processing and answering, so that interested parties could make use of locally stored and annotated video recordings.

3   CityWatcher Concepts

In this paper we propose, develop, discuss and evaluate the "CityWatcher" Android-based app. First of all, we discuss the use case "Road accident investigation" as illustrated in Fig.2.

Let's assume the following scenario. A road accident happens on a street. Road police has to decide who was at fault, but it is not always clear what happened. But if there was no road camera watching exactly that very spot, police will have to use other evidence which could be much weaker than the video recording. There is a good chance that the accident may have been recorded by someone's VMSC. Surely, the owner could stop the car and give his/her phone number to everyone involved in the accident and then they can get the video recording and present to investigation unit. In real life this is not always so. First of all – a car could be parked, while the driver was absent, if we assume that a VMSC continues recording while a car is parked. Or the driver hadn't seen the accident, or just had no time to stop. Anyway, there is a need to get a video. But how to get it ? The policeman opens a web browser, navigates to a special web site and makes a request to find all video recordings from that location by marking it on a map and specifying the time. Request is passed to a cloud service and linked to the account of a policeman. IoT middleware sets up a new task, which can last for several days or even longer. We can't assume that all smartphones that act as a VMSCs will be on and available via Internet just at the moment. So, the middleware gathers information as smartphones become available. When an available smartphone is discovered, middleware makes a request like "Have you been in 123.456/ 678.678 (lon/lat) on 03.03.14 12.14:45. The smartphone checks its own database. It sends no answers to the server. All the answers are connected with private data, so have to be handled very carefully. If there is no recording from that time and location – nothing happens. If there is – the application asks the user to check, if there is an accident on the record. If the users finds that there is, than he/she can share the fragment with the authorities. The user can also wipe out the sound, so that personal talks are not transferred.

All such answers are linked to the policeman's request, so when he/she gets back after some time he/she finds that he/she can look at the accident from different angles and make a fully qualified decision. We can also think about some incentives to make this service more attractive. People that provide significant help by sharing their video recordings can be rewarded somehow, for example by lowering the transport registration fee or by reducing fine points, or by giving some priorities in a queue for police service. This can depend on current country laws and traditions.

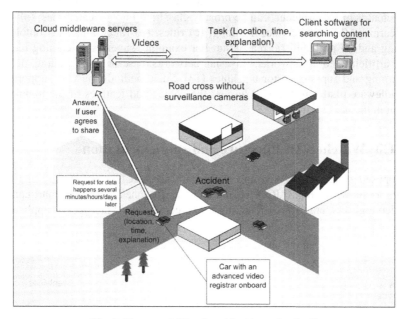

Fig. 2. Use case 1 "Road accident investigation"

As it was already mentioned, the main principle of data acquisition is crowdsourcing. Crowdsourcing seems to be a strong method of collecting data, but there are concerns about manual data processing. Though a lot of dispatching work can be automated, most of it has to be made manually and by professionals. There are also a number of challenges that would arise in any crowdsourcing solution:

- **Redundant data.** In our scenario all videos are complementary as there couldn't be identical videos from different cars.
- **False data.** Someone's joke can become a waste of time for the service team. One of the solutions of this problem can be in storing user's ratings in the database. These ratings can be made by giving marks to loaded videos by personnel which work with them.   Requests from users, who had already produced false data, can be blocked.
- **Feedback.** Users who send data want to understand what happens next. If they don't receive an acknowledgement that their request was processed and their work is needed, they will not send any requests any more. Citywatcher android application has a form with reports on all shared videos, so the user can check if his contribution was used.

Designing, running and automating a processing center is also a big challenge. At first, report processing can be done manually by professional operators. Later, when all the business logic becomes clear, many techniques can be used for process automation and reduction of the need in human resources. First of all, speech and video recognition techniques must be used for determining the context. When this is done, it becomes possible to apply different reasoning techniques for making a decision where to transfer the report. Context reasoning techniques, which will be used in future releases of Citywatcher, can be broadly classified into six categories: (a) supervised learning, (b) unsupervised learning, (c) rules, (d) fuzzy logic, (e) ontological reasoning and (f) probabilistic reasoning. For example, supervised learning methods include artificial neural networks, Bayesian networks, case-based reasoning, decision tree learning and support vector machines [13]. Such methods must be supported by the middleware platform. We will discuss the choice and features of the middleware platform in the next section.

4   CityWatcher Architecture and Implementation

The proposed system is divided into three main parts: ICOs, middleware and client software as shown in Fig.3. ICOs include the smartphones, VMSCs, smart cameras, cars video registers, etc. All participating ICOs record video via a special application.

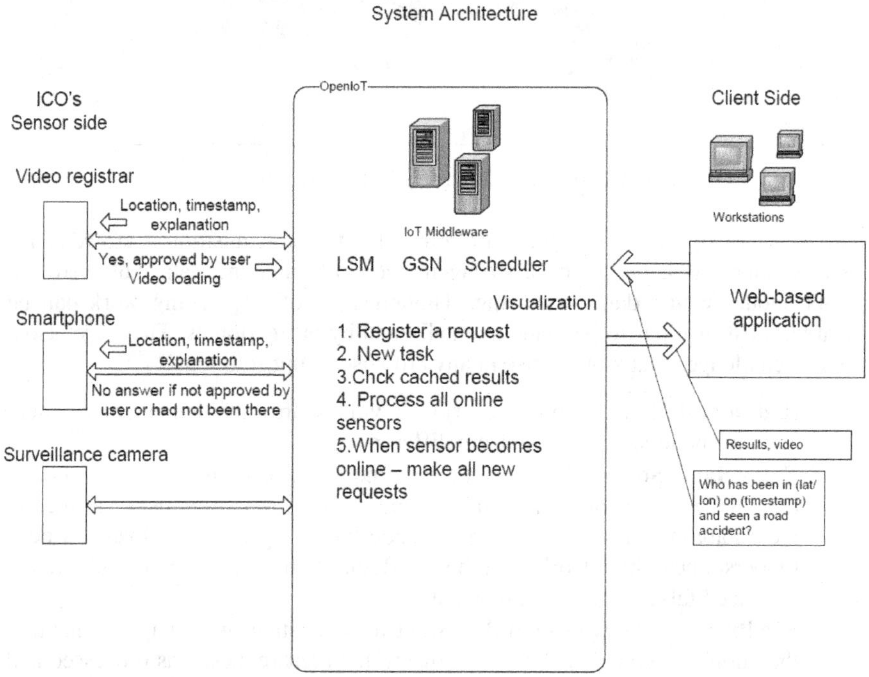

Fig. 3. CityWatcher system architecture

The prototype is an Android app. Application features include Video recording, Video annotating (gps coordinates, time), Self-discovery, auto register with middleware ,Request processing (like 'Have you been(time, location) ?'), Asking the owner to watch and share video, Loading part of the video to the cloud. Screenshots of

CityWatcher application for Android are presented in Fig.4. It is a working prototype app. The design of user-friendly interface will be included in future work.

CityWatcher Middleware is based on **OpenIoT** Platform [14]. OpenIoT project is a new open source middleware platform for intelligent IoT applications. It is an extension to cloud computing implementations and provides functions for managing IoT resources. In this way users can get IoT services including Sensing-as-a-Service.

OpenIoT platform is discussed in [16][17].

The CityWatcher project relates to urban crowdsensing. So we expect the city administrations to become service providers and a host to an OpenIoT platform. We use Global Sensor Network (GSN) to provide middleware functionality for registering and finding ICOs. Using an OpenIoT platform gives us an efficient and advanced way to fusing data sources. Not only VMSC can be used for getting video data. Streets in modern cities already have lots of surveillance cameras. Video streams from those cameras are stored in different systems, as cameras can have different vendors. It will be hard for users to make requests to check all such systems. OpenIoT platform can be used for efficient data fusing, so the user will have to make only one request to the middleware. The integration of CityWatcher application with OpenIoT middleware is shown in Fig.5. **Client software** located in Utility App Plane is a web-based application, which allows a user to make request to the system and view results.

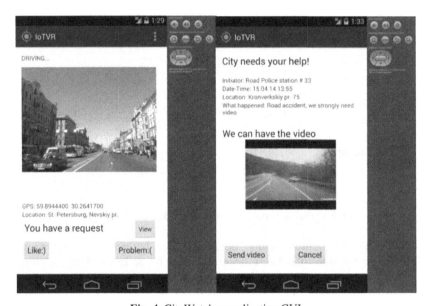

Fig. 4. CityWatcher application GUI

One of the technical problems needed to be solved is identifying the device, on which video is recorded and stored. Firstly, middleware has to be aware of all participating ICO's in the system, as it has to broadcast/multicast a request to all of them, register that the task has been sent and match the answer with the device IF. Secondly, middleware needs to have information about user personal preferences, as the user obtains some incentives for participating and sharing his/her videos. Besides, having information about the ICO owner makes the video fragment more legal in the court. We must also bear in mind that one user can have several devices and one device can be used by different users.

We propose to use the following scheme: every user gets an account in the CityWatcher system. It is a classical login/password pair. With this account a user can start the CityWatcher app. On the middleware side login is linked to full user description/profile, so the user can get his/her incentives from city services. As the video is stored on the users' devices and one user can have multiple devices, we need to uniquely identify every device. IP addresses and MAC addresses are poor candidates for that, as IP address can be changed when the device changes the network, and MAC address, in practice, can be not unique.

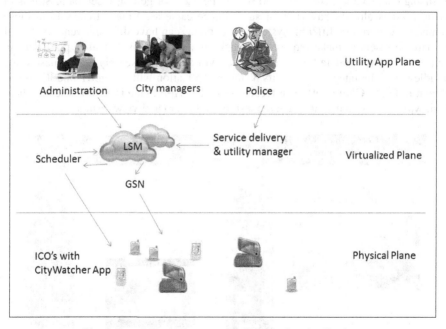

Fig. 5. OpenIoT middleware with CityWatcher Application

To solve this problem a unique device ID can be generated following the CityWatcher installation. It is like a license number for software. Then, all users of one device use this unique ID, but log in with their own accounts. This enables the middleware to distinguish devices for video searching possibilities and also, distinguish users for incentive purposes.

When the video fragment is loaded into the cloud, it should be stored and registered in the database. Additional information about the fragment includes task ID, on which it was found, user ID, device ID and obtaining time. Besides, some information about device like camera resolution and record format is also attached. This make possible to sort videos and firstly view the results with best resolution.

We use SQLite as an engine for storing data about timestamps and coordinates. The CityWatcher app instantly adds record to the database. When a task arrives and there is a need to understand if there is a video on local storage – a request to the database is made.

When the request is made, we search not exactly the same time and coordinates, but add delta values and perform range search. The middleware database structure is omitted in this paper due to space limitations.

As it was mentioned above, video and all the metadata about is stored on the device. This is more secure and gains more trust, than loading all the data to the cloud. On the other hand, searching in devices metadata depends on the online availability of the device. If a device is offline, searching can be slow, as the middleware will have to wait for the device to become available on the network to give it a search task. We can assume, that some users can agree to load the metadata about when and where they have been to the cloud immediately. That makes possible for the middleware to make search in a local database without spreading tasks among thousands of ICOs and waiting for their reply.

5   Conclusion and Future Work

We can think of a scenario, where people not only report problems by pressing the "red button", but, in contrast, they press a "green button", to appreciate things or changes they like. This "likes" can be marked on a map, aggregated and analyzed to understand, which locations in a city produce positive emotions. This can be a little step towards augmented reality. Another challenge is to measure the probability of the first scenario. We can expect only some percentage of cars to carry an ICO with CityWatcher app, so it will be challenging to find out, how often it will be possible to get the video of an accident. User study will be part of the future work. Measuring probability can be achieved by using modern modeling methods like traffic simulators. For example, open source projects like SUMO and Veins can be used [18],[19]. Making a precise model for prediction is a challenging task and can be considered for future work. We are confident that it is possible to extend the CityWatcher application with pervasive object search options. It is hard to imagine a location where ICOs cameras can't reach. ICOs can be deployed anywhere and anytime. This reinforces the motto of pervasive computing "anywhere, anytime, on any device and over any network". While it might still be hard and expensive to put surveillance cameras everywhere, with the advent of mobile computing we can assume that users carry their ICOs/smartphones everywhere they go. ICOs become more and more powerful and video recognition algorithms also are becoming more advanced. For example, we can include some library for car number plate recognition [20]. An ICO can receive a request from the police, searching for a stolen car. ICO analyzes all car number plates that it detects on the road. If a match is found – an alert can be sent back to police

As video recognition algorithms become mature, we foresee more and more scenarios of pervasive object search. Present day smartphones are able to record and store a large amount of video content with location and timestamp annotations, as well as other diverse context. The proper use of sharing this information can help in improving how the smart city runs and functions.

References

[1] Caragliu, A., Del Bo, C., Nijkamp, P.: Smart Cities in Europe. In: Proceedings of the 3rd Central European Conference on Regional Science, Košice (2009)

[2] Hu, M., Li, C.: Design Smart City Based on 3S, Internet of Things, Grid Computing and Cloud Computing Technology. In: Wang, Y., Zhang, X. (eds.) IOT Workshop 2012. CCIS, vol. 312, pp. 466–472. Springer, Heidelberg (2012)

[3] Smart city, http://en.wikipedia.org/wiki/Smart_city (accessed: April 18, 2014)

[4] Giffinger, R., Fertner, C., Kramar, H., Kalasek, R., Pichler-Milanovic, N., Meijers, E.: Smart cities – Ranking of European medium-sized cities. Smart Cities. Centre of Regional Science, Vienna (2007)

[5] Intelligent transportation system, http://en.wikipedia.org/wiki/Intelligent_transportation_system (accessed: April 24, 2014)

[6] Ki, Y.-K., Lee, D.-Y.: A Traffic Accident Recording and Reporting Model at Intersections. IEEE Transactions on Intelligent Transportation Sytems 8(2) (June 2017)

[7] Chae, K., Kim, D., Jung, S., Choi, J., Jung, S.: Evidence Collecting System from Car Black Boxes. School of Electronic Engineering Soongsil University Seoul, Korea

[8] Huang, T.: Surveillance Video: The Biggest Big Data. Computing Now 7(2) (February 2014), http://www.computer.org/portal/web/computingnow/archive/february2014?1f1=929719755f161316415589b16868896

[9] Schuurman, D., Baccarne, B., De Marez, L., Mechant, P.: Smart Ideas for Smart Cities: Investigating Crowdsourcing for Generating and Selecting Ideas for ICT Innovation in a City Context. Journal of Theoretical and Applied Electronic Commerce Research 7(3), 49–62 (2012)

[10] Georgakopoulos, D., Baker, M., Nodine, A.: Event-Driven Video Aware-ness Providing Physical Security. World Wide Web Journal (WWWJ) 10(1) (March 2007)

[11] Chatzimilioudis, G., Konstantinidis, A., Laoudias, C., Zeinalipour-Yazti, D.: Crowdsourcing with Smartphones. IEEE Internet Computing 16(5), 36–44

[12] Vondrick, C., Patterson, D., Ramanan, D.: Efficiently Scaling up Crowd-sourced Video Annotation. A Set of Best Practices for High Quality, Economical Video Labeling. Springer Scince+Business Media LLC (2012)

[13] Perera, C., Zaslavsky, A., Christen, P.: andD. Georgakopoulos, Context Aware Computing for Internet of Things: A Survey. IEEE Communications Surveys & Tutorials Journal (2013)

[14] OpenIoT Architecture Webpage, https://github.com/OpenIotOrg/openiot/wiki/OpenIoT-Architecture (accessed: April 12, 2014)

[15] OpenIoT Webpage, http://www.openiot.eu (accessed: March 12, 2014)

[16] PodnarZarko, I., Antonic, A., Pripužic, K.: Publish/Subscribe Middleware for Energy-Effcient Mobile Crowdsensing. In: UbiComp 2013 Adjunct Proceedings of the 2013 ACM Conference on Pervasive and Ubiquitous Computing Adjunct Publication, pp. 1099–1110 (2013)

[17] What is OpenIoT and what does it can do for you?, http://academics.openiot.eu/ (accessed: April 24, 2014)

[18] Sommer, C., German, R., Dressler, F.: Bidirectionally Coupled Network and Road Traffic Simulation for Improved IVC Analysis. IEEE Transactions on Mobile Computing 10(1), 3–15 (2011)

[19] Arellano, W., Raton, B., Mahgoub, I.: TrafficModeler extensions: A case for rapid VANET simulation using, OMNET++, SUMO, and VEINS. In: 2013 10th International Conference on High Capacity Optical Networks and Enabling Technologies (HONET-CNS), pp. 109–115 (2013)

[20] Buch, N., Kristl, V.S.A., Orwell, J.: A Review of Computer Vision Techniques for the Analysis of Urban Traffic. IEEE Transactions on Intelligent Transportation Systems 12(3)

Application of Face Recognition Methods
for Process Automation in Intelligent Meeting Room

Alexander L. Ronzhin

SPIIRAS, 39, 14th line, St. Petersburg, 199178, Russia
ronzhinal@iias.spb.su

Abstract. This paper describes the automatic registration technique based on face recognition of meeting participants, which has been implemented in the intelligent meeting room. This technique provides unobtrusive recognition and picture making of participant faces. Application of the developed technique makes it possible to reduce the work of secretaries and videographers; it also allows participants to concentrate on the discussed issues at the expense of automated control of sensory equipment. For the experimental evaluation of the developed methods and the technique about 52,000 photographs for 36 participants were accumulated from a high resolution camera. During the experiments three face recognition methods LBP, PCA and LDA were compared. The experimental results showed that method LBP has the highest recognition accuracy 79,3%, but the PCA method has the lowest percentage of the false positives 1,3%, which is important aspect in the participants identification.

Keywords: Multichannel signal processing, intelligent meeting room, computer vision, face recognition, processes automation.

1   Introduction

Application of intelligent information technologies in business and education, including at carrying out distributed events and for automation of the speaker's talk recording at the meeting, is important issue due to the increasing mobility of people and necessity to control the quality of decisions [1, 2]. Nowadays, the evolvement of a scientific paradigm of the intellectual space has shaped several models of intelligent environment that may serve users in a confined space: intelligent room, house, lecture hall, meeting room [3, 4, 5]. Development of the tools for capture and processing of audiovisual signals, which are capable to contactless evaluate the current situation in the room, is one of the main fields of research in this area.

When designing intelligent rooms for meetings, lectures, scientific and educational activities the following methods of audio and video signals processing are now most widely used: 1) detection and tracking of participants based on video monitoring [6]; 2) estimation of head orientation and face recognition [7]; 3) sound source localization [8, 9]; 4) speech recognition [10]; 5) speaker diarization [11]; 6) speech synthesis [12]. Application of these methods and their combination makes it possible to

S. Balandin et al. (Eds.): NEW2AN/ruSMART 2014, LNCS 8638, pp. 156–163, 2014.

develop tools for automatic recording of the speakers' talks, organizing of television broadcasts, journaling and archiving of audiovisual recordings after the event.

Let us to consider SPIIRAS intelligent meeting room. For its design ergonomic aspects of multimedia, audio-visual recording equipment location were taken into account to provide coverage and service of the greatest possible number of participants. Functionality of the intelligent meeting room includes its equipment as well as methods needed to the implementation of information support services and automation of events. At implementation of the participants system monitoring in the intelligent room, which based on distributed audiovisual signals processing were used as the existing methods of digital data processing (image segmentation, calculation and comparisons of the histograms, etc.), and developed own proprietary methods, such as method for meeting participants registration, the method for audiovisual recording of their performances. Detailed description of the equipment and audiovisual data processing methods used in the intelligent room is presented in [3, 8, 13, 14, 15].

This paper is organized as follows. The second section discusses methods of biometric identification based on face recognition. The third section describes the specifics of the developed technique and methods of automatic registration of meeting participants based on face recognition. The fourth section presents the experiments, conditions and results of the evaluation of face recognition methods.

2   Biometric Identification Methods Based on Face Recognition

Biometric identification is the automatic identification or user verification technology on the basis of physical characteristics and personal traits. Biometric characteristics and traits are divided into two categories: behavioral and physical. Behavioral characteristics include events such as signature and typing rhythm. Biometric systems for physical characteristics used to identify eyes, fingers, hands, voice and face.

Face recognition system is a computer application for the automatic identification or verification of the digital image part, with video frame read from the video source [16, 17]. Face recognition system allows user to be identified just by walking past the surveillance camera. People often recognize each other by unique facial features which is the most successful form of human observation. On this phenomenon the biometric face recognition technology is based. Currently, the interest in face recognition technology has been excited by the availability and low cost of video equipment as well as their unobtrusive presence. Such systems are already available on the market and are claimed to accurately, efficiently and rapidly recognize the human face. But there are some restrictions: 1) recognition errors occur due to the external conditions, such as ambient light and change of the position of the face relative to the camera, etc., and 2) the lack of accuracy appears during the processing of frames with a gestures variety in them.

In most cases, face recognition algorithm may be divided into two stages: 1) determining the position of the user face in the picture with a simple or complex background [18]; 2) recognizing face to identify a user. At both stages feature extraction procedure is performed, which converts pixels of the face region on the image into vector. Moreover, at these stages, a block for forms recognition is used. This block performs a search for and comparison of characteristics in the pre-arranged database to determine the best resemblance with the received face image of the user.

Further we consider in more detail the processes performed at each stage of image processing. Detecting the presence of a face on the image is a simple task for humans, but not for the computer [19]. Computer needs to determine the image pixels belonging to a face region and to the background. On the classic passport photo where the background image is clean, i.e. uniform and solid, it is not difficult to identify the face region. But when the background consists of several layers, where other objects are present, this problem becomes quite complex. Typically, methods for the face region detection are based on the face key point's determination such as eyes, or on the analysis of the color space of the image, as well as other face features, are used.

When the face region is determined (separated from the background image), normalization procedure should be performed. This means that the image must be standardized in terms of size, orientation, brightness and other parameters relative to the images in the database. In order to perform the normalization, the face key points must be accurately identified. The use of these key points in the normalization algorithm allows modification of images based on statistical conclusions or approximation. Recognition can be done properly only if the original image and the images in the database have the same parameters such as orientation, rotation, scale, size, etc. Only after the normalization features extraction and face recognition can be performed. At features extraction mathematical representation is generated. This representation is called biometric template or biometric reference [20], it is stored in the database and is the basis for any recognition.

The most common face recognition technologies such as PCA [20], LDA [21] and LBP [22] are applied with different success. In addition, in the Kanade and Yamada paper [23] authors describe the advantages of rigid components application, where the weigh coefficients for each of face fragments are predefined during processing of set of prepared photos. These coefficients depend on face position on the image. Methods based on holistic and rigid components of face representation are similar to the applied classification mechanism, because both of them compare image set with region "points" of feature space. In the approaches based on the use of rigid components several points are used, where each point is located in a distinct and to a considerable degree independent feature space. Until now, the advantage of application of free components to automatic face recognition with the presents of pose mismatches hasn't been fully investigated.

3   The Developed Technique for Automatic Registration of Meeting Participants

At the development of technique for automatic registration of meeting participants three cameras were implemented. So, this technique can be divided into three stages, as it is shown on Figure 2.

At the beginning of the first stage, frame from ceiling panoramic camera is received. Then in a cycle on total number of chairs located in the room, the procedure for cropping of a chair region from video stream is performed. Each chair region has predetermined size and position. After that, a histogram of color distribution on a received frame region is composed. Next, the created histogram is compared with prearranged template histogram of current chair region for calculation of the correlation coefficient. All numbers of occupied chairs, which correlation coefficient was more than threshold value, were added to list of chairs for processing at the next stage.

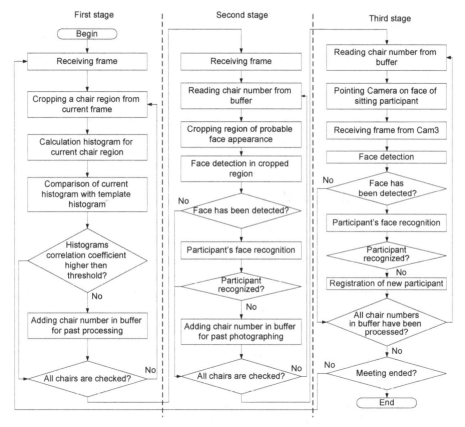

Fig. 1. Scheme of the technique for automatic identification and registration of participants in the intelligent meeting room

At the end of processing for all chairs, in the second stage a frame from the high resolution camera is received and processed by procedure for searching participant faces in areas of possible appearance of face corresponding to occupied chairs. Further, all zones in which faces have been found are processed using the face recognition method. Then a list of chair numbers is formed with unidentified participants, as well as a set of control commands, which are used in the for camera pointing on faces of the participants.

At the third stage, PTZ camera is pointing in close-up to the face of each previously unidentified participant. After checking the presence faces in the frame will be carried participant re-identification. If face hasn't been founded then number of chair with sitting participant passes into the end of the queue of unidentified participants. In case of absence a participant in the database new participant registration process is launched.

4   Experimental Results

First, we consider the dependence between the participant face size in the frame and the increasing of the distance between him/her and the stationary high resolution camera, as shown in Table 1. Distance from the camera to the near chair is 3m, and to the farthest – 9.5 m.

Table 1. Maximum participant's face size for chairs of the room

	Columns of chairs					
	66	64	65	59	55	43
Chair rows	56	54	50	43	44	38
	50	43	40	36	36	36
	38	36	36	34	33	30

Data presented in Table 1 shows that when the distance from the camera to chairs changes then the maximum detected participant face size gradually decreases, and at the farthest chair it becomes more than two times smaller.

For the experimental evaluation of a method of automatic identification of participants during the events in the intelligent room the accumulation of participant photos was produced only at the second stage of the method. As a result, the number of accumulated photos was more than 40,000 for 30 participants. In the database for training face recognition models, photos of each participant were added in order to have a difference between participant's head orientation and the direction of view from the Cam4 on every photo. As a result the created database contains 20 photos for each participant.

At the preliminary stage of experiments have been decided to determine the threshold for the three face recognition methods: 1) recognition of monolithic face representation with PCA method (PCA) [20]; 2) Linear Discriminant Analysis (LDA) [21]; 3) local binary patterns (LBP) method [22]. During this experiment threshold was calculated for each participant added to the recognition model, a maximum value of a correct recognition hypothesis for the LBPH method ranged from 60 to 100, for the PCA method from 1656 to 3576, for the LDA method from 281 to 858. As a consequence, for the further experiments were selected minimum threshold value - 60, 1656 and 281, respectively, for these methods. Table 2 presents average values of face recognition accuracy, as well as first (False Alarm (FA) rate) and second (Miss rate (MR)) errors type for each selected method.

Table 2. First experiment results of face recognition for selected methods

Method	FA, %	MR, %	Accuracy, %
LBP	10,5	10,1	79,3
PCA	0,8	25	74,1
LDA	7,6	15,3	77,1

For the training of recognition model during the second experiment some photos from the set of photos for 3 participants were replaced in such a way, that new photos of the equidistant from the camera chairs were added. Average values of the second experiment results with same parameters as in the first experiment are shown in Table 3.The presented results in Table 3 that the model trained on the photos of equidistant from the camera chairs improves the recognition accuracy and reduce the number of false positives for methods LBP and PCA. However, for the LDA results decreased significantly.

Table 3. Second experiment results of face recognition for selected methods

Method	FA, %	MR, %	Accuracy, %
LBP	10,3	10,3	79,4
PCA	0,8	24,5	74,7
LDA	9,8	15,6	74,6

In the third experiment photos for 13 participants were added to trained model in such a way as in the second experiment. The total amount of photos for each participant was equal to 20. In addition, training data base was increased to 36 participants as well as test data base includes more than 52 thousands photos. Average results values of the third experiment are shown in Table 4.

Table 4. Third experiment results of face recognition for selected methods

Method	FA, %	MR, %	Accuracy, %
LBP	12	8,5	79,5
PCA	1,3	23,5	75,2
LDA	19,2	7,8	73

The high value of false positives and miss rate errors due to the fact that the photos were stored in the course of actual of events, without a prepared scenario and focusing participants on a single object. Hereupon at the time of photographing participants can move freely, according to their face in the photos could be blurred or partially hidden. The experimental results showed that for systems of meetings process automation for face recognition PCA method should be used, this conclusion is based on the fact that this method as it is inferior in accuracy to LBPH method by 3-5%, but it has the lowest value of false alarm errors, which is an important aspect in the identification of meetings participants.

5   Conclusion

Information-control services provision based on human behavior and situation analysis is the main idea of intelligent space concept. An example of such space is intelligent meeting room, which is equipped by network of program modules, activation devices, multimedia facilities, and audiovisual sensors. Application of biometric identification technology based on face recognition methods provides automation of registration processes of meeting participants, thus reducing the work of secretaries and videographers; it also allows participants to concentrate on the discussing issues at the expense of automation control of sensory equipment.

During the research the technique for automatic registration of meeting participants was developed. It provides unobtrusive recognition and picture making of participants faces. At this stage of research for the experimental evaluation of the technique 52 thousands of photos for 36 participants were used. During experiments three face recognition methods LBP, PCA and LDA were compared. The results shows that LBP method has highest recognition accuracy (79,5%) as well as PCA method has the lowest false alarm rate (1,3%).

Acknowledgments. This work is partially supported by the Scholarship of the President of Russian Federation (Project № СП-1805.2013.5) and by the Committee on Science and Higher Education of the Administration of St. Petersburg.

References

1. Fillinger, A., Hamchi, I., Degré, S., Diduch, L., Rose, T., Fiscus, J., Stanford, V.: Middleware and Metrology for the Pervasive Future. IEEE Pervasive Computing Mobile and Ubiquitous Systems 8(3), 74–83 (2009)
2. Nakashima, H., Aghajan, H.K., Augusto, J.C.: Handbook of Ambient Intelligence and Smart Environments, 1294 p. Springer (2010)
3. Yusupov, R.M., Ronzhin, A.L., Prischepa, M., Ronzhin, A.L.: Models and Hardware-Software Solutions for Automatic Control of Intelligent Hall. Automation and Remote Control 72(7), 1389–1397 (2011)
4. Aldrich, F.: Smart Homes: Past, Present and Future. In: Harper, R. (ed.) Inside the Smart Home, pp. 17–39. Springer, London (2003)
5. Lampi, F.: Automatic Lecture Recording, 229 p. The University of Mannheim, Germany (2010)
6. Calonder, M., Lepetit, V., Strecha, C., Fua, P.: BRIEF: Binary Robust Independent Elementary Features. In: Daniilidis, K., Maragos, P., Paragios, N. (eds.) ECCV 2010, Part IV. LNCS, vol. 6314, pp. 778–792. Springer, Heidelberg (2010)
7. Ekenel, H.K., Fischer, M., Jin, Q., Stiefelhagen, R.: Multi-modal Person Identification in a Smart Environment. In: Proceedings of the Conference on Computer Vision and Pattern Recognition, CVPR 2007, pp. 1–8 (2007)
8. Ronzhin, A., Budkov, V.: Speaker Turn Detection Based on Multimodal Situation Analysis. In: Železný, M., Habernal, I., Ronzhin, A. (eds.) SPECOM 2013. LNCS (LLNAI), vol. 8113, pp. 302–309. Springer, Heidelberg (2013)

9. Zhang, C., Yin, P., Rui, Y., Cutler, R., Viola, P., Sun, X., Pinto, N., Zhang, Z.: Boosting-Based Multimodal Speaker Detection for Distributed Meeting Videos. IEEE Transactions on Multimedia 10(8), 1541–1552 (2008)

10. Karpov, A., Markov, K., Kipyatkova, I., Vazhenina, D., Ronzhin, A.: Large vocabulary Russian speech recognition using syntactico-statistical language modeling. Speech Communication 56, 213–228 (2014)

11. Imseng, D., Friedland, G.: Tuning-Robust Initialization Methods for Speaker Diarization. IEEE Transactions on Audio, Speech, and Language Processing 18(8), 2028–2037 (2010)

12. Lobanov, B., Tsirulnik, L., Ronzhin, A., Karpov, A.: A Model of Personalized Audio-Visual TTS-synthesis for Russian. In: Proceedings of International Conference on Speech Analysis, Synthesis and Recognition, Applications in Systems for Homeland Security, SASR-2008, Poland, pp. 25–32 (2008)

13. Ronzhin, A. L.: An audiovisual system of monitoring of participants in the smart meeting room. In: Proceedings of the 9th Conference of Open Innovations Framework Program FRUCT – Russia, Petrozavodsk, pp. 127–132 (2011)

14. Ronzhin, A.L., Budkov, V.Y., Karpov, A.A.: Multichannel System of Audio-Visual Support of Remote Mobile Participant at E-Meeting. In: Balandin, S., Dunaytsev, R., Koucheryavy, Y. (eds.) ruSMART/NEW2AN 2010. LNCS, vol. 6294, pp. 62–71. Springer, Heidelberg (2010)

15. Budkov, V.Y., Ronzhin, A.L., Glazkov, S.V., Ronzhin, A.L.: Event-Driven Content Management System for Smart Meeting Room. In: Balandin, S., Koucheryavy, Y., Hu, H. (eds.) NEW2AN 2011 and ruSMART 2011. LNCS, vol. 6869, pp. 550–560. Springer, Heidelberg (2011)

16. Lin, S.-H.: An Introduction to Face Recognition Technology. Information special issue on Multimedia Informing Technologies Part-2, 3(1), 1–7 (2000)

17. Rajeshwari, J., Karibasappa, K.: Face Recognition in Video Streams on Homogeneous Distributed Systems. International Journal of Advanced Computer and Mathematical Sciences 4(1), 143–147 (2013)

18. Gorodnichy, M.D.: Video-Based Framework for Face Recognition in Video. In: Proceedings of Second Canadian Conference on Computer and Robot Vision (CRV 2005), pp. 330–338 (2005)

19. Castrillon-Santana, M., Deniz-Suarez, O., Guerra-Artal, C., Hernandez-Tejera, M.: Real-time Detection of Faces in Video Streams. In: Proceedings of Second Canadian Conference on Computer and Robot Vision (CRV 2005), pp. 298–305 (2005)

20. Georgescu, D.: A Real-Time Face Recognition System Using Eigenfaces. Journal of Mobile, Embedded and Distributed Systems 3(4), 193–204 (2011)

21. Belhumeur, P.N., Hespanha, J., Kriegman, D.: Eigenfaces vs. Fisherfaces: Recognition Using Class Specific Linear Projection. IEEE Transactions on Pattern Analysis and Machine Intelligence. 19(7), 711–720 (1997)

22. Ahonen, T., Hadid, A., Pietikäinen, M.: Face Recognition with Local Binary Patterns. In: Pajdla, T., Matas, J(G.) (eds.) ECCV 2004. LNCS, vol. 3021, pp. 469–481. Springer, Heidelberg (2004)

23. Kanade, T., Yamada, A.: Multi-subregion based probabilistic approach toward pose-invariant face recognition. In: Proceedings of IEEE International Symposium on Computational Intelligence in Robotics and Automation, pp. 954–958 (2003)

Architecture of Data Exchange
with Minimal Client-Server Interaction
at Multipoint Video Conferencing

Anton I. Saveliev[1], Irina V. Vatamaniuk[2], and Andrey L. Ronzhin[1,2,3]

[1] SPIIRAS, 39, 14th line, St. Petersburg, 199178, Russia
[2] SUAI, 67, Bolshaya Morskaya, St. Petersburg, 190000, Russia
[3] ITMO University, 49 Kronverkskiy av., St. Petersburg, 197101, Russia
{saveliev,vatamaniuk,ronzhin}@iias.spb.su

Abstract. The problems of serverless connections between clients and audio-visual data transmission in peer-to-peer videoconferencing web application are discussed. The serverless connections based on the WebRTC protocol allow applications to transfer any data between clients directly. The partial or complete loss of service data for the WebRTC protocol, which hampers the client connection, occurs during interaction of several clients through the server of the videoconferencing application. The proposed architecture of data transmission and storage on client and server provides buffering and following processing of the signal data with the exception of their loss and support of interaction between client groups.

Keywords: peer-to-peer videoconferencing, peer-to-peer protocols, audio/video stream distribution, media transfer protocols, data exchange in web applications, group video chat.

1 Introduction

Nowadays, there are a lot of problems connected with streaming of video and audio data despite the great rate of development of Internet technologies. These problems are largely related to insufficient capacity of communication channels. Because video systems require greater channel bandwidth for video transmission, even for two participants, that support multiuser video conferencing is extremely challenging task. Currently hundreds of thousands of simultaneous users can use sharing peer-to-peer (P2P) networks [1, 2]. A common practice in P2P streaming systems today is a grouping of users, who looking the same content in a "swarm" and the redistribution of parts of video exclusively between members of the swarm. The P2P video systems with such a channel-isolated structure are characterized by channel change latency and lag of content playing related to outflow of channel clients and imbalance of the number of transmitting and host units. In general, the global P2P networks with channel-isolated structure have serious performance problems now that will become more serious with increasing the number of channels users.

S. Balandin et al. (Eds.): NEW2AN/ruSMART 2014, LNCS 8638, pp. 164–174, 2014.

Some related works are discussed in next section. In Section 3 the novel architecture of data exchange based on serverless connection control was described. Section 4 is devoted to the formation a web page and establishment of connection between client and server by WebSocket protocol. The algorithm for creation of client connections via WebRTC protocol is presented in Section 5.

2   Related Works

In addition, in streaming P2P networks there are problems associated with delays channel switching and low-productivity systems for channels with small number of clients. In [3] to solve these problems a multi-channel P2P streaming platform VUD (View-Upload Decoupling), which separates the users performing the file download and data view, and shares the resources by cross channels, was proposed. It is provides stability for multi-channel systems and qualitative distribution of network resources. Also to improve the performance data transmission the geographic location of users of the network is taken into account in order to establish the connection between the most closely spaced users with the necessary data.

In [4] the architecture of a distributed P2P multiplayer video conferencing system, which provides each participant to create, send and receive video at any time, is proposed. During a video conference participant transfers the video created by own device or received from another participant. Thus, a participant who sends his own video cannot act as an intermediate node for video stream from another participant. This distributed architecture can be used to implement the P2P video conferencing system and allows each participant to see other participants. Management of the system is based on chain configuration during the session, which is controlled by the configuration (service) messages sent between applications video conferencing participants.

To reduce the amount of transmitted data during video conferencing in [5] the detection of active speaker and priority transmission of his/her stream is proposed. The speaker identification, diarization and other speech processing methods is often implemented for automation of the telecommunication services [6, 7, 8].

In [9] the serverless endpoint video combiner architecture was proposed for high definition multiparty video conferencing. It is achieved by integration of the multipoint control unit into the client endpoints, which establishes a multipoint channel connection, encodes and decodes video from a client, distributes video to other client endpoints.

Performance methods video streaming in P2P networks also depends on the configuration of the network, its topology and heterogeneity of network resources of end users, the capacity of their communication channels [10]. In contrast to the joint file download, where low bandwidth channel leads to slow loading, in transmission of streaming video slow connection becomes a real problem. Also the issues of video compression, which can reduce the traffic without significant increasing of calculation on the end-user device during signal encoding/decoding.

The previous results obtained by the authors while researching the problems of multipoint video conferencing are presented in [11]. The optimization of main functioning stages of cross-platform video conferencing application including creation and deletion of audio and video streams of data, their transmission from the server to the client and back, creation chains of streams and their search on the server, was carried out. The simplification of the client part of the application and the reorganization of the structure of the server-side application was done during the research.

3   Main Components of the Video Conferencing Architecture

The developed architecture, which prevents loss of 'configuration message' (service) during connecting three or more participants video conferencing, is shown in Figure 1. The main structural components of the architecture are: 1 - the client part of the application, 2 - the server part, 3 - data transfer protocols. The client part is divided into two independent components - the user's device and the web page. The user device in video conferencing is required for creation of an audio and video stream from the camera and the microphone, which are connected to the user device or embedded in it. The web page consists of classes written in the programming language JavaScript, which are necessary to create the connections to the server and other clients using different protocols and data processing. The CSS and the HTML tools are used to build a graphical interface, to display and control the client part of the application. The JavaScript tools used on the video chat web page include three different types of instructions, which can be applied to transfer data using three protocols: WebRTC, WebSocket and HTTP. Also the JavaScript tools serve to capture and process data streams from the microphone and the video camera.

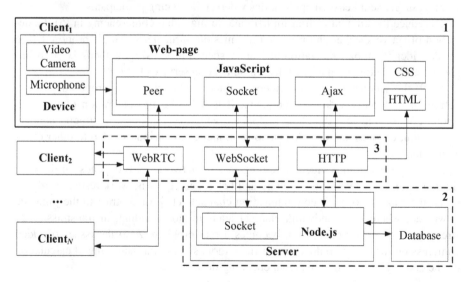

Fig. 1. The developed architecture of data exchange in the videoconferencing application

The second main component of the developed architecture is server side. It performs several functions: building client-side applications, client registration, client authorization, exchanging 'configuration message' between clients, creation of chat rooms and work with the database. The server is powered by Node.js, which transforms the JavaScript into a machine code and has a similar asynchronous architecture as the client side developed by means of the JavaScript. The MongoDB database based on NoSQL is located in the server-side application and that is suitable for simplifying the implementation of the server side. The interaction between the server and the MongoDB database is executed via the JavaScript and special library drivers designed for this database.

The third component of the architecture shown in Figure 1 consists of the HTTP, WebSocket and WebRTC protocols. These protocols provide communication at different stages of the application running. The client connection establishment is performed by WebRTC protocol for streaming audio and video data between the clients. There are some problems related to connection establishing. These problems appear owing to the asynchronous application architecture and the WebRTC protocol, which does not support the standard implementation of connecting a great number of clients. The difficulty of establishment connection procedure via WebRTC with exchange of 'configuration message' data between clients should also be mentioned. This requires special attention to the connection establishment.

In this paper we propose a solution of the problem described above by implementing algorithms of interaction between the client side and the server one and using of different data exchange protocols. Such architecture provides creation of a full-fledged peering application for video conferences that can operate as a group chat.

Further the implementation of the HTTP, WebSocket and WebRTC protocols will be considered, their capabilities and the specific tasks to create a connection and transfer of data in the application of video conferencing will be discussed. The next section is dedicated to the tasks of HTTP protocol, which perform data transmission related with logic operating and graphical representation of the web page. Also the WebSocket protocol will be considered. Its main task is to transfer the 'configuration message' to create connections between clients via WebRTC. Section 5 will describe the problems associated with the implementation of the WebRTC protocol, the asynchronous structure of the application and problems associated with the transfer of audio and video streams via WebRTC protocol.

4   Web Page Formation and Client-Server Connection Establishment by Means of WebSocket Protocol

Firstly let us to consider the main stages of the client and server parts operation of the developed application of video conferencing. The work of the application's client side starts with formation of a registration or authorization web page, which allows the client to interact with the server by sending or receiving data via the HTTP protocol.

The HTTP is an application layer protocol for transmission of arbitrary data. This protocol is used in the application to transfer the client GUI written in the HTML and the CSS and the logic of client-side application written in the JavaScript. Also it is used for client registration and authorization by the Ajax technology, which allows data exchanging between the client and the server via the HTTP without reloading the web page.

The client authorization allows user to get access to his/her personal data and the video conferencing page. To authorize, a user enters the login and the password in the appropriate forms. Then the typed data are collected and sent to the server using the Ajax technology. Next, the server processes the received data. It checks them for compliance with a specific set of characters and for exceeding the permissible size of a query, performs searching of the login – password couples in the database. After successful completion of all operations the video conferencing page with users' data is generated and sent to the client by the server via the HTTP. In case of mismatch of data or any another error, the server sends to the client an informative message, which offers possible solutions to the problem, using the HTTP protocol. The same procedure is executed during the client's registration. In case of the positive processing result, the server saves the user's data in the database, register the client in automatic mode, generate and send a response to a client request as a page of video conferencing. In case of incorrect data input, the server returns an error notification to the client by the HTTP.

Therefore, the application uses the HTTP protocol for secure transmission of the HTML, the CSS and the JavaScript data between the server and the client. The advantage of using this protocol is that it is specifically designed to transmit web pages and their logic. Also it is well supported by all existing browsers. The HTTP protocol has a set of standard commands including two basic commands: 'GET' and 'POST' allowing to carry out requests for the issuance of pages to the server and request for exchanged a varied type of data between client and server for authorization or client registration.

After the client receives the video conferencing web page by means of the JavaScript, a socket is created on the client. It connects the client to the server using the WebSocket protocol. The WebSocket is a protocol for full-duplex communication over the TCP-connection for real time message exchanging between the browser and the web server. The WebSocket protocol opens the sockets on the client and the server, which allow exchanging of data of any type. If a connection via the WebSocket is successful, then server creates a socket for client's data and starts its authorization: tries to get the client's HTTP cookie, which stores information, necessary for authorization, to unpack the cookies, to load the session, appropriate with the HTTP cookie from the database, to determine the user by the loaded session, to bind the user data with the socket, to create an unique number for the socket, to generate and send event 'connection' inside the server. If one of the socket authorization steps generates an error, the socket on the server will be disconnected and removed and the client socket will receive a disconnect message. Event 'connection', which occurs on the server, binds the server socket with the event handlers sent by the client socket. The client socket during its creation generates a set of event handlers sent by the server socket. Thereby the connection between the client and the server is established using the WebSocket protocol. It allows them to exchange quickly of messages of different types, which do not require identification due to presence of separate event handlers for each message.

In the construction of architecture of communication this protocol helps to achieve a high rate of information exchange and reducing the load on the client and the server due to absence of outlay data for identification. The WebSocket protocol is essential in the developed video conferencing application: it is engaged in the transfer of client browsers 'signaling' data. This allows creating a connection by the WebRTC protocol. Thus, this protocol is the basis of connection creating via the WebRTC. Also it simplifies the data transfer process, which is required for peering connections.

5   Creation of Client Connections via the WebRTC Protocol

After establishing the connection between clients and the server via the WebSocket protocol, users have to enable a camera and a microphone and to tell the browser to access them for video calls. Getting access to the camera and the microphone the browser can generate data media streams from the connected devices using JavaScript tools. The received audio and video streams can be transmitted via the WebRTC protocol directly between client browsers. The WebRTC is an Internet protocol designed to organize streaming data based on peer-to-peer technology between browsers or other applications supporting it. The following set of JavaScript instructions is required to connect two clients via the WebRTC protocol: 1) to create a peer for each of the clients, 2) to appoint one of the clients as a 'caller', 3) to appoint another client as a 'respondent', 4) to form a 'configuration message', 5) to exchange the 'configuration message', 6) to finish the connection.

The WebSocket protocol is used to transfer 'configuration messages' between the clients and the server. The sockets created before on the client side can transmit 'configuration message' data on specific channels to the server. By-turn, the server transmits these data to other clients they are intended for. There are three types of 'configuration message' required for client connections via the WebRTC protocol: the 'call offer', the 'call answer' and the 'candidate'. The 'Call offer' serves to initialize a WebRTC session. It is formed on one of the clients and sent to the server using the WebSocket protocol. By-turn the server sends it to the 'respondent' client. The 'Call offer' has an SDP (Session Description Protocol) format. An SDP message, transmitted from one node to another, may indicate: 1) destination address serving for media streams multicast addresses, 2) UDP port numbers for the sender and the recipient, 3)media formats (for example, codecs) used during the session, 4) start and stop time. The SDP message is used in broadcast sessions, for example, in television, radio or video conferencing. The client which received the 'call offer' generates and sends the 'call answer' data by the WebSocket protocol. It also has the SDP format. If the client, who sent the 'call offer', receives SDP message of the 'call answer' type, the clients will exchange by data of 'candidate' type via the WebSocket protocol. The 'candidate' data type has the ICE (Interactive Connectivity Establishment) format. The ICE is a technique used in computer networking, which involves network address translators (NATs) in such Internet applications as Voice over Internet Protocol

(VoIP), peer-to-peer communications, video, instant messaging and other interactive media. The 'candidate' data type is used for connecting clients, setting a path between them through which the media streams are transmitted. If the exchange of 'candidate' data type is successfully ended, each of the clients will open a channel for transmitting various types of data via the WebRTC protocol, including audio and video streams.

The WebRTC protocol has features creating difficulties during the user connection: to create a connection between the clients it is required to perform the operation 'handshake', which consists of the exchange of different types of 'configuration message' between browsers, but at the same time the client can establish only one connection by the WebRTC protocol.

This protocol specificity entails a number of problems when creating a full-fledged video conferencing. The problems occur owing to the asynchronous application architecture in following connection situations: one client to many, many clients to one, many clients to many. Such situations lead to destruction of the connection algorithm and total or partial loss of data required for establishing communication between clients. Several complementary approaches to the data exchange architecture construction in the application were proposed to solve this problem: buffering 'configuration message' of the WebRTC protocol on the client, association of connected clients socket in a 'chat room' on the server. These approaches help to exclude data loss, allow creating all the necessary connections between the clients and managing the processes of client connections at various stages of the application. Next, we consider algorithms based on the developed approaches which allow to create a connection via the WebRTC protocol, to supervise the handling of 'configuration message', to perform buffering of 'configuration message' and to group clients' sockets into separate 'chat rooms'. The algorithm presented in Figure 2 describes the step of connecting clients before the formation of 'configuration message'.

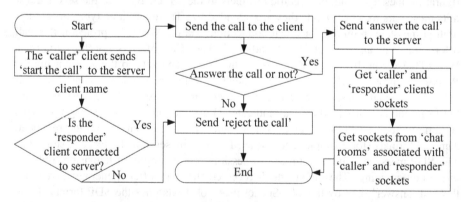

Fig. 2. Algorithm of preparation of client parts before forming the 'configuration message'

Firstly the 'caller' client submits a request to the server to connect to the 'respondent' client by the WebSocket protocol. Next, the server searches for the 'respondent' client among all clients connected to the server. If the 'respondent' client is not connected to the server, the server will end the call of 'caller' client. If server finds the 'respondent' client, it will send a connection request via the WebSocket protocol. The 'responder' client generates and sends a response to the request. If the response is negative, the server will end the call of the 'caller' client. In case of positive response the server receives the socket id of 'caller' and 'responder' clients, finds the 'chat room' in which the sockets are at the moment by socket id. After ending the algorithm socket buffers are created and processed on the server by the algorithm shown in Figure 3.

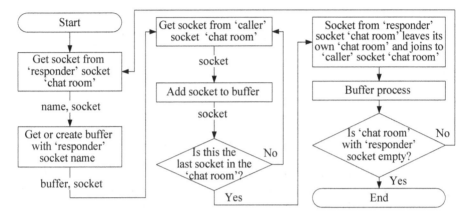

Fig. 3. The algorithm of allocation of sockets and processing their buffers on the server

After server has received the 'chat room' which contains the sockets of each client, the algorithm shown in Figure 3 starts. At first the 'chat room' socket of the 'responder' client is extracted, a buffer to store sockets waiting for the connection via the WebRTC protocol is created for that socket. Next, the 'chat room' socket of the 'responder' client adds the socket taken from the 'caller' client 'chat room' in the already existing buffer. If the socket taken from the 'chat room' of the 'caller' client is not the last, the operation of the socket extraction from the 'chat room' of the 'caller' client and adding it to the buffer will be repeated for the next socket from this room. After the server has taken the last socket from the 'chat room' of 'caller' client, the socket from the 'chat room' of 'responder' client is disconnected from its 'chat room' and is added to the 'caller' client 'chat room'. Next a function of processing socket buffer, which executes the query on the formation of the 'configuration message' for all clients queued this buffer is called. At the end of the algorithm the 'chat room' of the 'responder' client is checked for emptiness. If the 'chat room' is not empty, the algorithm will repeat all actions described above, otherwise the algorithm is complete. As can be seen, the algorithm adds the clients from 'chat room' of the 'responder' client to the 'caller' client 'chat room' and each time the 'chat room' of the 'caller' client is supposed to be increased. But it isn't, due to the fact that all users of the 'caller' client 'chat room' are duplicated in a separate array, before the algorithm starts, so it works correctly and it does not use the main 'chat room' every time, it uses the pre-prepared array.

The algorithm presented in Figure 4 is similar for two data types processing. The first data type is the request for data formation of the 'call offer' 'configuration message'. The second one is the 'call answer' 'configuration message' came from another client.

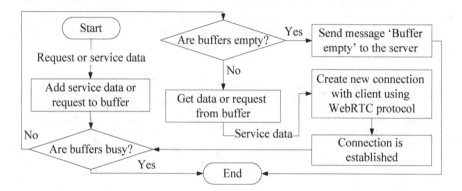

Fig. 4. The algorithm of working of data buffer on the client

There are two separate buffers for each data type in the video conferencing client-side application. Firstly, the new coming request or 'configuration message' is added to the related buffer. Then, both buffers are checked for being processed at this moment. If one buffer is busy by a processing function, the algorithm will end and the new data will be kept by the buffer and processed later. If none of the buffers are busy, they are checked for being empty. If both buffers are empty, the client will send the message that he is ready for receiving new data for establishing connections with other clients to the server via the WebSocket protocol. If one of the buffers contains data, the first data in the buffer queue will be extracted and treated appropriately to establish the connection. Once the connection is established, the algorithm will continue to work from the buffer occupancy survey. It is worth noting that the 'configuration messages' of 'candidate' type have no special buffer for storage either on the client or the server. These 'configuration messages' will be processed immediately as soon as the client receives them. During the processing of 'configuration message' of 'candidate' type buffers belonging to the 'call answer' or 'call offer' data may continue to be busy by the processing function. So, the rest of the data of the 'call answer' or 'call offer' type may continue to be added to buffers without destroying the algorithm of client connections via the WebRTC protocol.

Three algorithms mentioned above are one of the main parts of the application that implements the creation of video conferencing with other users by the peer-to-peer WebRTC protocol. The algorithms allow managing the asynchronous architecture of the client and the server parts of the application, performing data processing when it is necessary, preventing the emergence of situations of mixing data and interrupting performing connections via the WebRTC protocol. Thus, the ability to create and to control group video conferencing is achieved in the application by using the clients association in the 'chat rooms' on the server.

6   Experiments

For evaluation of performance of the developed architecture several comparative tests with the "Skype" application and the developed client's parts was made. The estimates of used RAM and CPU by applications in different work modes, which are obtained during several experiments, are shown in Table 1.

Table 1. Comparison of performance of the developed application with Skype

Modes of work of application client's part	RAM usage by client's part of the developed application	RAM usage by client part of the application "Skype"	CPU usage by client part of the developed application	CPU usage by client part of the "Skype" application
Verifying username and password mode	~ 8 000 Kb	~ 10 000 Kb	0%	1-2%
Standby mode	~ 11 000 Kb	~ 15 000 Kb	2-3%	2-4%
Mode of sending and receiving one audio and video stream	~ 17 000 Kb	~ 60 000 Kb	10-14%	12-15%
Mode of receiving four audio-video data streams and sending one audio-video data stream	~ 80 000 Kb	~ 250 000 Kb	25-40%	30-60%

Table 1 shows that the developed video conferencing application under high load consumes in three times less RAM and in twelve percent less CPU than "Skype" application. In other modes the difference between consumption of RAM and CPU is insignificant for the applications. Despite the problems restricting the use of WebRTC protocol on devices, the developed video conferencing application could compete by efficiency with "Skype" application.

7   Conclusion

The developed architecture of data exchange in video conferencing allows the client applications to distribute the certain data by the relevant protocols and to create serverless connections for them. As the initially architecture of video conferencing application is asynchronous so new algorithms of data control required for the connection via WebRTC protocol were developed. The created application for video

conferencing is capable to support group calls and monitor the status of clients connected to the group - 'room' on the server during the group chat. It should be noted that currently there are still problems restricting the use of WebRTC protocol on devices: 1) just three browsers (Opera, Mozilla Firefox, Google Chrome) support

WebRTC protocol; 2) powerful CPU and sufficient memory to handle audio and video data streams are required. Furthermore, these browsers do not work with graphical co-processor as a result the main processor is loaded. During experiments the developed video conferencing application showed that it could compete by efficiency with "Skype" application. Further research will be focused on simplifying and improving the audio and video processing and transmission using peer-to-peer connections in order to optimize the data processing load between clients.

Acknowledgments. This work is partially supported by the Russian Foundation for Basic Research (grant № 13-08-0741-a), as well as by the Government of Russian Federation (Grant 074-U01).

References

1. Hei, X., Liang, J., Liu, Y., Ross, K.W.: Inferring network-wide quality in P2P live streaming systems. IEEE Journal on Selected Areas in Communications 25(9), 1640–1654 (2007)
2. Paul, S., Pan, J., Jain, R.: Architectures for the future networks and the next generation Internet: A survey. Computer Communications 34, 2–42 (2011)
3. Wu, D., Liang, C., Liu, Y., Ross, K.W.: Redesigning multi-channel P2P live video systems with View-Upload Decoupling. Computer Networks 54, 2007–2018 (2010)
4. Civanlar, M.R., et al.: Peer-to-peer multipoint videoconferencing on the Internet. Signal Processing: Image Communication 20, 743–754 (2005)
5. Volfin, I., Cohen, I.: Dominant speaker identification for multipoint videoconferencing. Computer Speech and Language 27, 895–910 (2013)
6. Ronzhin, A., Budkov, V.: Speaker Turn Detection Based on Multimodal Situation Analysis. In: Železný, M., Habernal, I., Ronzhin, A. (eds.) SPECOM 2013. LNCS (LNAI), vol. 8113, pp. 302–309. Springer, Heidelberg (2013)
7. Ronzhin, A., Budkov, V., Kipyatkova, I.: PARAD-R: Speech Analysis Software for Meeting Support. In: Proc. of the 9th International Conference on Information, Communications and Signal Processing, ICICS 2013, Tainan, Taiwan (2013)
8. Ronzhin, A.L., Saveliev, A.I., Budkov, V.Y.: Context-Aware Mobile Applications for Communication in Intelligent Environment. In: Andreev, S., Balandin, S., Koucheryavy, Y. (eds.) NEW2AN/ruSMART 2012. LNCS, vol. 7469, pp. 307–315. Springer, Heidelberg (2012)
9. Baskaran, V.M., Chang, Y.C., Loo, J., Wong, K.: Software-based serverless endpoint video combiner architecture for high-definition multiparty video conferencing. Journal of Network and Computer Applications 36, 336–352 (2013)
10. Ramzan, N., et al.: Video streaming over P2P networks: Challenges and opportunities. Signal Processing: Image Communication 27, 401–411 (2012)
11. Saveliev, A.I.: Optimization algorithms distribution streams of multimedia data between server and client in videoconferencing application. SPIIRAS Proceedings 31, 61–79 (2013) (in Russ.)

NEW2AN

Joint Spatial Relay Distribution and Resource Allocation & ICIC Strategies for Performance Enhancement for Inband Relay LTE–A Systems

Miguel Eguizábal-Alonso and Angela Hernández-Solana

Aragon Institute for Engineering Research (I3A), University of Zaragoza,
50018 Zaragoza, Spain
{meguizab,anhersol}@unizar.es

Abstract. Relaying is seen as a cost-effective solution to increase the coverage and improve the capacity for LTE-A networks. However, the resource allocation among different nodes and links and inter-cell interference coordination (ICIC) are important challenges in order to make the most of the potential benefits of relay deployments. This paper focuses on Type 1 half-duplex inband relaying. Several strategies related to radio resource partition and frequency reuse between nodes and links are proposed for various relay deployments. Modified FFR-based resource sharing schemes for ICIC are proposed for both access and backhaul subframes, jointly considering the impact of the number and location of relays on the sector. The real impact of the capacity of the backhaul link on system performance is explicitly taken into account. In addition, relay cell range expansion with several bias offsets is analyzed.

1 Introduction

Relaying is seen as a cost-effective way to improve the system performance and solve the lacks of traditional macrocellular deployments in next generation mobile networks, such as LTE-A (Long Term Evolution-Advanced) [1]. Relaying implies that users (UEs: User Equipments) communicate with the network via relay nodes (RNs), which are wirelessly connected to a DeNB (Donor evolved NodeB). Communications through RNs imply the definition of two new links: DeNB-RN and RN-UE links are referred to as backhaul link and access link, respectively. The link between the DeNB and its connected UEs (mUEs) is referred to as direct link.

In this work, we focus on non-transparent Type 1 inband RNs over the FDD frame structure and on downlink (DL) transmissions [2]. Inband relaying involves that the backhaul link shares the same carrier frequency with access and direct links and RNs are assumed to operate in half-duplex fashion. Backhaul transmissions are time division (TD) multiplexed with access transmissions. Thus, based on 3GPP LTE frame structure, the DL radio frame is further partitioned into access and backhaul subframes as shown in Fig. 1. In backhaul subframes, RNs receive data from their DeNB and do not transmit to their connected UEs (rUEs). On the contrary, in access subframes, RNs do not listen to their DeNB and transmit data to their rUEs. Direct links can be allocated in both kinds of subframes and compete for resources with backhaul and access links.

S. Balandin et al. (Eds.): NEW2AN/ruSMART 2014, LNCS 8638, pp. 177–190, 2014.
© Springer International Publishing Switzerland 2014

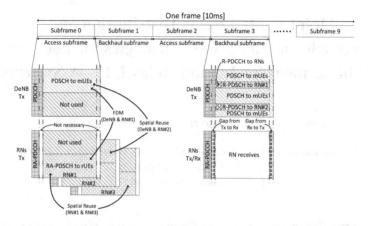

Fig. 1. Frame structure. Partition and reuse of resources among RNs and links.

In this context, it is important to investigate the resource sharing among and within the various links. Resource allocation (RA) for relay enhanced cellular systems has been widely studied [3-14]. In some works [3-5], RA schemes are based on the assumption that direct and access links can take place simultaneously with full frequency reuse in access subframes, whereas in backhaul subframes the DeNB allocates orthogonal resources to different RNs and mUEs. In this case, the studies are focused on the impact of the number and/or the spatial RN distribution on the system performance, and particularly, on the cell-edge performance, on scheduled based backhaul resource partition among RNs, and on resource partition between backhaul and access subframes [3-5]. However, in a multi-cell environment, the interference generated between nodes, both belonging to the same layer (co-layer, e.g. RN to RN) or to different layers (cross-layer, e.g. DeNB to RN or RN to DeNB), is a challenging issue in order to obtain a real benefit of HetNet deployments. Because RNs may create new cell-edges and introduce more interference, inter-cell interference management or coordination (ICIC) is needed. The radio resource management (RRM) is now more complex and conventional ICIC schemes for macrocellular systems (e.g. widely accepted schemes as soft frequency reuse (SFR) and fractional frequency reuse (FFR)) need to be revised in order to consider the new interference scenarios and guarantee a real performance gain linked to relay deployments. It is also important to take into account the different interference conditions and thus, different ICIC schemes, during access and backhaul subframes. A number of approaches for access subframes are based on the assumption that from the DeNB and its associated RNs perspective, the objective is to allocate resources statically or dynamically, in such a way that interfering links become orthogonal [6-7,10-11]. However, in many cases [6-7], the backhaul link overhead is not considered in system level evaluations. In [8-13], SFR-based ICIC schemes are proposed. In [11], the outer subband of each DeNB is reserved for its RNs, whereas in [12] RNs use the outer subband of one adjacent macrocell, so they reuse a portion of the inner subband of their DeNB with a low control over interference. In [13], where tri-sectorized cells are considered, frequency resources in access subframes are adaptively allocated according to the target data rate of users. In [14] the RA scheme is based on FFR and in a similar way to [11], the outer subband of each DeNB is used by RNs to allocate their rUEs, being the size of the outer subband adjusted adaptively. Note that, most proposals are only focused on access subframes and, on the other hand, a strict ICIC scheme can reduce interference, however, decreasing the availability of resources may degrade the network throughput. So, it is necessary to analyze the tradeoff between the restrictions of ICIC schemes and the throughput performance.

Several strategies can be contemplated as a result of a combination of frequency partition (the size of the parts depends on load conditions, traffic characteristics and QoS requirements) and frequency reuse. The goal should be to improve system efficiency by increasing the frequency reuse, but always ensuring that the interference levels between nodes and links are below the allowable thresholds. Relaying has to overcome the extra consumption of backhaul links, so RNs must intend to reuse the same frequency resources that are used by the DeNB for direct communications to its mUEs. It is required coordination between RNs and DeNBs. Links that do not interfere each other can use the same resources, whereas links that are interfering should be allocated orthogonally.

In this paper, we propose different radio resource partition and reuse schemes for several RN deployments, including configurations with two and four RNs per sector. Modified FFR-based resource sharing schemes for ICIC are proposed for both access and backhaul subframes, jointly considering the impact of the number and location of RNs on the sector. Thus, the impact of the capacity of the DeNB-RN backhaul link on system performance is explicitly taken into account. In addition, relay cell range expansion with several bias offsets is analyzed. The aim is to quantify the extent that RNs increase capacity and/or improve fairness.

The rest of the paper is organized as follows. Section 2 presents the system model and the proposed RA schemes linked to geometric relay deployments. Simulation results and the corresponding analysis are provided in section 3. Finally, conclusions are drawn in section 4.

2   System Model and Proposed Resource Allocation Schemes

In this section we describe the proposed RRM and ICIC schemes for Type 1 half-duplex inband relaying. We consider tri-sectorized cells, where the DeNB is located at the center (depicted as a blue star in Fig. 2) and is surrounded by RNs, depicted as black circles. 2-RNs per sector and 4-RNs per sector configurations are considered (Fig. 2(b) and Fig. 2(c) respectively). Each RN is positioned at the location (d, θ), where d is the distance between the RN and the DeNB and θ is the angle between the DeNB-RN direction and the closest sector-edge. Concerning the system model, half-duplex relay deployments, where backhaul and access links are TD multiplexed, are considered. In this work, backhaul and access subframes are alternated for each RN. Following the LTE recommendations, system frequency resources are divided into N_{RB} resource blocks (RBs), which will be the minimum RA unit. With the objective of distributing frequency resources clearly in the different RA strategies which will be presented in this section, the system bandwidth is divided into 9 groups of resources as shown in Fig. 2(a), denoted as GrRB.

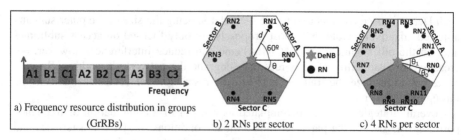

| A1 | B1 | C1 | A2 | B2 | C2 | A3 | B3 | C3 |
Frequency
a) Frequency resource distribution in groups
(GrRBs)

b) 2 RNs per sector

c) 4 RNs per sector

Fig. 2. Network configurations

RRM schemes are supported by a FFR scheme in which some variations have been introduced. As seen in Fig. 3, the inner subband, composed of 6 GrRBs, is reused in all DeNBs, but it is divided into 3 parts, each one mainly intended for one sector of the cell. In this way, users are preferably allocated in the RBs intended for their serving sector, although all RBs are really accessible in all sectors. In the outer subband, which is power boosted, the same procedure is followed but outer subbands are strictly sectorized. Note that, the same RB is not allowed to be simultaneously allocated in two sectors belonging to the same cell. This frequency occupation ordering reduces ICI variability in low load conditions and thus improves system performance as proved in [15], where a frequency resource occupation ordering, denoted as FFRopa, is applied to a conventional FFR.

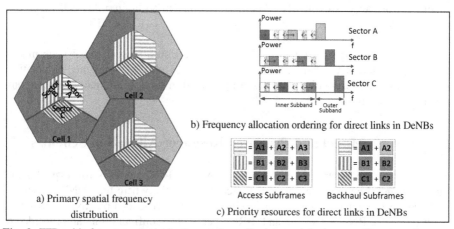

b) Frequency allocation ordering for direct links in DeNBs

a) Primary spatial frequency distribution

Access Subframes   Backhaul Subframes

c) Priority resources for direct links in DeNBs

Fig. 3. FFR with frequency sectorization at the cell-edge, and frequency allocation close to adaptive sectorization in the inner subband

In order to compare our RRM schemes with some strategies found in the literature, the SFR-based schemes for access subframes, inspired by the ideas proposed in [11](SFR-op1) and [12](SFR-op2), are shown in Fig. 4(a) and Fig. 4(b) respectively and are considered as a benchmark. In the SFR-op1, the outer subband of each DeNB is divided into three parts and these parts are allocated to its RNs alternatively. On the contrary, in the SFR-op2 the outer subband is used by DeNBs for their direct links and RNs use the outer subband of one adjacent DeNB, so RNs reuse a portion of the inner subband of their DeNB. Fig. 4(c) and Fig. 4(d) show the distribution of resources among sectors, and the RA for backhaul subframes is shown in Fig. 4(e).

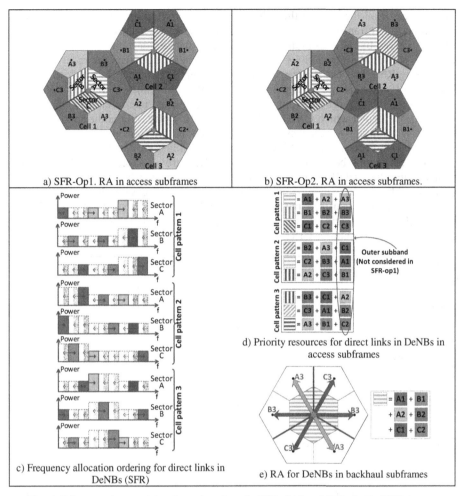

Fig. 4. RA strategies considered as a benchmark (SFR-Op1 and SFR-Op2). 2RNs/sector.

Starting from here, the main concept of the proposed scheme shown in Fig. 5(a1) (with Fig. 3(b) and Fig. 3 (c)), is that each RN reuses in its access links the GrRBs belonging to the inner subband of its DeNB, which will not be potentially used by its own sector, allocating as long as possible different GrRBs to adjacent RNs. Several RA options are considered to further increase the frequency reuse of the network. Thus, each RN can allocate its access links in two, three or four GrRBs (Fig. 5(a1) also depicts which group is chosen firstly to increase the RA). When two GrRBs are allocated to each RN, we choose one intended for the closest sector (adjacent sector) and the other one intended for the furthest sector (opposite sector). I.e. if we focus on the RN0 of the cell 1, the group C1 is intended for the sector C (adjacent sector), whereas the group B1 is intended for the sector B (opposite sector). Note that, the interference received by a RN in the RBs reused in the adjacent sector will be stronger than the received in the RBs reused in the opposite sector of its DeNB, due to the radiation pattern of antennas. The goal of our choice is to achieve that the performance of all the RNs of the cell is similar by providing resources with similar interference conditions to them. Note that, in heterogeneous load conditions, the interference received by rUEs from direct links of their own sector of the DeNB (using the resources potentially intended for adjacent sectors) is stronger than the received from neighboring RNs, because the transmit power of DeNBs is higher. In low load conditions, if only one GrRB is needed for each RN, the best choice is to allocate the resources potentially used by the opposite sector because the interference is weaker (in Fig. 5(a1), we depict which GrRB is chosen in the first place). In addition, the neighboring RNs belonging to adjacent macrocells would use different groups, reducing the co-layer interference as in the SFR-based schemes illustrated in Fig. 4(a) and Fig. 4(b), considered as a benchmark. When two GrRBs are allocated, no co-layer interference within the DeNB's sector coverage exists. However, a higher frequency reuse among neighboring RNs may be interesting because the increase in the co-layer interference may be compensated by the higher availability of resources in RNs. When three GrRBs are allocated, the reuse among RNs belonging to the same DeNB's sector increases, and when four GrRBs are allocated, a full reuse is achieved.

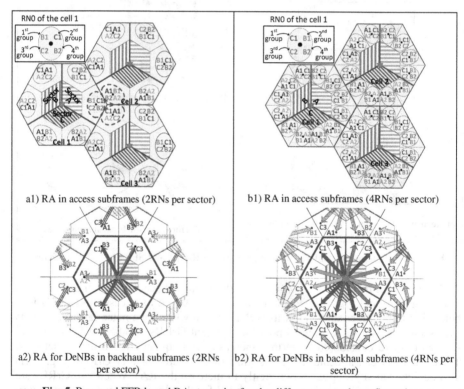

Fig. 5. Proposed FFR based RA strategies for the different network configurations

As seen in this scheme, only one cell pattern is considered and RNs use the resources intended for the other sectors of their DeNB, so these resources will be always interfered by the dominant interfering sectors of adjacent cells. In fact, a RN cannot use the frequency resources reserved for the outer subband due to the strong interference that rUEs would receive in these resources from the own and adjacent cells.

Regarding backhaul subframes (Fig. 5(a2)), the main concept is that each GrRB allocated to backhaul links, can be reused a maximum of two times in the same DeNB. Three groups are used twice by opposite backhaul links (i.e. A3, B3 and C3) and then we choose the GrRBs potentially allocated to mUEs by the opposite sector, with the aim of minimizing the intra-cell interference. RNs use a directive antenna to communicate with their DeNB, so the interference received from neighboring cells is greatly reduced.

In Fig. 5(b) (with Fig. 3(b) and Fig. 3 (c)), the 4-RNs per sector configuration is considered. Following the same philosophy as in the previous scheme, each RN reuses in its access links the GrRBs belonging to the inner subband of its DeNB, which will not be potentially used by its own sector and it can allocate its access links in two, three or four GrRBs (in Fig. 5(b1), we depict which GrRB is chosen firstly). There are four eligible groups for RNs in each sector (e.g. in the Sector A, the groups are B1, B2, C1 and C2). If only one GrRB is needed for each RN, neighboring RNs belonging to the own and adjacent cells use different GrRBs, reducing the co-layer interference. When two GrRBs are allocated to each RN, we allocate these groups in rotation in order to reduce the co-layer interference. A higher frequency reuse among neighboring RNs (we add a third and a fourth GrRB to access links of RNs) may be interesting because the increase in the co-layer interference may be compensated by the higher availability of resources in RNs. With regard to backhaul subframes, the RA is depicted in Fig. 5(b2) following the same idea as in the 2-RNs per sector configuration.

3   Simulation Results

3.1   Simulation Parameters

The evaluation of the RA algorithms has been carried out by system simulation with the set of parameters collected in Table 1 (following the system level parameters and the FDD frame structure of LTE [2]). We consider 21 tri-sectorized macrocells, and the wrap-around technique is applied to avoid border effects. In order to analyze the user's performance in different zones of the cell separately, each macrocell is divided into three concentric circles with the same area ($z0$, $z1$ and $z2$, where $z0$ is the innermost and $z2$ is the outermost). 180 users are homogeneously distributed on each cell and move at 3km/h within their respective sector to avoid handovers. A full buffer traffic model is applied.

The system bandwidth is divided into 27 RBs, so each of the 9 GrRBs contains 3 RBs. RRM entities (including power and frequency-time RA and adaptive MCS selection) are located at DeNB and RN. Scheduling decisions are independently performed by each node every Transmission Time Interval (TTI). RA is seen as a multidimensional problem. The ICIC scheme determines what part of frequency resources can be used in each node, sector or type of link. Subsequently, the scheduler allocates the resources to each UE or RN. The scheduler is composed of two separated components which are sequentially applied. First, a time domain (TD) scheduling policy provides the prioritized list of users which may be allocated. PF (Proportional Fair) algorithm [16] is applied. Afterwards, a frequency domain (FD) scheduling algorithm, according to this TD prioritization, chooses the resources (RB and power) for each user and selects the data rate (MCS), according to CQIs (Channel Quality Indicators). FD scheduling of DeNBs and RNs allocates to each user the RB where it experiences the best channel conditions, taking into account the constraints forced by the ICIC schemes. In order to avoid the monopolization of resources, each user only can be allocated in one RB and the scheduler selects the highest possible MCS, always satisfying the SINR requirements. The cell selection method is based on RSSI (Received Signal Strength Indicator).

Table 1. Simulation Parameters

Parameter		Value		
Cell layout		21 tri-sectorized cells and 6 RNs per cell		
Inter-Site Distance (ISD)		1732m		
System bandwidth (BW)		5MHz		
Carrier Frequency		2.5GHz		
Scheduling period (TTI)		1ms (14 OFDM symbols)		
Number of RBs/ RB size		27/12 subcarriers for 1 TTI (1ms)		
Path loss model [d(km)]		DeNB→UE(NLOS): $131.1+42.8\log_{10}d$		
		RN→UE(NLOS): $145.4+37.5\log_{10}d$		
		DeNB→RN(LOS): $100.7+23.5\log_{10}d$		
		eNB/RN→RN(NLOS): $125.2+36.3\log_{10}d$		
Shadowing standard deviation		DeNB→UE: 8dB		
		RN→UE: 10dB		
		DeNB/RN→RN: 2dB		
Shadowing correlation/distance		0.5/50m		
Channel/Doppler model (3Km/h)		Extended Pedestrian-A / Jakes		
Maximum output power	DeNB	43dBm		
	RN	37dBm		
Antenna gain	DeNB	14dB		
	RN (Tx/Rx)	5dB/7dB		
	UE	0dB		
Tx/ Rx diversity gain		3dB/3dB		
Antenna horizon-tal pattern	DeNB	70°(-3dB) with 25dB front-to-back ratio		
	RN (Tx)	Omnidirectional		
	RN (Rx)	70°(-3dB) with 20dB front-to-back ratio		
	UE	Omnidirectional		
AWGN density		-174 dBm/Hz		
Receiver noise figure	RN	5dB		
	UE	9dB		
Time between CQIs (T_{CQI})		1ms		
CQI averaging time (W_{CQI})		4ms		
CQI delay (Δ_{CQI})		4ms		
Scheduling algorithm (DeNB & RN)		Proportional Fair (PF)		
Traffic model		Full Buffer		
UE speed		3Km/h		
Link Adaptation/ MCSs considered				
Index	Modulation	Code rate	SINR$_{req}$	Info-bits/RB
0	QPSK	1/2	10.2dB	120
1		3/4	14dB	180
2	16QAM	1/2	17.5dB	240
3		3/4	21.4dB	360
4	64QAM	2/3	27.4dB	480
5		3/4	29.5dB	540

Interference is explicitly modelled, taking into account the effects of wireless channel. Channel State Information (CSI) is reported by each user or RN to its serving node, using CQI messages. Errors in both channel and interference estimation are explicitly modeled. CQI parameters and the set of MCSs, including their minimum SINR threshold (computed to achieve a BLER of 10^{-2}) and their information data rate, are collected in Table 1. In relay deployments, two types of subframes are defined and the interference conditions suffered by users in each type of subframe are different. Therefore, users compute different CQIs for each type of subframe. RNs also have to report their CSI but they calculate only one CQI in the type of subframe in which they receive their backhaul links. The maximum output power is equally distributed across all allocable RBs in each node and fixed power allocation is considered for both DeNBs and RNs.

When two transmit power levels (TPLs) are considered in DeNBs, the transmit power per RB in the outer subband has been defined as an increment of 3dB from the TPL in the inner subband. The two TPLs per RB are computed so that the total transmit power in the node never exceeds the maximum output power, even when it assigns all its allocable RBs. The same transmit power and MCS are applied to all subcarriers within a RB. Note that, relay site planning is realized by adding a bonus (5dB) to the path loss formula of the DeNB→RN link [2].

3.2   Numerical Results

The performance evaluation of the proposed ICIC schemes is mainly presented in terms of global throughput. The average cell throughput of the different RA strategies is shown in Table 2 (including the throughput gain with respect to a macrocellular deployment). RNs are located 2/3 cell radius away from DeNBs (cell radius is denoted as R, so $d=2/3R$). Note that, the global throughput only includes the throughput received by final users and it can be evaluated as shown in (1), where T_mUEs and T_rUEs is the throughput received by mUEs and rUEs respectively, and T_RNs is the maximum achievable capacity of backhaul links.

$$throughput=T\_mUEs(access)+T\_mUEs(backhaul)$$
$$+min(T\_rUEs(access),T\_RNs(backhaul)) \tag{1}$$

The results obtained in a macrocellular network (without relay deployment), employing FFR with frequency sectorization at the cell-edge are included in Table 2, in order to prove the performance gain achieved by relaying. In addition, the results obtained by the strategies based on the literature (SFR-op1 and SFR-op2) are also included. If we compare the SFR-based schemes, in SFR-op2 the whole bandwidth is employed for direct links in access subframes, increasing the throughput achieved by mUEs. With regard to rUEs, the GrRB reserved for each RN in SFR-op2 is used by its own DeNB, but it is not potentially used by its own sector and by its dominant interfering sectors of adjacent cells, so the throughput achieved by rUEs in SFR-op2 is higher.

Now we focus on our proposed RA schemes, described in Section 2, in order to check if they can increase frequency reuse and improve network capacity efficiently (see Table 2). Firstly, we analyze the effects of increasing the number of GrRBs allocated to access links of RNs in the 2-RNs per sector configuration. If we compare our proposed scheme, when two GrRBs (6RBs) are allocated to RNs, with the SFR-based schemes, we can see that increasing the number of allocable RBs of RNs the aggregate throughput achieved by rUEs (5.76Mbps) is greatly increased, whereas the performance of mUEs is not degraded, improving the global capacity. The next step is to increase the number of allocable RBs of RNs to 9, but in this case, the throughput gain obtained by rUEs is not as significant as when the resources are increased from 3 to 6RBs. If we further increase the resources allocated to RNs up to 12RBs, the aggregate throughput achieved by rUEs (6.28Mbps) is degraded, when it is compared with that achieved when only 9RBs are allocated to RNs (6.57Mbps).

Table 2. Cell throughput [Mbps] achieved in the macrocellular deployment and in relay deployments ($d=2/3R$). In the 2-RNs per sector configuration, $\theta=30°$ (Fig. 2(b)). In the 4-RNs per sector configuration, $\theta_0=15°$ and $\theta_1=45°$ (Fig. 2(c)). Table collects the achieved throughput in SFR-op1 (Fig. 4(a)) and SFR-op2 (Fig. 4(b)) strategies (considered as a benchmark) and in the proposed RA-ICIC schemes (Fig. 5).

Scheme	Nº RNs per sector	RBs/ RN	Bias [dB]	Access Subframe mUEs	Access Subframe rUEs	Backhaul Subframe mUEs	Backhaul Subframe Backhaul	Global [Mbps]	Zone 0 [Mbps]	Zone 1 [Mbps]	Zone 2 [Mbps]
Macro Deployment								10.65	5.13	3.25	2.27
Relays are located 2/3 cell radius away from DeNBs ($d=2/3R$)											
SFR Op1	2	3	0	4.16	3.49	3.81	4.26	11.46 (7.6%)	5.55 (8.2%)	3.45 (6.2%)	2.46 (8.4%)
SFR Op2	2	3	0	6.16	3.66	3.49	4.26	13.31 (25.0%)	6.39 (24.6%)	3.91 (20.3%)	3.01 (32.6%)
Proposed		6	0	5.39	5.76	4.70	5.67	15.76 (48.0%)	6.73 (31.2%)	5.78 (77.8%)	3.25 (43.2%)
RA-ICIC Relay Deployment	2	9	0	5.30	6.57	4.06	7.09	15.93 (49.6%)	6.13 (19.5%)	6.37 (96.0%)	3.43 (51.1%)
			2	5.37	6.75	4.14	7.09	16.26 (52.7%)	6.49 (26.5%)	6.35 (95.4%)	3.42 (50.7%)
			4	5.45	6.92	4.21	7.09	16.58 (55.7%)	6.91 (34.7%)	6.32 (94.5%)	3.35 (47.6%)
FFR-Based		12	0	5.25	6.28	4.06	7.09	15.59 (46.4%)	5.95 (16.0%)	6.27 (92.9%)	3.37 (48.5%)
			2	5.31	6.54	4.12	7.09	15.97 (50.0%)	6.28 (22.4%)	6.33 (94.8%)	3.36 (48.0%)
			4	5.40	6.80	4.16	7.09	16.36 (53.6%)	6.62 (29.0%)	6.39 (96.6%)	3.35 (47.6%)
Proposed		6	0	5.77	10.19	1.15	11.01	17.11 (60.7%)	5.60 (9.2%)	7.82 (140.6%)	3.69 (62.6%)
			2	5.85	10.32	1.19	11.01	17.36 (63.0%)	6.02 (17.3%)	7.71 (137.2%)	3.63 (59.9%)
			4	5.94	10.33	1.23	11.01	17.50 (64.3%)	6.38 (24.4%)	7.57 (132.9%)	3.55 (56.4%)
RA-ICIC Relay Deployment	4	9	0	5.70	10.19	1.14	11.01	17.03 (59.9%)	5.30 (3.3%)	8.08 (148.6%)	3.65 (60.8%)
			2	5.77	10.42	1.18	11.01	17.37 (63.1%)	5.67 (10.5%)	8.10 (149.2%)	3.60 (58.6%)
			4	5.86	10.57	1.22	11.01	17.65 (65.7%)	6.06 (18.1%)	8.02 (146.8%)	3.57 (57.3%)
FFR-Based		12	0	5.63	9.34	1.15	11.01	16.12 (51.4%)	5.04 (-1.8%)	7.69 (136.6%)	3.39 (49.3%)
			2	5.73	9.64	1.18	11.01	16.55 (55.4%)	5.38 (4.9%)	7.75 (138.5%)	3.42 (50.7%)
			4	5.82	9.93	1.22	11.01	16.97 (59.3%)	5.79 (12.9%)	7.75 (138.5%)	3.43 (51.1%)

Note that, as the amount of RBs allocated to RNs increases, the transmit power per RB is reduced, so rUEs are less protected against interference. In addition, we are increasing the frequency reuse and neighboring RNs use the same resources, so the co-layer interference is stronger. From the point of view of mUEs, when the number of allocable RBs of access links of RNs is increased, the amount of resources reserved for backhaul links also has to be increased. When 6RBs are allocated to RNs, each backhaul link only needs 4RBs, so we remove two RBs from the GrRBs reserved for backhaul links (i.e. A3, B3 and C3 in Fig. 5(a2)) and we add the six extra RBs to direct links. However, when 9 and 12RBs are allocated to RNs, each backhaul link uses 5RBs and only three extra RBs are added to direct links, reducing the throughput achieved by mUEs in backhaul subframes.

With the aim of enlarging the coverage area of RNs and increasing the number of rUEs, a positive bias is added to the RSSI from RNs. We can see that applying a bias we can increase the throughput achieved by rUEs, because the extra rUEs take advantage of free RBs. On the other hand, when the bias is applied, a portion of outer mUEs connects to RNs, so inner mUEs receive more assignments and the achieved throughput is increased because inner mUEs observe better channel conditions. When 6RBs are allocated to access links of RNs, applying a bias does not vary the aggregate throughput of rUEs, so only the results obtained when the bias is equal to 0dB are included in Table 2. We can conclude that increasing the allocable RBs of RNs from 6 or 9 to 12 do not provide a throughput gain. In heterogeneous load conditions, is more probable that one sector uses resources intended for other sectors of the cell, so it is proper to select the resources which are not allocated to RNs. In this situation, may be more beneficial to limit the amount of resources allocated to RNs, in order to facilitate the coordination with their DeNB. Therefore, if the system interference is not carefully controlled, a higher frequency reuse will not achieve a rise in the throughput performance.

If now we analyze in Table 2 the performance of the 4-RNs per sector configuration, we can see that the global throughput is increased. When four RNs are deployed in each sector, more users connect to RNs, taking advantage of the better channel conditions. In this configuration, increasing the allocable RBs of RNs to 9 do not provide a significant gain and when 12RBs are allocated to RNs, the throughput performance is degraded. Allocating 6RBs to RNs in the 4-RNs per sector configuration, the achieved frequency reuse is the same as in the 2-RNs per sector configuration when 12RBs are allocated to RNs, however is more beneficial to deploy more RNs than to increase excessively the amount of resources reused by RNs in their access links (co-layer interference among RNs is reduced). Note that, the throughput achieved by mUEs in backhaul subframes in the 4-RNs per sector configuration is reduced, because only 6RBs are reserved for direct links (i.e. if we focus on the sector A in Fig. 5(b2), only one RB of A1 and other RB of A2 are used by direct links, whereas the remaining RBs are reserved for backhaul links, using 4RBs for each RN). Note that, Table 2 shows in green when the throughput is lower than in the macrocellular network.

In Table 3, the results obtained when RNs are located on the borderline between the zones 2 and 3 (approximately 0.76*R) are collected (note that, in the 2-RNs per sector configuration, $\theta = 30°$ (Fig. 2(b)) and in the 4-RNs per sector configuration, $\theta_0 = 15°$ and $\theta_1 = 45°$ (Fig. 2(c))). The same conclusions explained above can be considered and the 4-RNs per sector configuration is the most suitable choice. When RNs are located on the outer region, most of the cell-edge users are connected to RNs, improving their achieved throughput (users of zone 2) and obtaining a fairer throughput distribution among zones. In this case, the 4-RNs per sector configuration increases the global throughput, greatly improves the performance of cell-edge users (throughput gains in zone 2 of up to 160%) and also achieves the fairest distribution. Fig. 6 shows that the 4-RNs per sector configuration improves the throughput distribution, particularly in the zones 1 and 2,

Table 3. Cell throughput [Mbps] achieved in relay deployments in other spatial RN distribution (RNs are located on the borderline between zones 2 and 3 ($d=0.76*R$)). Table also shows the achieved throughput in the 4-RNs per sector configuration, when $\theta_0 =12°$ and $\theta_l =36°$ (Fig. 2(c)). Table collects the achieved throughput in the proposed RA-ICIC schemes (Fig. 5).

Nº RNs per sector	RBs/ RN	Bias [dB]	Access Subframe		Backhaul Subframe		Global [Mbps]	Zone 0 [Mbps]	Zone 1 [Mbps]	Zone 2 [Mbps]
			mUEs	rUEs	mUEs	Backhaul				
Relays are located on the borderline between zones 2 and 3 ($d=0.76*R$)										
2 RNs	6	0	5.47	5.52	4.70	5.67	15.69 (47.3%)	6.56 (27.9%)	5.10 (56.9%)	4.03 (77.5%)
	9	0	5.38	6.41	4.07	7.09	15.86 (48.9%)	5.98 (16.6%)	5.26 (61.8%)	4.62 (103.5%)
		2	5.45	6.52	4.13	7.09	16.10 (51.2%)	6.27 (22.2%)	5.31 (63.4%)	4.52 (99.1%)
		4	5.51	6.69	4.19	7.09	16.39 (53.9%)	6.59 (28.5%)	5.33 (64.0%)	4.47 (96.9%)
	12	0	5.33	6.13	4.05	7.09	15.51 (45.6%)	5.80 (13.1%)	5.18 (59.4%)	4.53 (99.6%)
		2	5.40	6.35	4.11	7.09	15.86 (48.9%)	6.06 (18.1%)	5.28 (62.5%)	4.52 (99.1%)
		4	5.46	6.56	4.18	7.09	16.20 (52.1%)	6.40 (24.8%)	5.35 (64.6%)	4.45 (96.0%)
4 RNs	6	0	5.83	10.11	1.16	11.01	17.10 (60.6%)	5.29 (3.1%)	6.01 (84.9%)	5.80 (155.5%)
		2	5.90	10.26	1.19	11.01	17.35 (62.9%)	5.63 (9.7%)	5.99 (84.3%)	5.73 (152.4%)
		4	5.97	10.31	1.24	11.01	17.52 (64.5%)	5.96 (16.2%)	5.94 (82.8%)	5.62 (147.6%)
	9	0	5.78	10.11	1.15	11.01	17.04 (60.0%)	5.04 (-1.8%)	6.08 (87.1%)	5.92 (160.8%)
		2	5.86	10.30	1.19	11.01	17.35 (62.9%)	5.34 (4.1%)	6.16 (89.5%)	5.85 (157.7%)
		4	5.92	10.47	1.22	11.01	17.61 (65.4%)	5.67 (10.5%)	6.16 (89.5%)	5.78 (154.6%)
	12	0	5.72	9.30	1.15	11.01	16.17 (51.8%)	4.84 (-5.7%)	5.74 (76.6%)	5.59 (146.3%)
		2	5.80	9.60	1.18	11.01	16.58 (55.7%)	5.12 (-0.2%)	5.87 (80.6%)	5.59 (146.3%)
		4	5.88	9.79	1.22	11.01	16.89 (58.6%)	5.47 (6.6%)	5.89 (81.2%)	5.53 (143.6%)
Relays are located 2/3 cell radius away from DeNBs ($d=2/3R$)($\theta_0 =12º$ and $\theta_l =36º$)										
4 RNs	6	0	5.78	10.01	1.11	10.82	16.90 (58.7%)	5.45 (6.2%)	7.76 (138.8%)	3.69 (62.6%)
		2	5.84	10.12	1.15	10.82	17.11 (60.7%)	5.79 (12.9%)	7.73 (137.8%)	3.59 (58.1%)
		4	5.92	10.17	1.20	10.82	17.29 (62.3%)	6.17 (20.3%)	7.55 (132.3%)	3.57 (57.3%)
Relays are located on the borderline between zones 2 and 3($d=0.76*R$)($\theta_0 =12º$and $\theta_l =36º$)										
4 RNs	6	0	5.83	9.98	1.11	10.82	16.92 (58.9%)	5.12 (-0.2%)	5.97 (83.7%)	5.83 (156.8%)
		2	5.89	10.00	1.15	10.82	17.04 (60.0%)	5.42 (5.7%)	5.97 (83.7%)	5.65 (148.9%)
		4	5.96	10.15	1.18	10.82	17.29 (62.3%)	5.74 (11.9%)	5.99 (84.3%)	5.56 (144.9%)

whereas in the zone 0 it is degraded due to a lower availability of resources for direct links.

In macrocellular deployments, the users located on the frontier regions between sectors obtain poor channel conditions due to the radiation pattern of antennas. So, in order to stimulate the improvement in the performance of these users, we concentrate the RNs close to these regions (i.e. $\theta_0 = 12°$ and $\theta_1 = 36°$ (Fig. 2(c))). As we can see in Table 3, the performance of rUEs is slightly degraded due to a stronger interference from interfering sectors, and the capacity achieved by backhaul links is also reduced because RNs are now deployed closer to sector-edges and the signal received from their DeNB is weaker. After all, the achieved throughput performance is a bit lower than in the previous configuration (i.e. $\theta_0 = 15°$ and $\theta_1 = 45°$ (Fig. 2(c))).

From the above results we can conclude that four RNs deployed on the outer regions of cells, provide not only higher throughput (global throughput gains of up to 65%) but also ensure fair service to both cell-edge and inner users (gains in zone 2 of up to 160%). Table 2 and Table 3 show in red the most suitable options.

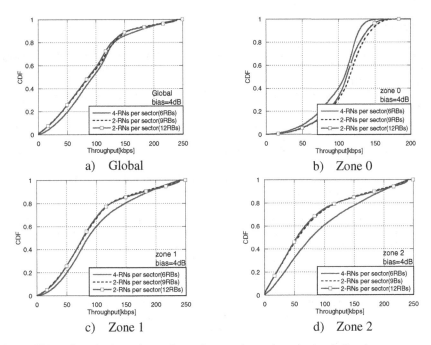

Fig. 6. CDFs of user's throughput when relays are located on the borderline between zones 2 and 3 ($d = 0.76 * R$)

4   Conclusions

In this work, various RA strategies and frequency reuse alternatives for relay deployments, in several spatial distributions, have been proposed and evaluated. Planning the frequency resources for each sector of the cell and coordinating the resources allocated to support the relaying operation with those used by DeNBs, the extra consumption of relay deployments can be compensated. Our proposed RA-ICIC schemes clearly outperform the proposals found in the literature, even when only 6RBs are allocated to access links of RNs in the 2-RNs per sector configuration. We have seen that if the interference is not carefully controlled, a higher frequency reuse (12RBs allocated to access links of RNs instead of 9RBs) will not achieve a rise in the throughput performance. However, deploying four relays per sector instead of two, allows increasing the global performance (gains of up to 65%) because it fits better to the spatial distribution of the users with bad channel conditions and a higher number of these users may connect to relays, taking advantage of relaying benefits. When RNs are deployed on the outer regions of cells, the 4-RNs per sector configuration achieves a fairer throughput distribution, providing similar throughput to the users of the three zones of cells (gains in zone 2 of up to 160%).

Acknowledgment. This work has been supported by the Spanish Government through the grant TEC2011-23037 from the Ministerio de Ciencia e Innovación (MICINN) and by the Diputación General de Aragón (DGA) through its fellowship program.

References

1. Saleh, A.B., et al.: On the Coverage Extension and Capacity Enhancement of Inband Relay Deployments in LTE-Advanced Networks. Journal of Electrical and Computer Egineering 2010, 1–12 (2010)
2. 3GPP, TR 36.814 (V9.0.0), Further Advancements for E-UTRA Physical Layer Aspects (March 2010)
3. Saleh, A.B., Haemaelaeinen, J., et al.: Resource Sharing in Relay-enhanced 4G Networks. In: Proc. EW 2011 (April 2011)
4. de Moraes, T.M., et al.: Resource allocation in relay enhanced LTE-advanced networks. EURASIP Jour. on Wireless Comm. and Networking 2012(1), 364 (2012)
5. de Moraes, T.M., et al.: QoS-aware Scheduling for In-Band Relays in LTE-Advanced. In: Proc. of 2013 ITG Conference on Systems, Communication and Coding (SCC) (January 2013)
6. Wang, L., et al.: Performance Improvement through Relay-Channel Partitioning and Reuse in OFDMA Multihop Cellular Networks. In: Proc. IWCMC 2008 (August 2008)
7. Park, W., Bahk, S.: Resource management policies for fixed relays in cellular networks. Computer Communications 32(4), 703–711 (2009)
8. Yu, Y., et al.: Inter-Cell Interference Coordination for Type I Relay Networks in LTE Systems. In: Proc. IEEE GLOBECOM 2011 (December 2011)
9. Yu, Y., et al.: Interference Coordination and Performance Enhancement for Shared Relay Networks in LTE-Advanced Systems. In: Proc. IEEE ISCIT 2011 (October 2011)
10. Yi, S., Lei, M.: Inter-Cell Interference Coordination in LTE-Advanced Inband Relaying Systems. In: Proc. IEEE PIMRC 2013 (September 2013)
11. Ren, P., et al.: A Novel Inter-Cell Interference Coordination Scheme for Relay Enhanced Cellular Networks. In: Proc. IEEE VTC 2011-Fall (September 2011)
12. Liu, J., et al.: Inter-Cell Interference Coordination based on Soft Frequency Reuse for Relay Enhanced Cellular Network. In: Proc. IEEE PIMRC 2010 (September 2010)
13. Cao, L., et al.: Joint Adaptive Soft Frequency Reuse and Virtual Cell Power Control in Relay Enhanced Cellular System. In: Proc. IEEE CNMT 2009 (January 2009)
14. Wang, J., et al.: Load Balance Based Dynamic Inter-Cell Interference Coordination for Relay Enhanced Cellular Network. In: Proc. IEEE VTC 2012-Spring (May 2012)
15. Hernández, A., et al.: Downlink Scheduling for Intercell Interference Fluctuation Mitigation in Partial-Loaded Broadband Cellular OFDMA Systems. In: Proc. IEEE ICUMT 2009 (October 2009)
16. Pokhariyal, A., et al.: HARQ Aware Frequency Domain Packet Scheduler with Different Degrees of Fairness for the UTRAN Long Term Evolution. In: Proc. VTC 2007-Spring (April 2007)

LTE Base Stations Localization

Aleksandr Gelgor[1], Igor Pavlenko[1], Grigoriy Fokin[2], Anton Gorlov[1],
Evgenii Popov[1], and Vladimir Lavrukhin[2]

[1] St. Petersburg State Polytechnical University, St. Petersburg, Russia
`a_gelgor@mail.ru`, `{ipavlenko.mail,eugapop}@gmail.com`,
`anton.gorlov@yandex.ru`
[2] The Bonch-Bruevich St. Petersburg State University of Telecommunications,
St. Petersburg, Russia
`{grihafokin,vladimir.lavrukhin}@gmail.com`

Abstract. Direction-of-arrival (DoA) estimation algorithm for LTE base stations (BSs) is proposed in the paper. The algorithm is assumed using of mobile
receiver station (MRS) with antenna array and is based on the analysis of peaks
of cell-specific reference signal (CRS) autocorrelation functions (ACFs). Line
of sight (LoS) identification is also provided by the algorithm. As a comparison,
DoA estimation algorithms for raw signal and for primary synchronization
signal (PSS) ACFs peaks are observed. It is shown, that proposed algorithm
provides the best accuracy in DoA estimation.

Keywords: mobile receiver station, antenna array, DoA estimation, LTE,
primary synchronization signal, cell-specific reference signal.

1 Introduction

Appearance of new wireless communication systems stimulates the development of
controlling services, whose objectives include estimation of parameters of signals and
localization of their sources.

Operators of mobile communication systems, such as GSM, UMTS, LTE, frequently admit illegal installation of BSs; in such case, one of the objectives of controlling services is to localize such BSs.

Most of localization methods are based on joint analysis of several DoAs of one
source. In its turn, DoA is usually got by using an antenna with variable diagram. For
example, antenna arrays are widely used.

*The objective of the present paper is development and approbation of DoA estima-
tion algorithm for LTE BSs using eight-element antenna array and LTE BS's primary
synchronization module, installed on MRS.*

Analysis of publications about this area has shown us that similar problems had
been solved before [1, 2, 3]. However, in these works DoA estimation was the com-
posite task of algorithm of improvement of signal reception by MRS with antenna
array, rather than BS localization. They have solved the problem of diagram forming
for antenna array with the direction of maximal power for the one BS and with zeros

S. Balandin et al. (Eds.): NEW2AN/ruSMART 2014, LNCS 8638, pp. 191–204, 2014.
© Springer International Publishing Switzerland 2014

in the directions for other BSs. The problem of presence or absence of LoS, which is important for BS localization, was not solved. Also, only algorithm based on processing of synchronization signals ACFs was observed.

In the present paper several DoA estimation algorithms are observed. Algorithms are based on detection of the direction, from which the signal of maximal power comes. The selection of direction is provided by observing a linear combination of signals coming to the antenna array.

The first DoA estimation algorithm does not perform a pre-processing of signals. Such method, although being simple in implement, and even does not require the presence of LTE BS's primary synchronization module, nevertheless, does not allow to separate signals from different sources and define the presence of LoS. The first algorithm is invariant with respect to the mobile communication system.

The second algorithm is based on processing of PSS ACFs peak values. This algorithm allows to separate signals from different sources, however, as the first algorithm, it does not allow to define the presence of LoS.

The third algorithm is based on processing of CRS ACFs peak values. Cell-specific reference signals are transmitted in wider bandwidths, than primary synchronization signal, therefore ACFs of CRSs has better time resolution than ACFs of PSSs. Nevertheless, the third algorithm does not allow to define the presence of LoS without additional processing either.

For LoS identification the following approach is proposed in the paper. CRS ACF is taken as an estimation of impulse response (IR) of channel between transmitting antenna of BS and receiving element of antenna array. Then, relation of CRS ACF maximal value to the sum of values of CRS ACF elements, exceeding the preset level (modified K-factor K_m), is calculated. Decision in favor of LoS is delivered in excess of K_m value of some threshold. Thus, the fourth algorithm differs from the third one by the fact that DoA estimation is performed only in case of LoS identified.

To compare the quality of proposed algorithms, environmental tests were conducted in Saint-Petersburg. The route of MRS was designed to include different conditions of signal propagation from BS to antenna array.

The material in the paper is stated in the following order. Correlation properties of PSSs are examined in the second part; properties of CRSs are examined in the third part. DoA estimation algorithm using antenna array is examined in the fourth part. The results of proposed algorithms' comparison taken during environmental tests are stated in the fifth part.

2   PSS Correlation Properties

In LTE standard there defined three different primary synchronization signals, which are used for detecting signal of BS. Regardless of bandwidth, given to the communications provider, PSS is always transmitted in the most narrow band of all possible ones – 1.08 MHz. Thus, if the initial record of signal is made with sampling rate 30.72 MHz, it is enough to examine decimated in $C_d = 16$ times signal (C_d is a decimation coefficient) while searching PSS, i.e. sampling rate is 1.92 MHz. The selection of C_d value which equals integer power of a two is conditioned by the wish to keep the possibility of using the fast Fourier transform algorithm for signal processing, e.g., while reading data of master information block (MIB) of physical broadcast channel (PBCH).

For both variants of duplex, FDD and TDD, PSS is always transmitted with normal cyclic prefix of 144 samples length for sampling rate 30.72 MHz and of 9 samples length for sampling rate 1.92 MHz.

Let us examine correlation properties of PSSs. For that, we will calculate correlation function (CF)

$$CorPSS_{i,j}[k,r] = \left| \sum_{n=0}^{127} TS\_PSS_j^{(1)}[k+16n]\exp(j\varphi(k,r,n))PSS^{(16)}{}^*[i,n] \right|, \qquad (1)$$

$$i, j = 0, 1, 2, \ k = -(2047 + 144), \ldots, 2047,$$

$$r = -NF, \ldots, NF, \ \varphi(k, r, n) = 2\pi\Delta fr(k + 16n)\Delta t,$$

where i, j are the numbers of PSSs we examine; k, r are shift variables along a time and frequency dimensions respectively; $TS\_PSS_j^{(1)}[k]$ denote test signal, calculated for sampling rate 30.72 MHz (non-decimated), consisting of PSS with number j, its cyclic prefix of 144 samples length and 2047 zero samples on the edges for computation non-periodical correlation function (table 1); $PSS^{(16)}$ is 3×128 matrix of primary synchronization signals time domain samples for sampling rate 1.92 MHz; symbol * denote complex conjugation; 128 is a sample-length of OFDM-symbol for $C_d = 16$; Δf is frequency increment in Hz for calculating CF in terms of non-nil frequency offset between central frequency of received signal and central frequency of receiver's oscillator; $\Delta t = 1/(15000 \times 2048)$ sec is sampling interval for sampling rate 30.72 MHz; $\varphi(k, r, n)$ denote the value of necessary phase shift of $(k + 16n)$ element of test signal $TS\_PSS_j^{(1)}[k]$ for providing the frequency shift Δfr; NF denote the half of observed values of frequency shifts.

Table 1. Values of test signal $TS\_PSS_j^{(1)}[k]$

k	$TS\_PSS_j^{(1)}[k]$	Number of values
−2191, …, −145	0	2047
−144, …, −1	$PSS^{(1)}[j, 2048 + k]$	144
0, …, 2047	$PSS^{(1)}[j, k]$	2048
2048, …, 4094	0	2047

Let us define possible frequency offsets between central frequency of received signal and central frequency of receiver's oscillator. The requirement for BS and mobile station (MS) oscillator accuracy is defined in [4, 5]: it is 0,1 ppm (excluding Home BS). For maximal central frequency 3800 MHz, defined in [4], this is equivalent to

$$\Delta f_{BS} = \Delta f_{MS} = 380 \text{ Hz}.$$

Frequency Doppler offset, e.g., for the speed of 60 km/h, is

$$\Delta f_D = 211 \text{ Hz}.$$

Thus, total offset can reach $(2 \times 380 + 211) = 971$ Hz, what is a bit less than 7% of subcarriers spacing of OFDM-symbol $1/T = 15$ kHz, where $T \approx 66.7$ μsec is OFDM-symbol duration.

Equation (1) describes a set of CFs for all possible combinations of i and j –9 functions in total. For $i = j$ we get the subset of three ACF. For $i \neq j$ we get the subset of six cross-correlation functions (CCF). For more convenient analysis of PSSs correlation properties we will examine the following functions:

$$MinCorPSS_{AC}[k,r] = \min_{i=j}\{CorPSS_{i,j}[k,r]\},$$

$$MaxCorPSS_{AC}[k,r] = \max_{i=j}\{CorPSS_{i,j}[k,r]\},$$

$$MinCorPSS_{CC}[k,r] = \min_{i \neq j}\{CorPSS_{i,j}[k,r]\},$$

$$MaxCorPSS_{CC}[k,r] = \max_{i \neq j}\{CorPSS_{i,j}[k,r]\}.$$

In the paper, correlation functions (1) were calculated for $\Delta f = 0.01 / T = 150$ Hz and $NF = 7$. In the fig. 1 sections $MinCorPSS_{AC}$ and $MaxCorPSS_{AC}$ are shown, sections $MinCorPSS_{CC}$, $MaxCorPSS_{CC}$ are shown in the fig. 2. In both cases $r = 0$, i.e. the presented results correspond to zero frequency shift; zero time shift PSS ACF value $MaxCorPSS_{AC}[0, 0]$ is taken as 0 dB.

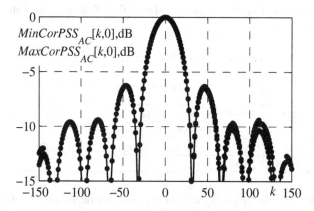

Fig. 1. ACFs of PSSs without frequency shift

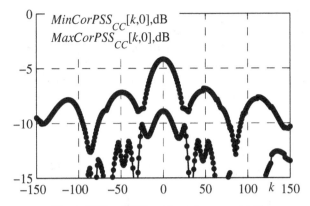

Fig. 2. CCFs of PSSs without frequency shift

From fig. 1-2 we can see that, firstly, PSSs ACFs have enough wide main lobe – 39 samples for the level of –3 dB and high side lobes. Hence, in the multipath channel PSS ACF peak will be extended, and its value will depend on several signal paths. Secondly, PSSs CCFs are strongly uneven and have a high peak at the zero time and frequency shift (from –9 to –4 dB). Thus, signal of neighbor BS sector can also have a high influence at the value of PSS ACF peak. Analysis of PSSs CFs in the presence of frequency shift ($-7 \leq r \leq 7$) has shown that the results obtained for $r = 0$ almost do not change and, consequently, there is no necessity to use frequency synchronization algorithms.

When processing real signals the value of PSS number and its position in the signal will be known (because BS localization station includes primary synchronization module), that is why in fact it will be necessary to make calculations of ACF of PSS for every element of antenna array:

$$IRE\_PSS_{Ar}[k] = \sum_{n=0}^{127} RS_{Ar}^{(16)}[Shift_{PSS} + n + k]PSS^{(16)} * [i, n],$$

$$Ar = 0, 1, ..., M - 1,$$

where $RS_{Ar}^{(16)}[k]$ denote signal from Ar element of antenna array filtered and decimated by 16 times; M denote number of elements of antenna array; $Shift_{PSS}$ is time synchronization with the beginning of PSS, common for all antenna elements; i is the number of PSS, defined during the primary synchronization.

3   CRS Correlation Properties

Taking into account the fact, that PSSs do not provide good time resolution of signal paths, it is reasonable to use wider-band CRSs. The advantage of CRSs relatively to PSSs is that resource elements of CRSs are evenly distributed within the whole band of OFDM-symbol, and, at the same time, 6 orthogonal patterns of resource elements allocation for CRSs depending on the value of cell physical identifier are defined. CRSs are transmitted 40 times per frame, whilst PSSs are transmitted only twice. These facts can provide improvement of IR estimation quality for CRSs ACFs, in comparison with PSSs ACFs. There another reference signals are defined in LTE: MBSFN reference signals, UE-specific reference signals and positioning reference signals. However they are all optional, hence the selection of CRSs for IR estimation is justified.

Let us examine correlation properties of CRSs:

$$CorCRS_{\mathbf{P1},\mathbf{P2}}[k,r] = \left| \sum_{n=0}^{2047} TS\_CRS_{\mathbf{P1}}[n+k] \exp(j\varphi(k,r,n)) CRS^{*}_{\mathbf{P2}}[n] \right|, \tag{2}$$

$$\mathbf{P} = [N_{ID}^{\text{cell}}, RB, At, Slot, Symbol, CP],$$

$$N_{ID}^{\text{cell}} = 0, 1, ..., 503; RB = 6, 15, 25, 50, 75, 100; At = 0, 1, 2, 3;$$

$$Slot = 0, 1, ..., 19; Symbol = 0, ..., 6; CP = 0, 1,$$

where N_{ID}^{cell} denote a cell physical identifier; RB is a number of resource blocks allocated to BS; At is a number of transmitting antenna of BS; $Slot$ is a number of slot in the frame; $Symbol$ is a number of OFDM-symbol in the slot; CP is a type of cyclic prefix of OFDM-symbol: 0 – normal, 1 – extended; CRS is 1×2048 matrix of CRS time domain samples, defined by the vector of parameters $\mathbf{P}$; $TS\_CRS_{P1}[k]$ is a test signal, computed in the same way to $TS\_PSS^{(1)}_{j}[k]$ in (1), but for CRS. Used values of At depend on the BS configuration; used values of $Symbol$ depend on values of At and CP. Everywhere below we will be examining the case of normal cyclic prefix ($CP = 0$), because the extended cyclic prefix is used only in single-frequency network mode in some subframes, and, consequently, this exclusion do not lead to loss of generality.

For analysis of CRSs autocorrelation properties it is necessary for every combination of RB and N_{ID}^{cell} values to observe 40 combinations of $Slot$ and $Symbol$ values for $At = 0$, 1, and 20 combinations of $Slot$ and $Symbol$ values for $At = 2$, 3. At the same time, in the first case it is enough to examine only $At = 0$, and only $At = 2$ for the second case, since the change of number of transmitting antenna only causes the frequency shift of CRS. In sum, for every RB value it is necessary to examine $504 \times (40 + 20) = 30\,240$ ACFs. It is obvious, that same to synchronization signals it is convenient to examine the function

$$MaxCorCRS_{AC}[k,r] = \max_{\mathbf{P1}=\mathbf{P2}} \{ CorCRS_{\mathbf{P1},\mathbf{P2}}[k,r] \}.$$

All CRSs ACFs for $-160 \leq k \leq 160$ and $NF = 7$ have been calculated in the paper. In the fig. 3 the sections $MaxCorCRS_{AC}$ for the case of $r = 0$ are shown, i.e. without frequency shift, and in table 2 the values of section width of $MaxCorCRS_{AC}$ for the case $r = 0$ up to preset level are represented.

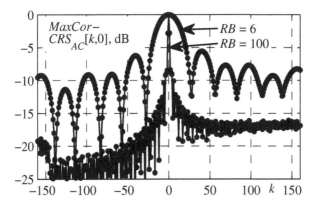

Fig. 3. CRS ACF without frequency offset

Table 2. Values of section width of $MaxCorCRS_{AC}$ for the case $r = 0$

$MaxCorCRS_{AC}[k, 0]$, dB	RB					
	6	15	25	50	75	100
−3	33.8	13.6	8.2	4.1	2.4	2.1
−5	41.4	16.6	10.0	4.7	3.0	2.8
−10	−	72.2	43.0	21.1	12.2	8.3
−15	−	−	−	−	63.2	39.1

From fig. 3 and table 2 it follows that, firstly, the main lobe of CRSs ACFs gets thinner with the increasing of number of resource blocks, allocated for the network, i.e. the time resolution of IR estimation gets better. For example, in passing from $RB = 6$ to $RB = 100$ the ACF width, calculated at the level of −3 dB, decreases from 33.8 to 2.1 samples, i.e. from 1.1 μsec to 67 nsec. Secondly, the level of side lobe of ACFs decreases with increasing of number of resource blocks, i.e. the magnitude resolution of IR estimation enhances. The research of $MaxCorCRS_{AC}$ for $r \neq 0$ has shown, that for $-7 \leq r \leq 7$, so for actual frequency offset values, central ACFs lobe for all values of RB reduce no more than 0.04 dB. Thus, the presence of frequency offset does not influent on the results of calculation of CRSs ACFs.

Let us examine the influence of additive white Gaussian noise (AWGN) on the quality of IR estimation for CRS ACF. When calculating CRS ACF, the signal-to-noise ratio (SNR) increases in the number of times, equal to the number of CRS resource elements, because CRS samples added to each other are correlated, and noise samples are not correlated. In table 3 the number values of CRS resource elements for one OFDM-symbol (in which CRSs are transmitted) N_{CRS} are shown, and ΔSNR values of increasing the SNR when calculating CRS ACF depending on the number of resource blocks.

Table 3. Gain of SNR when calculating CRS ACF

RB	6	15	25	50	75	100
N_{CSR}	12	30	50	100	150	200
ΔSNR, dB	10.8	14.8	17.0	20.0	21.8	23.0

From joint analysis of tables 2-3 it follows that constituent of AWGN in fact can be out of examination when analyzing CRS ACF, because SNR will always be lower than the level of CRS ACF side lobes peaks.

Finally, we want to estimate the influence of other BSs signals, i.e. examine cross-correlation properties of CRSs. The number of CCFs in (2) is extremely large, so we decided to make some simplifications to shorten the calculations. Firstly, we will examine only values of $CorCRS_{CC}(\mathbf{P1}, \mathbf{P2}) = CorCRS_{\mathbf{P1},\mathbf{P2}}[0, 0]$, i.e. without time and frequency shift. In this case, calculation of cross-correlation between OFDM-symbols comes to the calculation of cross-correlation between modulating sequences of CRSs. Secondly, taking into account the necessity of coincidence of resource elements of different reference signals, it is enough to examine only such combinations of $N_{ID}^{cell}{}_1$ and $N_{ID}^{cell}{}_2$, for which

$$\mathrm{mod}(N_{ID}^{cell}{}_1, 6) = \mathrm{mod}(N_{ID}^{cell}{}_2, 6).$$

Thirdly, we will examine only CRSs of eighth (for $Ar = 0$) and ninth (for $Ar = 3$) OFDM-symbols of subframes with numbers 0 and 5, i.e. those CRSs, which are transmitted at once after primary and secondary synchronization signals in FDD mode. In table 4 average and maximal values for calculated samples of $CorCRS_{CC}$ for different values of RB are shown.

Table 4. Average and maximal values of CRSs CCFs for two equipollent signals

RB	6	15	25	50	75	100
$\mathbf{E}\{CorCRS_{CC}\}$	–5.9	–7.9	–9.0	–10.5	–11.4	–12.0
$\max\{CorCRS_{CC}\}$	–0.7	–1.5	–2.8	–3.7	–4.6	–5.2

Data represented in table 4 can be used as IR magnitude resolution. Having the estimations of relative potency of signals of different BSs, it is possible to do corrections for values from table 4. Moreover, if we consider only signal of the most powerful BS, we can use values from table 4 directly.

When processing real signals, synchronization with the beginning of OFDM-symbol containing CRS is known; also all parameters defining CRS are known, so it is necessary to calculate CRS ACF for every element of antenna array:

$$IRE\_CRS_{Ar,At}[k] = \sum_{n=0}^{2047} RS_{Ar}^{(1)}[Shift_{Symbol} + n + k]CRS_{\mathbf{P}}^{*}[n],$$

where $RS_{Ar}^{(1)}[k]$ is the non-decimated signal from Ar antenna array element, i.e. recorded with sampling rate of 30.72 MHz; $Shift_{Symbol}$ is the time synchronization with the beginning of OFDM-symbol that contains CRS, common for all antenna array elements; the value of $\mathbf{P}$ is defined with parameters of CRS.

In the fig. 4-5 curves of absolute IRE_CRS function values for two BSs are presented: for more powerful BS at fig. 4 and for less powerful BS at fig. 5. These curves were obtained from real signal. In both cases $RB = 75$.

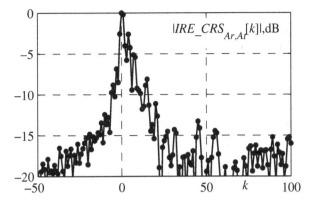

Fig. 4. Absolute CRS ACF values for strong BS

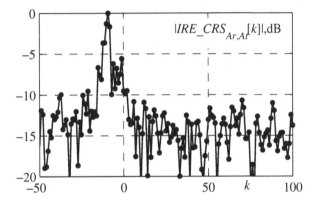

Fig. 5. Absolute CRS ACF values for weak BS

From fig. 4-5 it follows that the noise level of *IRE_CRS* function is lower for the signal of more powerful BS (≈ -15 dB) than for the signal of more weak BS (≈ -10 dB).

Calculation of modified K-factor will be made in the following way:

$$K_m = \frac{\left|IRE\_CRS_{Ar,At}[0]\right|}{\sum\limits_{k \in Q}\left|IRE\_CRS_{Ar,At}[k]\right|},$$

$$Q = \{k : IRE\_CRS_{Ar,At}[k] \geq b\}$$

i.e. by examining the relation of maximal value of |*IRE_CRS*| to the sum of all values of |*IRE_CRS*|, exceeding the threshold *b*. The question of calculation of corrections for values from table 4 (i.e. calculation of relative powers of signals from different BSs) was not examined in the present paper, therefore the value $\mathbf{E}\{CorCRS_{CC}\}$ from table 4 was chosen as a value of *b* for all BSs.

It should be pointed out that the best way of defining the number and the parameters of the signal paths is to solve the problem of superposition of CRS time shifted

ACFs. Then, the relation of powers of corresponding signal paths could be examined when calculating K-factor. However, in the present paper such approach was not implemented due to limited computing power of the MRS equipment.

4   Algorithm of DoA Estimation Using Antenna Array

In the present paper conventional beamformer is used for DoA estimation based on processing the signals coming to the antenna array [6]. Hereby we give the main computations.

Without loss of generality we believe that the emitting source (BS) is set in the distance which is long enough to provide the plain wave front. Antenna array elements are set on the plain parallel to the ground.

Depending on the DoA estimation algorithm, we will have either sequences of signal samples from antenna array (RS_{Ar}), or sequences of PSS ACFs peaks

$$IRE\_PSS_{Ar}[\hat{k}],$$

$$\hat{k} = \arg\{\max_{k}(IRE\_PSS_{Ar}[k])\},$$

or CRS ACFs peaks

$$IRE\_CRS_{Ar,At}[\hat{k}],$$

$$\hat{k} = \arg\{\max_{k}(IRE\_CRS_{Ar,At}[k])\}.$$

In any case, we will consider those sequences in the following way:

$$\mathbf{x(t)} = (\mathbf{x}(t_0), \mathbf{x}(t_1),...,\mathbf{x}(t_{N-1})) = \begin{pmatrix} x_0(t_0) & x_0(t_1) & \cdots & x_0(t_{N-1}) \\ x_1(t_0) & x_1(t_1) & \cdots & x_1(t_{N-1}) \\ \vdots & \vdots & \ddots & \vdots \\ x_{M-1}(t_0) & x_{M-1}(t_1) & \cdots & x_{M-1}(t_{N-1}) \end{pmatrix},$$

where M is the number of elements of the antenna array; N is the number of elements in the sequence of signal's samples, or in the sequence of peaks values of PSS ACFs or CRS ACFs.

For the antenna diagram synthesis with antenna array we examine the linear combination of signals from the elements of the antenna array:

$$\mathbf{y(t,\theta)} = \mathbf{a}^H(\theta)\mathbf{x(t)},$$

$$\mathbf{a}(\theta) = [1,\ \exp(j\varphi_1(\theta)),\ \exp(j\varphi_2(\theta)),\ ...,\ \exp(j\varphi_{M-1}(\theta))]^T,$$

where $\mathbf{a}(\theta)$ is a steering vector, and phase values $\varphi_1(\theta)$, $\varphi_2(\theta)$, ..., $\varphi_{M-1}(\theta)$ are chosen so that the resulting antenna diagram would have the direction of the θ angle.

Let us calculate the average power of the signal, coming from the θ direction:

$$P(\theta) = \frac{1}{N} \sum_{n=0}^{N-1} |\mathbf{y}(t_n, \theta)|^2 = \frac{1}{N} \sum_{i=0}^{N-1} \mathbf{a}^H(\theta) \mathbf{x}(t_n) \mathbf{x}^H(t_n) \mathbf{a}(\theta) = \mathbf{a}^H(\theta) \mathbf{R}_{xx} \mathbf{a}(\theta),$$

$$\mathbf{R}_{xx} = \frac{1}{N} \sum_{i=0}^{N-1} \mathbf{x}(t_n) \mathbf{x}^H(t_n).$$

Look-up steering vectors for θ values with increment of $2\pi/360 = 1°$ were calculated from geometric location of antenna array elements. Thus, calculation of DoA estimation $\hat{\alpha}$ reduces to the search of the argument of the maximum of signal's power:

$$\hat{\alpha} = \arg \{ \max_{\theta} (P(\theta)) \}.$$

Let us discuss some peculiarities of DoA estimation using different algorithms, which are conditioned by the capabilities of hardware components. Signal recording was made periodically, synchronous from all antenna array elements. Length of each record is so that at least 17 entire LTE frames are presented in it. Algorithm #1 involves 17 frames signal processing in the bandwidth of 1.08 MHz. Algorithm #2 involves processing of sequence of PSS ACFs peaks, obtained for every first PSS in 17 frames, i.e. sequence of 17 elements length. Algorithms #3 and #4 involve processing of sequence of CRS ACFs peaks, obtained for every first CRS in every slot in 17 frames, i.e. sequence of 340 elements length. For algorithms #3 and #4 only signal of first antenna of BS was taken into account, since the presence of another BS antennas is optional.

5   Comparison of DoA Estimation Algorithms

During the experiment all BSs locations and all directions of their antennas were known. The possibility of continuous calculation of coordinates, speed and course of the MRS was implemented. Thus the true DoA for every BS was obtained for every record.

To describe the algorithm effectiveness, it was decided to use the value of probability of getting the DoA estimation $\hat{\alpha}$ into the preset angle interval $\Delta\alpha$ around the true value of α (further noted as the probability of obtaining the correct DoA):

$$P_L = P\{|\alpha - \hat{\alpha}| \le \Delta\alpha\}.$$

Before going to the comparison of effectiveness of different DoA estimation algorithms, let us examine the algorithm #4 in detail, since it is necessary to choose threshold value K_m in it.

Comparison of the K_m value with the threshold K_0 value is needed to identify the LoS. This identification allows to exclude cases of DoA estimation from reflected signal, and consequently increase the probability of obtaining the correct DoA. In terms of environmental experiment it is hard to define the presence or absence of LoS, therefore probabilistic characteristics of LoS identification algorithm were not observed. Instead of this, threshold K_0 was chosen on the base of analysis of probability of false LoS identification, i.e. probability of deciding in favor of presence of LoS in its absence. In its turn, absence of LoS will be defined by not-getting of the DoA estimation into the preset angle interval of DoA's true value:

$$P_{FA}(K_0) = P\{K_m \geq K_0 / |\alpha - \hat{\alpha}| > \Delta\alpha\}.$$

The dependences $P_{FA}(K_0)$ and $P_D(K_0)$ are shown in the fig. 6, where $P_D(K_0)$ is probability of correct LoS identification:

$$P_D(K_0) = P\{K_m \geq K_0 / |\alpha - \hat{\alpha}| \leq \Delta\alpha\}.$$

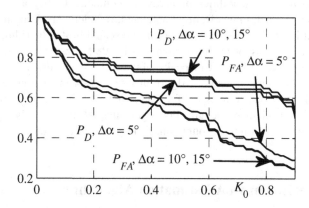

Fig. 6. Dependences $P_{FA}(K_0)$ and $P_D(K_0)$

From fig. 6 it follows that both probabilities reduce while K_0 is increasing, but the probability of false LoS identification reduces faster, than the probability of correct identification, what testifies for the reason of using K_m value for LoS identifying. The value of threshold K_0 was chosen according to the level of tolerance probability of the false LoS identification:

$$K_0 = \min_K\{K : P_{FA}(K) \leq 0.3\}.$$

Dependence of correct DoA probability on threshold K_0 for the algorithm #4 is shown in the fig. 7. It is obvious, that the curve in the fig. 7 increases while K_0 increases.

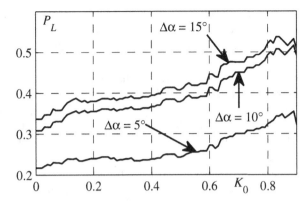

Fig. 7. Dependences of correct DoA probability on threshold K_0

In table 5 the values of correct DoA probability for different algorithms and different values of $\Delta\alpha$ are represented. Also, the results are separated by types of observed BSs. In first case only the most powerful BSs are examined, i.e. one BS for one record. In second case all BSs detected by primary synchronization module are examined – from 2 to 5 BSs. Although algorithm #1 does not provide the separation of different BSs signals, nevertheless, the following assumption was made to make it comparable in 2nd case: DoA estimated using algorithm #1 was assigned to all BSs detected by primary synchronization module.

Although the first algorithm was considered, nevertheless, its usage is out of interest in fact, since it does not allow to identify BS.

Table 5. P_L values for different algorithms and different cases

Case	Algorithm	P_L		
		$\Delta\alpha = 5°$	$\Delta\alpha = 10°$	$\Delta\alpha = 15°$
#1	#1	0.23	0.36	0.41
	#2	0.18	0.29	0.33
	#3	0.23	0.34	0.38
	#4	0.37	0.54	0.57
#2	#1	0.17	0.26	0.32
	#2	0.16	0.24	0.28
	#3	0.22	0.31	0.34
	#4	0.35	0.49	0.52

From table 5 we can conclude the following statements.

1. Increasing of $\Delta\alpha$ leads to the increasing of P_L for all algorithms.
2. Results of all algorithms become worse after going from case #1 to case #2. This is conditioned by the growing of power of structural interference while examining less powerful BSs with respect to the first one.

3. Algorithm #3 always shows better results than algorithm #2 does. Evidently, this is conditioned by better time resolution of CRSs ACFs with respect to PSSs ACFs and, therefore, less influence of reflected signal paths into the value of CRSs ACFs peaks.

4. In case #1, algorithm #1 shows better results than algorithms #2 and #3. Evidently, such phenomenon is conditioned by taking into account much higher energy of signal during the implementation of algorithm #1. Really, during the implementation of algorithm #1 the signal of all the frames in the band of 1,08 MHz is taken into account while during the implementation of algorithm #2 only the energy of one PSS per one frame is taken into account, and during the implementation of algorithm #3 – energy of twenty CRSs per one frame.

5. In all cases algorithm #4 shows noticeably better results, than all another algorithms.

References

1. Ko, Y.-H., Park, C.-H., Cho, Y.-S.: Joint Methods of Cell Searching and DoA Estimation for a Mobile Relay Station with Multiple Antennas. IEEE GLOBECOM 2008, 1–4 (2008)
2. Pec, R., Cho, Y.S.: A Parameter Estimation Technique for an LTE-based Mobile Relay Station with Antenna Array. In: 2013 International Conference on ICT Convergance (ICTC), pp. 134–135 (2013)
3. Ko, Y.H., Kim, Y.J., Yoo, H.I., Yang, W.Y., Cho, Y.S.: 2-D DoA Estimation with Cell Searching for a Mobile Relay Station with Uniform Circular Array. IEEE Transaction on Communications 58(10), 2805–2809 (2010)
4. 3GPP, "3GPP TS 36.104 V11.8.2. 3rd Generation Partnership Project; Technical Specification Group Radio Access Network; Evolved Universal Terrestrial Radio Access (E-UTRA); Base Station (BS) radio transmission and reception (Release 11)", 3rd Generation Partnership Project, Tech. Rep. (April 2014)
5. 3GPP, "3GPP TS 36.101 V11.8.0. 3rd Generation Partnership Project; Technical Specification Group Radio Access Network; Evolved Universal Terrestrial Radio Access (E-UTRA); User Equipment (UE) radio transmission and reception (Release 11)", 3rd Generation Partnership Project, Tech. Rep. (March 2014)
6. Chen, Z., Gokeda, G., Yu, Y.: Introduction to Direction-of-Arrival Estimation. Artech House Publishers (2010)

Delay Based Handover Algorithm Design for Femtocell Networks

Piri Kaymakçıoğlu, Kamil Şenel, and Mehmet Akar

Bogazici University, Turkey
Department of Electrical and Electronics Engineering
pkaymakcioglu@gmail.com, {kamil.senel,mehmet.akar}@boun.edu.tr

Abstract. Mobile communications systems have limited frequency resources to operate with. In traditional macrocell architectures, this problem is addressed by utilizing smaller cells to employ frequency reuse among cells. A new promising solution is to use two-tier networks that employs smaller cells called femtocells which are plug-and-play devices that use the broadband internet connection to connect to the operator's core network. However, successful deployment of femtocells requires modifications on the existing network structures. An important adjustment is required on the handoff protocols to provide quality of service requirements that the users need. The traditional network approaches that does not consider the broadband internet backhaul for quality of service is inadequate in two-tier networks. This is due to the fact that users connect to operators core network via broadband internet connection. In this paper we focus on the handover/handoff problem in two-tier networks and propose a novel handoff/handover algorithm that considers both wired and wireless medium requirements for service quality. The performance of the proposed algorithm is evaluated using the handover rate and signal degradation rate as key indicators. The simulation studies demonstrate that better results are achieved with the proposed algorithms compared to existing ones in the literature.

1 Introduction

Over the last decade, there has been an enormous growth in both the number of users and the demand for mobile services in cellular networks. Network operators have been providing service to increasing number of users while facing the challenge to cope with the incoming mobile standards. As the demand of a typical user has changed especially with the new mobile multimedia services, supplying different and higher data rates is becoming mandatory. Unfortunately, the existing mobile network setups are not equipped to deal with these demands, especially when one takes into account that % 90 percent of data communications is taking place indoors while indoor coverage is supplied with the outdoor macrocells in traditional cellular networks [1] [2]. Since most of the wireless transmission occurs indoors, the mobile infrastructure needs smaller cells such as microcells and nanocells to increase indoor capacities. Heterogeneous network configurations consisting of femto/pico/micro/macro-cells have been shown to increase system capacity and overall service quality in addition to reducing the load on macrocell

S. Balandin et al. (Eds.): NEW2AN/ruSMART 2014, LNCS 8638, pp. 205–218, 2014.

base stations. The most common examples of such successful heterogeneous mobile networks are the use of picocells and microcells in hotspots such as airports, shopping malls and train stations. Since picocells and microcells are small base stations deployed by the operator, they are not quite suitable for small businesses and home users due to their cost and difficult setup process. Femtocells on the other hand are low-powered wireless access points that are designed to operate in the licensed spectrum, and are suitable for small businesses or home users. As the demand for multi-media services tends to increase rapidly, small cell networks co-existing with the traditional macrocells promise to be a viable solution to meet the required capacity and service quality.

Heterogeneous network topology including femtocells, although very similar to single tier cellular systems, has very distinct properties. A single-tier network compromises of neighboring cells of same tier, whereas heterogeneous networks two-layer networks consist of smaller cells within macrocells. This multi-layer approach introduces challenges such as interference mitigation, self-organization and seamless mobility by handovers. Although these problems are considered for single tier networks, there are important differences in heterogeneous case such as unknown femtocell locations.

As a user moves through the coverage area of cellular network, the connection may be transfered to another cell. This procedure is called handover; it is used to maintain the required signal quality to user. Handover algorithms aim to optimize handover decisions in order to minimize the signal degradation rate and the number of unnecessary handovers. The wireless link quality between a user equipment and a base station is evaluated by the following parameters: Bit error rate, signal to noise ratio, distance, traffic load, received signal strength (RSS), and different combinations of these parameters [3–11]. Received signal strength based algorithms are simple to simulate and give satisfactory results, therefore they have received considerable attention [3–11]. The simplest handover algorithm based on the received signal strength is the one that the mobile equipment selects the base station with the highest RSS. However, this approach leads to handovers in cases where the serving base station signal can provide call continuity. These unnecessary can be eliminated by the threshold test [3], or the hysteresis test [4], or the threshold/hysteresis algorithms [5–7].

Macrocells are interconnected via a backbone providing high quality in the one tier macrocellular networks. However, the backhaul line of femtocell base station is generally connected to the core network via a connection of low rate, possibly incapable of maintaining quality of service for real time voice communications, therefore handover decision algorithms should consider femtocell backhaul quality in order to maximize cellular capacity. Call Admission Control (CAC) based methods are proposed in consideration for handovers from a macrocell to a femtocell [12–14]. However, if the Quality of Service (QoS) degrades during a call, it does not initiate handover. Becvar and Mach proposed a handover decision algorithm based on the delay at femtocell backbone. Although the proposed model brings an important enhancement for handovers from macrocell to femtocell, it does not provide handover decisions from femtocell to macrocell [15].

Real-time voice communications over IP networks needs measurable quantities to guarantee the quality of the communication [16]. There are many parameters that can be used to measure the quality of service, but the most important ones are the one-way delay, delay variance (jitter), packet loss and minimum bandwidth. Real time voice communications can tolerate delays up to 150 ms, packet loss up to 10^{-4}, and jitter should be less than 20 ms for good quality [17]. In this study, the one way delay parameter affecting the real time voice communications is used as a parameter in handover decision. Since one way delay parameter has been crucial to the real time applications, the delay dynamics of the Internet networks have been widely researched at different studies with different data sets [18–23]. The processing delay in the Internet networks is modeled as an Erlang Probability Distribution Function (PDF) by Wee et al [23]. The model is also similar to the Internet delay at Oboe and Fiorini's work that uses a shifted exponential distribution for a 150-km connection and an Erlang like probability density for a 10000-km connection [22]. In the delay model for femtocell backhaul line, two different delay probability distributions are used in order to classify the subscribers with "stable" delay at femtocell broadband lines and the subscribers with "unstable" delay at femtocell backhaul for comparison purposes.

The rest of the paper is organized as follows. In Section 2, the system description is given, path loss and one way delay models are introduced. The performance criteria used for the handoff algorithms in two-tier networks are defined in Section 3. Two new algorithms are proposed in Section 4. The simulation environment is described and the results are given in Section 5. Finally, Section 6 concludes this paper.

2   System Description

The experimental setup used in the design and comparison of the traditional handover algorithm and the proposed algorithms contains one femtocell base station and one macrocell base station similar to the setups considered in the literature [1,3,4,7]. The mobile station walks with a constant speed from the start point to the end point. The system setup is depicted in Figure 1. The distance between the macrocell base station (MBS) and the femtocell (FBS) base station is chosen as 200 meters and the mobile station(MS) is located in the middle of these two base stations. The MS starts to move from its start point towards the end point with a constant velocity taken to be 1 m/sec in the simulations.

2.1   Path Loss Model

Received signal strengths (in dBm) from macrocell and femtocell base stations are modeled as the output powers of MBS and FBS minus the pathloss of macrocell and femtocell signals, respectively; that are described as

$$s_m = P_m - PL_m, \tag{1}$$

$$s_f = P_f - PL_f, \tag{2}$$

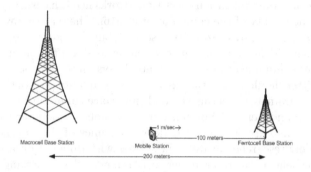

Fig. 1. Experimental Setup

where s_m and s_f denote the RSS from MBS and FBS, respectively; P_m and P_f represent the fixed output powers of the MBS and FBS, finally PL_m and PL_f refer to the pathloss for macrocell and femtocell signals, respectively. The pathloss models of macrocell and femtocell signals are given as:

$$PL_m = PL_{0m} + \eta_m 10log(l_m/l_{0m}) + PL_{sm}, \tag{3}$$

$$PL_f = PL_{0f} + \eta_f 10log(l_f/l_{0f}) + PL_{sf}, \tag{4}$$

where PL_{0m} and PL_{0f} refer to the pathloss of the macrocell and femtocell signal at reference distances, l_{0m} and l_{0f}, respectively. The second term in the model is logarithmic scaling with the distance from the mobile station to the macro and femto base stations, l_m and l_f; factored with the so-called the pathloss exponents, η_m and η_f for macro and femto base stations. PL_{sm} and PL_{sf} comprise the log-normal shadowing for macrocell signal and femtocell signal, respectively. PL_{sm} and PL_{sf} are auto-correlated Gaussian random variables with zero mean in decibel unit for macrocell signal and femtocell signal [24–26]. The mobile station moving from x_i to x_j has the shadowing correlation coefficient $C(x_i, x_j)$ described as

$$C(x_i, x_j) = \rho_{l_{ij}} = \sigma^2 e^{-\frac{l_{ij}}{D_c}} \text{ where } l_{ij} = \| x_i - x_j \|, \tag{5}$$

$\rho_{l_{ij}}$ denotes the correlation coefficient, and σ is the standard deviation of the Gaussian random variables and D_c is the effective correlation distance at which correlation coefficient reduces to 0.5 [27, 28]. The effective correlation distance for suburban areas is higher compared to urban areas. We can represent the correlation coefficients stated in equation (5) in matrix form as

$$C = \sigma^2 \begin{bmatrix} 1 & \rho_1 & \rho_2 & \cdots & \rho_{l-1} \\ \rho_1 & 1 & \rho_1 & \cdots & \rho_{l-2} \\ \rho_2 & \rho_1 & 1 & \cdots & \rho_{l-3} \\ \vdots & \vdots & \vdots & \ddots & \vdots \\ \rho_{l-1} & \rho_{l-2} & \rho_{l-3} & \cdots & 1 \end{bmatrix}, \tag{6}$$

where C denotes the autocovariance matrix. Since C is a positive definite symmetric matrix, it can be decomposed using Cholesky decomposition as $C = C_L C_U$, where $C_L = C_U{}^T$ is the lower triangular matrix and C_U is the upper triangular matrix. The correlated log-normal shadowing is $PL_s = \alpha C_U$, where PL_s denotes the correlated log-normal shadowing part of path-loss for macrocell signal in (3) and femtocell signal in (4); α is an $1 \times l$ array that is composed of uncorrelated, normally distributed random numbers with zero mean and unit variance [27].

2.2 Delay Models

In the delay model for femtocell backhaul line, two different delay probability distributions are used in order to classify the subscribers with good Internet connection that has a "stable" delay at femtocell broadband lines and the subscribers with a bad Internet connection that has an "unstable" delay at femtocell backhaul. For the subscribers who own a good Internet broadband line, the one way delay is modeled as a shifted exponential function [22]. The delay in this backhaul has a low variance and therefore the line does not suffer from jitter. For this good Internet connection, delay is modeled via random variables that are distributed according to the shifted exponential probability distribution (P_{gd}) expressed as

$$
P_{gd}(x, \lambda) = \begin{cases} \lambda e^{-\lambda(x - \triangle_d)} & x \geq \triangle_d \\ 0 & x < \triangle_d \end{cases}, \tag{7}
$$

where λ and $\triangle_d$ denote the rate parameter and the shift of delay, respectively. The probability of the occurrence of the event that the delay at stable femtocell connection (d_f) exceeds the maximum acceptable delay level for voice over IP communication (T_d) can be computed as a

$$
P_{gd}[x \geq T_d] = \begin{cases} e^{-\lambda(T_d - \triangle_d)} & T_d \geq \triangle_d \\ 0 & T_d < \triangle_d \end{cases}. \tag{8}
$$

The delay at "unstable" Internet connection is modeled as an Erlang-n probability distribution function given as

$$
P_{bd}(x, n, \lambda) = \frac{\lambda^n x^{n-1} e^{-\lambda x}}{(n-1)!} \text{ for } x, \lambda > 0, \tag{9}
$$

where n is the number of routers the communication packets pass through [23]. The delay at femtocell backbone can be unstable, i.e., it can change rapidly in the communication duration. The real-time communications suffer from rapid increase in delay which creates jitter. The probability of the occurrence of the event that the delay (d_f) at unstable femtocell backhaul exceeds the maximum acceptable delay level (T_d) for voice over IP communication can be defined as

$$
P_{bd}[x \geq T_d] = \sum_{k=0}^{n-1} \frac{1}{k!} e^{-\lambda T_d} (\lambda T_d)^k. \tag{10}
$$

3 Performance Criteria

The performance of handoff algorithms can be evaluated based on many different criteria such as total number of handoffs, handover delay, call quality and the number of unnecessary handoffs [29, 30]. The objective of the proposed handoff algorithms is to minimize the number of handoffs and the number of signal degradations as performance indicators. These performance indicators depend on the received signal strength and delay of the backhaul femtocell station broadband wired line. The tradeoff between the performance criteria leads us to achieve optimal algorithm parameters in order to minimize the performance indicators: Total number of signal degradations N_{SD} and total number of handoffs N_H [11]. The total number of signal degradations can be expressed as

$$N_{SD} = \sum_{k=1}^{K} \mathbb{I}\{(d_k > T_d) \lor (s_k < T_s)\}, \tag{11}$$

where $\mathbb{I}\{.\}$ denotes the indicator function and K is the total number of steps. A signal degradation is said to occur when the delay at step k (d_k) is more than the delay threshold (T_d) or if RSS at iteration k (s_k) is less than the RSS threshold (T_s). The total number of signal degradations is obtained by counting each signal degradation from the start point to the end point. A signal degradation does not necessarily imply that the call is dropped since T_s can be chosen higher than the minimum acceptable signal level and T_d can be picked lower than the maximum acceptable delay level. Hence, N_{SD} can measure the call quality from the point of human user perspective. The number of degradations that occur when the delay exceeds T_d, shows the quality of service level for wired backhaul line of femtocell base station, whereas the number of degradations that occur when the RSS falls below T_s, assesses the quality of service of the wireless link. Total number of handovers can be described as

$$N_H = \sum_{k=1}^{K} \mathbb{I}\{M(k) = 1\}, \tag{12}$$

where $M(k)$ denotes the handover function that can be defined as

$$M(k) = \begin{cases} 1 & \text{if } |r(k) - r(k-1)| = 1, \\ 0 & \text{otherwise.} \end{cases} \tag{13}$$

In equation (13), "1" is the handover state of mobile station from FBS to MBS or from MBS to FBS and "0" is the state that the mobile station stays at the serving station and the state function $r(k)$ is 1 (resp. 0) if MS is served by MBS (resp. FBS) at time t. The total number of handovers N_H criterion indicates the burden of the handovers to the cellular network since it is crucial to prevent excessive switching between the femtocell and the macrocell base stations.

4   Delay Based Handover Design

The objective of a handover algorithm is to minimize a weighted sum of N_{SD} and N_H. Two different approaches are introduced in this section, the first one is an ad-hoc approach and the second one tries to minimize the probability of signal degradation while avoiding excessive handovers.

4.1   Delay Based Ad-hoc Handover Algorithm

The first handover algorithm proposed in this study tends to achieve a tradeoff between N_{SD} and N_H, where N_{SD} and N_H are described in equations (11) and (12), respectively. The proposed handover function is described as:

$$M(t) = \begin{cases} 1 & \text{if } r(t) = 1 \text{ and } \mathbb{I}\{(s_m + h < s_f) \wedge (s_m < T_s) \wedge (d_f < T_d)\} \\ & \text{or } r(t) = 0 \text{ and } \mathbb{I}\{((s_f + h < s_m) \wedge (s_f < T_s)) \vee \\ & \quad\quad ((s_m > T_s) \wedge (d_f > T_d))\}, \\ 0 & \text{otherwise.} \end{cases} \quad (14)$$

This algorithm considers both received signal strength and delay in femtocell to decide for handover which is only performed if switching will result in a better service for the user. Based on the serving cell the algorithm works as follows:

(i) If the serving cell is a **macrocell** base station, the handover decision from macrocell to femtocell occurs only if all of these 3 conditions are satisfied
   (a) The RSS of femtocell (s_f) is higher than the received signal strength of macrocell (s_m) with a hysteresis (h) value
   (b) The RSS strength of the serving macrocell base station (s_m) falls below the RSS threshold (T_s)
   (c) The femtocell backhaul delay (d_f) is less than the delay threshold (T_d). Note that handoff decision is not taken unless both received signal from macrocell is degraded and femtocell base station has an acceptable wired and wireless connection. This prevents unnecessary handoffs and provides connection to femtocell when macrocell connection starts to deteriorate. Since this usually happens when a user goes from outdoor to indoor, it is natural for user to change serving base station.
(ii) If the serving cell is a **femtocell** base station, handover is performed only if received signal strength of macrocell (s_m) is higher than the received signal strength of femtocell (s_f) with a hysteresis (h) and if one of the following two conditions holds:
   (a) The RSS of femtocell (s_f) falls below the RSS threshold (T_s)
   (b) The delay of femtocell backhaul (d_f) exceeds the delay threshold (T_d). Femtocell to macrocell handover occurs when femtocell base station is not able to provide the required signal strength when mactocell base station can, or when femtocell backbone has a delay above the acceptable threshold and macrocell is able to provide the required RSS.

4.2   One-Step Optimized Algorithm

The second proposed algorithm considers the probability of signal degradation $Pr[SD]$, which can be defined as the probability of occurrence of the event that the delay at serving station backbone (d) exceeds T_d or the received signal strength (s) falls below T_s. Under the assumption that the delay of the wired backhaul is independent of the received signal level of the mobile station, this probability can be formulated as

$$Pr[SD] = Pr[(d > T_d) \parallel (s < T_s)] \tag{15}$$
$$= 1 - Pr[d \leq T_d] \, Pr[s \geq T_s]. \tag{16}$$

Two different cases are considered based on the serving base station.

(i) If the serving cell is a **macrocell** base station, $Pr[SD_m]$ is described as

$$Pr[SD_m] = 1 - Pr[d_m \leq T_d] \, Pr[s_m \geq T_s]. \tag{17}$$

Since the backhaul connection to the core network of the macrocell is capable of guaranteed quality of service for real time voice communications and the delay at macrocell backbone (d_m) is very low compared to T_d, the signal degradation originating from macrocell backhaul is negligible. Therefore, (17) can be simplified as

$$Pr[SD_m] = Pr[s_m < T_s]. \tag{18}$$

If the received signal strength of macrocell signal (s_m) defined in (1) is combined with (18), $Pr[SD_m]$ can be written as

$$Pr[SD_m] = Pr[P_m - PL_m < T_s]. \tag{19}$$

The pathloss of macrocell signal (PL_m) in (3) combined with (19) gives:

$$Pr[SD_m] = Pr[P_m - PL_{0m} - \eta_m 10 log(\frac{l_m}{l_{0m}}) - PL_{sm} < T_s] \tag{20}$$
$$Pr[SD_m] = Q\left(\frac{-T_s + m(l_m)}{\sigma_m}\right), \tag{21}$$

where $m(l_m)$, $\bar{\sigma}$ and Q are described as:

$$m(l_m) = P_m - PL_{0m} - \eta_m 10 log(l_m/l_{0m}), \tag{22}$$

$$\bar{\sigma_m}^2 = \sigma_m^2 (1 - a^2), \text{ where } a = e^{-d_s/D_c}, \tag{23}$$

$$Q(x) = \frac{1}{\sqrt{2\pi}} \int_x^\infty e^{-t^2/2} \, dt. \tag{24}$$

(ii) If the serving cell is a **femtocell** base station, using (16) $Pr[SD_f]$ can be described as:

$$Pr[SD_f] = 1 - Pr[d_f \leq T_d] \, Pr[s_f \geq T_s]. \tag{25}$$

If the received signal strength of femtocell signal (s_f) defined in (2) is combined with (25), $Pr[SD_f]$ can be written as

$$Pr[SD_f] = 1 - Pr[d_f \leq T_d] \, Pr[P_f - PL_f \geq T_s]. \tag{26}$$

The pathloss of femtocell signal (PL_f) in (4) combined with (26) gives

$$Pr[SD_f] = 1 - Pr[d_f \leq T_d] \, Pr[P_f - PL_{0f} - \eta_f 10log(\frac{l_f}{l_{0f}}) - PL_{sf} \geq T_s]. \tag{27}$$

Using cumulative distribution functions,

$$Pr[SD_f] = 1 - Q\left(\frac{T_s - m(l_f)}{\bar{\sigma}_f}\right) F(T_d), \tag{28}$$

where the complementary cumulative distribution functions $F(T_d)$ for stable delay and unstable delay at femtocell backbone are defined in Equations (8) and (10), respectively; $Q(\cdot)$ is defined in (24); and $m(l_f)$ and $\bar{\sigma}_f$ are given as

$$m(l_f) = P_f - PL_{0f} - \eta_f 10log(l_f/l_{0f}), \tag{29}$$

$$\bar{\sigma}_f{}^2 = \sigma_f^2(1 - a^2). \tag{30}$$

The handover function $M(t)$ for One-Step Optimized Algorithm is given as:

$$M(t) = \begin{cases} 1 & \text{if } r(t) = 1 \text{ and } Pr(SD_m) > c + Pr(SD_f), \\ & \text{or } r(t) = 0 \text{ and } Pr(SD_f) > c + Pr(SD_m), \\ 0 & \text{otherwise.} \end{cases} \tag{31}$$

Here, c is a nonnegative tradeoff parameter representing the cost of handover.

4.3 Hysteresis-Threshold Algorithm

A conventional handoff algorithm is used to the compare performances of the proposed algorithms. The handover function $M(t)$ is described as:

$$M(t) = \begin{cases} 1 & \text{if } r(t) = 1 \text{ and } \mathbb{I}\{(s_m + h < s_f) \wedge (s_m < T_s)\}, \\ & \text{or } r(t) = 0 \text{ and } \mathbb{I}\{(s_f + h < s_m) \wedge (s_f < T_s)\}, \\ 0 & \text{otherwise.} \end{cases} \tag{32}$$

This algorithm does not consider the delay in the backbone for femtocells, which may result in signal degradation. This is similar to the first proposed handoff algorithm with an ad-hoc approach. The handoff decision is based on only received signal strengths and a hysteresis (h) term is used to avoid ping-pong effect.

5   Numerical Analysis

In this section, the simulation results of the proposed algorithms and the traditional algorithm is presented. Two different setups based on the quality of femtocell backbone connection is considered. The parameters used for the simulations are shown in Table 1. Numerical analysis is carried out via Matlab simulations with given parameters. However optimal parameters for decision algorithms such as h, T_s and T_d are not known. Because of these uncertain parameters, the Monte Carlo simulation approach is used with the following values:

Table 1. Simulation Parameters

P_m	43 dBm	P_f	10 dBm	PL_{0m}	28	PL_{0f}	38.5
η_m	3.5	η_f	2	σ_m	8	σ_f	6
l_{mf}	200 m	l_{m0}	100 m	l_{f0}	100 m	v	1 m/s
T_f	150 ms	T_s	-70 dBm	D_c	5 m	$\triangle_d$	10 ms
n	5	λ	0.05				

The values h= [1, 2, 3, 4, 5, 6, 7, 8, 9, 10, 11, 12, 13, 14, 15](dB) for the hysteresis, T_s= [-90, -86, -82, -78, -74, -70, -66, -62](dBm) for the RSS threshold (T_s), T_d= [150, 165, 180, 195, 210, 225, 240, 255, 270, 285, 300](ms) for the delay threshold and c= [0:0.05:1] for the cost of handover are used to evaluate the performance of algorithms based on the total number of handovers and the total number of signal degradations as described in Section 3. Two different cases are considered based on the quality of the femtocell backbone: (i) stable delay, (ii) unstable delay.

5.1   Stable Delay at Femtocell Backhaul

The one way delay is modeled as a shifted exponential function in (7) for stable delay conditions at femtocell backbone [22]. The delay in femtocell backhaul resists change and therefore the line does not suffer from jitter. The simulation results of the hysteresis-threshold algorithm and the proposed algorithms under stable delay at femtocell backbone are depicted in Figure 2.

As expected, the simulation results of the conventional algorithm and the first of the two proposed algorithms do not show a significant difference. Signal degradations described in (11) occur when the received signal strength of the serving base station falls below the minimum acceptable RSS level or when the delay at femtocell backbone exceeds the maximum acceptable delay level. Since the probability of the occurrence of the event that the delay at stable femtocell backbone exceeds the maximum acceptable delay level for voice over IP communication is very low, most of the signal degradations in Figure 2 occur when wireless signal quality gets poor at femtocell and macrocell boundaries,

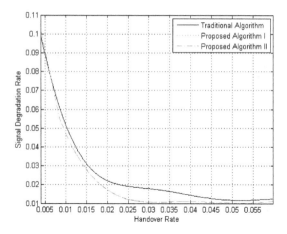

Fig. 2. Comparison of Handover Rates vs Degradation Rates of the Algorithms under Stable Delay

i.e., when the received signal strength of the serving base station falls below the minimum acceptable RSS level. This behavior of the proposed algorithms shows that the subscribers with clear channel and enough bandwidth on Internet broadband lines, can use femtocell architecture.

The tradeoff between the handover and signal degradation rates is achieved as shown in Figure 2. As hysteresis increases, the number of handovers is lowered and the number of signal degradations is raised as expected. The hysteresis parameter is introduced to act as an agent that blocks the ping-pong effect. If the hysteresis is kept too low, unnecessary handovers occur resulting in system signaling overload and undesired use of the system resources. If hysteresis is appointed too high, unnecessary handovers are not performed, however the number of signal degradations increases causing subscriber dissatisfaction and possible call drops. The RSS threshold parameter affects the performance of the handover algorithms by introducing a tradeoff as RSS threshold increases; the number of signal degradations decreases however the number of handovers raises. If the RSS threshold is chosen too high, the number of unnecessary handovers increases resulting in system signaling overload. If RSS threshold is picked too low, the number of signal degradations raises ending up user discontent on service quality.

The minimum value of delay threshold in the simulations is chosen as the recommended delay level for voice over IP communications [17]. The delay level of femtocell backhaul is inconsiderable to the minimum delay threshold value; therefore delay threshold parameter becomes irrelevant for subscribers with stable femtocell backbone. The cost of handover parameter c used in the second algorithm acts in similar way as hysteresis. The cost of handover parameter should be picked as high to decrease the number of handovers in the areas where system signaling is overloaded; whereas it should be chosen as low as possible in areas where the cost of call drops is high.

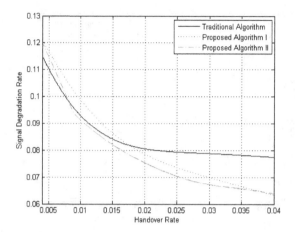

Fig. 3. Comparison of Handover Rates vs Degradation Rates of the Algorithms under Unstable Delay

5.2   Unstable Delay at Femtocell Backhaul

The unstable delay at femtocell backbone is modeled as an Erlang-n probability distribution function given in (9). The delay at femtocell backbone can be unstable, i.e., it can change rapidly during communication. The real-time communications suffer from rapid change in delay which creates jitter. The probability of the occurrence of the event that the delay at unstable femtocell backhaul exceeds the maximum acceptable delay level for voice over IP communication is given in (10). Compared to the stable delay case, this probability is significant and the effect of delay on number of signal degradation is higher. This can be observed in Figure 3.

Note that even though hysteresis-testing algorithm is comparable at low handover, high signal degradation region, for lower signal degradation rates its performance falls off drastically. This is due to the fact that hysteresis-testing algorithm does not consider delay which results in a high number of signal degradations. Between proposed algorithms, the second proposed algorithm performs better generally.

6   Conclusion

This paper has focused on handoff algorithm design in a two-tier macro-femto cellular environment. The tradeoff between performance criteria including call quality, the number of handovers has been studied for a traditional hysteresis-testing algorithm and two proposed algorithms. Two different setups are used in simulations based on the quality of the backbone of femtocells. The performance of the proposed algorithms has shown better results in terms of the average number of signal degradations at a fixed number of handovers especially in lower signal degradation regions. It is shown that by taking the quality of femtocell backhaul into consideration, an improvement can be achieved for macro-femto cellular network handover algorithms.

References

1. Zhang, J., Roche, G.: Femtocell: Technologies and Deployment. John Wiley & Sons, Inc., West Sussex (2010)
2. Cullen, J.: Radioframe Presentation. Femtocells Europe (2008)
3. Gudmundson, M.: Analysis of Handover Algorithms. In: Proceedings of Vehicular Technology Conference, pp. 537–542 (1991)
4. Vijayan, R., Holtzman, J.M.: A Model for Analyzing Handoff Algorithms. In: Proceedings of Vehicular Technology Conference, vol. 42, pp. 351–356 (1993)
5. Zhang, N., Holtzman, J.M.: Analysis of Handoff Algorithms Using Both Absolute and Relative Measurements. In: Proceedings of Vehicular Technology Conference, vol. 45, pp. 174–179 (1996)
6. Zonoozi, M., Dassanayake, P., Faulkner, M.: Optimum Hysteresis Level, Signal Averaging Time and Handover Delay. In: Proceedings of IEEE Vehicular Technology Conference, vol. 1, pp. 310–313 (1997)
7. Moghaddam, S., Vakili, V.T., Falahati, A.: New Handoff Initiation Algorithm (Optimum Combination of Hysteresis & Threshold Based Methods. In: IEEE Vehicular Technology Conference, pp. 1567–1574 (2000)
8. Kumar, P., Holtzman, J.M.: Analysis of Handoff Algorithms Using Both Bit Error Rate and Relative Signal Strength. In: International Conference on Universal Personal Communications, pp. 1–5 (1994)
9. Wang, S., Rajendran, A., Wylie-Green, M.: Adaptive Handoff Method Using Location Information. In: Proceedings of IEEE Personal, Indoor and Mobile Radio Communications, pp. D43–D47 (2001)
10. Itoh, K., Watanabe, S., Shih, J., Sato, T.: Performance of Handoff Algorithm Based on Distance and RSSI Measurements. IEEE Transactions on Vehicular Technology 51, 1460–1468 (2002)
11. Akar, M.: Integrated Power and Handoff Control for Next Generation Wireless Networks. Wireless Networks 15, 691–708 (2007)
12. Taleb, T., Ksentini, A.: QoS/QoE Predictions-based Admission Control for Femto Communications. In: Proceedings of IEEE International Conference of Communications, pp. 5146–5150 (2012)
13. Mase, K., Toyama, Y.: End-to-end Measurement Based Admission Control for VoIP Networks. In: Proceedings of IEEE International Conference on Communications, vol. 2, pp. 1194–1198 (2002)
14. Olariu, C., Fitzpatrick, J., Perry, P., Murphy, L.: A QoS Based Call Admission Control and Resource Allocation Mechanism for LTE Femtocell Deployment. In: Proceedings of Consumer Communications and Networking Conference, pp. 884–888 (2012)
15. Becvar, Z., Mach, P.: On Enhancement of Handover Decision in Femtocells. In: Proceedings of Wireless Days, vol. 2011, pp. 1–3 (2011)
16. Cole, R.G., Rosenbluth, J.H.: Voice over IP Performance Monitoring. ACM SIGCOMM Computer Communication Review 31, 9–24 (2001)
17. Cicconetti, C., de Blas, G.G., Masip, X., Silva, J.S., Santoro, G., Stea, G., Tarasiuk, H.: Simulation Model for End-to-End QoS across Heterogeneous Networks. In: Proceedings of 3rd IPS MoMe, pp. 79–89 (2005)

18. Bolot, J.: Characterizing End-to-End Packet Delay and Loss in the Internet. Journal of High Speed Networks 2, 305–323 (1993)
19. Paxson, V.: Measurements and Analysis of End-to-End Internet Dynamics. Ph.D. Thesis, University of California, Berkeley (1997)
20. Fei, A., Pei, G., Liu, R., Zhang, L.: Measurements on Delay and Hop-Count of the Internet. In: Proceedings of the IEEE Global Internet, pp. 1–8 (1998)
21. Hooghiemstra, G., Mieghem, P.V.: Delay Distributions on Fixed Internet Paths (2001), http://www.nas.ewi.tudelft.nl
22. Fiorini, P., Oboe, R.: A Design and Control Environment for Internet-Based Telerobotics. The International Journal of Robotics Research 17, 433–449 (1998)
23. Wee, S., Tan, W., Apostolopoulos, J., Etoh, M.: Optimized Video Streaming for Networks with Varying Delay. In: Proceedings of IEEE International Conference on Multimedia and Expo., vol. 2, pp. 89–92 (2002)
24. Gudmundson, M.: Correlation Model for Shadow Fading in Mobile Radio Systems. Electronics Letters 27, 2145–2146 (1991)
25. Baum, D.S., Hansen, J., Salo, J., Galdo, G.D., Milojevic, M., Kyösti, P.: An Interim Channel Model for Beyond-3G Systems. In: Proceedings of Vehicular Technology Conference, vol. 5, pp. 3132–3136 (2005)
26. Graziosi, F., Santucci, F.: A General Correlation Model for Shadow Fading in Mobile Radio Systems. IEEE Communication Letters 6, 102–104 (2002)
27. Khan, A., Constantinou, C., Stojmenovic, I.: Realistic Physical Layer Modelling for Georouting Protocols in Wireless Ad-Hoc and Sensor Networks. In: Proceedings of International Conference on Ultra Modern Telecommunications & Workshops, pp. 1–8 (2009)
28. Agrawal, P., Patwari, N.: Correlated Link Shadow Fading in Multi-hop Wireless Networks. IEEE Transactions on Wireless Communications 8, 4024–4036 (2009)
29. Akar, M., Mitra, U.: Variations on Optimal and Suboptimal Handoff Control for Wireless Communication Systems. IEEE Journal on Selected Areas in Communications, 1173–1185 (2001)
30. Akar, M., Mitra, U.: Soft Handoff Algorithms for CDMA cellular Networks. IEEE Transactions on Wireless Communications, 1259–1274 (2003)

Network Resource Control System for HPC Based on SDN

Petr Polezhaev[1], Alexander Shukhman[1], and Yuri Ushakov[2]

[1] Department of Mathematics, Orenburg State University, Orenburg, Russia
newblackpit@mail.ru, shukhman@gmail.com
[2] Information Technology Center, Orenburg State University, Orenburg, Russia
ushakov@unpk.osu.ru

Abstract. This article describes network resource control system for HPC based on software defined network (SDN). For job scheduling this system uses Back-Fill algorithm modifications with job assignment algorithms Summed Distance Minimization and Maximum Distance Minimization we have developed. We also offer data flow control methods for SDN in high-performance systems such as reactive and proactive algorithms. Our experiment has shown that BackFill scheduling algorithm in combination with Summed Distance Minimization and the proactive routing algorithm demonstrates a significant decrease in execution time for the reference communication-intensive job flow on a cluster.

Keywords: software defined networks, resource control, computing job scheduling, data flow control.

1 Introduction

At present there is a need for time-consuming task solutions in different fields of science, technology and economics, e.g. climate change modeling, construction of virtual test stands, search for optimal chemical structures. Such tasks can be solved with the use of HPC on clusters as well as grid systems.

Network resource control systems (NRCS) for HPC manage computing and network resources shared by multiple users. These systems are responsible for scheduling, starting, performance monitoring and storing the data of computing jobs.

The existing systems are not efficient enough. As a rule, they use job scheduling algorithms combining BackFill [1] with one of simple assignment methods such as First Fit or Best Fit [2,3]. On the average NRCSs provide computing resource load at 60-80%.

There is a possibility to improve job scheduling algorithm efficiency through the use of communication patterns for computing jobs and data transfer routes. In fact, if the communication pattern for the computing job is known alongside with the network actual state (its topology and delays on network ports), then the NRCS can assign jobs on computing nodes with maximal efficiency. It allows to localize network traffic so achieving less network contention.

S. Balandin et al. (Eds.): NEW2AN/ruSMART 2014, LNCS 8638, pp. 219–230, 2014.

SDNs provide additional possibility to improve NRCS job scheduling algorithms [4]. Before the job is started the NRCS can determine the data transfer routes between each pair of job processes aimed at less network contention. This approach results in shorter execution time of network operations which leads to a significant decrease in execution time for communication-intensive jobs and an increase of computing resource load in high-performance system (HPS).

The SDN concept was first brought up at the scientific laboratories in Stanford and Berkley and now is being developed within the Open Network Foundation and the European project OFELIA. The SDN approach is based on the possibility to dynamically control the network data flows using the OpenFlow protocol [5]. All of the network switches supporting OpenFlow are controlled by the OpenFlow controller that provides an access to network control for the applications.

At present there are no NRCS based on SDN for HPC aimed at the same time to control the data flows between job processes and to localize their assignment. The above makes our work innovative.

The idea to transmit the information on job communication patterns to the network level with the aim to optimize the transmission of data flow has been reflected in papers [6,7]. Their main disadvantages are orientation to specific network operations (for example, shuffle of Hadoop library or broadcast), which are implemented through modification of distributed computing frameworks or through creation of original data transmission protocols. The routing approaches suggested in our paper consider the communication patterns of arbitrary network operations.

Papers [8,9] describe application of SDN for optimization of network operations specific for MapReduce. In the papers [10,11] SDN is suggested for data flow control of arbitrary network operations, but this idea has not been implemented.

The rest of this paper is organized as follows. Section 2 suggests a structural model of HPS through weighted oriented multigraph and a computing job model considering its requirements for computational resources. Section 3 describes the suggested job scheduling algorithms which are combinations of Backfill and the developed job assignment methods (Summed Distance Minimization and Maximum Distance Minimization). Section 4 describes the suggested reactive and proactive network traffic routing approaches based on SDN. Section 5 presents the experimental HPS and the efficiency metrics for NRCS used for research of the suggested solutions. We present the NRCS efficiency experiment results in Section 6 and the conclusion in Section 7.

2   Structural Model of High-Performance System

Structural model of HPS can be described through multigraph:

$$DataCenter = (Devices, Links, Flows(t)). \qquad (1)$$

Its vertices are elements of a set:

$$Devices = Nodes \cup Switches \cup OFSwitches \cup Controllers, \qquad (2)$$

where $\text{Nodes} = \{\text{Node}_1, ..., \text{Node}_n\}$ means   a   set   of   computing   nodes, $\text{Switches} = \{\text{Switch}_i\}$ stands   for   the   switches   not   supporting   OpenFlow, $\text{OFSwitches} = \{\text{OFSwitch}_i\}$ stands   for   the   OpenFlow   switches, $\text{Controllers} = \{\text{Controller}_i\}$ are   correspondingly   the   OpenFlow   controllers.   The multigraph   edges $\text{Links} = \{(p_i, p_j)\}$ are   bidirectional   links   between   the   network ports; several parallel links are allowable between the same devices through different pairs of ports. $\text{Flows}(t) = \{\text{Flow}_i(t)\}$ is the data flows at a certain time t. Network state is described by this multigraph and weights of its vertices.

Each computing node Node_i has the following parameters and dynamic characteristics:

$$\text{Node}_i = (M_i, D_i, C_i, S_i, P_i^{\text{node}}, m_i(t), d_i(t), u_i(t), s_i^{\text{node}}(t)), \tag{3}$$

with $M_i \in N$ and $D_i \in N$ being respectively its RAM size (Mb) and disk capacity (Mb); $C_i \in N$ is the number of its computing cores, $S_i \in R^+$ is the relative performance of the cores; $P_i^{\text{node}} = \{p_{ij}^{\text{node}}\}$ is the network port set; $m_i(t) \in [0,1]$ and $d_i(t) \in [0,1]$ are RAM and disk relative loads for a computing node at time t; $u_i(t) = (u_{i1}(t), ..., u_{iC_i}(t))$ ($u_{ij}(t) \in [0,1]$) is the load vector for each computing node core at time t; $s_i^{\text{node}}(t) \in \{\text{"online", "offline"}\}$ is the node state at time t.

Computing job J_j from NRCS queue has the following parameters:

$$J_j = (cr_j, mr_j, dr_j, \tilde{t}_j, CP_j), \tag{4}$$

where $cr_j \in N$ is the number of required cores, $mr_j \in N$ and $dr_j \in N$ are correspondingly the RAM size (kb) and the disk capacity (kb) required for each process of the job, $\tilde{t}_j \in N$ is the user estimation for the job execution time on a node with the relative performance of cores equal to one, CP_j is the communication pattern of the job. Generally, a computing job J_j is a parallel non-interactive program using MPI, PVM, DVM or OpenMP. In simplest case the job can be single-process nonparallel program.

The communication pattern of the job can be described through the following weighted oriented graph:

$$CP_j = (\Pi_j, \lambda_j), \tag{5}$$

where $\Pi_j = \{\pi_{j1}, \pi_{j2}, ..., \pi_{jcr_j}\}$ is the set of job processes, $\lambda_j : \Pi_j \times \Pi_j \to R_+ \cup \{0\}$ is the function determining the weight for each arc equal to its communicational intensity.

The above described model allows to reflect the detail structure of modern HPS supporting SDN.

3   Job Scheduling Methods

NRCS scheduler builds the job J_j start schedule using the dynamic job scheduling algorithm, which schedules jobs as they arrive in the queue Q_{jobs}. This algorithm includes two sub-algorithms:

1. the job selection algorithm selects a next job J_j from the queue Q_{jobs} to assign it to computing resources, based on the available core information;
2. the core assignment algorithm for the selected job J_j determines the most suitable cores within the system.

Starting each scheduling cycle the job selection algorithm forms the scheduling window list WL at time t based on the current schedule Schedule and the queue of waiting jobs Q_{jobs}. Then the cycle is performed, each iteration of which implies the next job selection from Q_{jobs} based on certain criteria. The list of suitable windows WL′ is formed based on WL for the next selected job J and this list is then sent to the core assignment algorithm. The core assignment algorithm determines launch window R for job J, then R is used to update WL and Schedule.

The algorithm work results in the updated schedule Schedule and the selected job list U . The resulting schedule after all the jobs from the queue Q_{jobs} have been scheduled can be presented as a set of ordered pairs $Schedule = \{(J_j, W_j)\}_j$, where W_j is the launch window for the job J_j.

Let $dist(d_1, d_2)$ be the topological distance between the network devices $d_1 \in$ Devices and $d_2 \in$ Devices in high-performance system. As a rule, the topological distance is estimated as the total delay of packet transmission time, the current network state considered.

We suggest using the combination of the Backfill and the developed assignment algorithms Maximum Distance Minimization (MDM) or Summed Distance Minimization (SDM) as the job scheduling algorithms.

According to Backfill the jobs of the queue are ordered by priority. In each scheduling cycle the jobs are assigned to resources according to their priorities. The detailed algorithm description can be found in [1, 2].

Backfill has two conflicting goals 1) to improve the efficiency of computing resource load through filling empty windows in the schedule, 2) to avoid the job starvation due to reservations.

We described the suggested SDM algorithm as follows.

Algorithm 1. "Summed Distance Minimization".

Input parameters: J_j is the assigned computing job, WL' is the current list of scheduling windows suitable for J_j.

Output parameter: R is the launch window assigned for this job.

Step 1. Use the Floyd-Warshall algorithm to calculate the topological distances $dist_{cloud}(d_1, d_2)$ between all the pairs of network devices $d_1, d_2 \in$ Devices within HPS.

Step 2. Select the next suitable scheduling window W from WL'.

Step 3. Form the list L of the computing nodes, whose cores are included in scheduling window W and meet the resource requirements of computing job J_j. Each node is included in this list so many times as its cores are found in W.

Step 4. Select the next network device $d \in$ Devices and perform the following:

a) sort all the computing nodes $Node_i \in L$ in ascending order by the value of $dist(d, Node_i)$, if the distances are equal then the nodes are ordered by efficiency;

b) assign the first cr_j of computing nodes for job J_j, save their cores to launch window R_{Wd}.

c) set starting time for launch window R_{Wd} as starting time of scheduling window W and finishing time for R_{Wd} as starting time summed with $\tilde{t}_j$;

d) calculate the value of integral criterion I_{SDM}^{Wd} as the total of pairwise distances between the nodes containing the launch window R_{Wd} cores assigned for job J_j;

e) go to the next network device $d \in$ Devices.

Step 5. Go to the next scheduling window W.

Step 6. The result R of the algorithm is launch window R_{Wd}^{min}, yielding the minimum of integral criterion I_{SDM}^{Wd}:

$$I_{SDM}^{min} = \min_{\substack{W \in WL' \\ d \in Devices}} I_{SDM}^{Wd}. \tag{6}$$

This assignment algorithm generalizes the Manhattan Median method for random topologies [12] and topological distances between adjacent devices not equal to one.

MDM is different from SDM only in the calculation method for integral criterion I_{MDM}^{Wd} that is maximum distance of the pairwise distances between the nodes including the cores assigned for the job. Also MDM is analog of MC1x1 for random topologies [13].

The both developed methods provide significantly shorter job execution times and increase the whole system load due to localized assignment of computing job processes.

4   Network Traffic Routing Approaches

SDN has a limited scaling implying the quantity limits of flow rules at different software and hardware switches. The upper limit of flow rules approached, the delay becomes longer which leads to packet loss and QoS parameter violation. Flow rules can be used for a separate flow as well as for a group of flows which decreases the number of rules.

The reactive method for creating the rules in flow tables implies that the controller has no beforehand information on transmitted data flows. If incoming flow lacks a rule in the flow table then the controller creates corresponding rules and sends them to the switch. The advantage of this approach is its flexibility; its drawback is the additional delay during processing of each flow's first packet. Besides a rule retains its position in the flow table long enough after the suitable packets have been delivered.

The proactive method implies that the controller has beforehand information on the data flows scheduled for transmission and adds corresponding rules to flow tables before transmission starts. The advantage of this approach is processing of each flow's first packet without delay.

When a controller is requested to provide QoS parameters for certain flows it must create a set of proactive rules for the distribution of all network flows into the switch port queues. Those rules provide the minimal guaranteed bandwidth for each flow group. The flows without QoS requirements are routed by different routes or sent to the lowest priority queue. To calculate optimal route for each flow we use OSPF for residual bandwidth based on the statistics on flows and links.

5   Experimental HPS

For the experiment we have used the cluster system (Fig. 1) implemented at Orenburg State University. It includes 4 OpenFlow switches (2 HP 3500yl, 2 Netgear GSM7200), 8 computing nodes (32Gb RAM, 4 cores), 1 server (32Gb RAM, 8 cores) with OpenFlow controller and 1 server (32Gb RAM, 4 cores) with Torque.

NRCS was using Torque with implemented additional modules such as BackFill combined with SDM, MDM and the standard algorithm First Fit (FF).

For SDN we have used OpenFlow controller NOX [14]. We have developed the NOX application controlling the data flows for computing jobs according to the above proactive and reactive routing methods. This application receives the information on the communication pattern of each scheduled job and on the computing nodes assigned for the job processes. Before the job is started, this information is used by the proactive method routing the data between the nodes assigned for the job.

The principal parameter for a computing job reflecting network load during the job execution is the ratio of the network transmission time to the total job execution time. The value of this parameter affects directly the efficiency of our job scheduling algorithms and routing methods. The higher this value the more job execution time depends on the network interprocess communication time. Compact and localized process assignment by scheduling algorithm and the lower network contention through optimal route distribution by the routing method lead to a shorter execution time.

For experiment the benchmark tool CBench [15] was used due to its adequate set of computing jobs and suitable job management system Torque.

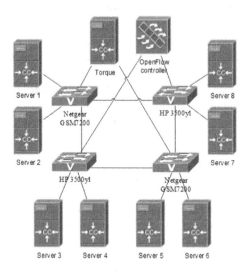

Fig. 1. Experimental HPS

All computing jobs can be classified into: computation-intensive (Table 1) and communication-intensive (Table 2).

Table 1. Computation-intensive jobs

Number	Job name	Job purpose
1	NAS Parallel Benchmark. Embarrassing Parallel	Generation of independent random numbers with normal distribution
2	NAS Parallel Benchmark. Fast-Fourier Transform	Solution of three-dimensional partial differential equations using Fast Fourier Transformation
3	NAS Parallel Benchmark. Lower-Upper Gauss-Seidel solver	Solution of linear equation systems using LU decomposition
4	High Performance Linpack	Benchmark for performance measure of HPS
5	CBench IRS	Solution of equation of radiation spread
6	Integral computing	Calculation of triple integral

All experiment stages have been carried out using the two reference job flows, each of them including jobs from only one of the above classes. These jobs solve the most frequently found problems pertaining to HPC.

Table 2. Communication-intensive jobs

Number	Job name	Job purpose
1	NAS Parallel Benchmark. MultiGrid benchmark	Approximate solution of three-dimensional Poisson equation Du=f on grid NxNxN with periodic boundary conditions
2	NAS Parallel Benchmark. Conjugate Gradient	Calculation of minimal eigenvalue for positive-definite matrix using Conjugate Gradients method
3	NAS Parallel Benchmark. Integer Sorting	Parallel integer sorting
4	CBench rotate	Synthetic test of CBench for network bandwidth
5	CBench AMG	Multi-grid algorithm for linear system solutions
6	QSS	Queuing system simulation

Each class of jobs uses a respective reference job flow including the jobs from tables 1 or 2 respectively repeated 10 times in the same order. The benchmark CBench is used to automatically send jobs to Torque at certain time intervals so simulating user behavior in HPS.

The existing researches on job flows in HPS [16,17,18] show the exponential distribution of time intervals between the sent jobs with a certain rate parameter λ. In our research $\lambda = 0.0003$ was chosen as the most adequate for an average job flow.

For efficiency evaluation of NRCS we have developed 3 metrics:

1. ΔT - the reference job flow execution time. It is determined as the difference between all jobs finish time T_1 and jobs flow starting time T_0 :

$$\Delta T = T_1 - T_0. \tag{7}$$

2. $\overline{U}$ - the average load (%) of HPS computational cores. It is determined as the mean value of average loads $\overline{U}_{ij}$ for each computational core of each node in HPS:

$$\overline{U} = \frac{\sum_{i=1}^{n} \sum_{j=1}^{C_i} \overline{U}_{ij}}{\sum_{i=1}^{n} C_i}. \tag{8}$$

3. σ - the load balance of computational cores. It can be calculated:

$$\sigma = \sqrt{\frac{\sum_{i=1}^{n} (\overline{U}_i - \overline{U})^2}{n-1}}, \tag{9}$$

where $\overline{U}_i$ is the average load (%) of the i-th computing node. This metric allows to evaluate how even computational resource load is. The resource load in HPS is highly balanced if $\sigma \leq 10\%$.

6   NRCS Efficiency Experiment Results

The NRCS efficiency experiment included two stages:

1. Efficiency evaluation for the developed scheduling algorithms compared to the standard Backfill and First Fit combination without using the SDN options;
2. Study of different algorithm combinations: the first stage most effective scheduling algorithm plus one of the routing methods (standard, reactive SDN and proactive SDN).

The first stage results are presented in Table 3. In the following tables we use abbreviations: Cmp is computation-intensive, Com is communication-intensive, S is standard, R is reactive SDN, P is proactive SDN. In addition, we evaluated Backfill combined with Best Fit, but its results were excluded due to their intermediate position between Backfill FF and Backfill MDM.

Table 3. Efficiency evaluation for the developed scheduling algorithms

Scheduling algorithms	Backfill FF		Backfill SDM		Backfill MDM	
Routing method	S	S	S	S	S	S
Reference job flow	Cmp	Com	Cmp	Com	Cmp	Com
Reference job flow execution time ΔT, s	19885	18748	19531	17630	19641	18077
Average load of HPS computational cores $\overline{U}$, %	87.3	71.6	87.5	75.4	87.4	74.1
Load balance of computational cores σ , %	10	11.1	8.1	5.7	8.5	6.1
Relative execution time reduction for the reference job flow compared to Backfill FF, %	0	0	1.8	6.0	1.2	3.6

Standard algorithm Backfill FF shows 19885 s execution time for the computation-intensive job flow and 18748 s execution time for the communication-intensive job flow. These values are the maximal of all the scheduling algorithms here.

The maximal relative execution time reduction compared to Backfill FF has been shown by Backfill SDM: 1.8% for the reference computation- intensive job flow and 6.0% for the reference communication-intensive job flow respectively. The reduction of execution time is more for communication-intensive jobs than for computation-intensive jobs which is due to the shorter time of transmissions between processes assigned to the localized computing nodes. The Backfill SDM also provides the best core load balance with excellent average load values. So this algorithm has been chosen as the most efficient for the second stage of our research.

The second stage results are presented in Table 4.

Table 4. Efficiency evaluation for the developed scheduling algorithms

Scheduling algorithms	Backfill SDM					
Routing method	S	S	R	R	P	P
Reference job flow	Cmp	Com	Cmp	Com	Cmp	Com
Reference job flow execution time ΔT, s	19531	17630	19583	17786	19457	17373
Average load of HPS computational cores $\overline{U}$, %	87.5	75.4	87.5	74.1	87.7	77.2
Load balance of computational cores σ, %	8.1	5.7	7.9	5.8	7.9	5.6
Relative execution time reduction for the reference job flow compared to Standard routing method, %	0	0	-0.3	-0.9	0.4	1.5

The standard routing method yielded a 19531 s time for the reference job flow (computation-intensive) and 17630 s (communication-intensive). The reactive method data were worse with respective 19583 s and 17786 s. The reason is the route setting delay processing the first packet during the job execution. The most efficient is the proactive method considering the communication patterns of the jobs and routing the interprocess data flows before the job has started. The proactive method also yields the best average load values and the best load balance of the computational cores.

According to the experiment results the Backfill SDM combined with the proactive routing method against the Backfill FF combined with the standard routing method shows 2.2% and 7.3% relative time reductions for computation-intensive and communication-intensive reference job flows respectively. Thus the Backfill SDM combined with the proactive routing method provides a significant increase in NRCS performance for communication-intensive jobs and a minor one for computation-intensive jobs on HPS using SDN.

7   Conclusions

Our research has resulted in the structural model of a high-performance system describing it through a weighted oriented multigraph. The computing job model has been developed considering the requirements for computational resources. These models reflect all the basic characteristics of modern HPS as well as the use of SDN. Job scheduling algorithms have been suggested combining Backfill with our SDM and MDM assignment methods. For data flow routing in SDNs we have developed the reactive and proactive methods.

The experiment was carried out on the cluster system of Orenburg State University. The Backfill SDM combined with proactive routing method has proved the most efficient for communication-intensive jobs in HPS using SDN. We plan a more detailed research for these algorithms on computing clusters using SDN with different job flows.

Acknowledgments. This work is supported by Russian Foundation for Basic Research (grants 13-07-97046 and 14-07-97034).

References

1. Feitelson, D., Weil, A.: Utilization and predictability in scheduling the IBM SP2 with backfilling. In: Proceedings of the First Merged International Parallel Processing Symposium and Symposium on Parallel and Distributed Processing, pp. 542–546 (1998)
2. Kovalenko, V.N., Semyachkin, D.A.: Using BackFill in grid-systems. In: Proceedings of the International Conference on Distributing Computing and Grid-technologies in Science and Education, pp. 139–144 (2004)
3. Polezhaev, P.N.: Simulator of cluster and its management system used for research of job scheduling algorithms. Bulletin of SUSU – Mathematical Modeling and Programming 35(6,211), 79–90 (2010)
4. McKeown, N., Anderson, T., Balakrishnan, H., Parulkar, G., Peterson, L., Rexford, J., Shenker, S., Turner, J.: Openflow: enabling innovation in campus networks. ACM SIGCOMM Computer Communication Review 38, 69–74 (2008)
5. Intro to OpenFlow, `https://www.opennetworking.org/standards/intro-to-openflow`
6. Chowdhury, M.: Managing data transfers in computer clusters with Orchestra. In: Proceedings of the ACM SIGCOMM 2011, pp. 98–109 (2011)
7. Hong, C.-Y., Caesar, M., Godfrey, P.B.: Finishing flows quickly with preemptive scheduling. ACM SIGCOMM Computer Communication Review, Special October Issue SIGCOMM 2012 42(4), 127–138 (2012)
8. Narayan, S., Bailey, S., Daga, A.: Hadoop acceleration in an OpenFlow-based cluster. In: Proceedings of 2012 SC Companion: High Performance Computing, Networking Storage and Analysis, pp. 535–538 (2012)
9. Narayan, S., Bailey, S., Daga, A., et al.: Openflow enabled Hadoop over local and wide area clusters. In: Proceedings of High Performance Computing, Networking, Storage and Analysis, pp. 1625–1628 (2012)
10. Chowdhury, M., Stoica, I.: Coflow: a networking abstraction for cluster applications. In: Proceedings of the 11th ACM Workshop on Hot Topics in Networks, pp. 31–36 (2012)
11. Chowdhury, M., Stoica, I.: Coflow: an application layer abstraction for cluster networking. Technical report EECS-2012-184 (2012)
12. Bender, M.A., Bunde, D.P., Fekete, S.P., Leung, V.J., Meijer, H., Phillips, C.A.: Communication-aware processor allocation for supercomputers: finding point sets of small average distance. Algorithmica 50(2), 279–298 (2008)
13. Mache, J., Lo, V., Windisch, K.: Minimizing message-passing contention in fragmentation-free processor allocation (2012), `http://citeseerx.ist.psu.edu/viewdoc/download?doi=10.1.1.55.390&rep=rep1&type=pdf`

14. NOX-Classic wiki, `https://github.com/noxrepo/nox-classic/wiki`
15. Cbench - Scalable Cluster Benchmarking and Testing,
    `http://sourceforge.net/apps/trac/cbench/`
16. Aida, K., Kasahara, H., Narita, S.: Job Scheduling Scheme for Pure Space Sharing among
    Rigid Jobs. In: Feitelson, D.G., Rudolph, L. (eds.) JSSPP 1998. LNCS, vol. 1459, pp. 98–
    121. Springer, Heidelberg (1998)
17. Feitelson, G.: Workload Modeling for Computer Systems Performance Evaluation. Work-
    load modeling book draft, Ver. 0.37, 508 p. (2012)
18. Lublin, U., Feitelson, G.: The workload on parallel supercomputers: modeling the charac-
    teristics of rigid job. Journal of Parallel and Distributed Computing 63(11), 542–546
    (2003)

Towards End-User Driven Power Saving Control in Android Devices

Vitor Bernardo[1], Bruno Correia[1], Marilia Curado[1], and Torsten Ingo Braun[2]

[1] Center for Informatics and Systems, University of Coimbra, Coimbra, Portugal
{vmbern,bcorreia,marilia}@dei.uc.pt
[2] Institute for Computer Science and Applied Mathematics,
University of Bern, Bern, Switzerland
braun@iam.unibe.ch

Abstract. During the last decade mobile communications increasingly became part of people's daily routine. Such usage raises new challenges regarding devices' battery lifetime management when using most popular wireless access technologies, such as IEEE 802.11. This paper investigates the energy/delay trade-off of using an end-user driven power saving approach, when compared with the standard IEEE 802.11 power saving algorithms. The assessment was conducted in a real testbed using an Android mobile phone and high-precision energy measurement hardware. The results show clear energy benefits of employing user-driven power saving techniques, when compared with other standard approaches.

Keywords: Energy Efficiency, Power Saving, IEEE 802.11, Android, Testbed.

1 Introduction

Nowadays, wirelessly connected mobile devices are present almost everywhere at any time. Apart from other available wireless technologies, IEEE 802.11 [1] seems to be the *de-facto* standard for wireless communications, being supported in millions of devices. Although several mobile devices have Internet access through mobile operator networks, performance limitations on the support of highly demanding multimedia applications enabled novel and hybrid communication paradigms (e.g. offloading) where IEEE 802.11 plays an important role [2].

In this context, energy consumption issues of battery-supported devices need to be addressed. In particular, since the Android [3] platform is responsible for a large part of the mobile device market growth, IEEE 802.11 energy management mechanisms in this platform should be carefully investigated.

This work studies and compares, using a real testbed, the most popular IEEE 802.11 power saving techniques implemented on the Android platform. Additionally, an Android framework for Extending Power Saving control to End-users (EXPoSE) is proposed, aiming at improving the devices' energy efficiency by considering end-users demands.

The rest of this paper is structured as follows: Section 2 introduces the technology background and discusses the most significant related work. An overview

S. Balandin et al. (Eds.): NEW2AN/ruSMART 2014, LNCS 8638, pp. 231–244, 2014.

of the Android platform architecture, followed by the presentation of the EX-PoSE framework is given in Section 3. Section 4 describes the evaluation testbed and discusses the obtained results. Finally, Section 5 presents the conclusions.

2 Related Work

This section presents the background concerning the standard power saving mechanisms of the IEEE 802.11 standard, followed by the discussion of the most relevant literature concerning IEEE 802.11 energy efficiency mechanisms in mobile devices.

The main goal of mobile device energy management is to keep as long as possible the network interface in a low energy consumption state, usually called sleep mode. Unlike in awake mode, a mobile device in sleep mode cannot receive or transmit data. The most popular power saving mechanism for IEEE 802.11 network interfaces is the Power Save Mode (PSM) [1], usually referred to as Legacy-PSM. When operating in Legacy-PSM, all the transmitted data to a certain device in sleep mode is queued at the Access Point (AP). Later, after being notified via *Beacon* messages (usually sent every 100 ms) the device must wake-up to perform pending data polling (sending a *PS-Poll* message for each pending frame). As this mechanism is usually associated with higher delays [4], the last generation of mobile devices addressed the problem by implementing an adaptive mechanism to switch faster between awake and sleep modes, commonly named Adaptive-PSM. In Android, the Adaptive-PSM implementation switches between awake and sleep modes depending on the network traffic, allowing the IEEE 802.11 interface to stay awake only when there is traffic.

The Adaptive-PSM implementation in Android devices does not consider traffic type and importance when switching between awake and sleep modes, leading to several unnecessary switches to awake mode. Trying to overcome this limitation, Pyles et al. [5] proposed the Smart Adaptive PSM (SAPSM). SAPSM's main goal is to avoid that low priority applications switch the interface to awake, which results in energy savings for this type of applications, while the high priority ones still have good performance. The application priority is given based on the statistical information collected in the device using the proposed Application Priority Manager service. Although the authors argue that such approach might have benefits for non-technical users, it does not allow the application or the end-user to fully control the decision process. Furthermore, if a continuous media application is classified as high priority, the SAPSM mechanism will not be able to go back into sleep mode.

An extension to the common Android's Adaptive-PSM, aiming at improving the VoIP energy efficiency, was proposed by Pyles et al. The silence prediction based WiFi energy adaptation mechanism [6], SiFi, manages the device energy states according to the VoIP application characteristics. The proposed mechanism predicts the silence periods in the VoIP call and uses this information to keep the wireless interface in sleep mode for a longer time. The results show 40% of energy savings, while keeping high voice quality. However, this approach is limited to VoIP applications, and the energy savings are closely related with the existence of silence periods.

A framework to reduce energy consumption of video streaming has been proposed by Csernai and Guly [7], aiming to dynamically adjust the awake interval according to the estimated video quality. The proposed mechanism was implemented in Android, but no accurate energy consumption study on the mobile equipment has been performed. An optimized power save algorithm for continuous media applications (OPAMA) was proposed by Bernardo et al. [8]. Although the OPAMA algorithm takes the end-user feedback into the process, it is limited by the AP configurations. The results show that OPAMA can save up to 44% of energy when compared with Legacy-PSM, but no real testbed validations have been performed. Korhonen and Wang [9] also proposed a mechanism to reduce energy consumption of multimedia streaming for UDP based applications. The mechanism dynamically adapts the burst intervals by analyzing network congestion. Despite of energy savings by sending the data into bursts, this proposal does not allow that the power control is driven by the end-users.

Ding et al. [10] introduce Percy, a mechanism that aims at reducing the energy consumption while keeping a low transfer delay for Web 2.0 flows. To achieve this goal a local proxy behind the AP was implemented. Energy savings are possible, since the device can go back into sleep mode when the proxy is queuing data to it. The assessment conducted in a mobile testbed shows energy savings between 44% and 67%. Nevertheless, as in other proxy-based approaches (e.g., [11]) the deployment of the solution is hard, since it requires changes in the access points. Additionally, these approaches do not consider end-users feedback.

To the best of our knowledge, this work advances the current state of the art by specifying an Android-based framework, which allows power saving mechanisms to be controlled by end-users. Such tool enables the possibility of including end-users and/or application preferences within the IEEE 802.11 interface power saving management.

3   EXPoSE: An Android Framework for Extending Power Saving Control to End-Users

This section discusses the Android platform internal design in Section 3.1, followed by the presentation of the EXPoSE framework in Section 3.2.

3.1   Android Overview and Motivation

Android [3] is an open source operating system (OS), based on Linux, which is being widely used in mobile devices. As Android OS is mainly used in mobile devices, the battery management is a critical issue to be addressed. Recent studies [12] have shown that wireless interfaces of mobile devices represent a non-negligible part of the total energy consumed. Therefore, aiming at saving energy, it is important to perform a proper management of the Android IEEE 802.11 interface. Figure 1 illustrates the Android IEEE 802.11 architecture.

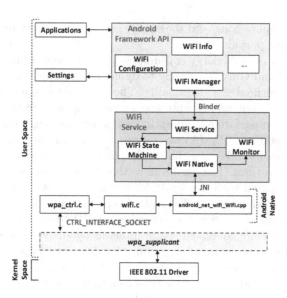

Fig. 1. Architecture of IEEE 802.11 in Android. (Based on [13])

To manage the configurations of the IEEE 802.11 interface (WiFi) applications must interact with the *"WiFi Manager"*, which handles the communication with *"WiFi Service"*. *"WiFi Service"* controls WiFi related communication between end-user and kernel spaces, being responsible for managing (*"WiFi State Machine"*) and translating (*"WiFi Native"*) all the requests. *"WiFi Native"* interfaces with the WiFi library, available as Android native code, through Java Native Interface (JNI). Finally, all the lower level calls to the IEEE 802.11 driver are performed through *"wpa_supplicant"*.

Although the IEEE 802.11 architecture of Android is clear and well defined, it does not expose any IEEE 802.11 sleep related feature in the "Application Framework API" nor in the "WiFi Service". Therefore, the defined architecture does not allow the management of the IEEE 802.11 sleep functions at higher-layers (e.g., application). This paper addresses this issue by proposing enhancements to the current IEEE 802.11 architecture in Android, allowing power saving to be controlled by end-users, as described in the next section.

3.2   EXPoSE Design

This section presents the Android framework for Extending Power Saving control to End-users (EXPoSE) design.

Figure 2 illustrates the architecture of the EXPoSE framework. The EXPoSE framework was implemented as an Android service (EXPoSE Service), plus a lower level control module included in the Android kernel. This module allows the IEEE 802.11 power saving functions to be exposed to higher level layers, enabling better control of power states.

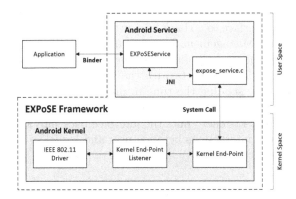

Fig. 2. EXPoSE framework architecture

To enable generic communication with the IEEE 802.11 driver, the developed Android kernel module is composed of two distinct components: the "Kernel End-Point" and the "Kernel End-Point Listener". The communication between the kernel end-point and the driver is performed through the proposed kernel end-point listener. Such abstraction plays an important role concerning energy efficiency, since, although the listener is always active, it is waiting in a semaphore and does not perform any additional processing (with extra energy costs).

Concerning the communication with the applications, the EXPoSE service can be configured in two distinct ways:

– **Pattern-based:** allows the application to configure the awake/sleep pattern over time. For instance, an application can specify that it must be awake for a certain period, α milliseconds, and it must be in sleep mode for a period of β milliseconds;
– **Maximum Allow Delay (MAD) definition:** enables the application to indicate that a maximum delay of γ milliseconds will be allowed. The control of the awake/sleep pattern over time will be performed by the EXPoSE service, taking into account the delay bound restriction.

To use the pattern-based approach, an application should indicate, at least, three distinct values. The first value is a flag to indicate if the specified pattern should be repeated over time. Such flag should be followed by two parameters, α and β, respectively, the awake and sleep periods in milliseconds. When using the MAD approach the application should only indicate a single value, γ, representing the maximum allowed delay in milliseconds.

As default Android Adaptive-PSM, the EXPoSE service also performs regular switching between sleep and awake modes. However, unlike Adaptive-PSM, the power modes switches are not based on the traffic load, but rather on application or end-users requirements. To change the IEEE 802.11 network interface to sleep mode for a certain period, EXPoSE changes the power mode on the IEEE 802.11 driver, forcing a NULL data frame with the Power Management flag enabled to be sent to the AP. Such action informs the AP that incoming data for that station should be queued, as in Legacy-PSM. Once the defined sleep period expires, the EXPoSE forces the interface to go back into awake mode and a NULL data frame to the AP with Power Management flag set to 0 is sent, allowing the queued data to be transmitted without any polling message.

The "EXPoSE service" interacts with the IEEE 802.11 driver through the "Kernel End-Point" by sending the time in milliseconds that the IEEE 802.11 network interface must be in sleep mode. Once the configured time expires, the "Kernel End-Point Listener" puts the interface back into awake mode. This scheme minimizes the interactions between user and kernel spaces, leading to a higher system level performance.

4   Experimental Evaluation

This section describes the experimental assessment performed in the testbed. Section 4.1 describes the evaluation goals, followed by the testbed presentation in Section 4.2. Finally, in Section 4.3, the results obtained in the testbed are presented and discussed.

4.1   Objectives

The experimental evaluation has two main goals. First, it aims to study the standard IEEE 802.11 power saving schemes, implemented in Android, in the presence of continuous media applications. The study includes the analysis of both energy efficiency and network-level performance metrics (e.g. delay) and the assessment of different application design options (e.g. packet size).

The second goal is to evaluate the EXPoSE approach effectiveness and to compare its performance with standard mechanisms.

4.2   Testbed Setup

This subsection presents the IEEE 802.11 testbed and the energy measurement methodology to assess energy consumption of mobile phones.

Figure 3 illustrates the IEEE 802.11 and energy measurement testbed. Figure 3a depicts the IEEE 802.11 architecture, and the energy measurement components are detailed in Figure 3b.

The "Mobile Node" used in this setup was a LG P990 mobile phone, running Android 4.2.2 and the "Access Point" was a Cisco Linksys E4200. The machines in the *Core Network* ("Server", "NTP Server", and "DNS+DHCP" entities) were virtualized and run in a HP ProLiant DL320 G5p server. Besides the "Mobile Node", all the other machines were running Debian 7.0. All the traffic generated in the following tests has "Server" as source and "Mobile Node" as destination. Traffic generation was performed using the D-ITG version 2.8.1 [14].

The mobile phone energy consumption assessment was addressed by extending a high-precision methodology [15] to support mobile devices. Figure 3b depicts the employed energy assessment testbed. The "Digital Multimeter" is a Rigol

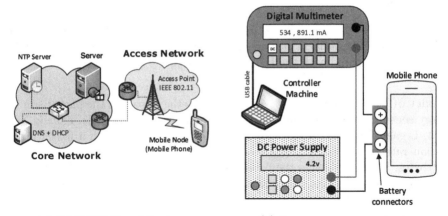

(a) IEEE 802.11 architecture.   (b) Energy measurement setup.

Fig. 3. IEEE 802.11 and energy measurement testbed

DM3061 unit, and supports up to 50.000 samples per second. This high-precision tool ensures the correct measurement of mobile phone energy consumption, since it will be possible to measure all the slight energy variations. The multimeter is managed by the "Controller Machine", which is a central control point for all the energy-related measurements. The communication between the "Control Machine" and the "Digital Multimeter" is performed using the Standard Commands for Programmable Instruments syntax (IEEE 488.2).

Rice et al. [16] were one of the first to explore the energy consumption in mobile phones. They proposed a methodology where a plastic battery holder replaces the battery, allowing the battery drop to be measured by using a high-precision measurement resistor in series between the holder cables and the battery. Although the accuracy behind this approach might be enough to measure mobile phone energy consumption, it still depends on the battery discharging pattern. Therefore, in the methodology used in this paper, an external "DC Power Supply" was employed. Nevertheless, as the mobile phone is expecting to receive battery status information via the smart battery system, the external power supply can not be directly used. The solution used in the measurements presented in this paper was to employ a specific voltage to allow the RT9524 unit of LG P990 (unit that controls the phone charging process) to be changed to "Factory Mode". This mode allows to supply the system using an external power supply and without connecting a battery.

The energy consumption of the IEEE 802.11 interface described in the next subsections is given by the difference between the mobile phone total energy consumption and the baseline energy consumption with the device operating in airplane mode (all radios off) with the display brightness at 100%. Each performed test has a duration of 60 seconds, and all the results presented include 20 distinct runs using with a confidence interval of 95%.

4.3 Results

This section discusses the obtained results regarding the No-PSM, Legacy-PSM and Adaptive-PSM performance when receiving data from a continuous media application, compared with the EXPoSE approach.

Impact of Packet Size: This section discusses the impact of the packet size in energy consumption of the three power saving approaches in study, namely No-PSM, Legacy-PSM and Adaptive-PSM. The data was sent with a fixed transmission rate of 100 packets per second. The packet size ranges from 200 to 1400 bytes. Figure 4 depicts the total energy consumed by the IEEE 802.11 interface (in Joule) during 60 s by each power saving algorithm, according to the employed packet size in bytes (x-axis).

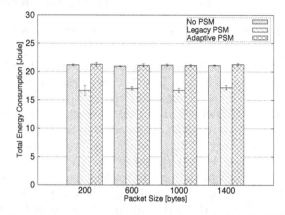

Fig. 4. Total energy consumed by the IEEE 802.11 interface with different packet sizes

The obtained results depict the Adaptive-PSM limitations in the presence of continuous media applications. Since in these applications there will be always data being transmitted from the core network to the mobile phone, the Adaptive-PSM algorithm does not have enough opportunities to sleep. Furthermore, due to the energy costs of multiple transitions between awake and sleep modes, this dynamic approach might consume more energy than No-PSM. When employing the Legacy-PSM the energy savings are between 18% and 21% compared to No-PSM and Adaptive-PSM schemes, respectively.

Concerning the relationship between packet size and energy consumption, it is possible to see that the packet size has a negligible impact on the total energy consumed. Such results highlight the energy benefits of using larger packets, since the energy cost per transmitted byte will be much lower.

Besides the energy consumption behavior, the impact of power saving algorithms on application performance should also be considered. Figure 5 shows a *boxplot* representing the one way delay, in milliseconds, for all the packets transmitted using the different algorithms. Figure 5a depicts the delay for all

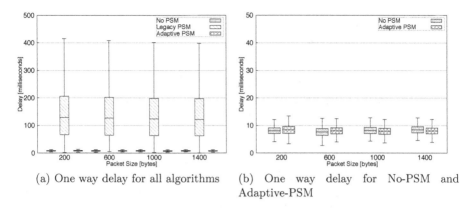

(a) One way delay for all algorithms

(b) One way delay for No-PSM and Adaptive-PSM

Fig. 5. One way delay for different packet sizes

the algorithms, while the Figure 5b zooms the same data only for No-PSM and Adaptive-PSM schemes.

Legacy-PSM introduces considerably more delay, when compared to No-PSM and Adaptive-PSM. The mean delay (second quartile) obtained for the Legacy-PSM is around 125 ms, while for No-PSM and Adaptive-PSM the mean delay is between 7 and 9 ms. Additionally, the No-PSM and Adaptive-PSM maximum delay never exceeds 14 ms and the Legacy-PSM has delays up to 400 ms. Again, the packet size does not reveal any impact on the results.

These results show that the energy savings (between 18% and 21%) obtained with the Legacy-PSM do not establish a good energy / performance trade-off, since there is a high impact on the application delay. Due to the polling phase, Legacy-PSM also generates packet loss, but always lower than 0.2%. Furthermore, the impact of packet size on the energy consumption is almost absent. Concerning the application design, the depicted data showed that using larger packets is highly preferable to improve overall energy consumption.

Impact of Transmission Rate: This section investigates the impact of the transmission rate on total energy consumption of No-PSM, Legacy-PSM and Adaptive-PSM. In this evaluation the packet size was fixed to 1000 bytes, with the transmission rate varying between 50 and 250 packets per second. Figure 6 presents the total energy consumed by the IEEE 802.11 interface (in Joules) during 60 s with the different transmission rates (x-axis). The results show that in both No-PSM and Adaptive-PSM it is possible to establish a linear relationship between the energy consumption over time and the transmission rate. When using Legacy-PSM, the energy consumption also increases with transmission rates, but only for rates up to 180 packets per second. With transmission rates above 180 packets per second, Legacy-PSM energy consumption is almost constant. Furthermore, Legacy-PSM only outperforms No-PSM and Adaptive-PSM for the lowest studied transmission rates.

Thus, in order to properly investigate the Legacy-PSM behavior, the one way delay and packet loss metrics were analyzed. No packet loss was detected

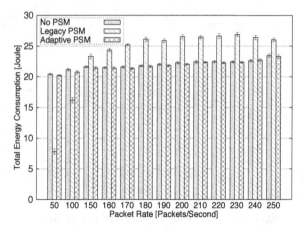

Fig. 6. Total energy consumed by the IEEE 802.11 interface with distinct transmission rates

with the No-PSM and the Adaptive-PSM schemes, and the mean delay ranges between 7 and 9 ms. The maximum delay observed was always lower than 14 ms. The *boxplot* depicting the delay and the packet loss rate for the Legacy-PSM are illustrated, respectively, in Figures 7a and 7b.

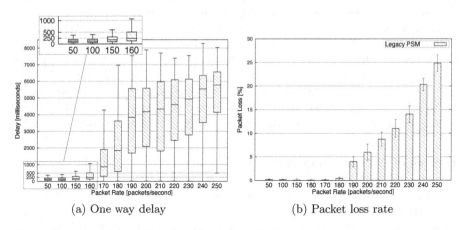

(a) One way delay   (b) Packet loss rate

Fig. 7. One way delay and loss rate for Legacy-PSM with distinct transmission rates

The Quality of Service (QoS) attained using Legacy-PSM is strongly affected by the transmission rate, as depicted by the mean delay for rates greater or equal than 160 packets per second. Apart from the unacceptable delay, Legacy-PSM also affects application performance by introducing a non negligible packet loss for rates above or equal to 190 packets per second. Such behavior is related to the Legacy-PSM protocol design, where the device must send one *PS-Poll* to the access point to request each pending frame [4]. The long delays resulting from the protocol polling mechanism also explain the depicted packet loss, since various packets are dropped in the access point queues due to time constraint violation.

In short, when analyzing the behavior of Legacy-PSM and Adaptive-PSM it is possible to conclude that they are not able to establish a proper energy / performance trade-off for continuous media applications. Legacy-PSM strongly affects the application performance, whereas the Adaptive-PSM can keep the application requirements, but without achieving significant energy savings.

EXPoSE Pattern-Based Sleep Approach: This section explores the employment of EXPoSE using the pattern-based sleep approach configured by the application, as described in Section 3.2. All the results presented next were performed using a fixed packet size of 1000 bytes with a constant transmission rate of 200 packets per second. This configuration was selected, since it represents a scenario where the Legacy-PSM performance is already worse than both No-PSM and Adaptive-PSM (see Figure 6).

As the goal of this assessment is to study the EXPoSE impact on the energy consumption and network performance, 9 distinct sleep patterns were selected as illustrated in Table 1. Apart from the parameters required to configure the EXPoSE pattern-based solution, the table also depicts a constant, κ, associated with each test. This constant allows to establish a relationship between the configured periods, that means $AwakePeriod = \kappa \times SleepPeriod$.

Table 1. EXPoSE pattern-based configurations

Test ID	T1	T2	T3	T4	T5	T6	T7	T8	T9
Loop Flag	1	1	1	1	1	1	1	1	1
SleepPeriod (ms)	30	30	30	60	60	60	120	120	120
AwakePeriod (ms)	90	30	10	120	60	20	360	120	40
κ	3	1	1/3	3	1	1/3	3	1	1/3

Figure 8a shows the EXPoSE pattern-based solution energy savings compared to Adaptive-PSM for the different configurations. As expected, the results show a direct relationship between the energy savings and the total time in sleep mode. Even for the scenarios with $\kappa =3$, where the time in sleep mode is 25% of the total time, the energy savings are up to 17.93% for the scenario with a sleep period equal to 120 ms. When reducing the awake period, an improvement in energy savings can be observed. With $\kappa =1$, where the awake and sleep periods are equal, the savings are up to 26.90%. If the awake period is reduced to 1/3 of the sleep period (i.e., $\kappa =1/3$) energy savings are 23.45%, 39.24% and 44.00% for configured sleep periods of 30, 60 and 120 ms, respectively.

The delay results, depicted in Figure 8b, show that EXPoSE impact on the delay is not negligible, such as with for Adaptive-PSM. For a similar scenario (with a rate of 200 packets per second and using packet size of 1000 bytes) the Legacy-PSM mean delay is around 4 s, with a maximum delay higher than 7 s. Moreover, packet loss with EXPoSE pattern-based approach was always lower than 0.02%, against 6.00% with Legacy-PSM.

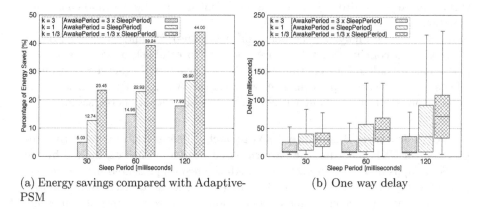

(a) Energy savings compared with Adaptive-PSM

(b) One way delay

Fig. 8. Energy savings and one way delay for EXPoSE pattern-based approach

In the studied scenarios, the EXPoSE pattern-based approach shows mean delays bellow 100 ms, enabling the possibility to be employed with continuous media application, as for instance, video streaming.

EXPoSE Maximum Allowed Delay Approach: This section studies the EXPoSE maximum allowed delay approach. In this evaluation, two applications were selected: one with a maximum allowed delay equal to 100 ms, and another allowing delays up to 200 ms. Both applications were emulated using a transmission rate of 200 packets per second, with packets of 1000 bytes length.

The main objective of EXPoSE's maximum allowed delay approach is to keep the application QoS requirements within specified bounds. Therefore, this section investigates the minimum awake period required to not exceed the defined maximum allowed delay. As defined in the EXPoSE maximum allowed delay mechanism, the sleep period will be equal to the configured maximum allowed delay.

Figure 9a shows the energy savings percentage compared with the Adaptive-PSM algorithm on the y-axis, while the x-axis depicts the tested awake periods, ranging from 50 ms to 500 ms. The one way delay for the same assessment is illustrated in Figure 9b.

The results show that to keep the application delay within the defined bounds, the awake period should be defined as 300 ms and 450 ms for maximum allowed delays of 100 ms and 200 ms, respectively. Energy savings for the lowest maximum allowed delay are 13.77% and 23.53% when the application supports delays up to 200 ms, respectively. Nonetheless, if the application allows that only 75% of the packets (*boxplot* third quartile) arrive within the configured limits, it is enough to be awake during 50 ms and the energy savings for 100 ms and 200 ms maximum allowed delay are 38.50% and 54.44%, respectively, compared to Adaptive-PSM.

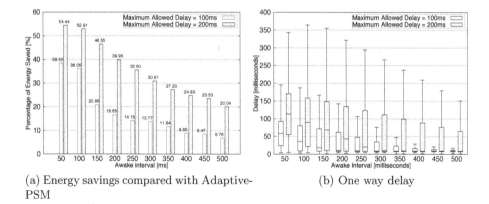

(a) Energy savings compared with Adaptive-PSM

(b) One way delay

Fig. 9. Energy savings and delay for EXPoSE maximum allowed delay scenarios

5 Conclusions

The fast growth of mobile devices deployment created new demands concerning the energy efficiency of the IEEE 802.11 network interfaces, which is one of the core access technologies supporting the communication of those devices. As Android is one of the most used platforms in portable equipment, the IEEE 802.11 power saving mechanisms in this system should be scrutinized.

Besides investigating the performance of IEEE 802.11 power saving techniques available in Android, this paper proposes an Android framework for Extending Power Saving control to End-users (EXPoSE), which enables an end-user and application based control of the IEEE 802.11 network interface power management. The achieved results showed that EXPoSE approaches, namely the pattern-based and the maximum allowed delay, are more energy efficient than both Legacy-PSM and Adaptive-PSM schemes. Depending on the scenarios and applications requirements, the EXPoSE energy savings, compared to Adaptive-PSM, can go up to 23.53% without violating the application delay constraints. Moreover, if some additional delay is acceptable (e.g., only 75% of the packets arriving on time), the energy savings can be more than 50%, compared to Adaptive-PSM.

Furthermore, the obtained results depicted the EXPoSE capabilities to improve continuous media applications energy efficiency, which is not well supported by Legacy-PSM and Adaptive-PSM strategies.

Acknowledgments. This work was partially supported by the COST Action IC0906, as well as by the iCIS project (CENTRO-07-ST24-FEDER-002003), co-financed by QREN, in the scope of the Mais Centro Program and European Union's FEDER. The first author was also supported by the Portuguese National Foundation for Science and Technology (FCT) through a Doctoral Grant (SFRH/BD/66181/2009).

References

1. IEEE: IEEE std 802.11-2012 (revision of ieee std 802.11-2007), 1–2793 (2012)
2. Costa-Pérez, X., Festag, A., Kolbe, H.J., Quittek, J., Schmid, S., Stiemerling, M., Swetina, J., van der Veen, H.: Latest trends in telecommunication standards. SIG-COMM Comput. Commun. Rev. 43(2), 64–71 (2013)
3. Android Open Source Project: Android open-soruce software stack (2014)
4. Tsao, S.L., Huang, C.H.: A survey of energy efficient MAC protocols for IEEE 802.11 {WLAN}. Computer Communications 34(1), 54–67 (2011)
5. Pyles, A.J., Qi, X., Zhou, G., Keally, M., Liu, X.: SAPSM: smart adaptive 802.11 PSM for smartphones. In: Proceedings of the 2012 ACM Conference on Ubiquitous Computing, UbiComp 2012, pp. 11–20. ACM, New York (2012)
6. Pyles, A.J., Ren, Z., Zhou, G., Liu, X.: Sifi: Exploiting voip silence for wifi energy savings insmart phones. In: Proceedings of the 13th International Conference on Ubiquitous Computing, UbiComp 2011, pp. 325–334. ACM, New York (2011)
7. Csernai, M., Gulyas, A.: Wireless adapter sleep scheduling based on video qoe: How to improve battery life when watching streaming video? In: ICCCN 2011, pp. 1–6 (July 2011)
8. Bernardo, V., Curado, M., Braun, T.: An IEEE 802.11 energy efficient mechanism for continuous media applications. Sustainable Computing: Informatics and Systems 4(2), 106–117 (2014)
9. Korhonen, J., Wang, Y.: Power-efficient streaming for mobile terminals. In: Proceedings of the International Workshop on Network and Operating Systems Support for Digital Audio and Video, NOSSDAV 2005, pp. 39–44. ACM (2005)
10. Ding, N., Pathak, A., Koutsonikolas, D., Shepard, C., Hu, Y., Zhong, L.: Realizing the full potential of psm using proxying. In: 2012 Proceedings IEEE INFOCOM, pp. 2821–2825 (March 2012)
11. Dogar, F.R., Steenkiste, P., Papagiannaki, K.: Catnap: Exploiting high bandwidth wireless interfaces to save energy for mobile devices. In: Proceedings of the 8th International Conference on Mobile Systems, Applications, and Services, MobiSys 2010, pp. 107–122. ACM, New York (2010)
12. Cui, Y., Ma, X., Wang, H., Stojmenovic, I., Liu, J.: A survey of energy efficient wireless transmission and modeling in mobile cloud computing. Mobile Networks and Applications 18(1), 148–155 (2013)
13. Amuhong: IEEE 802.11 architecture in Android (2014)
14. Botta, A., Dainotti, A., Pescapè, A.: A tool for the generation of realistic network workload for emerging networking scenarios. Computer Networks 56(15) (2012)
15. Bernardo, V., Curado, M., Staub, T., Braun, T.: Towards energy consumption measurement in a cloud computing wireless testbed. In: International Symposium on Network Cloud Computing and Applications (NCCA), pp. 91–98 (2011)
16. Rice, A., Hay, S.: Measuring mobile phone energy consumption for 802.11 wireless networking. Pervasive Mob. Comput. 6(6), 593–606 (2010)

Simulation-Based Comparison of AODV, OLSR and HWMP Protocols for Flying Ad Hoc Networks

Danil S. Vasiliev, Daniil S. Meitis, and Albert Abilov

Izhevsk State Technical University,
Department of Communication Networks and Systems,
ul. Studencheskaya, 7, 426069 Izhevsk, Russia
{danil.s.vasilyev,daniil.meitis,albert.abilov}@istu.ru
http://www.istu.ru/

Abstract. In this paper, we analyze Quality of Service (QoS) metrics for AODV, OLSR and HWMP routing protocols in Flying Ad Hoc Networks (FANETs) with the help of an NS-3 simulation tool. We compare proactive, reactive, and hybrid approaches to search and maintain paths in FANET based on hop count, PDR (Packet Delivery Ratio), and overheads metrics in source-destination transmission through the swarm of UAVs (Unmanned Aerial Vehicles). In the article, swarms of 10, 15 and 20 nodes were considered. The Gauss-Markov Mobility Model is used to simulate the UAV behavior in a swarm. The size of a simulated area is variable and changes from 250 to 750 meters. Average metrics were calculated in all cases. In addition, we calculate the Goodput metric and compare it with correspondent overheads. Results show that using HWMP in the considered mobile scenario grants higher PDR in trade-off, increased overheads.

Keywords: Mobile Ad Hoc Networks, Unmanned Aerial Vehicles, Quality of Service, Routing protocols, Computer simulation.

1 Introduction

Unmanned Aerial Vehicles (UAVs or drones) have become more functional since the last decade. They found new applications in search operations, pollution monitoring, wildfire management, border surveillance, etc. Drones communicate with each other in UAV systems during these operations, e.g., they transmit live video or sensor data.

Swarms of small drones are a cheap and fast way to provide a wide selection of services in a disaster area. UAVs maintain ad hoc connections in the swarm to deliver data safe and sound. Therefore, they could be considered as a set of nodes in a Flying Ad Hoc Network (FANET) [1], [2].

UAV-node velocities cause many challenges for MANET (Mobile Ad Hoc Networks) deployment. Mobility factors have an influence on QoS (Quality of Service) parameters in the network. Constant movement of nodes leads to frequent

S. Balandin et al. (Eds.): NEW2AN/ruSMART 2014, LNCS 8638, pp. 245–252, 2014.
© Springer International Publishing Switzerland 2014

link outages and packet loss. Thus, FANETs need special approaches for data delivery and routing [3], [4], [5], [6], [7]. Routing protocols are critical for live streaming from on-board cameras in the swarm.

Intraswarm communications impose new challenges for researchers. New simulation models and tools have been proposed to investigate routing and data delivery in FANETs [8], [9]. NS-3 provides routing protocols [10], signal propagation and mobility models [11]. Therefore, this simulation environment allows comparing ad hoc routing protocols in a FANET mobile scenario. In this paper, we analyze QoS metrics for AODV, OLSR and HWMP protocols in the case of source-destination transmission through a swarm of drones.

The remainder of this paper is organized as follows: Section 2, overview of routing protocols for FANET; Section 3, description of chosen simulation scenario; Section 4, QoS metrics used; Section 5, results; Section 6, conclusion.

2   Routing Protocols for FANET

FANET needs an efficient way to organize a swarm of nodes. Internode communication requires a routing mechanism to deliver information from one node to another through complicated mesh topology. Routing protocols for mesh networks use reactive, proactive, or hybrid approaches. In this paper, AODV, OLSR and HWMP protocols were considered. Each protocol presents a unique way to provide routing.

AODV (Ad hoc On-demand Distance Vector) protocol uses a reactive approach. This protocol constructs new routes as a user need them to transmit data through ad hoc network and maintains them until they exist.

OLSR (Optimized Link State Routing) is a proactive protocol, and, therefore, it tracks the network topology. Every OLSR node sends HELLO messages with regular intervals for its 1-hop neighbors. MPR (Multipoint Relays) reduce OLSR control overhead.

HWMP (Hybrid Wireless Mesh Protocol) is described in 802.11s draft and allows using reactive and proactive approaches within one network. In this protocol, AODV-like reactive routing competes with the root-centric proactive mode in a search for the best path through the ad hoc network with help of PREQ (Path Request) messages. In the NS-3 802.11s model, the node knows the root path but also tries to find new reactive paths to provide the best route based on an ALM (Air Time Link) metric following the hybrid nature of the protocol [8].

AODV and OLSR depend upon L3 and IP-addresses but HWMP is an L2.5 protocol and uses MACs to route data. Each of the protocols constructs routing tables (with MACs in HWMP case) on each node.

3   Simulation Scenario

Highly mobile nodes propose many challenges to researches. We have used the Gauss-Markov Mobility Model implemented in NS-3 to analyze routing in the swarm. Node position is always dictated by its previous position due to high moving speed [9]. The path of a drone is determined by the memory of the model.

We simulated FANET with AODV, OLSR and HWMP routing protocols in NS-3 environment. Fig. 1 illustrates a simulated network that consists of two ground stations and a variable number of nodes in the swarm between them. The area of simulation is constrained by the imaginary square box with variable side A (Fig. 1). Source and destination nodes are stationary and located in top-right and bottom-left corners of the box, correspondingly. The remaining nodes represent drones: they move following the Gauss-Markov Mobility Model and form the swarm. Swarm nodes have velocities from 25 to 30 meters per second. They are bound by the box and reflect from its borders without any speed reduction.

Swarms of 10, 15, and 20 nodes were simulated. Nodes used 802.11n on 5 GHz and 54 Mbps bandwidth. Signals were simulated with the Friis Signal Propagation Model, and transmission range of each node was about 250 meters. During the simulation, the source node transmitted 1406 bytes UDP datagrams (real-time video streaming) to the destination with a speed of 1 Mbps. Swarm nodes tried to deliver these datagrams to the destination through paths selected by AODV, OLSR or HWMP routing protocols.

Value A (Fig. 1) defined a box side. It was variable and changed from 250 to 750 meters with 50 m step. Therefore, the node density gradually dropped down and nodes were forced to find new paths in the swarm with help of routing protocols.

Stations and swarm nodes used one of the above mentioned routing protocols. The most important control messages for each protocol (HELLO or PREQ) were transmitted with 0,5 sec intervals.

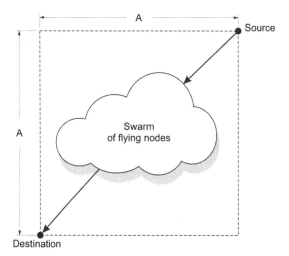

Fig. 1. Simulated network. Value A defines square box side

4 Quality of Service Metrics

We measured QoS metrics to compare effectiveness of protocols in this mobile scenario.

Average Packet Delivery Ratio (PDR_{ave}) shows a ratio between the number of datagrams received by the destination and the number of datagrams transmitted by the source. This metric is measured based on application layer sequence numbers added in the simulated UDP datagrams:

$$PDR_{ave} = \frac{Rx}{Tx} , \qquad (1)$$

where Rx – number of datagrams received by the destination, Tx – number of datagrams transmitted by the source.

Average Goodput is all UDP payload received by the destination during the simulation divided by the simulation time:

$$Goodput_{ave} = \frac{\sum_{i=1}^{n} S_i}{T} , \qquad (2)$$

where S_i – each received datagram payload size in kbits, n – number of received datagrams, T – simulation time in seconds.

Average Hop Count was found based on Time-to-Live field in the IP-header (for AODV and OLSR) or Mesh-header (for HWMP) in the datagrams received by the destination, summarized and divided by the number of received datagrams:

$$Hopcount_{ave} = \frac{\sum_{i=1}^{n} H_i}{n} , \qquad (3)$$

where H_i – each received datagram hop-count metric, n – number of received datagrams.

Overheads were measured as all control messages of each routing protocol in bytes divided by the simulation time:

$$Overheads_{ave} = \frac{\sum_{i=1}^{n} OH_i}{T} , \qquad (4)$$

where OH_i – protocol control message size in kbits, n – number of control messages in the simulation, T – simulation time in seconds.

5 Results

QoS metrics were calculated based on information in pcap-files collected during the simulations. Figures show average values of the 10 simulation runs. Each run was 5 minutes long.

Results for hop-count metric and PDR_{ave} are illustrated in Fig. 2 for the swarm of 20 nodes in the box with variable sides. Hop count metric (Fig. 2a) increases linearly as box size grows. HWMP demonstrated longer paths due to

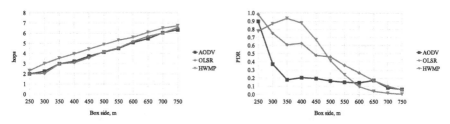

(a) Average hop count for each routing protocol

(b) Average packet delivery ratio for each routing protocol

Fig. 2. Measurement results for the swarm of 20 nodes

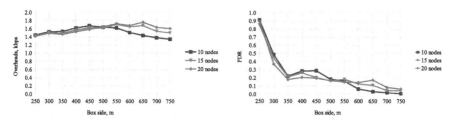

(a) Average overhead bandwidth for each swarm size

(b) Average packet delivery ratio for each swarm size

Fig. 3. Measurement results for AODV routing protocol

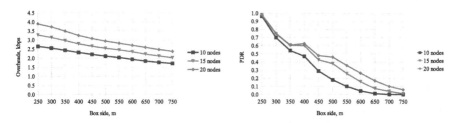

(a) Average overhead bandwidth for each swarm size

(b) Average packet delivery ratio for each swarm size

Fig. 4. Measurement results for OLSR routing protocol

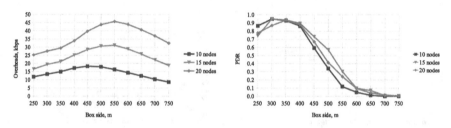

(a) Average overhead bandwidth for each   (b) Average packet delivery ratio for each
swarm size   swarm size

Fig. 5. Measurement results for HWMP routing protocol

using the ALM routing metric. While, AODV and OLSR use a hop-count metric
to find new paths.

The plot (Fig. 2b) for PDR_{ave} presents an advantage of HWMP for box sizes
from 300 to 500 meters. After this value, all protocols perform PDR lower than
0.5. PDR_{ave} for OLSR decreases linearly as paths lengthen and demonstrates
the second result for a packet delivery metric. PDR_{ave} for AODV rapidly drops
down and a path could not be established for the highly mobile scenario.

Figs. 3a, 4a and 5a show overheads for each protocol and box size. Reactive
AODV protocol has minimal overheads for all swarm sizes. As we showed in
Fig. 2, bigger box size causes more hop counts for the path between the source
and the destination. Overheads of AODV do not depend on hop count in the
chosen path. Proactive OLSR overheads are bigger for larger swarms, but slightly
decrease as node density lowers. HWMP has an extremum for each swarm size
and fully correlates to box size. This behavior could be explained by frequent
error message generation during recovery of lost paths. This process is intensive
in the middle part of the simulation.

Better results for PDR_{ave} metric for HWMP could be explained by higher
overheads. But an overhead metric is dependent on the maximum number of
control message retransmissions (PREQs). In the NS-3 model of HWMP, one
additional retransmission is allowed. Thus, this feature of HWMP routing algorithm gives it an advantage over AODV and OLSR in PDR_{ave} metric and a
disadvantage in overheads.

Average Goodput metric mimics PDR_{ave} curves. This metric helps comparing
payload and overheads. AODV and OLSR protocols Goodput metrics were much
higher than correspondent overheads in all cases; e.g., the worst case for both
protocols was the Goodput of 60.54 Kbps against 2.38 Kbps overheads (with
OLSR in the swarm of 20 nodes and 750 meters box size). HWMP demonstrated
overheads higher than the Goodput for the boxes with 700 and 750 meters sides.

Figs. 3b, 4b, and 5b allow comparing AODV, OLSR and HWMP by PDR_{ave}
metric. AODV does not show any dependence on swarm size; e.g., 10 nodes
or 20 nodes. OLSR slightly increases for bigger swarms and, therefore, it gets
benefits from higher node density. HWMP tries to hold PDR_{ave} high (around

0.8–0.9) but after 400 meters, it falls exponentially. PDR_{ave} curves for HWMP demonstrate similar behavior for all swarm sizes.

6 Conclusions

In this paper, we have examined AODV, OLSR and HWMP routing protocols in order to choose the best of them for live streaming through a highly mobile ad hoc network of UAVs. We compared effectiveness of protocols based on hop count, PDR, and overheads metrics. HWMP showed the highest PDR and the highest overheads in source-destination transmission through the swarm of drones. OLSR earned the second place and AODV was unpredictable in such mobile environments due to its pure reactive nature. PDR metric for AODV and HWMP protocols does not depend on swarm size. This behavior is defined by the reactive component of both protocols. Simulation results show that a hybrid approach and HWMP can significantly improve the QoS for the video transmission in mobile ad hoc networks of flying robots.

In the simulated scenario, overheads for OLSR are lower than for HWMP, but we could manage time intervals between control messages in order to improve PDR for OLSR. Nevertheless, control overhead reduction needs additional research due to special conditions for transmission in a wireless medium.

Node movement and link outage affect on QoS metrics in any case. Mobility prediction algorithms, opportunistic data delivery techniques, and new coding schemes are needed to organize live video streaming in such networks. Moreover, further improvement of routing protocols are vital to provide high quality communication in FANETs.

References

1. Bekmezci, I., Sahingoz, O.K., Temel, S.: Flying Ad-Hoc Networks (FANETs): A survey. Ad Hoc Networks 11(3), 1254–1270 (2013)
2. Bok, P.-B., Tuchelmann, Y.: Context-Aware QoS Control for Wireless Mesh Networks of UAVs. In: 2011 Proceedings of 20th International Conference on Computer Communications and Networks (ICCCN), pp. 1–6 (2011)
3. Beard, R., McLain, T., Nelson, D., Kingston, D., Johanson, D.: Decentralized Cooperative Aerial Surveillance Using Fixed-Wing Miniature UAVs. Proceedings of the IEEE 94(7), 1306–1324 (2006)
4. Robinson, W., Lauf, A.: Resilient and efficient MANET aerial communications for search and rescue applications. In: 2013 International Conference Computing, Networking and Communications (ICNC), pp. 845–849 (2013)
5. Rosati, S., Kruzelecki, K., Traynard, L.F., Rimoldi, B.: Speed-Aware Routing for UAV Ad-Hoc Networks. Report, GLOBECOM 2013, Atlanta, GA, USA (2013)
6. Guo, Y., Li, X., Yousefi'zadeh, H., Jafarkhani, H.: UAV-aided cross-layer routing for MANETs. In: Proceedings of Wireless Communications and Networking Conference (WCNC), pp. 2928–2933 (2012)
7. Li, Y., Shirani, R., St-Hilaire, M., Kunz, T.: Improving routing in networks of Unmanned Aerial Vehicles: Reactive-Greedy-Reactive. Wireless Communications and Mobile Computing 12(18), 1608–1619 (2012)

252 D.S. Vasiliev, D.S. Meitis, and A. Abilov

8. Wei, Y., Blake, M.B., Madey, G.R.: An Operation-time Simulation Framework for UAV Swarm Configuration and Mission Planning. In: 2013 International Conference Computational Science (ICCS), pp. 1949–1959 (2013)

9. Ho, D., Grotli, E.I., Shimamoto, S., Jahansen, T.A.: Optimal Relay Path Selection and Cooperatice Communication Protocol for a Swarm of UAVs. In: The 3rd International Workshop on Wireless Networking & Control for Unmanned Autonomous Vehicles: Architectures, Protocols and Applications, pp. 1585–1590 (2012)

10. Andreev, K., Boyko, P.: IEEE 802.11s Mesh Networking NS-3 Model. Report, Workshop on NS-3, Malaga, Spain (2010)

11. Broyles, D., Jabbar, A., Sterbenz, J.: Design and Analysis of a 3-D Gauss-Markov Mobility Model for Highly Dynamic Airborne Networks. In: Proceedings of the International Telemetering Conference 2010, San Diego, CA (2010)

A Comparative Analysis of Beaconless Opportunistic Routing Protocols for Video Dissemination over Flying Ad-Hoc Networks

Denis Rosário[1,2], Zhongliang Zhao[2], Torsten Braun[2],
Eduardo Cerqueira[1,3], and Aldri Santos[4]

[1] Federal University of Pará, Belém, Brazil
[2] University of Bern, Bern, Switzerland
[3] University of California Los Angeles, Los Angeles, USA
[4] Federal University of Paraná, Curitiba, Brazil
{denis,cerqueira}@ufpa.br, {zhao,braun}@iam.unibe.ch, aldri@inf.ufpr.br

Abstract. A reliable and robust routing service for Flying Ad-Hoc Networks (FANETs) must be able to adapt to topology changes, and also to recover the quality level of the delivered multiple video flows under dynamic network topologies. The user experience on watching live videos must also be satisfactory even in scenarios with network congestion, buffer overflow, and packet loss ratio, as experienced in many FANET multimedia applications. In this paper, we perform a comparative simulation study to assess the robustness, reliability, and quality level of videos transmitted via well-known beaconless opportunistic routing protocols. Simulation results shows that our developed protocol XLinGO achieves multimedia dissemination with Quality of Experience (QoE) support and robustness in a multi-hop, multi-flow, and mobile networks, as required in many multimedia FANET scenarios.

Keywords: FANETs, Multiple video flows, Opportunistic Routing, QoE support, and Robustness.

1 Introduction

Multi-flow video transmission over Flying Ad-Hoc Networks (FANETs) enable a large class of multimedia applications, such as safety & security, natural disaster recovery, and others [1]. However, transmission of video flows over FANETs becomes a hard task due to topology changes mainly caused by node failures and mobility, as well as channel conditions. Particularly, topology changes have impact on both network performance and video quality [2], where the latter is essential to provide visual support for humans about the best action to be taken in the monitored area. Moreover, multimedia transmission over FANETs usually involves multiple mobile nodes transmitting multiple video flows simultaneously, leading to higher network congestion, buffer overflow, and packet loss ratio. Multimedia services require real-time video transmissions with low frame loss

S. Balandin et al. (Eds.): NEW2AN/ruSMART 2014, LNCS 8638, pp. 253–265, 2014.

rate, tolerable end-to-end delay, and jitter to support multimedia dissemination with Quality of Experience (QoE) support [3].

Several routing protocols have been proposed to meet the requirement to simultaneously deliver multiple video flows with robustness and QoE support over FANET scenarios [1]. Those protocols are based on flat, hierarchical, or geographical approaches, and rely on end-to-end routes to forward packets [4]. However, end-to-end routes might be subject to frequent interruptions or may not always exist, which is not desirable for FANET multimedia applications. In this context, beaconless Opportunistic Routing (OR) protocols enable packet transmission even in case of continuous topology changes by acting in a completely distributed manner to pick up one of the possible relay nodes to forward packets [5]. The routing service should also detect and recover from route failures, enabling a smoother operation in mobile networks.

To address the issues discussed above, our previous work introduced the Cross-layer Link quality and Geographical-aware beacon-less OR protocol (XLinGO) [6], which improves the human experience when watching live video sequences. More specifically, it combines a set of cross-layer, video, and human-related parameters for routing decisions, namely packet delivery ratio, QoE, queue length, link quality, geographical location, and residual energy. XLinGO also relies on a recovery mechanism to deal with route failures, providing a smoother operation in harsh environments and mobile networks, such as experienced in FANET scenarios.

In this paper, we introduce a comparative simulation analysis of well-known beaconless OR protocols for video dissemination over FANETs, in terms of robustness and quality level of transmitted videos. We perform simulations to analyse the impact of node speed, different number of video flows, and videos with different motion and complexity levels on the final video quality level, frame loss ratio, and signalling overhead. Based on simulation results, we identify that in contrast to well-known beacon-less OR protocols, namely BLR [7], BOSS [8], and MRR [9], XLinGO provides multimedia transmission with reliability, robustness, and QoE support, such as required in many multimedia FANET applications.

The remainder of this paper is structured as follows. Section 2 outlines beacon-less OR protocols under evaluation in this paper. Section 3 discusses simulation results. Section 4 concludes this paper.

2   Evaluated Beaconless Opportunistic Routing

Multimedia dissemination over FANETs provide visual information, as soon as the standard fixed network infrastructure is unavailable due to a natural disaster, such as an earthquake or hurricane (as illustrated in Figure 1). Hence, multimedia content plays an important role in enabling humans in the control center to take actions to explore a hazardous area, which rescuers are unable to reach easily and quickly. Moreover, it enables image processing by systems, and also allows the end-users to be aware of what is happening in the environment.

Routing protocols that rely on end-to-end routes, such as DSR, AODV or OLSR, might not be a feasible approach to forward packets. This is because end-to-end routes might be subject to frequent interruptions or may not always exist in case of node mobility or wireless channel variations, such as experienced in multimedia FANET scenarios. In this context, OR improves communication performance by exploiting the broadcast nature and spatial diversity of the wireless medium. In beaconless OR protocols, a sender sends a packet not only to a single next-hop, but also to multiple neighbours simultaneously. Afterwards, one or more receiving nodes forward the packet towards the destination, based on a coordination method to select the best candidate to forward packets. Hence, the routing mechanism forwards the packet towards the destination based on a hop-by-hop routing decision at the receiver side. In other words, in beaconless OR protocols forwarding decisions are performed by the receiver of a packet based only on information contained in the packet, as well as in local information, such as node position and direction. Table 1 presents the most important characteristics of beaconless OR protocols under evaluation in this paper.

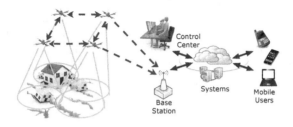

Fig. 1. Multimedia FANET deployed in an emergency situation

Heissenbüttel et al. [7] introduced the concept of Dynamic Forwarding Delay (DFD) as forwarding decision in the Beacon-Less Routing protocol (BLR). In its operation, the source node broadcasts a data packet. Before forwarding the received packet, possible relays within a forwarding area compute a DFD value in the interval $[0, DFD_{Max}]$ based only on location information, start a timer according to the DFD value, and wait for the expiration of this timer to forward the received packet. In this way, the relay node that generates the smallest DFD value, replaces the source node location with its own locations in the packet header, and forwards the packet first, becoming a forwarder node. Neighbour nodes recognize the occurrence of relaying, and cancel their scheduled transmissions for the same packet that they overhear. At the same time, BLR uses the transmitted packet as passive acknowledgement, and thus the source node knows which relay forwarded the packet first, in order to unicast subsequent packets. The algorithm continues until the packet reaches the destination, which sends an explicit acknowledgement.

Sanchez et al. [8] proposed the Beacon-less On-demand geographic routing Strategy (BOSS), which extends BLR by introducing a different forwarding area and applying a three-way handshake mechanism. In addition, BOSS assumes a full data payload for the Request to Send (RTS) message size, since selecting a forwarder using small control messages may lead to the choice of a forwarder unable to receive larger data packets. Al-Otaibi et al. [9] proposed a Multipath Routeless Routing protocol (MRR), which defines a forwarding area as a rectangle and combines multiple metrics to compute the DFD. Based on the proposed DFD function MRR selects forwarder nodes closer to boundaries of the rectangle, which does not mean that the forwarder provides geographical advance towards the Destination Node (DN), i.e., MRR does not select a forwarder closer to the DN. In addition, MRR prefers a forwarding node that receives a packet with a weak signal.

LinGO follows our initial XLinGO definition [10], which relies on periodic route reconstruction, and combines link quality, geographical information, and energy to compute the DFD. On the other hand, XLinGO [6] combined a set of cross-layer parameters for routing decisions, and also relies on a recovery mechanism. More specifically, it consider queue length, link quality, geographical information to compute the DFD, and thus it selects forwarders closer to the destination with a reliable link, as well as sufficient queue and energy capacity to forward packets from simultaneous multi-flows and mobile nodes with a reduced packet loss ratio. XLinGO also relies on a recovery mechanism, which considers link quality and packet received ratio to detect and quickly react to route failures, providing smoother operation in harsh and mobile networks, such as experienced in multimedia FANET scenarios.

Table 1. Characteristics of Beaconless OR Protocols under Evaluation

Beaconless OR protocols	Metrics to compute the DFD	Route reconstruction
BLR [7]	Distance	Periodic
BOSS [8]	Distance	Periodic
MRR [9]	Energy, distance, and Link quality	Periodic
LinGO [10]	Energy, distance, and link quality	Periodic
XLinGO [6]	Distance, link quality, and queue length	Recovery mechanism

3   A Comparative Simulation Study

This section describes the methodology and metrics used to evaluate the quality level of transmitted videos by different beaconless OR protocols. Afterwards, we evaluate the impact of different moving speeds and the impact of the number of multimedia flows on the video quality, overhead, and number of dropped packets.

3.1   Simulation Description and Evaluation Metrics

We used the Mobile MultiMedia Wireless Sensor Network (M3WSN) OMNeT++ framework [11] for our evaluations. The simulations last for 200 seconds (s) and run with the lognormal shadowing path loss model. We set the simulation parameters to allow dynamic wireless channel variations, link asymmetry, and irregular radio ranges, as expected in a multimedia FANETs environment. Results are averaged over 33 simulation runs with different randomly generated seeds to provide a confidence interval of 95%. In our simulations, we deployed 40 nodes with a Destination Node (DN) located at $(50, 0)$. Other nodes are moving following the Random Waypoint mobility model over the entire flat terrain of 100×100 m. Nodes are equipped with IEEE 802.11 radio, using a transmission power of 12dBm. Nodes rely on a CSMA/CA MAC protocol without RTS/CTS messages and retransmissions. Nodes also considered the QoE-aware redundancy mechanism [12] to add redundant packets to priority frames at the application layer. To compute the DFD value, XLinGO requires coefficients (α, β, and γ) to give different importance to each metric, i.e. link quality, geographical information, and queue length. It also considers DFD_{Max} to specify the maximum DFD value. In this way, for the following simulation, XLinGO considers $DFD_{Max} = 100$ ms, $\alpha = 0.4$, $\beta = 0.4$, and $\gamma = 0.2$, since these values provide the best trade-off between progress, link quality, and queue length [6, 10]. The XLinGO recovery mechanism considers $timeout = 0.3$s and $PRR_{th} = 30\%$ to detect topology or channel changes.

In our simulations, the Source Node (SN) transmitted a set of videos, i.e., Hall, UAV$_1$, and UAV$_2$ video sequences, with different characteristics. These videos are downloaded from the YUV video trace library and YouTube [13]. More specifically, the Hall video sequence has similar characteristics as if a mobile node stopped in a certain area to capture a video. UAV$_1$ and UAV$_2$ videos have similar motion and complexity levels, compared to a UAV capturing video flows while it is flying, but UAV$_2$ has a higher motion level than UAV$_1$ caused by UAV instability during the flight. We encoded those videos with a H.264 codec at 300 kbps, 30 frames per second, Group of Picture (GoP) size of 18 frames, and common intermediate format (352 x 288). The decoder uses Frame-Copy for error concealment to replace each lost frame with the last received one in order to reduce frame loss on the video quality. We selected a set of videos transmitted via XLinGO, LinGO, BLR, BOSS, and MRR together with original videos and make them available at [14].

QoE metrics overcome the Quality of Service (QoS) scheme limitations to assess the quality level of multimedia applications. Hence, we used a well-known objective QoE metric, namely Structural Similarity (SSIM). SSIM is based on frame-by-frame assessment of three video components, i.e., luminance, contrast, and structural similarity. It ranges from 0 to 1, and a higher SSIM value means better video quality. We used the MSU Video Quality Measurement Tool (VQMT) to measure the SSIM value for each transmitted video [15].

3.2 Impact of the Mechanism to Recover from Route Failures

We deployed 40 static nodes with one DN, one Source Node (SN), and 38 possible relay nodes. We created the worst-case scenario for topology changes, where SN established the persistent route ($P_{SN,DN}$) and 10% of its neighbour have individual node failures. We defined three configurations for XLinGO: i) XLinGO does not experience node failures and does not use the recovery mechanism; ii) XLinGO – Failure relies on periodic route reconstruction and experience node failures; and iii) XLinGO – Recovery considers the recovery mechanism and experiences node failures. The original video represents an error free video transmission to serve as a benchmark video quality level, since the video coding and decoding process also introduces impairments on video quality even in the absence of packet loss.

Figure 2 shows the SSIM for all frames that compose the Hall video sequence for those three XLinGO configurations. XLinGO without node failures delivers the video with high and constant quality, i.e., $SSIM$ around 0.94, which is similar to the benchmark video quality level. This is because XLinGO builds a reliable path $P_{SN,DN}$, which protects the video frames during link error periods. Hence, XLinGO enables video dissemination with QoE support in scenarios with dynamic topologies caused by channel quality variations. Frames 0 – 17 have a good quality when transmitted by XLinGO – Failure and XLinGO – Recovery. This is because the destination node received frame 0, and the decoder relies on frame-copy for error concealment to replace each lost frame with the last received one, enabling to reconstruct those frames 0 – 17 with a good quality. XLinGO – Failure transmitted video frames 18 – 89 with bad quality, since bursts of packets were lost until the SN re-establishes the path $P_{SN,DN}$ as soon as one of the nodes from the $P_{SN,DN}$ is no longer available to forward packets. In the worst case, packet loss lasts during the interval for route reconstruction. In our experiments, XLinGO – Failure reconstructs routes every 3s, and this explains the poor video quality for frames 18 – 89. Afterwards, frames 90 – 299 have similar quality compared the benchmark video quality level, since XLinGO enables source node to reconstruct another reliable $P_{SN,DN}$. XLinGO – Recovery transmitted only frames 18 – 35 with poor quality. Due to the proposed recovery mechanism considers a timeout value of 0.3s to detect that one of the forwarders is no longer available to forward packets. As soon as the time-out expires, the node must re-establish route $P_{SN,DN}$. Frames 36 – 299 transmitted via XLinGO – Recovery have similar quality compared to the benchmark video quality.

Aguiar et al. classified the Hall video sequence as a low motion video flow, which means that there is a small moving region on a static background [15]. Videos with static background make frame loss less severe than videos with high motion. Hence, we also carried out simulations with a source node transmitting the UAV_1 video sequence to analyse how the proposed recovery mechanism performs with more dynamic video sequences. Let us summarize the results to analyse the impact of the proposed mechanism to recover from route failures, as shown by results of Figure 3. In a scenario without any node failure, XLinGO transmits the entire Hall video with SSIM equal to 0.94. On the other hand, XLinGO – Failure transmits the entire video with an SSIM value of 0.84. Finally, the entire video transmitted by XLinGO – Recovery has a SSIM value of 0.91. Based on simulation results, we found out that for XLinGO with periodic route reconstruction, as soon as one of the forwarding nodes from a given route is no longer available to forward packets, a burst of packets might be lost until XLinGO re-establishes the route, reducing video quality for longer periods. Route reconstruction interval should be adjusted according to the desired degree of robustness and energy consumption. More specifically, from an energy consumption point-of-view, route reconstruction must occur with low frequency. On the other hand, based on a robustness point-of-view, high frequency of route reconstruction provides better robustness. Hence, it is not trivial to find the route reconstruction interval to provide robustness, QoE support, and reduced overhead simultaneously. In this way, XLinGO with recovery mechanism provides smoother operation in harsh environments and mobile networks, as shown by results of Figure 3.

We conclude that the proposed recovery mechanism reduces the video quality level less than periodic route reconstruction schemes. This is because the proposed mechanism enables XLinGO to quickly detect and respond to topology changes, and thus it enables multimedia dissemination with robustness and QoE assurance. Moreover, Hall and UAV_1 video sequences have similar quality regardless of the transmission scheme.

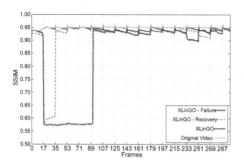
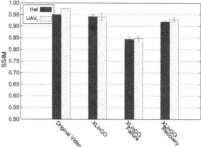

Fig. 2. SSIM for All Frames for the Hall Video Sequence

Fig. 3. Overall Video Quality for Hall and UAV_1 Video Sequences

3.3 Impact of Node Mobility and Number of Video Flows

We defined two configurations, where the first has one static DN, two mobile SNs transmitting simultaneous video flows, and 37 mobile RN_i. The second configuration has one DN, three mobile SNs, and 36 mobile RN_i. For both configurations, each SN transmits a different video sequence with six different maximum speed limits. Based on the simulation results introduced in Figures 4(a) and 4(b), we conclude that XLinGO provides multimedia dissemination with higher quality than LinGO, BLR, MRR, and BOSS regardless of the moving speed. This is because BLR only considers geographical information to compute the DFD, and due to the unreliability of the wireless channel, the most distant node might suffer from a bad connection, increasing the packet loss ratio for BLR. MRR selects a forwarder that receives a packet with a weak signal, reducing reliability and quality of transmitted videos. Videos transmitted via BOSS have higher quality compared to BLR and MRR, even if BOSS only considers geographical information for routing decisions like BLR, since BOSS considers a three-way handshake mechanism to select forwarder. However, videos transmitted via BOSS have lower video quality than LinGO and XLinGO. XLinGO increases video quality compared to LinGO, because XLinGO considers queue length for routing decisions, and also a mechanism to recover from route failures. For videos transmitted through LinGO, BLR, MRR, and BOSS, as soon as the maximum speed (s_{max}) increases, the video quality decreases. This is because those protocols rely on periodic route reconstruction and do not consider queue length to compute the DFD, which increases the packet loss ratio.

Figure 5 shows the frame loss ratio to explain the video quality level for results of Figure 4(a). In particular, XLinGO has an overall frame loss rate of 17.8%, which is 48%, 61%, 57%, 70% lower than for LinGO, BLR, BOSS, and MRR, respectively. From the I-frame loss perspective, XLinGO only lost 13.6% of I-frames, which is 55%, 70%, 62%, 75% lower than LinGO, BLR, BOSS, and MRR, respectively. XLinGO lost 19.4% of P-frames, which is 60%, 63%, 62%, 75% lower than for LinGO, BLR, BOSS, and MRR, respectively. We conclude that XLinGO transmits video packets with reduced frame loss rate, and thus it protects priority frames (I- and P-frames) in periods of congestion and link errors. Frame loss ratio for other values of s_{max} also confirms that XLinGO protects priority frames, but due to restricted space of this paper they are not shown.

Figure 4(c) shows the number of packets dropped at intermediate queues, where BLR has the lowest number of packets dropped. This is because BLR only considers geographical information to compute the DFD, and due to the unreliability of wireless channels, the most distant node might suffer from a bad connection. MRR has the higher number of packets dropped at intermediate buffers, since it considers a rectangle as forwarding area. In particular, MRR selects forwarders closer to boundaries of the rectangle, which does not mean that a forwarder provides geographical advance towards DN, i.e., MRR does not select forwarders closer to DN. Hence, MRR increases the number of hops, interferences, and buffer overflow. Finally, XLinGO has less buffer overflow than LinGO, BOSS, and MRR. This is because XLinGO avoids the selection of a forwarder node with heavy traffic load, which prevents buffer overflow, minimizes packet loss, delay and jitter, and also provides load balancing. The number of dropped packets increases, as soon as the number of SN increases, as shown in Figures 4(c) and 4(d). This is because more packets are transmitted, which increases interference.

3.4   Signalling Overhead Evaluation

For the signalling overhead evaluation, we counted the number of control messages transmitted to deliver a given video flow, as shown by results of Figures 4(e) and 4(f). BLR defines a recovery strategy to deal when route creation fails, where SN broadcasts a control message and all of its neighbours reply with another message indicating their positions. In this way, SN chooses the RN_i closer to the DN as forwarder node. LinGO and MRR do not include any control message, since they define a simple recovery strategy, where the SN repeats the contention-based forwarding mode. It is important to mention that the original

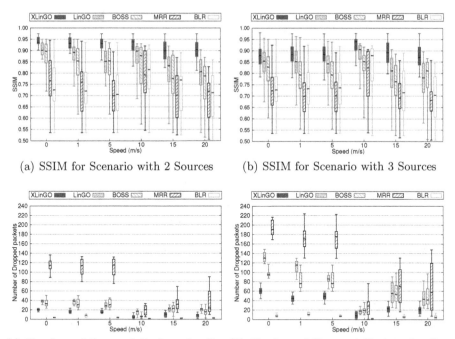

(a) SSIM for Scenario with 2 Sources   (b) SSIM for Scenario with 3 Sources

(c) Number of Dropped Packets for Sce-(d) Number of Dropped Packets for Scenario with 2 Sources   nario with 3 Sources

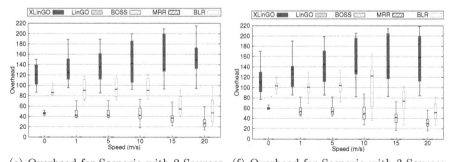

(e) Overhead for Scenario with 2 Sources (f) Overhead for Scenario with 3 Sources

Fig. 4. Video Quality, Number of Dropped Packets, and Overhead for a Scenario with
Different Number of Multimedia SN and Moving Speed

MRR includes extra overhead and delay for the location update mechanism,
since nodes transmit control messages to find a DN's location. However, we
implemented only the routing algorithm, because we consider a static DN.

BOSS includes overhead by the tree-way handshake mechanisms to help the
forwarding selection. Finally, XLinGO adds more control packets, because the
persistent route mode considers a recovery mechanism to detect and quickly
react to route failures, where every node should continually assess whether the

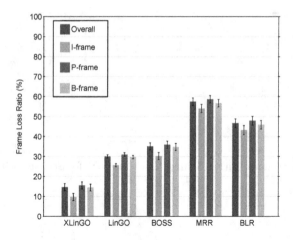

Fig. 5. Frame Loss Ratio for Videos Transmitted with s_{max} equals to 20m/s and 2 Simultaneous Video Flows Transmissions

persistent route $P_{SN,DN}$ is still reliable or available to transmit packets. It is important to highlight that XLinGO provides the best trade-off between the video quality and signalling overhead.

3.5   Video Quality Evaluation

We selected a random frame (i.e., Frame 143) from the UAV_1 video sequence transmitted by each protocol to show the impact of transmitting video streams from the standpoint of the end-user, as displayed in Figure 6. Frame number 143 is the moment when the church tower appears in the scene, as shown in Figure 6(a). This frame transmitted via XLinGO has the same quality compared to the original frame, as can be seen in Figure 6(b), and makes the benefits of XLinGO for video transmission evident. Apart from distortions at frames transmitted via LinGO, BLR, MRR, and BOSS, the buildings do not appear in the same position compared to original frame, as it can be seen in Figures 6(c), 6(d), 6(e), and 6(f), respectively. This is because this frame was lost, and the decoder reconstructed it based on previously received frames.

Figure 7 shows SSIM values for each frame of the UAV_1 video sequence for videos of Figure 6. For example, frame 143 transmitted via XLinGO, LinGO, BLR, BOSS, and MRR has SSIM values equal to 0.97, 0.69, 0.58, 0.61, and 0.29, respectively. First, these SSIM results can be attributed to the fact that DN received the frame correctly when transmitted via XLinGO. Second, this is an I-frame, which was lost when transmitted via LinGO, BLR, BOSS, and MRR. Hence, there is a higher distortion for a longer period, since an I-frame has more information to update the scene in case of scene changes. Moreover, distortion propagates in subsequent frames, because the decoder uses an I-frame as a reference for remaining frames within the GoP [15]. Error propagation explains the poor quality level for videos transmitted via LinGO, BLR, BOSS, and MRR. Third, the decoder uses Frame-Copy as an error concealment method, which means that the decoder must replace each lost frame with the last one that was correctly received. Hence, frame 143 was reconstructed based on frame numbers 130, 120, 131, and 73 when transmitted via LinGO, BLR, BOSS, and MRR, respectively. This is because they are the last frames received correctly by the DN.

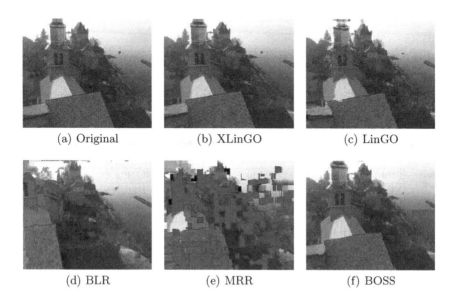

(a) Original   (b) XLinGO   (c) LinGO

(d) BLR   (e) MRR   (f) BOSS

Fig. 6. UAV Frame #143 transmitted via different beacon-less OR

This result also helps us to illustrate the ability of XLinGO to deliver all video frames with quality level support. We can see that frames transmitted via XLinGO have similar and constant quality compared to the baseline video quality level (original video), since XLinGO builds a reliable persistent route by taking into account cross-layer multiple metrics for routing decision, and also provides a recovery mechanism to deal with route failures. The video quality of frames transmitted via LinGO, BLR, BOSS, and MRR have a higher distortion compared to the baseline video quality. XLinGO, LinGO, BLR, BOSS, and MRR reduced the SSIM for entire video in 3%, 16%, 25%, 16, and 45%, respectively, compared to the benchmark video quality level.

Table 2 summarizes the results achieved by this study, where we conclude that LinGO, BLR, BOSS, and MRR perform poorly compared to XLinGO in scenarios composed of mobile nodes, simultaneous multiple flows video transmission, and videos with different motion and complexity levels.

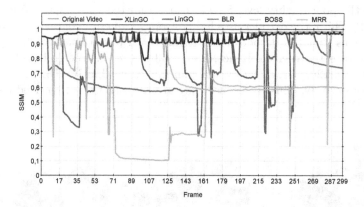

Fig. 7. SSIM for All Frames for the Compose the UAV_1 Video Sequence

Table 2. Performance Comparison Among Different Beaconless OR Protocols

Protocols	Reliability	Robustness	Overhead	QoE assurance
XLinGO	High	High	High	High
LinGO	High	Medium	None	Medium
BLR	Medium	Medium	Medium	Medium
BOSS	Medium	Medium	Medium	Medium
MRR	Low	Low	None	Low

4   Conclusions

In this paper, we introduced a comparative simulation study of different beaconless OR protocols in terms of video quality level, frame loss ratio, and signalling overhead. We performed simulations to analyse the impact of node mobility, different numbers of video flows, and videos with different motion and complexity levels, such as expected in many multimedia FANET applications. Based on our evaluation analysis, we identified that LinGO, BLR, BOSS, and MRR perform poorly compared to XLinGO in scenarios composed of mobile nodes, multiple flows, and videos with different motion and complexity levels. This is because XLinGO builds a reliable persistent route $P_{SN,DN}$ by combining link quality, geographical information, and queue length. Hence, it selects forwarder nodes closer to DN with a reliable link, and sufficient queue capacity to forward packets with a reduced packet loss ratio. It also considers a mechanism to detect topology changes, providing smoother operation in mobile networks. This performance is desirable for many multimedia FANET applications, such as safety & security, environmental monitoring, and natural disaster recovery.

References

1. Bekmezci, İ., Sahingoz, O.K., Temel, Ş.: Flying Ad-Hoc Networks (FANETs): A survey. Ad Hoc Networks 11, 1254–1270 (2013)
2. Huang, P., Chen, H., Xing, G., Tan, Y.: SGF: a State-free Gradient-based Forwarding Protocol for Wireless Sensor Networks. ACM Transactions on Sensor Networks (TOSN) 5(2), 14 (2009)
3. Ickin, S., Wac, K., Fiedler, M., Janowski, L., Hong, J.-H., Dey, A.: Factors Influencing Quality of Experience of Commonly Used Mobile Applications. Communications Magazine 50, 48–56 (2012)
4. Ehsan, S., Hamdaoui, B.: A survey on energy-efficient routing techniques with QoS assurances for wireless multimedia sensor networks. Communications Surveys Tutorials 14(2), 265–278 (2012)
5. Hsu, C.-J., Liu, H.-I., Seah, W.K.G.: Survey Paper: Opportunistic Routing - A Review and the Challenges Ahead. Computer Network 55(15), 3592–3603 (2011)
6. Rosário, D., Zhao, Z., Braun, T.I., Cerqueira, E., Santos, A., Alyafawi, I.: Opportunistic Routing for Multi-flow Video Dissemination over Flying Ad-Hoc Networks. In: 15th IEEE International Symposium on a World of Wireless, Mobile and Multimedia Networks. IEEE, Sydney (2014)
7. Heissenbüttel, M., Braun, T., Bernoulli, T., Wälchli, M.: BLR: Beacon-less Routing Algorithm for Mobile Ad hoc Networks. Computer Communications 27(11), 1076–1086 (2004)
8. Sanchez, J., Ruiz, P., Marin-Perez, R.: Beacon-less geographic routing made practical: challenges, design guidelines, and protocols. IEEE Communications Magazine 47(8), 85–91 (2009)
9. Al-Otaibi, M., Soliman, H., Zheng, J.: A multipath routeless routing protocol with an efficient location update mechanism. International Journal of Internet Protocol Technology 6(1/2), 75–82 (2011)
10. Rosário, D., Zhao, Z., Santos, A., Braun, T., Cerqueira, E.: A beaconless Opportunistic Routing based on a Cross-layer Approach for Efficient Video Dissemination in Mobile Multimedia IoT Applications. Computer Communications 45(1), 21–31 (2014)
11. Rosario, D., Zhao, Z., Silva, C., Cerqueira, E., Braun, T.: An omnet++ framework to evaluate video transmission in mobile wmsn. In: 6th International Workshop on OMNeT++, Cannes, pp. 277–284 (2013)
12. Zhao, Z., Braun, T., Rosário, D., Cerqueira, E., Immich, R., Curado, M.: QoE-aware FEC mechanism for intrusion detection in multi-tier WMSNs. In: Inter. Workshop on Wireless Multimedia Sensor Networks, Barcelona, pp. 697–704 (2012)
13. Library, V.: YUV Video Sequences (June 2014), http://trace.eas.asu.edu/yuv/
14. A Set of Videos Transmitted via XLinGO, LinGO, BLR, BOSS, and MRR together with Original Videos (April 2014), https://plus.google.com/u/0/b/102508553201652207043/102508553201652207043/posts
15. Aguiar, E., Riker, A., Cerqueira, E., Abelém, A., Mu, M., Braun, T., Curado, M., Zeadally, S.: A Real-time Video Quality Estimator for Emerging Wireless Multimedia Systems. Wireless Networks, 1–18 (2014)

A Balanced Battery Usage Routing Protocol to Maximize Network Lifetime of MANET Based on AODV

Esubalew Yitayal[1], Jean-Marc Pierson[2], and Dejene Ejigu[1]

[1] IT Doctoral Program, Addis Ababa University,
Addis Ababa, Ethiopia
[2] Laboratoire IRIT UMR 5505, Universite Paul Sabatier,
Toulouse, France

Abstract. Energy efficiency is a critical issue for battery-powered mobile devices in ad hoc networks. Failure of node or link allows re-routing and establishing a new path from source to destination which creates extra energy consumption of nodes, sparse network connectivity and a more likelihood occurrences of network partition. Routing based on energy related parameters is one of the important solutions to extend the lifetime of the network. In this paper, we are designing and evaluating a novel energy aware routing protocol called a balanced battery usage routing protocol (BBU) which uses residual energy, hop count and energy threshold as a cost metric to maximize network life time and distribute energy consumption of Mobile Ad hoc Network (MANET) based on Ad hoc on-demand Distance Vector (AODV). The new protocol is simulated using Network Simulator-2.34 and comparisons are made to analyze its performance based on network lifetime, delivery ratio, normalized routing overhead, standard deviation of residual energy of all Nodes and average end to end delay for different network scenarios. The results show that the new energy aware algorithm makes the network active for longer interval of time once it is established and fairly distribute energy consumption across nodes on the network.

Keywords: AODV, BBU-AODV, MANET, NS-2.34, Network Lifetime, energy consumption, residual energy.

1 Introduction

Recent advances in wireless communication technologies and availability of less expensive computer processing power have led to an interest in mobile computing applications. Mobile Ad hoc Network (MANET) is a special type of wireless network in which a collection of mobile entities form a temporary network without the aid of any established infrastructure or centralized administration [1]. Therefore, dynamic topology, unstable links, limited energy capacity and absence of fixed infrastructure are special features for MANET when compared to wired networks. These characteristics put special challenges in routing protocol design.

The key challenge in the design of wireless ad hoc networks is the limited availability of the energy resources. Each of the mobile nodes is operated by a limited energy battery and usually it is impossible to recharge or replace the batteries during a mission such as in battlefields and emergency relief scenarios [2, 3]. Since each mobile node in a MANET acts both as a router and host and most of the mobile nodes rely on other nodes

S. Balandin et al. (Eds.): NEW2AN/ruSMART 2014, LNCS 8638, pp. 266–279, 2014.

to forward their packets, the failure of a few nodes, due to energy exhaustion, might cause the disruption of service in the entire network. Thus, researchers have focused on design of power-aware network protocols for the ad hoc networking environment to extend network lifetime and balance energy usage among mobile nodes.

In recent years, many researchers have focused on the optimization of energy consumption of mobile nodes, from different points of view. Some of the proposed solutions try to adjust the transmission power of wireless nodes; other proposals tend to efficiently manage a sleep state for the nodes [4]. Finally, there are many proposals which try to design an energy efficient routing protocol by means of an energy efficient routing metric instead of the minimum-hop count.

In this paper a new energy efficient algorithm called BBU-AODV, which maximizes the life time of a MANET by avoiding routing of packets through nodes with low residual energy and balance the total energy consumption among all nodes in the network while selecting a route to the desired destination, is proposed. BBU routing protocol is developed on top of the popular AODV routing protocol.

The remainder of the paper is organized as follows. In section 2, we describe MANET routing protocols. In Section 3, we review some of the proposed energy-aware routing protocols for MANETs. We explain in detail our proposed work and its integration with AODV in Section 4. In Section 5, we compare the performance of our protocol with that of AODV via Network simulator NS-2.34 simulations for a variety of network scenarios, and finally, we conclude the paper in Section 6.

2   MANET Routing Protocols

In this section we describe routing protocols in MANET and the basic operation of the reactive AODV routing protocol. MANET routing protocols could be classified into three categories based on the routing information update mechanism: proactive (table-driven), reactive (on-demand) and hybrid [5].

Proactive routing protocols require nodes to exchange routing information periodically and compute routes continuously between any nodes in the network, regardless of using the routes or not. This means a lot of network resources such as energy and bandwidth may be wasted, which is not enviable in MANETs where the resources are constrained. On the other hand, reactive routing protocols do not exchange routing information periodically. Instead, they discover a route only when it is needed for the communication between two nodes. Proactive protocols inherently consume more energy than the Reactive ones; hence most of the research works involve modifications to reactive protocols. The last category which is Hybrid routing protocols combine the basic properties of the first two classes of protocols. That is, they are both reactive and proactive in nature. It uses the route discovery mechanism of reactive protocol to determine routes to far away nodes and the table maintenance mechanism of proactive protocol to maintain routes to nearby nodes.

Among reactive protocols, AODV is considered potentially the most energy efficient routing protocol. Hence many research studies have focused on making AODV routing protocol more energy efficient [6].

Ad hoc On-demand Distance Vector (AODV): When a node wants to find a route to a destination and does not have a valid route to that destination, it will initiate a path discovery process. Path discovery process is initiated by broadcasting a route request packet (RREQ) to its neighbors. When a node receives RREQ in case it has routing information to the destination, it sends a route reply (RREP) packet back to the source. Otherwise, it rebroadcasts RREQ packet further to its neighbors till either the destination is reached or another node is found with a fresh enough route to the destination. Nodes that are part of an active route keep its connectivity by broadcasting periodically local Hello messages to its neighbors. If Hello messages stop arriving from a neighbor beyond some time threshold, the connection is assumed to be lost. When a node detects that a route to a neighbor node is not valid it removes the routing entry and sends a route error (RERR) message to neighbors that are active and use the route. This procedure is repeated at nodes that receive RERR messages till it reaches to the source node. A source that receives an RERR can reinitiate a route discovery by sending a RREQ Packet. In AODV, the routing process will not consider about the energy of the node rather it considers only minimum hop-count along the paths [7]. Hence AODV algorithm may result in a quick depletion of nodes battery along the most heavily used routes in the network.

3   Related Works

Routing is one of the important solutions to the problem of energy efficiency in Mobile ad hoc network. In the recent past years energy efficient routing in Ad hoc network has been addressed by many research works which has produced so much innovation and novel ideas in this field. The majority of energy efficient routing protocols for MANET try to reduce energy consumption by means of an energy efficient routing metric instead of the minimum-hop metric. Each and every protocol has some advantages and shortcomings. None of them can perform better in every condition. It depends upon the network parameters which decide the protocol to be used. This section reviews some of the many energy efficient schemes based on AODV developed by researchers in the field.

In Zhaoxiao et al. [8], to mitigate the energy saving problem, an energy-aware routing named EAODV for Ad Hoc networks is proposed. The algorithm selects routing according to the dynamic priority-weight (β) and takes the hop count as an optimization condition. The dynamic priority weight is determined using the square of the ratio of residual battery energy(R) and consumed energy(C) of a node at time t as shown below.

$$\beta_i(t) = \left(\frac{R_i(t)}{C_i(t)} \right)^2$$

The destination node selects two maximum summation of priority-weight which spends less energy and owns larger capacity based on synthetic analysis among possible routes and propagates the route reply (RREP) messages to the source node. The second path will be used when the primary path fails. Since the work considered the summation of priority-weight, the selected path for data transmission might contain a node which has less remaining energy.

Jie et al [9] propose a PS-AODV routing protocol based on load conditions of a node to balance uneven nodes energy consumption of the traditional AODV. They made an improvement in route discovery process. Node checks its load value when received a RREQ packet before forwarding RREQ packets. If the node load is too high, it refuses to forward the RREQ packet until the load is reduced. However queue load condition could not give guarantee to protect nodes with little battery capacity which decreases network life time.

Lei and Xiaoqing [10] propose an improved energy aware AODV strategy to extend network life time. The improvements made by the authors are on route request packets and hello mechanisms of AODV. However the algorithm did not consider fair distribution of energy usage across nodes on the network.

Patil et al. [11] introduce an algorithm which combines Transmission Power and Remaining Energy Capacity and integrates these metrics into AODV so that the Ad hoc network has a greater life time and the energy consumption across the nodes is reduced. During route discovery from source to destination the transmission and remaining energy values along the route are accumulated in the RREQ packets. At the destination or intermediate node (which has a fresh enough route to the destination) these values are copied into the RREP packet which is transmitted back to the source. The source alternates between the maximum remaining energy capacity route and minimum transmission route every time it performs route discovery. Since hop count did not consider as a cost metric and transmission power is used for route selection, the selected link for data transmission might be frequently broken which creates more energy consumption and shorten life time of network.

Kim and Jang [12] propose an enhanced AODV routing protocol to maximize networks lifetime in MANET using an Energy Mean Value algorithms. Here, energy remaining of each node in the path between source and destination is accumulated and delivered to the destination by adding a field on a RREQ message. The destination node does not give a RREP reply immediately to the first RREQ, rather it waits for 3 * NODE_TRAVERSAL_TIME to receive additional RREQ packets destined for the node. Then the destination node adds the accumulated residual energy of each path and divides by the number of hops along the paths to obtain the mean energy of network. Finally the destination node unicasts RREP messages along the reverse path of the RREQ message received first and nodes hearing the RREP message store the mean energy. When a new path is discovered, the mean energy stored in each node is compared with the residual energy in the node. If the residual energy is less than the mean energy, the delay time of RREQ message is set to be 0.5ms otherwise the delay time of RREQ message is set to 0.05ms.Since the nodes are mobile, cumulative delay of each node affect the relay node out its position during data transmission which minimizes network lifetime and consumes battery.

Liu et al. and Sara et al. [13,14], propose a multipath mobile ad hoc routing protocol which extends the Ad Hoc On-demand Multipath Distance Vector routing protocol . The protocol finds the minimum remaining energy of each route and sort multi-route by descending nodal residual energy. Once a new route with higher nodal residual energy is emerging, it is reselected to forward rest of the data packets.

The work done in Tie et al. [15] proposes ALMEL-AODV which considers node remaining energy as a routing metric to balance and extend the survival time of the nodes in the network. The proposed algorithm chooses two highest summations of residual energy routes for data transmission. The second route will be used as a backup. Although the metric used is important, a node which has very low residual energy might be included during message transmission as they centered on maximum summation of remaining energy irrespective of nodal residual energy. Hence the remaining capacity of each host should be consider as a metric to prolong the life time of the network.

Kim et al. [16] introduce an energy drain rate metric, which represents the rate of battery consumption. It estimates the lifetime of a node; therefore, if the estimated value is below a threshold, the traffic passing through it can be diverted in order to avoid node failure due to battery exhaustion. The cost of a node i is calculated as the ratio between the Remaining Battery Power (RBC) and the Drain Rate (DR):

$$C = RBC / DR$$

C-K Toh [17] proposes a routing algorithm called Minimum Total Transmission Power Routing (MTPR) based on minimizing the amount of energy required to get a packet from source to destination. The problem is mathematically stated as:

$$\underset{\Pi}{Min} \left\{ \sum_{(i,j) \in \Pi} T_{ij} \right\}$$

Where Tij denotes the energy consumed in transmitting between two consecutive nodes i and j in route Π. Although the MTPR can reduce the total energy consumption of the overall network, it does not reflect the lifetime of each mobile entity.

Singh et al. [18] propose the Minimum Battery Cost Routing (MBCR) which used the remaining energy capacity as a cost metric, and the cost function is defined as:

$$C_R = \sum_{i=1}^{k-1} f(E_r^i(t))$$

where $f(E_r^i(t)) = \dfrac{1}{E_r^i(t)}$; $E_r^i(t)$ – remaining energy of node i at time t

MBCR selects routes with a minimum cost value to choose the route with the maximum remaining energy capacity. However, MBCR only considers the summation of the inverse of residual battery capacities for all nodes along the path. Thus, routes containing small energy capacity nodes can still be chosen.

Local Energy-Aware Routing Protocol is proposed by works in [19][20][21] [22][25]. When a node receives a RREQ message at time t, it compares its current remaining energy capacity with the predefined threshold value or computed value. If the residual energy is less than the threshold or computed value, the RREQ message is dropped. Otherwise, the message is processed and forwarded. However, the destination will receive a route request message only when all intermediate nodes along the route have enough battery levels. If all the paths to destination have small residual energy, the RREQ message will not be reached at the destination.

Kumar and Banu [23] present an E2AODV scheme to balance load distribution of nodes. A threshold value is used to judge if intermediate node was overloaded or not. Here, an intermediate node receiving the RREQ will compare its current queue length with its threshold before rebroadcasting it. If queue length is greater than the threshold, the RREQ will be dropped. Otherwise, the node will broadcast it. In their scheme, the threshold value plays the key role in selecting nodes whether or not to forward RREQ. Every time an intermediate node receives a RREQ, it will recalculate the threshold according to the average queue length of all the nodes along the path to the node itself. Therefore, the threshold is variable and changing adaptively with the current load status of network.

4   The Proposed Work

The main aim of this work is to propose a routing protocol that increase the life time of network and fairly distribute an energy consumption of hosts in MANET. The algorithm which we propose combines threshold, summation of residual energy, min residual energy and hop count as a cost metric and integrates these metrics into AODV in an efficient way. This metrics ensure that all the nodes in the network remain up and running together for as long as possible.

4.1   Modification on RREQ Packet

The proposed energy aware AODV modifies route request (RREQ) packet for route discovery process as shown in Figure 1. We modified the fields in the RREQ packet by adding minimum residual energy (MRE) and sum of residual energy (SRE) which keeps the minimum remaining energy and sum of remaining energy along the path respectively. An EnergyDifference (D) field, which stores the difference between either average minimum residual energy (AME) and threshold (Th) or average sum of residual energy (ASE) and threshold (Th), is also added on the routing table at a destination node.

In BBU-AODV, when all nodes in some possible routes between a source-destination pair have large remaining energy than the threshold then a route with maximum of the difference of average sum of residual energy and threshold among the routes is selected. Otherwise the maximum difference of the average minimum residual energy and threshold among the routes is selected.

TYPE	Reserved	Hop Count
Broadcast ID		
Destination IP Address		
Destination Sequence Number		
Source IP Address		
Source Sequence Number		
Minimum Residual Energy(MRE)		
Sum Residual Energy(SRE)		

Fig. 1. Modified RREQ Packet format

4.2   Mathematical Model of BBU-AODV

If we consider a generic route $r_j = n_0 , n_1 , n_2 , \dots , n_d$, where n_0 is the source node and n_d is the destination node , h is the number of hop between n_0 and n_d and a function $r(n_i)$ denotes the residual energy of node n_i then the average minimum residual energy(AME) and average summation of residual energy(ASE) for the route r_j is calculated as:

$$AMR(r_j)= (\min_{\forall n_j \in r_j} r(n_j))/h$$

$$ASR(r_j) = (\sum_{\forall n_j \in r_j} r(n_j))/h$$

The BBU-AODV algorithm selects an optimal route k O_k (D) which verifies the following condition:

If (minimum residual energy along a path is greater than or equal to the threshold i.e.

$$\underset{r_j \in A}{AMR}(r_j) \times h >= Th)$$

Choose a route which has the maximum of the difference of average residual summation and threshold i.e.

$$O_k(D) = \max_{r_j \in A}(ASR(r_j)) - Th$$

Else

Choose a route with maximum difference of average minimum residual energy and threshold i.e.

$$O_k (D)= \max_{r_j \in A}(AMR(r_j)) - Th$$

Where A is the set of all routes under consideration and Th is a predefined energy threshold.

4.3   Algorithm for RREQ Handling

The pseudo code in Figure 2 shows the algorithm used to search for the desired path and the flow chart of RREQ handling at the intermediate and destination node for BBU-AODV is as shown in Figure 3.

The intermediate nodes process RREQ as follows:

Step 1: It checks whether RREQ is new by looking up the source node id and broadcast ID in a routing table

Step 2: If RREQ is the first or greater Destination Sequence Number, a node updates additional MRE and SRE fields of RREQ as follow, then rebroadcast RREQ.
   MRE=min (residual energy of current node, MRE of RREQ received)

   SRE= (residual energy of current node + SRE of RREQ received)

Step 3: If RREQ is not the first or Destination Sequence Number is not greater than the sequence number in the routing table, then the coming RREQs is discarded.

The destination node processes RREQ as follows:

Step 1: The node checks whether RREQ is first arrived by looking up the source node id and broadcast ID in a routing table.

Step 2: If RREQ is first arrived, it calculates an EnergyDifference(D) value as shown below and waiting time (δ) for additional RREQ's packet and keeps it on a routing table for additional RREQ.

Let threshold (Th) = some constant energy E

If (MRE >= Th)

$$D= ((SRE/hopcount) - Th)$$

Else

$$D= ((MRE/hopcount) - Th)$$

Step 3: If RREQ is not the first, then the node checks its waiting time δ.

Step 4: If RREQ is not expired, then the algorithm calculates an energyDifference (routing cost) for the new RREQ and compares it with an EnergyDifference value on the the routing table. If the route cost (D) of the incoming RREQ is greater than an EnergyDifference (D) in the routing table, then the destination node replaces the routing table entry of an existing RREQ by the incoming copies of RREQ otherwise the incoming RREQ is discarded.

Step 5: If the node receives another copy of RREQ, it executes step 4 till its waiting time expires.

Step 6: If waiting time expires a destination node sends an RREP on the reverse path which has large value of EnergyDifference to a source node.

Fig. 2. Pseudo code on how node process RREQ

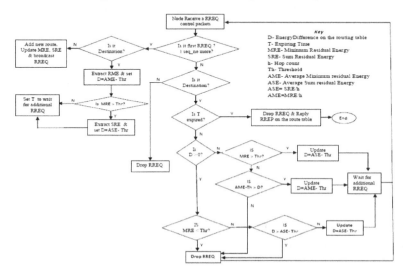

Fig. 3. The flow chart of RREQ handling by BBU-AODV

4.4   Comparison of Routing Protocols

To understand the operations of the proposed protocol, we consider three different routing protocols namely AODV, ALMEL-AODV and BBU-AODV. In Fig.4, the number written above a node corresponds to the value of residual node energy during RREQ received and inside a node indentifies a particular node. We have also used 10 joule as an energy threshold for the network.

Case 1: Choose a route with minimum hop count between source and destination (AODV routing protocol). AODV selects route < S-6-7-8-D > which has the smallest hop count of 4.

Case 2: Choose a route with largest Summation of residual energy. (Max_Sum Energy (ALMEL-AODV) routing protocol. The ALMEL-AODV algorithm selects route <S-1-2-3-4-5-D> which is the largest summation of residual energy.

Case 3: Choose a route with large summation of residual energy and less hop count if possible; otherwise choose a route with largest minimum residual energy and less hop count (proposed routing protocol i.e. BBU-AODV). Our proposed model selects a route with largest value of EnergyDifference (D). Thus route <S-9-10-11-12-D> which has largest D value of 6.6 is selected.

Case 1 selects the shortest path without considering remaining energy of nodes. Thus, case 1 does not give guarantee for long network lifetime. Case 2 selects a route with largest summation of residual energy but it has serious problem in terms of life time and hop count as it may still choose a route with nodes containing small remaining battery capacity as shown in Figure 4. Case-3 improves the drawbacks of Case 1 and Case-2 by considering both residual energy and hop count as a cost metric. Based on our algorithm the cost function (D) for path S-1-2-3-4-5-D, S-6-7-8-D, S-9-10-11-12-D and S-13-14-15-16-17-18-D is -9.5, -9.25, 6.6 and 5.4 respectively as shown in Figure 4. Hence BBU-AODV selects path S-9-10-11-12-D which is the largest value of D i.e. 6.6 for data transmissions. So the proposed algorithm always chooses a route which extends network lifetime by taking energy capable nodes and distributes load among mobile nodes as well by taking either large summation of residual energy or maximum residual energy.

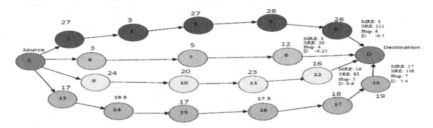

Fig. 4. BBU-AODV route setup from node S to node D

5   Simulation and Results

5.1   Simulation Environment

In this paper the simulations are carried out using Network Simulators-2 version 2.34 [24] to evaluate the performance of the proposed energy efficient routing protocol against AODV and ALMEL_AODV. We used Wireless Channel/Wireless Physical, Propagation model is Free Space Propagation Model, Queuing model is Drop Tail/Priority Queue, Mobility model is Random Waypoint model and MAC protocol is 802.11. The simulation setup consists of an area of 500m X 500m with different number of nodes ranging from 100 to 200 for each simulation. Each packet starts travelling from a random location to a random destination with a randomly chosen speed. When a node reaches a destination, it moves to another randomly chosen destination after a pause. To emulate the dynamic environment, all nodes move around in the entire region with maximum speed of 20m/sec. Constant Bit Rate (CBR) traffic source with packet size of 512 bytes is used. Traffic scenarios with 15 source-destination pairs were used to establish the routes. All the simulations were run for a period of 500 sec. The initial energy of each node was set as 100 Joule with transmission and reception power of 1 W and 0.5W respectively. The Energy threshold value for the simulation is set to 30 Joule and the expiration time of D at the destination routing table is set to 0.1sec. Identical movement and traffic scenarios are used across all protocols.

Once the trace file is generated, a Perl and AWK scripts are used to analyze the information from the trace file. Based on the output of these scripts, graphs are plotted for network lifetime v/s number of nodes, delivery rate v/s number of nodes, normalized routing overhead v/s number of nodes, standard deviation of residual energy of all nodes v/s number of nodes and average end to end delay v/s number of nodes. The parameters used in the simulation are listed in Table 1.

Table 1. Simulation Parameters

Simulation Parameters	Values
Number of Nodes	100 to 200
Geographical areas(m²)	500mX500m
Packet Sizes(Bytes)	512
Traffic Type	CBR
Pause time(sec)	40
Mobility Model	Random way Point
Simulation Time(sec)	500
Initial Energy(Joule)	100
Transmission Energy(Watt)	1
Reception Energy(Watt)	0.5
Traffic Sources	15
Maximum Speed(m/s)	20
Threshold(Joule)	30
Expire-Time(T) of D(sec)	0.1

5.2   Performance Metrics

The following performance metrics are used to evaluate our algorithm against AODV routing protocol.

Network lifetime: the time it takes for the first node to deplete its energy.

Delivery ratio: the ratio of data packets reaching the destination node to the total data packets generated at the source node.

Standard deviation of residual energy of all Nodes: shows how much remaining energy variation or "dispersion" exists from the average.

Normalized routing overhead: the total number of routing packets transmitted per data packets delivered at the destination.

Average End-to-End Delay: the interval time between sending by the source node and receiving by the destination node, which includes the processing time and queuing time.

5.3   Simulation Results

The following results show that how BBU-AODV improves the performance of a MANET.

Network lifetime

Our modified energy aware algorithm outperforms both AODV and ALMEL-AODV by achieving long duration of time for the first node exhausts its energy on the network during the simulations as illustrated by Fig. 5. The improvement in network life time is due to the fact that our enhanced energy aware AODV prevents small residual energy nodes as a relay node or selects a path which has long duration than shorter path between source and destination.

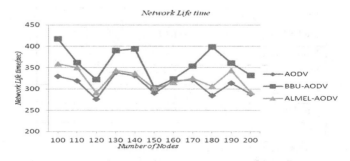

Fig. 5. Network life time with different number of nodes

Standard deviation of residual energy of all Nodes

In our energy aware AODV as the value of residual energy of each node on the path is above the threshold maximum summation of energy path with less hop count will be used. This technique protects nodes at the early stage before they exhaust their battery capacity (i.e. the battery will be used more fairly) and thus the standard deviation of residual energy of the enhanced energy aware AODV algorithm is smaller than AODV and ALMEL-AODV as shown in Fig.6.

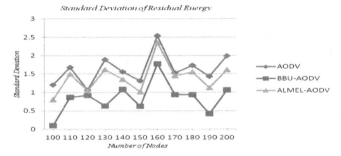

Fig. 6. Standard deviation with different number of nodes

Delivery Ratio

Fig.7 shows that the proposed scheme provides higher delivery ratio than AODV and ALMEL-AODV. This is due to the fact that our algorithm selects energy capable path which is active for longer duration of time.

Fig. 7. Delivery ratio with different number of nodes

Normalized Routing Overhead

Fig.8 shows the normalized control packet overhead required by the transmission of the data packets. AODV and BBU-AODV has less normalized control packet overhead than ALMEL-AODV. The reason is that both schemes consider hop count as a cost metric unlike ALMEL-AODV. Thus ALMEL-AODV uses more number of hops as a relay node which creates additional number of route reply message and link failure.

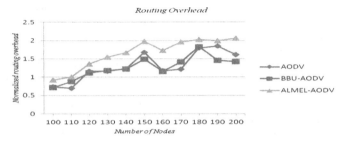

Fig. 8. Normalized routing overhead with different number of nodes

Average End to End Delay

In Fig. 9, we observe that the average end to end delay of AODV is smaller than both BBU-AODV and ALMEL-AODV. The reason is that AODV only finds shorter routes during route discovery irrespective of other parameters. Furthermore, the average end to end delay of ALMEL-AODV is higher than BBU-AODV. This is because ALMEL-AODV doesn't consider hop count as a cost metric. Hence the probability that longer paths will be selected increases.

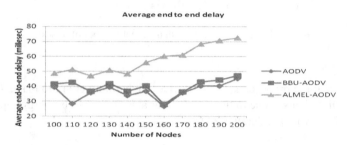

Fig. 9. Average end to end delay with different number of nodes

6   Conclusion

Maximum network life time and fair utilization of energy usage is very important in a MANET as a new route discovery process has to be reinitiated if less energy capable path is selected which creates extra energy consumption of nodes and affects delivery ratio of the network. To overcome these problems, a BBU-AODV algorithm is proposed. The simulation results show that BBU-AODV has better network lifetime, delivery ratio and reasonable distribution of energy consumption across nodes than both ALMEL_AODV and classical AODV. But AODV has relatively low normalized overhead and average end to end delay than the modified one. Now we are working on to examine the effect of various mobility and traffic models on BBU-AODV algorithm. We are also looking forward to minimize routing overhead of both AODV and BBU-AODV.

References

1. Royer, E.M., Toh, C.K.: A Review of Current Routing Protocols for Ad Hoc Mobile Wireless Networks. IEEE Personal Communications 6(2), 46–55 (1999)
2. Meshram, A., Rizvi, M.A.: Issues and Challenges of Energy Consumption in MANET Protocols. International Journal of Networking and Parallel Computing, 56–60 (2013)
3. Shivashankar, H., Suresh, N., Golla, V., Jayanthi, G.: Designing Energy Routing Protocol with Power Consumption Optimization in MANET. IEEE Transactions on Emerging Topics in Computing (2013)
4. Prakash, S., Saini, J.P., Gupta, S.C., Vijay, S.: Design and Implementation of Variable Range Energy Aware Dynamic Source Routing Protocol for Mobile Ad Hoc Networks. International Journal of Computer Engineering and Technology (IJCET), 105–123 (2013)
5. Abolhasana, M., Wysockia, T., Dutkiewicz, E.: A review of routing protocols for mobile ad hoc networks. Elsevier Ad Hoc Networks 2 (2004)

6. Patil, A.P., Kanth, K.R., Sharanya, B., Kumar, M.P.D., Malavika, J.: Design of an Energy Efficient Routing Protocol for MANETs based on AODV. IJCSI International Journal of Computer Science, 215–220 (2011)
7. Perkins, C.E., Belding-Royer, E.M., Das, S.: Ad hoc on-demand distance vector (AODV) routing. Network Working Group, RFC 3561 (2003)
8. Zhaoxiao, Z., Tingrui, P., Wenli, Z.: Modified Energy-Aware AODV Routing for Ad hoc Networks. In: IEEE Global Congress on Intelligent Systems, pp. 338–342 (2009)
9. Jie, T., Yu, W., Jianxing, L.: Researching on AODV and PS-AODV Routing Protocols of Ad Hoc Network for Streaming Media. In: CSAM The 2nd International Conference on Computer Application and System Modeling (2012)
10. Lei, Q., Xiaoqing, W.: Improved energy-aware AODV routing protocol. In: IEEE WASE International Conference on Information Engineering, pp. 18–21 (2009)
11. Patil, A.P., Kanth, K.R., Sharanya, B., Kumar, M.P.D., Malavika, J.: Design of an Energy Efficient Routing Protocol for MANETs based on AODV. IJCSI International Journal of Computer Science Issues 8(4(1)), 215–220 (2011)
12. Kim, J.M., Jang, J.W.: AODV based Energy Efficient Routing Protocol for Maximum Lifetime in MANET. In: IEEE Proceedings of the Advanced International Conference on Telecommunications and International Conference on Internet and Web Applications and Services (2006)
13. Liu, Y., Guo, L., Ma, H., Jiang, T.: Energy Efficient on demand Multipath Routing Protocol for Multi-hop Ad Hoc Networks. IEEE (2008)
14. Sara, G.S., Pari, N.S., Sridharan, D.: Energy Efficient Ad Hoc on Demand Multipath Distance Vector Routing Protocol. ACEE International Journal of Recent Trends in Engineering 2(3), 10–12 (2009)
15. Tie, T.H., Tan, C.E., Lau, S.P.: Alternate Link Maximum Energy Level Ad Hoc Distance Vector Scheme for Energy Efficient AdHoc Networks Routing. In: IEEE International Conference on Computer and Communication Engineering (2010)
16. Kim, D., Aceves, G.L., Obraczka, K., Cano, J.C., Manzoni, P.: Power-aware routing based on the energy drain rate for mobile ad hoc networks. In: IEEE in Proc. 14th ICCCN (2002)
17. Toh, C.-K.: Maximum battery life routing to support ubiquitous mobile computing in wireless adhoc networks. IEEE Communications Magazine 39(6), 138–147 (2001)
18. Singh, S., Woo, M., Raghavendra, C.S.: Power-aware with Routing in Mobile Ad Hoc Networks. In: Proceedings of Mobicom 1998, Dallas, TX (1998)
19. Wang, K., Xu, Y.L., Chen, G.L., Wu, Y.F.: Power-aware on-demand routing protocol for manet. In: IEEE in ICDCSW 2004: Proceedings of the 24th International Conference on Distributed Computing Systems Workshops- W7: EC, ICDCSW 2004, pp. 723–728 (2004)
20. Senouci, S.-M., Naimi, M.: New routing for balanced energy consumption in mobile ad hoc networks. In: ACM in PE-WASUN 2005: Proceedings of the 2nd ACM International Workshop on Performance Evaluation of Wireless Ad hoc, Sensor, and Ubiquitous Networks, pp. 238–241 (2005)
21. Wang, X., Li, L., Ran, C.: An energy-aware probability routing in manets. In: IEEE Workshop on IP Operations and Management, pp. 146–151 (2004)
22. Dwivedi, D.K., Kosta, A., Yadav, A.: Implementation and Performance Evaluation of an Energy Constraint AODV Routing. International Journal of Science and Modern Engineering, IJISME (2013)
23. Kumar, R.V., Banu, R.S.D.W.: E2AODV Protocol for Load Balancing in Ad-Hoc Networks. Journal of Computer Science 8 (7) (2012)
24. Network Simulator, ns-2, http://www.isi.edu/nsnam/-ns/
25. Malek, A.G., LIb, C., Yang, Z., Hasan, A.H.N., Zhang, X.: Improved the Energy of Ad Hoc On-Demand Distance Vector Routing Protocol. In: Elsevier IERI Procedia 2 International Conference on Future Computer Supported Education, pp. 355–361 (2012)

Testbed Evaluation
of Sensor Node Overlay Multicast

Gerald Wagenknecht and Torsten Ingo Braun

Institute of Computer Science and Applied Mathematics,
University of Bern, Switzerland
{wagen,braun}@iam.unibe.ch
http://cds.unibe.ch

Abstract. The Sensor Node Overlay Multicast (SNOMC) protocol supports reliable, time-efficient and energy-efficient dissemination of data from one sender node to multiple receivers as it is needed for configuration, code update, and management operations in wireless sensor networks. SNOMC supports end-to-end reliability using negative acknowledgements. The mechanism is simple and easy to implement and can significantly reduce the number of transmissions. SNOMC supports three different caching strategies namely caching on each intermediate node, caching on branching nodes, or caching on the sender node only. SNOMC was evaluated in our in-house real-world testbed and compared to a number of common data dissemination protocols. It outperforms the selected protocols in terms of transmission time, number of transmitted packets, and energy-consumption.

1 Introduction

Typical sensor data traffic in wireless sensor networks is from many-to-one communication, e.g., many sensor nodes transmitting their data to a single sink. Code updates and other management data require one-to-many communication and data dissemination should be reliable. Distributing data from one sender node to many receivers can be done in different ways. The simplest one is flooding, which is inherently very inefficient, energy-consuming and unreliable. Another way is to rely on multiple unicast flows between source and desired receivers, which is also not efficient for many receivers. In wireless sensor networks (WSNs), however, redundant transmissions might cause more collisions, which leads to inefficient and unreliable data dissemination.

A more efficient data dissemination scheme is multicast, which is able to propagate data from a single sender to many receivers and avoid redundant transmissions. Currently, to our knowledge, there is no multicast protocol able to meet the requirements for reliability and time/energy efficiency simultaneously. Multicast communication should be IP-based to support connectivity between WSN and Internet and the current trend of using IP in WSNs as part of the Internet of Things. One arising question is: How can we design an IP-based multicast dissemination scheme in a wireless sensor network? How can we support end-to-end

S. Balandin et al. (Eds.): NEW2AN/ruSMART 2014, LNCS 8638, pp. 280–293, 2014.
© Springer International Publishing Switzerland 2014

reliability (necessary for code updates) while still keeping energy consumption and delays low? To address these research questions, we developed the Sensor Node Overlay Multicast (SNOMC) scheme. SNOMC's novelty is the combination of various mechanisms such as overlay networks, multicasting, caching, and reliability support.

After discussing related work in Section 2, Section 3 describes the design choices forSNOMC. Section 4 presents in detail SNOMC's performance evaluation results based on measurements in an in-house testbed. Section 5 concludes the paper.

2   Related Work on Multicast Routing and Data Dissemination Protocols

In this section, we present different WSN multicast routing and data dissemination protocols related to our work. Directed Diffusion [1] follows a data-centric routing approach. Data generated by sensor nodes is named by an attribute-value pair. Directed diffusion consists of naming, interest propagation, gradient establishment, data propagation, path reinforcement and repair, but does not support reliability.

Trickle [7] is an early mechanism for code propagation in WSNs. Since it is based on flooding of metadata, all nodes in a WSN will be affected, while a multicast approach only affects nodes in a group, e.g., based on hardware characteristics and software functionality. Multicast can then be used to distribute code updates only to a subset of nodes.

Multipoint Relay (MPR) [3] is also broadcast-based like Flooding and tries to reduce the number of forwarding intermediate nodes. In MPR only a subset of sensor nodes (so called multipoint relays) rebroadcast messages. Relays are chosen based on local knowledge at each node from its two-hop neighbourhood. MPR does not provide reliability mechanisms.

PIM-WSN [9] (PIM: Protocol Independent Multicast) aims to provide energy-efficient multicast routing in WSNs. Data are broadcast but only designated nodes acknowledge reception of a packet. This approach does not guarantee delivery as it is needed in code updates.

Stateless Multicast RPL Forwarding (SMRF) [10] builds a destination-oriented directed acyclic graph and populates sensor nodes' routing tables.

uCast [6] intends to support many multicast sessions, in particular for small numbers of destinations. As with other multicast routing protocols, additional reliability mechanisms are needed to support reliability as required by code dissemination and network management operations.

PSFQ [2] addresses one-sink-to-many-sensors communication. Since reliability is generally more important than transmission time, the main idea of PSFQ's pump operation is to slowly inject packets into the network to avoid congestion. In case of packet errors, PSFQ's fetch operation performs aggressive hop-by-hop recovery at each sensor node to fetch the lost packets from neighbour nodes. To prevent message implosion, PSFQ limits negative acknowledgements (NACK) message transmission to the one-hop neighbourhood. The report operation supports end-to-end error control.

TinyCubus [4], [5] is an adaptive cross-layer framework for wireless sensor networks. It is also broadcast-based and deploys a role-based code distribution algorithm using cross-layer information such as role assignments to decrease the number of messages needed for code distribution to specific nodes. The data dissemination protocol is an integral part of role-based code distribution. As PSFQ, TinyCubus is compared to SNOMC and it turns out that these approaches are less efficient than SNOMC.

Low-Power Wireless Bus [8] is a simple communication protocol turning a multi-hop WSN into a network infrastructure similar to a shared bus. As with other broadcast-based approaches, code update messages can not be limited to a subset of nodes, but affect all nodes in a WSN.

Splash [21] is another dissemination protocols flooding the whole network with the same issues as discussed for Low-Power Wireless Bus. It combines several concepts such as tree pipelining, opportunistic overhearing, XOR encoding, channel cycling etc.

Banerjee [16] et al. discuss energy-efficient broadcast and multicast schemes for reliable communication in multi-hop wireless networks. The choice of neighbors in the broadcast and multicast trees in these schemes are based on link distance and link error rates. The proposed schemes can be implemented using both positive and negative acknowledgment based reliable broadcast techniques in the link layer. The proposed protocols have been evaluated by simulating networks of 100 nodes, but no real-world implementation is provided.

An energy-efficient retransmission scheme based on network coding for wireless sensor networks is discussed by Wang et al. [15]. Intermediate nodes can recover lost packets requested by the nodes having better channel condition. Nodes do not need to keep their radio on during the whole retransmission process and thus get to resulting in consuming less energy. Simulation results show that the proposed approach can reduce energy consumption.

3   Sensor Node Overlay Multicast (SNOMC)

We aimed to design a protocol called Sensor Node Overlay Multicast (SNOMC), which supports multicast in wireless sensor networks in a time and energy-efficient way. SNOMC has been introduced in a previous publication [17], but only preliminary simulation results have been presented [18]. This paper evaluates SNOMC by implementation expirements in a real-world WSN testbed.

There are several possible approaches towards the design of multicast in WSNs such as overlay multicast [14] and IP Multicast. IP Multicast builds its distribution tree between IP routers, while overlay multicast is implemented on the application layer, and therefore, its distribution tree is built between the participating nodes. In case of overlay multicast, different transport protocols (UDP or TCP) can be used for overlay links. Due to its stateless character UDP does not provide reliability, but benefits from low complexity. TCP can support end-to-end reliability but on the cost of certain communication overhead.

We can choose between sender-driven and receiver-driven construction of the multicast tree. In the former, the sender decides on the receiving nodes (the participants in the multicast group) while in the latter the receiving nodes decide themselves whether they want to be part of the multicast group.

Caching is a way to have data closer to the requesting side and hence decreases delays. There are three possibilities on where to cache data: on sender nodes, on branching nodes or on all intermediate nodes (forwarding and branching nodes). Furthermore, nodes that cache data fragments can also detect gaps in the fragment sequence and pro-actively request missed fragments.

To meet our design requirements we choose an overlay multicast approach. The main argument for an overlay approach is that it is easier to implement it at the application layer than at the network layer. We are using UDP as transport protocol and a simple mechanism based on negative acknowledgements (NACK) on application layer (end-to-end) to support reliability. This is option is called UDP-E2E.

3.1 Multicast Distribution Tree Construction

To distribute data from one sender node to many receiver nodes we need to build a distribution tree. A distribution tree can be built only after the receiver nodes have joined a multicast group. A multicast group can be formed by nodes with similar characteristics such as the same hardware type. Joining to a multicast group can happen in two different modes, namely, sender-driven or receiver-driven. In a sender-driven mode the sender decides which receiver nodes should be in the multicast group. The result of the joining phase is a distribution tree including sender (**S**), forwarding nodes (**F**), branching nodes (**B**), and receivers (**Rx**). The joining phase for the sender-driven mode is shown in Fig. 1. In the receiver-driven join approach the receivers themselves decide whether they want to be in a multicast group. In this case they have to know which node is the sender node. The joining phase is shown in Fig. 2. The distribution tree is built through the path of the join messages prior to data transmission.

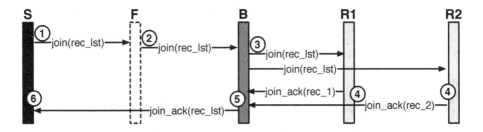

Fig. 1. SNOMC: Distribution tree establishment for sender-driven mode

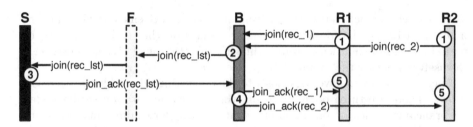

Fig. 2. SNOMC: Distribution tree establishment for receiver-driven mode

3.2 Data Transmission Phase and Caching

Propagating data from sender to receivers is done using overlay links established as result of the tree construction phase. The overlay links are established between the sender, caching nodes, branching nodes, and receivers.

Fig. 3 shows the data transmission for the case of caching at all intermediate nodes. On the branching nodes the data messages have to be duplicated and transmitted to two or more successor nodes. Instead, we can use a broadcast transmission instead of two or more unicast transmissions. Fig. 4 shows the broadcast optimization, where the branching node broadcasts a single packet to receivers 1 and 2 instead of transmitting separate unicast packets as in Fig. 3.

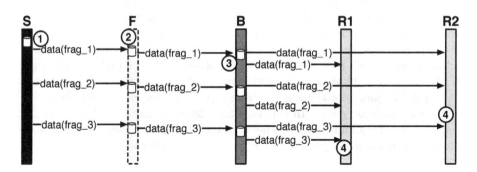

Fig. 3. SNOMC data transmission with caching on forwarding nodes

3.3 End-to-End Reliability

End-to-end reliability is ensured using a mechanism based on negative acknowledgments. If a **data** fragment gets lost a receiver recognizes it and requests the lost **data** fragment by transmitting backwards a **nack** message. A **nack** message, arriving at a sensor node that has the missing **data** fragment cached, triggers retransmission of the fragment. The success of a transmission is indicated using a **data_ack** message transmitted by the receivers towards the sender node. The nodes that cache the **data** fragment also can pro-actively request fragments, if they detect a gap in the fragment sequence. In this case a missed fragment is

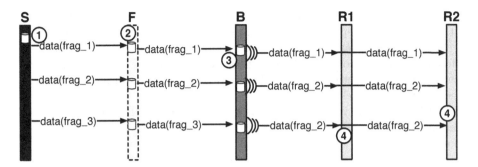

Fig. 4. SNOMC data transmission using broadcast optimization

detected earlier and not only at arrival at receiver nodes. Fig. 5 shows the pro-active mode with caching on each intermediate node, where fragment 2 gets lost between sender and forwarding node.

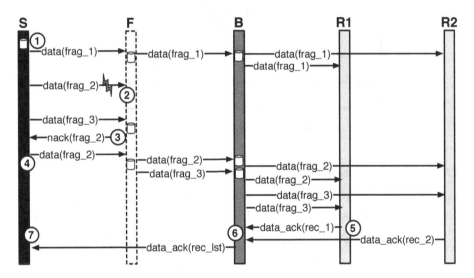

Fig. 5. SNOMC: Reliability, caching on each intermediate node, pro-active retransmission requests

4   SNOMC Evaluation in Real-World Testbed

4.1   Implemented Protocol Stack

The performance of SNOMC is compared to other protocols such as Flooding, MPR and TinyCubus, PSFQ, and Directed Diffusion. We also compare SNOMC to the unicast-based distribution using UDP and TCP. We implemented these protocols in Contiki OS to evaluate SNOMC under real-world conditions running testbed experiments. To ensure a fair evaluation we implemented on top of each of the protocols (except for TCP) a simple NACK-based reliability mechanism including different caching strategies as used in SNOMC. Furthermore, we used different MAC protocols, namely NullMAC and ContikiMAC [19] [20], which are integrated in the Contiki OS [13]. The protocol stack is shown in Fig. 6 and is based on the μIP stack from Contiki OS.

ContikiMAC is a very energy-efficient MAC protocol. In the unicast case ContikiMAC sends a frame until the receiver node is wake up and returns an acknowledgement. In the broadcast case there are no acknowledgements. ContikiMAC transmits for a time longer than the time between two wake-ups of any neighbours to ensure that every neighbour was awake during this time. In our evaluation scenarios, ContikiMAC rebroadcasts a packet for the full wake-up interval of 125 ms.

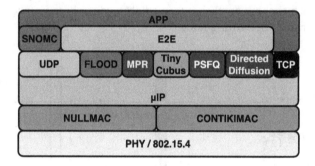

Fig. 6. Implemented protocol stack

4.2   Experimentation Scenarios

For the evaluation of SNOMC we used three different scenarios in an in-house testbed based on Wisebed [12] testbed technology. For all we configured static unicast routes. All experiment runs have been performed 20 times. The first scenario, shown in Fig. 7 left, consists of all 40 sensor nodes in the in-house testbed. We have one sender node (7), and three receiver nodes (16, 19, and 21). The SNOMC distribution tree spans over the nodes 4, 40, 23, and 1. The other nodes are passive in SNOMC but they play an active role while transmitting data using Flooding or MPR. Since most of the broadcast-based protocols did not terminate, i.e. did not deliver the application data reliably, and thus failed in scenario 1 with all 40 nodes, we choose a second scenario (Fig. 7 right). This scenario is similar to scenario 1, but it does not include all the passive nodes, which leads to much less traffic and transmitted packets, in particular for broadcast-based protocols. The third scenario, shown in Fig. 8, is used to compare SNOMC specifically with UDP-E2E. This scenario yields advantages of a multicast-based data distribution scheme compared to a unicast-based one.

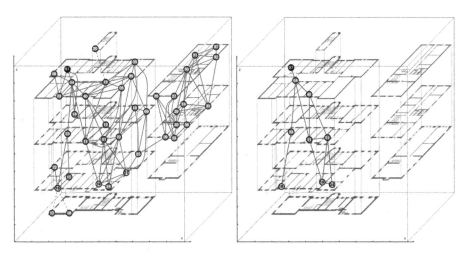

Fig. 7. Evaluation scenarios 1 and 2

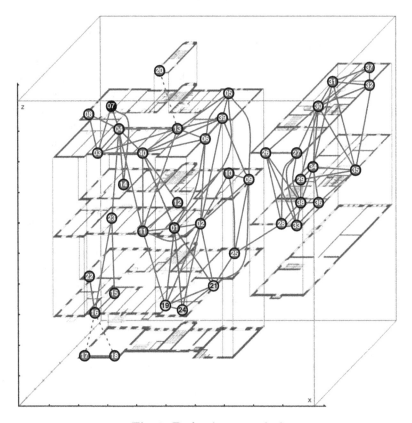

Fig. 8. Evaluation scenario 3

4.3 Transmission Times

In this section, we present the results for the transmission time, which indicates the time needed between the start of transmitting a certain amount of data at the sender and the completion of the reception at the receiver. This includes also any retransmissions caused by packet errors and loss. All figures show on the x-axis the combinations of data dissemination protocols with the different caching strategies. In our notation **s** stands for caching only on sender node, **b(-pa)** for caching on branching nodes (with pro-active requests), **i(-pa)** for caching on each intermediate sensor node (with pro-active requests), and **i-bc** for broadcast optimization of SNOMC.

Fig. 9 shows the transmission times transmitting 1000 bytes from one sender node to the three receiver nodes. SNOMC has the best performance for Null-MAC (left part of Fig. 9). Only the performance of UDP-E2E comes close to SNOMC. Broadcast-based protocols such as Flooding and MPR work generally worse than unicast-based protocols except compared to TCP. Broadcast of data and NACK messages leads to a huge amount of messages in the network and causing high interference. For TinyCubus only nodes 4, 40, 23, and 1 rebroadcast the packets. Thus, there is less traffic in the network and TinyCubus performs somewhat better. Directed Diffusion delivers poor results, because the found paths have been rather bad and thus there are no stable connections. TCP performs poorly, because each packet is acknowledged leading to higher traffic and interferences. Comparing the various caching strategies, we see that caching on each intermediate node shows the best results. Pro-actively requesting missed packets results in higher network traffic. The right part of Fig. 9 shows the results using ContikiMAC. We see that several protocols do not work in this scenario. Only SNOMC, UDP-E2E, TinyCubus and partly Flooding show meaningful results at all. The other protocols fail because of the way ContikiMAC handles broadcasts.

Since several protocols fail using ContikiMAC as MAC protocol in scenario 1, we run the same experiments on a smaller scenario, which only consists of a subset of nodes, i.e. 9 nodes instead of 40 nodes. Fig. 10 shows the results for SNOMC, UDP-E2E, and TCP. They are almost the same as in scenario 1, because the same set of nodes are involved in data transmission. Again, SNOMC outperforms all protocols. In scenario 2 broadcast-based protocols such as Flooding, MPR, and TinyCubus perform better than in scenario 1 but still rather poorly. Again, many packets in the network lead to high interferences and thus many retransmissions. The results in scenario 2 show that the caching strategy has significant influence. While most of the data dissemination protocols using ContikiMAC failed in scenario 1, we have meaningful results for all of them in scenario 2. Fig. 10 shows that SNOMC outperforms all other protocols, broadcast-based protocols perform worst, and the different caching strategies have a certain influence. If we look at the broadcast optimization of SNOMC we see the strong influence of the underlying MAC protocol. In combination with NullMAC the broadcast optimization performs very well. It shows the overall best performance of all protocol and caching combinations. In combination with

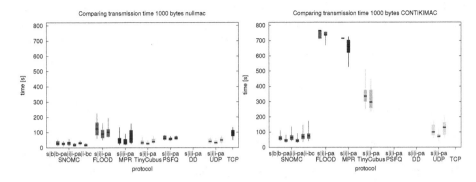

Fig. 9. Transmission time in scenario 1 using NullMac/ContikiMAC

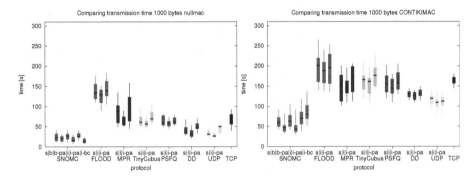

Fig. 10. Transmission time in scenario 2 using NullMac/ContikiMAC

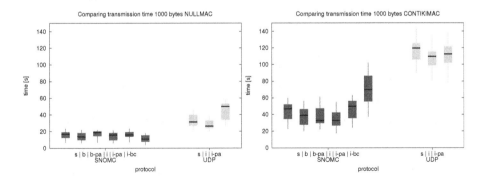

Fig. 11. Transmission time in scenario 3 using NullMac/ContikiMAC

ContikiMAC the broadcast optimization has an opposite impact, since Contiki-MAC handles broadcast transmissions somewhat inefficiently.

For comparison to the strongest competitor of SNOMC, UDP-E2E, we choose a third scenario. Fig. 11 shows that SNOMC outperforms UDP-E2E by factor of 2 to 2.5 with both MAC protocols due to the reduced number of hops. With

UDP-E2E at least 12 (= 3 * 4) hops are necessary to transmit a packet to all three receiver nodes. In case of using SNOMC just 6 (= 3 + 3 * 1) hops, or even 4 (= 3 + 1) hops with broadcast optimization, are needed. This demonstrates the advantage of the overlay multicast communication scheme compared to a unicast-based one.

4.4   Number of Transmissions

The results using scenario 2 are shown in Figure 12. SNOMC outperforms all other protocols such as Directed Diffusion, PSFQ, TCP, and partly TinyCubus, MPR and Flooding. SNOMC requires a somewhat lower number of transmissions than UDP-E2E. SNOMC requires additional packets for the tree establishment phase, but has some advantage in the number of hops due to the use of a distribution tree instead of using three parallel unicast flows as UDP-E2E.

The broadcast-based protocols use a high number of transmissions due to re-broadcasting of packets. Directed Diffusion requires many packets in the Interest phase, which uses broadcast transmissions to find adequate paths. Also PSFQ uses broadcast transmissions in its pump operation. ContikiMAC requires significantly more transmissions than NullMAC for the broadcast-based protocols. We also see this effect comparing the SNOMC broadcast optimization for NullMAC and ContikiMAC.

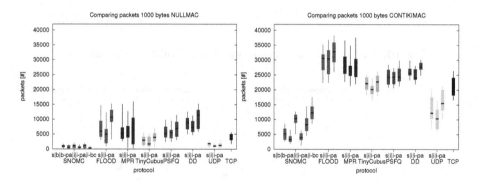

Fig. 12. Number of transmitted packets in scenario 2 using NullMac/ContikiMAC

4.5   Energy Consumption

Energy consumption has been measured using the software-based energy profiling tool powertrace [11], which is integrated into Contiki. The consumed energy includes the energy for transmitting and receiving packets, the operation of the micro-controller on the sensor node, and reading/writing flash memory. For evaluation of energy consumption we used scenario 1. Since NullMAC is an always on protocol (without any energy saving mechanisms) the measured energy values behave as the transmission times, cf. Fig. 13 left)

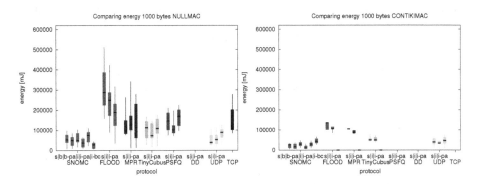

Fig. 13. Energy consumption in scenario 2 with NullMAC/ContikiMAC

For ContikiMAC (cf. Fig. 13 right), we again see that various protocol combinations failed, e.g., Directed Diffusion and PSFQ. Using ContikiMAC in scenario 2 leads to longer transmission time but also to lower energy consumption. Saving energy using ContikiMAC works fine when we have mainly unicast transmissions as in SNOMC, UDP, TCP, and partly Directed Diffusion and PSFQ. In these cases we save a lot of energy compared to NullMAC. For the broadcast-based protocols such as Flooding, MPR and TinyCubus we save less energy, which can be explained by the way how ContikiMAC handles broadcast transmissions.

5   Conclusions

We propose the Sensor Node Overlay Multicast (SNOMC) protocol to support reliable, time-efficient and energy-efficient dissemination of code and management data from one sender node to many receivers. To ensure end-to-end reliability we designed and implemented a NACK-based reliability mechanism. Further, to avoid costly end-to-end retransmissions we propose different caching strategies. We evaluated and compared SNOMC against a number of data dissemination protocols for wireless sensor networks, such as Flooding, MPR (Multipoint Relay), PSFQ (Pump Slowly, Fetch Quickly), TinyCubus, and Directed Diffusion as well as the unicast-based protocols UDP and TCP. To ensure a fair competition we implemented similar reliability mechanisms and caching strategies on top of the selected protocols.

SNOMC performs very well compared to the selected data dissemination protocols. Especially, broadcast-based data dissemination protocols, such as Flooding, MPR, and TinyCubus cause broadcast storms and thus perform very poorly. Data dissemination protocols especially designed for wireless sensor networks, such as Directed Diffusion or PSFQ, have some protocol phases also relying on broadcast transmissions. Thus, also these protocols perform worse than SNOMC. The strongest competitor of SNOMC is UDP-E2E. In this case the advantage of SNOMC depends on the topology of the network and the distribution tree. Further, we showed the influence of the different caching strategies. In general, caching on intermediate nodes improves the performance of each protocol, since expensive end-to-end retransmissions are avoided. Pro-actively requesting missed fragments does not improve performance. SNOMC can be used in different environments and in different systems, independently of the used protocol stack. We conclude that SNOMC can offer a robust solution for the efficient distribution of code updates and management information in wireless sensor networks.

References

1. Intanagonwiwat, C., Govindan, R., Estrin, D., Heidemann, J., Silva, F.: Directed diffusion for wireless sensor networking. IEEE/ACM Transactions on Networking 11(1) (February 2002)
2. Wan, C.-Y., Campbell, A.T., Krishnamurthy, L.: PSFQ: A reliable transport protocol for wireless sensor networks. In: Proceedings of the ACM International Workshop on Wireless Sensor Networks and Applications (WSNA 2002), Atlanta, GA, USA (September 2002)
3. Liang, O., Sekercioglu, Y., Mani, N.: A survey of multipoint relay based broadcast schemes in wireless ad hoc networks. IEEE Communications Surveys and Tutorials 8(4) Fourth Quarter (2006)
4. Marron, P.J., Lachenmann, A., Minder, D., Hhner, J., Sauter, R., Rothermel, K.: Tinycubus: A flexible and adaptive framework for sensor networks. In: Proceedings of the European Conference on Wireless Sensor Networks (EWSN 2005), Istanbul, Turkey (January 2005)
5. Marron, P.J., Minder, D., Lachenmann, A., Rothermel, K.: Tinycubus: An adaptive cross-layer framework for sensor networks. it - Information Technology 47(2) (February 2005)
6. Cao, Q., He, T., Abdelzaher, T.: uCast: Unified Connectionless Multicast for Energy Efficient Content Distribution in Sensor Networks. IEEE Transactions on Parallel and Distributed Systems 18(2) (February 2007)
7. Levis, P., Patel, N., Culler, D., Shenker, S.: Trickle: a self-regulating algorithm for code propagation and maintenance in wireless sensor networks. In: 1st Conference on Symposium on Networked Systems Design and Implementation, NSDI 2004, vol. 1. USENIX Association, Berkeley (2004)
8. Ferrari, F., Zimmerling, M., Mottola, L., Thiele, L.: Low-power wireless bus. In: 10th ACM Conference on Embedded Network Sensor Systems (SenSys 2012). ACM, New York (2012)
9. Marchiori, A., Han, Q.: PIM-WSN: Efficient multicast for IPv6 wireless sensor networks. In: IEEE International Symposium on a World of Wireless, Mobile and Multimedia Networks, June 20-24 (2011)
10. Oikonomou, G., Phillips, G.I.: Stateless multicast forwarding with RPL in 6LowPAN sensor networks. In: IEEE International Conference on Pervasive Computing and Communications Workshops (PERCOM Workshops), March 19-23 (2012)
11. Dunkels, A., Eriksson, J., Finne, N., Tsiftes, N.: Powertrace: Network-level power profiling for low-power wireless networks. Technical Report T2011:05, Swedish Institute of Computer Science (March 2011)
12. Coulson, G., Porter, B., Chatzigiannakis, I., Koninis, C., Fischer, S., Pfisterer, D., Bimschas, D., Braun, T., Hurni, P., Anwander, M., Wagenknecht, G., Fekete, S., Krller, A., Baumgartner, T.: Flexible experimentation in wireless sensor networks. Communications of the ACM 55(1) (January 2012)

13. Dunkels, A., Grönvall, B., Voigt, T.: Contiki - A Lightweight and Flexible Operating System for Tiny Networked Sensors. In: Proceedings of the 29th Annual IEEE International Conference on Local Computer Networks, LCN 2004. IEEE Computer Society, Washington, DC (2004)

14. Hosseini, M., Ahmed, D.T., Shirmohammadi, S., Georganas, N.D.: A survey of application-layer multicast protocols. Communications Surveys & Tutorials 9(3), 58–74 (2007)

15. Wang, X., Tan, Y., Zhou, Z., Zhou, L.: Reliable and energy-efficient multicast based on network coding for wireless sensor networks. In: 6th International ICST Conference on Communications and Networking in China (CHINACOM), pp. 1142–1145 (August 2011)

16. Banerjee, S., Misra, A., Jihwang, Y., Agrawala, A.: Energy-efficient broadcast and multicast trees for reliable wireless communication. Wireless Communications and Networking, 660–667 (March 20, 2003)

17. Wagenknecht, G., Anwander, M., Braun, T.: SNOMC: An overlay multicast protocol for Wireless Sensor Networks. In: 9th Annual Conference on Wireless on-demand Network Systems and Services, WONS 2012, Courmayeur, Italy, January 9-11 (2012)

18. Wagenknecht, G., Anwander, M., Braun, T.: Performance Evaluation of Reliable Overlay Multicast in Wireless Sensor Networks. In: Koucheryavy, Y., Mamatas, L., Matta, I., Tsaoussidis, V. (eds.) WWIC 2012. LNCS, vol. 7277, pp. 114–125. Springer, Heidelberg (2012)

19. Dunkels, A.: The ContikiMAC Radio Duty Cycling Protocol SICS Technical Report T2011:13

20. Dunkels, A., Mottola, L., Tsiftes, N., Österlind, F., Eriksson, J., Finne, N.: The Announcement Layer: Beacon Coordination for the Sensornet Stack. In: Marrón, P.J., Whitehouse, K. (eds.) EWSN 2011. LNCS, vol. 6567, pp. 211–226. Springer, Heidelberg (2011)

21. Doddavenkatappa, M., Chan, M.C., Leong, B.: Splash: Fast Data Dissemination with Constructive Interference in Wireless Sensor Networks. In: 10th USENIX conference on Networked Systems Design and Implementation, Lombard, IL, April 02-05 (2013)

M2M Traffic Models and Flow Types
in Case of Mass Event Detection

Alexander Paramonov and Andrey Koucheryavy

Saint-Petersburg State University of Telecommunications,
Pr. Bolshevikov, 22, Saint-Petersburg, Russia
`alex-in-spb@yandex.ru`, `akouch@mail.ru`

Abstract. The Internet of Things (IoT) is the new ITU-T concept for the network development [1, 2]. The IoT is based today on the Ubiquitous Sensor Network (USN) [3, 4] and M2M decisions [5, 6]. So the USN and M2M traffic models should be studied well. The USN traffic models were considered for telemetry applications in [7], for medical applications in [8], for image applications in [9]. There are many M2M traffic model investigation papers too [10, 11, 12]. The M2M traffic models and flow types definition in the case of mass event detection are the investigation goal of this paper. The anti-persistent flow type for M2M traffic in the case of mass event detection is identified. The results can be used for Recommendation Q.3925 "Traffic flow types for testing quality of service parameters on model networks" modification.

Keywords: Traffic, anti-persistent, self-similar, M2M, network.

1 Introduction

The ITU-T standardized IoT [1, 2] and USN [3, 4] during study period 2009-2012. The M2M Focus group was created too. The M2M Focus group prepared Recommendation set which will be approved at the nearest time. The USN and M2M is the technological base for IoT concept implementation. So the USN and M2M traffic models should be studied well. The ITU-T Recommendation Q.3925 "Traffic flow types for testing quality of service parameters on model networks" [13] defined the traffic flow types for varied networks, including USN. The traffic flow types for M2M are not defined in the Q.3925 up to now.

There are many traffic model investigation papers [10, 11, 12]. The M2M traffic applications via wireless networks was investigated at [10] under FP7 LOLA (Achieving Low-LAtency in wireless communications). The mutual M2M and online games traffic influence to network delays was studied. The M2M traffic was modeled as an on-off traffic with constant/uniform packet sizes. The M2M device categorization proposed in the [11]. The asset tracing, building security, fleet, generic communication modems, metering, telehealth are considered as M2M devices. The downlink and uplink and aggregated traffic volumes cumulative distribution functions are studied. The study was based on a week long traffic trace. The M2M and

S. Balandin et al. (Eds.): NEW2AN/ruSMART 2014, LNCS 8638, pp. 294–300, 2014.
© Springer International Publishing Switzerland 2014

smartphone traffic analysis was made together. The resource allocation mechanism for M2M traffic can be improved in accordance with paper conclusions. The M2M traffic is different to human-based traffic [12]. The logistic process M2M traffic is analyzed in [12]. The M2M messages can be varied from few bits in case of temperature measurements up to megabytes in case of video monitoring logistic process. The LTE (Long Term Evolution) network performance can degrade and network can blocked for human-based services in case of emergency events in according with avalanche-like M2M traffic growth. So, the LTE operators should change scheduling mechanisms for support guaranteed QoS to both traffic types: human-based and M2M.

The M2M network as a USN network performance analysis is shown that there are many cases than one or more events detected by mass devices or sensor nodes [14]. Furthermore, the mass events can advance due to coincidence many simple events. It can be cars alarm removing in the morning than people going to the work places, for example. So, the M2M traffic models and flow types definition for mass event detection are the investigation goal of this paper.

2   The Investigation Model

There are many M2M applications. The most important M2M traffic differences to H2H (Human-to-Human) traffic is a traffic initialization body: the devices are the transmitters in M2M traffic models. The three important initialization events can be considered for M2M traffic:

- The environment event. It can be physical parameter varies, for example,
- The scheduling time interval finished. The information should transmitted in accordance with scheduling,
- The management or operation information event. The signaling information should be transmitted.

So, we proposed classify the M2M traffic models to three big groups. First of all is the Random traffic in accordance with environment events. The second M2M traffic models group is the determination (scheduling) traffic. The SCADA (Supervisory Control and Data Acquisition) systems can be a good example this M2M applications. The third M2M traffic models group is the service M2M traffic models. The service M2M traffic model flows are studied in [7]. The traffic models and flow types in the case of mass event detection for Random traffic will investigate in the paper further. Moreover, the t new traffic models for scheduling traffic will propose and flow types in this case will investigate too.

3   The Random M2M Traffic Model

The M2M network as an USN network performance analysis is shown that there are many cases than one or more events detected by mass devices or sensor nodes [14, 15]. Furthermore, the mass events can advance due to coincidence many simple events. It can be cars alarm removing in the morning than people going to the work places, for example. The traffic flow should define by monitoring traffic data during whole network life-time. Moreover, ad hoc event can change the traffic structure. The ad hoc event is event than the mass events can advance. Let p is a probability of ad hoc events for m/n devices. The p can be defined the traffic types and flow parameters. We will prove this assumption further.

The next simulation model was investigated:

- The each from n devices can be passive or active at the time t,
- The passive device is the mode than device monitoring environment events periodically with T_i time interval. The T_i value is selected randomly in the initialization model moment in accordance with uniform distribution in the interval $\{T_{\min}, T_{\max}\}$.
- The active device is the mode than ad hoc event detects by this device. The active device will come back to the passive mode, than ad hoc event is finished. The traffic value from active device is random in accordance with uniform distribution in the interval $\{v_{\min}, v_{\max}\}$ with mean value $\bar{v}$.
- The ad hoc event occurs in the independent random time moment. The interval between ad hoc events is random in according with exponential distribution with mean value $\bar{t}_E$.

The simulation was made by ns-2. The next values are using: $n = 10$, $T_{\min} = 0$, $T_{\max} = 1$, $v_{\min} = 0$, $v_{\max} = 1000$ (messages), $\bar{t}_E = 100$.

The flow realization example is shown on the fig.1. As we see the peak traffic values are obtained during ad hoc events.h respect to

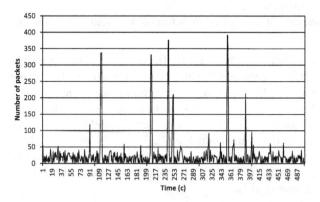

Fig. 1. The traffic flow realization in case of mass events

The Hurst parameters estimations by the method of changes in the variance function are shown on the fig.2. The variance was calculated in according with:

$$\left(\sigma^2\left(X_t^{(m)}\right)\right) \sim -\beta \cdot \log(m) + \log(a),$$

where $X_i^{(m)} = \dfrac{1}{m} \displaystyle\sum_{t=m(i-1)+1}^{mi} X(t)$;

the m is the number of data aggregation intervals.

There are three functions on the fig.2 for the different probability values $p = 0.1; 0.5; 1.0$. The Hurst parameters are $H = 0.451; 0.375; 0.292$ respectively.

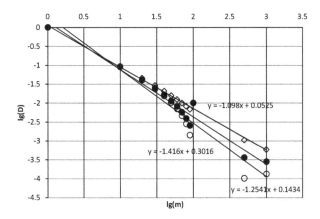

Fig. 2. The Hurst parameter estimation

So, the anti-persistent flow type for M2M traffic in the case of mass event detection is identified.

4   The Scheduling M2M Traffic Models

The SCADA (Supervisory Control and Data Acquisition) systems can be a good example the scheduling M2M applications. The master-slave concept can be used for SCADA model performance. The master is a data collection server and the slave is a sensor (device), for example. The n servers and m devices can create the network generally.

The traffic model for the scheduling M2M network is the next. The traffic between server i and device j is the deterministic flow including requests data flow from servers to devices and replies data flow from devices to servers. The flow parameters define by schedule in accordance with time period T_i.

The generally case traffic model for the scheduling M2M networks to n monitoring system has the deterministic flow too. The phase shift between requests occurs each monitoring system can be defined as φ_i $i = 1 \ldots n$. The generally traffic model can be defined as the deterministic flow too with request time period $T_S = lcm\{T_i\}$ $i = 1 \ldots n$. The is the minimal common multiple for whole request time periods . The generally model is shown on the fig.3.

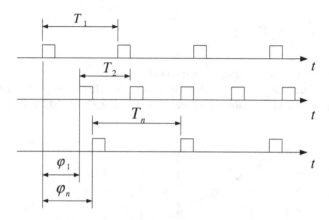

Fig. 3. The generally case traffic model for the scheduling M2M networks

There are many cases than the phase shifts are not the stable. It can be following to restart separate monitoring system, for example. The generally traffic can be not deterministic in this case.

The next simulation model was investigated:

- The number of monitoring system is n,

- The T_i $i = 1...n$ period selected randomly from interval [1, 60] s under the uniform distribution,

- The phase shift φ_i $i = 1...n$ can be change randomly in accordance with restart moment,

- The interval between restarts following to exponential low with mean value 0.6 restart per hour.

The traffic flow realization for the scheduling M2M network with restarts is shown on the fig.4.

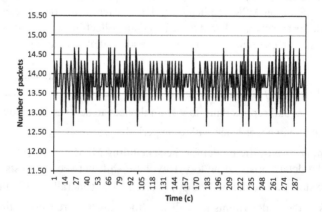

Fig. 4. The traffic flow realization for the scheduling M2M network with restarts

The Hurst parameters estimations by the method of changes in the variance function are shown on the fig.5.

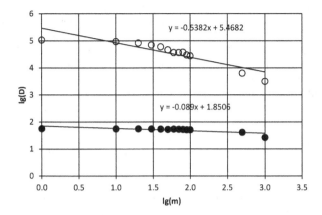

Fig. 5. The Hurst parameters estimations for the scheduling M2M network with restarts

So, the restart events influence to the deterministic flow structure for the scheduling M2M network. The flows change to the self-similar structure. The self-similarity depends from the restarts intensity. The Hurst parameter equal 0.73 than $p = 0.1$ and 0.96 than $p = 1$.

5   Conclusion

The M2M traffic models classification is proposed. The random M2M traffic model in the case of mass event detection was investigated. The anti-persistent flow type for M2M traffic in the case of mass event detection is identified. The Hurst parameter estimations are found.

The traffic model for the scheduling M2M networks under the restart conditions was investigated. The restart events influence to the deterministic flow structure for the scheduling M2M network and the flows change to the self-similar structure. The Hurst parameter estimations are found.

References

1. Recommendation, Y. 2060. Overview of Internet of Things. ITU-T, Geneva (February 2012)
2. Recommendation, Y. 2063. Framework of the WEB of Things. ITU-T, Geneva (July 2012)
3. Recommendation, Y. 2062. Framework of Object-to-Object Communication using Ubiquitous Networking in NGN. ITU-T, Geneva (February 2012)
4. Recommendation, Y. 2221. Requirements for Support of USN Applications and Services in the NGN Environment. ITU-T, Geneva (January 2010)

5. Carugi, M., Li, C., Ahn, J.-Y., Chen, H.: M2M enabled ecosystems: e-health. ITU-T, FG M2M, San Jose, November 13-15 (2012)
6. Andreev, S., Galinina, O., Koucheryavy, Y.: Energy-efficient client relay scheme for machine-to-machine communication. In: Proceedings of the Global Telecommunications Conference (GLOBECOM 2011), Houston, Texas, USA, December 5-9 (2011)
7. Koucheryavy, A., Prokopiev, A.: Ubiquitous Sensor Networks Traffic Models for Telemetry Applications. In: Balandin, S., Koucheryavy, Y., Hu, H. (eds.) NEW2AN/ruSMART 2011. LNCS, vol. 6869, pp. 287–294. Springer, Heidelberg (2011)
8. Vybornova, A., Koucheryavy, A.: Ubiquitous Sensor Networks Traffic Models for Medical and Tracking Applications. In: Andreev, S., Balandin, S., Koucheryavy, Y. (eds.) NEW2AN/ruSMART 2012. LNCS, vol. 7469, pp. 338–346. Springer, Heidelberg (2012)
9. Koucheryavy, A., Muthanna, A., Prokopiev, A.: Ubiquitous Sensor Networks Traffic Models for Image Applications. Internet of Things and its Enablers (INTHITEN). In: Proceedings of the Conference, State University of Telecommunication, St. Petersburg, Russia, June 3-4 (2013)
10. Drajic, D., et al.: Traffic Generation Application for Simulating Online Games and M2M applications via Wireless Networks. In: 9th Conference on Wireless on-demand Network Systems and Services WONS 2012, Courmayeur, Italy, January 9-11 (2012)
11. Shafig, M.Z., et al.: A First Look at Cellular Machine-to-Machine Traffic: Large Scale Measurement and Characterization. In: 12th ACM Sigmetrics Performance International Conference, London, England, UK, June 11-15 (2012)
12. Potsch, T., Marwat, S.N.K., Zaki, Y., Gorg, C.: Influence of Future M2M Communication on the LTE System. In: Wireless and Mobile Networking Conference, Dubai, United Arab Emirates, April 23-25 (2013)
13. Recommendation Q.3925. Traffic Flow Types for Testing Quality of Service Parameters on Model Networks. ITU-T, Geneva (March 2012)
14. Dashkova, E., Gurtov, A.: Survey on Congestion Control Mechanism for Wireless Sensor Networks. In: Andreev, S., Balandin, S., Koucheryavy, Y. (eds.) NEW2AN/ruSMART 2012. LNCS, vol. 7469, pp. 75–85. Springer, Heidelberg (2012)

Modelling a Random Access Channel with Collisions for M2M Traffic in LTE Networks[*]

Vladimir Y. Borodakiy[1], Konstantin E. Samouylov[2], Yuliya V. Gaidamaka[2], Pavel O. Abaev[2], Ivan A. Buturlin[2], and Shamil A. Etezov[2]

[1] JSC"VIVOSS and OI"
Nizhnaya Krasnoselskaya str. 13-1, 105066 Moscow, Russia
[2] Telecommunication Systems Department,
Peoples' Friendship University of Russia,
Ordzhonikidze str. 3, 115419 Moscow, Russia
bvu@systemprom.ru, {ksam,ygaidamaka,pabaev}@sci.pfu.edu.ru,
{ivan.buturlin,setezov}@gmail.com

Abstract. One of the main problems in LTE networks is distribution of a limited number of radio resources among Human-to-Human (H2H) users and increasing number of machine-type-communication (MTC) devices in machine-to-machine (M2M) communications. The radio resources allocation scheme of M2M traffic service in LTE networks discussed in the paper is adopted by 3GPP and implements Random Access Channel (RACH) mechanism for transmitting data units from a plurality of MTC devices. This mechanism determines the sequence of signaling messages transmitted between a MTC device and a ENodeB (eNB) without reserving radio resources. The mathematical model for calculation of the mean time from the instant of MTC device activation to the instant of data transition beginning via eNB is built. The Monte Carlo method for simulation the behavior of the system is used.

Keywords: LTE, M2M, MTC, random access network, RAN, random access procedure, RACH procedure, Markov chain, access delay, simulation.

1 Introduction

Over the past years, a large number of technological devices were launched in the market that support various applications involving the transfer of data automatically, without human intervention. Traditional schemes of the modern wireless communication including 3GPP LTE network, do not allow serving effectively M2M connections between a large number of interacting MTC devices. Extensive increase of network performance is not possible due to the high cost of this approach. One possible solution of the problem is based on the use of random access procedure (RACH procedure) [1,2]. The advantage of this method is that the MTC devices can obtain access

---

[*] The reported study was partially supported by the RFBR, research projects No. 13-07-00953, 14-07-00090.

S. Balandin et al. (Eds.): NEW2AN/ruSMART 2014, LNCS 8638, pp. 301–310, 2014.
© Springer International Publishing Switzerland 2014

to the radio channel for data transmission, regardless of their arrangement and centralized management.

RACH procedure overload could cause congestion in LTE networks. M2M traffic differs from traditional H2H traffic, and existing overload control mechanisms are not capable to overcome overload effectively. MTC devices such as fire detectors usually send small amounts of data periodically while operating in the normal mode. However, in the case of emergency MTC devices generate bursty traffic [3, 4]. So solving various optimization problem for radio resources allocation [5-8] other considerations are overload control problems [5, 9, 10].

With the increasing number of MTC devices, more radio resources are required to provide a service with proper quality. In case of high loaded network average data transfer time will greatly increase and that may be critical in various emergency situations [1]. In papers [5, 9] the authors proposed an overload control mechanisms based on Access Class Barring schemes. New mechanisms of M2M traffic service in papers [11, 12] are based on data transferring that used resources of Physical Uplink Shared Channel (PUSCH). Some features of M2M traffic service were considered in [13-15] by the same authors. The purpose of this paper is to investigate a setup procedure in LTE network taking into account collisions in RACH procedure because of mass of MTC devices. The mathematical model of setup procedure in the form of Markov chain is constructed and analyzed. The Markov chain model is the basis of simulation for estimation of access delay, i.e. the time interval from the instant of MTC device activation to the instant of data transition beginning via eNB.

The paper is organized as follows. In Section 2 we describe RACH signaling process. Section 3 details our mathematical model and introduces its core assumptions. Further, in Section 4 simulation methodologies are presented, and the main performance measure of the proposed model via a numerical example is illustrated. Finally, we conclude the paper in Section 5.

2   RACH Signaling

In this section we consider RACH procedure that is the initial synchronization process between user equipment (UE) and the base station eNB while data exchange performs over Physical Random Access Channel (PRACH) in LTE network [1]. Since UEs' attempts for data transmission can be performed randomly and the value of distance to the eNB is unknown, requests for synchronization from various UEs should come with different delay, which is estimated by the level of incoming PRACH signal by eNB.

In LTE FDD (Frequency Division Duplex) networks, data transmission over PRACH occurs only once during the time interval, that corresponds one subframe of radio carrier [2]. A transmission consists of a preamble and a cyclic prefix.

RACH procedure specified for M2M traffic service in LTE network defines the sequence of signaling messages transmitted between the UE and the eNB as shown in Fig. 1.

The procedure begins with a random access preamble transmission to the eNB (Msg 1) by means of one of available PRACH slots (RACH opportunity). The information about slots is broadcasted by the eNB in System Information Block messages. The number of RACH opportunities and the number of preambles depends on the particular LTE network configuration.

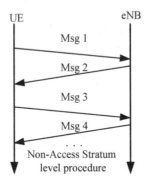

Fig. 1. Signaling flow in LTE 3GPP random access

After preamble sending the UE waits for a Random access response (RAR) (Msg 2) from the eNB within the time interval called a response window. RAR message transmitting over Physical Downlink Control Channel (PDCCH) contains a resource grant for transmission of the subsequent signaling messages. If after the response window is over the UE has not received Msg 2, it means that a collision occurs. For LTE FDD networks a collision of a preamble transmission may occur when two or more UEs select the same preamble and send it at same time slot [1]. According to RACH procedure in the case of a collision the UE should repeat preamble transmission attempt after a response window. If a preamble collision occurs, the eNB will not send RAR message to all UEs, which have chosen the same preamble. In that case, preambles will be resent after the time interval called the Backoff window. In the case of series of collisions for a UE after the number of failed attempts exceeds the preamble attempts limit, the RACH procedure is recognized failed.

In the case of successful preamble transmission after receiving Msg 2 from the eNB and RAR processing time, the UE sends a RRC connection request (Msg 3) to the eNB over Physical Uplink Shared Channel (PUSCH). RACH procedure is considered completed after the UE receiving a contention resolution message (Msg 4) from the eNB. Hybrid Automatic Repeat request (HARQ) procedure guarantees a successful transmission of Msg 3/Msg 4. HARQ procedure provides a limit in Msg 3/Msg 4 sequential transmission attempts. If the limit is reached UE should start a new RACH procedure by sending a preamble.

In the next section simplifying assumptions for RACH procedure are made and the mathematical model in the form of Markov chain is constructed.

3   Assumptions and Model

According to [1] the time interval for completing RACH procedure is one of the important KPIs (Key Performance Indicator) to estimate the effectiveness of functioning of the M2M traffic service scheme in LTE networks. We propose a mathematical model in the form of discrete Markov chain that follows the steps of RACH procedure. Here a state of the Markov chain determines the number of preamble attempt collisions and the number of sequential Msg 3/Msg 4 transmission attempts. With this model the access delay for each state of the Markov chain can be calculated by summing up the corresponding time intervals introduced below:

- Δ_1^1 – waiting time for a RACH opportunity to transmit a preamble;

- Δ_1^2 – preamble transmission time;

- Δ_1^3 – preamble processing time at the eNB;

- Δ_1^4 – RAR response window;

- $\Delta_1 := \Delta_1^1 + \Delta_1^2 + \Delta_1^3 + \Delta_1^4$ – the time interval from the beginning of RACH procedure until sending message Msg 3 or resending a preamble;

- Δ_2 – Backoff window;

- Δ_3 – RAR processing time;

- Δ_4 – time for Msg 3 transmission, waiting for Msg 4, and Msg 4 processing.

A call flow for successful and two fragments of call flows for successful unsuccessful session setup establishment based on RACH procedure are shown in Fig. 2.

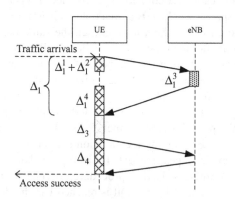

(a)   Access success without collisions

Fig. 2. Message sequence to transmit data with retransmissions in LTE random access

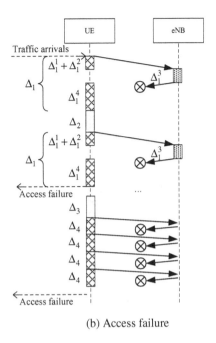

(b) Access failure

Fig. 2. (*continued*)

In the case of reliable connections a session setup delay is equal to the sum of the mentioned above variable $\Delta_i, i = 1,3,4$. If collision occurs or connection is unreliable the number of message retransmissions are limited by $N = 9$ for Msg 1 and $M = 4$ for Msg 3 [1]. Let p denote the probability that collision occurs while sending a preamble. This value depends on the number of MTC devices which are served by eNB, on intensity of incoming calls and on LTE network configuration. In [2] the probability is given by the formula $p := 1 - e^{-\gamma/L}$ for the case when a cell of LTE FDD network is serving a large number of UEs with RACH procedure, where γ - random access intensity, L - the total number of RACH opportunities. Let g denote the HARQ retransmission probability for Msg 3/Msg 4.

Let consider inhomogeneous discrete Markov chain $\{\xi_t\}$ over the state space $L = \{(0),(1),(2),(n,m),0 \le n \le N,0 \le m \le M\}$ which determines the process of transitions between states. We assume that state (0) is the start point of RACH procedure, state (1) is the absorbing state denoting access success, state (2) is another absorbing state denoting access failure, the pair of (n,m) denotes the state when n Msg 1 retransmissions and m Msg 3/Msg 4 retransmissions occurred. This Markov chain can be represented by the graph in Fig. 3.

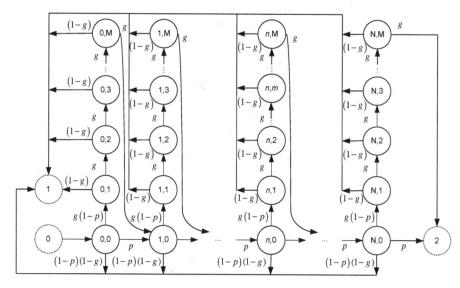

Fig. 3. State transition diagram of discrete Markov chain $\{\xi_i\}$

Consider a Markov chain path $l^k_{(n,m)} = \{(0),(0,0),...,(n,m),(1)\}$ with the state (n,m) just prior to the absorbing state (1). We assume that $L(n,m) = \{l_{(n,m)} : n = \overline{0,N}, m = \overline{0,M}\}, |L(n,m)| = 2^n$, is the set of all possible paths of the Markov chain. The path length (the number of intermediate states) depends on the total number of collisions that occur when sending message Msg 1 and Msg 3/Msg 4. For each state $(n,m) \in L$ the minimum $l^{\min}_{(n,m)}$ and maximum $l^{\max}_{(n,m)}$ paths for $L(n,m)$ are shown in Fig. 4 and corresponding path lengths are given by the following formulae

$$l^{\min}_{(n,m)} = \{(0),(0,0),(1,0),...,(n,0),(n,1),...,(n,m),(1)\},$$
$$|l^{\min}_{(n,m)}| = n + m + 1,$$
(1)

$$l^{\max}_{(n,m)} = \{(0),(0,0),(0,1),...,(0,M),...,(n,0),(n,1)...,(n,m),(1)\},$$
$$|l^{\max}_{(n,m)}| = (M+1)n + m + 1,$$
(2)

$$|l^{\min}_{(n,m)}| \le |l^k_{(n,m)}| \le |l^{\max}_{(n,m)}|, \ 1 \le k \le 2^n.$$
(3)

As mentioned above a state (n,m) of the Markov chain determines the number of preamble attempt collisions and the number of sequential Msg 3/Msg 4 transmission attempts. So each state (n,m) just prior to the absorbing state (1) determines the access delay of RACH procedure.

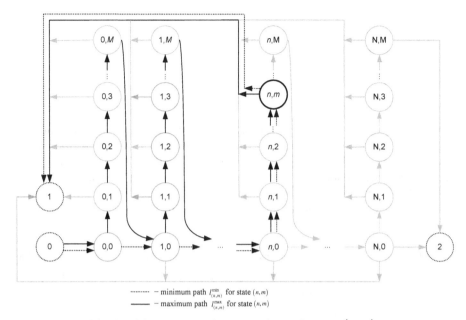

Fig. 4. Minimum and maximum path length for state (n, m)

The next section is devoted to the simulation of process of traffic service based on RACH procedure with collisions.

4 Simulation and Case Study

The simulation model was developed to find the above described characteristics of the Markov chain. The Monte Carlo method used to simulate the behavior of the system. A state-to-state transition process of the model developed is shown below.

Algorithm 1. State-to-state transition process

```
define p, g, M, N
n ← 0
f = true
while n ≤ N and f = true do
    if random[0;1] < p then
    n ← n + 1
    else
    m ← 0
    while m ≤ M and f = true do
            if random[0;1] < g then
                m ← m + 1
    else
            f = false
    if f = true then
```

```
    m ← 0
      n ← n + 1
  if f = true then
        fail
  else
      success
```

We present an example of a single LTE FDD cell supporting M2M communications to illustrate some performance measures for RACH. The initial data for simulation is presented in Table 1 [1, 3, 4]. We assume that the PRACH configuration index is equal to 6. Then the total number of available preamble for data transmission from multiple MTC devices is equal to 54. Given the above assumptions, the probability that collision occurs while sending a preamble is given by the formula [1]:

$$p = 1 - e^{-\gamma/(54*200)} \tag{4}$$

By changing the random access intensity γ and the HARQ retransmission probability g we compute statistic of mean access delay $E\tau$, average path length El for the case close to overload mode [1, 11].

Table 1. Parameter settings

Parameter	Setting
L	200
N	9
M	4
g	0,1; 0,7
Δ_1^1	2,5 ms
Δ_1^2	1 ms
Δ_1^3	2 ms
Δ_1^4	5 ms
Δ_2	20 ms
Δ_3	5 ms
Δ_4	6 ms

Figure 2 introduces plots illustrating statistic of mean access delay for two cases ($g = 0,1$ and $g = 0,7$) on increasing random access intensity from MTC devices. The figure indicates that $E\tau$ varies significantly with the changing of the collision probability p and can reach values exceeding 140 ms due to a significant number of Msg 1 retransmissions.

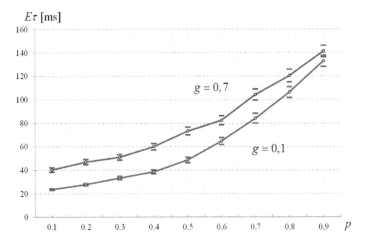

Fig. 3. Mean access delay and 95% confidence intervals

The plots for average path length El ($g = 0,1$, $g = 0,7$ and $g = 0,99$) are shown in Figure 6. With the increasing of the collision probability p the number of retransmissions is growing, hence the number of intermediate states before the absorbing state (1) also increases. The figure indicates that when $g = 0,99$ and $p = 0,9$ average path length El tends to minimum length to the absorbing state (2) denoting access failure.

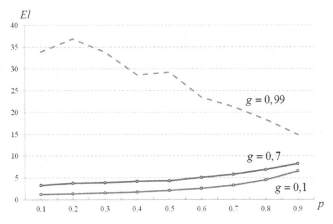

Fig. 4. Average path length

5   Conclusion

In this paper, we addressed a RACH procedure for service M2M traffic in LTE cell and give a model in the form of discrete Markov chain. The model of RACH procedure is based on the number of retransmissions of signaling message between the UE and eNB. An interesting task for future study is the analysis of various admission control schemes for M2M communication.

References

1. 3GPP TR 37.868 – Study on RAN Improvements for Machine-type Communications (Release 11) (2011)
2. 3GPP LTE Release 10 & beyond (LTE-Advanced)
3. Ericsson, R1-061369, LTE random-access capacity and collision probability
4. Beale, M.: Future challenges in efficiently supporting M2M in the LTE standards. In: Proceedings of the 10th Wireless Communications and Networking Conference, WCNCW 2012, Paris, France, pp. 186–190. IEEE (2012)
5. Hossain, E., Niyato, D., Han, Z.: Dynamic Spectrum Access andManagement in Cognitive Radio Networks. Cambridge University Press (2009)
6. Borodakiy, V.Y., Buturlin, I.A., Gudkova, I.A., Samouylov, K.E.: Modelling and analysing a dynamic resource allocation scheme for M2M traffic in LTE networks. In: Balandin, S., Andreev, S., Koucheryavy, Y. (eds.) NEW2AN/ruSMART 2013. LNCS, vol. 8121, pp. 420–426. Springer, Heidelberg (2013)
7. Buturlin, I.A., Gaidamaka, Y.V., Samuylov, A.K.: Utility function maximization problems for two cross-layer optimization algorithms in OFDM wireless networks. In: Proc. of the 4th International Congress on Ultra Modern Telecommunications and Control Systems, ICUMT 2012, October 3-5, pp. 63–65. IEEE, St. Petersburg (2012)
8. Andreev, S., Koucheryavy, Y., Himayat, N., Gonchukov, P., Turlikov, A.: Activemode power optimization in OFDMA-based wireless networks. In: Proc. of the IEEE Global Telecommunications Conference Workshops, pp. 799–803 (2010)
9. Cheng, M., Lin, G., Wei, H.: Overload control for machine-type-communications in LTE-advanced system. IEEE Communications Magazine 50(6), 38–45 (2012)
10. Dementev, O., Galinina, O., Gerasimenko, M., Tirronen, T., Torsner, J., Andreev, S., Koucheryavy, Y.: Analyzing the overload of 3GPP LTE system by diverse classes of connected-mode MTC devices. In: Proc. of the IEEE World Forum on Internet of Things, pp. 309–312 (2014)
11. Gerasimenko, M., Petrov, V., Galinina, O., Andreev, S., Koucheryavy, Y.: Impact of MTC on energy and delay performance of random-access channel inLTE-Advanced. Transactions on Emerging Telecommunications Technologies 24(4), 366–377 (2013)
12. Andreev, S., Larmo, A., Gerasimenko, M., Petrov, V., Galinina, O., Tirronen, T., Torsner, J., Koucheryavy, Y.: Efficient small data access for machine-type communications in LTE. In: Proc. of the IEEE International Conference on Communications, pp. 3569–3574 (2013)
13. Gudkova, I., Samouylov, K., Buturlin, I., Borodakiy, V., Gerasimenko, M., Galinina, O., Andreev, S.: Analyzing Impacts of Coexistence between M2M and H2H Communication on 3GPP LTE System. In: Mellouk, A., Fowler, S., Hoceini, S., Daachi, B. (eds.) WWIC 2014. LNCS, vol. 8458, pp. 162–174. Springer, Heidelberg (2014)
14. Andreev, S., Galinina, O., Koucheryavy, Y.: Energy-efficient client relay schemefor machine-to-machine communication. In: Proc. of the IEEE GlobalTelecommunications Conference, pp. 1–5 (2011)
15. Andreev, S., Pyattaev, A., Johnsson, K., Galinina, O., Koucheryavy, Y.: Cellulartraffic offloading onto network-assisted device-to-device connections. IEEE Communications Magazine 52(4), 20–31 (2014)

The Quality of Experience Subjective Estimations and the Hurst Parameters Values Interdependence

Maria Makolkina, Andrey Prokopiev, Alexandr Paramonov,
and Andrey Koucheryavy

State University of Telecommunication, Pr. Bolshevikov, 22, St. Petersburg, Russia
akouch@mail.ru

Abstract. The Hurst parameter estimation methods likely R/S, Higushi method, Wittle method, HEAF2 are considered in the paper for multiservice traffic (voice and video). The Hurst parameter values are obtained during simulation for differences network technical conditions. The Quality for Experience (QoE) subjective estimation method Single Stimulus Continuous Quality Evaluation (SSCQE) is considered. The interdependence between SSCQE values and Hurst parameter values are obtained during simulation. So, the Hurst parameter can be objective metrics for QoE estimation.

Keywords: Hurst parameter, Quality of Experience, Single Stimulus Continuous Quality Evaluation, subjective estimation, objective estimation.

1   Introduction

The QoE definitions and characteristics are done in the ITU-T Recommendations [1,2,]. The QoE estimation is a very difficult task especially during operation. There are many subjective and objective methods for QoE estimation. The subjective methods based on the well known Mean Opinion Score(MOC) methodic [3]. We use the SSCQE subjective method further as the standard method which is recommended by ITU-R [4]. The most famous objective method is Video Quality Metrics (VQM) [5]. This method gives very good results at the laboratory or testing center but it using in the network during operation is difficult. The R-factor [6] is a suitable metric for voice quality estimation. We would like found the similar metric for video QoE.

First of all we investigated the Hurst parameter varies from network traffic flows set (video and voice), flow rate and so on. The results show that the Hurst parameter value depends from the network performance. So, the next investigation step is the Hurst parameter values and SSCQE values interdependence likely R-factor and MOS curve for voice quality.

S. Balandin et al. (Eds.): NEW2AN/ruSMART 2014, LNCS 8638, pp. 311–318, 2014.

2   The Hurst Parameter Value Dependences from the Network Performance

For the Hurst parameter estimation was created the model with the characteristics, based on the recommendations of the International Telecommunication Union [7,8].The studied IP-based network is presented on Figure 1. Modeling was performed using the package Network Simulator 2 (ns2), which allows to analyze in detail the processes occurring in the network, as well as evaluate and analyze network performance in a wide range of variable parameters [9].

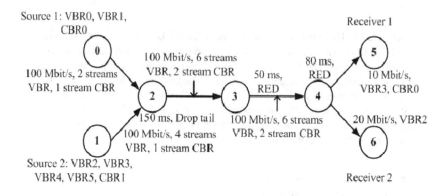

Fig. 1. Block diagram of the simulated network

Figure 1 shows a simplified network configuration, which consists of seven nodes: 2 sources, 3 network nodes, 2 recipients. Node "0" and the node "1" are the sources of video and voice traffic. In the link between nodes "0" and "2" delay is set to 150 ms, the method of traffic management is «DropTail», the transmission rate is 100 Mbit/s. In the link between nodes "1" and "2" delay also set to 150 ms, the method of traffic management is «DropTail», the transmission rate is 100 Mbit/s. Links 2-3 and 3-4 are the core elements of the transport network, the delay is set to 50 ms and 80 ms, the used method of traffic management is «RED» (Random Early Detection algorithm), the transmission rate is 100 Mbit/s. Nodes "5" and "6" are the recipient nodes, so the speed on the links from node "4" to the recipient nodes is 10 and 20 Mbit/s, delay - 200 ms, «DropTail» selected as a traffic control method. Sources generate 6 video streams (VBR0, VBR1, VBR2, VBR3, VBR4, VBR5) and 2 voice stream (CBR0 and CBR1). Video streams have different speeds, VBR0 simulates broadcasting of film encoded with MPEG 2 standard, it's transmitted on the speed 5 Mbit/s, a packet size of 1500 bytes is selected because of using Ethernet technology on the link layer. VBR1 stream simulates video traffic with a low degree of motion, so transmitted at a speed of 3.5 Mbit/s.VBR2 generates a rate of 2.5 Mbit/s, VBR3 - 5 Mbit/s, VBR4 - 10 Mbit/s, VBR5 - 5 Mbit/s. During simulation flow rates changes on the range from 1.5 to 100 Mbit/s. Examined various simulation periods, but regarding video traffic features, was chosen the most interesting simulation period - 1000 seconds.

Aggregate flow is considered as the superposition of several separate independent ON/OFF sources, which transmit at the same rate, but with duration distributed according to an exponential distribution and Pareto distribution.

Hurst parameter estimate was made using the method of analysis of variance changes, rescaled range analysis, Whittle's method, the method of Higuchi, HEAF method [10,11,12].

Figure 2 shows the simulation results for 6 video streams and 2 voice streams. In this experiment, for two video streams were set Pareto law, the remaining four used exponential. During work authors estimate Hurst parameter changes depending on increase of flow rate of a CBR, ranged from 5 to 95 Mbit/sec.

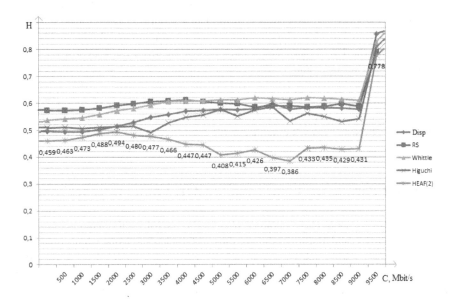

Fig. 2. Hurst parameter dependence on the flow rate CBR

As seen on Figure 2 with the presence of multiple video streams in network when the rate of the voice stream increased the Hurst parameter of the total flow does not increase significantly. Significant increase of the Hurst parameter observed when the reached maximum bandwidth of the researched network that rather due to a change in the status of the network connected with communication overload.

Conclusions from the simulation. During the simulation conducted a series of experiments with different initial conditions, including influence of the packet size to the value of Hurst parameter. With the packet size equal to 1000 bytes than the vaerage value of Hurst parameter is around of 0.724 (by the method of "dispersion changes plot analysis") and 0.684 (Higuchi method).

For the analyzing of dependence on the increasing of the flow rate step, authors turn off one VBR stream, thereby reducing the load on the network, so the network include 5 VBR and 2 CBR streams, flow rate varies in 10,000 kbit/s step. Hurst parameter is around of 0.456 (by the method of "dispersion changes plot analysis") and 0.442 (Higuchi method).

When reducing step of the flow rate of the simulation was changed from 10 000 kbit/s to 1000 kbit/s. Hurst parameter values are in range from 0,496 to 0,520.

In the experiment, which was not take care of the packet size of CBR stream, Hurst parameter values are in the range 0.606 - 0.709. During a burst at the 60 Mbit/s rate, Hurst parameter estimation was 0.7 (by the method of "dispersion changes plot analysis") and 0.9 (Higuchi method).

When authors disable the video streams with the Pareto distribution, and varying only the CBR rate, there was decrease of the Hurst parameter values to 0,456 - 0,442.Also investigated the dependence of the Hurst parameter from queue length at node 4, ranged from 5 to 25%. Hurst parameter values were around 0.684 - 0.724.

Prooved, that the Hurst parameter values depending from the following parameters: a packet size, a queue length at a node, a flow rate increase step, an observation period.

Found that the aggregated load in the network with the presence of a large number of different types of traffic flows is self-similar with an average level of self-similarity, which is the basis for inclusion the Hurst parameter to the list of factors affecting the assessment of the quality of perception.

3   Hurst Parameter Values and SSCQE Values Interdependence

In present times big popularity obtained a video broadcast by its manufacturers, for example, many TV and Radio Company have Internet sites, from which programs, belonging to them are broadcasted "online". Thus, a larger share of network traffic begins to be "online" video traffic, thereby influencing the characteristics of Internet traffic. In order to maintain the quality of the video and other types of traffic at the current level and improve it in the future, much interest rise in the research of the features of the traffic occurring in the network when users access service broadcasting channels in the "online" via the Internet, as well as for timely accounting of traffic increasing in the network.

To investigate the features of video traffic specified above considered traffic from the "First Channel" online broadcasting on the official site «stream.1tv.ru».

Results. For the experiments was used internet access channel with bandwidth 20 Mbit/s, the number of hops was 9, the delay on the route is in the range of 9-12 ms.For investigation was used video, presented in different formats: standard resolution (SD - Standard Definition) and high definition (HD - HighDefinition). The most popular standard for video resolution: 640×360 (corresponding to the average quality) and 640×480 (corresponds to a good quality), HD-video, usually has a resolution of 1280×720 (720p) and assumes the highest quality.

On the figures 3 and 4 show the traffic flow of the "First Channel" broadcast.

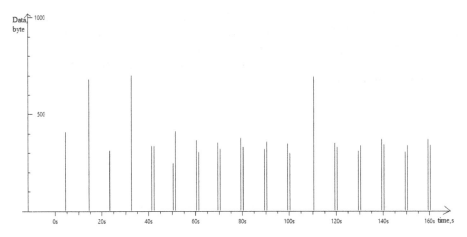

Fig. 3. The traffic flow ("First Channel", measurement interval 1 s)

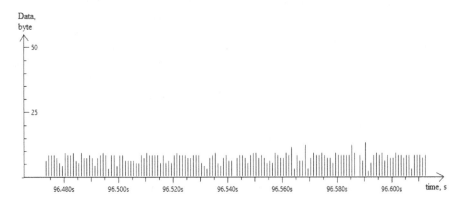

Fig. 4. The traffic flow ("First Channel", measurement interval 10 ms)

Analyzing figures above shows that the traffic on figure 3 has a distinct burst character, the burst nature of traffic on figure 4 is less pronounced.

Data transmitted in packets of maximum length (1514 bytes) using the TCP protocol. Hurst parameter estimation was done by the method of dispersion changes plot analysis [7]. Simultaneously was performed the subjective evaluation of the quality of the transmitted video by the expert group.

As a method of subjective evaluation of video transmission was chosen SSCQE (Single Stimulus Continuous Quality Evaluation) described in ITU BT.500-13 [4]. This method is used to assess the quality of video transmission under conditions close to home video viewing users when there is not possible to compare with the reference test video, obtained directly from the source. Because this method is continuous assessment video, it allows to evaluate each piece of test material. Observers viewed the video once without the reference image. To assess the quality used a continuous scale. Observers offer to evaluate the overall quality of the video image by putting a mark on the vertical scale. Scale is a system of continuous evaluation to avoid quantization errors, but it is divided into five segments of equal length, corresponding to the usual five-point quality scale ITU-R. Terms that define various levels are the same as those usually used: excellent, good, fair, poor, unacceptable.

Table 1 shows the Hurst parameter values for aggregated traffic with different number of video streams.

Table 1. Hurst parameter estimation and SSCQE for aggregated traffic

Number	SSCQE	Hurst parameter
1	1.55	0.733
2	2.73	0.758
3	3.26	0.796
4	3.55	0.812
5	3.98	0.895

The results shown in Table 1 suggest that there is a fairly clear correlation between the values of subjective evaluations of the quality and Hurst parameter, so the value of Hurst parameter can be regarded as an objective quality metric for video perception.Moreover, one can quite reasonably assume that the maximum value of subjective quality assessments will be observed when the Hurst parameter tends to 1.

Also, were performed series of experiments with different initial conditions, but the relationship of the Hurst parameter estimation and the SSCQE method was observed in all results. The results are shown in tables 2 and 3. The data presented in Table 2 were obtained by assessing the aggregate flow on link with bandwidth of 10 Mbit/s. Table 3 shows the values for the Hurst parameter aggregated traffic streams with different numbers of video data streams, additional data traffic and bandwidth shortage.

Table 2. Hurst parameter estimation and SSCQE for aggregated traffic on link with bandwidth of 10 Mbit/s

Number	SSCQE	Hurst parameter
1	1.92	0.642
2	2.83	0.635
3	4.08	0.677
4	3.96	0.689
5	4.41	0.692

Table 3. Hurst parameter estimation and SSCQE for aggregated traffic in case of bandwidth shortage

Number	SSCQE	Hurst parameter
1	1.92	0.635
2	2.51	0.715
3	3.74	0.767
4	3.52	0.741
5	4.16	0.804

Figure 5 shows the relationship between the subjective assessments of the quality of video transmission and Hurst parameter, which is approximated by a logistic curve.

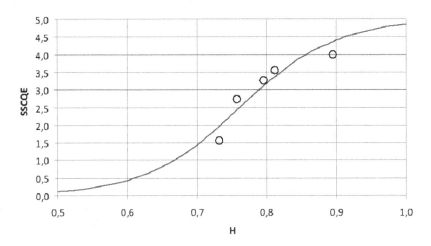

Fig. 5. Relationship between Hurst parameter and subjective assessments SSCQE

Relationship between Hurst parameter and subjective assessments of the quality of video, obtained by the method of valuation SSCQE, can be described by the following expression:

$$F(H) = \frac{M}{1 + e^{-\frac{h - h_0}{b_0}}}$$

where, M - the maximum ratings of the subjective method (in the case of method SSCQE - 5);

h - Hurst parameter value;

h0 - relative value of Hurst parameter;

b0 - parameter which consideret network characteristics.

As you can see, the relationship between Hurst parameter and subjective assessments of the quality of the video is similar to the relationship of perception of subjective evaluations MoS and R-factor used to estimate the voice quality.

4   Conclusions

Shown correlation between the subjective perception of video quality assessments recommended by the International Telecommunication Union, and the value of the Hurst parameter for aggregated video streams.

Defined the dependence of the subjective evaluations of the quality of video experience from Hurst parameter values.

Proposes the use of the Hurst parameter as an objective video quality metric by analogy with the R-factor for the voice quality. On the basis of the relationship between Hurst parameter and subjective video quality assessment can be developed a method for objective evaluation of video quality.

References

1. Recommendation, G. 1011. Reference Guide to Quality of Experience Assessment Methodologies. ITU-T, Geneva (May 2013)
2. Recommendation G.1080. Quality of Experience requirements for IPTV Services. ITU-T, Geneva (December 2008)
3. Recommendation P.800. Methods for Subjective Determination of Trans-mission Quality. ITU-T, Geneva (August 1996)
4. Recommendation ITU-R BT.500-13. Methodology for the Subjective Assessment of the Quality of Television Pictures. ITU-R, Geneva (January 2012)
5. Recommendation ITU-R BT.1683. Objective Perceptual Video Quality Measurement Techniques for Standard Definition Digital Broadcast Television in the Presence of a Full Reference. ITU-R, Geneva (June 2004)
6. Recommendation G.107. The E-model: a Computational Model for use in Transmission Planning. ITU-T, Geneva (February 2014)
7. Recommendation Y.1540. IP Packet Transfer and Availability Performance. ITU-T, Geneva (March 2011)
8. Recommendation Y.1541. Network Performance Objectives for IP-based Services. ITU-T, Geneva (December 2011)
9. Koucheryavy, A., Prokopiev, A.: Ubiquitous Sensor Networks Traffic Models for Telemetry Applications. In: Balandin, S., Koucheryavy, Y., Hu, H. (eds.) NEW2AN/ruSMART 2011. LNCS, vol. 6869, pp. 287–294. Springer, Heidelberg (2011)
10. Sheluhin, O.I., Smolskiy, S.M., Osin, A.V.: Self-Similar Processes in Telecommunications. John Wiley & Sons (2007)
11. Vybornova, A., Koucheryavy, A.: Ubiquitous Sensor Networks Traffic Models for Medical and Tracking Applications. In: Andreev, S., Balandin, S., Koucheryavy, Y. (eds.) NEW2AN/ruSMART 2012. LNCS, vol. 7469, pp. 338–346. Springer, Heidelberg (2012)
12. Kettani, H., Gubner, J.A.: A Novel Approach to the Estimation of the Hurst Parameter in Self-Similar Traffic. In: Proceedings of the 27th Annual IEEE Conference on Local Computer Networks (LCN 2002), Tampa, Florida (November 2002)
13. Dunaytsev, R., Koucheryavy, Y., Harju, J.T.: NewReno throughput in the presence of correlated losses: The slow-but-steady variant. In: proceedings of the INFOCOM 2006: 25th IEEE International Conference on Computer Communications, Barcelona, Spain (April 2006)

Fuzzy Logic and Voronoi Diagram Using for Cluster Head Selection in Ubiquitous Sensor Networks

Yahya Al-Naggar and Andrey Koucheryavy

Saint-Petersburg State University of Telecommunications,
Pr. Bolshevikov, 22, Saint-Petersburg, Russia
yahya_alnaggar@yahoo.com, akouch@mail.ru

Abstract. In this paper, we propose a new algorithm called Cluster-Head selection using Fuzzy Logic with Voronoi diagram in USNs (CHS-FL-VD) that is a distributed algorithm which makes local decisions to select cluster head using a fuzzy inference system based on two parameters which are remaining energy and centrality by Voronoi diagram. Cluster formation using Voronoi diagram. Disseminating data from cluster heads to the base station using gateway and direct connection to properly distribute energy and ensure maximum network life time. A comparison with LEACH and Fuzzy C-Means algorithms has been done. Simulation results show that the proposed algorithm consumes less energy and prolongs the network life time compared with these algorithms.

Keywords: Ubiquitous Sensor Networks, residual energy, lifetime, clustering, Voronoi diagram, Cluster Head Selection, Fuzzy Logic.

1   Introduction

Sensor networks currently represent a new area in full development and emerging innovations in communication technologies.

In general, Ubiquitous Sensor Networks (USNs) are composed of a large number of small and cheap sensor nodes and one base station deployed in geographical area or in special environment and designed to collect and transmit autonomously different types of data from the USNs and to make it accessible for end users over the internet.

Sensors nodes have very limited energy, processing power and storage. These sensors collect data and send them to the base station (BS) via radio transmitter [5], [6], [7], [12].

Wireless sensor networks are used in health, army, hygiene, education, industry, agriculture and so on. The potential applications of sensor networks are highly varied such as environmental monitoring, target tracking, battle field surveillance, monitoring the enemy territory, detection of attacks and security etiquette. Other applications of these sensors are in the health sectors where patients can wear small sensors for physiological data and in deployment in disaster prone areas for environmental monitoring [8], [11].

S. Balandin et al. (Eds.): NEW2AN/ruSMART 2014, LNCS 8638, pp. 319–330, 2014.

Clustering network is an efficient and scalable way to organize USNs. Appropriate cluster-head selection can significantly reduce energy consumption and prolong the lifetime of USNs. In clustering the sensor nodes are divided into some clusters and one node is selected as cluster head in each cluster. Cluster heads receive data from other sensor nodes and send them to base station. Selecting a suitable cluster head decreases energy consumption to a great extent and as a result increases networks lifetime [4], [9], [11].

In this paper, we propose a new algorithm called Cluster-Head selection using Fuzzy Logic with Voronoi diagram in heterogeneous USNs (CHS-FL-VD) that is a distributed algorithm which makes local decisions to select cluster head using a fuzzy inference system based on two parameters which are residual energy and centrality in Voronoi diagram, cluster formation using Voronoi diagram and disseminating data from cluster heads to the base station using gateway and direct connection to properly distribute energy and ensure maximum network life time.

The rest of this paper is organized as follows. Section 2 gives an overview of some related works, previous studies and summarizes network radio model used. Section 3 describes our proposed algorithm. In section 4, we evaluate by simulation our algorithm compared with LEACH and Fuzzy C-Means algorithms. Finally, section 5 concludes the paper.

2   Related Works

There are many clustering algorithms used in USNs. In this section, we review some algorithms.

• The most famous algorithm is Low-Energy Adaptive Clustering Hierarchy or LEACH. W.R. Heinzelman & al. proposed Low Energy Adaptive Clustering Hierarchy (LEACH), the LEACH is divided into rounds. Each round consists of a set-up phase and a steady state phase. The set-up phase consists of CH election and cluster formation. The steady state phase consists of sensing, and transmission of the sensed data to the CH and then to the BS. At the start of the first set-up phase and in every round, each sensor node generates a random number between 0 and 1 to determine it will become a cluster-head or not, then compares this number with threshold [1].

• The Gupta algorithm uses a Fuzzy Logic approach to select CHs. The Fuzzy Inference System (FIS) designer considered three descriptors: energy level, concentration, and centrality, each divided into three levels, and one output which is chance, divided into seven levels. The system also uses 27 IF-THEN rules. In this protocol there are two phases (set-up and steady-state) as in LEACH. The difference between the two protocols lies in the set-up phase where the BS needs to collect energy level and location information for each node, and evaluate them in the designed FIS to calculate the chance for each node to become a CH. The BS then chooses the node that has the maximum chance of becoming a CH. After the CH selection, everything (advertising message, join CH message, and the steady-state phase) will be the same as in LEACH [2].

• In [3], Kim offers CHEF the CHs are selected based on a fuzzy logic. The difference is that in this approach more than one cluster head is selected locally in each round. The fuzzy set includes nodes' energy and their local distances. CHEF also generates a random number for each sensor and if it is less than a predefined threshold, Popt, then the node's chance is determined. Thus, there may be some qualified nodes that lose their chance on a random manner.

- N. A. Torghabeh & al. adopt in [7] a two-level fuzzy logic to evaluate the probability of sensor node to elect itself as a cluster head. In the first level (Local Level), node's energy and number of neighbors are used as parameters to decide node as cluster-head. In the second level (Global Level), node's overall cooperation is considered to the whole network with three fuzzy parameters.

- In [12] Mourad Hadjila & al. propose Fuzzy C-Means algorithm that combine cluster formation and cluster head selection in wireless sensor networks. They have used Fuzzy C-Means technique to create clusters and fuzzy logic system based on two parameters, which are energy level and centrality to elect cluster heads then the data is sent in multi-hop path from cluster head to cluster head until reaching the base station. Simulation results show that proposed algorithm extends the network lifetime about 34.13% when the last node dies comparing with the LEACH algorithm.

In this paper, we assume S sensors which are deployed randomly in a field to monitor environment. We represent the *i-th* sensor by s_i and consequent sensor node set $S = s_1, s_2, \ldots, s_n$.

- Sensor field is divided into 2 logical areas (first area 0-50m and second area 50-100m) for effective communication between nodes and BS. Nodes in closer vicinity of BS and gateway use direct transmission (in single hop), nodes in farther the 50 m from the BS use gateway to transmission (in multi-hop).

- We deploy the BS far away from the sensing field. Sensor nodes and the BS are stationary after deployment. A gateway node is deployed in the same network field at the centre of the network. Gateway node is stationary after deployment and rechargeable. Gateway is used to enhance the communication time of sensor nodes.

- We use homogeneous sensor nodes with same computational and sensing capabilities. All sensor nodes have the same initial energy. The radio channel is symmetric such that energy consumption of transmitting data from node X to node Y is the same as that of transmission from node Y to node X.

- We use first order radio model as used in [1] and [10]. This model represents the energy dissipation of sensor nodes for transmitting, receiving and aggregating data.

3   Proposed Algorithm

3.1   Cluster Formation Using Voronoi Diagram

The Voronoi diagram [4], also known as Dirichlet tessellation, represents one of the most fundamental data structures in computational geometry. It has interesting mathematical and algorithmic properties and potential applications.

In our algorithm, the formation of clusters is done by Voronoi diagram. In every round, sensing field is divided randomly. After cluster head selection compute the distance between the sensor node and the cluster head in each cluster as shown in equation (1).

$$\text{distance}\left(S_i, C_j\right) = \sum_{i=1}^{m}\sqrt{\left(S_i - C_j\right)^2} \tag{1}$$

where S_i is the sensor node in cluster (i=1,…,m) and C_j is the cluster head (j=1,…,k).

3.2   Selection of Cluster Heads Using Fuzzy Logic System

The concept of fuzzy set and fuzzy logic was introduced by Zadeh in 1965. The structure of a fuzzy logic system is shown in Fig.1. When an input is applied to a fuzzy logic system, the inference engine computes the output set corresponding to each rule. The defuzzifier then computes a crisp output from these rule output sets.

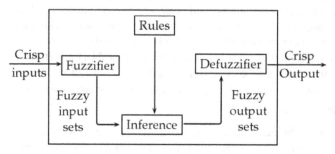

Fig. 1. The structure of a fuzzy logic system

We use Mamdani-style fuzzy inference in this paper. The process is performed four steps:

➢ Fuzzification of the input variables. Take the above input variables and determine the degree to which these inputs belong to each of the appropriate fuzzy sets.
➢ Rule evaluation. Take the fuzzified inputs, and apply them to the antecedents of the fuzzy rules. Because the given fuzzy rule has multiple antecedents, the fuzzy operator (AND) is used to obtain a single number that represents the result of the antecedent evaluation.
➢ Aggregation of the rule outputs. Take the membership functions of all rule consequent previously clipped or scaled and combine them into a single fuzzy set.
➢ Defuzzification. Evaluate the rules, but the final output of a fuzzy system has to be a crisp number. We use the centre of gravity (COG) as a defuzzification method is expressed as shown in equation (2):

$$\text{Output (PCHS)} = \frac{\int x\,\mu_A\,(x)dx}{\int \mu_A\,(x)dx} \tag{2}$$

where $\mu_A\,(x)$ is membership function of set A.

Once clusters formation is complete, we proceed to the selection of cluster heads using a system based on fuzzy logic [14]. The common decision of the nodes in the cluster is selecting the cluster heads. The table 1 consists of the Fuzzy Logic system linguistic variables. For the primary fuzzy system of this paper, two Input linguistic variables are defined, representing the remaining energy and the centrality by Voronoi diagram. For remaining energy and centrality by Voronoi diagram, the term set is defined three labels, so the fuzzy inference rule has 3×3 = 9 rules.

Moreover, the system has one output linguistic variable named Probability of Cluster Head Selection (PCHS). The term set of output linguistic variable is divided into nine levels.

Table 1. Fuzzy Logic system linguistic variables

Input variables	x_1	variable	Remaining energy
		Term-fuzzy set	{low, medium, high}
		Value limits	[0, 0.1] J
	x_2	variable	Centrality by Voronoi diagram
		Term-fuzzy set	{far, middle, close}
		Value limits	[0, 100] %
Output variable		variable	Probability of Cluster Head Selection
		Term-fuzzy set	{very small, small, rather small, med small, medium, med large, rather large, large, very large}
		Value Limits	[0, 100] %

The Cluster Head Selection system is depicted in fig.2 and the membership functions of these linguistic variables are depicted in fig.3.

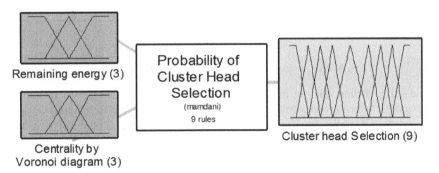

Remaining energy (3)

Centrality by Voronoi diagram (3)

Probability of Cluster Head Selection
(mamdani)
9 rules

Cluster head Selection (9)

System PCHS: 2 inputs, 1 outputs, 9 rules

Fig. 2. The Cluster Head Selection system

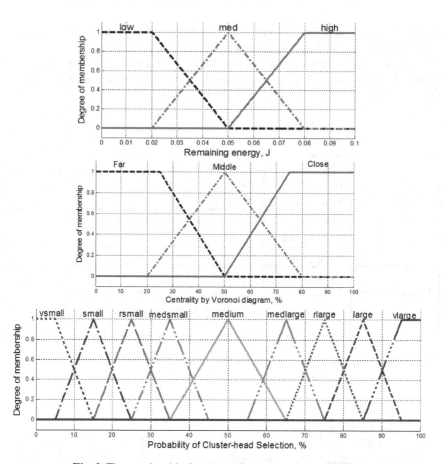

Fig. 3. The membership functions of these linguistic variables

The fuzzy if-then rules in the Cluster Head Selection are also shown in Table 2. For example, we can read a rule in the following manner: if the Remaining energy is high and the Centrality by Voronoi diagram is close then the Probability of Cluster Head Selection is very large.

Table 2. The fuzzy if-then rules in the Cluster Head Selection

Rule No	Remaining energy	Centrality by Voronoi diagram	Probability of Cluster Head Selection
1	Low	Far	Very small
2	Low	Middle	Small
3	Low	Close	Rather small
4	Medium	Far	Medium small
5	Medium	Middle	Medium
6	Medium	Close	Medium large
7	High	Far	Rather large
8	High	Middle	Large
9	High	Close	Very large

The rules are created using the Fuzzy Inference System (FIS) editor contained in the Matlab Fuzzy Toolbox [13]. Fig.4 shows a sample fuzzy calculation of a probability of cluster head selection based on the amount of remaining energy and centrality by Voronoi diagram. Fig.5 shows a control surface.

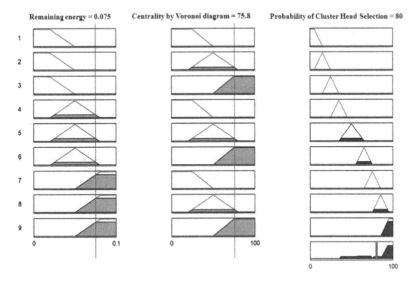

Fig. 4. Sample fuzzy calculation of a probability of cluster head selection

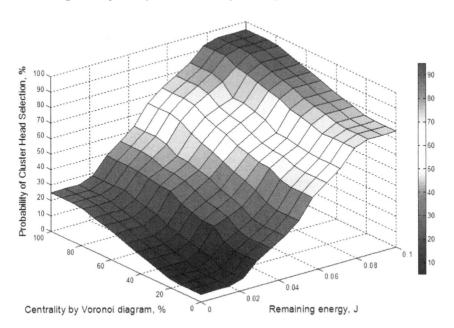

Fig. 5. Surface view of probability of cluster head selection with respect to remaining energy and centrality by Voronoi diagram

4   Simulation Results and Analysis

In this section, we present the simulation result which is done in Matlab. The simulation parameters are shown in table 3.

Table 3. Simulation parameters

Type	Parameter	Value
Network topology	Number of nodes	100
	Probability P_{opt}	0.1
	Network coverage	(0, 0) (100, 100) m
	Base station location	at (50, 105) m
	Gateway location	at (50, 50) m
Radio model	Initial energy E_0	0.1 Joules
	Energy for data aggregation E_{DA}	5 n J/bit/signal
	Transmitting and receiving energy E_{elec}	50 n J/bit
	Amplification energy for short distance E_{fs}	10 p J/bit/m^2
	Amplification energy for long distance E_{amp}	0.0013 p J/bit/m^2
	Data packet size	1000 bits

Fig.6 shows sensor nodes deployment randomly. Once the nodes are deployed, we proceed to the formation of clusters using Voronoi diagram as depicted in fig.7.

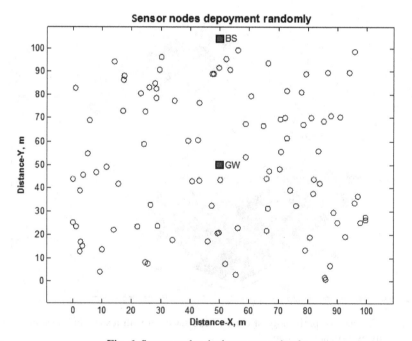

Fig. 6. Sensor nodes deployment randomly

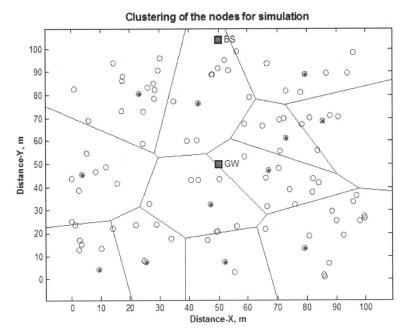

Fig. 7. Clusters formation using Voronoi diagram

Fig.8 shows simulation results of cluster head selection using Fuzzy Logic with Voronoi diagram.

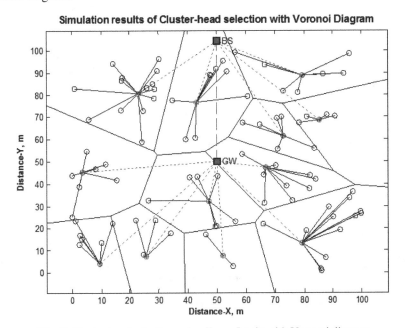

Fig. 8. Cluster head selection using Fuzzy Logic with Voronoi diagram

In order to evaluate our algorithm, we take two metrics, which are the residual energy of the network and network lifetime. Fig.9 shows respectively the residual energy of the network and fig.10 shows the number of alive nodes in the network for proposed algorithm compared with LEACH and Fuzzy C-Means algorithms.

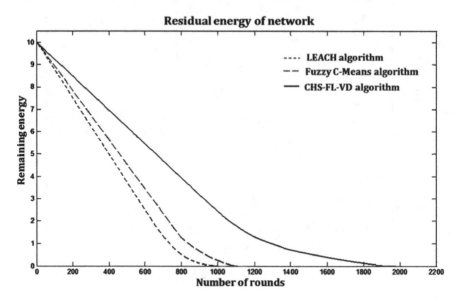

Fig. 9. Remaining energy vs. number of rounds

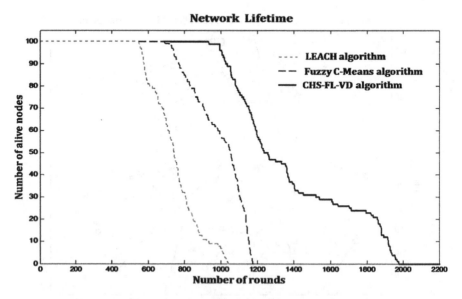

Fig. 10. Number of alive nodes vs. number of rounds

Fig.11 shows duration of the first node, the half node and the last node die in such algorithms. According to fig.11, our proposed algorithm has larger FND, HND and LND.

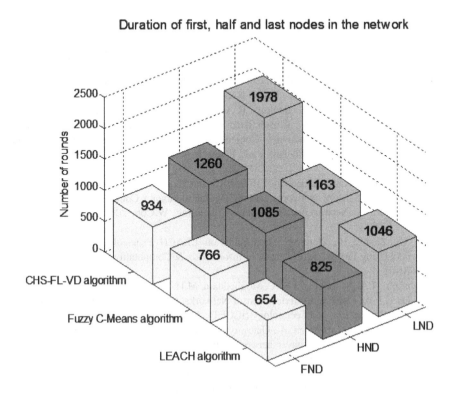

Fig. 11. Node death comparison algorithms

5 Conclusion

We have proposed in this paper a new algorithm that combine cluster formation and cluster head selection in ubiquitous sensor networks. We have used Voronoi diagram to create clusters and fuzzy logic system based on two parameters, which are remaining energy and centrality by Voronoi diagram to select cluster heads. The nodes collect data and send it to their cluster heads than from cluster heads to the base station using gateway and direct connection. Simulation results show that proposed algorithm extends network lifetime about 89 % when the last node dies comparing with the LEACH algorithm and about 70 % when the last node dies comparing with the Fuzzy C-Means algorithm.

References

1. Heinzelman, W.R., Chandrakasan, A.P., Balakrishnan, H.: An application-specific protocol architecture for wireless microsensor networks. IEEE Transactions on Wireless Communications 1(4), 660–670 (2002)
2. Gupta, I., Riordan, D., Sampalli, S.: Cluster-head Election using Fuzzy Logic for Wireless Sensor Networks. In: Proceedings of the 3rd Annual Communication Networks and Services Research Conference (CNSR 2005), pp. 255–260 (May 2005)
3. Myoung Kim, J., Park, S., Han, Y., Chung, T.: CHEF: Cluster Head Election mechanism using Fuzzy logic in Wireless Sensor Networks. In: 10th International Conference on Advanced Communication Technology (ICACT), pp. 654–659 (February 2008)
4. Salim, A., Koucheryavy, A.: Cluster head selection for homogeneous Wireless Sensor Networks. In: 11th International Conference on IEEE Advanced Communication Technology, ICACT 2009, Phoenix Park, Korea, February 15-18, vol. 03, pp. 2141–2146 (2009)
5. Hu, Y., Shen, X., Kang, Z.: Energy-Efficient Cluster Head Selection in Clustering Routing for Wireless Sensor Networks. In: 5th International Conference on Wireless Communications, Networking and Mobile Computing, WiCom 2009, September 24-26, pp. 1–4 (2009)
6. Ran, G., Zhang, H., Gong, S.: Improving on LEACH Protocol of Wireless Sensor Networks Using Fuzzy Logic. Journal of Information & Computational Science 7(3), 767–775 (2010)
7. Torghabeh, N.A., Totonchi, M.R.A., Moghaddam, M.H.Y.: Cluster Head Selection using a Two-Level Fuzzy Logic in Wireless Sensor Networks. In: 2nd International Conference on Computer Engineering and Technology (ICCET), April 16-18, vol. 2, pp. 357–361 (2010)
8. Alla, S.B., Ezzati, A., Mohsen, A.: Gateway and Cluster Head Election using Fuzzy Logic in heterogeneous wireless sensor networks. In: International Conference on Multimedia Computing and Systems (ICMCS), May 10-12, pp. 761–766 (2012)
9. Lee, J.-S., Member, S., Cheng, W.-L.: Fuzzy-Logic-Based Clustering Approach for Wireless Sensor Networks Using Energy Predication. IEEE Sensors Journal 12(9), 2891–2897 (2012)
10. Siew, Z.W., Chin, Y.K., Kiring, A., Yoong, H.P., Teo, K.T.K.: Comparative Study of Various Cluster Formation Algorithms in Wireless Sensor Networks. In: 7th International Conference on Computing and Convergence Technology (ICCCT), December 3-5, pp. 772–777 (2012)
11. Karimi, A., Abedini, S.M., Zarafshan, F., Al-Haddad, S.A.R.: Cluster Head Selection Using Fuzzy Logic and Chaotic Based Genetic Algorithm in Wireless Sensor Network. Journal of Basic and Applied Scientific Research 3(4), 694–703 (2013)
12. Hadjila, M., Guyennet, H., Feham, M.: A Routing Algorithm based on Fuzzy Logic Approach to Prolong the Lifetime of Wireless Sensor Networks. International Journal of Open Scientific Research IJOSR 1(5), 24–35 (2013)
13. MathWorks - MATLAB and Simulink for Technical Computing, Fuzzy Logic Toolbox, Documentation Center,
   http://www.mathworks.com/help/fuzzy/index.html
14. Fuzzy Logic Toolbox™ User's Guide © COPYRIGHT 1995–2012 The MathWorks, Inc.

Power-Aware Architecture to Combat Intelligent Adaptive Attacks in Ad-Hoc Wireless Networks

Joseph Soryal, Fuad A. Alnajjar, and Tarek Saadawi

Electrical Engineering Department,
The City University of New York, City College,
New York 10031, USA
{jsoryal00,fuad,saadawi}@ccny.cuny.edu

Abstract. The paper presents a novel technique to combat smart adaptive attacks in ad-hoc wireless networks where battery power and bandwidth are scarce. Nodes in ad-hoc wireless networks can assume the roles of transmitters, receivers, and/or routers in case they fall geographically between two communicating nodes. Malicious nodes deviate from the standard communications protocols to illegally maximize their share of the bandwidth and save their battery power for their own communication. When the malicious node has data packets to transmit, it increases the power of the transmitted signal to increase the throughput and reduce the delay and retransmission attempts. This behavior adversely affects the rest of the nodes that follow the standard communication protocols. When the malicious node receives a packet that is not destined to it, the malicious node forwards the packet with very low transmission power level to save power and in this case, the packet reaches the next hop without enough signal strength to be decoded correctly. In this type of sophisticated attacks, malicious nodes deal with the control packets according to the standards so it keeps itself visible inside the network in case there are packets destined to it. Our end-to-end algorithm detects and isolates those types of attackers to maintain the resiliency of the network against the malicious behavior.

Keywords: Network Security, Routing Protocols, Wireless Networks, DSR.

1 Introduction

Ad-hoc wireless networks are self-organizing wireless networks that do not require fixed assets during the communication session. Each node has the resources to be a workstation and/or a router. The source and destination can communicate directly if the receiver is within the signal coverage of the sender's transmission range, otherwise other hops within the coverage function as routers to relay the data. Ad-hoc wireless Networks is an attractive solution for many groups specially those who do not have the resources to build an infrastructure like disaster recovery teams and army units.

The most significant constraints that challenge the use of Ad-hoc wireless networks are the bandwidth and power supply limited lifetime (batteries). Due to the lack of centralized authority to manage routing and communication sessions, malicious nodes can take unfair advantage over the other innocent nodes to maximize their share of the bandwidth and reduce their power consumption on the expense of

S. Balandin et al. (Eds.): NEW2AN/ruSMART 2014, LNCS 8638, pp. 331–344, 2014.
© Springer International Publishing Switzerland 2014

the innocent nodes. This greedy behavior only benefits the attacker and can go unnoticed or detected by current technologies. Due to the distributed nature of ad-hoc wireless networks, the technology is vulnerable to attacks.

One of the most common routing protocols in ad-hoc wireless networks is DSR (Dynamic Source Routing) protocol [1].

DSR is an on-demand and self-configuring routing protocol that enables nodes that have data to send to reach out to their respective destinations traversing the network through other nodes in the network. DSR consists of two main phases: Routing Discovery and Routing Maintenance. All the routing information is cached in the nodes and is continuously updated to reflect any changes in the network. During the routing discovery phase, a source creates a "RouteRequest" packet and then floods the network with this packet. Every node gets this packet; it forwards it to its neighbours, as long as the packet time to live is still larger than zero, until the packet reaches the destination. Each "RouteRequest" has a sequence number initiated by the source node and the path it has traversed and is used to prevent loop formations and multiple transmissions of the same "RouteRequest" by an intermediate node. Nodes respond to "RouteRequest" by generating "RouteReply" packets that traverses the network back to the source node. Nodes gather paths and routing data from these control packets. The Route Maintenance phase is triggered upon a route change indicated by Route Error packets. The words "attacker" and "malicious" are used interchangeably in this paper.

Malicious nodes that forward packets with very low transmission power emulate the black hole attacks that drop the packets since the packet transmitted with low transmission power is not received properly by the next hop due to its weakened signal. Intelligent attackers resort to lower the transmission power of the forwarded packets rather than wholly dropping the packets, as in the black hole attack, to evade detection, which is studied extensively in the literature. Nodes that drop packets are considered a very serious threat to the integrity of the network [2]. Various methods to detect nodes that refuse to forward the packets and drop them are presented in the literature [3] and [4]. Authors in [5], [6], [7]. [8], [9], [10], and [11] present good detection schemes for nodes dropping the packets but have not dealt with intelligent attacker nodes that selectively drop the packets to dodge detection, or those attackers that purposely lower the transmission levels. Authors in [12] propose multiple simultaneous routes, which may consume the bandwidth unnecessarily. Authors in [13] present a technique to detect nodes that selectively drop packets; the technique depends on the routing information such as number of hops and expiration time for packets that could be a challenging task in a distributed changeable environment like ad-hoc wireless networks.

Authors in [14] proposes a watchdog that keeps a copy of every single packet transmitted in its buffer until it hears that other nodes forward the same packet then it discards the copy, however the paper assumes unlimited buffer size and it does not account for collisions and innocent nodes that may be experiencing harsh environment.

Mostly these researches ignore the fact that the node might be just an innocent node experiencing environmental changes that impacts its wireless capabilities. In our approach, we account for those innocent nodes so they are not falsely flagged as attackers. Researches dealing with the detection and combating of greedy nodes that increase the transmission power level to gain extra throughout illegally are scarce in the literature. Although, some researches adopted the idea of manipulation of the transmitted power levels for routing algorithms but not for a malicious purpose as presented in [15] and [16].

The rest of the paper is organized as follows: Section 2 describes the attack algorithm and the impact on the rest of the network. Section 3 presents the algorithm. The conclusion is in Section 4.

2   Attack Algorithm and Impact

The attacking node is a selfish and greedy node that deviates from the DSR routing protocol [17] and IEEE 802.11 standards [18] to increase its share of the bandwidth in terms of transmitted packets and to save power and resources when it is positioned as an intermediate node that should forward packets. The attacker in this case is sophisticated and adapts to the environment rules and to its own needs to maximize the gain. Current standards for IEEE 802.11b and DSR do not have mechanisms to detect and deal with these types of intelligent attacks [1] and [18]. We modeled the attacks using OPNET [19] and chose a simple topology shown in Fig.1 to present the concept of the attack, gains to the attacker, and the negative impact on the innocent nodes. In Fig.1, node_1 is the attacker. The simulation shows the severity of the attack on the network and details the benefits of the attacker. In Fig.1, all nodes are using DSR protocol to send data to node_3 with constant load of 0.001seconds inter-arrival time between two consecutive packets. The packet size is 5000 bits. The load was selected to ensure that all nodes are under saturation whereas every node has always a packet to transmit. The physical layer is using DSSS (Direct Sequence Spread Spectrum) techniques with channel data rate of 11 Mbps. The goals of the attacker are:

1. Illegally Increase the throughput: The IEEE 802.11 standard [18] and the FCC (Federal Communications Commission) rules [20] determine the transmission power levels for nodes and expect that all nodes would comply with the specifications to achieve harmony and fair channel access to all nodes inside a coverage area. The rules in [20] provide details on limitations of EIRP (Equivalent Isotropically Radiated Power) which state that the omni-directional antennas shall have less than 6 dB gain. The FCC rules require EIRP to be 1 watt (1,000 milliwatts) or less. In our simulations, the default transmission power for all rule-abiding nodes is 0.005 Watt and the receiver's sensitivity is (-95) dBm. A malicious node can unilaterally increase its transmission power to increase its throughput on the expense of other innocent nodes. Since the attacker in this case is intelligent and adaptive, it increases its power gradually to achieve at least double its normal throughput whenever it has data to transmit. Subsequently this behavior increases the noise level in the communication channels and decreases the number of packets transmitted by legitimate nodes. The throughput is proportionate with the transmission power level [21].

For the simplicity, and to present the concept, the attacker here aims to at least double the throughput. However the attacker can aim to significantly increase its throughout by raising the transmission power exponentially but this might affect its own battery life and requires detailed calculations to determine the relationship

Attacker's Algorithm

```
Packet to Transmit (PK_TX)
If   PK_TX = Control Packet
Then    Power_TX = (normal level)
Else
//data packet to be sent out
    {
{
While (Initial_Throghput< (2*Initial_Throghput))
{
Power_TX   = 2 * (normal level)
// the power is doubled in each iteration
}}}
```

between loss and gain in terms of power spent and throughput achieved respectively. In Fig.2, a comparison between the rates of traffic sent when node_1 was operating in normal mode (blue line) and when it activated the attack mode (red line). The throughput almost increased three folds when the attack mode is active. In this case, the attacker doubled its transmission power four times. The difference between packets dropped when node_1 is behaving normally (blue line) and when the attack mode is active (red line) is shown in Fig.3.When the attack mode is active; the total packets dropped by the attacker were reduced to almost one third. The innocent nodes in the network suffer degradation of the bandwidth usage and increased delay. Fig.4 shows the number of route error message inside the network is increased about two and half times when the attack is present.

2. Save power and processing resources: As per the DSR protocol, intermediate nodes function as hops to forward packets between sources and destinations. When the malicious node is an intermediate hop that should be forwarding packets as per the DSR standards, it does not follow the standards [20] regarding the minimum transmission power and instead it purposely lowers transmission power, which eventually resembles a black hole attack. This behavior causes network disruption, increased delay, and excessive packet loss.

In Fig.5, the total traffic exchanged among the nodes inside the network is negatively affected when the attacker refused to follow the transmission power level specification. Also, in Figures (6 and 7), the individual traffic sent by innocent nodes dropped when the attacker activated its attack mode. While the innocent nodes suffer from packet loss and extra power consumption related to the retransmission attempts and longer route discovery times, the attacker is saving its power that constitutes greedy behavior.

Attacker's Algorithm

```
Packet Received (PK_RX)
If      PK_RX = Control Packet
Then    Process&Respond using normal power level
Else
// data packets received
        {
If      PK_RX => (Destined to me)
Then    Process&Respond
Else    Forward the packet with (0.1*Normal Power level)
}
```

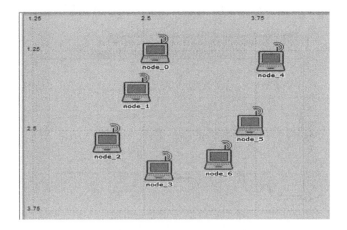

Fig. 1. Network Topology

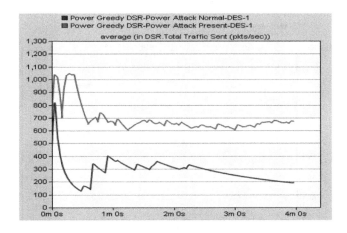

Fig. 2. Throughput (Packets/Second) for the innocent node

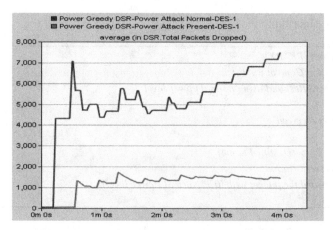

Fig. 3. Packets Dropped (Packets) for the Attack Node

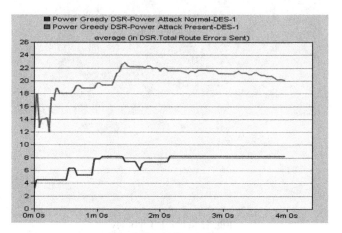

Fig. 4. Route Errors Sent

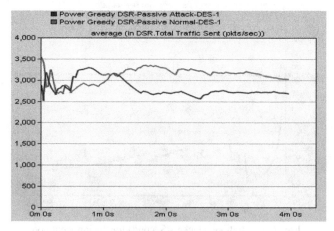

Fig. 5. Total Traffic Sent (Packets)

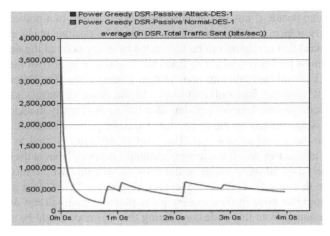

Fig. 6. Traffic Sent by Innocent Node (bits/sec)

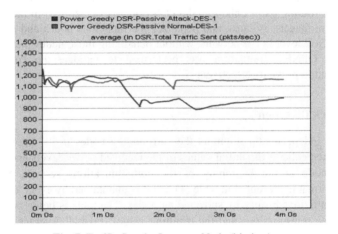

Fig. 7. Traffic Sent by Innocent Node (bits/sec)

3   Algorithm and Simulation

The algorithm resides in every node and runs independently from other copies of the algorithm residing in other nodes in the network. The algorithm enables every node to police other nodes in its wireless range. The algorithm consists of two modules. The first module is the "Attack Detection" module which enables the nodes to consistently monitor other nodes inside the network. Once the Attack Detection module detects a malicious behaviour then the second module, "Resiliency Module", gets activated to reduce the negative impact on the network. To implement the algorithm, some modifications to the DSR firmware will be included. Each node monitors and tracks two parameters in the received data packets, as shown in Table (1); the power level received and the SNR (Signal to Noise Ratio) with the assumption that all nodes are stationary. Also, we assume that all nodes start the communication session with the same battery level Based on the history of observed power levels and SNR values, every node can decide if any of the adjacent nodes is acting in a malicious way with high certainty. Once a malicious node is detected, a custom made control packet (Detect_Packet) that is required to be forwarded by every node in the network similar to RouteRequest packet is sent out to flood the network to inform all the nodes inside the network to avoid the malicious node in their routing paths. The Detect_Packet will include the IP (Internet Protocol) and MAC (Media Access Control) addresses of the suspicious node and the level of certainty that this node is really malicious. The level of certainty is determined by the nodes that detect the potential malicious node based on the matrix presented in table (2). The level of certainty is used as a measure to penalize the attacker or not. If the level of certainty is "High" that is the detected node is an attacker, then all other nodes in the network will discard packets destined and coming from this specific node. If the level of certainty is "Low" then all other nodes will only avoid this node in their routing paths planning and will not discard packets destined or coming from this node considering that this node might be suffering from environmental changes that affect the wireless signal strength and quality. The importance of flagging the nodes even with "Low" certainty is that avoiding such nodes in the routing paths enhances the QoS (Quality of Service) in the network. So our algorithm detects, isolates, and penalizes the malicious nodes and also enhances the overall QoS level in the network without penalizing those nodes that are affected. Preliminary OPNET simulation was performed to validate the concept of the algorithm and to prove its effectiveness.

3.1   Attack Detection

Every node independently monitors the surrounding nodes in its wireless coverage range and then collects the initial values of the received power levels and SNR values for every node at the beginning of the communications session. Sample data collection format is presented in table (1) and it is maintained by every node to collect initial transmissions history by surrounding nodes. The table is being populated after the first three transmissions and the resulting values (Overall Average PL/SNR) are used as the baseline for the rest of the communication session. Table (1) shows a sample log maintained by node_1 which has node_0 and node_2 in its coverage area.

Table 1. Sample table maintained by every node

Node IP	Node MAC	PL /SNR (1)	PL/SNR (2)	PL/SNR (3)	Overall Average PL /SNR
IP_0	MAC_0	X1/Y1	X2/Y2	X3/Y3	X/Y
IP_2	MAC_2	L1/M1	L2/M2	L3/M3	L/M

Node_1 collects the data for each node in a separate row. IP_0 and MAC_0 are the IP and MAC addresses for Node_0. X1/Y1 are the first readings for the received PL (Power Level) and SNR values. Second set of readings goes to the second column and so on. Each set of readings in every column is an average value of a single data transmission event which is defined herein as a continuous stream of packets sent by these specific nodes regardless of the number of packets sent or duration and taking into account the of the propagation delay   between two consecutive packets.

Table 2. Decision Matrix

Baseline (Power Received/SNR)	Monitored Node is (Source or Intermediate)	Power Received	SNR	Result**
X/Y	Source	Within range*	Within range*	A
X/Y	Source	> baseline	>baseline	D-1
X/Y	Source	> baseline	<baseline	C
X/Y	Source	<baseline	>baseline	C
X/Y	Source	< baseline	<baseline	B
X/Y	Intermediate	Within range*	Within range*	A
X/Y	Intermediate	> baseline	>baseline	A
X/Y	Intermediate	> baseline	<baseline	B
X/Y	Intermediate	<baseline	>baseline	C
X/Y	Intermediate	< baseline	<baseline	D-2

* Within range = +/- 10% of baseline value and it is based on experiment observations.

**   Results description is listed below:

A= Node is behaving normally and no action is required.

B= Node is behaving normally however the wireless environment might be degrading around the node and avoiding this node in the path selection may increase the throughput

C= Node might be a suspect or the wireless environment might be degrading. Avoiding this node in the path selection may increase the throughput

D-1= Node is an Attacker (Increasing Throughput illegally) and it should be penalized.

D-2= Node is an Attacker (Energy Saver) and it should be penalized.

Detection Module Algorithm

```
// The module kicks in when the communication session starts
Populate Table (1) for specific node is complete and Baselines obtained
No        Do Nothing
Yes
{
Monitor transmissions from this node
      && Compare values to Baselines
Use Decision Matrix (Table (2)) to obtain Result
If   {
Result = A        Then    Do Nothing
Result = B // C
      Then
      Activate Resiliency Module && Severity Level = Low
Result = D
      Then
      Activate Resiliency Module && Severity Level = High
      }
```

3.2   Resiliency Module

This portion of the algorithm is triggered by the detection module. Nodes that detect a potential malicious node create a Detect_Packet with the IP and MAC addresses of the potential malicious node and the level of severity based on the criteria outlined in the detection algorithm. The Detect_Packet floods the network end-to-end following the RouteRequest packet behaviour in terms of being forwarded by every node and looping avoidance. Source nodes select new routing paths to avoid the node in question; this is on the contrary of the original DSR algorithm that selects the shortest path regardless of the conditions of the intermediate nodes. Whenever there is a node that is reported as potential malicious with severity level of "High", all other nodes in the network drop packets destined or coming from this node to penalize its behaviour.

One of the strengths in the resiliency module is that, besides isolating and penalizing the malicious node, it also enhances the QoS of the network based on the observed parameters without penalizing nodes with poor performance.

The resiliency mode is implemented in OPNET manually to provide the proof of concept of our algorithm and the results are presented in the next section. The algorithm can be integrated to actual DSR fir products to provide protection against real-life greedy users that manipulate the standards for illegal benefits. The modified firmware would be interoperable with the market-available DSR firmware.

Resiliency Module Algorithm

// The node is actively monitoring its physical range
Detection module activated Resiliency module
No   Do Nothing
Yes
{
(Create "Detect_Packet")
(Broadcast "Detect_Packet"&&Avoid this node in routing paths selections)
If   Severity Level = High
Then   Discard all packets destined to and from the malicious node //penalizing the attacker
Else   Do Nothing
}
// The node is not in the attacker's physical coverage
Received "Detect_Packet"   // from at least 3 different nodes reporting the same attacker to prevent bad mouthing attacks
No   Do Nothing
Yes
{
Avoid the reported node in routing paths selections
&&
{
If   Severity Level = High
Then   Discard all packets destined to and from the malicious node
Else   Do Nothing
}}

3.3   Results

The results presented below show improvements in the achieved throughput of innocent nodes where the malicious node purposely manipulates its power level during data transmissions to save power. Fig.8 shows the comparison of the achieved throughput by one innocent node under three conditions. The first is normal condition (blue line) where all nodes follow the standards and there is no malicious behaviour implemented by any of the nodes. The (green line) presents the throughput by an innocent node with the presence of an attacker in the network and the attacker is manipulating the transmission power level whenever it sends a stream of data packets to illegally gain higher throughput when it is acting as a data source and save power when it acts as a router. The graph shows that the throughput of the innocent node has declined due to the effect of the attacker. The (red line) presents the throughput of the innocent node when the proposed algorithm is in effect. The result shows improvement in the throughput with higher rate than the one when the malicious node is operating freely. The result presents the throughput after the baselines are determined and the malicious node is detected and flagged as an attacker, then the route is changed to avoid this malicious node. Also, innocent nodes discard all packets destined or originating from the malicious node as a penalty. Although, the new route does not include the malicious node, the throughput of the innocent nodes do not get restored to the previous level where the attacker did not exist because the attempts of data transmission by the malicious node consumes some bandwidth especially they are transmitted with high transmission power which creates unnecessary interference for other nodes and overpower their transmissions. Fig.9 presents the delays (in seconds) encountered by innocent node. The blue line presents the delay where all nodes follow the standards. The red and green lines present the network when the malicious node acts freely and when the proposed algorithm is in effect, respectively. The algorithm reduced the delay for the innocent node but however it did not restore the delay levels as in the case of normal conditions because of the lasting effect of the malicious node.

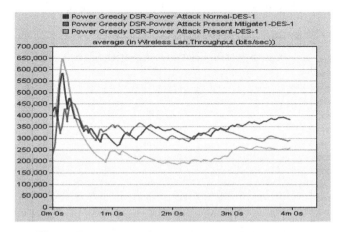

Fig. 8. Throughput Achieved by Innocent Node (bits/sec)

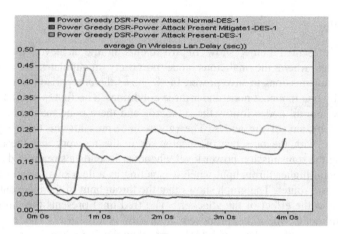

Fig. 9. Delay Encountered by Innocent Node (sec)

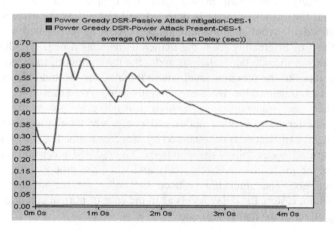

Fig. 10. Delay Encountered by Innocent Node (sec)

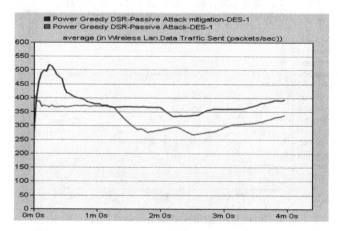

Fig. 11. Data Traffic Sent by innocent node (packets/sec)

Figures (10 and 11) show the event when an innocent node detects a strange behavior according to Table (2) but the observed node is not verified as an attacker, so the algorithm would choose a different path away from this node to improve the QoS. Figures (10 and 11) show improvement in the delay encountered by an innocent node and the throughput, respectively, before and after selecting a different routing path.

4   Conclusion and Future Work

The paper presents a novel approach to detect, isolate, and penalize intelleigent attacksin ad-hoc wireless networks. The intelligent attackers aim to save power and increase their throughput on the expense of other innocent users. The developed algorithm enables innocent nodes to detect the malicious behavior, isolates, and penalizes the attackers to help maintain the network operable. As a future step, the algorithm will be enhanced to deal with the false postives so it could distinguish the malicious behavior from the imperfections of the wireless enviroment. Some enhancements will be added to the algorithm to react to multiple attackers simultenously. Also, the algorithm will be expanded so it can be used in a mobile environment.

References

[1]   http://tools.ietf.org/html/rfc4728
[2]   Deng, H., Li, W., Dharma, P.: Agrawal: Routing Security in Ad Hoc Networks. IEEE Communications Magazine, Special Topics on Security in Telecommunication Networks 40(10), 70–75 (2002)
[3]   Tseng, et al.: A survey of black hole attacks in wireless mobile ad hoc networks. Human-centric Computing and Information Sciences 1, 4 (2011)
[4]   Balakrishnan, K., Deng, J., Varshney, P.K.: TWOACK: Preventing Selfishness in Mobile Ad Hoc Networks. In: Proc. IEEE Wireless Comm. and Networking Conf. (WCNC 2005) (March 2005)
[5]   Sun, B., Guan, Y., Chen, J., Pooch, U.W.: Detecting Black-hole Attack in Mobile Ad Hoc Networks. Paper presented at the 5th European Personal Mobile Communications Conference, April 22-25, United Kingdom, Glasgow (2003)
[6]   Tamilselvan, L., Sankaranarayanan, V.: Prevention of Blackhole Attack in MANET. Paper presented at the 2nd International Conference on Wireless Broadband and Ultra Wideband Communications, Sydney, Australia, August 27-30 (2007)
[7]   Djenouri, D., Badache, N.: Struggling Against Selfishness and Black Hole Attacks in MANETs. Wireless Communications & Mobile Computing 8(6), 689–704 (2008), doi:10.1002/wcm.v8:6
[8]   Kozma, W., Lazos, L.: REAct: Resource-Efficient Accountability for Node Misbehavior in Ad Hoc Networks based on Random Audits. Paper presented at the Second ACM Conference on Wireless Network Security, Zurich, Switzerland, March 16-18 (2009)
[9]   Raj, P.N., Swadas, P.B.: DPRAODV: A Dynamic Learning System Against Blackhole Attack in AODV based MANET. International Journal of Computer Science 2, 54–59 (2009), doi: abs/0909.2371

[10] Jaisankar, N., Saravanan, R., Swamy, K.D.: A Novel Security Approach for Detecting Black Hole Attack in MANET. Paper presented at the International Conference on Recent Trends in Business Administration and Information Processing, Thiruvananthapuram, India, March 26-27 (2010)

[11] Mistry, N., Jinwala, D.C., Iaeng, Zaveri, M.: Improving AODV Protocol Against Blackhole Attacks. Paper presented at the International MultiConference of Engineers and Computer Scientists, Hong Kong, March 17-19 (2010)

[12] Al-Shurman, M., Yoo, S.-M., Park, S.: Black Hole Attack in Mobile Ad Hoc Networks. Paper presented at the 42nd Annual ACM Southeast Regional Conference (ACM-SE 42), Huntsville, Alabama, April 2-3 (2004)

[13] Su, M.-Y.: Prevention of Selective Black Hole Attacks on Mobile Ad Hoc Networks Through Intrusion Detection Systems. IEEE Computer Communications 34(1), 107–117 (2011), doi:10.1016/j.comcom.2010.08.007

[14] Marti, S., Giuli, T., Lai, K., Baker, M.: Mitigating Routing Misbehavior in Mobile Ad Hoc Networks. In: Proc. of the Sixth Annual International Conference on Mobile Computing and Networking, MobiCom (August 2000)

[15] Sharma, S., Gupta, H.M., Dharmaraja, S.: EAGR: Energy Aware Greedy Routing scheme for wireless ad hoc networks. In: International Symposium on Performance Evaluation of Computer and Telecommunication Systems, SPECTS 2008, June 16-18, pp. 122–129 (2008)

[16] Singh, S., Woo, M., Raghavendra, C.: Power-Aware Routing in Mobile Ad Hoc Networks. In: MobiCom, pp. 181–190 (1998)

[17] Johnson, D.B., Maltz, D.A.: Dynamic Source Routing in Ad Hoc WirelessNetworks. In: Imielinski, T., Korth, H. (eds.) Mobile Computing, ch. 5, pp. 153–181. Kluwer Academic Publishers (1996)

[18] IEEE Standard 802.11 - Part 11: Wireless LAN Medium Access Control(MAC) and Physical Layer (PHY) specifications (1999)

[19] http://www.OPNET.com

[20] US (Federal Communications Commission) FCC Part 15.247

[21] Adapting Transmission Power for Optimal Energy Reliable Multi-hop Wireless Communication Suman Banerjee, Archan Misra. Wireless Optimization Workshop, WiOpt 2003, Sophia-Antipolis, France (March 2003)

Cross-Layer Rate Control for Streaming Services over Wired-Wireless Networks

Yu-ning Dong[1], Yu-jue Peng[1], and Hai-xian Shi[2]

[1] College of Telecommunication and Information Engineering,
Nanjing University of Posts and Telecommunications, Nanjing 210003, China
[2] School of Horticulture, Nanjing Agricultural University, Nanjing 210095, China
dongyn@njupt.edu.cn

Abstract. Many advanced technologies including ARQ (Automatic Repeat Request), HARQ (Hybrid ARQ), and AMC (Adaptive Modulation and Coding) have been introduced in modern mobile communication systems, which target to support high-speed streaming services. These services consume a considerable part of resources in core networks and wireless channels. Therefore, an efficient rate control mechanism is necessary. This paper presents a novel cross-layer rate control algorithm which makes use of the aforementioned advanced technologies. The receiver estimates the quality of wireless channels by ARQ information and improves the measurement accuracy of delay with cross-layer design. Packet losses are classified into wireless and congestion losses based on the quality of wireless channels. Following the idea of Delay-Based Congestion Control, the sender adjusts the sending rate according to the tendency of delay variation. Simulation results show that the proposed algorithm outperforms existing TFRC-like schemes in the terms of loss rate control and throughput.

Keywords: rate control, cross-layer design, wired-wireless heterogeneous networks.

1 Introduction

Modern mobile communication systems such as 3G/LTE/4G [1] are designed to support not only the applications like voice services and low data rate services, but also numerous high-speed applications. These applications, including videophone, video conferencing and other multimedia applications, require a higher level of QoS (Quality of Service). As a result, efficient end-to-end rate/delay/error control is necessary.

In wired-wireless heterogeneous networks, a considerable part of packet losses is due to rapid time-variation and high bit error rate of wireless channels. Conventional TFRC (TCP-Friendly Rate Control) algorithm [2] assumes that all packet losses are caused by network congestions and slows down the sending rate unnecessarily when packet losses are actually caused by wireless errors. Consequently it cannot make full use of network resources. To overcome this problem, some loss discrimination algorithms [3,4] are introduced in TFRC. Meanwhile, many advanced functionalities including ARQ, HARQ and AMC are introduced in current mobile communication

S. Balandin et al. (Eds.): NEW2AN/ruSMART 2014, LNCS 8638, pp. 345–355, 2014.

systems. These functionalities enhance the performance of air interface considerably, but make the nature of packet loss more sophisticated, which results in the performance degradation of TFRC-based algorithms.

Some schemes [5-10] take advantage of the characteristic of these functionalities. A cross-layer adaption of TFRC proposed in [5] classifies packet losses and delays observed at transport layer as capacity-related and erroneous-channel-related, and distinguishes them by layer 2 ARQ information. TFWSRC (TCP-Friendly Rate Control for Wireless Streaming) [6] minimizes the wireless losses from the total packet losses by the explicit SDU (Service Data Unit) assembly failure notification offered by RLC (Radio Link Control) sublayer. MARC (Mobile-aware Adaptive Rate Control) [7] estimates the available bandwidth by TFRC equation, and the congestion degree by comparing the available bandwidth and current sending rate. MARC uses AIHD (Additive Increase Heuristic Decrease) algorithm for congestion control. Reference [8] adds two parameters to TFRC equation which provide flexible and smooth transmission rate and adaptability to unpredictable wireless channel condition. RBRC (Round trip time Based Rate Control) [9] classifies congestions in UMTS as long-term and short-term and distinguishes them by RTT (Round Trip Time) value. When the network is in long-term congestion, TFRC is used for rate control, otherwise a modified MIMD (Multiplicative Increase Multiplicative Decrease) algorithm is used. Reference [10] calculates a user's bandwidth by CQI (Channel Quality Identifier) and the number of TTIs (Transmission Time Interval) assigned in a second.

Some other related schemes [11,12] are not end-to-end approaches. I-TFRC (Improved TFRC) [11] needs to set a RTP proxy at the base station. VCP-BE (Variable-structure Congestion control Protocol using Bandwidth Estimation) [12] requires the cooperation of routers configured with ECN (Explicit Congestion Notification).

This paper proposes a novel cross-layer wireless loss aware rate control algorithm called WATC (Wireless-loss Aware Transmission Control). The main contributions of this research are three folds: 1) The algorithm estimates the quality of wireless channels by ARQ (i.e. ACK/NACK) information and distinguishes wireless losses from congestion losses more precisely than the existing end-to-end loss differentiation algorithms; 2) WATC removes the congestion-unrelated variation in delay measurement brought by RLC retransmission mechanism and estimates the level of congestion by the variation of delay; 3) It adopts the idea of Delay-Based Congestion Control which makes congestion decisions based on delay variations, and adjusts the sending rate in advance of severe congestion.

The rest of this paper is organized as follows. Section 2 describes the details of the proposed algorithm. In Section 3, WATC is compared with some existing schemes via simulation. Finally, Section 4 concludes the paper.

2   WATC Algorithm

2.1   Related Wireless Technology

In this sub-section, we briefly introduce related 3G/UMTS technology [1]. The Data Link Layer of UMTS air interface can be divided into two sublayers in Control Plane: MAC (Media Access Control) sublayer and RLC sublayer. The RLC sublayer operates in three modes: Transparent Mode (TM), Unacknowledged Mode (UM) and Acknowledged Mode (AM), according to the type of applications.

TM is specially designed for real-time voice applications. No protocol header is added to the data from higher layer. SDU is transmitted with or without segmentation depending on the type of data.

UM provides segmentation, concatenation and padding functionalities by means of header added to the data. No retransmission mechanism is in use and the delivery is not guaranteed. The corrupt PDU is discarded or marked erroneous depending on the configuration. UM is used for VoIP and broadcasting in the cell.

AM provides segmentation, concatenation, padding functionalities and duplicate detection. An ARQ mechanism is used for error correction. In case the RLC fails to deliver the data successfully within a predefined transmission time or a given number of retransmission attempts, the SDU is discarded and the peer AM entity is informed.

Streaming applications are bandwidth demanding real-time applications and have lower delay limit requirements compared with conversational applications, but their requirement of packet loss rate is stricter [13]. RLC in AM offers more reliable data transmission by retransmitting the lost or erroneous packets. In consequence, delay and jitter will increase but the requirement of packet loss rate may be met in severe wireless condition. For this reason, AM with high reliability is suitable for streaming services and the proposed WATC is designed on the basis of AM.

2.2   Estimation of Wireless Channel Quality

WATC estimates the wireless channel quality by ACK/NACK information in ARQ message, which is a cross-layer approach across Data Link Layer and Transport Layer. By exploiting the close and simple interaction between Data Link layer and Transport Layer, it can provide optimal performance of a sophisticated wireless network [13]. Each AM sender maintains two buffers, a transmission buffer called TX-buff and a retransmission buffer called RTX-buff. When a packet (i.e. an RLC PDU) is sent from TX-buff, it will be put into RTX-buff while waiting for its acknowledgement in the ARQ messages. An ARQ message is a combination of ACK/NACK flags. When a PLC PDU is successfully received, there will be a corresponding ACK flag in the ARQ message sent by AM receiver, otherwise a NACK flag. AM sender discards the RLC PDUs with ACK flag in the RTX-buff and retransmits the RLC PDUs with NACK flag. Note that packets in RTX-buff are given higher priority to those in TX-buff in the sending process. The worse the wireless channel quality is, the more frequently the RLC PDU transmission failure occurs. Consequently, an ARQ message contains higher proportion of NACK flags and AM sender sends higher proportion of RLC PDUs from RTX-buff. Due to the bi-directional connection, either AM sender or AM receiver is able to estimate the wireless channel quality in the following way.

AM sender keeps a count of RLC PDUs that were sent from RTX-buff (N_{RTX}) and those from TX-buff (NTX) in a recent period of time or several recent TTIs, and calculates the *retransmission ratio* which is defined below:

$$RTX\_ratio = N_{RTX} / (N_{TX} + N_{RTX})   \tag{1}$$

AM receiver keeps a count of NACK flags (N_{NACK}) and ACK flags (N_{ACK}) in several recent ARQ messages, and calculates the *negative acknowledgement ratio* as defined below:

$$NACK\_ratio = N_{NACK} / (N_{ACK} + N_{NACK}) \tag{2}$$

Either *retransmission ratio* or *negative acknowledgement ratio* can be an estimate of packet loss rate of RLC PDU (PLR_{RLC}). However, *NACK_ratio* outperforms *RTX_ratio* in timeliness. AM sender allows the ARQ messages to be piggybacked on the RLC PDU with user data whenever possible. Piggybacking enhances the radio resource utilization at the expense of increasing the delay of ARQ messages. That is to say, the calculated *NACK_ratio* reflects the PLR_{RLC} during the past τ_1 seconds , where τ_1 is the additional delay of status message caused by Piggybacking, i.e. *NACK_ratio(t)=$PLR_{RLC}(t-\tau_1)$*. Then AM sender receives the ARQ messages and executes the retransmission mechanism, which generates another additional delay τ_2. That is to say, the calculated *RTX_ratio* reflects the PLR_{RLC} at $\tau_1+\tau_2$ seconds ago, i.e. *RTX_ratio(t)=$PLR_{RLC}(t-\tau_1-\tau_2)$*. As a result, WATC takes *NACK_ratio* as the estimate of PLR_{RLC} and then converts PLR_{RLC} into the *packet error rate* (PER) for the Transport Layer by (3):

$$PER = 1 - (1 - PLR_{RLC})^N \tag{3}$$

where $N = \lceil S / S_{RLC} \rceil$ is the number of RLC PDUs that contain a complete packet in Transport Layer (i.e. RLC SDU), S_{RLC} is the size of RLC PDU and S is the size of RLC SDU.

2.3   Loss Differentiation Algorithm

WATC receiver maintains two loss event queues similar to that in TFRC, a modified loss event queue $\{fix_i\}, i=0,1,2,...,n$ and a congestion loss event queue $\{cg_i\}, i=0,1,2,...,n$, where fix_i is the number of packets received between the ith and (i+1)th loss events, cg_i is the number of packets received between the ith and (i+1)th congestion loss events. The former queue amends its update according to the estimated wireless channel quality. If there is no packet losses and $fix_0 > 1/PER$, WATC receiver inserts an additional loss event and updates $\{fix_i\}$. Let p_{fix} be the average loss rate calculated from $\{fix_i\}$. When the network is congested, packets are lost much more frequently. Consequently, the value of each fix_i decreases and p_{fix} increases remarkably. When the network is idle, packet losses happen less frequently and are more likely to be caused by SDU discard function adopted by AM sender. The amplitudes of fix_i decrease and p_{fix} increase are much smaller than those when the network is congested. As a result, one can discriminate the packet loss by p_{fix} and PER: if the ratio of p_{fix} and PER is greater than a certain threshold TH_{fix}, the loss is judged to be a congestion loss. Otherwise, the loss is judged to be a wireless loss. The congestion loss event queue $\{cg_i\}$ ignores the wireless loss and updates when a packet loss is judged to be a congestion loss. The whole loss differentiation algorithm is described below:

Algorithm 1 Loss differentiation

On every packet received:

01: **if** a packet loss is detected

02:    discard fix_n, fix_{i-1} replaces fix_i, set fix_0 to one

03:    **if** p_{fix} / $PER > TH_{fix}$

04:        the loss is judged to be a congestion loss

05: discard cg_n, cg_{i-1} replaces cg_i, set cg_0 to one

06:    **else**

07:        the loss is judged to be a wireless loss

08:        cg_0 increases by one

09:    **end**

10: **else** // no packet is lost

11:    cg_0 increases by one

12:    **if** $fix_0 > 1 / PER$

13:        discard fix_n, fix_{i-1} replaces fix_i, set fix_0 to one

14:    **else**

15:        fix_0 increases by one

16:    **end**

17: **end**

2.4   Rate Control Mechanism

WATC sender controls sending rate mainly on the basis of TFRC equation [2], as defined in (4).

$$T = \frac{s}{RTT \cdot \sqrt{\dfrac{2p}{3}} + RTO \cdot (3\sqrt{\dfrac{3p}{8}}) \cdot p \cdot (1 + 32p^2)} \tag{4}$$

where T is the sending rate, s is the packet size, RTT is Round Trip Time, RTO is TCP retransmission timeout value, p is the average congestion loss rate. A widely accepted value [2] of RTO is four times RTT. But the sender takes action to smooth the sending rate according to the congestion level when the sending rate changes rapidly.

Stabilization of Sending Rate

The WATC sender controls sending rate mainly on the basis of (4). The average loss rate p in (4) is equal to the average congestion loss rate pcg calculated from {cgi}. When the sending rate calculated by (4) is much lower than the current sending rate, i.e. the ratio of the calculated rate and current rate is smaller than a threshold THrate, WATC will take actions to stabilize the sending rate according to the congestion level as follows: 1) If pfix/PER>THhigh (the congestion level is regarded as high), WATC will decrease the sending rate by 1/6 every other feedback packet. 2) If pfix/PER<THlow and no packet is lost in the recent RTT interval (the network is getting idle), WATC will increase the sending rate by 1/10 every three feedback packets, in order to enhance the network utilization. 3) If neither condition 1) nor 2) is met (the congestion level is regarded as relatively low), WATC will decrease the sending rate by 1/8 for every other feedback packet.

Pre-adjustment of Sending Rate

Following the idea of Delay-Based Congestion Control schemes [14], the WATC sender adjusts the sending rate in advance of severe congestion by monitoring the tendency of ROTT (Relative One-way Trip Time) variation.

It is worth noting that the retransmission and sequential delivery strategies exert significant influence on ROTT. If an RLC PDU of the jth SDU(j) is lost or corrupt, this RLC PDU will be retransmitted later and the delivery of SDU(j) is delayed. As SDU(j) is blocked, SDU(j+1) will not be delivered to the upper layer even if it is successfully received. When SDU(j) is successfully received, SDU(j) and SDU(j+1) will be delivered simultaneously. This process adds an additional delay to both SDU(j) and SDU(j+1), which has nothing to do with the network congestion [15]. Our solution is accomplished by the assistance of RLC sublayer. When AM receiver receives the last RLC PDU of SDU(i), it records the arrival time $recvTime_{AM}(i)$. ROTT of SDU(i) is the difference between the sending time $sendTime(i)$ and $recvTime_{AM}(i)$:

$$ROTT(i) = recvTime_{AM}(i) - sendTime(i) \qquad (5)$$

WATC uses PCT (Pairwise Comparison Test) and PDT (Pairwise Difference Test) [16] to determine the tendency of ROTT variation:

$$ROTT_{PCT} = \frac{1}{n-1}\sum_{i=2}^{n} I[ROTT(i-1) - ROTT(i)] \qquad (6)$$

$$\text{where } I(x) = \begin{cases} 1, & x > 0 \\ 0, & x \le 0 \end{cases}$$

$$ROTT_{PDT} = \frac{[ROTT(1) - ROTT(n)]}{\sum_{i=2}^{n}|ROTT(i-1) - ROTT(i)|} \qquad (7)$$

where n equals 16, that is the recent 16 samples of ROTT to be taken into consideration. The value of $ROTT_{PCT}$ is 1 for the fastest increase of ROTT and 0 for the fastest decrease. The value of $ROTT_{PDT}$ is 1 for the fastest increase of ROTT and -1 for the fastest decrease.

Meanwhile, WATC keeps on measuring the maximum and minimum values of $ROTT$ (i.e. $ROTT_{max}$ and $ROTT_{min}$) which are relatively stable based on the assumption that the network topology is unchanged. One more index of ROTT called $ROTT$ normalized value ($ROTT_{nv}$) is defined as:

$$ROTT_{nv}(i) = [ROTT(i) - ROTT_{min}(i)] / ROTT_{diff}(i) \qquad (8)$$

$$ROTT_{diff}(i) = ROTT_{max}(i) - ROTT_{min}(i) \qquad (9)$$

where the value of $ROTT_{nv}(i)$ is normalized by $ROTT_{diff}$.

WATC sender estimates the congestion level by $ROTT_{PCT}$, $ROTT_{PDT}$ and $ROTT_{nv}$, and makes an additional adjustment to the sending rate T calculated by (4) in two cases: 1) If $ROTT_{nv}>TH_1^{nv}$ and $ROTT_{PCT}>TH_1^{PCT}$, T will be adjusted as: $T' = T / [1 + \gamma \cdot (ROTT_{nv} - TH_1^{nv}) / (1 - TH_1^{nv})]$. 2) If $TH_2^{nv\_low}<ROTT_{nv}<TH_2^{nv\_high}$ and $ROTT_{PCT}>TH_2^{PCT}$ and $ROTT_{PDT}>TH_2^{PDT}$, T will be adjusted as: $T' = T / [1 + \gamma \cdot (ROTT_{nv} - TH_2^{nv\_low}) / (TH_2^{nv\_high} - TH_2^{nv\_low})]$. The values of the thresholds are determined by experiment: $TH_1^{nv}=0.9$, $TH_1^{PCT}=0.52$, $TH_2^{nv\_low}=0.3$, $TH_2^{nv\_high}=0.9$, $TH_2^{PCT}=0.8$, $TH_2^{PDT}=0.75$, $\gamma=0.05$.

3   Performance Evaluation

To assess the performance of the proposed algorithm by simulation, we implemented WATC with NS-2 (Network Simulator 2) and the 3G/UMTS technology. The simulation platform consists of NS-2 and the UMTS extension offered by EURANE Project [17]. WATC is compared with TFRC [2], SPLD [4], AIHD [9] and RBRC [11] in terms of packet loss rate and throughput.

3.1   Simulation Parameters and Performance Metrics

Three scenarios are taken into consideration: 1) no background flow, 2) background pulse flow in wireless link, 3) background pulse flow in wired link. The parameters shared by all scenarios are shown in Table 1.

Table 1. Simulation parameters

Parameters	Value and comment
SDU size	762bytes
period of background pulse flow	60s, turn on/off every 60s
bandwidth and delay between RNC and BS	622Mb / 10ms
bandwidth of DCH uplink/downlink	256Kbps / 2Mbps
RLC PDU size	40bytes
RLC buffer size	15Kbytes
TTI	10ms
acknowledge scheme of AM	bitmap scheme
error model in wireless link	Gilbert-Elliot error model
wireless byte error rate	1×10^{-5} - 2.5×10^{-4}
simulation duration	800s

Network topologies are shown in Fig. 1 and Fig. 2. Topology in Fig. 1 is used in Scenario 1 and Scenario 2. Source node sends streaming data to UE (User Entity). GGSN is Gateway GPRS Support Node and SGSN is Serving GPRS Support Node. RNC is Radio Network Controller and BS is Base Station. Topology in Fig. 2 is used in Scenario 3.

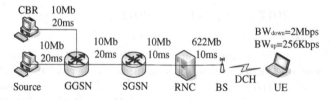

Fig. 1. Network topology of Scenario 1 and Scenario 2

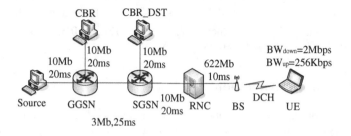

Fig. 2. Network topology of Scenario 3

Packet loss rate is the ratio of the number of lost packets (N_{loss}) to the total number of packets sent (N_{send}):

$$lossrate = N_{loss} / N_{send} \qquad (10)$$

Throughput is normalized by the available bandwidth BW:

$$T_{normalize} = T / BW \qquad (11)$$

where T is the throughput and the initial 25 seconds is ignored when calculating the throughput. The available bandwidth BW is different in different scenarios and its value will be given in next sub-section.

3.2   Simulation Results

Scenario 1

CBR node does not generate any flows so the available bandwidth in (11) is the bandwidth of wireless downlink, i.e. $BW_1=BW_{down}=2$Mbps. Fig. 3 shows the packet loss rate and the normalized throughput versus wireless byte error rate.

The packet loss rate of WATC is the lowest and stable around 0.3%. The throughput of WATC decreases slightly when the byte error rate increases. However, WATC outperforms other protocols (higher than 97% and close to 100% when byte error rate is low). This is because WATC distinguishes packet loss reasons more precisely by estimating the wireless channel quality with ARQ information. The stabilization strategy adopted by WATC prevents the sending rate from changing rapidly and enhances the network utilization. The standard TFRC algorithm without any loss differentiation mechanism performs the worst.

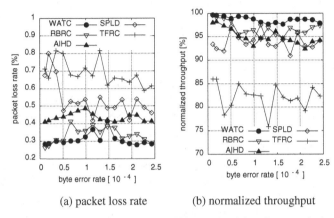

(a) packet loss rate   (b) normalized throughput

Fig. 3. Performance comparisons of different algorithms in scenario 1

Scenario 2

In Scenario 2, CBR node generates a periodic CBR (constant bit rate) on/off flow and sends it to UE. The CBR node turns on or off every 60 seconds and the sending rate is R_{cbr}=512Kbps when it turns on. Wireless link is the bottleneck so the average available bandwidth in (11) is $BW_2=BW_{down}-0.5\times R_{cbr}$=1.792Mbps. Fig. 4 shows the simulation results.

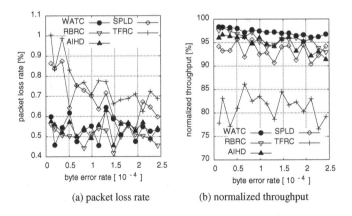

(a) packet loss rate   (b) normalized throughput

Fig. 4. Performance comparisons of different algorithms in Scenario 2

Packet loss rates of all protocols increase due to the existence of CBR on/off flow. WATC, AIHD and RBRC performs similarly in terms of loss rate. But WATC maintains the highest throughput (average 97%). The throughput of AIHD and RBRC are similar (around 95%). SPLD is worse (around 93%) and TFRC is the worst. This is because WATC adopts the idea of Delay-Based Congestion Control strategy and adjusts the sending rate appropriately in advance of severe congestion. The pre-adjustment strategy adopted by WATC avoids severe congestions. In addition, the stabilization strategy enhances the network utilization by adapting the sending rate to reach the steady state faster after the CBR pulse flow turns on or off.

Scenario 3
CBR node generates a periodic CBR pulse flow and sends it to CBR_DST. When the CBR flow is off, the wireless link is the bottleneck. When the CBR pulse flow is on (at the rate of R_{cbr}=1.5Mbps), the link between GGSN and SGSN becomes the bottleneck. So the average available bandwidth in (11) is BW_3=0.5×(BW_{down}+ $(3–R_{cbr})$)=1.81Mbps. Fig. 5 shows the results.

WATC outperforms other protocols in terms of loss rate and throughput, as shown in Fig. 5 WATC maintains the lowest packet loss rate (0.2%~0.4%) and highest throughput (97%~99%). This is because that the loss differentiation algorithm and the pre-adjustment and stabilization strategy still work in Scenario 3, in which the network state is more complicated.

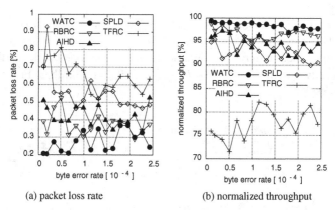

(a) packet loss rate   (b) normalized throughput

Fig. 5. Performance comparisons of different algorithms in scenario 3

4   Conclusion

In this paper, a novel cross-layer rate control algorithm is proposed for streaming services in wired-wireless heterogeneous networks. The proposed algorithm is implemented on the basis of ARQ information and adopts the idea of Delay-Based Congestion Control. Through simulations in NS-2, a performance comparison with AIHD, RBRC, SPLD and TFRC validates that the proposed algorithm can reduce the packet loss rate and enhance throughput effectively in such heterogeneous networks.

Although the current version of the proposed algorithm is developed upon UMTS technology, the basic idea can be extended to the utilization of advanced functionalities in LTE/4G systems, and this will be our future work.

Acknowledgement. This work was supported in part by National Natural Science Foundation of China (No. 61271233, 60972038), the Ministry of Education (China) Ph.D. Programs Foundation (No.20103223110001).

References

1. Molisch, A.F.: Wireless Communications, 2nd edn., pp. 635–645. John Wiley & Sons Ltd., New York (2011)
2. Handley, M., Floyd, S., Padhye, J., Widmer, J.: TCP friendly rate control. IETF RFC 3448 (January 2003)
3. Nguyen, N., Yang, E.: End-to-end loss discrimination for improved throughput performance in heterogeneous networks. In: Proc. IEEE-CCNC, 3rd edn., pp. 538–542 (2006)
4. Cen, S., Cosman, P.C., Voelker, G.M.: End-to-End differentiation of congestion and wireless losses. IEEE/ACM Trans. on Networking 11(9), 703–717 (2003)
5. Jung, I.M., Karayiannis, N.B., Pei, S., Benhaddou, D.: The cross-layer adaptation of TCP-friendly rate control to 3G wireless links. In: Proc. CSNDSP 7th, pp. 283–288 (2010)
6. Fu, Y., Hu, R., Tian, G., Wang, Z.: TCP-friendly rate control for streaming service over 3G network. In: Proc. WiCOM-WCNMC 2006, pp. 1–4 (2006)
7. Koo, J., Chung, K.: MARC: Adaptive rate control scheme for improving the QoE of streaming services in mobile broadband networks. In: Proc. ISCIT 2010, pp. 105–110 (2010)
8. Dehkordi, A.M., Vakili, V.T.: Equation based rate control and multiple connections for adaptive video streaming over cellular networks. In: Proc. SoftCOM-STCN 17th, pp. 176–180 (2009)
9. Zhao, X., Dong, Y., Zhao, H., Hui, Z., Li, J., Sheng, C.: A real-time congestion control mechanism for multimedia transmission over 3G wireless networks. In: Proc. IEEE-ICCT 12th, pp. 1236–1239 (2010)
10. Huang, X., Jiang, N., Yang, J.: A cross-layer flow control algorithm over HSDPA networks. In: Proc. IEEE-WCNIS 2010, pp. 317–312 (2010)
11. Zheng, T., Hua, G., Yu, F.: An improved TFRC mechanism of mobile streaming media based on RTCP. In: Proc. TMEE 2011, pp. 2134–2137 (2011)
12. Wang, J., Chen, J., Zhang, S., Wang, W.: An explicit congestion control protocol based on bandwidth estimation. In: Proc. IEEE-GLOBECOM 2011, pp. 1–5 (2011)
13. Mehmood, M.A., Sengul, C., Sarrar, N., Feldmann, A.: Understanding cross-layer effects on Quality of Experience for video over NGMN. In: Proc. IEEE-ICC 2011, pp. 1–5 (2011)
14. Budzisz, L., Stanojevic, R., Shorten, R., Schlote, A., Baker, F., Shorten, R.: On the fair coexistence of loss- and delay-based TCP. IEEE/ACM Trans. on Networking 19(6), 1811–1824 (2011)
15. Yuan, H., Du, M., Li, C.: Improve TFRC mechanism for streaming media in wireless network. Journal of Harbin University of Science and Technology 16(2), 40–42 (2011)
16. Han, T., Huang, Y., Qu, L., Shi, M.: Research on improving the performance of TCP for heterogeneous networks. Computer Science 38(10A), 279–281 (2011)
17. SEACORN, Enhanced UMTS radio access network extensions for NS-2 (October 2006), http://eurane.ti-wmc.nl/

A Relevant Equilibrium in Open Spectrum Sharing: Lorenz Equilibrium in Discrete Games

Ligia C. Cremene[1] and Dumitru Dumitrescu[2]

[1] Technical University of Cluj-Napoca, Romania
[2] Babes-Bolyai University, Cluj-Napoca, Romania
ligia.cremene@com.utcluj.ro, ddumitr@cs.ubbcluj.ro

Abstract. A new game theoretical solution concept for open spectrum sharing in cognitive radio (CR) environments is highlighted – the Lorenz equilibrium (LE). Both Nash and Pareto solution concepts have limitations when applied to real world problems. Nash equilibrium (NE) rarely ensures maximal payoff and it is frequently Pareto inefficient. The Pareto set is usually a large set of solutions, often too hard to process. The Lorenz equilibrium is a subset of Pareto efficient solutions that are equitable for all players and ensures a higher payoff than the Nash equilibrium. LE induces a selection criterion of NE, when several are present in a game (e.g. many-player discrete games) and when fairness is an issue. Besides being an effective NE selection criterion, the LE is an interesting game theoretical situation *per se*, useful for CR interaction analysis.

Keywords: open spectrum sharing, cognitive radio environments, non-cooperative games, Lorenz equilibrium, fairness.

1 Introduction

The problem of finding an appealing solution concept for Open Spectrum Sharing (OSS) is addressed from a game theoretical perspective. *Lorenz equilibrium (LE)* [1] ensures both a Pareto optimal solution [2] and a technique/criterion for selecting a Nash equilibrium (NE) when many are present.

OSS refers to spectrum sharing among secondary users (cognitive radios) in unlicensed spectrum bands [3]. Cognitive radio (CR) interactions are strategic interactions [4], [5]: the utility of one CR depends on the actions of all the other CRs in the environment. Game Theory (GT) provides a fertile framework and the tools for CR interaction analysis [4], [6], [7], [8], [9], [11], [12]. Insight may be gained on unanticipated situations that may arise in spectrum sharing, by devising GT simulations.

Usually, the outcome of a non-cooperative game is the Nash equilibrium - the most common solution concept. In OSS fairness is an issue [4], [11], [12] and usually, NE does not provide it.

Standard GT analysis of spectrum sharing is performed based on the continuous forms of the games. Yet, discrete modeling seems more realistic for spectrum access as

S. Balandin et al. (Eds.): NEW2AN/ruSMART 2014, LNCS 8638, pp. 356–363, 2014.

users get to choose discrete quantities of the radio resources (number of channels, power levels, etc.). Discrete GT models usually yield multiple Nash equilibria. The number of NEa increases with the number of users. Equilibrium selection thus becomes an issue.

The basic assumptions of our approach are: *(i)* CRs (i.e. CR access points or CR Tx-Rx pairs) are modelled as self-regarding players, *(ii)* CRs do not know in advance what actions the other CRs will choose, and *(iii)* CRs have channel sensing and RF reconfiguration capabilities [9], [10]. Given these assumptions, one-shot, non-cooperative, discrete game analysis is considered relevant.

A well known game theoretical model – Cournot oligopoly [5] – is chosen as support for simulation, due to its simple and intuitive form and suitability for resource access modelling. Although it has been intensively used for spectrum trading modelling, the Cournot model has also been reformulated in terms of spectrum access [6], [11], [13] capturing general channel access scenarios.

Considering the limitations of Nash and Pareto equilibria, a new solution concept is considered - the Lorenz equilibrium (a subset of Pareto optimal solutions) [1]. The Lorenz solution concept is a transformation that is applied to the payoffs (outcomes) of the game. As Lorenz equilibrium is both equitable and Pareto efficient, it induces a NE selection criterion. Besides being an effective NE selection criterion, the LE is an interesting GT situation *per se*.

Numerical simulations reveal equilibrium situations that may be reached in simultaneous, open spectrum access scenarios. Lorenz equilibrium is detected and analyzed along with four other types of equilibria: Nash, Pareto, and the joint Nash-Pareto and Pareto-Nash equilibria capturing the heterogeneity of players. The later become more distinct for n-player interactions.

2   Lorenz Equilibrium

The Lorenz equilibrium is a new GT solution concept [1]. This solution concept is inspired by multicriteria optimization (MCO) where it is known as Lorenz dominance (LD) [14], or equitable dominance relation.

The standard solution concept in MCO is the Pareto set. Informally, the Pareto optimality (or Pareto efficiency) is a strategy profile so that no strategy can increase one player's payoff without decreasing any other player's payoff [2], [14]. A refinement of the Pareto dominance, LD is used in decision theory and fair optimization problems. The set of equitable efficient solutions is contained within the set of efficient solutions (Pareto). *Lorenz equilibrium is the set of the most balanced and equitable Pareto efficient solutions so that no payoff can be improved without degradation of other payoffs.*

In GT the Lorenz dominance relation is applied to address some limitations of standard GT solution concepts. Both Nash and Pareto equilibria have limitations when applied to real world problems. Nash equilibrium rarely ensures maximal payoff – it suffers from excessive competition among selfish players in a non-cooperative game, and the outcome is inefficient [4]. The Pareto equilibrium is a set of solutions that is often too hard to process. However, the LE provides a small subset of efficient solutions that are equitable for all players.

Generally, a game is defined as a system $G = (N, A_i, u_i, i = 1,\ldots, n)$ [5] where:

(i) N represents the set of n players, $N = \{1,\ldots, n\}$.

(ii) for each player $i \in N$, A_i represents the set of actions $A_i = \{a_{i1}, a_{i2}, \ldots, a_{im}\}$; $A = A_1 \times A_2 \times \ldots A_n$ is the set of all possible game situations;

(iii) for each player $i \in N$, $u_i : A \to R$ represents the utility function (payoff).

A strategy profile is a vector $a = (a_1,\ldots,a_n) \in A$, where $a_i \in A_i$ is an action of player i.

By (a_i, a_{-i}^*) we denote the strategy profile obtained from a^* by replacing the action of player i with a_i, i.e. $(a_i, a_{-i}^*) = (a_1^*, a_2^*,\ldots, a_{i-1}^*, a_i, a_{i+1}^*,\ldots, a_1^*)$.

A strategy profile is said to be a Nash equilibrium if no player can improve her payoff by unilateral deviation [5].

Let us consider the payoffs ordered in an ascending order

$$u_{(1)}(a) \le u_{(2)}(a) \le \ldots \le u_{(n)}(a), a \in A \tag{1}$$

and define the quantities:

$$l_1(a) = u_{(1)}(a),$$
$$\ldots$$
$$l_n(a) = \sum_{i=1}^{n} u_{(i)}(a). \tag{2}$$

Strategy x is said to Lorenz dominate strategy y (and we write $x \succ_L y$) if and only if [1], [14]:

$$l_i(x) \ge l_i(y), i = 1,\ldots, n,$$
$$\exists j : l_j(x) > l_j(y). \tag{3}$$

Lorenz equilibrium of the game is the set of non-dominated strategies with respect to relation $\succ_L$, that is considered the generative relation for the Lorenz equilibrium [1]. Generative relations represent an algebraic tool for characterizing game equilibria [15].

Similarly, generative relations for Nash, Pareto, and joint Nash-Pareto equilibria may be defined [15]. The joint Nash-Pareto and Pareto-Nash equilibria applied to CR interaction analysis are discussed in [13], [17] for different game models.

Lorenz equilibrium is characterized by equitable payoffs. In addition to the basic objective, aiming to maximize individual utilities, fairness refers to the idea of favouring well-balanced utility profiles.

Where multiple equilibria exist (in discrete games, for instance), the need for equilibrium selection arises. The main equilibrium selection criteria, as discussed in [4], are: Pareto optimality, equilibrium refinement, and evolutionary equilibrium.

If multiple Nash equilibria exist in a game (e.g. discrete many-player game), the closest ones to the Pareto optimal are usually chosen. But, usually, a large set (sometimes infinite) of Pareto optimal solutions exists. Considering the distance to the Lorenz set of optimal solutions, a more specific condition (criterion) is introduced. The selected NE will be closest to Pareto optimal and also equitable.

3 Open Spectrum Sharing Scenario

In order to illustrate the detection and usefulness of the Lorenz equilibrium, a general open spectrum access scenario may be considered [6], [13]. The scenario is modelled as a non-cooperative, one-shot game. The discrete form of the game is analyzed, as it exhibits multiple Nash equilibria. Standard game models, in their continuous form ([4], [5], [8], [9], [11], [12]), do not capture the discrete nature of choices made by CRs. Therefore, we consider the discrete form games.

The players are n CRs (i.e. CR access points or CR Tx-Rx pairs) attempting to access a certain set of available channels (or whitespace) W. Each CR i is free to implement a number of frequency hopping channels $a_i \in \{0, 1, ..., |W|\}$, where a_i is a CR i individual strategy. A strategy profile is vector $a = (a_1, ..., a_n)$. Let us consider K_i to be the rate of interfered symbols on each channel, and let $K_i \in [0, 10]$. This rate may also account for a range of factors that generally cause unused symbols (interference, noise, etc.) and it actually reflects the link-level performance on each channel.

Each CR is attempting to maximize its payoff u_i given as:

$$u_i(a) = \left(|W| - \sum_{k=1}^{n} a_k \right) a_i - K_i a_i, i = 1, ..., n , \tag{4}$$

The form of the chosen payoff function u_i accounts for the difference between a function of goodput $\left(|W| - \sum_{k=1}^{n} a_k \right) a_i$ (a linear approximation of the number of non-interfered symbols per channel × number of frequency hopped channels) and the cost of simultaneously supporting a_i channels: $K_i a_i$ (rate of interfered symbols per channel × number of channels). Other payoff functions may be considered (e.g. the ones in [11], [12]), as well as asymmetric costs.

The question for this general scenario is: how many simultaneous frequency hopping channels should each CR access in order to maximize its payoff in a stable game situation (equilibrium)?

Challenges of discrete games are related to: *(i)* computing Nash equilibrium, *(ii)* existence of multiple NE, and *(iii)* selection of an efficient NE (close to Pareto optimal).

4 Equilibrium Detection – Numerical Experiments

The effectiveness of the Lorenz criterion for the equilibrium selection problem becomes clearer for many-player games (n-dimensional space). For the sake of accuracy and simplicity in illustrating the presence of LE, we have chosen to represent the two- and three-dimensional cases. Two and three CR simultaneous spectrum access scenarios are therefore considered. As the continuous modelling captures only partially the variety of possible equilibrium situations, challenging discrete instances of the game are considered.

Lorenz equilibrium is detected and analyzed along with four other types of equilibria: Nash, Pareto, and the joint Nash-Pareto, and Pareto-Nash equilibria.

An evolutionary method, based on generative relations [16], is used for equilibrium detection. An adaptation of the state-of-the-art Differential Evolution [18] is the underlying evolutionary technique. Other choices are also possible. The method is robust with respect to the nature of the game (continuous, discrete) and scalable to the number of players and the number of available channels. It allows comparison of strategies and payoffs of several equilibria.

Let us consider the following simulation parameters: $|W| = 10$ (number of available channels) and $K_i = 1$ (equal unitary rate of interfered symbols per channel).

The reported results represent a sub-set of more extensive simulations. A population of 100 strategies has been evolved using a rank-based fitness assignment technique. In all experiments the process converges in less than 20 generations.

Fig. 1 illustrates the equilibrium strategies achieved by two CRs simultaneously trying to access the same set of ten available channels ($|W| = 10$).

The computation of the NE strategy for the standard, continuous-form game – taken as a reference point – is straightforward and yields (3,3), the NE being unique [5]. NE is a stable strategy from which no CR has any incentive to individually deviate.

The discrete instance of the 2-player game reveals three Nash equilibria: (2,4), (3,3), (4,2) and one LE: (2,2) (Fig. 1 and Fig. 2). For a larger number of players even more NEa exist and LE is also a multi-element set.

The existence of multiple NEa indicates a certain degree of flexibility in choosing the number of channels to access – there are several situations from which the CRs have no incentive to unilaterally deviate.

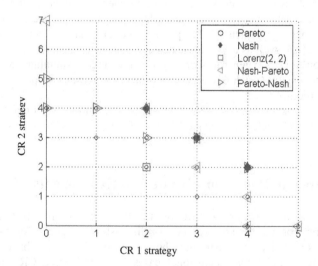

Fig. 1. Discrete modelling. – two CRs (card W=10, K=1). Evolutionary detected equilibrium **strategies**: Nash: (2,4), (3,3), (4,2) Pareto, Nash-Pareto, Pareto-Nash, and Lorenz: (2,2).

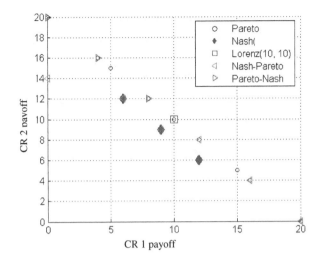

Fig. 2. Discrete modelling – two CRs (card W=10, K=1). **Payoffs** of the evolutionary detected equilibria: Nash: (6,12), (9,9), (12,6), Pareto, N-P, P-N, and Lorenz: (10,10).

The (3,3) NE strategy is the most stable game situation as it maintains even for the joint N-P and P-N strategies (they overlap, Fig.1). It is also the closest one to LE (2,2). As expected, the corresponding payoffs – NE: (9,9) – (Fig. 2), are the most equitable of the three NEa. The other two NEa, (2,4) and (4,2), are also stable and are maintained for one of the joint strategies (N-P or P-N), but are not equitable: payoffs are (6,12) and (12,6).

Fig. 2 illustrates the payoffs of the two players (CRs): $u_1(a_1, a_2)$ and $u_2(a_1, a_2)$. The three NE payoffs (6,12), (9,9), (12,6) offer a diversity of utilities among which one equitable solution: (9,9). LE payoff (10,10) is slightly higher than NE payoff (9,9). Yet, the corresponding number of accessed channels is smaller for LE: (2,2) than for NE: (3,3).

Fig. 3 and Fig. 4 capture the equilibrium situations (strategies and payoffs, respectively) for the discrete modelling of the 3-CRs simultaneous sharing of the same set of channels.

Seven Nash equilibria) are detected (Fig. 3): (2,2,2), (2,2,3), (2,3,2), (3,2,2), (1,3,3), (3,1,3), and (3,3,1). This indicates an even higher flexibility in choosing the number of accessed channels for each CR. Also the range of available payoffs is increased (Fig. 4). NE payoffs are (6, 6, 6), (4, 4, 6), (4, 6, 4) , (6, 4, 4), (2, 6, 6), (6, 2, 6), (6, 6, 2). We may even notice one NE strategy (2,2,2) overlapping a LE and yielding identical payoffs (6,6,6). Obviously this is the preferred NE among the seven detected.

The effectiveness of LE criterion in NE selection is even more evident for n-player $(n \geq 3)$ games. In the n-player case the number of NEa increases polynomially. Moreover, the closeness to LE can no longer be indicated by visual inspection. The NE that is computationally detected as closest to LE may then be selected.

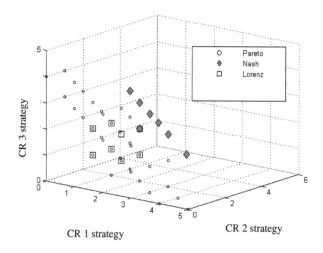

Fig. 3. Three CR simultaneous access (card W=10, K=1). Discrete resource access modelling. **Strategies**: Nash: (2,2,2), (2,2,3), (2,3,2), (3,2,2), (1,3,3), (3,1,3), (3,3,1), Pareto, N-N-P, N-P-P, Lorenz: (2,2,2), (1,1,2), (1,2,1), (2,1,1), (2,2,1), (2,1,2), (1,2,2).

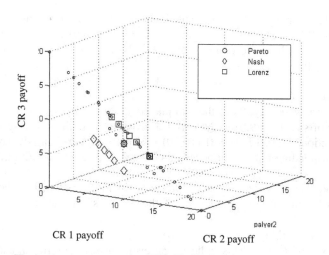

Fig. 4. Three CR simultaneous access (card W=10, K=1). Discrete resource access modelling. **Payoffs**: Nash: (6, 6, 6), (4, 4, 6), (4, 6, 4) , (6, 4, 4), (2, 6, 6), (6, 2, 6), (6, 6, 2), Pareto, N-N-P, N-P-P, and Lorenz: (6, 6, 6), (5, 5, 10), (5, 10, 5), (10, 5, 5), (8, 8, 4), (8, 4, 8),   (4, 8, 8).

5   Conclusions

A new equilibrium concept that may prove useful for modelling CR open spectrum sharing is highlighted: the Lorenz equilibrium. Its usefulness is illustrated on a general channel access scenario, yet other radio resource access models may be analyzed in the same manner. LE is a subset of Pareto optimal solutions that proves useful in selecting a NE when multiple ones exist (e.g. in many-player discrete games). LE is an appealing solution concept for spectrum sharing games as it is both fair and profitable (ensures a higher payoff than NE and is Pareto optimal).

Acknowledgment. Thanks are due to dr. Réka Nagy for her kind help in preparing the experiments. This work was partly funded by the Technical University of Cluj-Napoca, Romania, research project no. 24311.

References

[1] Nagy, R., Suciu, M., Dumitrescu, D.: Lorenz Equilibrium: Evolutionary Detection. In: GECCO 2012, pp. 489–496 (2012)
[2] Fudenberg, D., Tirole, J.: Multiple Nash Equilibria, Focal Points, and Pareto Optimality. In: Game Theory. MIT Press (1983)
[3] Akyildiz, I., Lee, W., Vuran, M., Mohanty, S.: NeXt generation/dynamic spectrum access/cognitive radio wireless networks: A survey. Computer Networks: Int. J. of Comput. Telecomm. Netw. 50(13), 2127–2159 (2006)
[4] Wang, B., Wu, Y., Liu, K.J.R.: Game theory for cognitive radio networks: An overview. Computer Networks, Int. J. of Omput. Telecomm. Netw. 54(14), 2537–2561 (2010)
[5] Osborne, M.J.: An Introduction to Game Theory. Oxford, U.P.(2004)
[6] Neel, J.O.: Analysis and Design of Cognitive Radio Networks and Distributed Radio Resource Management Algorithms, PhD Thesis (2006)
[7] MacKenzie, A., Wicker, S.: Game Theory in Communications: Motivation, Explanation, and Application to Power Control. In: GLOBECOM 2001, pp. 821–825 (2001)
[8] Huang, J.W., Krishnamurthy, V.: Game Theoretic Issues in Cognitive Radio Systems. J. of Comm. 4(10), 790–802 (2009)
[9] Niyato, D., Hossain, E.: Microeconomic models for dynamic spectrum management in cognitive radio networks. In: Hossain, E., Bhargava, V.K. (eds.) Cognitive Wireless Communication Networks, pp. 391–423. Springer, NY (2007)
[10] Cordeiro, C., Challapali, K., Birru, D.: IEEE 802.22: An Introduction to the first wireless standard based on cognitive radios. J. of Comm. 1(1), 38–47 (2006)
[11] da Costa, G.W.O., Cattoni, A.F., Kovacs, I.Z., Mogensen, P.E.: A scalable spectrum-sharing mechanism for local area network deployment. IEEE T. Veh. Technol. 59(4) (May 2010)
[12] Nie, N., Comaniciu, C.: Adaptive channel allocation spectrum etiquette for cognitive radio networks. Mobile Netw. App. 11(6), 779–797 (2006)
[13] Cremene, L.C., Dumitrescu, D., Nagy, R., Cremene, M.: Game theoretic modelling for dynamic spectrum access in TV whitespace. In: CROWNCOM 2011, Osaka, pp. 336–340 (2011)
[14] Kostreva, M.M., Ogryczak, W.: Linear Optimiation with multiple equitable criteria. RAIRO Op. Research 33, 275–297 (1999)
[15] Dumitrescu, D., Lung, R.I., Mihoc, T.D.: Generative relations for evolutionary equilibria detection. In: GECCO 2009, pp. 1507–1512 (2009)
[16] Lung, R.I., Dumitrescu, D.: Computing Nash Equilibria by Means of Evolutionary Computation. IJCCC 3, 364–368 (2008)
[17] Cremene, L.C., Dumitrescu, D., Nagy, R., Gasko, N.: Cognitive Radio Simultaneous Spectrum Access/ One-shot Game Modelling. In: IEEE, IET CSNDSP 2012, Poznan, pp. 1–6 (2012)
[18] Storn, R., Price, K.: Differential evolution – simple and efficient heuristic for global optimization over continuous spaces. J. of Global Optimization 11, 341–359 (1997)

A Stochastic Game in Cognitive Radio Networks for Providing QoS Guarantees

Jerzy Martyna

Institute of Computer Science, Faculty of Mathematics and Computer Science
Jagiellonian University, ul. Prof. S. Lojasiewicza 6, 30-348 Cracow, Poland

Abstract. Providing quality of service (QoS) guarantees is an important objective in the design of the cognitive radio (CR) networks. In this paper, we studied the stochastic game in CR networks with multiple classes of secondary users (SUs). Considering the spectrum environment as time-varying and that each group of SUs is able to use an adaptive strategy, the providing QoS guarantees is identified by finding the effective capacity radio channel. The performance of the proposed stochastic game for providing QoS guarantees has been studied through computer simulations and the results are quite satisfactory.

1 Introduction

The key enabling technology of dynamic spectrum access is cognitive (CR) technology. The term 'cognitive radio' defined [5] means that CR is an intelligent wireless communication system, which is able to change its transmission or reception parameters to improve certain operating parameters (e.g., transmit power, carrier-frequency, and modulation strategy). A cognitive radio shall sense the environment (cognitive capability), analyse and learn sensed information (self-organised capability) and adapt to the environment (reconfigurable capabilities).

Cognitive radio network is a networking system of cognitive radios that make use of cutting-edge technology from computer networks to solve the problems in traditional wireless networks. In the paper [12] the cognitive network is described as a network with the cognitive radios which can perceive current network conditions, plan, decide, act on those conditions, learns from consequences of its actions, and follow end-to-end goals. Cognitive radio network can be also defined as a network that can utilize both radio spectrum and wireless resources opportunistically, based on the knowledge of availability such resources. Therefore, the CR network can opportunistically organize cognitive radios.

We use Fig. 1 to illustrate cognitive radio network, in which there are two types of users sharing a common spectrum portion but with different rules: Primary Users (PUs) have the priority in spectrum utilization within the band they have licensed, and the Secondary Users (SUs) must opportunistically access the spectrum without interfering with PUs. The resource allocation problem in downlink (Cognitive Radio Base Station/Core Station to SUs) and uplink (SUs to Cognitive Radio Base Station/Core Station) is without synchronization between the primary system and cognitive radio network.

S. Balandin et al. (Eds.): NEW2AN/ruSMART 2014, LNCS 8638, pp. 364–374, 2014.

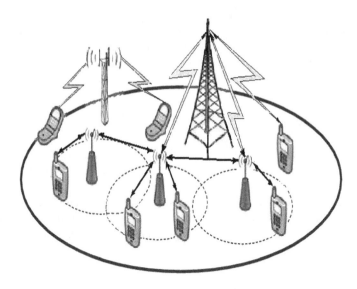

Fig. 1. An example of downlink/uplink cognitive radio

The research in CR network is mainly devoted to spectrum sensing techniques and spectrum sharing approaches to determine which portion of the spectrum is available and detect the presence of primary users when a SU operates in a licensed band in order to avoid any harmful interference to PUs [2], [9], [10],. There is very little research on the providing quality of service (QoS) guarantees over cognitive radio channels. Multimedia services such as video and audio transmission require bounded delays or guaranteed bandwidth. A hard delay bound guarantee is infeasible due to the impact of the time varying fading cognitive channels.

The spectrum management technologies including spectrum sensing, spectrum decision, spectrum sharing, and spectrum handoff schemes with QoS provisioning in CR networks have been studied in [13]. In this paper to evaluate performance of different spectrum management schemes between the primary and the secondary users the preemptive resume priority (PRP) M/G/1 queuing model was suggested. An analysis of QoS provisioning in cognitive radio networks was presented in [6], in which the QoS-oriented CR system together with the underlying QoS-provisioned DSA protocol (called QPDP) has been simulated. Unfortunately, the presented QoS provisioning in the CR networks are not based on the *effective capacity* context. We recall that in the paper [15] has been developed a powerful concept defined as *effective capacity*, which provides the statistical QoS guarantee in general wireless communication.

Two key contributions of this paper are the following: (a) formulating a state-transition model for QoS service provisioning over cognitive radio channels, (b) providing a stochastic game for QoS guarantees. Our simulations demonstrate that the proposed approach provides efficient solution to implementing QoS provisioning in CR networks.

The rest of the paper is organized as follows. Section 2 describes the network model, extracts a cognitive channel model and presents concept of effective bandwidth and effective capacity. Section 3 provides a stochastic game for QoS guarantees. Section 4 details the minimax-Q learning algorithm to obtain the optimal policy that maximizes the expected sum of payoffs of the stochastic game. Section 5 gives a simulation-based performance evaluation of the proposed method. We finish by drawing some conclusions in section 6.

2   Network Model

In this section, we present the network model of cognitive radio network. We also formulate the radio channel model and provide the effective capacity term of the cognitive radio channel.

Consider a cognitive radio network with the secondary users and free radio channels available for use by multiple secondary users. Each channel can be used simultaneously by multiple secondary users. Moreover, a single secondary user can use several channels at the same time to achieve their requirements.

2.1   Cognitive Channel Model

Cognitive radio channel model allows the sending of information by a secondary transmitter to a secondary user, possibly in the presence of primary users. The cognitive radio channel will be tested by secondary users. If the secondary transmitter selects its transmission when the channel is busy, the average power is $\overline{P}_1$ and the rate is r_1. When the channel is idle, the average power is $\overline{P}_2$ and the rate is r_2. We assume that $\overline{P}_1 = 0$ denotes the stoppage of the secondary transmission in the presence of an active primary user. Both transmission rates, r_1 and r_2, can be fixed or time-variant depending on whether the transmitter has channel side information or not. In general, we assume that $\overline{P}_1 < \overline{P}_2$. In the above model, the discrete-time channel input-output relation in the absence in the channel of the primary users is given by

$$y(i) = h(i)x(i) + n(i), \quad i = 1, 2, \ldots \tag{1}$$

where $h(i)$ is the channel coefficient, i is the symbol duration. If primary users are present in the channel, the discrete-time channel input-output relation is given by

$$y(i) = h(i)x(i) + s_p(i) + n(i), \quad i = 1, 2, \ldots \tag{2}$$

where $s_p(i)$ represents the sum of the active primary users' faded signals arriving at the secondary receiver $n(i)$ is the additive thermal noise at the receiver and is zero-mean, circularly symmetric, complex Gaussian random variable with variance $E\{|n(i)|^2\} = \sigma_n^2$ for all i.

We assume that the receiver knows the instantaneous values $\{h(i)\}$, while the transmitter has no such knowledge. We construct a state-transition model for cognitive transmission by considering the cases in which the fixed transmission rates are lesser or greater than the instantaneous channel capacity values. In particular, the ON state is achieved if the fixed rate is smaller than the instantaneous channel capacity. Otherwise, the OFF state occurs.

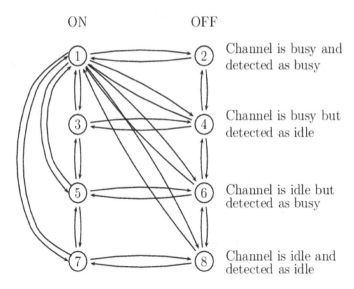

Fig. 2. State transition model for the cognitive radio channel

We assume that the maximum throughput can be obtained in the state-throughput model [1], which is given in Fig. 2. Four possible scenarios are associated with the model, namely:

1) channel is busy, detected as busy (correct detection),
2) channel is busy, detected as idle (miss-detection),
3) channel is idle, detected as busy (false alarm),
4) channel is idle, detected as idle (correct detection).

If the channel is detected as busy, the secondary transmitter sends with power $\overline{P}_1$. Otherwise, it transmits with a larger power, $\overline{P}_2$. In the above four scenarios, we have the instantaneous channel capacity, namely

$$C_1 = B \log_2(1 + SNR_1 \cdot z(i)) \quad \text{channel is busy, detected as busy} \qquad (3)$$

$$C_2 = B \log_2(1 + SNR_2 \cdot z(i)) \quad \text{channel is busy, detected as idle} \qquad (4)$$

$$C_3 = B \log_2(1 + SNR_3 \cdot z(i)) \quad \text{channal is idle, detected as busy} \qquad (5)$$

$$C_4 = B \log_2(1 + SNR_4 \cdot z(i)) \quad \text{channel is idle, detected as idle} \qquad (6)$$

where B is the bandwidth available in the system, $z(i) = [h(i)]^2$, SNR_i for $i = 1, \ldots, 4$ denotes the average signal-to-noise ratio (SNR) values in each possible scenario.

The cognitive transmission is associated with the ON state in scenarios 1 and 3, when the fixed rates are below the instantaneous capacity values ($r_1 < C_1$ or $r_2 < C_2$). Otherwise, reliable communication is not obtained when the transmission is in the OFF state in scenarios 2 and 4. Thus, the fixed rates above are the instantaneous capacity values ($r_1 \geq C_1$ or $r_2 \geq C_2$). The above channel model has 8 states and is depicted in Fig. 2. In states 1, 3, 5 and 7, the transmission is in the ON state and is successfully realised. In the states 2, 4, 6 and 8 the transmission is in the OFF state and fails.

2.2   Effective Capacity

The statistical QoS constraints in cognitive radio networks can be identified through effective capacity. Effective capacity was introduced by Wu and Negi [15] as the maximum constant arrival rate that a given time-varying service process can support while meeting the QoS requirements.

In the paper [3] has been proved that for queueing system with a stationary ergodic arrival and service process, the queue length process $Q(t)$ converges to a random variable $Q(\infty)$ such that

$$- \lim_{x \to \infty} \frac{\log(Pr\{Q(\infty) > x\})}{x} = \theta \tag{7}$$

The parameter $\theta(\theta > 0)$ indicates the xponential decade rate of the QoS violation probabilities. A small θ corresponds to a slow decay rate, which implies that the system provides a looser QoS guarantee. On the other hand, a larger θ indicates a fast decay rate, and thus the system provides a strict QoS guarantee.

We assume that the maximum throughput can be obtained in the state-transition model [1]. Thus, the maximum The effective capacity is expressed by:

$$E_c(\theta) = -\frac{1}{\theta} \log(E[e^{-\theta R}]) \tag{8}$$

where R is the independent identical distributed (i.i.d.) service process and $E[y]$ is taking expectation over y. Specifically, if $\theta \geq -\log \epsilon / D_{max}$, then:

$$\sup_t Pr\{D(t) \geq D_{max}\} \leq \epsilon \tag{9}$$

where D_{max} is the maximum tolerable delay of the traffic rate and $D(t)$ is the delay at time t. The Eq. (8) indicating that the probability that the traffic rate delay exceeds the maximum tolerable delay is below ϵ.

By using effective capacity Tang and Zhang [11] have determined that the optimal power and rate adaptation techniques that maximise the system throughput under QoS constraints. The effective capacity and resource allocation strategies for Markov wireless channel models were studied by Liu et al. [8]. In this study, the continuous Gilbert-Elliot channel model with ON and OFF states was used. The energy efficiency under QoS constraints was analysed by Gursoy et al. [4] in low power and wideband regions. Unfortunately, none of the above-mentioned papers have been not considered in the application of stochastic game to solve the formulated problem.

3   Stochastic Game Formulation

There is a stochastic game among a number of players belonging to various classes. The stochastic game $\mathcal{G}$ is defined as a set of states denoted by $\mathcal{S}$, a set of actions described as $\mathcal{A}_1, \mathcal{A}_2, \ldots, \mathcal{A}_n$ - one for each player in the game. Each player selects a new state with a transmission probability determined by the current state and one action from each player, $\mathcal{T} : \mathcal{S} \times \mathcal{A}_1 \times \ldots \times \mathcal{A}_n \to PD(\mathcal{S})$. At each stage, each player attempts to maximise his expected sum of payoffs, namely: $E\{\sum_{j=0}^{\infty} \beta^j \eta_{i,t+j}\}$, where $\eta_{i,t+j}$ is the reward received j steps into the future by player i and β is the discount factor.

A secondary user can utilise unused spectrum bands belonging to L primary users. We assume that the bandwidth of licensed bands may be different, and each licensed band is partitioned into a set of adjacent channels with the same bandwidth. Thus, we can denote N_l channels in the primary user l's band. In our approach, when the primary user is active at time t in the l-th band, this is denoted by $P_l^t = 1$. Otherwise, the state is defined as $P_l^t = 0$. We assume that the data are packetized with an average packet length l_p.

Since the channel is modelled as a finite-state Markov chain (FSMC), the channel quality in terms of SNR of the lst band can be expressed by FSMC. Thus, the achievable gain of the licensed band depends on the primary users' status ($P_l^t = 1$ when the primary user uses the lst band at any time t, otherwise $P_l^t = 0$). Thus, each state of the FSMC is jointly modelled by the pair (P^t, g_l^t), where g_l^t is the channel quality. The channel quality can take any value from a set of discrete values, i.e. $g_l^t \in \{SNR_1, \ldots, SNR_8\}$.

Consider the scenario with a two type of secondary users belonging to two classes. The actions of the secondary users from the first class of the secondary users can be defined as $\mathbf{a}^t = (a_{l,D_1}^t, a_{l,C_1}^t, a_{l,D_2}^t, a_{l,C_2}^t)$. The action a_{l,D_1}^t (or a_{l,C_1}^t) denotes that the secondary network will transmit data (control) messages at channels uniformly selected at time slot t. Next, the action a_{l,D_2}^t (or a_{l,C_2}^t) indicates that the secondary network will transmit data (control) messages in the a_{l,D_2}^t (or a_{l,C_2}^t) channel selected from previously used channels without success.

Similarly, the action of the secondary users belonging to the second type of the secondary users is defined $\mathbf{a}_S^t = (a_{S,l,D_1}^t, a_{S,l,C_1}^t, a_{S,l,D_2}^t, a_{S,l,C_2}^t)$, where the action a_{S,l,D_1}^t (or a_{S,l,C_1}^t) denotes the secondary network will transmit data (control) messages at channels uniformly distributed at time slot t. Analogous, the action a_{S,l,S_2}^t (or a_{S,l,C_2}^t) denotes that the secondary users will transmit the data (control) messages in the a_{S,l,D_2}^t (or a_{S,l,C_2}^t) channel from previously used channel without success.

After defining the state at each stage, we may provide the state transition rule, namely assuming that secondary users should observe which channel has been occupied by secondary users. Based on these observations, the secondary users can define the pair $\{S_{l,D}^t, S_{l,C}^t\}$, where $S_{l,D}^t$ and $S_{l,C}^t$ denote data and control channel numbers being used by secondary users of the second class in the lst band observed at time slot t. We assume that the secondary users cannot be informed as to whether an idle channel is occupied or not by the secondary users from the second class. Thus, the number of idle channels that are not being engaged by the secondary users of the second class is not an observation by the secondary users from the first class.

Thus, at every time slot time t, the state of the stochastic game $\mathcal{G}$ is defined by $\mathbf{s}^t = \{s_1^t, s_2^t, \ldots, s_L^t\}$ where $a_l^t = (P_l^t, g_l^t, S_{l,D}^t, S_{l,C}^t)$ indicates the state associated with band l ($l \in \{1, \ldots, L\}$).

After defining the state at each stage, we may provide the state transition rule, namely

$$p(\mathbf{s}^t \mid \mathbf{s}^t, \mathbf{a}^t, \mathbf{a}_S^t) = \prod_{l=1}^{L} p(s_l^{t+1} \mid s_l^t, a_l^t, a_{l,S}^t) \tag{10}$$

The transition probability $p(s_l^{t+1} \mid s_l^t, a_l^t, a_{l,S}^t)$ can be further expressed by

$$p(s_l^{t+1} \mid s_l^t, a_l^t, a_{l,S}^t) = p(S_{l,D}^{t+1}, S_{l,C}^{t+1} \mid S_{l,D}^t, S_{l,C}^t, a_l^t, a_{l,S}^t)$$
$$\times p(P_l^{t+1}, g_l^{t+1} \mid P_l^t, g_l^t) \tag{11}$$

where the first term on the right side represents the transition probability of the number of secondary users of second type and data channels, and the second term denotes the transition of the primary user status and the channel conditions.

In particular, in the game the all players choose their actions. The secondary users will transmit data and control messages in the selected channels and the secondary users will intercept their channels. We assume that the same control messages are transmitted in all the control channels, and one correct copy of control information at time t is sufficient for coordinating the spectrum management in the next time slot.

We assume that the stage payoff of the secondary users maximizes the spectrum gain, namely

$$r(\mathbf{s}^t, \mathbf{a}^t, \mathbf{a}_S^t) = T(\mathbf{s}^t, \mathbf{a}^t, \mathbf{a}_S^t) \times (1 - p^{block}(\mathbf{s}^t, \mathbf{a}^t, \mathbf{a}_S^t)) \tag{12}$$

where $T(\mathbf{s}^t, \mathbf{a}^t, \mathbf{a}_S^t)$ indicates the expected spectrum gain when not all control channels get intercept and $p^{block}(\mathbf{s}^t, \mathbf{a}^t, \mathbf{a}_S^t)$ denotes the probability that all control channels in all L bands are intercepted.

4 The Minimax-Q Learning to Obtain the Optimal Policy of the Stochastic Game

In this section, the minimax-Q learning for the secondary users to obtain the optimal policy of the stochastic game is presented.

In general, the secondary users treat the payoff in different stages differently. Then, the secondary users' objective is find an optimal policy that maximizes the expected sum of payoffs

$$\max E\{\sum_{t=0}^{\infty} \beta^t r(\mathbf{s}^t, \mathbf{a}^t, \mathbf{a}_S^t)\} \tag{13}$$

where β is the discount factor of the secondary user. In our approach, the policy of the secondary network is expressed by $\pi : \mathcal{S} \to \mathcal{PD}(\mathcal{A})$ and the policy of the secondary

for $\forall\, s \in \mathcal{S}$ **and** $\forall\, a \in \mathcal{A}$ **do**
  begin
    After receiving reward $r(\mathbf{s}^t, a^t, \mathbf{a}_S^t)$ for moving from $\mathbf{s}^t$ to $\mathbf{s}^{t+1}$ by taking action $\mathbf{a}^t$
    Compute $Q(\mathbf{s}^t, \mathbf{a}^t, \mathbf{a}_S^t)$
    Update the optimal strategy $\pi^*(\mathbf{s}^t, \mathbf{a})$ by
      $\pi^*(\mathbf{s}^t) := \arg\max_{\pi(\mathbf{s}^t)} \min_{\pi(\mathbf{s}^t)} \sum_a \pi(\mathbf{s}^t, a^t) Q(\mathbf{s}^t, \mathbf{a}^t, \mathbf{a}_S^t)$
    Compute $V(\mathbf{s}^t) := \min_{\mathbf{s}_S(\mathbf{s}^t)} \sum_a \pi^*(\mathbf{s}^t, \mathbf{a}) Q(\mathbf{s}^t, \mathbf{a}, \mathbf{a}_S)$
      $\alpha^{t+1} := \alpha^t \cdot \mu;$
  end;

Fig. 3. The learning phase of the minimax-Q learning algorithm for two groups of secondary users

users of second type $\pi_S : \mathcal{S} \to \mathcal{PD}(\mathcal{A}_S)$, where $\mathbf{s}^t \in \mathcal{S}, \mathbf{a}^t \in \mathcal{A}, \mathbf{a}_S^t \in \mathcal{A}_S$. It is noticeable that the policy π^t at time t is independent of the states and actions in all previous states and actions. Then, the policy π is said to be Markov. If the policy is independent of time, the policy is said to be stationary.

In the stochastic game between the secondary users and the secondary users of second type is a zero-sum game, the equilibrium of each stage is the minimax equilibrium. To solve the game, we can use the minimax-Q learning method [7,14]. The Q-function of stage t is defined as the expected discounted payoffs when the secondary users take action $\mathbf{a}^t$ and the secondary users of second type take the action $\mathbf{a}_S^t$. Then the Q-value in the minimax-Q learning of the game can be expressed as

$$Q(\mathbf{S}^t, \mathbf{a}^t, \mathbf{a}_S^t) = r(\mathbf{s}^t, \mathbf{a}^t, \mathbf{a}_S^t) + \beta \sum_{\mathbf{s}}^{t+1} p(\mathbf{s}^{t+1} \mid \mathbf{s}^t, \mathbf{a}^t, \mathbf{a}_S^t) V(\mathbf{s}^{t+1}) \qquad (14)$$

where $r(\mathbf{s}^t, \mathbf{a}^t, \mathbf{a}_S^t)$ is reward when making a transition from $\mathbf{s}^t$ to $\mathbf{s}^{t+1}$, $V(\mathbf{s}^{t+1})$ is the value of a state in the game of secondary users of second type. In order to obtain the state transition probability, we can modify the value of iteration and the Q-function according to [7], namely

$$Q(\mathbf{s}^t, \mathbf{a}^t, \mathbf{a}_S^t) = (1 - \alpha^t) Q(\mathbf{s}^t, \mathbf{a}^t, \mathbf{a}_S^t) + \alpha^t \{r(\mathbf{s}^t, \mathbf{a}^t, \mathbf{a}_S^t) + \beta \cdot V(\mathbf{s}^{t+1})\} \qquad (15)$$

where α^t denotes the learning rate decaying over time by $\alpha^{t+1} = \mu \cdot \alpha^t$ and $0 < \mu < 1$. Then, the learning phase of the used minimax-Q algorithm for the two groups of secondary users is given in Fig. 3.

5   Simulation Results

We simulated the cognitive framework as an extension of wireless LANs with cognitive radio capability. All simulations are executed from the MATLAB environment. We used an arrangement of 12 secondary transmitters and receivers randomly distributed

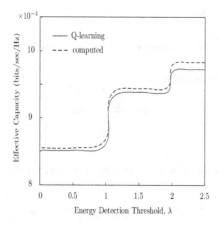

Fig. 4. Effective capacity as a function of the detection threshold value for the secondary users

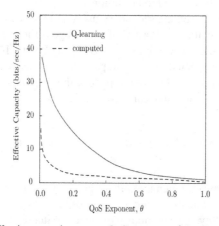

Fig. 5. Effective capacity versus QoS exponent θ for the secondary users

on a square equal to 200 m $\times$ 200 m. The maximum power each of the secondary users is equal to 20 dBm. A primary unity with a maximum power equal to 30 dBm is located at the central point of the simulated area.

For a CR network, Q-value is initially unknown. Thus, an initial estimation is made by initial flooding from the primary unit. Furthermore, all neighbours estimate their Q values according to the neighbours' attributes.

In this illustrative simulation, we assume that the highlight of the impact on the multimedia quality of all six pairs of secondary transmitters stream the multimedia data to their receivers. The primary user randomly disturbs their transmission. The length of the time slot is here equal to 10^{-3} s. For each pair of secondary users, we simulated 3000 samples of Rayleigh faded received signals. Analogously, for the primary user, we generated 3000 randomly distristributed Rayleigh faded signal samples. We assume that the radio transmitting range for secondary users is equal to 50 m. The channel

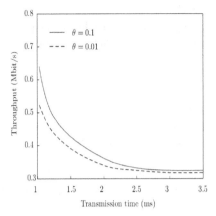

Fig. 6. The throughput of the CR channel under different transmission times of data for various values of θ

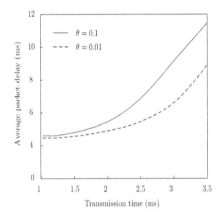

Fig. 7. The average packet delays in dependence on the transmission times of data for various values of θ

bandwidth is equal to 100 kHz. The QoS exponent is equal to 0.001. The average SNR values when the channels are correctly detected are $SNR_1 = 0$ dBm and $SNR_4 = 1$ dBm for busy and idle channels respectively.

The results obtained by the simulation have been compared with the computational results achieved using the effective capacity method applied to the streaming multimedia data. Fig. 4 shows the effective capacity as a function of the detection threshold value λ. As we see in Fig. 4 the effective capacity is increasing with increasing the detection threshold value λ.

Fig. 5 plots the effective capacity as a function of the QoS exponent obtained for both classes of the secondary users under the assumption that the probability of false alarm is equal to 0 and the probability of detection is equal to 1.

Fig. 6 shows the throughput of the CR channel with proposed values of the QoS exponent, $\theta = 0.01$ and $\theta = 0.1$, respectively, under different transmission times of data. Fig. 7 plots the average packet delays in dependence on the transmission times of data. It can be seen that increasing the data rate (reducing the packet transmission time) increases the throughput and decreases the average packet delay.

6 Conclusions

In this paper, we have presented a stochastic game for providing QoS guarantees in cognitive radio networks. At each stage of the game, the proposed mechanism selects the QoS exponent to allocate the available resource. To improve the performance activity of our stochastic game, we have proposed a learning algorithm to predict the possible future reward at each state. Through simulations, we demonstrated the effectiveness of the proposed method. In most cases we have observed up to 20% improvement in QoS provisioning percentage.

References

[1]   Akin, S., Gursoy, M.C.: Effective Capacity of Cognitive Radio Channels for Quality of Service Provisioning. IEEE Trans. Wireless Comm. 8(11), 3354–3364 (2010)

[2]   Akyildiz, I.F., Lo, B.F., Balakrishnan, R.: Cooperative Spectrum Sensing in Cognitive Radio Networks: A Survey. Physical Communication, 40–62 (2011)

[3]   Chang, C.-S.: Stability, Queue Length, and Delay of Deterministic and Stochastic Queueing Networks. IEEE Trans. on Automat. Contr. 39(5), 913–931 (1994)

[4]   Gursoy, M.C., et al.: Analysis of Energy Efficiency in Fading Channel under QoS Constraints. IEEE Trans. Wireless Comm. 8(8), 4252–4263 (2009)

[5]   Haykin, S.: Cognitive Radio: Brain-empowered Wireless Communications. IEEE Journal on Selected Areas in Communications 23(2), 201–220 (2005)

[6]   Kumar, A., Wang, J., Challapali, K., Sin, K.G.: Analysis of QoS Provisioning in Cognitive Radio Networks: A Case Study. Wireless Personal Communications 57(1), 53–71 (2010)

[7]   Littmann, M.L.: Markov Games as a Framework for Multi-Agent Reinforcement Learning. In: Proc. 11th International Conference on Machine Learning, pp. 157–163 (1994)

[8]   Liu, L., Parag, P., Tang, J., Chen, W.-Y., Chamberland, J.-F.: Resource Allocation and Quality of Service Evaluation for Wireless Communication Systems Using Fluid Models. IEEE Trans. on Information Theory 53(5), 1767–1777 (2007)

[9]   Ma, L., Han, X., Shen, C.: Dynamic Open Spectrum Sharing MAC Protocol for Wireless Ad Hoc Network. In: IEEE DySPAN, pp. 88–93 (2005)

[10] Sankaranarayanan, S., Papadimitratos, P., Mishra, A., Hershey, S.: A Bandwidth Sharing Approach to Improve Licensed Spectrum Utilization. In: IEEE DySPAN, pp. 279–288 (2005)

[11] Tang, J., Zhang, X.: Quality of Service Driven Power and Rate Adaptation Over Wireless Links. IEEE Trans. Wireless Comm. 6(8), 3058–3068 (2007)

[12] Thomas, R.W., Da Silva, L.A., MacKenzie, A.B.: Cognitive Networks. In: IEEE Int. Symp. on Dynamic Spectrum Access Networks (DySPAN), pp. 352–360 (2005)

[13] Wang, L.-C., Wang, C.-W.: Spectrum management techniques with QoS provisioning in cognitive radio networks. In: IEEE Int. Symp. on Wireless Pervasive Computing (ISWPC), pp. 116–121 (2010)

[14] Watkins, C.J.C.H., Dayan, P.: Q-learning. Machine Learning 8, 279–292 (1992)

[15] Wu, D., Negi, R.: Effective Capacity: A Wireless Link Model for Support Quality of Service. IEEE Trans. on Wireless Communications 2(4), 630–643 (2003)

Short-Term Forecasting: Simple Methods to Predict Network Traffic Behavior

Anton Dort-Golts*

Saint-Petersburg State University of Telecommunications,
Saint-Petersburg, Russia
dortgolts@gmail.com

Abstract. In the article we evaluate the accuracy of simple linear fore-casting methods applied to short-term prediction of network traffic be-havior, namely the traffic intensity. Such investigation is carried out in order to determine the possibility of such methods employment in net-work management systems and various TE-implementations. Also time series extracted from real network traffic are statistically analysed to obtain general properties of aggregated network traffic behavior.

Keywords: Traffic forecasting, traffic engineering, short-term predic-tion, time series analysis.

1 Introduction and Motivation

Forecasting of some stochastic process based on some previous observations is one of the most general applied science problems, therefore numerous approaches and methods were suggested in this field. Simple and at the same time robust forecasting methods are also critical for the various network traffic engineering solutions. Today there are two basic network management approaches [1]: offline and online. Both of these methods are in need of traffic behavior prediction. The difference could be found in duration of prediction intervals: offline-methods are based on a long-term forecasting (hours and days), while online-methods, relying on rapid changes occurred in network, operate with short-term forecasts (from seconds to tens of minutes). In this paper we concentrate on methods suitable for online network management. Novelty of this research is in the investigation of simple forecasting methods using real network traffic intensity time series. All considered methods could be implemented in a hardware and don't require any operator's participation during the process.

The rest of the paper is organized as follows. In section 2 we take a look at some related works, in section 3 specify forecasting methods to be investigated, while experiment methodology is expounded in section 4. Section 5 is dedicated to preliminary statistical analysis of test time series, used for forecasting accuracy evaluation. Results of numerical experiments are presented in section 6 and final section 7 contains conclusions.

---

* I would like to express sincere gratitude to Andrey Koucheryavy, Alexander Para-monov, Olga Simonina and Oleg Vinogradov for their helpful comments and advices.

S. Balandin et al. (Eds.): NEW2AN/ruSMART 2014, LNCS 8638, pp. 375–388, 2014.

2   Related Works

A lot of network traffic forecasting investigations for the last two decades could be found, but the majority of them is dedicated to long-term prediction. Examples of the early approaches are papers [2] and [3], in which FARIMA (autoregressive fractionally integrated moving average) models were considered for long-term network traffic intensity forecasting. Another polular model is ARIMA (autoregressive integrated moving average), and its analysis results could be found e.g. in [4] and [5]. The third general approach is the usage of neural networks for forecasting [5], [6].

All mentioned forecasting methods display high accuracy even for the far prediction horizon. But at the same time they have some grave shortcomings when try to use it for the purpose of online traffic management. At first any ARIMA/FARIMA model requires a time series expert to evaluate adequate model parameters. Neural networks are free of such restrictions, but they need comparatively long time for training, while real traffic behavior could change rapidly and this sort of change will result in the new training cycle. And finally, such methods rather hard to implement in real network hardware, because of significant computational complexity, especially in the case of neural networks.

Long-term forecasting methods mentioned above are redundant for the purposes of short-term forecasting. The main requirements here are comparative simplicity, full process automation and ability to react fast on occurred changes. Simple linear methods considered further, such as approximation, smoothing, autoregressive models etc., meet all of these requirements. Moreover, it was stated in the paper [7] that in most cases the differences in quality between short-term forecasts of network traffic produced by complex (such as ARIMA-like and neural networks) and linear models (namely exponential smoothing) are not statistically significant.

3   Forecasting Methods

Generally speaking, any observed time series could be represented by sum (additive model) or product (multiplicative model) of several unobservable components [8]. In this paper it is reasonable to consider only additive type of models, because of multiplicative model instability in the case of pikes and bursts occurred in real traffic time series.

$$X(t) = \mu(t) + S(t) + e(t) \tag{1}$$

Here $\mu(t)$ – trend model, $S(t)$ – seasonal component, $e(t)$ – stationary remainder. In the case of short-term forecasting we can neglect seasonal part and concentrate on smooth trend and residual centred random component. Further we will consider some simple linear forecasting methods.

Polynomial approximation. Method based on representation of time series segment in a form of defined power polynomial, approximating points of the former.

$$P_k(x_t) = a_0 + \sum_{i=1}^{k} a_i x_{t-i}^{i} \tag{2}$$

Here $a_0, \ldots, a_n$ – coefficients, $x_{t-1}, \ldots, x_{t-n}$ – values of time series segment, k – polynomial power. Coefficients could be obtained with least squares method, minimising cumulative squared error of prediction.

$$\sum_{i=1}^{n} (f(x_i) - P_k(x_i))^2 \longrightarrow \min; n \geq k+1 \tag{3}$$

Here $f(x_i)$ – observed i-th value of time series segment, $P_k(x_i)$ – approximated i-th value and n – length of segment.

Polynomial extrapolation. The principal difference between polynomial extrapolation and approximation is a former's neccessity to match exactly all values of the segment. Corollary of such condition is rigid restriction for segment size: it should be exactly $n = k + 1$ points.

Smoothing with Spencer's formulae. Other well-known methods of obtaining trend from observed data are Spencer's formulas [9]. In this paper we used smoothing procedures based on 5- and 7-points formulas, which refer to weighted moving averaged process. Original Spencer's formulas were modified to extrapolate trend value outside of considered time series segment.

$$x_t[5] = \frac{1}{35} \left[-28x_{t-5} + 77x_{t-4} - 28x_{t-3} - 98x_{t-2} + 112x_{t-1} \right] \tag{4}$$

$$x_t[7] = \frac{1}{21} \left[-12x_{t-7} + 18x_{t-6} + 12x_{t-5} - 9x_{t-4} - 24x_{t-3} - 12x_{t-2} + 48x_{t-1} \right] \tag{5}$$

Here $x_t[5]$ – forecast based on 5-point formula, and $x_t[7]$ – consequently, on 7-point formula. Practically, Spencer's formulae are low-frequency filters, rejecting high-frequency noise of stationary residual.

Linear prediction. Linear prediction is an autoregressive method, based on representation of random process as a time-invariant linear system. Unknown parameters could be obtained by analysis of inputs and outputs of such system [10].

$$\hat{y}(n) = -\sum_{k=1}^{p} a_k y(n-k) \tag{6}$$

Here $\hat{y}(n)$ – prediction of the n-th value of time series. Weight parameters $a_1, \ldots, a_k$ could be obtained by minimization of cumulative squared prediction error on the segment of length n.

$$\sum_{n} [y(n) - \hat{y}(n)]^2 \longrightarrow \min_{\{a_k\}} \tag{7}$$

This method is more sophisticated than previously mentioned ones, because it takes into account autocorrelation of time series values.

Exponential smoothing. Exponential smoothing (ES) could be considered as a particular case of weighted average, taking into account all previous values of the series with exponentially decaying coefficients. There are a lot of exponential smoothing model types: additive and multiplicative, using trend, seasonality, etc. In this paper we considered three following additive models:

- Simple exponential smoothing (N-N model)
- Exponential smoothing with additive trend (Holt's double ES, A-N model)
- Exponential smoothing with damped additive trend (DA-N model)

Table 1. Used exponential smoothing models

Model N-N	Model A-N	Model DA-N
$S_t = \alpha X_t + (1-\alpha)S_{t-1}$	$S_t = \alpha X_t + (1-\alpha)(S_{t-1} + T_{t-1})$	$S_t = \alpha X_t + (1-\alpha)(S_{t-1} + \phi T_{t-1})$
$\hat{X}_t(m) = S_t$	$T_t = \gamma(S_t - S_{t-1}) + (1-\gamma)T_{t-1}$	$T_t = \gamma(S_t - S_{t-1}) + (1-\gamma)\phi T_{t-1}$
	$\hat{X}_t(m) = S_t + mT_t$	$\hat{X}_t(m) = S_t + \sum_{i=1}^{m} \phi^i T_t$

Corresponding formulas could be found in Table 1. Also, it is notable, that exponential smoothing models are equivalent to a whole number of statistical models including particular cases of more complex ARIMA models [11].

4   Experiment Methodology

To evaluate accuracy of all methods mentioned above for the purpose of short-term forecasting we performed number of numerical experiments. In each experiment iterative procedure of one-step-ahead forecasting carried out: each new time series point prediction is based on a part of previous observations – segment of length n. This segment moves along the whole time series of length N, emulating real iterative forecasting procedure and performing $(N - n)$ single predictions.

Time series of traffic intensity were obtained from real network traces of aggregated traffic, captured in two data-centers. All time series show various behavior, thus allowing to evaluate forecasting methods precision in a diverse conditions. Detailed analysis of used traces could be found in [12]. All original time series (with step 1 second) were additionally transformed into smoothed series with different step of averaging: 5, 11 and 21 seconds.

To estimate forecasting accuracy we used mean absolute percentage error value (MAPE).

$$MAPE = \frac{1}{N-k} \sum_{i=k}^{N} \left| \frac{X_i - \hat{X}_i}{X_i} \right| \tag{8}$$

Here N – time series length, k – length of segment on which single prediction is based, $\hat{X}_i$ – predicted value and X_i – observed value.

In this paper we assumed forecasting accuracy as follows [4]:

- Excellent forecast. $MAPE \leq 10\%$
- Good forecast. $10\% \leq MAPE \leq 20\%$
- Acceptable forecast. $20\% \leq MAPE \leq 50\%$
- Unaccepatable forecast. $MAPE \geq 50\%$

5   Time Series Analysis

All real network traffic intensity time series used for forecasting accuracy estimation were previously analysed to obtain statistical properties and reveal the inner structure of processes.

Probability distribution law. All series were fitted to obtain parameters of common distribution laws set. After that results were sorted by maximum likelihood estimator (MLE) value and most likely ones were chosen.

Performed experiments show that the most probable distribution laws are lognormal and Weibull. Also, the growth of data averaging step leads to increased likelihood of normal distribution.[1]

Trend detection. To detect presence of trend in data two high-performance tests were used:

- Autocorrelation test [13].
- Bartels test [14].

All tests have shown trend presence in the original data, though averaging led to positive results only in cases, when the obvious trend was presented. So we can conclude, that time averaging helps to get rid of fluctuations influence.

Time series stationarity. Stationarity of all series were inspected with augmented Dicky-Fuller test [15]. The majority of the original series were recognized stationary with 95% confidence. It was observed that the growth of averaging step led to decrease probability of stationarity null-hypoteze in the case of aggregated traffic of different types, while the network traffic intensity process in the case of predominated UDP-based filesharing remains stationary.

Hurst parameter estimation. To estimate self-similarity degree of network traffic intensity processes we used Hurst parameter value [16], obtained by two different procedures: variance plot and R/S-statistics. For the best reliability [17] only 30% of lowest resulting points were taken into consideration, during estimation procedure and in the case of obvious trend presence in the data it was removed too.

---

[1] Such behavior seems to correspond the central limit theorem.

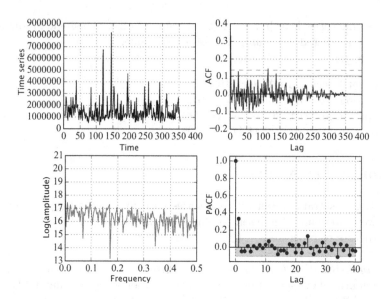

Fig. 1. Analysis of original time series. Time series example, its autocorrelation and partial autocorrelation functions and spectrum of the series are presented.

Hurst parameter values belonging to the interval $(0; 0.5)$ correspond to antipersistent process, in the interval $(0.5; 1)$ – to persistent process, while $H = 0.5$ indicates white noise random process with no autocorrelation and no dependence between sequential values.

Results of performed tests show that in most cases Hurst parameter value obtained from original time series exceeds 0.5 boundary and we can suppose it to be self-similar ones with average Hurst parameter value about $H \approx 0.75$. At the same time, increase of averaging step leads to scattering of estimates, though most frequently it still hits the range $[0.7, 0.8]$.

Spectrum analysis. Analysis of time series spectrum, its autocorrelation and partial autocorrelation functions was performed in order to indicate most significant harmonic components in the data and reveal former's inner correlation structure. This investigation has shown lack of observable splashes in series spectrums (Fig. 1).

At the same time autocorrelation (ACF) and partial autocorrelation (PACF) function shapes indicate presence of periodic components in data, but it is hard to determine what oscillation periods are most significant.

Especial oscillations could be observed in the case of dominant UDP-filesharing traffic, and it may be caused by application level mechanism's operation.

Increase of averaging time step doesn't change appreciably spectrum characteristics and corresponding correlation functions, just smoothing it slightly. So we can conclude that time series of real network traffic intensity, both original and smoothed ones, doesn't have any evident oscillation periods to take into account during short-term, namely one-step-ahead, forecasting. This observation justifies our assumption about unreasonable consideration of models with seasonality for the particular case of short-term traffic prediction.

6 Forecasting Accuracy Evaluation

Here we present results of experimental investigations of forecasting accuracy using simple linear methods mentioned above. Because of bursts occurred in processes, the median was choosen as a main statistical characteristic to describe multiple expriments results, due to its robustness in such conditions. Main results of performed experiments therefore are medians of MAPE and its standard deviation.

6.1 Approximation Polynomial Power Test

In this experiment we studied forecasting using approximation polynomials of different power: from 2 to 8. It is evident (Fig. 2) that forecasting MAPE and its standard deviation increase at exponential rate when approximation polynomial power is augmented. All obtained results are in the area of unacceptable prediction accuracy.

6.2 Approximation vs. Extrapolation Test

Here we investigated the prediction accuracy of low-power approximation polynomials (linear, quadratic and cubic) compared to the same power extrapolation polynomials. It can be seen (Fig. 3), that highest forecasting precision was

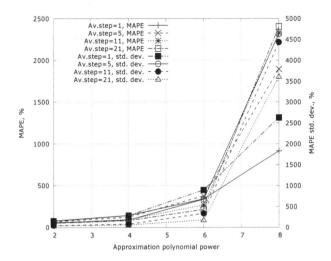

Fig. 2. Accuracy of polynomial approximation forecasting with high powers

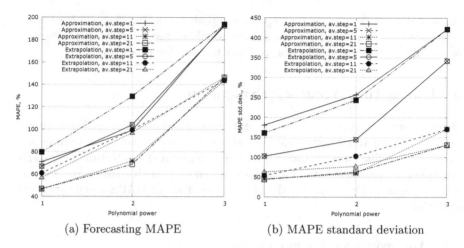

(a) Forecasting MAPE   (b) MAPE standard deviation

Fig. 3. Accuracy of forecasting using approximation and extrapolation low-power polynomials (linear, quadratic, cubic)

achieved with linear polynomials operating time series, smoothed as much as possible (time step 21 second). Also, approximation methods show better forecasting quality than extrapolation ones, but even the best results are near the unacceptable area boundary.

6.3   Approximation Segment Length Test

In this experiment we examined the influence of segment's length, on which prediction is based (i.e. number of previously observed points taken into account for the next prediction), upon forecasting accuracy with a linear approximation polynomial. As anyone can see (Fig. 4) forecasting accuracy tends to increase with enlargement of segment size until some threshold is reached. After that accuracy remains roughly the same or even come down (as in the case of time series averaged with step 21 seconds). So we can observe some optimal value of segment size about 5-10 points.

6.4   Spencer's Formulae Extrapolation Test

In this experiment we evaluated forecasting accuracy of a method based on trend extrapolation, obtained with Spencer's formulae. Here smoothing procedures for 5 and 7 points were tested. Derived dependence (Fig. 5) displays significant relative accuracy improvement while increasing averaging step of time series. Unfortunately, all obtained results are even beyond the area of acceptable forecasting quality. Actually, such results are expected, because Spencer's formulae are particular cases of cubic polynomial approximation.

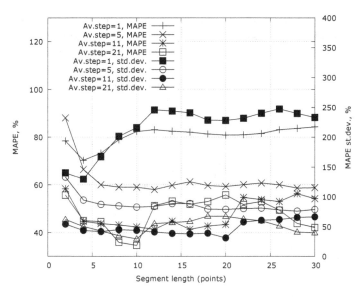

Fig. 4. Segment's length impact on precision of linear approximation forecasting

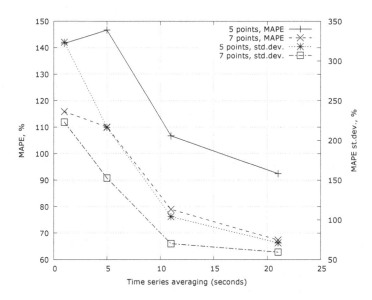

Fig. 5. Forecasting accuracy of extrapolated trend, obtained with Spencer's smoothing formulae

6.5   Linear Prediction Test

Here we investigated forecasting accuracy of the linear prediction method, referred to a class of autoregressive models. Influence of two parameters was examined: analysed segment length, used to obtain correlation matrix, and prediction
filter order. Consequently, two sets of experiments were performed with one of
the mentioned parameters fixed and another one varying.

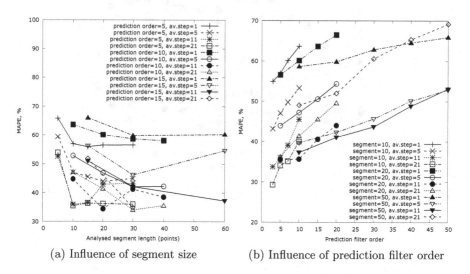

(a) Influence of segment size   (b) Influence of prediction filter order

Fig. 6. Linear prediction accuracy

Analysing the results of experiments we can conclude that forecasting accuracy of prediction filter increases until some threshold, and after that remains
roughly the same (Fig. 6a). The highest forecasting quality (with MAPE about
30-35%) could be reached operating highly smoothed time series, i.e. preliminary
data averaging is of primary importance.

Prediction filter order influences the quality of forecasting as well (Fig. 6b):
best results could be reached by low-order filter with small (here 10 points) analysed segment size. But averaging step still remains dominant factor of accuracy
– the larger you set averaging step, the better will be prediction quality.[2] At the
same time, obtained MAPE standard deviation dependence shows that alter-
ation of prediction filter's order doesn't matter considerably on forecast errors
dispersion.

Also we should mention, that concerning implementation of linear prediction,
it has some restrictions (though non-critical): algorithm is sensitive to the long
zero-sequences, leading to appearance of singular correlation matrix.

---

[2] We can assert this statement only in examined averaging step boudaries from 1 to
21 seconds.

6.6   Exponential Smoothing Test

In this set of experiments we examined forecasting accuracy of previously considered (see p. 378) exponential smoothing models. As it was mentioned above the models are:

- N-N, simple exponential smoothing, considering only smoothed value level (one parameter of the model to be evaluated),
- A-N, exponential smoothing with additional trend (two parameters),
- DA-N, exponential smoothing model with additive damped trend (three parameters).

First of all for our iterative one-step-ahead prediction procedure we have chosen two strategies for adaptive evaluation of the model's parameters:

- parameters are evaluated on every forecasting step (step-by-step evaluation)
- parameters are evaluated for some number of steps (per-interval evaluation)

In both strategies parameters are evaluated, minimizing cumulative squared error on some segment of time series. Exponential smoothing with fixed parameters was discarded, because of its inability to adjust to changing conditions (i.e. traffic patterns). It was asserted in research [18], that better results could be obtained with parameters optimization after each forecast, but such procedure has greater computational complexity compared to per-interval one.

According to the results of additional experiments there is minimum threshold value of segment size, on which parameter values are evaluated: about 10 points for volatile time series (averaging steps 1-5 seconds), and about 5 points for smoothed (11-21 seconds averaging) series. With segment size below this threshold, prediction process becomes unstable. For the same reason we compulsory restricted [11], [19] possible values of smoothing parameter to a range [0; 1].

Also some researchers (e.g. [20]) convince that initial parameter values impact on further forecasting quality, so we set it the way it is recommended. Initial smoothing parameter value was $\alpha = 0.2$, initial level was equal to the mean of observed values along the initial segment, and initial trend value – as a mean trend on the segment.

In these experiments forecasting accuracy was investigated depending on segment length (on which parameter values were evaluated) and averaging of given time series. Results of experiments show that averaging step of the series influence dominantly upon forecasting accuracy in all considered models (Fig. 7-14). The best quality forecasting results (near 20% MAPE) could be obtained by simple N-N model operating highly averaged time series (21 seconds averaging). And also this type of model has least computational complexity among all considered.

Change of segment length in 5-20 points range doesn't influence tangibly on prediction quality. And also, the difference in accuracy given by per-interval and step-by-step types of N-N model practically doesn't matter, whereas the computational complexity of the former is lower.

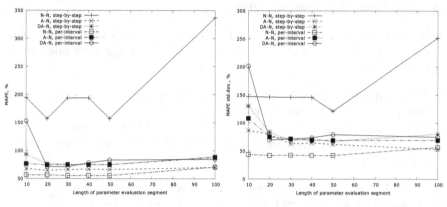

Fig. 7. MAPE, step – 1 sec   **Fig. 8.** Std.dev., step – 1 sec

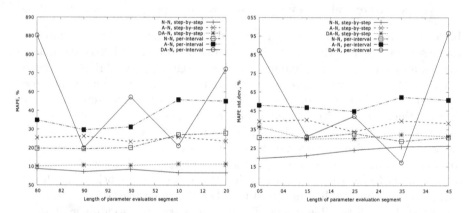

Fig. 9. MAPE, step – 5 sec   **Fig. 10.** Std.dev., step – 5 sec

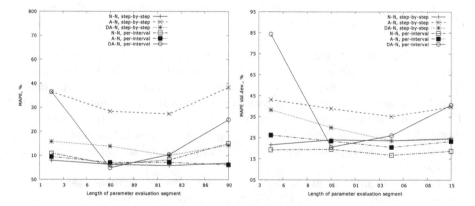

Fig. 11. MAPE, step – 11 sec   **Fig. 12.** Std.dev., step – 11 sec

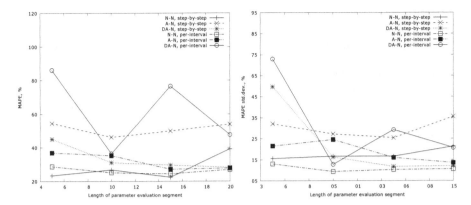

Fig. 13. MAPE, step – 21 sec   **Fig. 14.** Std.dev., step – 21 sec

7   Conclusion

In this paper we analysed the ability of simple linear forecasting methods to predict short-term real network traffic behavior. To evaluate such ability we experimentally estimate forecasting accuracy of some basic methods operating time series based on a real network traffic.

The results of performed experiments show that some of examined forecasting methods could predict the behavior of traffic intensity with acceptable and even high accuracy. Considering these results we could conclude that the most convenient method for autonomous short-term network traffic forecasting is a simple exponential smoothing. It combines implementation and computational simplicity, universality and at the same time robustness and the highest accuracy among all the examined methods. Simple exponential smoothing with adaptive smoothing parameter and evaluation interval size about 10-15 points, operating on highly averaged time series with 21-seconds timestep, was recognized as the best prediction method in the whole set of merits. Its mean averaged percentage error proved to be near 20%, on the border dividing high and acceptable forecasting quality.

Main contribution of the paper is accurate estimation of short-term forecasting methods performed on the samples of the real network traffic. We believe that obtained results will help researchers and engineers developing network management solutions, especially load-balancing systems, that are in need of such methods, to choose the most appropriate ones. Statistical analysis of used real network traffic intensity time series could be regarded as an additional contribution of the paper, revealing the inner structure of the processes and giving better understanding of its inherent properties and characteristics.

References

1. Awduche, D., et al.: Overview and principles of Internet traffic engeeniring, RFC 3272 (May 2002)
2. Shu, Y., et al.: Traffic prediction using FARIMA models. In: IEEE International Conference on Communications, ICC 1999, pp. 891–895. IEEE (1999)
3. Xue, F., Lee, T.T.: Modeling and predicting long-range dependent traffic with FARIMA processes. In: Proc. International Symposium on Communication, Kaohsiung, Taiwan (1999)
4. Grebennikov, A., Krukov, Y., Chernyagin, D.: A prediction method of network traffic using time series models, pp. 1–10 (2011)
5. Rutka, G.: Network Traffic Prediction using ARIMA and Neural Networks Models. Electronics and Electrical Engineering, 47–52 (2008)
6. Guang, S.: Network Traffic Prediction Based on The Wavelet Analysis and Hopfield Neural Network. International Journal of Future Computer and Communication 2(2), 101–105 (2013)
7. Klevecka, I.: Forecasting network traffic: A comparison of neural networks and linear models. In: Proceedings of the 9th International Conference: Reliability and Statistics in Transportation and Communication, RelStat 2009, pp. 21–24 (2009)
8. Koopman, S.J., Shephard, N.: State space and unobserved component models. Cambridge University Press (2004)
9. Kendall, M.G., Stuart, A.: The advanced theory of statistics, vol. II and III (1961)
10. Makhoul, J.: Linear prediction: A tutorial review. Proceedings of the IEEE, 561–580 (1975)
11. Gardner Jr., E.S.: Exponential smoothing: The state of the art. Part II. International Journal of Forecasting, 637–666 (2006)
12. Benson, T., Akella, A., Maltz, D.A.: Network traffic characteristics of data centers in the wild. In: Proceedings of the 10th ACM SIGCOMM Conference on Internet Measurement, pp. 267–280. ACM (2010)
13. Knoke, J.D.: Testing for randomness against autocorrelation: Alternative tests. Biometrika, 523–529 (1977)
14. Bartels, R.: The rank version of von Neumann's ratio test for randomness. Journal of the American Statistical Association, 40–46 (1982)
15. Fuller, W.A.: Introduction to statistical time series. John Wiley and Sons (2009)
16. Rose, O.: Estimation of the Hurst parameter of long-range dependent time series. University of Wurzburg, Institute of Computer Science Research Report Series (February 1996)
17. Clegg, R.G.: A practical guide to measuring the Hurst parameter, arXiv preprint math/0610756 (2006)
18. Fildes, R., et al.: Generalising about univariate forecasting methods: further empirical evidence. International Journal of Forecasting, 339–358 (1998)
19. Taylor, J.W.: Smooth transition exponential smoothing. Journal of Forecasting, 385–404 (2004)
20. Kalekar, P.S.: Time series forecasting using Holt-Winters exponential smoothing. Kanwal Rekhi School of Information Technology, 1–13 (2004)

Traffic Analysis in Target Tracking Ubiquitous Sensor Networks

Anastasia Vybornova and Andrey Koucheryavy

Saint-Petersburg State University of Telecommunications,
Pr. Bolshevikov, 22, Staint-Petersburg, Russia
a.vybornova@gmail.com, akouch@mail.ru

Abstract. With an increase of number of Ubiquitous Sensor Networks (USN) relevance of the traffic analysis in these networks also grows. In this research we tried to describe the traffic produced by USN performing one of the specific tasks target tracking. We used ON-OFF model for the source traffic simulating. The purpose of the research was to determine how parameters of source traffic model (such as ON and OFF periods average length, shape parameter of the Pareto distribution and packet rate during ON period) influence on the traffic at the sink. The research showed that ON periods length and packet rate during ON periods take strong effect on the aggregated traffic in the target tracking USN. On the other hand weight of the ON and OFF times distribution tail and average OFF periods length do not cause statistically significant influence on the traffic characteristics.

Keywords: Internet of Things, Ubiquitous Sensor Network, self-similar traffic, Hurst parameter, target tracking.

1   Introduction

According to the latest forecasts [1], [2] future telecommunication networks will mainly consist of wireless units dealing with Machine-to-machine (M2M) communication. Different types of Ubiquitous Sensor Networks (USN) therefore will constitute a significant part of future global telecommunication networks and will produce significant part of overall traffic in the telecommunication networks [3]. For this reason researching of traffic produced by all types of USN are required for the purpose of prediction the global traffic growth due to USN expansion.

Unfortunately, now there is no much research activity in this area. The Poisson arrival process was assumed as traffic model for each individual sensor node in [4]. The ON/OFF method for USN traffic models was proposed for source traffic modelling in [5]. The pseudo long range dependent (LRD) traffic model was proposed in [6] for mobile sensor networks. The USN traffic models for telemetry applications was studied in [7], where self-similarity of USN traffic was discovered for telemetry applications.

However the main problem of the USN traffic research is that the different USN applications produce types of traffic with different characteristics. In this

S. Balandin et al. (Eds.): NEW2AN/ruSMART 2014, LNCS 8638, pp. 389–398, 2014.

paper we will try to describe the traffic produced by USN performing one of the specific tasks target tracking.

The goal of the target tracking is to trace the paths of moving objects (i.e. targets). At present target tracking is usually done using Global Positioning System (GPS). However, GPS has some limitations coming from Line of Sight requirement, size, weight and cost of GPS receivers, etc. Sensor nodes in contrary are small, cheap and can be used indoors or on high-relief terrains. Therefore target tracking is one of the most promising applications for USN, especially for military purposes.

Target tracking applications at sensor node in general case produce non-regular traffic because a node sends traffic to the sink only when there is a target in its sensing area. According to [5], the source traffic of the sensor node in case of target tracking scenario may be captured by ON-OFF model with the length of ON and OFF periods following the Pareto distribution. On that ground we used the source model represented on [5] as a application-layer traffic generator of each node in the USN model proposed in Section 2.

The purpose of the research was to determine how parameters of source traffic model (as ON and OFF periods average length, shape parameter of the Pareto distribution and packet rate during ON period) influence on the traffic at the sink.

In this research we estimated an average value of packets per second and variance of packets per second obtained during the simulation. As self-similarity of the traffic produced by some types of USN were shown in [7], we also calculated Hurst parameter to estimate the degree of self-similarity.

Two types of traffic at the sink were analyzed: data traffic and signaling traffic (e.g routing packets, acknowledgments etc).

Importance of the data traffic analysis in the USN sinks lies in fact that in the nearest future sensor networks will produce significant part of overall traffic in the telecommunication networks and understanding of USN influence to the global networks is needed.

Research of not only data but also signaling and total traffic in the USN has more practical purposes. Understanding of traffic flows in USN allows us to design networks and nodes in the optimal way (e.g. choose optimal length of buffer, make decision on sensor nodes energy supply type, etc).

2   Analysis Model and Simulation Parameters

For the simulation of the target tracking USN we used Network Simulator 2.

24 sensor nodes of modeling network were regularly placed at the 30x30 meters field; a sink was placed at the center of the sensing area as shown in the Fig. 1. All the nodes were stationary. On a physical and data link layers we used IEEE 802.15.4 standard operated on 2.4 GHz as the most popular standard for the physical and data link layers of USN. Transmission range of of the sensor nodes was about 15-20 meters therefore separate sensing areas form dense coverage of the territory. Internet protocol (IP) was used on a network layer and

Transmission Control Protocol (TCP) was used on a transport layer. We chose Ad hoc On-Demand Distance Vector Routing (AODV) as a routing algorithm because of it wide usage in sensor ad-hoc networks.

Presented model gives a reasonably accurate picture of the USN with minimum number of multi-hop transmission and a single node for a data collection. Designed that way, this USN may be used as one cluster of a multi-cluster network in case one need to track a target on a wider territory.

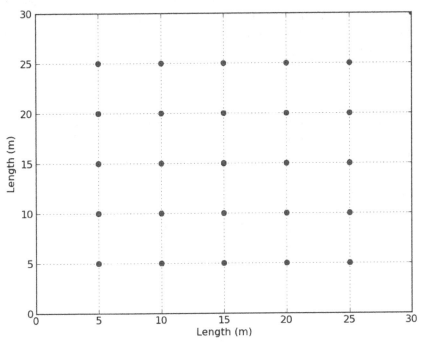

Fig. 1. Sensing area with sensor nodes deployed

3   Self-similarity Estimation Methods

As the self-similarity of target tracking USN traffic was expected, methods of self-similarity degree estimation were needed to perform the analysis. In this research we decided to use a Rescaled range analysis as the most widely-accepted method [8] and a local Whittle estimator as a relatively simple method which brings reasonably accurate results [9]. The rescaled range analysis (R/S analysis) was defined by H. E. Hurst in 1951. It was shown by Hurst that some natural phenomena (e.g. breathing of the Nile river) meet the following relationship:

$$E[(R/S)_t] \sim cn^H \tag{1}$$

where $E[]$ means an average value, $(R/S)_t$ — rescaled range, i.e. ratio of values range exhibited in a portion of a time series to the standard deviation of the same portion, c — constant, n — length of the time series, H — Hurst parameter i.e. measure of self-similarity of the time series.

The local Whittle estimator in contrast to the Rescaled range analysis analyzes an frequency domain of the initial time series [9].

In general case Whittle estimator gives the value of the vector η which minimizes the function of the difference between the signal spectral density $f(v; \eta)$ (the formula for the signal spectral density is assumed known) and the periodogram of the time series $I(v)$:

$$Q(\eta) = \int_{-\pi}^{\pi} \frac{I(v)}{f(v; \eta)} dv + \int_{-\pi}^{\pi} \log f(v; \eta) dv \rightarrow \min \tag{2}$$

The local Whittle estimator gives a semiparametric approach to the selfsemilarity estimation based on the same likelihood principle. Under the condition:

$$f(v) \sim G(H)|v|^{1-2H}, v \rightarrow 0 \tag{3}$$

And for discrete v formula 2 takes the form:

$$Q(G, H) = \frac{1}{m} \sum_{j=1}^{m} \left(\frac{I(v_j)}{G(v_j)^{1-2H}} + \log G(v_j)^{1-2H} \right) \rightarrow \min \tag{4}$$

Minimization of the function above gives the estimated value of Hurst Parameter H.

4   Simulation Results

4.1   Traffic Dependence on ON Periods Average Time

The state ON of the ON-OFF model corresponds to an active state when node sends packet to the sink with a constant bit rate. Contrary to ON state, OFF periods correspond to an idle state when node sends no data to the sink. According to that, traffic flow (number of packets per second) has to grow with increase of average ON state time. Simulation confirms this expectation, as shown in the Fig. 2: both data traffic and signaling traffic grow with the growth of average ON periods length, but data traffic grows faster. Variance of the number of packets per second also have positive dependence on average ON periods length.

Simulation also shows that traffic flow become more self-similar with increase of average ON time (see Fig. 3). Explanation of this observation lies in fact that with the growth of average time of ON period probability of two or more nodes send data packets to the sink in the same moment also increases. So data traffic distribution becomes more bursty and causes burstiness of signal traffic (this connection of data and signaling traffic is caused mostly by acknowledgment packets).

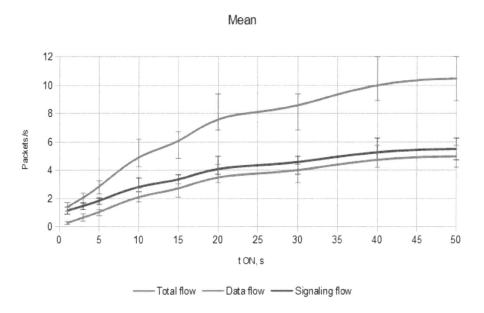

Fig. 2. Dependence of traffic flows on average ON periods length

4.2   Traffic Dependence on OFF Intervals Average Time

As OFF periods in ON-OFF model correspond to an idle state when node does not send any data to the sink, extension of the average OFF period time theoretically has to cause decrease of average number of packets per second. Simulation results meet this theoretical speculation, as shown in the Fig. 4.

Simulation shows that variance and Hurst parameter values show no significant dependence on the average OFF time.

4.3   Traffic Dependence on Shape Parameter of Pareto Distribution

Shape parameter of the Pareto distribution is also known as tail index and describes the weight of the distribution tail: small values of shape index characterize distribution with heavy tail and large values of shape index represent distributions with light tail.

The ON-OFF model we used for source traffic modeling supposes that the lengths of ON and OFF periods distributed according the Pareto law. By means of the simulation we tried to analyze dependence of traffic parameters on the

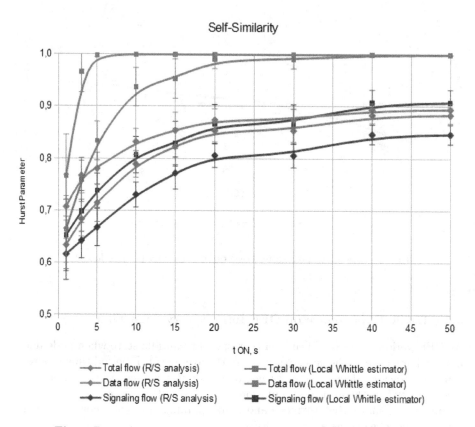

Fig. 3. Dependence of Hurst parameter on average ON periods length

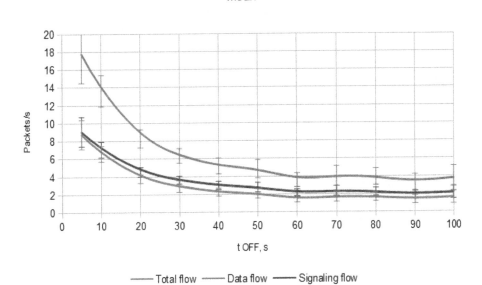

Fig. 4. Dependence of traffic flows on average OFF periods length

shape parameter of Pareto distribution. Simulation shows that mean value of packets per second is independent on the shape parameter value. Self-similarity shows some positive dependence but not statistically significant. On the contrary variance of the number of packets per second shows positive dependence on shape parameter (see Fig. 5), but this observation needs further investigation.

4.4   Traffic Dependence on Packet Rate during ON Periods

Simulation shows close to linear positive dependence of average number of packets in the sink on the packet rate during ON period (see Fig. 6). Variance as well shows approximately linear positive dependence.

Self-semilarity degree of the flow on the sink also shows positive dependence on the packet rate, especially for data traffic, but Hurst parameter significantly faster for the values 0,25 1 packet/s then for the values more than 1 packet/s, especially tor data flow as shown in Fig. 7.

Explanation of this observation (as well as for the dependencies on ON period length) lies in superposition of several traffic sources that in case of target tracking USN gives a bursty traffic on the sink.

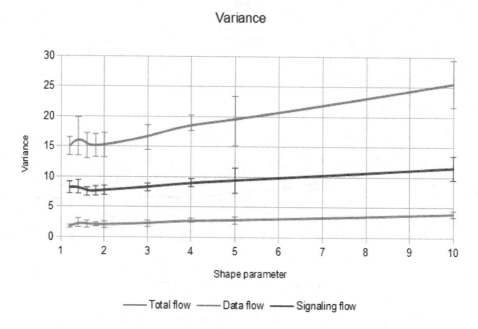

Fig. 5. Dependence of traffic flow variance on shape parameter of Pareto distribution

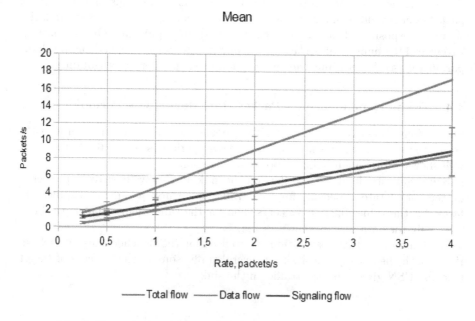

Fig. 6. Dependence of traffic flows on packet rate during ON periods

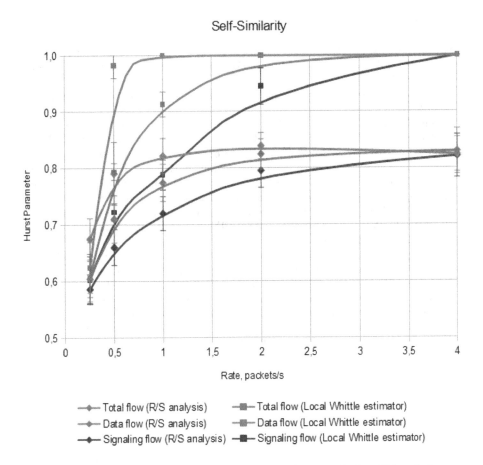

Fig. 7. Dependence of Hurst parameter on packet rate during ON periods

5   Conclusion

The research showed that ON periods length and packet rate during ON periods take strong effect on the traffic in the target tracking USN. In case source traffic of a sensor node is captured well with the ON-OFF model with average ON periods length more than 5-10 seconds (e.g if there is more than 1 target in the sensing area), significant growth of self-similarity degree of aggregate traffic of the USN should be taken into consideration when designing USN. Packet rate more than 1 packet per second also caused the growth of Hurst parameter to the values grater than 0.8, therefore the packet rate of the sensor node when it observes the target should be less than 1 packet/s.

On the other hand weight of the ON and OFF times distribution tail and average OFF periods length do not cause statistically significant influence on the traffic flow characteristic except the positive dependence of packet variance on shape parameter that needs further investigation.

References

1. Sorensen, L., Skouby, K.E.: Use scenarios 2020 - a worldwide wireless future. Visions and research directions for the Wireless World. Outlook. Wireless World Research Forum 4 (2009)
2. Waldner, J.-B.: Nanocomputers and Swarm Intelligence. ISTE. Wiley&Sons, London (2008)
3. Recommendation Y.: Overview of Internet of Things. ITU-T 2060, Geneva (2012)
4. Willinger, W., Taqqu, M., Sherman, R., Wilson, D.: Self-similarity through High-variability. IEEE/ACM Transaction on Networking 15(1), 71–86 (1997)
5. Wang, Q., Zhang, T.: Source Traffic Modelling in Wireless Sensor Networks for Target Tracking. In: 5th ACM International Simposium on Performance Evaluation of Wireless Ad Hoc Sensor and Ubiquitous Networks, pp. 96–100 (2008)
6. Wang, P., Akyildiz, I.F.: Spatial Correlation and Mobility Aware Traffic Modelling for Wireless Sensor Networks. In: IEEE Global Communications Conference, GLOBECOM 2009 (2009)
7. Koucheryavy, A., Prokopiev, A.: Ubiquitous Sensor Networks Traffic Models for Telemetry Applications. In: Balandin, S., Koucheryavy, Y., Hu, H. (eds.) NEW2AN 2011 and ruSMART 2011. LNCS, vol. 6869, pp. 287–294. Springer, Heidelberg (2011)
8. Hurst, H.E.: Long-term storage of reservoirs: An experimental study. Transactions of the American Society of Civil Engineers 116, 770–799 (1951)
9. Robinson, P.M.: Gaussian Semiparametric Estimation of Long Range Dependence. Annals of Statistics 23, 1630–1661 (1995)

LAN Traffic Forecasting Using
a Multi Layer Perceptron Model

Octavio J. Salcedo Parra[1], Gustavo Garcia[2], and Brayan S. Reyes Daza[3]

[1] Intelligent Internet Research Group
[2] Universidad Distrital "Francisco José de Caldas" Facultad de Ingeniería
[3] Universidad Nacional De Colombia. Facultad de Ingeniería
Bogotá DC, Colombia
osalcedo@udistrital.edu.co, gustavoandresg@gmail.com,
bsreyesd@correo.udistrital.edu.co

Abstract. The main idea in the failures forecasting is to predict catastrophic faults in the network, doing that it is possible to guarantee reliability and quality (QoS) in real time to maintain the network availability and reliability and to initiate appropriate actions of restoration of "normality". The following article describes the process performed for implementing failures prediction system in LAN using artificial neuronal networks multi-layer Perceptron. It describes the system, the tests made for the selection of the own parameters of the neuronal network like the training algorithm and the obtained results.

Keywords: Failures, MIB, artificial neuronal Network multi-layer Perceptron, back-propagation.

1 Introduction

The main idea in the failures forecasting is to predict catastrophic faults in the network, doing that it is possible to guarantee reliability and quality (QoS) in real time to maintain the network availability and reliability and to initiate appropriate actions of restoration of "normality". The necessity arises to implement systems that by means of analysis of the traffic of the network could predict the failures in file servers that could be presented/displayed, there are several different techniques for prediction which they will be mentioned in the following section, but used in the developed system the this cradle in artificial neuronal networks, to which it is due to determine of experimental and no theoretical form the architecture and the learning algorithms which the neuronal network will train. Next a brief introduction will occur to the prediction tools, later one was the implemented system of prediction, its parts and the different made tests to find the parameters of the prediction system that offer a better performance in the prediction of failures.

S. Balandin et al. (Eds.): NEW2AN/ruSMART 2014, LNCS 8638, pp. 399–407, 2014.

2   Description

2.1   Tools Used in the Prediction of Failures

There are different types from tools used in the prediction such as:

Artificial Neuronal Networks
According to Simon Haykin "a neuronal network is a processor massively parallel distributed that is flat by nature to store experimental knowledge and to make it available for its use". This mechanism is looked like the brain in two aspects:

- The knowledge is acquired by the network to traverse a process that denominates learning.
- The knowledge is stored by means of the modification of the force or synaptic weight of the different unions between neurons".

The Artificial Neurons are also known like process units, and its operation is simple, because it consists of receiving in the input the neighboring cells output and to calculate a value of output, which is sent to all the remaining cells. Three types of cells or units exist [1]:

- Input Neurons: They receive signals from the surroundings; these inputs (that are simultaneously   input to the network) come generally from a series of time with data previous to which is tried to predict, result of pre-processing such as normalizations, derived generally, threshold levels among others.
- Output Neurons: The output units send a signal outside the network; in the application of prediction the exit would correspond to the future or considered value.
- Hidden neurons: Those are whose the input and output are within the system; that is to say, they do not have contact with the outside. The neuronal networks can learn of experiences that are provided like entrance-exit of the network with no need to express the exact relation between (s) the entrance (s) and the exit, these can generalize the learned experience and obtain the correct exit when new 4 situations are found [3].

Auto-regressive Models (AR)
They are models commonly used to describe signals of non stationary stochastic time series, and its basic characteristics is that the average and 2 the 5 variance go beyond statistical measures like [2] [4], a autoregressive model as Proakis mentions, "is a single process of poles whose function of transference in Z is in the equation 1 which is denominated autoregressive process of order p" [3].

Machine Learning
In agreement to Kyriakakos (ML the machine learning, are adaptive systems of finite states that interact continuously with a general atmosphere. Through the answer of a probabilistic process of test and error, they learn it to choose or to adapt to a behavior that generates the best answer. [6]

Circulant Markov Modulated Poisson Process (CMMP)

This tool captures not only the statistics of second order since they but that make the processes auto-regressive of average (ARMA) also the statistics of first order whose distribution can be different from the Gaussian one, the technique to construct this process is explained in detail in [7].

3   Methodologic Development

In graph 1, it is the proposed system of prediction to set of variables MIB enters to him (Management Information Base): IpInReceives, IpInDelivers, IpOutRequests, tcpActiveOpens, TcpRetransSec. These variables taken from the FTP server, which is found in the network shown in figure 2, are used in the preprocessing stage and those statistical calculations are made passing them to the input neurons of the neuronal network.

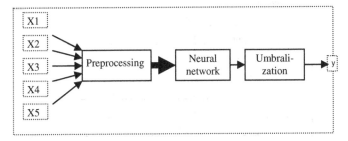

Fig. 1. Prediction System Failure proposed

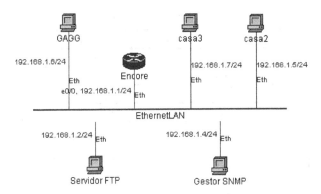

Fig. 2. Test Lan

In order to find the configuration of the system that gave the best performance in the failure prediction different tests to find were performed:

- Number of Stages of training
- Input constraints to the neuronal network
- Neuronal network Architecture

- Learning Algorithm
- Learning algorithm Parameters
- Selection of the threshold level

3.1 Selection of the Number of Times of Training of the Neuronal Network

The neuronal network configuration is found in table 1. This parameter was varied from a value from 500 to 4500 in passages of 500 times. By every time 10 tests were made to verify the consistency of the results. It was selected to 2000 times (Table 2.) like the value which better performance presented. As it bases for this selection considered that the idea is not to have the smaller error of training but that is small and that it does not require many times since this affects the times of training considerably.

Table 1. Test selection number of times of training

Hidden Neurons	8
Training algorithm	backPropagation (traingd)
Learning rate	0.05
Output neurons	1
Input neurons	5
Type	MultiLayer Perceptron
Stages	500 - 4500
Test by Stages	10
Total tests	90

Table 2. Testing results stages

Stagess	MseMin Avg	UltMse Avg
500	0,1143	0,114323222
1000	0,112908222	0,112908222
1500	0,112372556	0,112372556
2000	0,111893556	0,111893556
2500	0,110850333	0,110850333
3000	0,110573111	0,110573111
3500	0,110519444	0,110519444
4000	0,110476333	0,110476333
4500	0,110439889	0,110439889

3.2 Selection of the Incoming Number to the Neuronal Network

In this stage, different tests with different input to the network were done to determine under what conditions they generated a better performance in the prediction system. The proven inputs to the neuronal network are: variables MIB IpInReceives, IpInDelivers, IpOutRequests, tcpActiveOpens, TcpRetransSec; the average values, deviations standard and previous values of the variables IpInReceives, IpInDelivers, IpOutRequests. The conformation of the diverse entrances to the neuronal network is generated in the pre-processing stage (figure 1).

In table 3 is shown the configuration of the used neuronal network in the tests for the different entrances. In the process of results verification or being one proves very important in the prediction system; the visual verification of the logouts was done also. Figure 2 shows these exits in where the green line when it takes the value from one, it represents the points in which the system would have to indicate that it is going away to present a failure in the servant, and the blue lines are those in which the system made the prediction. In this figure corresponding DatosOriMediaDesv the average image when coming out of the system when the entrances are variables MIB, values and the standard deviations of last the twenty minutes of acquisition, are had is the one that better behavior presents taking as it bases the amount of times that the system predicted correctly versus the amount of missed predictions.

Table 3. Neuronal network Configuration, input test

Hidden Neurons	8
Training algorithm	backPropagation with moment (traingd)
Learning rate	0.05
Moment	0.04
Output neurons	1
Input neurons	5 to 26
type	Perceptron multi-layer
Times	2000
Tests by input set	10
Total tests	50

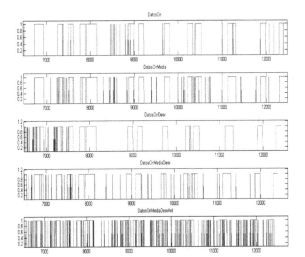

Fig. 3. Neural network input results

3.3   Selection of the Number of Hidden Neurons in the Neuronal Network

The configuration used in the selection of the number of hidden neurons one is in table 4 where it is observed that the results obtained until the moment were including in the tests. Tests were made varying the number of hidden neurons from two to fifteen obtaining the results in table 5. In this table they are also counted on parameters that allow to identify as of the tests I present/display better behavior in the prediction of failures, not only considering the amount of guessed right predictions (nOk) but the been mistaken ones (nErradas).

Table 4. Neuronal network Configuration, Hidden neuron test

Hidden Neurons	2-15
Training algorithm	backPropagation with moment (traingd)
Learning rate	0.05
Moment	0.04
Output neurons	1
Input neurons	11
type	Perceptron multi-layer
Times	2000
Tests by input set	10
Total tests	140

Configuration of the neuronal network with 11 neurons in the layer is chosen since the value average of the relation between the number of guessed right predictions versus the number of missed predictions is the quarter more upper, but the standard deviation is but low of these, which represents a greater homogeneity of the predictions, additionally reviewing the amount of guessed right predictions versus the mistaken ones.

3.4   Selection of the Training Algorithm

Tests with the following algorithms of training were made:

- traingd: Backpropagation of descendent gradient
- traingdm: Backpropagation of descendent gradient and moment
- traingda: Backpropagation of descendent gradient with adaptive rate of training
- trainrp: Resilient Backpropagation

To each one of the algorithms ten test were made to determine which one of the four proven presented better behavior in the prediction of failures. In table 6 is the configuration of the neuronal network and the table 7 contains the results summary. In that is possible to be observed that the selected algorithm is the algorithm traingdm (Backpropagation of descending gradient and momentum) although the one that greater index of correct predictions versus missed predictions was trainrp. The reason

Table 5. Configuration neuronal network, tests algorithm learning

Hidden Neurons	11
Training algorithm	traingd, traingdm, traingda, trainrp
Learning rate	0.05
Moment	0.04
Output neurons	1
Entrance neurons	11
type of network	Perceptron MultiLayer
Stages	2000
tests by algorithm	10
Total tests	40

Table 6. Hidden Neural Networks Test Results

Test	Neur ons	Mse In	nErradas	nOk	nOk Prom	(nOk/nErradas)	(nOk/n Erradas) prom	Dev Est nOk
9	2	0,110531	198	117	91,3	0,59	0,47	0,21
3	3	0,112086	133	93	82,8	0,70	0,44	0,21
3	4	0,110273	161	107	91	0,66	0,59	0,16
9	5	0,110105	233	122	125,5	0,52	0,44	0,16
3	6	0,111664	186	133	70,3	0,72	0,49	0,24
1	7	0,109914	227	144	63,2	0,63	0,52	0,27
3	8	0,110727	181	114	68,7	0,63	0,48	0,19
4	9	0,109966	94	69	38,5	0,73	0,32	0,31
2	10	0,109988	180	111	85,6	0,62	0,41	0,23
6	11	0,110726	219	134	96,2	0,61	0,51	0,07
3	12	0,109578	158	105	107,4	0,66	0,45	0,14
9	13	0,112075	233	139	90,4	0,60	0,62	0,16
4	14	0,110857	247	149	95,3	0,60	0,47	0,15
9	15	0,110076	171	103	58,9	0,60	0,37	0,23

Table 7. Test results of the selection of training algoritms

Test	Algorithm	MseEntMin	nErradas	nOk	nOk/nErradas	PROM Ind	PROM Nok	DesvEst Ind
2	traingd	0,1103	315,00	162,00	0,51	0,37	56,60	0,15
1	traingdm	0,1107	204,00	124,00	0,61	0,54	100,20	0,16
4	traingda	0,1074	306,00	161,00	0,53	0,44	86,80	0,05
5	trainrp	0,1035	389,00	241,00	0,62	0,54	234,90	0,06

obeys to that this last one presents in average an elevated number of erroneous predictions as 389 missed predictions versus are it the 204 that presented the selected algorithm.

3.5   Selection of the Moment in the Training Algorithm

After having found the training algorithm of the neuronal network "traingdm", their parameters are: the momentum and the rate of learning (learning rate). Continuation in table 8 the configuration of the neuronal network can be seen with which moment was proven the parameter and in table 9 they are observed the summary of test results. The parameter selected momentum was the one of 0,10; already has a high index of prediction (of the 0,68), the number of predictions is much greater to the one of 0,72 that single had 72 guessed right predictions and the number of failures was not very elevated (it was smaller to 250 failures than actually, it is observed that it was a little most efficient for the functionality of the system).

Table 8. Momentum test

Hidden Neurons	11
algorithm training	traingdm
Learning rate	0.05
Moment	0.01 - 0.15
Output neurons	1
Input neurons	11
Type	Perceptron Multi-Layer
Stages	2000
tests by moment	10
Total tests	150

4   Conclusions

The system development for failures in networks requires low computational complexity so that time used in the prediction of failures allows that the system is implementable.

The neuronal networks Perceptron are a useful tool in the prediction of failures; although other algorithms are due to prove of training to improve the performance of the prediction system obtained.

Variables MIB IpInreceives, IpIndelivers, IpOutRequests, TcpActiveOpens, tcpRetranSec allows you determine the faults of a network LAN, using its values average and deviations Standard to the entrance of the neuronal network.

With the input to the neuronal network IpInreceives, IpIndelivers, IpOutRequests, TcpActiveOpens, tcpRetranSec, its values average and standard deviations, the architecture that better behaves in the prediction is the one that has eleven neurons in the hidden layer.

From the Backpropagation algorithms of descendent gradient, Backpropagation of descendent gradient and moment, Backpropagation of descendent gradient with adaptive rate of training and Resilient Backpropagation, the best one for the prediction of failures of a network LAN using neuronal network multi-layer Perceptron of eleven neurons in the layer it hides, is backpropagation of descendent gradient and moment, with parameters rate of training of a 0,04 and momentum of 0.01.

For the determination of the failure of network LAN, the best parameter for the umbralization of the neuronal network output is that with 2.3 multiplied by Standard deviation of the last 120 outputs.

The performance of the proposed system could be improved using different a neuronal network architecture or algorithms that allow better learning to the neuronal network.

References

[1] Neuronal networks, Alfaomega, 2000 Box and Jenkins, Time Series Analysis, Forecasting and Control, Holden Day Series (1976)

[2] Box, G.E.P., Jenkins, G.M.: Time Series Analysis, Forecasting and Control, Holden Day Series (1976)

[3] Proakis, J.G., Manolakis, D.G.: Digital treatment of Signals. Pretince-Hall (1998)

[4] Ouyang, Y., Yeh, L.-B.: Predictive bandwidth for control MPEG video: To wavelet approach for self-similar parameters estimation. In: IEEE International Conference on Bowl Communications, ICC 2001, vol. 5, pp. 1551–1555 (2001)

[5] Thottan, M., Ji, C.: Fault prediction AT the network to layer using intelligent agents. In: IFIP/IEEE Eighth International Symposium on Integrated Management Network, 2003, pp. 547–588 (1965), Lucky, R. W.: Automatic equalization for digitalis communication. Bell. Syst. Tech. J., Bowl. 44(4), pp. 547–588, (1965)

[6] Frangiadakis, N., Kyriakakos, M., Merakos, R.B.: Enhanced path prediction for network resources management in wireless LANs. IEEE Wireless Communications, 620–69 (2003)

[7] Li, S.Q., Hwang, C.L.: On the convergence of traffic measurement and queueing analysis: To statistical-matching and queueing (SMAQ) tool. IEEE/ACM Transactions on Networking, Bowl 5(1), 95–110 (1997), G. R. Faulhaber, Design of service systems with priority reservation, In: IEEE Int. Conf. Communications on Conf. Rec., pp. 3–8(1995)

[8] Sang, A., Li, S.: To predictability analysis of network traffic. In: Doyle, W.D. (ed.) IEEE INFOCOM (2000), Biaxial reversal Magnetization in films with anisotropy. In: 1987 Proc. INTERMAG Conf., pp. 2.2–1-2.2–6

Session Setup Delay Estimation Methods for IMS-Based IPTV Services[*]

Yuliya V. Gaidamaka and Elvira R. Zaripova

Telecommunication Systems Department,
Peoples' Friendship University of Russia,
Ordzhonikidze str. 3, 115419 Moscow, Russia
{ygaidamaka,ezarip}@sci.pfu.edu.ru

Abstract. This paper proposes an IMS architecture for delivering IPTV services like Content on Demand and analytical methods of signaling delay estimation. Three methods of session setup delay estimation are proposed - the multiclass BCMP queueing networks method, the method based on homogeneous queueing networks with given variation coefficients for the probability distribution of service time, and the method with regard for background traffic. A numerical example for initial data of Iskratel SI3000 Multi Service Control Plane is provided. In conclusion advantages and area of application for the methods are discussed.

Keywords: IMS, SIP, IPTV, Content on Demand, session setup signaling flow, approximate method, session setup delay.

1 Introduction

According to Cisco's Visual Networking Index Forecast Internet video-to-TV traffic will increase nearly 5-fold between 2012 (1.3 exabytes per month) and 2017 (6.5 exabytes per month), while video applications traffic growth in global mobile data traffic will be 69 percent over 2013-2018 [1]. Delivering of IPTV services, such as Broadcast TV, Video on Demand, Time Shifted TV, will be promoted via IP Multimedia Subsystem (IMS) [2, 3] designed by 3rd Generation Partnership Project (3GPP). The IMS is an overlay architecture that provides session establishment, authentication, required quality of service (QoS), so it suites well for session signaling concerning IPTV scenario. The general media signaling for an IPTV scenario is based on Session Initiation Protocol (SIP) [4]. The description of the IMS-based IPTV procedures and functional entities taking part in the procedures related to session signaling and media signaling are provided in ETSI technical specifications [5,6].

With increased global use of video services/applications the quality of service (QoS) and the quality of experience (QoE) aspects remain still important [7]. One of

---

[*] The reported study was partially supported by the RFBR, research projects No. 12-07-00108, 14-07-00090.

S. Balandin et al. (Eds.): NEW2AN/ruSMART 2014, LNCS 8638, pp. 408–418, 2014.

QoE parameters that effects subscriber satisfaction is signaling delay or session setup delay, i.e. the time required to the session initiation. In this paper three methods for session setup delay estimation are proposed. The methods are described on the example of the Video on Demand (VoD) service which is a special case of the Content on Demand (CoD) service [8]. Unlike CoD the media signaling for VoD can be realized by Real-Time Transport Control Protocol (RTSP) [9]. The first method (Method 1) for session setup delay estimation is based on open multiclass BCMP queueing network [10]. The second method (Method 2) is built upon approximate method from [11] and concerns homogeneous open queueing network with given variation coefficients for the probability distribution of customer arrivals and service time for the network nodes. The third method (Method 3) is also approximate and considers multiphase queuing system with background traffic based on an open queueing network [12]. The purpose of the paper is a comparative analysis of the methods in the context of accuracy and application area.

The rest of the paper is organized as follows. In Section 2 the IMS-based IPTV architecture is designated and the signaling messages flow for Content on Demand setup is shown. The three methods for session setup delay estimation are proposed in Section 3. A numerical example for initial data of Iskratel SI3000 Multi Service Control Plane is provided in Section 4. Advantages and application area for the methods are discussed in Section 5.

2   IMS-Based IPTV Architecture

IPTV includes three main groups of services: Broadcast Television in real-time, Time-Shifted TV (replays a TV program that was broadcast days or months ago) and Video on Demand for unicast communications. With the help of electronic program guide the user can choose their own content. Besides video services the internet protocol's abilities allow to transmit a lot of other services, including integrated and interactive services. The Content on Demand service includes all above mentioned services.

In [5, 13] relations between the functional entities involved into CoD scenario based on SIP and RTSP are represented. The basis of the IMS Core is Call Session Control Function (CSCF), which consists of three functional blocks: Proxy CSCF (P-CSCF) identifies the subscriber, Interrogating CSCF (I-CSCF) is responsible for the selection of the appropriate application server and access to it, and the Serving CSCF (S-CSCF) handles all SIP-messages between entities. The CoD architecture includes the User's Equipment (UE), for example, Set Top Box (STB), backbone IP/MPLS, CoD Application Server (CoD AS) and CoD Media Function (CoD MF). CoD AS carries out service discovery function and service selection function. Media servers (CoD Media Functions) are responsible for managing the interaction with the user's equipment (Media Control Function, MCF) and delivering media streams to the user's equipment (Media Delivery Function, MDF).

By definition of ETSI Content on Demand program is provided at the request of the end user for direct consumption (real-time streaming). Fig. 1 shows the flow of

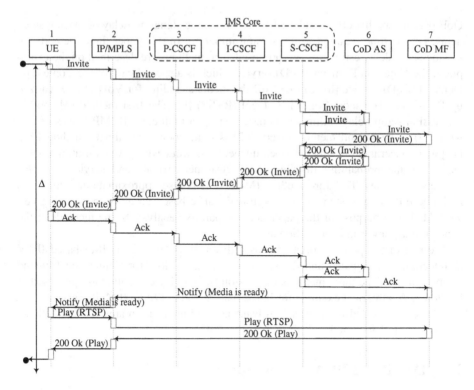

Fig. 1. Signaling messages flow for Content on Demand setup

signaling messages between the functional entities involved in a session initiation. The registration procedure is not included in Fig. 1, because we assume that the user has registered in IMS Core heretofore.

The user's request to get CoD initiates a session. The user selects the content of interest in the electronic program guide within the offered services, and after he pressed a button the connection request starts from his UE through the IP/MPLS network. UE supports content protocols and coding used by CoDs. IMS Core configured to forward all CoD related SIP requests to CoD AS. The Invite message of SIP is an initiating message, and passing through the IMS Core it reaches CoD AS and CoD MF. Session initiation procedure comes with 200 Ok, Ack and Notify messages of SIP as well as the message Play of RTSP. The last 200 Ok message from CoD MF completes the procedure, and the session is admitted to be initiated. The total signaling delay for this procedure including the reservation of channel for transmitting data is a critical component of IPTV QoE.

In the next section we develop three analytical methods for a session setup delay estimation. All methods are based on open queueing networks models with nodes corresponding to the functional entities involved into CoD scenario and customers corresponding to the signaling messages. This approach allows modeling the session setup delay as a sojourn time of a customer in the queueing network.

3   Mathematical Modelling and Methods for Session Setup Delay Estimation

In this section we built three mathematical models to analyze the setup procedure shown in Fig. 1. Common to the three models is that they all are seven-node open queueing networks. We number the functional entities in Fig. 1 sequentially and get the set of nodes $\mathbf{M} = \{1,2,3,4,5,6,7\}$, $|\mathbf{M}| = 7$. External arrivals in all networks are Poisson process with a given rate λ. The service rates on the nodes μ_i, $i \in \mathbf{M}$, are fixed. Customers' behavior between nodes of the network is described by routing matrix Θ. Throughout the paper we use the following notation: λ_i is the arrival rate of the aggregated flow of all customers to i-node; $\rho_i = \lambda_i / \mu_i$ is the offered load on i-node, $i \in \mathbf{M}$. We denote Δ a sojourn time of a customer in the queueing network, which corresponds to a session setup delay.

The model in Section 3.1 is a BCMP network [10] with nodes of two types - an infinite server and an infinite queue with service discipline First Come First Served (FCFS) and fixed service rates μ_i, $i \in \mathbf{M}$. The model in Section 3.2 is a homogeneous open queueing network with given variation coefficients for the probability distribution of service times for the network nodes [11]. The model in Section 3.3 is a multiphase queuing system with background traffic based on an open queueing network [12].

3.1   The BCMP Queueing Network Model

The BCMP queueing network is the well-known class of queueing networks for open, closed or mixed models with multiple classes of customers and various service disciplines and service time distributions. The most noted result concerning product form solution was presented by Baskett, Chandy, Muntz and Palacios in [10]. This apparatus was used for telecommunication network analysis for example in [14, 15]. To provide Method 1, we split the set of the queueing network nodes into two subsets $\mathbf{M} = \mathbf{M}_1 \bigcup \mathbf{M}_2$. The set $\mathbf{M}_1 = \{1,2\}$ includes the nodes of a type $M \mid M \mid \inf$, and the set $\mathbf{M}_2 = \{3,4,5,6,7\}$ includes the nodes of a type $M \mid M \mid 1 \mid \inf$ with service discipline FCFS. There is a set $\mathbf{R}$ of classes of customers numbered according to Table 1, $|\mathbf{R}| = 5$.

Table 1. Classes of customers

Messages	Class of customers
Invite	1
200 Ok	2
Ack	3
Notify	4
Play	5

The arrival rates to i-node of r-class customers can be received from the balance equations concerning routing matrix $\Theta = \left(\theta_{ir,js}\right)$, $i, j \in \mathbf{M}$, $r, s \in \mathbf{R}$. The average sojourn time in the FCFS nodes is class independent and equals $\left(\mu_i - \lambda_i\right)^{-1}$, $i \in \mathbf{M}_2$. The steady-state condition for queuing network is the following:

$$\lambda < \min\left(\frac{\mu_3}{3}; \frac{\mu_4}{3}; \frac{\mu_5}{6}; \frac{\mu_6}{3}; \frac{\mu_7}{3}\right). \tag{1}$$

Applying the approach of [16] an average sojourn time Δ of a customer in the queueing network can be estimated by the following formula:

$$\Delta = 4\mu_1^{-1} + 6\mu_2^{-1} + \frac{3}{\mu_3 - 3\lambda} + \frac{3}{\mu_4 - 3\lambda} + \frac{6}{\mu_5 - 6\lambda} + \frac{3}{\mu_6 - 3\lambda} + \frac{3}{\mu_7 - 3\lambda}. \tag{2}$$

3.2   Queueing Network with Given Variation Coefficients for Service Times

Method 2 concerns homogeneous open queueing network with given variation coefficients for the probability distribution of service times for the network nodes. This method first published in [11], where it was called "Peoples' Friendship University (PFU) method". Modeling each node as an infinite queue with service discipline FCFS or an infinite server we suppose service times at each node are independent and identically distributed random variables with the cumulative distribution function (CDF) $B_i(x)$, $i \in \mathbf{M}$, mean b_i, and coefficient of variation (CV) $C_B(i)$, $i \in \mathbf{M}$. The elements θ_{ij} of stochastic routing matrix $\Theta = \left(\theta_{ij}\right)$, $i, j \in \mathbf{M}$, denotes the probability that a customer leaving i-node immediately goes to j-node, whereas θ_{i0} is the probability that it leaves the network, $\theta_{i0} = 1 - \sum_{j=1}^{7} \theta_{ij}$. The routing matrix Θ is shown in Table 2.

Table 2. The routing matrix Θ

Θ	0	1	2	3	4	5	6	7	Σ
0	0	1	0	0	0	0	0	0	1
1	1/4	0	3/4	0	0	0	0	0	1
2	0	1/2	0	1/3	0	0	0	1/6	1
3	0	0	1/3	0	2/3	0	0	0	1
4	0	0	0	1/3	0	2/3	0	0	1
5	0	0	0	0	1/6	0	1/2	1/3	1
6	0	0	0	0	0	1	0	0	1
7	0	0	2/3	0	0	1/3	0	0	1

We introduce the following notation:

λ_{ij} - the flow rate from i-node to j-node;

$A_{ij}(x)$, $C_A(i,j)$ - the CDF and the CV of random time intervals between arrivals of customers from i-node to j-node;

$A_i(x)$, $C_A(i)$ - the CDF and the CV of random time intervals between arrivals of customers to i-node, $i,j \in \mathbf{M}$.

The steady-state condition necessary for each node is $\rho_i < 1$, $i \in \mathbf{M}$. In this case $\sum_{j=0}^{7} \lambda_{ij} = \sum_{j=0}^{7} \lambda_{ji} = \lambda_i$, $i \in \mathbf{M}$, hence the flow rates λ_{ij} can be obtained from the system of linear equations $\lambda_{ij} = \lambda_i \theta_{ij}$, $i,j \in \mathbf{M}$, where $\lambda = \sum_{i=1}^{7} \lambda_{0i}$.

We can obtain CV $C_A(i)$ from the following system of linear equations:

$$\lambda_i \left[C_A^2(i) - 1 \right] - \sum_{k=0}^{M} \lambda_k \left[C_A^2(k) - 1 \right] \left(1 - \rho_k^2 \right) \theta_{ij}^2 =$$
$$= \lambda_{0i} \left[C_A^2(0,i) - 1 \right] + \sum_{k=0}^{M} \mu_k \left[C_B^2(k) - 1 \right] \rho_k^2 \theta_{ki}^2, \quad i \in \mathbf{M}. \tag{3}$$

For instance, the CV $C_A(i)$ for the queueing network with the nodes $M|M|1|\inf$ and $M|M|\inf$, are the solutions of the equations (3a), and for the network with the nodes $M|D|1|\inf$ and $M|D|\inf$ - of the equations (3b):

$$\lambda_i \left[C_A^2(i) - 1 \right] - \sum_{k=0}^{M} \lambda_k \left[C_A^2(k) - 1 \right] \left(1 - \rho_k^2 \right) \theta_{ij}^2 = 0, \quad i \in \mathbf{M}, \tag{3a}$$

$$\lambda_i \left[C_A^2(i) - 1 \right] - \sum_{k=0}^{M} \lambda_k \left[C_A^2(k) - 1 \right] \left(1 - \rho_k^2 \right) \theta_{ij}^2 = -\sum_{k=0}^{M} \mu_k \rho_k^2 \theta_{ki}^2, \quad i \in \mathbf{M}. \tag{3b}$$

Knowing the coefficients of variation $C_A(i)$ and arrival rates λ_i for i-node we determine the average waiting time of a customer in i-node using the approximate formula Kramer and Langenbach-Beltz [11, 17]:

$$\omega_i = \frac{1}{\mu_i} \frac{\rho_i}{2(1-\rho_i)} \left[C_A^2(i) + C_B^2(i) \right] \cdot g\left(\rho_i, C_A(i), C_B(i) \right), \quad i \in \mathbf{M}, \tag{4}$$

where

$$g\left(\rho_i, C_A, C_B \right) = \begin{cases} \exp\left\{ -\dfrac{1(1-\rho)}{3\rho} \dfrac{(1-C_A^2)^2}{C_A^2 + C_B^2} \right\}, & C_A \leq 1, \\[4mm] \exp\left\{ -(1-\rho) \dfrac{C_A^2 - 1}{C_A^2 + 4C_B^2} \right\}, & C_A > 1. \end{cases} \tag{5}$$

The average sojourn time of a customer in i-node can be determined by the formula

$$v_i = \omega_i + b_i, \quad i \in \mathbf{M},$$ (6)

and an average sojourn time Δ of a customer in the queueing network - by the formula

$$\Delta = \sum_{i=1}^{7} v_i.$$ (7)

3.3   Multiphase Queuing System with Background Traffic

Method 3 is approximate like Method 2 and considers background traffic [12]. Method 3 involves dividing incoming flow at each node into foreground and background traffic. To illustrate this separation we describe the two types of traffic on the example of the first node (UE, 1-node). Let us describe the setup procedure for a customer as a routing chain from 1-node to 1-node passing multiple times every of the rest transitionary nodes. Fig. 1 shows that the chain runs through the 1-node four times. Thus, when servicing the Invite message as foreground traffic, three other messages (2 messages 200 Ok, Notify) form background traffic. After that, when servicing the 200 Ok message as foreground traffic, three other messages (Invite, 200 Ok, Notify) form background traffic and so on. Applying this approach we should take into account different signaling message lengths, which affect service times for foreground and background customers. Signaling message lengths are listed in Table 3 [18].

Table 3. Signaling messages length

Messages	Length of message, byte
Invite	976
200 Ok	1036
Ack	676
Notify	596
Play	90

The routing chain consists of $K = 28$ steps, from the processing of Invite message in 1-node (UE) till the processing of 200 Ok message in the same node. Denote the arrival rate and the average service time of the foreground customers at k-step as λ_0 and d_k, and the arrival rate and the average service time of the background customers as λ_k and b_k. Note that $\lambda_0 = \lambda$. The sojourn time Δ of a customer in the queueing network is equal to the sum of the average waiting times and the average service times on each node in the context of the foreground customers:

$$\Delta = \sum_{k=1}^{K} (w_k + d_k).$$ (8)

Here w_k is the average waiting time in the corresponding node at k-step, obtained from the Pollaczek-Khinchin's formula

$$w_k = \frac{\rho_k d_k}{1-\rho_k}\left(\frac{1+C_k^2}{2}\right),\tag{9}$$

where $\rho_k = \lambda_0 d_k + \lambda_k b_k$ is the offered load on the corresponding node at k-step. The squared CV of the service time at k-step is shown by the formula

$$C_k^2 = \frac{(\lambda_0 + \lambda_k)\left(\lambda_0 d_k^{(2)} + \lambda_k b_k^{(2)}\right)}{\left(\lambda_0 d_k^{(1)} + \lambda_k b_k^{(1)}\right)^2} - 1.\tag{10}$$

Note, that for this method we can find the quantile level ψ of a sojourn time of a customer in the queueing network by the formula

$$Q_\psi = q_\psi + \sum_{k=1}^{K}\left(\frac{\ln(\gamma_k w_k)}{\gamma_k} + d_k\right).\tag{11}$$

In formula (11) q_ψ is the one positive root of equation

$$1 - \psi = \sum_{k=1}^{K} e^{-\gamma_k q_\psi}\frac{(\gamma_k q_\psi)^k}{k!}.\tag{12}$$

The attenuation parameter γ_k of the distribution functions of a sojourn time of a customer at k-step is the only positive root of the equation $\alpha(\gamma_k)\beta(-\gamma_k) = 1$, where $\alpha(s) = \dfrac{\lambda_0 + \lambda_k}{\lambda_0 + \lambda_k + s}$ is a Laplace-Stieltjes transform (LST) CDF of interarrival time at k-step, $\beta(s) = \dfrac{\lambda_0}{\lambda_0 + \lambda_k}e^{-sd_k} + \dfrac{\lambda_k}{\lambda_0 + \lambda_k}e^{-sb_k}$ is LST CDF of service time of a customer at k-step, $k = 1,...,K$.

4   Numerical Example

For the numerical experiment we choose the input data from [16, 18] for Iskratel SI3000 Multi Service Control Plane. We also assume that the average processing time of UE (1-node of the queueing network) is 1 ms, the propagation delay through the network modelled by 2-node of the queueing network is 50 ms. The average processing time of a message by the P-CSCF, the I-CSCF and the S-CSCF (3-, 4-, 5-nodes) equals 0,4 ms, and the average processing time of a message by the CoD AS and the CoD MF (6-, 7- nodes) equals 0,5 ms.

The results of a session setup delay calculations are shown in Fig. 2. We represent five curves: the full line for Mehod 1 (BCMP) calculated by the formula (2), two dashed lines for Mehod 2 (PFU) calculated by the formula (7) for exponential distributed and constant service times in queueing network nodes, the dashed-dotted line for Mehod 3 (background traffic) calculated by the formula (8), in addition we show as dotted line 95% quintile of a session setup delay calculated by the formula (12).

Upper limit $\lambda = 414$ requests/s of the intensity of users' requests for Content on Demand services is determined by the steady-state condition (1). All the plots except Method 2 (PFU) calculated by the formula (7) for exponential distributed, lie below 95% quintile curve, hence, can be used for of a session setup delay estimation.

Each method has its own advantages and disadvantages, which will differ for each case of application depending on purpose of the analysis. Method 1 can be classed among exact methods. Another advantage of this method is its simplicity. The disadvantage of this method is that the method is applicable only for queueing network with exponentially distributed service times in FCFS nodes. Method 2 and Method 3, are approximate methods. Their advantages are that the methods are applicable to any service times distribution with given mean and variation coefficients. The disadvantage of Method 2 is that for exponential distributed service times in queueing network nodes the results are out of the confidence interval even in the medium and high load area – see the dashed curve for Method 2 (PFU) calculated by the formula (7) for exponential distributed service times in Fig. 2. Method 3 devoid of this disadvantage. The 5% confidence interval covers the results obtained by Method 3 with respect to the simulation for all the test load range.

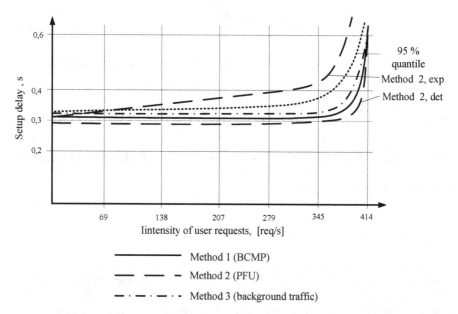

Fig. 2. Session setup delay vs intensity of user requests

5   Conclusion and Outlook

The results of calculation using the three methods shows that a session setup delay is not more than one second. That confirms the findings of [16] that Iskratel SI3000 Multi Service Control Plane, which is designed to service up to one million subscribers with a peak load of two million requests per hour, in busy hours is ready to provide IPTV service with proper quality to 15% of its subscribers, while its request is usually not more than 10% of subscribers.

As a results of a session setup delay estimation for CoD service using the three proposed methods we concluded that Method 3 considering multiphase queuing system with background traffic based on an open queueing network is applicable to a wider range of load and allows to calculate not only the mean session setup delay, but quantile of this random variable, i.e. to prescribe a probability distribution of a session setup delay. The above advantages of Method 3 allow recommending it to the analysis of a session setup for other IPTV services. This is the aim of our further research.

The authors thank the head of the Telecommunication Systems Department of the Peoples' Friendship University of Russia professor K. Samouylov for useful notes concerning the problem's formulation.

References

1. Cisco 2014-2018 Cisco Visual Networking Index: Global Mobile Data Traffic Forecast Update, 2013-2018, FLGD 11446, USA, pp. 1–40. Cisco Systems Inc. (2014)
2. 3GPP TS 23.228 V7.7.0 (2007-03): IP Multimedia Subsystem (IMS), Stage 2 (Release 7)
3. ETSI TS 182 006 V2.0.4 (2008-05) Technical Specification. Telecommunications and Internet converged Services and Protocols for Advanced Networking (TISPAN); IP Multimedia Subsystem (IMS); Stage 2 description (3GPP TS 23.228 Release 7, modified)
4. IETF RFC 3261 (2002-06): SIP Session Initiation Protocol
5. ETSI TS 183 063 V3.5.2 (2011-03) Technical Specification. Telecommunications and Internet converged Services and Protocols for Advanced Networking (TISPAN); IMS-based IPTV stage 3 specifications
6. ETSI TS 186 020 V3.1.1 (2011-07) Technical Specification. Telecommunications and Internet converged Services and Protocols for Advanced Networking (TISPAN); IMS-based IPTV interoperability test specification
7. Paramonov, A., Tarasov, D., Koucheryavy, A.: The Video Streaming Monitoring in the Next Generation Networks. In: Balandin, S., Moltchanov, D., Koucheryavy, Y. (eds.) NEW2AN/ruSMART 2009. LNCS, vol. 5764, pp. 191–205. Springer, Heidelberg (2009)
8. ETSI TS 102 542-2 V1.3.1 (2010-01) Technical Specification. Digital Video Broadcasting (DVB);Guidelines for the implementation of DVB-IPTV Phase 1 specifications; Part 2: Broadband Content Guide (BCG) and Content on Demand
9. IETF RFC 2326 (1998-04): RTSP Real-Time Transport Control Protocol
10. Baskett, F., Chandy, K.M., Muntz, R.R., Palacios, F.G.: Open, Closed, and Mixed Networks of Queues with Different Classes of Customers. Journal of the ACM 22(2), 248–260 (1975)
11. Basharin, G.P., Bocharov, P.P., Kogan, Y.A.: Queueing analysis for computer networks. Theory and computational methods, 336 p. Nauka, Moscow (1989)
12. Naumov, V.A., Abaev, P.O.: Approximate Performance Analysis of Queues in Series Bulletin of Peoples' Friendship University of Russia. Mathematics. Information Sciences. Physics. (3-4), 64–70 (2007)
13. Wilson, P.R., Ventura, N.: A Direct Marketing Platform for IMS-Based IPTV. In: SATNAC 2009: Proceedings of the 12th Southern African Telecommunications Networks and Applications Conference (2009)

14. Samouylov, K.E., Luzgachev, M.V., Plaksina, O.N.: Modeling SIP Connections with Open Multiclass Queueing Networks. Bulletin of PFUR. Series "Mathematics. Information Sciences. Physics." (3-4), 16–26 (2007)
15. Buzyukova, I., Gaidamaka, Y., Yanovsky, G.: Estimation of GoS Parameters in Intelligent Network. In: Balandin, S., Moltchanov, D., Koucheryavy, Y. (eds.) NEW2AN/ruSMART 2009. LNCS, vol. 5764, pp. 143–153. Springer, Heidelberg (2009)
16. Ali Raad, A.M., Gaidamaka, Y.V., Pshenichnikov, A.P.: Session initiation model of IPTV service using IMS platform. Electrosvyaz (10), 46–51 (2013)
17. Kramer, W., Langenbach-Belz, M.: Approximation for the delay in the queueing systems GI|GI|1, Congressbook. In: 8th International Congress, Melbourne (1976)
18. Samouylov, K.E., Sopin, E.S., Chukarin, A.V.: Signalling Traffic Characteristics Measurement in IMS-based Network. T-Comm. Telecommunications and Transport (7), 8–13 (2010)

Wireless Sensor Network Based Smart Home System over BLE with Energy Harvesting Capability

Olga Galinina[1], Konstantin Mikhaylov[2],
Sergey Andreev[1], and Andrey Turlikov[3]

[1] Tampere University of Technology (TUT), Finland
{olga.galinina, sergey.andreev}@tut.fi
[2] University of Oulu (UO), Finland
konstantin.mikhaylov@ee.oulu.fi
[3] State University of Aerospace Instrumentation (SUAI), Russia
turlikov@vu.spb.ru

Abstract. In this paper, we study a smart home system, which incorporates an intelligent gateway serving a number of battery-powered wireless sensors. These communicate their measured data to the gateway by employing the novel Bluetooth Low Energy (BLE) radio technology. Additionally, sensors have capability to harvest wireless energy transmitted to them from the gateway over another dedicated radio interface and thus recharge their battery. We evaluate performance of the envisioned system in terms of overall delay and waiting time, as well as battery charge level and drain probability. To this end, we construct an optimistic estimate of energy harvesting capability, mindful of power distribution and consumption, which quantifies the potential limitations of wireless energy transfer. These results may be important to improve and optimize future smart home technology deployments.

Keywords: Smart home system, intelligent gateway, wireless sensors, Bluetooth Low Energy, energy harvesting capability, performance estimation.

1 Introduction and Motivation

With the advancement of Internet of Thing (IoT), more and more attractive end-user applications are enabled by the current technology. One such application is smart home [4], where human user has possibility to connect in-home wireless sensor network (composed of sensors, actuators, smart meters, etc.) to a smart home gateway and thus control the required functionality. However, as sensors are typically small-scale and battery-powered, their operation time without recharging becomes of paramount concern in smart home scenarios. To this end, a novel technology has recently been proposed, named Bluetooth Low Energy (BLE) protocol, to enable more energy efficient operation of a wireless sensor network. Another exciting innovation is the capability of modern devices

S. Balandin et al. (Eds.): NEW2AN/ruSMART 2014, LNCS 8638, pp. 419–432, 2014.
© Springer International Publishing Switzerland 2014

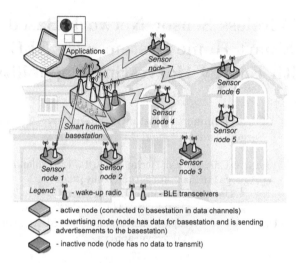

Fig. 1. Example network with seven sensor nodes and a gateway with four BLE transceivers

to harvest wireless energy from the ambient enwironment to recharge their batteries and thus extend operation lifetimes.

Inspired by these recent innovations, in this work we study a system, which comprises two major components: the energy-limited sensor nodes and the smart home gateway (see Figure 1). The gateway is equipped with multiple BLE transceivers that allow it to establish multiple active data sessions at a time. Additionally, it features a dedicated non-BLE radio interface which could transfer energy wirelessly to the sensor nodes and thus refill their energy supply.

The major purpose of developing the BLE protocol has been to enable transceivers with lower power consumption, lower complexity, and lower price than the ones possible with the conventional Bluetooth solution [3]. Although BLE operates in the same 2.4 GHz ISM band, uses same modulation and has inherited many features of its predecessor Bluetooth, the two technologies are not compatible [6]. The first studies (e.g., [6]) have shown that the BLE technology can provide 1.5-2 times higher throughput, is more energy efficient and has lower cost compared to other available solutions (e.g., IEEE 802.15.4). Among the major limitations of the BLE technology, however, are the restriction of the supported network topologies to the single-hop scenario, lower distance of communication compared to e.g., IEEE 802.15.4, and the use of stop-and-wait flow control mechanism based on cumulative acknowledgments, which might cause degradation of the protocol performance in the lossy environment.

In more detail, BLE divides the available frequency band into 40 2-MHz wide channels. The three channels are assigned specifically for advertising and discovery of the services and are named *advertising channels*. The remaining 37 *data channels* might be used for peer-to-peer data transfers. The data transmission between BLE devices is bound to time units known as advertising and connection events.

The periodic advertising events are used by the BLE transceivers to broadcast small blocks of data or to agree on the parameters of the connection to be established in the data channels. At the beginning of an advertising event the *advertiser*, i.e., the device that has some data to transmit, sends an advertising frame. Then, the advertiser might switch to receive mode and wait for the possible connection establishment requests. If the connection request from a device (which is referred to as an *initiator*) is received, the two devices start peer-to-peer connection in the data channels.

In the past, the legacy Bluetooth protocol has been intensively analyzed in the context of the smart home gateway scenario by use of queuing theory [9] and, in particular, with the polling models, which have been studied quite widely [12], [11]. By contrast, specific features of the *recently introduced* BLE protocol have been subject to much less research from the queuing theory community. To compensate for that, in this paper we formulate a new multi-channel discrete queuing model, which accounts for the processes of (i) file service and (ii) energy circulation, mindful of characteristic features of BLE protocol, such as existence of special advertising channels, which are separated from *orthogonal* transmission channels.

In what follows, we construct an *optimistic estimate* of wireless energy harvesting capability by the sensor nodes, which reveals the limits of any practical energy distribution mechanism. Particularly, we aim at evaluation of the key system parameters, such as overall data delay (or waiting time), i.e. the interval between file arrival and the end of its transmission (or beginning of transmission), as well as the probability of battery drain and the average level of remaining battery power.

More generally, there exist two possibilities for the analysis of our system, namely, *discrete* and *hybrid* continuous/discrete approaches, which differ in the properties of energy circulation process. Discrete techniques have been widely studied in the past literature, e.g. in [8], while hybrid techniques may be preferred when dealing with small granularity (which would otherwise result in prohibitive number of discrete states) and require solving a system of differential equations, often with intricate boundary conditions. A comprehensive overview of hybrid techniques (fluid models) may be found in [7]. However, given the specifics of the considered protocol (e.g., packet-based structure with small file sizes) and constant harvested energy consumption, we conclude that small granularity is not an issue here. Therefore, focusing on simpler discrete formulation, we leave hybrid fluid model for further study.

The rest of the text is organized as follows. Section 2 introduces general assumptions of the system model and Section 3 details our approach to obtain the stationary distribution and respective performance metrics. Importantly, we also provide relevant description of the BLE technology, its major characteristic features and application scenario considerations in Section 4.

2   Considered System Model

We focus on capturing dynamics of a centralized wireless sensor network observed in its stationary mode. The network consists of fixed number of sensors M spatially distributed around the connecting *gateway*. The network enables uplink and downlink data transmission channels between the sensors and the gateway. For the purpose of this research, we omit consideration of downlink channel, concentrating solely on uplink traffic, without loss of generality. Downlink channel operation could be easily taken into account by means of adding necessary auxiliary timings (see Section 4).

Sensor nodes are equipped with two radio interfaces, so that each node could communicate with the gateway using one of them (BLE) and might also harvest the energy of the received radio signal on another (non-BLE) and use it for refilling own energy battery (buffer). All devices in the network are synchronized and not interfering between each other. The system time is assumed to be *slotted*, so that the transmission of one data packet takes exactly one slot. We assume that at the beginning of any slot, each sensor may generate data files of various size with probability λ, so that the inter-arrival times follow geometrical distribution with parameter λ, whereas the file size in packets is geometrically distributed with average μ^{-1}.

For the sake of notation, we term the sensor *active*, if it has a data file to transmit, or *inactive* otherwise. Once awake, the sensor node will switch to inactive mode and start sensing the environment. After acquiring sensed data to be reported to the gateway, a sensor starts sending the advertisement packets in the BLE advertisement channels (which we do not consider in our model directly, assuming respective information exchange immediate). The gateway constructs the optimal schedule for receiving these files using the available transceivers, i.e., for each active sensor *a dynamic schedule* provided by the gateway may grant one out of $K < M$ orthogonal channels.

Following such optimal schedule, the gateway establishes the connections with different sensors using available channels and receives the sensed data at the fixed transmission rate r until the file is communicated completely. If there is no available channel for a new active sensor, it is *waiting* for service according to the FIFO discipline. Once all the data of a sensor are sent to gateway, the node switches back to inactive mode until a file arrives. It should be emphasized that the file transfer duration in slots, as well as the file size in packets, are defined completely by the file size in bits, the rate r in kbit/s, and the slot length $\Delta \tau$.

In our power-rate model, we let the transmit power p of a sensor and its corresponding transmit rate r be coupled by a well-known *Shannon's formula*:

$$r = w \log\left(1 + \text{SINR}\right) = w \log\left(1 + \frac{p\gamma}{N_0}\right), \tag{1}$$

where p is the output power of the RF power amplifier, w is the channel bandwidth, and N_0 is the noise power. The parameter γ stands for the channel gain and it is assumed to be proportional to some power function of the distance d

between the transmitter and the receiver. In order to limit the value of γ, we assume an upper bound, such that:

$$\gamma = \min \left(Gd^{-k}, \gamma_{max} \right), \tag{2}$$

where G is the propagation constant and k is the propagation exponent. The parameter γ_{max} defines the maximum level of SNR at the distance $d_0 = \left(\frac{G}{\gamma_{max}} \right)^{\frac{1}{k}}$, so that its further increase does not change the achievable rate.

Moreover, we employ realistic assumption regarding the sensor *power control* and assume L *discrete* levels of *power consumption* $p_{min} = p_1 < ... < p_L = p_{max}$. For the sake of analytical tractability, we also assume that there is a number of equal steps $\Delta p = (p_{max} - p_{min})/(L - 1)$.

We note that p_{max} defines the upper limit on the *radiated power* for a sensor. In other words, all sensors, which are able to maintain the predefined rate r, are located within the sphere of radius R:

$$R = \left[\frac{p_{max} G}{N_0 \left(e^{\frac{r}{w}} - 1 \right)} \right]^{\frac{1}{k}}. \tag{3}$$

Therefore, we may further assume that all connected sensors are uniformly distributed *within the sphere* of radius R. We also consider the following *three* levels of power consumption at the sensor:

(i) *transmit power* consumption $P_{tx} = \lceil \frac{p}{\Delta p} \rceil \Delta p + P_{tx_0}$ [power units] while the sensor is transmitting, where p is minimum required radiated power, P_{tx_0} is the circuit power, and the rest is power consumption of the amplifier;

(ii) *active power* P_a [power units], when the sensor has data and awaits service;

(iii) *sleep mode/inactive power*, when the sensor has no data to transmit. It is assumed to be zero, without loss of generality.

While sensor energy is being spent according to the levels P_{tx} and P_a, we employ the following model of *energy replenishment*. We recall that sensors are capable of *harvesting energy* from the gateway through the dedicated radio interface. In order to build *an optimistic estimate* of power consumption, we assume that our system is assisted by an *oracle*, which collects all energy flows into a common *rechargeable buffer* of capacity Z units and distributes it between the sources (see Figure 2). Energy arrived into the oracle's shared buffer follows Bernoulli distribution with probability q and the volume of each energy unit is assumed to be equal p_0. Further, the harvested energy is immediately stored in the buffer and may be spent starting from the following slot.

We note that any file transmission is possible *if and only if* there is enough energy in the buffer at the beginning of the slot, i.e. the required energy for transmission and waiting does not exceed the current energy level. Otherwise, we say that the energy resource is drained out and the system can neither transmit nor keep the sensors waiting. That is, all the sensors switch to the sleep mode, while their files are considered *lost*.

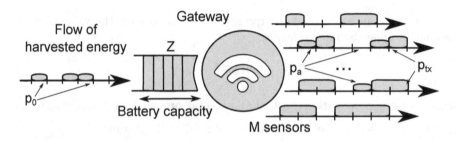

Fig. 2. Proposed discrete energy harvesting scheme

3 Our Mathematical Approach

3.1 Spatial Distribution

Continuous Case. Under the assumption that all sensors are uniformly distributed in a sphere of known radius R around the gateway, let us find the distribution of distances between the sensor and the gateway, that is effectively the distance between a random point within a sphere and its center. Due to the uniform distribution, the probability density function of sensor coordinates $f(x, y)$ is defined as:

$$f(x, y, z) = \frac{1}{\frac{4}{3}\pi R^3}, x^2 + y^2 + z^2 \leq R^2.$$

In terms of the spherical coordinates, we obtain

$$f_{r,\theta,\phi}(r, \theta, \phi) = f_{x,y,z}(r \sin\theta \cos\phi, r \sin\theta \sin\phi, r \cos\theta) \cdot J = \frac{3r^2 \sin\theta}{4\pi R^3}, 0 < r < R.$$

Therefore, we may conclude that

$$f_\phi(\phi) = \frac{1}{2\pi}, 0 \leq \phi \leq 2\pi, f_\theta(\theta) = \frac{1}{2}\sin\theta, 0 \leq \theta \leq \pi,$$

and, more importantly, the distribution of distances to the center of the sphere is $f_r(r) = \frac{3r^2}{R^3}, 0 < r < R$. It implies that the distribution of the distances between an arbitrary sensor and the gateway is defined as:

$$f_d(d) = \frac{3d^2}{R^3}, 0 < d < R, F_d(d) = \frac{d^3}{R^3}, 0 < d < R.$$

We recall that power consumption is limited by p_{min}. Without loss of generality, we assume that d_0 or γ_{max}, the limited SNR above, corresponds to p_{min} (otherwise, it can be done by updating e.g. d_0 to the maximum value). Consequently, if the transmit power is defined as

$$p_i = \max\left[\frac{N_0}{\gamma_i}\left(e^{\frac{r}{w}} - 1\right), p_{min}\right] = \max\left[d^k \frac{N_0}{G}\left(e^{\frac{r}{w}} - 1\right), p_{min}\right],$$

$$d = \left[\frac{pG}{N_0 \left(e^{\frac{r}{w}} - 1 \right)} \right]^{\frac{1}{k}}, \quad d'_p = \frac{1}{k} p^{\frac{1}{k} - 1} \left[\frac{G}{N_0 \left(e^{\frac{r}{w}} - 1 \right)} \right]^{\frac{1}{k}},$$

then, the distribution of continuous powers may be calculated as follows:

$$f_p(p) = \begin{cases} p^{\frac{3}{k} - 1} \cdot \frac{3}{k} (p_{max})^{-\frac{3}{k}}, p_{min} < p \leq p_{max} \\ F_p(p_{min}) = \left[\frac{p_{min}}{p_{max}} \right]^{\frac{3}{k}} \end{cases} \tag{4}$$

Discrete Case. We remind that a sensor is able to transmit on one of L different power levels. Let us establish the probability that an arbitrary sensor in the sphere has to radiate a particular power level. Since if the sensor always selects the closest higher power level $p_l = p_{min} + (l - 1)\Delta p$, then the probability mass function may be obtained as:

$$c_p(p_{min}) = c_p(p_1) = \Pr\{0 < p \leq p_1\} = F_p(p_1) = \left[\frac{p_{min}}{p_{max}} \right]^{\frac{3}{k}},$$

$$c_p(p_l) = \Pr\{p_l < p \leq p_{l+1}\} = F_p(p_{l+1}) - F_p(p_l) = \frac{[p_l]^{\frac{3}{k}} - [p_{l-1}]^{\frac{3}{k}}}{[p_{max}]^{\frac{3}{k}}}. \tag{5}$$

The summary power radiated by all sensors would depend on the sum of random variables p:

$$y_i = \sum_{n=1}^{i} p_n, 0 \leq i \leq K, \tag{6}$$

which may be calculated via consecutive convolutions of distributions or approximated by normal distribution in case of large i.

3.2   Power Consumption

Further, let us find the *random* power consumption P_i of the system at the state i, including radiated and circuit power of *random* sensors as well as active power of waiting sensors:

$$P_i = \frac{1}{K} \sum_{n=1}^{i_0} (p_n + P_c - P_a) + P_a i, 0 \leq i \leq M,$$

where $i_0 = \min(i, K)$ is the number of transmitting sensors and p_n is the radiated power of a random sensor n located at some point. We remind that during a particular file transmission by the tagged sensor the rest $(i - 1)$ sensors stay active and consume P_a power units.

Since the spatial distribution and service process are assumed to be independent, we may employ the assumption that all transmitting sensors at the state i are distributed according to (5).

Therefore, in a stationary state, the spatially averaged *total system* power consumption $\bar{p}_i$ may be found via averaging the following *random* power consumption:

$$\bar{p}_i = \frac{1}{K}E[y_i] + \frac{\min(i,K)}{K}(P_c - P_a) + P_a i, 0 \leq i \leq M,$$

where $y_i = \sum_{n=1}^{\min(i,K)} p_n$ is the sum of random variables (6).

3.3 Stationary Distribution of Energy Harvesting Operation

We remind that our system is equipped with a rechargeable battery of volume Z, which may either provide energy for sensors or be recharged according to the harvested energy. That all together may be represented as a two-dimensional discrete-time Markov chain with finite number of states $(N(t), X(t))$, where $N(t) \in [0, M]$ is the number of active sensors and $X(t) \in [0, Z]$ is the level of energy in the buffer. For the $N(t)$ process illustration, see Figure 3.

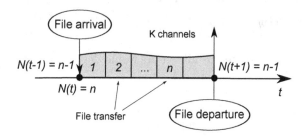

Fig. 3. File service example time diagram

Since the considered Markov chain is homogeneous, the elements of associated transition matrix $P \in \mathbf{R}^{s \times s}$ corresponding to the states (i, x) and (j, y) may be calculated straightforwardly accounting for the fact that $s = (M+1)(Z+1)$ is the number of possible states of process $(N(t), X(t))$. In more detail, consider the transition from (i, x) to (j, y). Let c out of $i_0 = \min(i, K)$ files be transmitted, while $j - i + c$ out of $M - i$ empty sensors generate new files with probability λ, so that $0 \leq j - i + c \leq M - i$. The energy level may change either at arrival of an energy block of volume p_H with probability q, or energy consumption of i active sensors, so that the new level of energy equals $y = \min(Z, x + [p_H] - P_{tx} i_0 - P_a(i - i_0))$. Therefore, we may exploit the following additive procedure to derive transition probabilities:

$$P_{(i,x),(j,y)} += \sum_{c=0}^{i_0} \binom{i_0}{c} \mu^c (1-\mu)^{i-c} \binom{M-i}{j-i+c} \lambda^{j-i+c} (1-\lambda)^{M-j-c} \cdot q,$$
$$y = \min(Z, x + p_H - P_{tx} i_0 - P_a(i - i_0)),$$

$$P_{(i,x),(j,y)} += \sum_{c=0}^{i_0} \binom{i_0}{c} \mu^c (1-\mu)^{i_0-c} \binom{M-i}{j-i+c} \lambda^{j-i+c} (1-\lambda)^{M-j-c} \cdot (1-q),$$
$$y = (x - P_{tx} i_0 - P_a(i - i_0))$$

In equation above we assume $x - P_{tx}i_0 - P_a(i - i_0) \geq 0$. In case when sensors drain the battery, all the files are considered lost:

$$p_{(i,x),(0,0)} = 1 - q, \text{ if } x < P_{tx}i_0 + P_a(i - i_0),$$
$$p_{(i,x),(0,p_H)} = q, \text{ if } x < P_{tx}i_0 + P_a(i - i_0).$$

Moreover, the probability to drain the battery at any state (i, x) up to the level when all the sensors lose their files and force the sleep mode/inactive state may be obtained as:

$$p_{(i,x)}^{loss} = 1, \text{ if } x < P_{tx}i_0 + P_a(i - i_0).$$

Further, let us obtain the stationary probability distribution:

$$\pi_{(i,x)} = \lim_{t \to \infty} \Pr\{N(t) = i, X(t) = x\}. \tag{7}$$

For that purpose, we solve the following system of linear equations:

$$\pi = P^T \pi, \sum_{i,x} \pi_{(i,x)} = 1, \tag{8}$$

where P is the transition probability matrix, and $\pi = \{\pi_{(i,x)}\}_{i,x} \in R^{1 \times (M+1)(Z+1)}$ is the stationary distribution vector.

Then, basing on the stationary distribution, we may calculate the average metrics in our system, such as the average number of active sensors $\bar{N}$, the average number of waiting sensors $\bar{N}_s$, the average level of buffer charge, and the probability p^{loss} to drain the battery:

$$\bar{N} = \sum_i \sum_{n=1}^{M} n\pi_{(i,x)}, \bar{N}_w = \sum_i \sum_{n=1}^{M-K} \pi_{(n+K,x)},$$
$$\bar{X} = \sum_{i,x} x\pi_{(i,x)}, p^{loss} = \sum_{i,x} p_{(i,x)}^{loss}\pi_{(i,x)}. \tag{9}$$

By following the same idea, since the vector of immediate throughput at the state (i, x) is $S_{(i,x)} = \mu \min(i, K), \forall x \geq P_{tx}i_0 + P_a(i-i_0)$, the average throughput may be obtained as:

$$S = \mu \sum_{i=0}^{K} \sum_{x=iP_{tx}}^{Z} i\pi_{(i,x)} + \mu K \sum_{i=K+1}^{M} \sum_{x=P_{tx}K+P_a(i-K)}^{Z} \pi_{(i,x)},$$

where the lower border of the summation by Z is the minimum available energy level for the operation of i sensors.

The average delay (i.e., time interval between the file arrival and the end of its transmission) and the average waiting time (i.e., time interval between the file arrival and the beginning of its transmission) may be derived by employing the Little's law as:

$$\bar{T} = \frac{\bar{N}}{S} \text{ and } \bar{T}_w = \frac{\bar{N}_w}{S}. \tag{10}$$

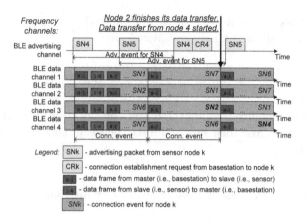

Fig. 4. Example resource management scenario for the case of seven sensor nodes and a gateway with four BLE transceivers

4 Technology and Application Scenario Discussion

As follows from Section 2, the proposed model is somewhat non-conventional for the wireless sensor networks. This is due to the consideration of the recently developed BLE (Bluetooth Low Energy) protocol [3], as well as because of the following factors:

1. The special advertising channels are assumed, which are separated from the channels where the data are transferred. We assume that those are used by the sensors to announce to the gateway the amount and the parameters of the available sensed data, as well as to agree on the employed data channels. Also those might be used by the sensors to inform the gateway of the amount of available and harvested energy.
2. The used time-slot based communication model for the data channels, according to which the two communicating nodes are periodically alternating on the used radio channels. In our case, we assume that all the nodes need to communicate with the same gateway, which enables the gateway to synchronize the time-slots for all the nodes. Additionally, the gateway may arrange the frequency hopping sequences for each of the sensor nodes in such a way, that the nodes will not interfere with each other (see Figure 4).
3. The stop-and-wait flow control mechanism based on cumulative acknowledgments used in BLE data channels allows us to take into account the parameters of the downlink through the parameters of the uplink.
4. We assume that data and energy are transferred using non-overlapping radio bands, as discussed below.

A state-of-the-art solution on the low and zero-power radios is well illustrated by Fig. 1 in [10]. There, as one can see, the minimum power of the received radio signal enabling energy harvesting is on the order of -30 dBm. The reason

Table 1. Input parameters for energy transfer efficiency calculation

Receive antennas:	halfwave dipole
Radio propagation model:	Friis equation
Minimum power of radio signal enabling wake-up radio and EH:	-30 dBm
Efficiency of the rectifier (RF-DC conversion):	30%
Energy storing efficiency (DC to capacitor):	90%

Table 2. Wireless energy transfer efficiency

	Frequency bands		
	2400-2483.5 MHz	869.4-869.65 MHz	433.05-434.79 MHz
Maximum duty cycle, % [1,2]	100	10	100
Maximum power, mW [1]	10 EIRP	500 ERP	10 ERP
Maximum distance to gateway enabling energy harvesting, m [3]	1.2	31.8	9
Amount of power transferred into energy and buffered, nW [3,4]	0	20	1.7
Harvesting time to send one BLE advertising packet with 31 byte payload, s [3,4,5]	-	1.8	22

[1] according to [2]

[2] relative to one hour, see [2]

[3] assumed that the gateway's transmitter uses maximum permitted transmit power

[4] distance between the gateway and the sensor is 5 m

[5] as for the TI CC2540 BLE transceiver the energy required for sending an ADV_NONCONN_IND packet is 39 μJ

behind this is that less powerful radio signals are unable to provide the minimum voltage required to fully commutate the stages of rectifiers, which are used for retrieving the energy of the received radio signals.

Keeping in mind this technology feature and accounting for the restrictions by the frequency regulation agencies (i.e., Finnish Communications Regulatory Authority, FICORA) on the use of the unlicensed ISM channels [2], we shortly summarize (a) which frequency bands might be used for wireless energy transfer, (b) what is the maximum distance for successful energy transfer, and (c) what is the amount of energy one can potentially transfer from the gateway to the sensors. The input data used for our calculations is given in Table 1 and the results are presented in Table 2.

Our results indicate that an attractive compromise between the maximum permitted transmit power, path losses, and antenna efficiency is achieved in 869.5 MHz band. Nonetheless, even there, the energy replenishment rate is quite small and it will take a while to harvest enough energy for system's operation. Meanwhile, due to the high losses and low maximum permitted transmit power, the energy transfer in 2.4 GHz band (where BLE operates) is hardly feasible without the use of strongly-directive antennas.

5   Numerical Results and Conclusions

In order to illustrate our analysis with numerical results, we adopt typical parameters from Table 3. Note that we express energy values, as well as battery capacity, in terms of energy blocks of $1mW$ to keep them integer. We vary the probability of energy arrival q and observe the key system parameters for different number of sensors $(10, 40, 50, 100)$.

Table 3. Primary system parameters

Notation	Parameters description	Value
$\Delta\tau$	Slot length	10ms
M	Number of sensors	var
λ	Average file arrival rate per sensor	$1/60\mathrm{s}^{-1}$
μ^{-1}	Average file size	512byte
μ^{-1}	Average file size in packets	2
K	Number of avaliable channels	37 [1]
r_0	Fixed channel rate	236.7kbit/s [6]
f	Frequency	2.4GHz[1]
w	Spectral bandwidth	2 MHz [1]
k	Propagation exponent	4 [5]
l	Wavelength	0.123m
p_{max}	Maximum output power of the RF power amplifier	10mW [1]
p_{min}	Minimum output power of the RF power amplifier	0.01mW [1]
P_a	Power consumption of waiting sensor	2mW
P_c	Circuit power consumption	60mW
Z	Energy buffer capacity	2mAh

Figure 5 illustrates the ratio between the average level of energy in the buffer and the total battery capacity Z. The less is the number of users, the sooner system achieves full battery level with the growing energy flow. For $M = 100$, even $q = 1$ is not enough and we need to increase the volume of energy portion p_0 instead. All plots begin from the point where energy is not enough, so that the buffer is almost empty.

Further, the probability to drop files due to insufficient battery power level is represented in Figure 6. Starting with some level of energy (different for various total loads), the energy resource is sufficient for transmission and the considered probability tends to zero. For $M = 100$, p_0 is clearly not enough as seen in Figure 5 as well.

We conclude that our model is helpful to characterize performance of a smart home scenario controlled by a gateway. With the proposed solution, the BLE/energy harvesting parameters may be optimized and the data loss probability can be controlled for a wide range of practical system parameters.

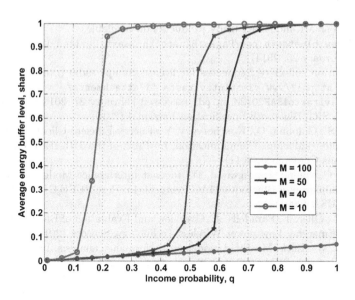

Fig. 5. Dependence of average battery level on energy arrival probability q

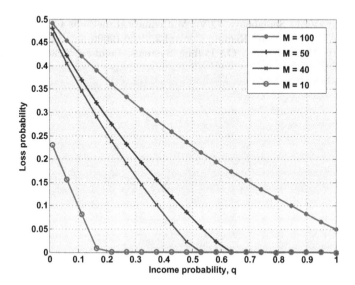

Fig. 6. Dependence of the probability of file dropping on the probability of energy arrival q

References

1. Bluetooth smart and smart ready products now available, http://www.bluetooth.com/Pages/Bluetooth-Smart-Devices-List.aspx (accessed February 26, 2014)
2. Regulation on collective frequences for license-exempt radio transmitters and on their use, https://www.viestintavirasto.fi/attachments/Viestintavirasto15AF2013M_en.pdf (accessed February 27, 2014)
3. Bluetooth SIG, Bluetooth Specification Version 4 (2010)
4. Andreev, S., Galinina, O., Koucheryavy, Y.: Energy-Efficient Client Relay Scheme for Machine-to-Machine Communication. In: Proc. of the 54th IEEE Global Communications Conference (2011)
5. Cordeiro, C., Sadok, D., Agrawal, D.: Piconet Interference Modeling and Performance Evaluation of Bluetooth MAC Protocol. In: Proc. IEEE GLOBECOM, San Antonio, USA (2001)
6. Gomez, C., Oller, J., Paradells, J.: Overview and Evaluation of Bluetooth Low Energy: An Emerging Low-power Wireless Technology. Sensors 12(9) (August 2012)
7. Gribaudo, M., Telek, M.: Fluid models in performance analysis. In: Bernardo, M., Hillston, J. (eds.) SFM 2007. LNCS, vol. 4486, pp. 271–317. Springer, Heidelberg (2007)
8. Kleinrock, L.: Queueing Systems. Theory, vol. I. Wiley Interscience (1975)
9. Miorandi, D., Zanella, A., Pierobon, G.: Performance Evaluation of Bluetooth Polling Schemes: An Analytical Approach. Mobile Networks and Applications 9(1), 63–72 (2004)
10. Roberts, N., Wentzloff, D.: A 98nW wake-up radio for wireless body area networks. In: Proc. 2012 IEEE Radio Frequency Integrated Circuits Symp., pp. 373–376 (2012)
11. Vishnevsky, V.M., Semyonova, O.V.: Mathematical methods to study the polling systems. Automation and Remote Control 2, 3–56 (2006)
12. Vishnevsky, V., Semenova, O.: Polling Systems. Lambert Academic Publishing (2012)

Further Investigations
of the Priority Queuing System with Preemptive Priority and Randomized Push-Out Mechanism

Alexander Ilyashenko, Oleg Zayats, Vladimir Muliukha, and Leonid Laboshin

St, Petersburg State Polytechnical University, Russia
{ilyashenko.alex,zay.oleg}@gmail.com, vladimir@mail.neva.ru,
laboshin1@neva.ru

Abstract. This article is written about a queuing theory models with limited buffer size and one service channel with two incoming flows. One of the flows is more important than another flow. In this article we prefer to call packets of these flows as priority and non-priority packets. This priority can be realized as a preemptive priority, which allows high-priority packets to take place in system queue closer to service channel and push-out low-priority packets out of service channel or as a randomized push-out, which allows to push out non-priority packets out of the system when it is full. Authors present in this article algorithm for computing statistical characteristics of the model for all values of push-out probability α. For getting solution is used generating functions method. This method reduces size of linear equations system from $k(k+1)/2$ to $(k+1)$. Using this method allowed authors to study model behavior for all load values from 0 to 4 by first and second incoming flows. In this article provided zones of model "closing" for non-priority packets. Also authors considered a relative deviation of loss probability and it's approximation by linear law depending on push-out probability α to get areas of possible using linear law for approximating results of changing this push-out probability.

Keywords: priority queuing, preemptive priority, randomized push-out mechanism.

1 Introduction

Queuing systems appear everywhere in our life, and corresponding mathematical models are very important for all applications from economics to telecommunications, IT and medicine. Studying such systems is very important for modern network technologies. Results can help to improve parameters of technical devices used for data transfer [13,14,15].

Priority queuing systems are standing out among all models of queuing theory. These models allow us to construct systems, where is important to transfer (service) data of

S. Balandin et al. (Eds.): NEW2AN/ruSMART 2014, LNCS 8638, pp. 433–443, 2014.

varying importance. But considering systems only with priorities is meaningful only when data streams have low intensities, because at high loads effect of prioritization will not be visible due to permanent filling of system buffer and consequently data loss of any types. In real life some systems have flows with high intensity, for example, computer networks. To resolve this problem usually researchers add push-out mechanism in model, which allows pushing non-priority packets out of the system to free up space, which could be taken by packets with a higher priority. In literature considered only one type of push-out mechanism – deterministic push-out. It this case all non-priority packets are pushed out by priority packets out of the system. Adding a deterministic push-out mechanism can have too powerful influence on model behavior.

In a first articles about randomized push-out mechanism [1, 2] N.O. Vilchevsky and co-authors considered it as a mechanism with one parameter - the push-out probability α. Main feature of this mechanism is if α equals to zero then the system has no push-out mechanism at all, but, on the contrary, if α equals one, the mechanism is deterministic. So getting solution for this mechanism allows to get solution for all possible types of push-out mechanisms.

Method which is used for getting solution firstly was applied to queuing model with unlimited buffer size by H. White and L.S. Christie [3] as well as F.F. Stephan [4]. It was a classical method of generating functions. Using this method in this article obtained analytical expressions for the model characteristics. This paper presents application of this method to the systems with the limited storage and randomized push-out mechanism. Main characteristics of the model depending on α were plotted. And this plots show how this α parameter can effectively control behavior of the model, even of heavy-loaded ones.

Presented model has a single channel, finite buffer size k ($1 \leq k < \infty$) and two independent elementary streams of packets with intensities $\lambda 1$ and $\lambda 2$ accordingly. Time interval between packets of ith type is distributed exponentially with the rate λi. Processing time for both types of packets is independently and equally distributed by the exponential law with the rate μ. Time intervals between incoming packets and all processing time intervals are independent too. Packets of the first type are priority packets and of the second type are non-priority packets. Priority in this model is preemptive.

Considered model is represented by scheme in Figure 1. Probability of push-out equals zero or one, realizes two limiting cases of pushing out, mentioned above. G.P. Basharin [5] considered such types of pushing-out and modified classification system for queuing systems created by D.Kendall [6]. Basharin [7] proposed to add one more symbol$\overrightarrow{M_2}/M/1/k/f_j^i$ to classification, where i – type of priority (0 – without priority, 1 – non-preemptive, 2 – preemptive priority), j – type of push-out mechanism (0 – without push-out, 2 – determined push-out). In the Basharin's notation the value j = 1 remained free, and can be assigned to mechanism described in [1,2]. So we can now classify the system, which is presented on the Figure 1, as of the $\overrightarrow{M_2}/M/1/k/f_2^1$ class.

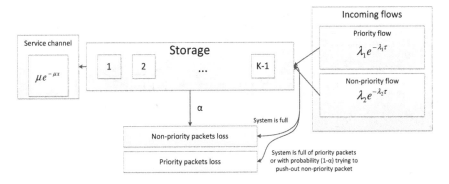

Fig. 1. Scheme of system $\overrightarrow{M_2}/M/1/k/f_2^1$

In this article was improved the algorithm of calculations from [8-10], and obtained more complete and detailed numerical results. Also studies "closing" area for non-priority packets and areas where loss probabilities of ith type packets can be approximated by linear law depending on push-out probability.

2   Building Kolmogorov's System of Equations

Let's consider a system $\overrightarrow{M_2}/M/1/k/f_2^1$ in the steady state. Process in this system is Markov process [6] and ergodic. Phase space of the system can be defined by the following equation:

$$\Omega = \left\{(i,j): i = \overline{0,k}, j = \overline{0,k}, 0 \leq i + j \leq k\right\}.$$

Probabilities of different states for this phase space:

$$P(i,j;t) = P\{N_1(t) = i, N_2(t) = j, 0 \leq i + j \leq k\},$$

where $N_q(t)$ means the q-th type packets number in the system at the moment t. Final probabilities are:

$$P_{i,j} = \lim_{t \to \infty} P(i,j;t), (i = \overline{0,k}, j = \overline{0,k}, 0 \leq i + j \leq k).$$

The state graph is used for building a Kolmogorov's system of equations for the final probabilities. As a result, using Kronecker delta symbol for unification of each specific state equation, we obtain:

$$-[\lambda_1(1-\delta_{j,k-i}) + \alpha\lambda_1(1-\delta_{i,k})\delta_{j,k-i} + (1-\alpha)\lambda_1\delta_{i,0}\delta_{j,k-i} + \lambda_2(1-\delta_{j,k-i}) +$$
$$+\mu(1-\delta_{i,0}\delta_{j,0})]P_{i,j} + \mu P_{i+1,j} + \mu\delta_{i,0}P_{i,j+1} + \lambda_2 P_{i,j-1} + \lambda_1 P_{i-1,j} + \tag{1}$$
$$+\alpha\lambda_1\delta_{j,k-i}P_{i-1,j+1} + (1-\alpha)\lambda_1\delta_{j,k-i}\delta_{i,1}P_{i-1,j+1} = 0, (0 \leq i \leq k; 0 \leq j \leq k-i).$$

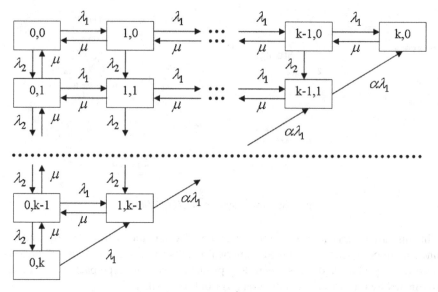

Fig. 2. The state graph of $\overrightarrow{M_2}/M/1/k/f_2^1$ system

System matrix is singular. But we have additional normalization condition (2) which can be replace any other equation from the system

$$\sum_{i=0}^{k}\sum_{j=0}^{k-i}P_{i,j}=1.\tag{2}$$

After replacing system of linear equations (1), (2) has a unique solution. This system has of $k\,(k+1)/2$ equations and the same number of unknown variables. System size grows rapidly with increasing k. In [1, 2] proposed method of reduction this size to k+1.

3   Generating Function Method Application

Generating function for probabilities $P_{i,j}$ is:

$$G(u,v)=\sum_{i=0}^{k}\sum_{j=0}^{k-i}P_{i,j}u^{i}v^{j}.$$

with normalization condition in terms of generating function

$$G(1,1)=\sum_{i=0}^{k}\sum_{j=0}^{k-i}P_{i,j}=1.$$

Multiply both sides of equation (1) on uivj and sum over all admissible values of (i,j). Final equation for generating function is:

$$[\lambda_1 u(1-u) + \lambda_2 u(1-v) + \mu(u-1)]vG(u,v) = \mu(u-v)G(0,v) + (1-\alpha)\lambda_1 P_{0,k} v^k u(u-v) +$$

$$+\mu u(v-1)G(0,0) + \alpha\lambda_1 u^{k+1}(v-u)P_{k,0} + [\alpha\lambda_1(u-v) + \lambda_1(1-u)v + \lambda_2(1-v)v]u\sum_{i=0}^{k} P_{i,k-i} u^i v^{k-i}. \tag{3}$$

Generating function includes $k\,(k+1)/2$ unknown variables $P_{i,j}$, expressed through the $(2k+1)$ state probabilities from first row and last diagonal of state graph. Technique from [1, 2] can be used to reduce required states to only $(k+1)$ unknown probabilities of the "diagonal" states. As a result all characteristics of such systems can be expressed using probabilities

$$p_i = P_{k-i,i}, (i = \overline{0,k}). \tag{4}$$

Obtaining expressions for the probability $P_{i,j}$ can be done by resolving (3) respectively to G:

$$G(u,v) = \frac{(u-v)G(0,v) + u(u-1)G(0,0) + \alpha\rho_1 u^{k+1}(v-u)P_{k,0}}{v\rho_1(u-u_1)(u-u_2)} +$$

$$+\frac{[\alpha\rho_1(u-v) + \rho_1(1-u)v + \rho_2(1-v)v]u\sum_{i=0}^{k} P_{i,k-i} u^i v^{k-i} + (1-\alpha)\rho_1 P_{0,k} v^k u(u-v)}{v\rho_1(u-u_1)(u-u_2)}. \tag{5}$$

Using roots of denominator we can require that the following two conditions of analyticity were satisfied:

$$\operatorname*{Re\,s}_{u=u_1} G(u,v) = 0, \operatorname*{Re\,s}_{u=u_2} G(u,v) = 0.$$

Doing simple transformations we obtain

$$(1-v)G(0,0) - G(0,v) +$$

$$+\sum_{i=0}^{k} P_{i,k-i} v^{k-i} \left[\rho_1(v-\alpha)\frac{u_2^{i+2} - u_1^{i+2}}{u_2 - u_1} + v(\rho_1(\alpha-1) + \rho_2(v-1))\frac{u_2^{i+1} - u_1^{i+1}}{u_2 - u_1} \right] +$$

$$+\alpha\rho_1 P_{k,0}\left(\frac{u_2^{k+2} - u_1^{k+2}}{u_2 - u_1} - v\frac{u_2^{k+1} - u_1^{k+1}}{u_2 - u_1} \right) + (1-\alpha)\rho_1 P_{0,k} v^k(u_2 + u_1) = 0.$$

Now consider a divided difference of the first order for function ui at u1 and u2 argument values [12]

$$Q_i = \frac{u_2^i - u_1^i}{u_2 - u_1}, \tag{6}$$

Decomposing divided differences into Taylor series using Gegenbauer polynomials as described in [8-10] and applying it to states from the first row of state graph obtain in a uniform manner for all $j = \overline{0, i-1}$

$$P_{k-i,j} = p_0 \rho_1^{-1} \rho_1^{(1-i)/2} \beta^j (C_{i-1-j}^{j+1} - \rho_1^{1/2} C_{i-2-j}^{j+1}) + p_i(\alpha-1)\delta_{i,k}\delta_{j,i-1} +$$

$$+ \sum_{s=1}^{j} p_s((\rho_1^{-1} + \alpha)\rho_1^{(1-i+s)/2} C_{i-1-j}^{j-s+1} - \rho_1^{-1}\rho_1^{(1-i+s+1)/2} C_{i-2-j}^{j-s+1})\beta^{j-s} - \tag{7}$$

$$- \alpha \sum_{s=0}^{j} \rho_1^{(1-i+s)/2} C_{i-1-j}^{j-s+1} p_{s+1}\beta^{j-s}, (0 \le j \le i-1).$$

All final probabilities were expressed in terms of "diagonal" probabilities p_i. It remains to build a system of equations for the p_i.

To do this, we use the distribution of the total number of packets in the system:

$$r_n = \sum_{i=0}^{n} P_{i,n-i} = \sum_{i=0}^{n} P_{n-i,i}. \tag{8}$$

Then we substitute expressions for final probabilities (7) in (8) and get system of linear equations for only "diagonal" states

$$r_z = \sum_{j=0}^{z} P_{z-j,j} = p_0 \rho_1^{-1}\zeta_z - \alpha\varphi_z p_{z+1} + \sum_{j=1}^{z} p_j\xi_{z,j} + p_k(\alpha-1)\delta_{z,k-1}, (0 \le z \le k-1), \tag{9}$$

where

$$\zeta_z = \sum_{j=0}^{z} \rho_1^{(1-k+z-j)/2} \beta^j (C_{k-z-1}^{j+1} - \rho_1^{1/2} C_{k-z-2}^{j+1}), \phi_z = \rho_1^{(z+1-k)/2} C_{k-1-z}^{1}, \tag{10}$$

$$\xi_{z,j} = \sum_{s=j}^{z} \rho_1^{(z-k-s+j)/2}(\rho_1^{-1/2} C_{k-1-z}^{s-j+1} - C_{k-2-z}^{s-j+1})\beta^{s-j} - \alpha\rho_1^{(j-k)/2}\beta^{z+1-j} C_{k-1-z}^{z-j+2}. \tag{11}$$

System has only k equations and k+1 unknown variables. To get unique solution we need to add one more equation to this system. This equation is:

$$r_k = \sum_{i=0}^{k} P_{k-i,i} = \sum_{i=0}^{k} p_i. \tag{12}$$

Solving system (9), (12) where the coefficients are given by (10) and (11) can be found all characteristics of the queuing model. Also it can be seen that this system has a specific almost triangular matrix. So solving process can be simplified more.

4   Computational Results

Most interesting characteristic of any queuing model is a probability of losing packet of any type. Each type losing probability can be obtained as

$$P_{loss}^{(1)} = q_k + (1-\alpha)\sum_{i=1}^{k-1} p_i, \quad P_{loss}^{(2)} = r_k + \alpha\frac{\rho_1}{\rho_2}\sum_{i=1}^{k-1} p_i + \frac{\rho_1}{\rho_2} p_k.$$

Figure 3 shows losing probability for changing incoming priority flow relative intensity from 0.2 to 2.6 for weak and strong non-priority packets flow. From figure 4 we can see that changing the push-out probability α of non-priority packets model can completely lock-up for non-priority packets.

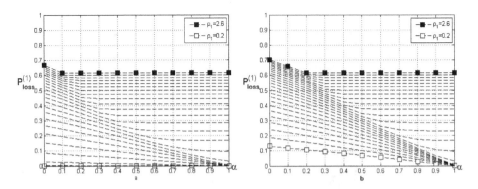

Fig. 3. a) Loss probability of high-priority packets with weak low-priority flow (ρ_2=0.5). b) Loss probability of high-priority packets with strong low-priority flow (ρ_2=1.5) with 0.1 step by ρ_1

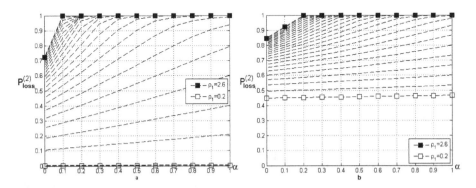

Fig. 4. a). Loss probability of low-priority packets with weak low-priority flow (ρ_2=0.5), b) Loss probability of low-priority packets with strong low-priority flow (ρ_2=1.5)

From Figure 4.a we can see that increasing α allows us to reduce the loss of priority packets almost to zero, while the number of lost non-priority packets rises slower. Also from Figure 4 it's seen that there is almost linear behavior of loss probability for low priority load. For a high load chart it is quite different.

Interesting results presented on Figure 4 attracted our attention and we decided to study systems deeper. Results where we can see linear behavior and "closing" of the system are useful for real time systems where this model can be used. So we decided to find areas of "closing" system for non-priority packets. We changed incoming flows intensities from 0 to 4.0 with a 0.01 step and computed losing probability of non-priority packets for push-out probability equal to 1. On Figure 5 presented a plot of areas built using levels [0, 0.5, 0.9, 0.95, 0.99, 0.999, 0.9999]. Level where losing probability less than 0.5 is black on this plot. Right area is area where losing probability more than 0.9999.

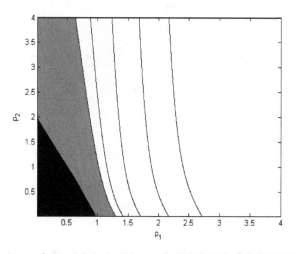

Fig. 5. Areas of system "closing" for levels [0,0.5,0.9,0.95,0.99,0.999,0.9999]

The second studied result is linear behavior of losing probabilities on weak loads of a model. We already have analytical solution for systems where push-out probability equal to 0 and 1. In real systems we don't have enough time for real0time calculations using provided expressions from this article, but for some systems we can find areas where we can use linear approximation for predicting losing probabilities values. For building linear approximation of losing probability curve we used two points where push-out probability points equal to 0 and 1. We used same results as for Figure 5 and built a relative deviation of losing probability curve and its linear approximation for all points from 0 to 4 for both flows with step 0.01.Figure 6 presents areas where relative deviation for priority packets changes from 0% to 40% with 1% step. Black areas are where deviation less than 1%.

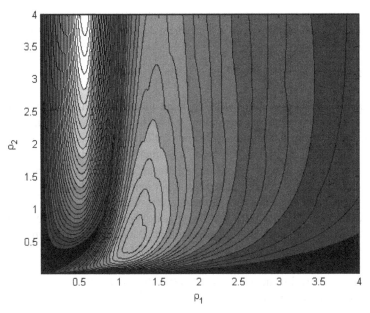

Fig. 6. Relative deviation of priority packets losing probability curve and its linear approximation from 0% to 40% with 1% step

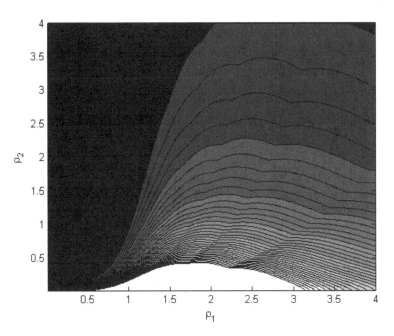

Fig. 7. Relative deviation of non-priority packets losing probability curve and its linear approximation from 0% to 40% with 1% step

On Figure 7 we can see results for non-priority packets. Black area is an area where deviation less than 1% and white area is an area where deviation more than 40%.

5   Conclusion

This article contains method of calculating queuing model characteristics with two incoming flows preemptive priority and randomized push-out mechanism. First chapter of paper describes main definitions and significations and algorithm of building Kolmogorov's system of equations. The second chapter of this article contains expressions for generating function and system of linear equations where used only "diagonal" states of state graph. Third chapter is about numerical computation results. There are detailed plots of characteristics of the queuing model.

As a main result it can be concluded that the provided mechanism if model control is highly effective and efficient tool for controlling data flows which are going through the system. One more important result is a model "closing" for non-priority packets. As a result was plotted Figure 5 where we can see areas where losing probability reaches specified level from array [0,0.5,0.9,0.95,0.99,0.999,0.9999].

One more result was shown on Figures 6 and 7 where we can see areas where we able to use linear approximation for losing probabilities with specified deviation level. These results can be used for improving performance of real-time technical systems and predict losing probabilities without additional computations.

References

1. Avrachenkov, K.E., Shevlyakov, G.L., Vilchevsky, N.O.: Randomized push-out disciplines in priority queueing. Journal of Mathematical Sciences 122(4), 3336–3342 (2004)
2. Avrachenkov, K.E., Vilchevsky, N.O., Shevlyakov, G.L.: Priority queueing with finite buffer size randomized push-out mechanism. Performance Evaluation 61(1), 1–16 (2005)
3. White, H., Christie, L.S.: Queueing with preemptive priorities or with breakdown. Operation Research 6(1), 79–95 (1958)
4. Stephan, F.F.: Two queues under preemptive priority with Poisson arrival and service rates. Operations Research 6(3), 399–418 (1958)
5. Basharin, G.P.: A single server with a finite queue and items of different types. Teoriyaveroyatnostey iee Primeneniya 10(2), 282–296 (1965) (in Russian)
6. Kleinrock, L.: Queueing systems. John Wiley and Sons, N.Y (1975)
7. Basharin, G.P.: Some results for priority systems. Massovoe Obsluzhivanie v Sistemakh Peredachi Informacii. M.: Nauka, 39–53 (1969) (in Russian)
8. Zaborovsky, V., Mulukha, V., Ilyashenko, A., Zayats, O.: Access control in a form of queueing management in multipurpose operation networks. International Journal on Advances in Networks and Services 4(3/4), 363–374 (2011)
9. ZayatsO., Z.V., Muliukha, V., Verbenko, A.S.: Control packet flows in telematics devices with limited buffer, absolute priority and probabilistic push-out mechanism. Part 1. Programmnaya Ingeneriya (2), 22–28 (2012) (in Russian)

10. ZayatsO., Z.V., Muliukha, V., Verbenko, A.S.: Control packet flows in telematics devices with limited buffer, absolute priority and probabilistic push-out mechanism. Part 2. Programmnaya Ingeneriya (3), 21–29 (2012) (in Russian)
11. Zaborovsky, V., Podgursky, U., Semenovsky, V.: International ATM-project involving Russia. Globalnie Seti i Telekommunikacii (10), 56–58 (1998) (in Russian)
12. Bateman, H., Erdelyi, A.: Higher transcendental functions. McGrow-Hill, NY (1953)
13. Shi, Z., Beard, C., Mitchell, K.: Analytical Models for Understanding Misbehavior and MAC Friendliness in CSMA Networks. Performance Evaluation 66(9-10), 469–487 (2009)
14. Shi, Z., Beard, C., Mitchell, K.: Analytical Models for Understanding Space, Backoff and Flow Correlation in CSMA Wireless Networks. Wireless Networks 19(3), 393–409
15. Shi, Z., Beard, C., Mitchell, K.: Competition, Cooperation, and Optimization in Multi-Hop CSMA Networks with Correlated Traffic. International Journal of Next-Generation Computing (IJNGC) 3(3), 228–246 (2012)

Low Temperature Ferroelectric and Magnetic Properties of Doped Multiferroic $Tb_{0.95}Bi_{0.05}MnO_3$

Natalia Vl. Andreeva[1], Victoria A. Sanina[2], Sergej B. Vakhrushev[1,2], Alexey Vl. Filimonov[1], Alexander E. Fotiadi[1], and Andrey I. Rudskoy[1]

[1] St.-Petersburg State Polytechnical University, St.-Petersburg, Russia
nvandr@gmail.com, {filimonov,fotiadi}@rphf.spbstu.ru
[2] Ioffe Physical Technical Institute, St.-Petersburg, Russia
{sanina,s.vakhrushev}@mail.ioffe.ru

Abstract. Multiferroics are very promising materials for application in RF/microwave electronic devices. Electrical tuning of magnetism due to strong magnetoelectric coupling in these materials gives an opportunity to use them in reconfigurable microwave devices, ultra-low power electronics and magnetoelectric random access memories (MERAMs). $Tb_{0.95}Bi_{0.05}MnO_3$ (TMNO) multiferroic is a solid solution of $TbMnO_3$ and $BiMnO_3$, the interest to investigation of this nanocomposite material is caused by the possibility to obtain the multiferroic with close temperatures of magnetic and ferroelectric ordering which are higher than in pure $TbMnO_3$. Results of TMNO crystal investigations obtained using cryogenic magnetic force and piezopresponse force microscopy techniques are presented. An existence of ferroelectric and ferromagnetic ordering in TMNO at low temperatures was observed.

Keywords: Multiferroics, Low temperature magnetic force microscopy, Low temperature piezoresponse force microscopy.

1 Introduction

A great potential of multiferroic materials in a voltage tunable and low power devices caused an increasing amount of works aimed at investigation of their properties and new material design. Strong magnetoelectric (ME) coupling in multiferroics provides opportunities for developing different multiferroic devices such as magnetic sensors, voltage and magnetic field tunable RF/microwave devices, including non-reciprocal tunable bandpass filters, low loss and high power handling phase shifters, etc. Nowadays ME coupling at room temperatures was demonstrated in multiferroic heterostructures [1]. Bulk multiferroics with strong ME coupling are considered as prospective materials for RF/microwave application, but nowadays the great challenge for material sciences is increasing temperatures of their magnetic and ferroelectric ordering.

$Tb_{0.95}Bi_{0.05}MnO_3$ is a solid solution of the initial compounds $TbMnO_3$ and $BiMnO_3$ and at room temperature has the rhombic perovskite structure with the space group Pnma (62) and the cell parameters a = 5.321 E, b = 5.858 E, and c = 7.429 E [2].

S. Balandin et al. (Eds.): NEW2AN/ruSMART 2014, LNCS 8638, pp. 444–450, 2014.
© Springer International Publishing Switzerland 2014

$TbMnO_3$ has the structure of a rhombically distorted perovskite and is a multiferroic with ferroelectric and magnetic orders occur near 30 and 40 K respectively [3]. It was demonstrated that $TbMnO_3$ has gigantic magnetoelectric and magnetocapacitance effects, which can be attributed to switching of the electric polarization induced by magnetic fields [herein]. Further theoretical and experimental studies reveal that the ferroelectricity in $TbMnO_3$ and many other cycloidal-spin magnets is induced by the noncollinear Mn spiral spin order with inverse Dzyaloshinskii–Moriya (DM) interaction, which is the driving force of oxygen atom displacements [4]. $BiMnO_3$ is a multiferroic with the ferromagnetic and ferroelectric Curie temperatures $T_C = 105$ and 750–800 K, respectively, and has the monoclinic symmetry with the space group C2 [5].

The interest to the investigations of the solid solution of $TbMnO_3$ and $BiMnO_3$ is caused by the possibility to obtain the multiferroic with close temperatures of magnetic and ferroelectric ordering which are higher than in $TbMnO_3$. Earlier it was shown that in the multiferroic compound, the doping and substituting with different types of magnetic ions could modify the properties of the compound, see, for example [6, 7, 8].

Studies of structural, magnetic and dielectric $Tb_{0.95}Bi_{0.05}MnO_3$ properties revealed the existence of a state with a high dielectric constant ($\varepsilon \sim 10^5$) for temperatures above 165 K in the $Tb_{0.95}Bi_{0.05}MnO_{3 + \delta}$ single crystals [9]. Relying on the obtained dispersion of the dielectric constant and the conductivity below 150 K in $Tb_{0.95}Bi_{0.05}MnO_{3 + \delta}$ the conclusion about the presence of the inhomogeneous state of the crystal and the coexistence of the local domains of dipole correlations of different scales was made. It was shown that external magnetic field effects on the resistivity and capacitance of the crystal in all temperatures range. Thus it was supposed that $Tb_{0.95}Bi_{0.05}MnO_{3 + \delta}$ crystal is an RF multiferroic at temperatures 5–500 K in which restricted domains of various sizes with simultaneous polar and magnetic correlations coexist.

In [10] properties of $Tb_{0.95}Bi_{0.05}MnO_3$ ceramics were investigated. It was shown that Bi occupies the Tb site and alters the multiferroicity of TBMO due to reduction in the exchange interactions J_{Tb-Tb} and J_{Mn-Tb}. The Bi partial substitution in $TbBiMnO_3$ ceramics suppresses the Tb-spin ordering point (T_{Tb}) and ferroelectric ordering point (T_C). Bi-doping significantly depressed the dielectric constant. At low temperature relaxorlike dielectric characteristics of $Tb_{0.95}Bi_{0.05}MnO_3$ can be related to the dipolar effect induced by charge-carrier hopping motions [10].

The aim of the present study was an elucidation of the low temperature ferroelectric and magnetic states in TBMO using atomic-force microscopy (AFM) technique. Single crystal perovskite manganese oxides $Tb_{1-x}Bi_xMnO_3$ with x=0.05 has been studied with magnetic force and atomic force microscopy techniques at low temperatures. The ferroelectric long-range order typical of TMO and the restricted domains with ferromagnetic correlations were revealed in TBMO at T = 8 K.

2 Material and Methods

All measurements were carried out on a TMBO single-crystal grown by the spontaneous crystallization technique, dimensions 2 Ч 1 Ч 0.5 mm3. Directions of crystallographic axes were determined by X-ray diffraction analysis using SuperNova diffractometer

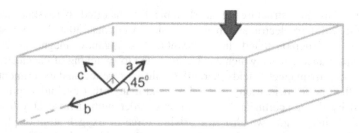

Fig. 1. TBMO crystal orientation. The surface for AFM measurements is pointed with the gray arrow.

(Oxford Diffraction, UK). The orientation of the crystal surface for ferroelectric and magnetic measurements is shown of a Fig.1.

Ferroelectric and magnetic states of the TBMO crystal were measured using the cryogenic atomic-force microscope AttoAFM I (Attocube Systems, Germany) according previously developed procedures of measurements [11]. Both for magnetic and ferroelectric measurements cantilevers with Co coating were taken. The conductivity of the soft magnetic Co coated cantilever ensures the possibility to register magnetic properties with magnetic force microscopy (MFM) and ferroelectric properties of the surface with piezoresponse force microscopy (PFM) at the same place of the sample. We used MAGT cantilevers (Applied NanoStructures Inc., USA), with the resonant frequency of 62 kHz, k constant of 3 N/m and tip radius curvature of 40 nm. Taking into account the low value of TMO polarization (8410^{-4} C·m^{-2} [3] at T ~10 K) comparing with the same value for classic ferroelectrics (for ex. 2.6410^{-2} C·m^{-2} for BaTiO$_3$ at 296 K [herein]), measurements of ferroelectric sample state were done under conditions of the tip-surface local contact resonance. This allowed to enhance the piezoresponse from the surface by the cantilever Q-factor times. All measurements were done in the temperature range of 8 – 30 K.

3   Experimental Results

The results of TMNO crystal measurements with MFM are shown on a Fig. 2 a,b. The distribution of magnetic properties on the TMNO surface revealed the existence of isolated ferromagnetic domains. The distribution of these domains doesn't correlate with the surface topography. The autocorrelation analysis of MFM image was done (Fig. 2c).

The shape of the spatial autocorrelation function corresponded to the isotropic spatial distribution of ferromagnetic properties. On this evidence the absence of the long-range ferromagnetic ordering in TBMO crystal was concluded. The absence of preferred direction in ferromagnetic domain distribution followed from the symmetry of the autocorrelation function peak. Determined autocorrelation length was 750 nm and evidenced of local ferromagnetic ordering in TBMO. The autocorrelation analysis of obtained topographic image didn't reveal any correlation in surface topography of TBMO. This confirmed the fact that ferromagnetic ordering was not caused by topographic features. It should be noticed that the real size of ferromagnetic domains could be differed from that was determined in autocorrelation analysis. It could happen that visualized on MFM image ferromagnetic domains have a fine structure, but due to the finite size of the tip, fine structure wasn't resolved.

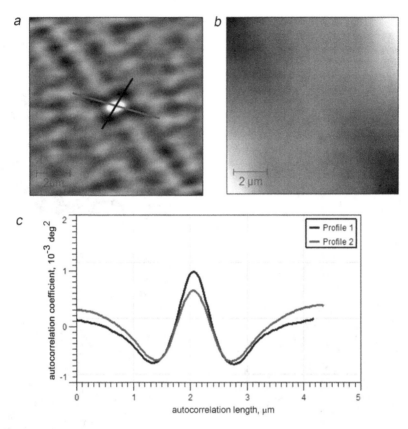

Fig. 2. An autocorrelation analysis of ferromagnetic properties of TMNO crystal. a – autocorrelation of image of ferromagnetic properties distribution, b – autocorrelation of topography, c – profiles from autocorrelation of image of ferromagnetic properties distribution.

According to the results of PFM measurements, polar domains with weak piezoresponse were found (Fig. 3 c, d) on the TBMO surface at low temperatures. Ferroelectric domains had linear shape with 250 nm average thickness and length up to several microns. On some parts of the piezoresponse image it could be seen the union of linear domains resulted in a thickening effect of linear structures. Taking into account the fact that in TMO crystal polarization P is parallel to the crystallographic axe c ($P_c \gg P_a, P_b$, $P_c = 8*10^{-4}$ C/m^2 [3]) it is reasonable to assume that obtained long-range ferroelectric ordering in TBMO originated from the long-range ordering in TMO. In the investigated TBMO crystal the axe c (Fig. 1) had a projection on the surface where out-of-plane PFM measurements were done, thus measured piezoresponse from polar domain structure in TBMO could have polarization in the direction of axe c. It means that the great part of TBMO crystal is mainly consist of TMO while the doped TMO occupied the small part of the crystal and didn't reveal itself on the background of long-range ferroelectric ordering of TMO.

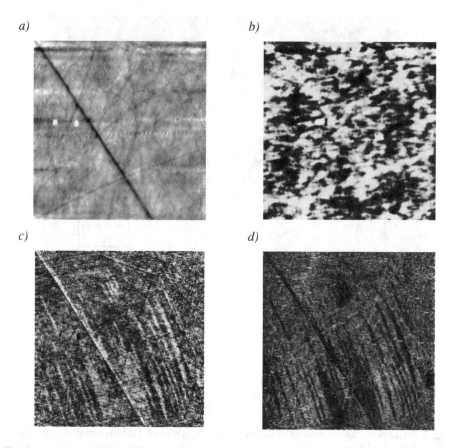

Fig. 3. Results of TBMO single crystal measurements using PFM and MFM techniques in the temperature range of 8 – 30 K. a – topography of TBMO crystal surface; b – ferromagnetic properties distribution; c – out-of-plane amplitude PFM signal from the TBMO; d – out-of-plane phase PFM signal from the TBMO.

A comparison of ferroelectric domain structure (Fig. 3 c, d) with ferromagnetic ordering (Fig. 3 b) at low temperatures reveals the absence of correlation between ferroelectric and ferromagnetic properties of TBMO crystal.

4   Discussion

According to our experimental results it was found that in the temperature range of 8 – 30 K there are local ferromagnetic and long-range ferroelectric ordering in TBMO single crystal. We suppose that the main contribution in the ferromagnetic properties of TBMO crystal arise from the isolated ferromagnetic domains. The initial spiral magnetic ordering of TMO crystal isn't revealed in the background of strong response from ferromagnetic domains. These isolated ferromagnetic domains could be formed in the process of crystal growth. TBMO single crystals were grown by the spontaneous crystallization method. Substitution of the bigger Bi^{3+} ions in Tb^{3+} positions causes local lattice distortions that bring to appearance of the smaller ions Mn^{4+} with changed valence. As a result of the TBMO contains both Mn^{3+} and Mn^{4+} ions, and free charge carries appear in it. Probably, charge carriers arise also due to the hybridization of the lone pair of $6s^2$-electrons of Bi^{3+} ions, 2p-oxygen orbitals and 3d-orbitals of Mn^{3+} ions, because their energies are close (see e.g. [12]), i.e. $Bi^{3+} = Bi^{5+} + 2 e_g$ and $2 e_g + 2 Mn^{4+} = 2Mn^{3+}$. Another explanation of ferromagnetic domain formation could be formation of the conducting ferromagnetic domains inside the volume of TMO due to phase separation. The charge carriers and ferromagnetic Mn^{3+} - Mn^{4+} ion pairs originating from doping are not distributed in TBMO statistically. Due to the double exchange interaction resulting in the phase separation [13] they form conducting ferromagnetic domains inside the volume of TMO.

As noted above, at low temperatures these domains occupy the small volume of TMO. But in accordance with our results it would be consider that isolated ferromagnetic domains have sufficiently large sizes to give rise to a long-range ferromagnetic ordering inside them. However, no long-range ordering occurs in the entire crystal.

The ferroelectric domains structure of TBMO crystal corresponds to the ferroelectric ordering in TMO crystal. It proves that the main volume of TBMO is occupied by the initial TMO crystal. Ferroelectric long-range ordering from the structure of TMO crystal isn't disturbed by doped parts of the TBMO crystal. Long range ferroelectric ordering doesn't correlate with local ferromagnetic ordering in TBMO crystal.

Acknowledgements. This work is supported by the Ministry of Education and Science Program "5-100-2020".

References

1. Sun, N.X., Liu, M.: Voltage Control of Magnetism in Multiferroic Heterostructures. Phil. Trans. R.Soc. A 372, 20120439 (2014)
2. Golovenchits, E.I., Sanina, V.A.: Dielectric and Magnetic Properties of the Multiferroic $Tb_{(1-x)}Bi_xMnO_3$: Electric Dipole Glass and Self-Organization of Charge Carries. JETP Lett. 81(10), 509–513 (2005)
3. Kimura, T., Goto, T., Shintani, H., Ishizaka, K., Arima, T., Tokura, Y.: Magnetic Control of Ferroelectric Polarization. Nature 426(6), 55–58 (2003)
4. Hur, N., Park, S., Sharma, P.A., Ahn, J.S., Guba, S., Cheong, S.-W.: Electric Polarization Reversal and Memory in a Multiferroic Material Induced by Magnetic Fields. Nature 392, 392–395 (2004)
5. Kimura, T., Kawamoto, S., Yamada, I., Azuma, M., Takano, M., Tokura, Y.: Magnetocapacitance Effect in Multiferroic $BiMnO_3$. Phys. Rev. B 67, 180401(R) (2003)
6. Cui, Y.: Decrease of Loss in Dielectric Properties of $TbMnO_3$ by Adding TiO_2. Physica B: Condensed Matter 403(18), 2963–2966 (2008)

7. Yang, C.C., Chung, M.K., Li, W.-H., Chan, T.S., Liu, R.S., Lien, Y.H., Huang, C.Y., Chan, Y.Y., Yao, Y.D., Lynn, J.W.: Magnetic Instability and Oxigen Defiency in Na-doped $TbMnO_3$. Phys. Rev. B. 74, 094409 (2006)

8. Mufti, N., Nugroho, A.A., Blake, G.R., Palstra, T.T.M.: Relaxor Ferroelectric Behaviour in Ca-doped $TbMnO_3$. Phys. Rev. B. 78, 024109 (2008)

9. Sanina, V.A., Golovenchits, E.I.: Magnetic-Field-Induced Phase Transition in $Tb_{0.95}Bi_{0.05}MnO_{3+\delta}$ Multiferroic. JETP Letters 84(4), 190–194 (2006)

10. Zhang, C., Yan, H., Wang, X., Kang, D., Li, L., Lu, X., Zhu, J.: Effect of A-site Bi-doping on the Megnetic and Electrical Properties in $TbMnO_3$. Materials Letters 111, 147–149 (2013)

11. Andreeva, N.V., Tyunina, M., Filimonov, A.V., Rudskoy, A.I., Pertsev, N.A., Vakhrushev, S.B.: Low-Temperature Evolution of Local Polarization Properties of $PbZr_{0.65}Ti_{0.35}O_3$ Thin Films Probed by Piezoresponse Force Microscopy. Appl. Phys. Lett. 104, 112905 (2014)

12. Wang, K.F., Liu, J.-M., Ren, Z.F.: Multiferroicity: the Coupling between Magnetic and Polarization Orders. Adv. Phys. 58(4), 321–448 (2009)

13. Kagan, M.Y., Kugel, K.I.: Inhomogeneous Charge Distributions and Phase Separation in Manganites. PHYS-USP 44(6), 553–570 (2001)

Reflectivity Properties of Graphene and Graphene-Coated Substrates

Galina L. Klimchitskaya[1,2], Vladimir M. Mostepanenko[1,2], and Viktor M. Petrov[2]

[1] Central Astronomical Observatory at Pulkovo of the Russian Academy of Sciences, St. Petersburg, 196140, Russia
[2] Institute of Physics, Nanotechnology and Telecommunications, St. Petersburg, State Polytechnical University, St. Petersburg, 195251, Russia

Abstract. The reflectivity properties of graphene and graphene-coated substrates, as prospective materials for nanocommunications, are calculated using the formalism of the polarization tensor. Simple analytic expressions for the transverse magnetic and transverse electric reflection coefficients are obtained at zero temperature and for sufficiently high frequencies at any temperature. The previously known results for the transverse magnetic case are reproduced. The transverse electric coefficients of graphene and graphene-coated plates are shown to depend on the angle of incidence.

Keywords: Graphene, polarization tensor, reflection coefficients, nanocommunications.

1 Introduction

During the last few years carbon nanostructures, such as graphene, carbon nanotubes and fullerenes, find diverse applications in nanotechnology due to their unique electrical, mechanical and optical properties [1]. Specifically, they can be used as components of high-frequency electronics. For one example, radio receivers based on a single carbon nanotube have been demonstrated [2, 3]. Because of this, it is important to investigate the physical properties of these nanostructures which are needed for their use in electromagnetic communications, e.g., the reflectances of incident electromagnetic waves.

In this work, we find the reflection coefficients of graphene and graphene-coated substrates for both independent polarizations of the electromagnetic field (transverse magnetic and transverse electric) at zero and nonzero temperature. Graphene is a two-dimensional sheet of carbon atoms having a hexagonal structure. At low energies, quasiparticles in graphene are described by the massless Dirac equation, but move with a Fermi velocity, rather than the velocity of light [4, 5]. Previously the transverse magnetic reflection coefficients on graphene and graphene-coated substrates were expressed via the longitudinal conductivity in the far-infrared and visible regions of the spectrum [6, 7]. Later both the transverse magnetic and transverse electric coefficients were found in terms of the longitudinal and transverse conductivities of graphene (or, equivalently, respective

S. Balandin et al. (Eds.): NEW2AN/ruSMART 2014, LNCS 8638, pp. 451–458, 2014.

density-density correlation functions), but it was underlined that the transverse versions are not available in the literature [8].

Recently, the longitudinal, as well as transverse, conductivities and respective density-density correlation functions were expressed [9] via the components of the polarization tensor of graphene in (2+1)-dimensional space-time. In so doing, explicit expressions for the polarization tensor have been used found in the literature at zero [10] and nonzero [11] temperature. Furthermore, the transverse magnetic and transverse electric coefficients for graphene-coated substrates were directly expressed in terms of the polarization tensor and the dielectric permittivity of substrate material [12]. It is pertinent to note that papers [8–12] are devoted to investigation of the Casimir interaction [13] between two graphene sheets, a graphene sheet and a material plate, or a graphene-coated plate and a sphere. For this reason, the reflection coefficients along the imaginary rather than the real frequency axis have been a major focus of attention. This also means that the wave vector and the frequency were considered as independent quantities and the mass-shell equation was not satisfied [8–12]. Finally, the experimental data of the first measurement of the Casimir interaction in graphene system [14] were found [12, 15] to be in a very good agreement with theory.

Below we use the results of Refs. [11, 12], but apply them to ordinary electromagnetic waves satisfying the mass-shell equation. At some conditions, we express both the transverse magnetic and transverse electric reflection coefficients on graphene and graphene-coated substrates in terms of the polarization tensor. The obtained simple analytic expressions hold at zero temperature. They are also valid at nonzero temperature provided that the wave frequency is sufficiently high. We reproduce some of the results obtained previously [7] for the transverse magnetic reflection coefficient. As to the transverse electric reflection coefficient, which was not explicitly calculated in Refs. [6, 7], we show that it is equal to the transverse magnetic one only at the normal incidence and depends on the incidence angle.

2   Reflection Coefficients for Graphene in Terms of the Polarization Tensor

We consider the electromagnetic wave of frequency ω incident under the angle θ_i to the normal of the graphene sheet kept at a temperature T in thermal equilibrium with an environment. In the following, it is assumed that the graphene under consideration is gapless and undoped. We suggest that our graphene has zero chemical potential and the applicability conditions of the Dirac model are satisfied (the generalization for the case of nonzero mass-gap parameter and chemical potential is straightforward). For the transverse magnetic (TM) or, in the other notation, p-polarized wave, the magnetic field is perpendicular to the plane containing the incident, reflected, and refracted waves. For the transverse electric (TE), or s-polarized wave, the electric field is perpendicular to this plane.

The amplitude reflection coefficients for graphene along the imaginary frequency axis $\xi = -i\omega$ were expressed in terms of the polarization tensor Π_{kl} in

(2+1)-dimensional space-time [9, 11, 16]

$$r_{\rm TM}(i\xi, k_\perp) = \frac{q\Pi_{00}(i\xi, k_\perp)}{2\hbar k_\perp^2 + q\Pi_{00}(i\xi, k_\perp)}, \tag{1}$$

$$r_{\rm TE}(i\xi, k_\perp) = -\frac{k_\perp^2\, \Pi_{\rm tr}(i\xi, k_\perp) - q^2\Pi_{00}(i\xi, k_\perp)}{2\hbar k_\perp^2 q + k_\perp^2\, \Pi_{\rm tr}(i\xi, k_\perp) - q^2\Pi_{00}(i\xi, k_\perp)}.$$

Here, $k_\perp = |\mathbf{k}_\perp|$ is the magnitude of the projection of the wave vector on the plane of graphene independent on the frequency ξ and

$$q \equiv q(\xi, k_\perp) = \left(k_\perp^2 + \frac{\xi}{c^2}\right)^{1/2}. \tag{2}$$

The 00-component of the polarization tensor is given by

$$\Pi_{00}(i\xi, k_\perp) = \frac{\pi\hbar\alpha k_\perp^2}{f} + \frac{8\hbar\alpha c^2}{v_F^2}\int_0^1 dx \left[\frac{k_B T}{\hbar c}\ln\left(1 + 2\cos\frac{\hbar\xi x}{k_B T}e^{-\Theta_T} + e^{-2\Theta_T}\right)\right.$$

$$\left. -\frac{\xi}{2c}(1-2x)\frac{\sin\frac{\hbar\xi x}{k_B T}}{\cosh\Theta_T + \cos\frac{\hbar\xi x}{k_B T}} + \frac{\xi^2\sqrt{x(1-x)}}{c^2 f}\frac{\cos\frac{\hbar\xi x}{k_B T} + e^{-\Theta_T}}{\cosh\Theta_T + \cos\frac{\hbar\xi x}{k_B T}}\right], \tag{3}$$

where $\alpha = e^2/(\hbar c)$ is the fine structure constant, $v_F = 8.74 \times 10^5\,{\rm m/s}$ [17, 18] is the Fermi velocity in graphene, and the following notations are introduced:

$$f \equiv f(\xi, k_\perp) = \left(\frac{v_F^2}{c^2}k_\perp^2 + \frac{\xi^2}{c^2}\right)^{1/2}, \tag{4}$$

$$\Theta_T \equiv \Theta_T(\xi, k_\perp, x) \equiv \frac{\hbar c}{k_B T}f(\xi, k_\perp)\sqrt{x(1-x)}.$$

The quantity $\Pi_{\rm tr}$ in Eq. (1) denotes the sum of spatial components Π_1^1 and Π_2^2. For the undoped gapless graphene with zero chemical potential it is given by [9, 11, 16]

$$\Pi_{\rm tr}(i\xi, k_\perp) = \Pi_{00}(i\xi, k_\perp) + \frac{\pi\hbar\alpha}{f}\left(f^2 + \frac{\xi^2}{c^2}\right) \tag{5}$$

$$+8\hbar\alpha\int_0^1 dx\left[\frac{\xi}{c}(1-2x)\frac{\sin\frac{\hbar\xi x}{k_B T}}{\cosh\Theta_T + \cos\frac{\hbar\xi x}{k_B T}}\right.$$

$$\left. -\frac{\sqrt{x(1-x)}}{f}\left(f^2 + \frac{\xi^2}{c^2}\right)\frac{\cos\frac{\hbar\xi x}{k_B T} + e^{-\Theta_T}}{\cosh\Theta_T + \cos\frac{\hbar\xi x}{k_B T}}\right].$$

It is easily seen [9, 11, 16] that under the condition

$$\frac{\hbar|\xi|}{k_B T} \gg 1 \tag{6}$$

all terms under the integrals in Eqs. (3) and (5) become exponentially small. Then one obtains much simpler expressions for the polarization tensor

$$\Pi_{00}(i\xi, k_\perp) = \frac{\pi \hbar \alpha k_\perp^2}{f}, \qquad \Pi_{\mathrm{tr}}(i\xi, k_\perp) = \frac{\pi \hbar \alpha (q^2 + f^2)}{f}. \tag{7}$$

At zero temperature, $T = 0\,\mathrm{K}$, these expressions are in fact the exact ones. At $T > 0\,\mathrm{K}$, Eq. (7) is the approximate one and can be used at sufficiently high frequencies satisfying Eq. (6).

Now we return to the real frequencies ω by putting $\omega = i\xi$ and consider the ordinary, rather than the fluctuating, electromagnetic fields satisfying the mass-shell equation

$$|k_\perp| = \frac{\omega}{c} \sin\theta_i. \tag{8}$$

Substituting this in Eq. (7), we obtain

$$\Pi_{00}(\omega, \theta_i) = \frac{i\pi\hbar\alpha\omega \sin^2\theta_i}{c\left(1 - \frac{v_F^2}{c^2}\sin^2\theta_i\right)^{1/2}} \approx i\pi\hbar\alpha\frac{\omega}{c}\sin^2\theta_i, \tag{9}$$

$$\Pi_{\mathrm{tr}}(\omega, \theta_i) = -\frac{i\pi\hbar\alpha\omega}{c\left(1 - \frac{v_F^2}{c^2}\sin^2\theta_i\right)^{1/2}} \left(\cos^2\theta_i + 1 - \frac{v_F^2}{c^2}\sin^2\theta_i\right)$$

$$\approx -\frac{i\pi\hbar\alpha\omega}{c}(\cos^2\theta_i + 1).$$

Note that in the last transformation we have neglected by the quantity of the order of 10^{-5}, as compared with unity. Using Eqs. (8) and (9), the following combination entering the transverse electric reflection coefficient in Eq. (1) takes the form

$$k_\perp^2 \Pi_{\mathrm{tr}} - q^2 \Pi_{00} = -\frac{i\omega^3 \pi \hbar \alpha}{c^3} \sin^2\theta_i. \tag{10}$$

Substituting Eqs. (9) and (10) in Eq. (1) considered along the real frequency axis, we finally obtain

$$r_{\mathrm{TM}}(\omega, \theta_i) = \frac{\pi\alpha \cos\theta_i}{2 + \pi\alpha \cos\theta_i}, \qquad r_{\mathrm{TE}}(\omega, \theta_i) = -\frac{\pi\alpha}{2\cos\theta_i + \pi\alpha}. \tag{11}$$

As can be seen in Eq. (11), both reflection coefficients do not depend on the frequency. This property is valid over the application region of Eq. (11), i.e., at $T = 0\,\mathrm{K}$ and at any T for sufficiently high frequencies satisfying the condition $\omega \gg \omega_T \equiv k_B T/\hbar$. For example, at room temperature $T = 300\,\mathrm{K}$ the thermal frequency $\omega_T \approx 3.9 \times 10^{13}\,\mathrm{rad/s}$. Thus, at room temperature Eq. (11) is well applicable to the optical frequencies. At lower temperatures Eq. (11) is applicable starting from much smaller frequencies.

Using Eq. (11), the reflectances are given by

$$|r_{\mathrm{TM}}(\theta_i)|^2 = \frac{\pi^2\alpha^2 \cos^2\theta_i}{(2 + \pi\alpha \cos\theta_i)^2}, \qquad |r_{\mathrm{TE}}(\theta_i)|^2 = \frac{\pi^2\alpha^2}{(2\cos\theta_i + \pi\alpha)^2}. \tag{12}$$

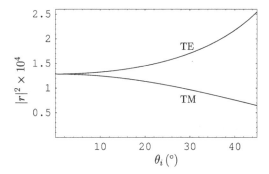

Fig. 1. Reflectances of the transverse magnetic and transverse electric electromagnetic waves of optical frequencies on graphene are shown as functions of the angle of incidence by the lower and upper lines, respectively.

In Fig. 1, we plot the reflectances multiplied by the factor 10^4 as functions of the incidence angle. As is seen in Fig. 1, at the normal incidence ($\theta_i = 0°$) the transverse magnetic and transverse electric reflectances take one and the same value

$$|r_{\mathrm{TM}}(0)|^2 = |r_{\mathrm{TE}}(0)|^2 = \frac{\pi^2\alpha^2}{(2+\pi\alpha)^2} \approx 1.28 \times 10^{-4}. \tag{13}$$

This is in agreement with the result of Ref. [7] obtained for a visible region of spectrum. However, with increasing angle of incidence, the trasverse magnetic and transverse electric reflectances behave differently. The quantity $|r_{\mathrm{TM}}|^2$ is a monotonously decreasing function of θ_i, whereas $|r_{\mathrm{TE}}|^2$ monotonously increases with the increase of θ_i. The latter is in disagreement with the qualitative conclusion [7] made with no explicit calculation that $|r_{\mathrm{TE}}(\theta_i)|^2 = |r_{\mathrm{TM}}(0)|^2$ in the framework of the Dirac model.

3   Reflection Coefficients for Graphene Deposited on a Substrate

In this section, we consider the reflection of the electromagnetic waves incident on a graphene sheet covering a material plate. In below equations it is assumed that the plate is infinitely thick. For dielectric materials at the optical frequencies this model works good for plates with thicknesses of about $2\,\mu\mathrm{m}$ and more (the presented results can be easily generalized for plates of smaller thicknesses using the formalism of the Lifshitz theory for layered structures [13]).

The reflectivity properties of graphene are described by the polarization tensor discussed in Sec. 2. The plate material is characterized by the frequency-dependent dielectric permittivity $\varepsilon(\omega)$. On the imaginary frequency axis, the reflection coefficients for a graphene-coated plate are given by [12]

$$r_{\text{TM}}(i\xi, k_\perp) = \frac{\varepsilon q + k\left(\frac{q}{\hbar k_\perp^2}\Pi_{00} - 1\right)}{\varepsilon q + k\left(\frac{q}{\hbar k_\perp^2}\Pi_{00} + 1\right)}, \tag{14}$$

$$r_{\text{TE}}(i\xi, k_\perp) = \frac{q - k - \frac{1}{\hbar k^2}\left(k_\perp^2 \Pi_{\text{tr}} - q^2\Pi_{00}\right)}{q + k + \frac{1}{\hbar k^2}\left(k_\perp^2 \Pi_{\text{tr}} - q^2\Pi_{00}\right)},$$

where $\varepsilon \equiv \varepsilon(i\xi)$ and the following notation is introduced:

$$k = \left[k_\perp^2 + \varepsilon(i\xi)\frac{\xi^2}{c^2}\right]^{1/2}. \tag{15}$$

Now we turn to the real frequency axis in Eq. (15) using Eqs. (8)–(10), and obtain

$$r_{\text{TM}}(\omega, \theta_i) = \frac{\varepsilon(\omega)\cos\theta_i + \sqrt{\varepsilon(\omega) - \sin^2\theta_i}\,(\pi\alpha\cos\theta_i - 1)}{\varepsilon(\omega)\cos\theta_i + \sqrt{\varepsilon(\omega) - \sin^2\theta_i}\,(\pi\alpha\cos\theta_i + 1)},$$

$$r_{\text{TE}}(\omega, \theta_i) = -\frac{\sqrt{\varepsilon(\omega) - \sin^2\theta_i} - \cos\theta_i + \pi\alpha}{\sqrt{\varepsilon(\omega) - \sin^2\theta_i} + \cos\theta_i + \pi\alpha}. \tag{16}$$

It is seen that from Eq. (16) we return back to Eq. (11) in the case $\varepsilon(\omega) = 1$, i.e., in the absence of a substrate. The obtained expressions are again applicable either at $T = 0\,\text{K}$ or at any T at sufficiently high frequencies $\omega \gg \omega_T$. In the case of graphene deposited on a substrate, both reflection coefficients depend on the angle of incidence and on the wave frequency. The latter dependence, however, is irrelevant to graphene. It arises from the frequency dependence of the dielectric permittivity of a substrate. For the normal incidence ($\theta_i = 0$) one has

$$r_{\text{TM}}(\omega, 0) = |r_{\text{TE}}(\omega, 0)| = \frac{\sqrt{\varepsilon(\omega)} - 1 + \pi\alpha}{\sqrt{\varepsilon(\omega)} + 1 + \pi\alpha}. \tag{17}$$

From this it is seen that the impact of graphene coating on the reflectance increases with decreasing dielectric permittivity of the plate material. Really, the difference between the reflection coefficients at the normal incidence with and without graphene coating is equal to

$$\Delta r_{\text{TM}}(\omega, 0) = \frac{2\pi\alpha}{(\sqrt{\varepsilon(\omega)} + 1)(\sqrt{\varepsilon(\omega)} + 1 + \pi\alpha)}, \tag{18}$$

which increases with the decrease of ε.

In Fig. 2, we present the the transverse magnetic and transverse electric reflectances at $\omega = 1\,\text{eV} = 1.52 \times 10^{15}\,\text{rad/s}$ as functions of the incidence angle for a graphene sheet deposited on a SiO$_2$ substrate. The value $\varepsilon = 2$ for the

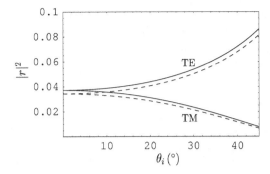

Fig. 2. Reflectances of the transverse magnetic and transverse electric electromagnetic waves of optical frequencies on a graphene-coated SiO$_2$ plate are shown as functions of the angle of incidence by the lower and upper solid lines, respectively. The dashed lines provide the same information for a uncoated SiO$_2$ plate.

dielectric permittivity of SiO$_2$ at a chosen frequency has been used in computations. The dashed lines show the respective reflectances for a SiO$_2$ plate with no graphene coating. For the normal incidence one has

$$|r_{\mathrm{TM}}(0)|^2 = |r_{\mathrm{TE}}(0)|^2 \approx 0.037 \tag{19}$$

in the presence of a graphene coating. The same quantity is equal to 0.034 for a uncoated SiO$_2$ plate. Our results at the normal incidence are in agreement with Ref. [7]. We have shown, however, that not only the transverse magnetic, but also the transverse electric reflectance depends on the angle of incidence similar to that of a freestanding graphene sheet considered in Sec. 2.

4   Conclusions and Discussion

In the foregoing, we have calculated the reflectivity properties of graphene and graphene-coated substrates using the formalism of the polarization tensor in (2+1)-dimensional space-time developed previously for applications to the Casimir effect. This allowed to obtain not only the transverse magnetic reflection coefficients, considered previously in the literature in terms of the longitudinal conductivity of graphene, but the transverse electric ones as well. We have found simple analytic expressions for both types of coefficients for a freestanding graphene and for graphene-coated material plates which are valid at zero temperature and for sufficiently high frequencies at any temperature. At normal incidence, our results are in agreement with those obtained previously in the literature. We have shown, however, that the transverse electric reflectances on both a freestanding graphene and graphene-coated substrates depend on the angle of incidence. This is in disagreement with what was qualitatively discussed in previous literature in the absence of explicit calculation. The obtained results are important for researchers working in nanocommunications.

In future it is interesting to find the reflection coefficients on graphene in terms of the polarization tensor at arbitrary temperature and any real frequency. This will be helpful for application of graphene and other carbon nanostructures in nanoscale networking and communications.

References

1. Dresselhaus, M.S.: On the past and Present of Carbon Nanostructures. Phys. Status Solidi B 248, 1566–1574 (2011)
2. Rutherglen, C., Burke, P.: Carbon Nanotube Radio. Nano Lett. 7, 3296–3299 (2007)
3. Jensen, K., Weldon, J., Garcia, H., Zettl, A.: Nanotube Radio. Nano Lett. 7, 3508–3511 (2007)
4. Castro Neto, A.H., Guinea, F., Peres, N.M.R., Novoselov, K.S., Geim, A.K.: The Electronic Properties of Graphene. Rev. Mod. Phys. 81, 109–162 (2009)
5. Katsnelson, M.I.: Graphene: Carbon in Two Dimensions. Cambridge University Press, Cambridge (2012)
6. Falkovsky, L.A., Pershoguba, S.S.: Optical Far-Infrared Properties of a Graphene Monolayer and Multylayer. Phys. Rev. B 76, 153410-1–4 (2007)
7. Stauber, T., Peres, N.M.R., Geim, A.K.: Optical Conductivity of Graphene in the Visible Region of the Spectrum. Phys. Rev. B 78, 085432-1–8 (2008)
8. Sernelius, B.E.: Retarded Interactions in Graphene Systems. Phys. Rev. B 85, 195427-1–10 (2012)
9. Klimchitskaya, G.L., Mostepanenko, V.M., Sernelius, B.E.: Two Approaches for Describing the Casimir Interaction in Graphene: Density-Density Correlation Function Versus Polarization Tensor. Phys. Rev. B 89, 125407-1–9 (2014)
10. Bordag, M., Fialkovsky, I.V., Gitman, D.M., Vassilevich, D.V.: Casimir Interaction Between a Perfect Conductor and Graphene Described by the Dirac Model. Phys. Rev. B 80, 245406-1–5 (2009)
11. Fialkovsky, I.V., Marachevsky, V.N., Vassilevich, D.V.: Finite-Temperature Casimir Effect for Graphene. Phys. Rev. B 84, 035446-1–10 (2011)
12. Klimchitskaya, G.L., Mohideen, U., Mostepanenko, V.M.: Theory of Casimir Interaction from Graphene-Coated Substrates Using the Polarization Tensor and Comparison with Experiment. Phys. Rev. B 89, 115419-1–8 (2014)
13. Bordag, M., Klimchitskaya, G.L., Mohideen, U., Mostepanenko, V.M.: Advances in the Casimir Effect. Oxford University Press, Oxford (2009)
14. Banishev, A.A., Wen, H., Xu, J., Kawakami, R.K., Klimchitskaya, G.L., Mostepanenko, V.M., Mohideen, U.: Measuring the Casimir Force Gradient from Graphene on a SiO$_2$ Substrate. Phys. Rev. B 87, 205433-1–5 (2013)
15. Klimchitskaya, G.L., Mostepanenko, V.M.: Observability of Thermal Effects in the Casimir Interaction from Graphene-Coated Substrates. Phys. Rev. A 89, 052512-1–7 (2014)
16. Chaichian, M., Klimchitskaya, G.L., Mostepanenko, V.M., Tureany, A.: Thermal Casimir-Polder Interaction of Different Atoms with Graphene. Phys. Rev. A 86, 012515-1–9 (2012)
17. Wunsch, B., Stauber, T., Sols, F., Guinea, F.: Dynamical Polarization of Graphene at Finite Doping. New J. Phys. 8, 318-1–16 (2006)
18. Peres, N.M.R., Guinea, F., Castro Neto, A.H.: Electronic Properties of Disordered Two-Dimensional Carbon. Phys. Rev. B 73, 125411-1–23 (2006)

Nanoporous Glasses with Magnetic Properties as a Base of High-Frequency Multifunctional Device Making

Alexander Naberezhnov[1,2], Andrey I. Rudskoy[1], Igor Golosovsky[4],
Viktor Nizhankovskii[3], Alexey Vl. Filimonov[1], and Bernard Nacke[5]

[1] St. Petersburg State Polytechnical University, 195251,
Polytechnicheskaya 29, St.-Petersburg, Russia
`filimonov@rphf.spbstu.ru`
[2] Ioffe Physico-Technical Institute, 194021, Polytechnicheskaya 26, St.-Petersburg, Russia
`alex.nabereznov@mail.ioffe.ru`
[3] International Laboratory of High Magnetic Fields and Low Temperatures,
Gajowicka 95, 53-421 Wroclaw, Poland
`nizhan@alpha.ml.pan.wroc.pl`
[4] Petersburg Nuclear Physics Institute, Gatchina, Leningrad District, 188300, Russia
`golosov@pnpi.spb.ru`
[5] Leibniz University of Hannover, ETP, 4 Wilhelm-Busch-Str., 30167 Hannover, Germany
`nacke@etp.uni-hannover.de`

Abstract. Two types (microporous - Fe20-MIP and macroporous Fe20-MAP) porous magnetic matrices were prepared on base of Fe20 (60% SiO_2–15% B_2O_3–5% Na_2O–20% Fe_2O_3) alkali borosilicate glass. The average pore diameters were 5 nm for Fe20-MIP and 5 and 50 nm (bimodal distribution) for Fe20-MAP. The morphology, magnetic properties and coefficients of volume and linear magnetostriction have been studied for Fe20-MIP glass.

Keywords: Nanoporous matrices, ferriferous glasses, iron oxides, magnetic properties, magnetostriction.

1   Introduction

At present time multifunctional materials (especially nanocomposites) attract the steadfast attraction not only from the point of view of fundamental science but for real application, for example as model materials for nanoferronics [1]. There are the different ways of nanocomposite material (NCM) production and one of them is the embedding of substances into the pores of artificial or/and natural dielectric matrices[2,3]. NCM with ferroelectric, magnetic and metal materials demonstrate very winning for practical application macroscopic properties: giant dielectric permittivity at low frequencies [4,5], large ion mobility [6-9], suppression of ferroelectric and magnetic ordering in the surface layers [10,11], increasing of critical magnetic fields in nanostructured superconductors [12,13] et cetera. Usually the porous matrix plays a passive role - it forms the conditions of nanoconfinement and does not play any role in a modification of NCM physical properties. We propose the

S. Balandin et al. (Eds.): NEW2AN/ruSMART 2014, LNCS 8638, pp. 459–466, 2014.

absolutely new approach to preparation of NCM with coexisting ferroelectric and magnetic properties at room temperature on base of porous magnetic glasses filled by ferroelectric materials.

2   Samples and Measurement Procedure

The microporous (Fe20-MIP) and macroporous (Fe20-MAP) ferriferous matrices were produced in Grebenshchikov Institute of Silicate Chemistry (Russian Academy of Sciences) by one- and two-stages etching of two-phase ferriferous glasses [14-16] with the initial (before melting) mixture 60 wt. % SiO_2–15 wt. % B_2O_3—5 wt. % Na_2O–20 wt. % Fe_2O_3 and were kindly given for study of their morphology and magnetic properties. The average pore diameter in Fe20-MIP sample was $D = 5 \pm 2$ nm, and the total porosity was $W \sim 15\%$; in the case of the Fe20-MAP sample the pores had a bimodal structure with $D1 \sim 5$ nm, $D2 \sim 50$ nm and $W \sim 60\%$ [15]. X-rays diffraction studies have been shown that at melting of this mixture hematite (α-Fe_2O_3) transforms into magnetite (Fe_3O_4) and there is no additional admixture of other iron oxides [15]. The diffraction sizes of magnetite particles were 168 ± 7 Å in Fe20-MIP and 180 ± 5 Å in Fe20-MAP glasses. For studies of morphology and surface magnetic properties we have used atomic-force microscope (AFM) Attotcube AFM1 including magnetic force microscopy (MCM) mode. The samples were the rectangular polished plates $10 \times 10 \times 1$ mm^3 and $4,8 \times 4,8 \times 0,54$ mm^3 for magnetostriction measurements. The magnetic properties have been studied by SQUID and the vibration magnetometer. The linear and volume magnetostriction coefficients were measured in the special dilatometer cell [17].

3   Results and Discussions

3.1   Morphology

In Fig. 1 the topography of Fe20-MIP (a), Fe20-MAP (b) and two-phase nonporous Fe20 (c) glasses are presented.

In contrast to two-phase nonporous Fe20 glass on surface of porous glasses there are some "defects" on place of ferriferous agglomerates (large spots in Fig.1a and Fig. 1b) due to partial (Fe20-MIP) or practically total etching (Fe20-MAP) ferriferous phase. Moreover for Fe20-MIP and Fe20-MAP we have observed the formation of branching dendrite system (or net) of channels (Fig. 2). The chemical compositions of Fe20-MIP and Fe20-MAP after one- or two-stage etching are presented in Table 1 [15]. So one can conclude that the most part of iron oxide is located in channels and at formation of porous glasses it is etched out of channel net.

The remnant iron oxide concentrates in the large agglomerates with visible lateral size (according to ACM data) $\sim 450 \pm 10$ nm. This lateral size is essentially larger than characteristic size of Fe_3O_4 nanoparticles determined from X-ray diffraction for Fe20-MIP (168 ± 7 Å) and Fe20-MAP (180 ± 5 Å) glasses [15].

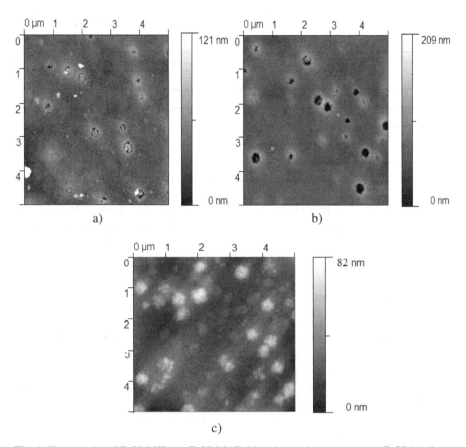

Fig. 1. Topography of Fe20-MIP (a), Fe20-MAP (b) and two-phase nonporous Fe20 (c) glasses

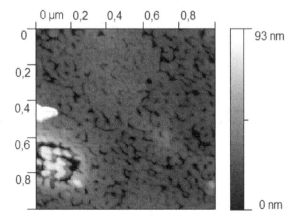

Fig. 2. The ACM image of Fe20-MIP surface with the size 1μm×1mμ

Table 1. Chemical compositions of Fe20-MIP and Fe20-MAP glasses according to the results of chemical analytical analysis – from the paper [15]

substance	quantity, wt. %	
	Fe20-MIP	Fe20-MAP
Na_2O	0.5	1.0
B_2O_3	3.9	5.8
SiO_2	86.8	86.8
Fe_2O_3	8.6	5.9
K_2O	0.2	0.5

Really it means that at melting and chemical phase separation of initial Fe20 glasses a self-organization of magnetite nanoparticles in the large agglomerates takes place. Earlier the magnetic properties of these agglomerates for nonporous Fe20 were confirmed by magnetic force microscopy measurements [18].

3.2   Magnetic Properties

In spite of etching of the most part of Fe_3O_4 the both glasses (Fe20-MIP and Fe20-MAP) demonstrate the magnetic properties. In Fig. 3 the field dependence of magnetization at room temperature for Fe20-MAP containing 5.9 wt. % of iron oxides is presented. It is easy to see the appearance of field hysteresis loop, the saturation in M(H) achieves at H ~ 3000 Oe, the coercive field is equal to 346 (11) Oe. The obtained dependence M(H) is typical for system of superparamagnetic particles and the hysterisis loop indicates that there are the large particles in this system. It coincides with our ACM data.

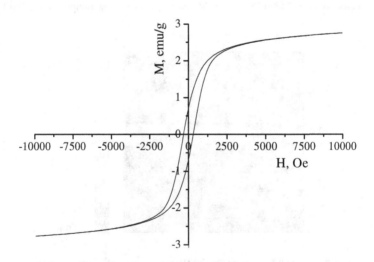

Fig. 3. Field dependence of Fe20-MAP magnetization

For confirmation of superparamagnetic properties of these glasses we have carried out the measurements of temperature dependence of magnetization by SQUID in two regimes (Fig. 4): initially the samples was cooled in the zero magnetic field down to 5 K, then it was heated in applied field 100 Oe (ZFC- branch) up to 300 K and was cooled at 100 Oe down to 5 K (FC-branch). At high temperature (~ 300 K) the curves FC and ZFC coincide practically. It means that above ~300 K all magnetic particles are in so-called "unblocked" state typical for systems of Stoner-Wohlfarth`s [19] magnetic particles. The vertical arrows indicate the anomalies in temperature dependences M(T) at ZFC and FC regimes. Their positions ( ~ 120 K) are in a good agreement with Verwey`s transition in the magnetite. These experimental facts can be considered as an additional argument in favour of magnetite presence in the matrix skeleton.

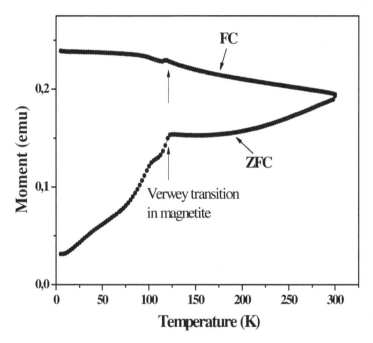

Fig. 4. Temperature dependences of magnetization of Fe20-MIP glass. ZFC (zero field cooling) – heating in applied field 100 Oe after initial cooling in zero field. FC (field cooling) – cooling in applied field 100 Oe. The arrows indicate the anomalies corresponding to Verwey`s transition in the bulk magnetite

3.3   Magnetostriction in Fe20-MIP Glasses

The field dependences of magnetostriction coefficients we have studied using the special measurement cell [17] at fields applied parallel and perpendicular sample surface. As sample we have used Fe20-MIP glass with embedded into the pores KNO_3 from melted state under pressure. The measurements have carried out at 4.2 K and in applied fields up to 14 T. The obtained results for longitudinal and perpendicular components of relative expansions are presented in Fig. 5 (the upper and the middle parts), the bottom part – calculated coefficient of volume magnetostriction.

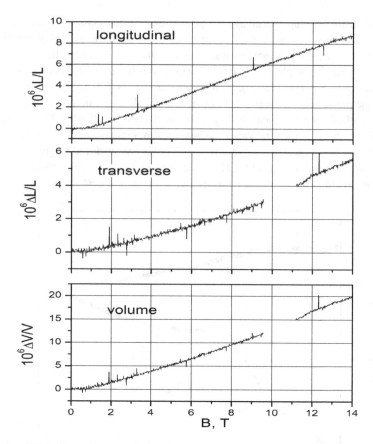

Fig. 5. Field dependences of longitudinal, transverse and volume coefficients of magnetostriction for nanocomposite Fe20-MIP+KNO$_3$

At low field (below 1 T), where the saturation of magnetization does not reach (Fig. 3), the manetostriction coefficients are very small. But at field increasing above 1 T (the region > 10000 Oe in Fig. 3) these coefficients depend on linearly from applied field and achieve $8{,}8 \cdot 10^{-6}$ и $5{,}6 \cdot 10^{-6}$ for longitudinal and transversal components and $2{,}0 \cdot 10^{-5}$ for volume magnetostriction. The last value is comparable with the coefficients of volume magnetostriction for the majority of materials: usually they are in the limits from $2*10^{-5}$ up to $9*10^{-5}$.

4   Conclusion

The ACM and SQUID data have shown that the magnetic agglomerates consisting of magnetite are contained in the porous Fe20-MIP and Fe20-MAP glasses after one- and two-stage etching. ZFC and FC measurements confirm the superparamagnetic origin of magnetic properties. The coercive field for Fe20-MAP glass (346 (11) Oe) have been found from the field dependence of magnetization at 4.2 K. The values of magnetostriction are determined for nanocomposite Fe20-MIP+KNO$_3$. It is shown that these glasses can be used for preparation of multifunctional nanocomposite materials with spatially separated magnetic and ferroelectric ordering.

Acknowledgements. This work is supported by RFBR (grant 12-02-00230) and DAAD grant "Strategic Partnership with St. Petersburg State Polytechnical University and Leibniz Universität Hannover" and by the Ministry of Education and Science Program "5-100-2020".

References

1. Bibes, M.: Nanoferronics is a winning combination. Nature Materials 11, 354–357 (2012)
2. Filimonov, A., Rudskoy, A., Naberezhnov, A., Vakhrushev, S., Koroleva, E., Golosovsky, I., Kumzerov, Y.: New Nanocomposite materials on the Basis of Dielectric Porous Matrices. Material Sciences and Applied Chemistry (Journal of Riga Technical University) 28, 73–84 (2013)
3. Kumzerov, Y., Vakhrushev, S.: Nanostructures within porous materials. Encyclopedia of Nanoscience and Nanotechnology 10, 1–39 (2003)
4. Pan'kova, S.V., Poborchii, V.V., Solov'ev, V.G.: The giant dielectric constant of opal containing sodium nitrate nanoparticles. J. Phys. Condens. Matter 8, L203 (1996)
5. Colla, E., Fokin, A.V., Kumzerov, Y.A.: Ferroelectrics properties of nanosize KDP particles. Solid State Commun. 103, 127 (1997)
6. Fokin, A.V., Kumzerov, Y.A., Naberezhnov, A.A., Okuneva, N.M., Vakhrushev, S.B., Golosovsky, I.V., Kurbakov, A.I.: Temperature Evolution of Sodium Nitrite Structure in a Restricted Geometry. Phys. Rev. Lett. 89, 175503-1–175503-4 (2002)
7. Golosovsky, I.V., Delaplane, R.G., Naberezhnov, A.A., Kumzerov, Y.A.: Thermal motions in lead confined within porous glass. Phys. Rev. B 69, 132301-1–132301-4 (2004)
8. Golosovsky, I.V., Smirnov, O.P., Delaplane, R.G., Wannberg, A., Kibalin, Y.A., Naberezhnov, A.A., Vakhrushev, S.B.: Atomic motion in Se nanoparticles embedded into a porous glass matrix. Eur. Phys. Jour. B 54, 211–216 (2006)
9. Vakhrushev, S.B., Kumzerov, Y.A., Fokin, A., Naberezhnov, A.A., Zalar, B., Lebar, A., Blinc, R.: ^{23}Na spin-lattice relaxation of sodium nitrite in confined geometry. Phys. Rev. B 70, 132102-1–132102-3 (2004)
10. Beskrovny, A.I., Vasilovskii, S.G., Vakhrushev, S.B., Kurdyukov, D.A., Zvorykina, O.I., Naberezhnov, A.A., Okuneva, N.M., Tovar, M., Rysiakiewicz-Pasek, E., Jagus, P.: Temperature Dependences of the Order Parameter for Sodium Nitrite Embedded into Porous Glasses and Opals. Physics of the Solid State 52(5), 1092–1097 (2010)
11. Golosovsky, I.V., Mirebeau, I., André, G., Kurdyukov, D.A., Kumzerov, Y.A., Vakhrushev, S.B.: Magnetic Ordering and Phase Transition in MnO Embedded in a Porous Glass. Phys. Rev. Lett. 86(25), 5783–5786 (2001)
12. Panova, G.K., Nikonov, A.A., Naberezhnov, A.A., Fokin, A.V.: Resistance and Magnetic Susceptibility of Superconducting Lead Embedded in Nanopores of Glass. Physics of the Solid State 51(11), 2225–2229 (2009)
13. Shikov, A.A., Zemlyanov, M.G., Parshin, P.P., Naberezhnov, A.A., Kumzerov, Y.A.: Superconducting Properties of Tin Embedded in Nanometer-Sized Pores of Glass. Physics of the Solid State 54(12), 2345–2350 (2012)

14. Stolyar, S.V., Anfimova, I.N., Drozdova, I.A., Antropova, T.V.: New two-phase ferriferous sodium borosilicate glasses for preparation of nanoporous materials with magnetic properties. Nanosystems, Nanomaterials, Nanotechnologies 9(2), 433–440 (2011) (in Russian)
15. Antropova, T.V., Anfimova, I.N., Golosovsky, I.V., Kibalin, Y.A., Naberezhnov, A.A., Porechnaya, N.I., Pshenko, O.A., Filimonov, A.V.: Structure of Magnetic Nanoclusters in Ferriferous Alkali Borosilicate Glasses. Physics of the Solid State 54(10), 2110–2115 (2012)
16. Koroleva, E., Burdin, D., Antropova, T., Porechnaya, N., Naberezhnov, A., Anfimova, I., Pshenko, O.: Dielectric properties of two-phase and porous ferriferous glasses. Optica Applicata 42(2), 287–294 (2012)
17. Nizhankovskii, V.I.: Magnetostriction of terbium molybdate in high magnetic field. Europ. Phys. Jour. B. 71, 55–57 (2009)
18. Porechnaya, N.I., Plyastcov, S.A., Naberezhnov, A.A., Filimonov, A.V.: Study of topology and magnetic response of ferriferous alkali borosilicate glasses by magnetic force microscopy. Journal Physics and Mathematics 4(109), 113–117 (2010) (in Russian)
19. Stoner, E.C., Wohlfarth, E.P.: A Mechanism of Magnetic Hysteresis in Heterogeneous Alloys. Philos. Trans. R. Soc. London, Ser. A. 240, 599–642 (1948)

Wavelet Analysis of Non-stationary Signals in Medical Cyber-Physical Systems (*MCPS*)

Sergey V. Bozhokin[1,3,*] and Irina B. Suslova[2]

[1] Department of Theoretical Physics, St. Petersburg State Polytechnic University
(National Research University), Russia
bsvjob@mail.ru
[2] Department of Mathematical Physics, St. Petersburg State Polytechnic University
(National Research University), Russia
ibsus@mail.ru
[3] Polytechnicheskaya 29, St. Petersburg, 195251 Russia
bsvjob@mail.ru
http:www.spbstu.ru

Abstract. The advantages of multichannel medical cyber-physical systems (*MCPS*), which are designed to receive and process signals of human biological rhythms (EEG, ECG, blood pressure) at the remote server and to issue diagnostic conclusions, are discussed. The paper presents new data processing algorithms for *MCPS* based on continuous wavelet transform (*CWT*). The proposed method provides an array of parameters characterizing frequency restructuring of medical signals in different frequency bands. Since frequency fluctuations in brain and heart rhythms are closely related to different processes in human organism, the obtained data can allow us: to identify the disease at its early stages; to test the adaptive capacity of the human organism; to give diagnostic reports on cardiovascular and nervous systems; to analyze changes in rhythm during biofeedback sessions. The techniques set forth in the paper can help in the creation of a unique "rhythmic portrait" of a person to diagnose his physiological state.

Keywords: medical cyber-physical systems, wavelet transform, spectral integral, biomedical signal processing.

1 Introduction

Currently we are witnessing a rapid development of the industry of medical devices as well as of communication networks that integrate these devices into a single complex. High-tech sensors and software development for analyzing various biomedical signals allow upgrading many techniques of medical diagnostics. At present, distributed systems of biomedical signal analysis are replacing stand-alone diagnostic devices related to the individual patient. Modern diagnostic systems can monitor simultaneously a large number of channels with biomedical signals coming from each patient. Highly

---

[*] Corresponding author.

S. Balandin et al. (Eds.): NEW2AN/ruSMART 2014, LNCS 8638, pp. 467–480, 2014.

sensitive equipment and control devices make it possible to exchange information between the diagnostic tool and a remote server over the Internet. Advances in software and algorithm development allow making a decision automatically taking account of many factors. All this turns modern medical diagnostic devices into a separate class of medical cyber-physical systems (*MCPS*) [1]-[4].

To process a multidimensional array of biomedical data (heart rate, electrocardiogram (ECG), blood pressure (BP), blood oxygen saturation, respiratory rate, electroencephalogram (EEG), circadian rhythms) we need to develop appropriate mathematical techniques. It is known that the majority of human medical signals are non-stationary. This means that their characteristics calculated for a sufficiently large time interval T change when taking any other time interval T arbitrarily displaced relative to the first time interval. The reason of the non-stationarity can be found not only in specific features of the origin and generation of bioelectric signals, but also in internal transients at various levels of integration, which appear to the "observer" as random. Various amplitude and frequency irregularities in the form of synchronization, desynchronization, temporal bursts and so on are observed even in normal (in the absence of any obvious external disturbing factors) and may be due to spontaneous fluctuations in the level of functional activity such as for example a mental and cognitive activity during the signal registration.

Non-stationarity of human rhythms demonstrates itself most clearly during functional tests. To diagnose cardio-vascular system dynamic tests (veloergometrics, tredmill, dynamometry), respiratory tests a (Valsalva tests, rhythmic breathing, breath-holding on inspiration or expiration), orthostatic, drugs and psycho-emotional tests are conducted. Such tests as photostimulation, hyperventilation and cognitive tests are used to study human central nervous system. As during these tests we register a lot of frequency restructuring in heart and brain activity, the urgent issue is to create new methods taking into account the essential non-stationarity of biological signals.

The aim of the present work is to provide *MSPS* with new algorithms for processing non-stationary medical signals and to calculate a set of quantitative parameters which can serve as the foundation for expert assessments and automatic diagnostic conclusions. The original technique based on continuous wavelet transform *(CWT)* together with skeleton and spectral integral analysis in the fixed spectral band is proposed to solve the problem. The technique has been tested on different specially constructed models of EEG and ECG signals and applied to real records of human rhythms. The approach lends to further development by increasing the dimension and informative quality of the calculated array of parameters that would expand the range of diagnostic and scientific problems. Finding new quantitative characteristics related to human rhythm rearrangements and interactions will form the data base for modern diagnostic complex *MCPS*.

2   Methodology

2.1   *MCPS* in Electroencephalography

Modern electroencephalographs enable us to conduct Holter monitoring of EEG signals read out from the surface of the brain. This is done through radio-telemetry devices placed directly on the EEG cap. The high quality of the EEG recording is due to the absence of artifacts which arise from the use of ordinary wired contacts. Noise reduction technology provides protecting against any external interference and permits daily EEG monitoring of a patient with the possibility of on-line translating the registered data to a remote server via the Internet [5]-[8].

The remote server provides data storage and mathematical processing of EEG signals. The high rate of data processing in *MSPS* is determined by efficient parallelization of computational algorithms and optimal use of all processors [9-11]. Thus, a strategy of cloud computing which provides ubiquitous access to computing resources (networks, servers, data storage devices, applications and services) is carried out. The central task for computing resources of *MCPS* while processing non-stationary EEG is to create mathematical algorithms for computing the values of diagnostic parameters describing brain activity. These algorithms should perform filtering of signals by removing artifacts, receive and analyze quantitative parameters characterizing the rhythmic patterns, calculate the statistical characteristics that detect synchronization and desynchronization in EEG channels. Such approach makes it possible to trace the participation, cooperation and contribution of different parts of the brain during its functioning, both at rest and in diagnostic procedures.

All the proposed algorithms are based on the application of the wavelet theory as the main tool to process non-stationary biological signals [12]-[16].

Define continuous wavelet transform (*CWT*) $V(v,t)$ of an arbitrary signal $Z(t)$ as follows

$$V(v,t) = v \int_{-\infty}^{\infty} Z(t') \psi^* (v(t'-t)) dt', \tag{1}$$

where $\psi(x)$ is the Morlet mother wavelet function. In [17]-[20] one can find the explicit form of the mother wavelet and the formula for value $\varepsilon(v,t)$

$$\varepsilon(v,t) = \frac{2}{C_\psi} \frac{|V(v,t)|^2}{v}, \tag{2}$$

which describes instantaneous distribution of signal energy at frequency V (local density of signal energy spectrum) and $C_\psi \approx 1.0132$. Spectral integral $E_\mu(t)$ determines temporal dynamics of frequencies

$$E_\mu(t) = \frac{1}{\Delta v} \int_{v_\mu - \Delta v/2}^{v_\mu + \Delta v/2} \varepsilon(v,t) dv. \tag{3}$$

$E_\mu(t)$ is the mean value of energy spectrum local density of signal $Z(t)$ integrated over frequency interval $\mu = [v_\mu - \Delta v/2; v_\mu + \Delta v/2]$, where v_μ denotes the center of the interval and Δv – its width. The time-dependence of $E_\mu(t)$ performs some kind of filtering by summing the contributions of local spectrum density $\varepsilon(v,t)$ only in the given frequency band μ. By analyzing CWT and calculating $E_\mu(t)$ in deferent spectral bands $\mu = \{\delta, \theta, \alpha, \beta\}$ we can follow the dynamics of the appearance and disappearance of frequencies in spectral band μ, where δ–rhythm (0.5–4 Hz), θ–rhythm (4–7.5 Hz), α–rhythm (7.5–14 Hz), β–rhythm (14–30 Hz).

The graphs of $Z(t)$ for occipital leads, CWT transform $V(v,t)$ and spectral integral $E_\alpha(t)$ of this EEG signal in the band of α–rhythm are shown in Fig.1-3.

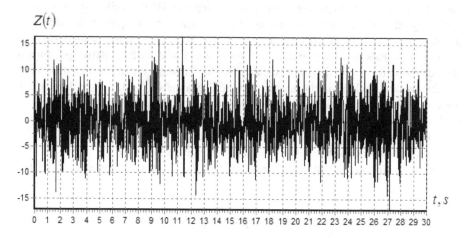

Fig. 1. EEG signal $Z(t)$ in the state of rest (eyes closed, occipital lead Oz) depending on time t, s

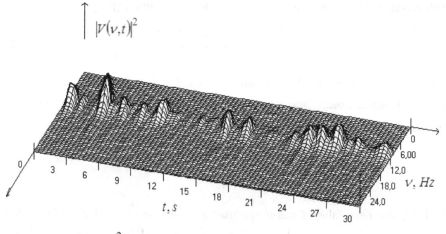

Fig. 2. Value $|V(v,t)|^2$ (1) for EEG signal $Z(t)$ (Fig.1.) depending on time t, s and frequency v, Hz

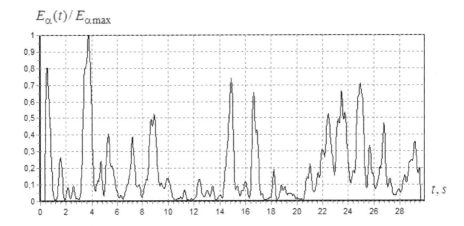

Fig. 3. Spectral integral $E_\alpha(t)$ (12) in α – rhythm band (7.5–14 *Hz*) for EEG signal $Z(t)$ (Fig.1) depending on time t, s

Next to each brain lead we study skeletons of *CWT*, which are the lines representing the position of maxima (ridges) of $|V(v,t)|^2$ surface. The time-dependence of instantaneous local frequencies $F_\mu(t)$ related to maximum values of $|V(v,t)|^2$ can be calculated within spectral band $\mu = \{\delta, \theta, \alpha, \beta\}$. Spectral integrals $E_\mu(t)$ are analyzed and the characteristic times of the increase and decrease of EEG burst activity are determined. Specifying the cut off level to 5% from the maximum value of $E_\mu(t)$, we can see that the whole observation period for spectral integral $E_\mu(t)$ is divided into many separate bursts. For each burst numbered as $J=0,1...N_F-1$, the automatic algorithm is implemented to calculate local frequency $F_\mu(t)$, the beginning of J-burst and its end - $t_B(J)$ and $t_E(J)$, duration of burst $\tau(J)$, the center of burst localization $t_C(J)$ and average frequency of the burst $F_\mu(J)$.

For record $Z(t)$ (Fig.1) in the band $\mu = \alpha$, we have $N_F =20$ bursts of α-activity. The dependences of $\tau(J)$ and $F_\alpha(J)$ on $t_C(J)$ are shown in Fig.4 and Fig.5. The maximum and minimum values for burst durations are $\tau_{min}(8) \approx 0.08$ s at $t_C(8) \approx 11.28$ and $\tau_{max}(17) \approx 6.8$ s at $t_C(17) \approx 24.04$ s.

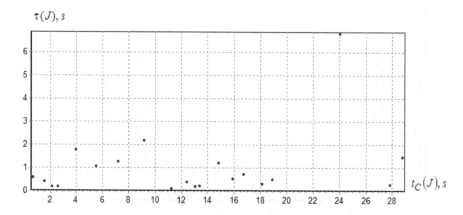

Fig. 4. The dependence of burst time-duration $\tau(J)$, s with number $J=0,1,\ldots N_F-1$ on time-value of its center $t_C(J)$, s

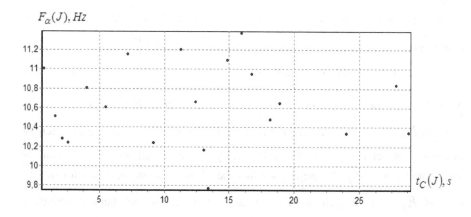

Fig. 5. The dependence of J- burst frequency mean value $F_\alpha(J)$, Hz on the time-value of its center $t_C(J)$, s

Statistical characteristics of the entire ensemble of bursts, i.e. mean values and standard deviations have the following values: $<\tau>\approx 0.99$ s, $<\Delta\tau>\approx 1.45$ s, $<F>\approx 10.63\ Hz$, $<\Delta F>\approx 0.4\ Hz$. Quantitative parameters of each burst and of the entire ensemble of bursts reflect the processes of restructuring, pattern formation, synchronization and desynchronization of bursts. The total set of the parameters can be the data base for identifying and classifying the EEG patterns that is a part of automatic diagnostic devices *MCPS*.

2.2   *MCPS* in Electrocardiography

In the recent years there has been a significant progress in the development of diagnostic methods in arrhythmology and in creation of compact antiarrhythmic implanted devices: implantable pacemakers, internal cardioverter defibrillators (*ICDs*) and devices for cardiac resynchronization therapy (*CRT*) - biventricular pacemakers. Production of artificial pacemakers and devices that restore sinus rhythm in the event of life-threatening arrhythmias requires development of automated medical diagnostic systems *MCPS* based on rhythm analysis [21]-[23].

MCPS provide automated collection and wireless transmission of statistical and diagnostic information about the patient under monitoring. Automatic methods for computer diagnostic of electrocardiogram (EEG) allow analyzing the patient's state, adjusting the therapy, anticipating and preventing possible complications while using implantable devices. An increasing number of patients with implantable devices require significant time even for a standard medical examination. Development of automated technologies for remote monitoring of ECG parameters, as well as a remote monitoring of entire implantable system, significantly reduces the number of scheduled visits to your doctor. We should emphasize that besides systematic monitoring, the traditional schemes of therapy using implants require thorough expert examination of long ECG records to detect the symptomatic events. In addition, the results are not always accurate. Unfortunately, the consequences of the late detection of such events are progressive arrhythmia, increased risk of stroke, progressive heart failure by developing methods for remote patient heart control we give the chance for the immediate identification and prevention of dangerous situations, in some cases helping to save the patient's life.

As a rule, an automated cardiological complex consists of the following components:

1) The implanted device (implantable pacemaker, internal cardioverter defibrillator (*ICDs*), biventricular pacemaker) telemetrically connected with a special transmitter which receives and processes the information from the implant located in the patient's body. Handheld diagnostic devices allow broadcasting ECG signals to the Internet server and accessing to the stored records and the results of processing.

2) The Internet server receiving an unlimited number of ECGs, which conducts processing, storage and access to saved data. The information is transmitted by means of GPRS-based connection with the identification of access rights for each remote user.

3) The Experts-cardiologists with computers connected to the server, who analyze the information and make recommendations on correcting the treatment process. Computing resources that solve the problems of diagnostics are located on the server. This gives access to a large number of users and ensures high reliability of automatic ECG processing. Notably, in principle the problem of diagnosing and suggesting recommendations can be solved without the participation of experts by using modern mathematical methods of data processing and decision making. Thus, we have a chain to transfer the information [patient- service center-doctor]. The decisions made by the doctor create a chain of feedback [doctor-service center- patient]. Such automated monitoring has allowed conducting the research works: TRUST [24], REFORM,

COMPAS [25], OEDIPE [26]. These studies have shown the advantages of remote monitoring, which is alternative to control visits to medical institutions. Remote monitoring gives the opportunity for early detection of device failure and of symptomatic events in the ECG of the patients. Early detection of ECG rhythm disturbances permits physicians to make quick decisions, minimizing the risk of stroke, heart failure, ventricular tachycardia, atria and ventricular fibrillation.

The Automated complex for remote monitoring can be regarded as highly effective, for it identifies the majority of violations at an early stage of their development. Moreover, it gives an economy of means on medical examinations under such heavy conditions as complex transport infrastructure and territory extension. All this makes the task of creating the automated diagnostic complex for remote monitoring quite urgent. .

While developing *MCPS* much attention should be paid to creation of software algorithms, which study the non-stationary heart rate variability (*HRV*). *HRV* analysis considers the rhythmogram –the sequence of *RR* -intervals between adjacent heartbeats to evaluate a cardiovascular system. However, the statistical parameters *(RRNN, SDNN, RMSSD)* describing *HRV*, spectral characteristics of cardio intervals obtained using Fourier transform (*VLF, LF, HF*) and histogram methods can be used only in stationary situations, when spectral properties of the signal do not change over time. This paper presents new methods of mathematical analysis of non-stationary heart rate variability (*HRV*) during functional tests on the basis of a general approach to the processing of signals with time-varying amplitude and frequency characteristics proposed in [27]-[29].

We simulate the rhythmogram by superposition of Gaussian peaks of equal amplitude. The centers of heart beats with number n are separated by time-intervals RR_n and located on irregular time-grid in time moments t_n, where $t_n = t_{n-1} + RR_n$, n =1,2,3..N–1, $t_0 = RR_0$. This new model takes into account the obvious fact that the heart rhythmogram is a frequency modulated signal. For the given model we can obtain analytical expressions for a continuous wavelet transform with the Morlet mother wavelet function [18]-[19],[28]. Considering maximum value of *CWT*, we have calculated the time-variation of local frequency $F_{max}(t)$. By applying *CWT* once again (*DCWT* method*)* to signal $F_{max}(t)$ we reveal the aperiodic and oscillatory motions of local frequency against the trend. We emphasize that it is the non-stationary behavior of $F_{max}(t)$ that requires *DCWT*. The procedure implies the modification of the ordinary wavelet transform. It enables us to obtain additional important information about time-frequency characteristics of heart rate variability (*HRV*).

To illustrate *DCWT* technique we consider the model of functional test with deep breathing (*DB*), when the rhythmogram consists of three stages. At stage of rest *A* (end time t_A =500 *s*, beat number n_A =652) the rhythmogram forms a frequency modulated signal as a superposition of many oscillations with different frequencies and modulation coefficients. Stage B (t_B =800 s, n_B =985) corresponding to deep breath is characterized by predominant frequency 0.1 Hz (6 cycles per minute) which complies with the operator commands. Stage of relaxation C differs from stage A in frequencies and modulation coefficients. The example of this rhythmogram RR_n is given in Fig.6.

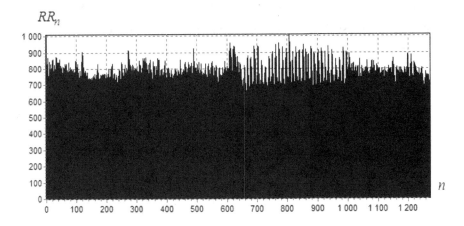

Fig. 6. Rhythmogram RR_n depending on heart beat number n

Secondary wavelet transformation $V_{DCWT}(v,t)$, when we apply *CWT* once again but to signal $F_{max}(t)$, is given in Fig.7.

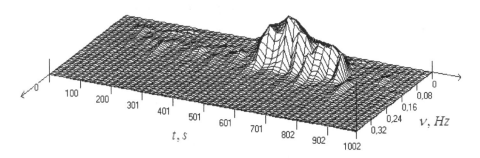

Fig. 7. Value $V_{DCWT}(v,t)$ depending on time t, s and frequency v, Hz

$V_{DCWT}(v,t)$ shows a sharp increase within time-interval t =[500; 800 s] related to the period of deep breath. The rhythmogram model allows to calculate the skeletons and spectral integrals $E_{\mu}(t)$ in subbands μ ={ULF,VLF,LF,HF} describing aperiodic and oscillating behavior of local frequency $F_{max}(t)$. The characteristic durations of time related to the increase and decrease of the frequencies have been calculated. In Fig. 8 we present the dependence on time t, S of spectral integral $E_{LF}(t)$ for subband LF=[0.04; 0.15 Hz] normalized on its maximum value $E_{LF\,max}(t)$.

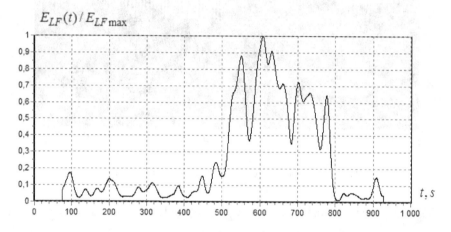

Fig. 8. The time dependence of spectral integral $E_{LF}(t)$ normalized on its maximum

For the rhythmogram record in Fig.6 we have external breath rhythm assimilation coefficient $K_M(LF)\approx7.73$ and retention rhythm factor $K_C(LF)\approx0.027$. The average frequency value and standard deviation at stage B are $< f_{LF}(B) >\approx0.113$ $Hz\ and\ < \Delta f_{LF}(B) > \approx0.017\ Hz$. Time τ_M of mastering the breath is the sum of time duration of silence τ_S and time duration of increase τ_I, i.e. $\tau_M = \tau_S + \tau_I$. Parameters τ_S и τ_I show different degrees of cardiovascular system adaptation to the imposed external breathing rhythm. The analysis of $E_{LF}(t)$ gives τ_S =17 s, τ_I =87 s. Similarly we can calculate the parameters at the stage C when operator stops commanding.

All quantitative parameters of HRV have been calculated for the model rhythmograms and for the real rhythmogram records at various functional tests. The presented technique can be applied to study cardiac arrhythmias, as well as for the analysis of heart rate turbulence associated with extrasystoles.

3   Results, Conclusions and Prospects

3.1   Application of *CWT*–Technique to *EEG* Processing in *MCPS*

Non-stationary EEG signal is represented as a superposition of brain activity bursts in various spectral bands $\mu=\{\delta,\theta,\alpha,\beta\}$. Brain activity bursts can be associated with different pathologies, influence of drugs, cognitive processes [17]-[18],[20]. The array of parameters calculated during the execution of the proposed algorithms, contains information on individual time-frequency properties of each burst and of their interaction. The study, comparison and classification of these data for different states of brain activity can help us recognize and, perhaps, enable to manage these states. EEG analysis is based on continuous wavelet transform (*CWT*) of EEG signal, which allows performing the procedure of exact calculation of frequencies and spectral integrals $E_\mu(t)$ in frequency bands μ. For each burst with number $J=0,1\ldots N_F-1$ an automatic algorithm is implemented to find the values of local frequency $F_\mu(t)$, time $t_B(J)$ and time $t_E(J)$ of J–burst beginning and end, J–burst duration $\tau(J)$, time of center localization $t_C(J)$ и average burst frequency $F_\mu(J)$. The procedure to calculate statistical characteristics of entire ensemble of bursts: $<\tau>$, $<\Delta\tau>$, $<F>$, $<\Delta F>$ is proposed.

The developed system of quantitative characteristics can be used to identify individual variability of alpha EEG activity; to study synchronous states of different brain regions and coherent activity associated with cognitive processes as well as with various neurological diseases (epilepsy, sensorimotor disorders, sleep disorders, Parkinson's disease, etc.). Further, the array of rhythm parameters can be supplemented with new data. This requires additional experimental study of EEG patterns in all frequency bands. The models of EEG bursts suggested as part of *MCPS* may be useful for testing the quality of learning processes in neural networks. The developed method can be applied during many functional tests (photostimulation, hyperventilation, psycho-emotional test) to diagnose the appearance and disappearance of specific frequencies in the EEG signal.

3.2   Application of *CWT* – Technique to *ECG* Processing in *MCPS*

Analytical expression $V(v,t)$ is obtained for *CWT* transform of frequency modulated non-stationary signal presented as superposition of equal Gaussian peaks on the irregular grid of time. Local frequency $F_{\max}(t)$ related to the extremum of $V(v,t)$ is determined. In real medical practice $F_{\max}(t)$ defines the input of respiration waves caused by the influence of sympathetic and parasympathetic divisions of an autonomic nervous system (*ANS*) in transient periods. To study $F_{\max}(t)$ we have applied double continuous wavelet transformation *DCWT* in calculating the skeletons and spectral integrals $E_\mu(t)$ in frequency bands $\mu = \{ULF,$ $VLF, LF, HF\}$. This algorithm performs some sort of signal filtering by summing the contributions of local spectrum density $\varepsilon(v,t)$ only in the given frequency band μ. The rigorous study of $F_{\max}(t)$ - oscillations against the trend allows detecting various restructuring scenarios in a complex rhythmogram signal. The important advantage of the proposed method is that it is a new tool to study the dynamics of *LF* spectral components, giving the opportunity to calculate transition periods during a great number of functional tests. The algorithm involves automatic computation of external breath rhythm assimilation coefficient $K_M(\mu)$, retention rhythm factor $K_C(\mu)$ and time-parameters τ_M, τ_S, τ_I, characterizing human adaptive capacities. This method eliminates long-term visual assessment of a long rhythmogram during Holter monitoring, which is the advantage of *MCPS* and of great importance in practice.

The algorithm of processing non-stationary rhythmogram (*DCWT*) as part of *MCPS* gives the opportunity to obtain a set of quantitative parameters associated with various cardiac arrhythmias during the functional tests, including the duration of transient restructuring in heart rhythm. The calculation of characteristic data array at functional tests holds the folowing opportunities:

1. Identifying the disease manifested in rhythm disturbances at early stages;

2. Testing the adaptive abilities of a person, that is an important task of training in many fields of human activity;

3. Studying the dynamics of interaction between the sympathetic and parasympathetic divisions of the autonomic nervous system (*ANS*).

4. Analyzing the rhythmogram during biofeedback sessions.

For applying the proposed technique in multichannel *MCPS* we should develop a unified algorithm to process non-stationary biological signals of different origin such as EEG, ECG, blood pressure fluctuations, circadian rhythms, etc. The technique is based on the theory of continuous wavelet transform (*CWT*). The method makes it possible to obtain a large number of signal parameters characterizing rhythm restructuring. The array of these parameters describes almost all physiological processes of human organism in dynamics. Thus, multichannel automatic signal processing in MCPS leads to the creation of a uniform "rhythmic portrait" of a person to diagnose his physiological state.

References

1. Sokolsky, O., Chen, S., Hatcliff, J., Jee, E., Kim, B., King, A., Mullen-Fortino, M., Park, S., Roederer, A., Venkatasubramanian, K.K.: Challenges and Research Directions in Medical Cyber–Physical Systems. Proceedings of the IEEE 100(1), 75–90 (2012)
2. Poovendran, R.: Cyber–Physical Systems: Close Encounters Between Two Parallel Worlds (Point of View). Proceedings of the IEEE 98(8), 1363–1366 (2010)

3. Lee, E.A.: Cyber Physical Systems: Design Challenges. In: 11th IEEE International Symposium on Object Oriented Real-Time Distributed Computing (ISORC), pp. 363–369 (2008)
4. Lee, E.A., Seshia, S.A.: Introduction to embedded systems: A cyber-physical systems approach, UC, Berkeley (2011)
5. Sutter, R., Fuhr, P., Grize, L., Marsch, S., Rüegg, S.: Continuous video-EEG monitoring increases detection rate of nonconvulsive status epilepticus in the ICU. Epilepsia 52(3), 453–457 (2011)
6. Authier, S., Paquette, D., Gauvin, D., Sammut, V., Fournier, S., Chaurand, F., Troncy, E.: Video-electroencephalography in conscious non human primate using radiotelemetry and computerized analysis: Refinement of a safety pharmacology model. Journal of Pharmacological and Toxicological Methods 60(1), 88–93 (2009)
7. Abubakr, A., Ifeayni, I., Wambacq, I.: The efficacy of routine hyperventilation for seizure activation during prolonged video-electroencephalography monitoring. Journal of Clinical Neuroscience 17(12), 1503–1505 (2010)
8. Kennett, R.: Modern electroencephalography. J. Neurol. 259, 783–789 (2012)
9. Migdalas, A., Pardalos, P., Sverre, S.: Story Parallel Computing in Optimization. Springer Publishing Company, Incorporated (2012)
10. Kontoghiorghes, E.J.: Handbook of Parallel Computing and Statistics. CRC Press (2005)
11. Khludova, M.: On-line parallelizable task scheduling on parallel processors. In: Malyshkin, V. (ed.) PaCT 2013. LNCS, vol. 7979, pp. 229–233. Springer, Heidelberg (2013)
12. Mallat, S.: A Wavelet Tour of Signal Processing, 3rd edn. Academic Press, New York (2008)
13. Koronovskii, A.A., Khramov, A.E.: Continuous wavelet analysis and its applications, Moscow, Fizmatlit (2003) (in Russian)
14. Chui, C.K.: An introduction to wavelets. Academic Press, New York (1992)
15. Daubechies, I.: Ten lectures on wavelet. Society for industrial and applied mathematics, Philadelphia (1992)
16. Pavlov, A.N., Hramov, A.E., Koronovskii, A.A., Sitnikova, E., Yu, M.V.A., Ovchinnikov, A.A.: Wavelet analysis in neurodynamics. Physics-Uspekhi. Advances in Physical Sciences 55(9), 845–875 (2012)
17. Bozhokin, S.V., Suvorov, N.B.: Wavelet analysis of transients of an electroencephalogram at photostimulation. Biomed. Radioelektron (3), 21–25 (2008)
18. Bozhokin, S.V.: Wavelet analysis of learning and forgetting of photostimulation rhythms for a nonstationary electroencephalogram. Technical Physics 55(9), 1248–1256 (2010)
19. Bozhokin, S.V.: Continuous wavelet transform and exactly solvable model of nonstationary signals. Technical Physics 57(7), 900–906 (2012)
20. Bozhokin, S.V., Suslova, I.B.: Wavelet analysis of a mathematical model of the brain bursts electroencephalogram. In: Proceedings of the XVI - Russian Scientific and Technical Conference, Neuroinformatics–2014, Moscow Engineering Physics Institute, January 27-30 (2014); Collection of scientific papers. In: 3 parts. CH.2.M.: MEPhI 2014. P.51-66. ISBN 978-5-7262-1899-1 (in Russian)
21. Boriani, G., Auricchio, A., Klersy, C., et al.: Healthcare personnel resource burden related to in-clinic follow-up of cardiovascular implantable electronic devices: a European Heart Rhythm Association and Eucomed joint survey. Europace 13(8), 1166–1173 (2011)
22. Lomidze, N.N., Revishvili, A.S., Kuptsov, V.V., Spiridonov, A.A.: Remote monitoring of patients - results of clinical studies. Journal of Arrhythmology 74, 71–77 (2013) (in Russian)

23. Ricci, R.P., Morichelli, L., Santini, M.: Home monitoring remote control of pacemaker and implantable cardioverter defibrillator patients in clinical practice: impact on medical management and health-care resource utilization. Europace 10(2), 164–170 (2008)

24. Varma, N., Epstein, A.E., Irimpen, A., et al.: Efficacy and safety of automatic remote monitoring for implantable cardioverter-defibrillator follow-up: the Lumos-T Safely Reduces Routine Office Device Follow-up (TRUST) trial. Circulation 122(4), 325–329 (2010)

25. Mabo, P.: Home monitoring for pacemaker follow-up: The first prospective randomised trial. Presentation at Cardiostim, Nice Acropolis, French Riviera, Jun 16-19 (2010)

26. Halimi, F., Clémenty, J., Patrick, A., et al.: Optimized post-operative surveillance of permanent pacemakers by home monitoring: the OEDIPE trial. Europace 10(12), 1392–1399 (2008)

27. Bozhokin, S.V., Shchenkova, I.M.: Analysis of the heart rate variability using stress tests. Human Physiology 34(4), 461–467 (2008)

28. Bozhokin, S.V., Suslova, I.M.: Double Wavelet Transform of Frequency-Modulated Nonstationary Signal. Technical Physics 58(12), 1730–1736 (2013)

29. Bozhokin, S.V., Suslova, I.B.: Analysis of non-stationary HRV as a frequency modulated signal by double continuous wavelet transformation method. Biomedical Signal Processing and Control 10(3), 34–40 (2014)

Generation and Reception of Spectral Efficient Signals Based on Finite Splines

Mikhail A. Kryachko[1], Alexander F. Kryachko[1], Sergey B. Makarov[1],
Victor I. Malyugin[1], Mikhail V. Silnikov[1], and Li Yanling[2]

[1] St. Petersburg State Polytechnical University, St. Petersburg, Russia
{mike_kr,alex_k34.ru}@mail.ru, makarov@cee.spbstu.ru,
director@mes.spbstu.ru
[2] Jiangsu Normal University, Xuzhou, China
ylli@jsnu.edu.cn

Abstract. Spectral efficient signals based on finite splines, which are obtained by repeated usage of discrete convolution procedure, is considered. Energy spectra of random sequence of signals is presented. Practical realization of device for generation and reception is proposed. Generation of SEPSK signals is done by matched filter with feedback line. Proposed method for reception of SEPSK signals is similar to method for reception of classical OFDM signals. High reduction rate of out-of-band emissions for random sequence of SEPSK signals is provided by proposed atomic functions.

Keywords: SEPSK, atomic functions, finite spline, out-of-band emissions.

1 Introduction

Application of spectrally efficient signals with phase shift keying (SEPSK), constructed on the basis of atomic functions or finite splines [1], [2], allows to obtain efficient compression of the energy spectrum of a random sequence of signals. Application compression method based on atomic functions allows the usage of their formation and receiving simple to implementation approach based on the calculation of convolution and deconvolution of functions [1].

The objective of this paper is to study possibility of realization of device for forming and receiving SEPSK signals, obtained based on atomic functions.

2 Basic Properties of Atomic Functions

Atomic functions has properties of polynomials and splines [1]. Splines of degree γ are functions which are "piecewise" polynomials of degree γ. Reduction of out-of-band emissions for signals on the basis of finite splines of degree γ is equal to $C/\omega\gamma + 1$ (if all derivatives of the envelope of signal up to $(\gamma - 1)$-th order have no discontinuities, and γ-th derivative is everywhere finite). Distinctive features of the atomic functions are:

S. Balandin et al. (Eds.): NEW2AN/ruSMART 2014, LNCS 8638, pp. 481–487, 2014.

- analyticity;
- combination of finitely of function and high decreasing rate of Fourier transformation (faster than any power);
- connection with the derivative of functions and explicit expression for the spectrum.

Let's consider generation of spectral efficient signals with duration T and based on application of finite splines. Repetition of convolution of basis functions is used for generation those signals. For rectangular form of envelope for PSK signals with amplitude A and centered about start of timing

$$a(t) = \begin{cases} A, & |t| \leq T/2 \\ 0, & |t| > T/2 \end{cases}.$$ (1)

N-fold convolution $(N+1)$ $a(t)$ may be represented in the form of finit splines $\Theta_N(x)$ [1]. Expression (1) may be rewritten as follows:

$$\varphi(t) = \frac{A}{2\pi} \int_{-\infty}^{\infty} e^{jut} \frac{\sin(u/2)}{u/2} du.$$ (2)

We obviously have for (2):

$$\varphi(t) = \mathbb{F}^{-1}[v(u)],$$

where $\mathbb{F}^{-1}[v(u)]$ – Fourier transform of function $v(u) = \dfrac{\sin(u/2)}{u/2}$. Convolution of functions can be written (using the theorem of Borel and the symmetry of the Fourier transform):

$$\varphi(t) * \varphi(t) = \mathbb{F}^{-1}[v^2(u)].$$ (3)

Then:

$$\varphi(t) * \varphi(t) = \frac{A^2}{2\pi} \int_{-\infty}^{+\infty} e^{jux} \left[\frac{\sin(u/2)}{u/2}\right]^2 du.$$ (4)

This process may be repeated N times for obtaining more smoother function, i.e. calculation of convolution procedure functions such $\varphi(t) * \varphi(t) * \ldots * \varphi(t) * \ldots$. Thus, the result of infinite convolution is a new finite function defined on the interval $[-NT/2; NT/2]$.

3   Formation of Spectrally Effective Signals on the Basis of Atomic Functions

In general, we can write the expression for the spline $\Theta_N(t)$ for any value of N by next expression:

$$\Theta_N(t) = \frac{A^{N+1}}{2\pi} \int_{-\infty}^{\infty} e^{jut} \left(\frac{\sin(u/2)}{u/2} \right)^{N+1} du. \tag{5}$$

Let's consider the form of the function $\Theta_N(t)$ for $N = 1...5$. For $N = 1$:

$$\Theta_1(t) = \frac{A^2}{2\pi} \int_{-\infty}^{+\infty} e^{jut} \left[\frac{\sin(u/2)}{u/2} \right]^2 du \ ;$$

$N = 2$:

$$\Theta_2(t) = \frac{A^3}{2\pi} \int_{-\infty}^{+\infty} e^{jut} \left[\frac{\sin(u/2)}{u/2} \right]^3 du \ ;$$

$N = 3$:

$$\Theta_3(t) = \frac{A^4}{2\pi} \int_{-\infty}^{+\infty} e^{jut} \left[\frac{\sin(u/2)}{u/2} \right]^4 du \ ;$$

$N = 4$:

$$\Theta_4(t) = \frac{A^5}{2\pi} \int_{-\infty}^{+\infty} e^{jut} \left[\frac{\sin(u/2)}{u/2} \right]^5 du \ ;$$

$N = 5$:

$$\Theta_5(t) = \frac{A^6}{2\pi} \int_{-\infty}^{+\infty} e^{jut} \left[\frac{\sin(u/2)}{u/2} \right]^6 du \ .$$

Form of envelope $a(t)$ of SEPSK signals for each value of N is determined by function $\Theta_N(t)$. $a(t)$ for $N = 1,...,5$ is shown on fig. 1. Normalized duration of SEPSK signals for $N = 1, ..., 5$ is increased from $T = 0.2$ to $T = 0.6$ at this figure. SEPSK signal for $N = 0$ has rectangular form of envelope (1) and duration $T = 0.1$. The degree of smoothness of the envelope of SEPSK signals with fixed energy increases with increasing of N (fig. 1).

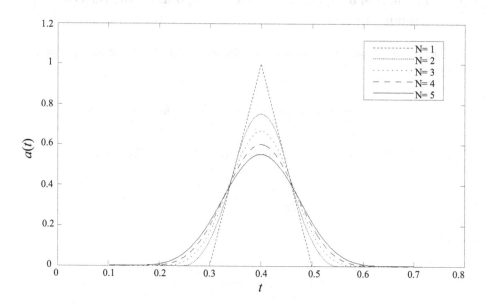

Fig. 1. Envelope $a(t)$ of SEPSK signals

Spectrum of random sequence of SEPSK signals, whose number is M, may be represented in the following form:

$$S(\omega) = A \int_0^{MT} \sum_{k=0}^{M-1} a(t-kT) d_r^{(k)} \cos \omega_0 t \cdot \exp(-i\omega t) dt = S_+(\omega) + S_-(\omega),$$

where $S_+(\omega) = \dfrac{A_0}{2} \int_0^{MT} \sum_{k=0}^{M-1} a(t-kT) d_r^{(k)} \exp(-i(\omega_0 + \omega)t) dt$, $S_-(\omega) = \dfrac{A_0}{2} \int_0^{MT} \sum_{k=0}^{M-1} a(t-kT)$

$\times d_r^{(k)} \exp(i(\omega_0 - \omega)t) dt$ and values of symbols of message depends on the location of symbol in the sequence and the index $r = 1, 2$. In particular, $d_1^{(k)} = 1$ and $d_2^{(k)} = 2$; $r = 1, 2$.

After the change of variable $x = t - kT$:

$$S_+(\omega) = F_a(\omega) \sum_{k=0}^{M-1} d_r^{(k)} \exp\left[-i(\omega_0 - \omega)t\right],$$

where $F_a(\omega) = \int_0^T a(t) \exp\left[-i(\omega_0\right] F_a(\omega) = \dfrac{A}{2} \int_0^T a(t) \exp\left[-i(\omega - \omega_0)\right] dt$ – spectrum of envelope $a(t)$.

Energy spectrum of random sequence of signals is calculated with tendency of M to infinity:

$$G(\omega) = \lim_{M \to \infty} \frac{1}{MT} m_1 \left\{ |S(\omega)|^2 \right\}$$

and the mathematical expectation $m_1 \left\{ |S(\omega)|^2 \right\}$ is determined by averaging over all possible finite sequences of symbols $d_r^{(k)}$.

Expression for energy spectrum (for narrowband signals) has the next form:

$$G(\omega) = G_+(\omega) + G_-(\omega),$$

where

$$G_+(\omega) = \lim_{M \to \infty} \frac{1}{MT} m_1 \left\{ |S_+(\omega)|^2 \right\}.$$

It is easy to show that:

$$m_1 \left\{ |S_+(\omega)^2| \right\} = |F_a(\Delta\omega)|^2 \sum_{k=0}^{M-1} \sum_{l=0}^{M-1} \exp\left[-i\Delta\omega(k-l) \right] m_1 \left\{ d_r^{(k)} d_q^{(l)} \right\},$$

where $\Delta\omega = \omega - \omega_0$.

For case of equally probable and independent symbols:

$$m_1 \left\{ d_r^{(k)} d_q^{(l)} \right\} = \begin{cases} 1, & k = l, \\ 0, & k \neq l. \end{cases}$$

The final expression for calculating the power spectrum of the random sequence of SEPSK signal in area of $\omega > 0$:

$$G_+(\omega) = \lim_{M \to \infty} \frac{1}{4MT} \left\{ M |F_a(\Delta\omega)|^2 \right\} = \frac{1}{4T} |F_a(\Delta\omega)|^2.$$

Thus, energy spectrum of random sequence of SEPSK signals is determined by Fourier transform of single signals and has same frequency bandwidth.

Energy spectra for random sequences of SEPSK signals, which form of real envelope is finite splines, is shown on fig. 2. Normalized energy spectrum are presented on Y-axis, relative frequency ($(f - f_0)T$ (where f_0 – central frequency) is shown on X-axis. As expected, the rate of out-of-band emissions increases with increasing of N.

We can see from analysis energy spectra of those signals that the reduction of out-of-band emissions is very high for large values of N (for example, $N = 3$-5).

The advantage of these signals is the principle of their generation and reception based on multiple repetitions of the convolution of functions.

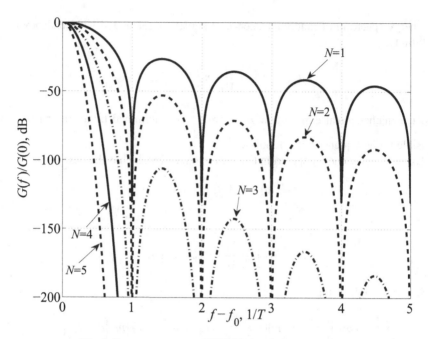

Fig. 2. Energy spectrum of SEPSK signals for $N = 1,\ldots,5$

Functional scheme of device for generation spectral effective signals based on finite splines is shown on fig. 3. This device is constructed by using matched filter with adjustable feedback. The impulse response of this filter has the envelope, which form is determined by (1). Signals from matched filter`s output come to input of delay line. Time delay of the filter response is determined by value of $N$. Delay of signal from matched filter`s output is equal to $2T$ for $N = 2$.

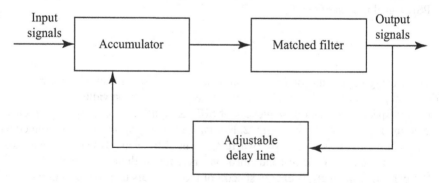

Fig. 3. Functional scheme of device for SEPSK signals generation

Voltage from matched filter`s output is transferred to its input again after calculating first procedure of convolution. Result of calculating second procedure convolution is response of matched filter as spline $\Theta_2(t)$. This spline is the same as form of envelope $a(t)$.

Reception of SEPSK signals may be done by several stage. On first stage, input process is converted to frequency domain by Fourier transform. On next stage, reception device must do nonlinear procedure of N-th degree root extraction from frequency function. Conversion to time domain by Fourier inversion is done on final stage. Output result is transmitted to classic demodulation algorithm, which based on application of correlation processing of PSK signals with rectangular form of envelope.

Of course, the proposed method of reception for SEPSK signal is not optimal. However, this method is simple (especially with usage of matched filter for realization). Proposed method for reception of SEPSK signals is similar to method for reception of classical OFDM signals.

4 Conclusions

High reduction rate of out-of-band emissions for random sequence of SEPSK signals is provided by proposed atomic functions (finite splines). In particular, reduction of out-of-band emissions is equal to $1/f^6$ for 2-3 iterations of convolution.

Generation of SEPSK signals is done by matched filter with feedback line. Reception of SEPSK signals is performed sequentially for 3 stage. On the first stage transformation of input mix of signals with additive noise to frequency domain is done. On next stage, reception device must do nonlinear procedure of N-th degree root extraction. Reverse operation (transformation to time domain) is performed on final stage.

References

1. Gulyaev, Y.V., Kravchenko, V.F., Pustovoit, V.I.: A New Class of WA-Systems of Kravchenko-Rvachev Functions // Doklady Mathematics 75(2), 325–332 (2007)
2. Ghavami, M., Michael, L.B., Kohno, R.: Ultra Wideband Signals and Systems in Communication Engineering. John Wiley & Sons (2004)

Generation of SEFDM-Signals Using FFT/IFFT

Alexandr B. Kislitsyn, Andrey V. Rashich, and Ngok Nuen Tan

St. Petersburg State Polytechnical University, St. Petersburg, Russia
andrey.rashich@gmail.com, kislitcyn@rambler.ru,
ngoctan1610@yahoo.com

Abstract. A study of Spectrally Efficient Frequency Division Multiplexing (SEFDM) signals generation is done. Three new algorithms for generation and reception of SEFDM-signals based on FFT/IFFT are proposed. Their complexity and conditions of applicability are analyzed. It is shown that the proposed schemes outperform the proposed earlier by 1,2-6 times. The BER performance of the developed methods for AWGN channel is analyzed, it is shown that the number of the guard subcarriers significantly affects the BER performance of SEFDM (up to 4 dB energy loss when doubling the guard part of the signal).

Keywords: OFDM, SEFDM, computational complexity, BER performance.

1   Introduction

Modern broadband wireless telecommunication systems (LTE, WiMAX, Wi-Fi) widely use orthogonal frequency division multiplexing (OFDM) on PHY level. The need of further increase of data rates in these systems brings a high interest to modifications of OFDM to achieve better spectral efficiency. As the most perspective alternative of OFDM comes the method of non-orthogonal frequency division multiplexing also known as spectrally efficient frequency division multiplexing (SEFDM). In [1], [2] and [4] it is shown that SEFDM-signals can provide spectral efficiency increase of up to 2-3 times in comparison with classic OFDM signals. With that SEFDM modulation shows the same performance as OFDM in fading channels.

One of the main drawbacks of SEFDM-signals is high computing complexity. In ([1], [2]) SEFDM transmitter consisted of a bank of N analogue modulators was proposed, where N is the number of subcarriers. Discrete version of the method is based on Inverse Discrete Fourier Transform (IDFT) and requires at least $N^2\alpha$ complex multiplications, where α is normalized frequency spacing between the subcarriers. In ([3], [4]) Fractional Fourier Transform (FrFT) was used for SEFDM generation with quadruple complexity of a conventional Inverse Fast Fourier Transform (IFFT) of length N. Ahmed and Darwazeh in [5] and [6] proposed the alternative method, based on IFFT/FFT operations, conventional for OFDM transceivers. However complexity of these algorithms seems to be substantially redundant.

S. Balandin et al. (Eds.): NEW2AN/ruSMART 2014, LNCS 8638, pp. 488–501, 2014.
© Springer International Publishing Switzerland 2014

That keeps SEFDM from application in 5G telecommunication systems. In this article 3 new algorithms for SEFDM generation based on IFFT are proposed. The corresponding reception schemes based on FFT are also offered. The proposed algorithms show considerably smaller computing complexity than methods in [1], [2], [3], [4], [5], [6] based on analog modulation, (Inverse) Discrete Fourier Transform, (Inverse) Fractional Fourier Transform and conventional (Inverse) Fast Fourier Transform. BER performance for BPSK-SEFDM is analyzed.

2   SEFDM Model Description

The frequency division multiplexing (FDM) signal $s(t)$ at zero frequency can be written as:

$$s(t) = \sum_{k=0}^{N-1} \left\{ C_k(t) e^{j2\pi\Delta fkt} \right\},$$

N – number of used subcarriers, Δf – frequency separation between adjacent subcarriers, $C_k(t)$ – complex manipulation function for k-th subcarrier. For classic OFDM-signals

$$C_k(t) = const(t) = C_N^{(n)}(k) \, \text{for} \, t \in \left[(n-1)T; nT \right], n \in \mathbb{N},$$

here $C_N^{(n)}(k)$ – complex amplitude for n-th OFDM-symbol, T – OFDM-symbol duration, $T = 1/\Delta f_{orth}$, Δf_{orth} – spacing between subcarriers, providing their orthogonallity in enhanced sense. In modern OFDM systems special guard subcarriers with zero amplitude are used on the both side of frequency band:

$$C_k^{(n)} = 0 \, \text{ for } \, k \in [0; N_{guard\_left} - 1] \, \text{ and } \, k \in [N - N_{guard\_right}; N-1],$$

N_{guard_left} – number of unused subcarriers in the negative frequencies band, N_{guard_right} – number of unused subcarriers in the positive frequencies band. The normalized frequency spacing α is $\alpha = \Delta fT = \Delta f / \Delta f_{orth}$.

For OFDM signals $\alpha = 1$, whilst for SEFDM $\alpha < 1$. The packing factor β is reciprocal to α: $\beta = 1/\alpha$.

In digital domain the signal is considered in the certain moments of time: $s(i\Delta t)$, $\Delta t = 1/F_s$ – sampling interval, F_s – sampling frequency. Usually $F_s = N\Delta f$, for that $\Delta t\Delta f = 1/N$ and $T / \Delta t = N\alpha = L$. We consider signals with natural L:

$$N\alpha = L \in \mathbb{N}. \tag{1}$$

Using upper notations $s(t)$ can be written in the following way:

$$s(i) = \sum_{k=0}^{N-1} \left\{ C_k(i) e^{j2\pi\Delta fki\Delta t} \right\} = \sum_{k=0}^{N-1} \left\{ C_k(i) e^{j2\pi\frac{ki}{N}} \right\}, \tag{2}$$

$$C_k(i) = const(i) = C_N^{(n)}(k) \text{ for } i \in \left[(n-1)L; nL \right], n \in \mathbb{N}.$$

3   DFT-Based Generation and Reception Schemes for SEFDM-Signals

3.1 N-Point DFT/IDFT-Based Generation and Reception Scheme

Let us consider a finite discrete signal $s_N^{(n)}(i)$, corresponding to n-th FDM-symbol of $s(i)$, continued on duration up to N samples:

$$s_N^{(n)}(i) = \sum_{k=0}^{N-1} \left\{ \psi_N(i) C_N^{(n)}(k) e^{j2\pi\frac{ki}{N}} \right\}, \ i = 0\ldots(N-1), \ \begin{cases} \psi_N(i) = 1, \ i = \overline{0\ldots(N-1)} \\ \psi_N(i) = 0, \ i \neq \overline{0\ldots(N-1)} \end{cases}.$$

Periodic extension $\zeta_N^{(n)}(i)$ of $s_N^{(n)}(i)$ with period equal to N samples is

$$\zeta_N^{(n)}(i) = \sum_{m=-\infty}^{m=+\infty} \left\{ s_N^{(n)}(i-mN) \right\} = \sum_{m=-\infty}^{m=+\infty} \left\{ \sum_{k=0}^{N-1} \left[\psi_N(i-mN) C_N^{(n)}(k) e^{j2\pi\frac{k(i-mN)}{N}} \right] \right\} =$$

$$= \sum_{k=0}^{N-1} \left\{ C_N^{(n)}(k) e^{j2\pi\frac{ki}{N}} \right\}. \tag{3}$$

Equation (3) is the Inverse Fourier Transform formula.

Thus, FDM signal $s(i)$ can be obtained by the successive generation of FDM-symbols. FDM-symbol with index n is generated using N-point IDFT on samples $C_N^{(n)}(k)$, after that the output sequence of IDFT with length N is reduced to length L sequence. For $\alpha = 1$ ($L = N$) OFDM signal is generated, for $\alpha < 1$ ($L < N$) SEFDM-signal is produced.

Using the N-point DFT on the receiver side the reception of SEFDM-signal becomes possible. To get the estimates of degradation for SEFDM reception one should use special intersymbol interference compensation algorithms which are another subject for research.

Assuming that the number of subcarriers in the FDM-system is a power of 2: $N = 2^n$, $n \in \mathbb{N}$ it becomes possible to use fast algorithms for DFT and IDFT, greatly reducing the computational complexity of the method. Regarding to (1) the conditions of applicability of the described scheme and FFT/IFFT algorithms can be written as

$$N = 2^n, \ n \in N; \ \alpha = l/p = l/2^m, \ \{l,m\} \in N; \ n \geq m. \tag{4}$$

The number of samples in each FDM-symbol is $L = N\alpha = Nl/p = l2^n/2^m = l2^{(n-m)}$.

Hereinafter, a method of generation and receiving SEFDM-signals using N-point DFT/IDFT is denoted as $(I)DFT_N$ and special case of the FFT/IFFT is denoted as $(I)FFT_N$. Equation (4) assigns to each value of N a finite set of possible values for $\alpha < 1$ with a number of values in each set $M_\alpha(N) = N - 1$. Examples of allowable coefficients α for $N = 2, 4, 8, 16$ are shown in Table 1. When the number of subcarriers is of practical interest ($N \geq 2^6 = 64$), a number of coefficients $M_\alpha(N)$ is more than sufficient in terms of use in real communication systems. In particular, for $N = 64$ there are 63 possible values for $\alpha, \alpha \in (0, 1)$.

Table 1. Sets of α when using the $(I)FFT_N$ for $N = 2, 4, 8, 16$

N	A	$M_\alpha(N)$
2	1/2	1
4	1/2, 1/4, 3/4	3
8	1/2, 1/4, 3/4, 1/8, 3/8, 5/8, 7/8	7
16	1/2, 1/4, 3/4, 1/8, 3/8, 5/8, 7/8, 1/16, 3/16, 5/16, 7/16, 9/16, 11/16, 13/16, 15/16	15

Computational complexity of $(I)FFT_N$ scheme for one SEFDM-symbol generation expressed in a number of pairs of elementary operations of addition and multiplication is determined by the following expression [7]:

$$Compl\left((I)FFT_N\right) = Nlog_2(N) = Nn. \tag{5}$$

3.2   L-Point DFT/IDFT-Based Generation and Reception Scheme

The obvious drawback of $(I)FFT_N$ scheme is the need for N-pint DFT/IDFT, while the number of transmitted samples for each FDM-symbol is $L \leq N$. The conversion to L-point transforms is possible if in eq. (3) replace signal $s_N^{(n)}(i)$ by $s_L^{(n)}(i)$:

$$s_L^{(n)}(i) = \sum_{k=0}^{N-1}\left\{\psi_L(i)C_N^{(n)}(k)e^{j2\pi\frac{ki}{N}}\right\}, i = 0\ldots(L-1), \quad \begin{cases} \psi_L(i) = 1, \ i = \overline{0\ldots(L-1)} \\ \psi_L(i) = 0, \ i \neq \overline{0\ldots(L-1)} \end{cases}.$$

Then

$$\zeta_L^{(n)}(i) = \sum_{m=-\infty}^{m=+\infty}\left\{s_L^{(n)}(i-mL)\right\} = \sum_{m=-\infty}^{m=+\infty}\left\{\sum_{k=0}^{N-1}\left[\psi_L(i-mL)C_N^{(n)}(k)e^{j2\pi\frac{k(i-mL)}{N}}\right]\right\}. \tag{6}$$

It is necessary to use L-point DFT to calculate spectral samples $S_L^{(n)}(l)$ of signal $\zeta_L^{(n)}(i)$:

$$S_L^{(n)}(l) = \frac{1}{L}\sum_{i=0}^{L-1}\zeta_L^{(n)}(i)e^{-j2\pi\frac{il}{L}} = \frac{1}{L}\sum_{i=0}^{L-1}\left(\sum_{m=-\infty}^{m=+\infty}\left\{\sum_{k=0}^{N-1}\left[\psi_L(i-mL)C_N^{(n)}(k)e^{j2\pi\frac{k(i-mL)}{N}}\right]\right\}\right)e^{-j2\pi\frac{il}{L}},$$

for $l = \overline{0,L-1}$:

$$S_L^{(n)}(l) = \frac{1}{L}\sum_{i=0}^{L-1}\left(\sum_{k=0}^{N-1}\left[C_N^{(n)}(k)e^{j2\pi\frac{ki}{N}}e^{-j2\pi\frac{il}{L}}\right]\right) = \sum_{k=0}^{N-1}\left(C_N^{(n)}(k)\frac{1}{L}\sum_{i=0}^{L-1}e^{j2\pi i\left(\frac{k}{N}-\frac{l}{L}\right)}\right) =$$

$$= \sum_{k=0}^{N-1}C_N^{(n)}(k)K(l,k), \quad \text{where } K(l,k) = \frac{1}{L}\sum_{i=0}^{L-1}e^{j2\pi i\left(\frac{k}{N}-\frac{l}{L}\right)}.$$

Thus, spectral samples $S_L^{(n)}(l)$ represent a linear combination of $C_N^{(n)}(k)$. It is similar to multiplication of column vector $\overline{C}_N$, with its elements $C_N^{(n)}(k)$ ($k = \overline{0,N-1}$), by matrix $\mathbf{K}_{L\times N}$ of dimension L x N, with its elements $K(l,k)$. The result is column vector $\overline{S}_L$, with elements $S_L^{(n)}(l)$ ($l = \overline{0,L-1}$):

$$\overline{S}_L = \mathbf{K}_{L\times N}\overline{C}_N. \tag{7}$$

For OFDM signals, $L = N$ and $\mathbf{K}_{L\times N} \to \mathbf{K}_{N\times N}$, with identity matrix $\mathbf{K}_{N\times N}$. Samples $S_L^{(n)}(l)$ and $C_N^{(n)}(l)$ are equal.

No we can say that SEFDM-signal $s(i)$ can be generated in the following way. For each n-th SEFDM-symbol spectral samples $S_L^{(n)}(l)$, $l = \overline{0,L-1}$ are calculated via linear combination of modulation symbols $C_N^{(n)}(l)$, $k = \overline{0,N-1}$. Then L-point IDFT is applied to samples $S_L^{(n)}(l)$, the output of IDFT is time domain samples $s_L^{(n)}(i)$ of n-th SEFDM-symbol.

Similarly, the use of L-point DFT at the receiver side provides an estimate of the spectral counts $\hat{S}_L^{(n)}(l)$. However, the calculation of estimates for modulation symbols $\hat{C}_N^{(n)}(k)$ is generally impossible because of the fact that matrix $\mathbf{K}_{L\times N}$ is not square. This fact can be considered as the interference between subcarriers in SEFDM-signal and interference compensation algorithms are needed. Considering the expression (7) as underdetermined system of linear equations, it is possible to obtain estimates of $\hat{C}_N^{(n)}(k)$, using various methods of approximating the exact solution.

For example, using the minimum mean square error optimality criterion, the approximate solution of system (7) can be obtained as follows:

$$\mathbf{K}_{L \times N} \mathbf{K}_{N \times L}^* \left(\mathbf{K}_{L \times N} \mathbf{K}_{N \times L}^* \right)^{-1} \overline{S}_L = \mathbf{K}_{L \times N} \overline{C}_N$$

$$\mathbf{K}_{N \times L}^* \left(\mathbf{K}_{L \times N} \mathbf{K}_{N \times L}^* \right)^{-1} \overline{S}_L = \overline{C}_N \qquad \mathbf{K}_{N \times L}^+ \overline{S}_L = \overline{C}_N ,$$

here $\mathbf{K}_{N \times L}^*$ – Hermitian conjugate matrix of $\mathbf{K}_{L \times N}$, $\mathbf{K}_{N \times L}^+$ – pseudoinverse matrix $\mathbf{K}_{L \times N}$.

When the number of transmitted L samples of each SEFDM-symbol is a power of 2: $L = N\alpha = 2^n$, it becomes possible to use fast DFT and IDFT algorithms to significantly reduce the computational complexity of the method. Furthermore, given that the number of subcarriers must be an natural $N \in \mathbb{N}$, we can write the conditions of using FFT/IFFT:

$$L = 2^q, \ q \in N; \ \alpha = l / p = 2^m / p, \ \{p,m\} \in N; \ n \ge q . \qquad (8)$$

The condition (8) includes the condition (1). Hereinafter, a method of generation and receiving SEFDM-signals using L-point DFT/IDFT is denoted as $(I)DFT_L$ and special case of the FFT/IFFT is denoted as $(I)FFT_L$. Condition (8) imposes restrictions on the possible values for α. For $\alpha < 1$, the $M_\alpha(N)$ is determined by the amount of numbers of the form 2^q ($q \in N$), less than or equal than N. Examples of allowable coefficients α for $N = 2, 4, 8, 16$ are shown in Table 2.

Table 2. Sets of α when using the $(I)FFT_L$ for $N = 2, 4, 8, 16$

N	α	$M_\alpha(N)$
2	-	0
4	1/2	1
8	1/2, 1/4	2
16	1/2, 1/4, 1/8	3

It is clear to see that the $(I)DFT_N$ scheme provides a much wider range of possible values for α than method $(I)DFT_L$ for any allowed N. At the same time, the number of possible values for N in $(I)DFT_N$ scheme is severely limited by the conditions (4), while $(I)DFT_L$ scheme is applicable for any N, starting with 3. Moreover, for values of N of practical interest ($N \ge 2^6 = 64$), $(I)DFT_L$ scheme allows selection of at least 5 different coefficients α in the range $\alpha \in (0,1)$, which is sufficient for most applications solutions.

Computational complexity of the $(I)DFT_L$ scheme for one SEFDM-symbol generation expressed in a number of pairs of elementary operations of addition and multiplication is determined by the following expression

$$Compl\left((I)FFT_L \right) = NL + L log_2 (L). \qquad (9)$$

It is considered in (9) that calculation of matrixes $\mathbf{K}_{L \times N}$ and $\mathbf{K}^{+}_{N \times L}$ is done while initializing SEFDM-modem and is not carried out during signal generation and reception.

3.3   Multiple-Block L-Point DFT/IDFT-Based Generation and Reception Scheme

The main drawback of $(I)DFT_L$ scheme is the need to perform computationally inten-sive linear transforms on the modulation symbols on the transmitting and receiving side. Alternative approach is to use several L-point DFT/IDFT blocks in parallel. For this algorithm to perform the migration from the sequence $\left\{C_N^{(n)}(k)\right\}_{k=0}^{N-1}$ of N modu-lation symbols to the sequence $\left\{C'^{(n)}_{Nl}(m)\right\}_{m=0}^{Nl-1}$ of Nl modulation symbols must be done. Here $l = \alpha p, \{l, p\} \in \mathbb{N}$:

$$\begin{cases} C'^{(n)}_{Nl}(m) = C_N^{(n)}(k) \, for \, m = kl \\ C'^{(n)}_{Nl}(m) = 0 \qquad for \, m \neq kl \end{cases}. \tag{10}$$

In this way the formal conversion to SEFDM-signal with Nl subcarriers is done. Expression for a finite signal $s_L^{(n)}(i)$ can be rewritten as

$$s_L^{(n)}(i) = \sum_{k=0}^{Nl-1}\left\{\psi_L(i)C'^{(n)}_{Nl}(k)e^{j2\pi\frac{ki}{lN}}\right\} = \sum_{b=0}^{p-1}\left[\sum_{k=0}^{Nl/p-1}\left\{\psi_L(i)C'^{(n)}_{Nl}(kp+b)e^{j2\pi\frac{(kp+b)i}{lN}}\right\}\right] =$$

$$= \sum_{b=0}^{p-1}\left[\sum_{k=0}^{\frac{Nl}{p}-1}\left\{\psi_L(i)C'^{(n)}_{Nl}(kp+b)e^{j2\pi\frac{kpi}{lN}}\right\}e^{j2\pi\frac{bi}{lN}}\right] =$$

$$= \sum_{b=0}^{p-1}\left[\sum_{k=0}^{L-1}\left\{\psi_L(i)C'^{(n)}_{Nl}(kp+b)e^{j2\pi\frac{ki}{L}}\right\}e^{j2\pi\frac{bi}{lN}}\right].$$

Turning to the periodic continuation of the signal $s_L^{(n)}(i)$ with period of L sam-ples, $= \overline{0, L-1}$, we can obtain the following expression:

$$\zeta_L^{(n)}(i) = \sum_{b=0}^{p-1}\left[\sum_{k=0}^{L-1}\left\{C'^{(n)}_{Nl}(kp+b)e^{j2\pi\frac{ki}{L}}\right\}e^{j2\pi\frac{bi}{lN}}\right], i = \overline{0, L-1}$$

The latter equation is the sum of p vectors $\overline{\zeta_L^{(n)}(b)}$:

$$\zeta_L^{(n)}(i,b) = \sum_{k=0}^{L-1}\left\{C_L^{'(n)}(k,b)e^{j2\pi\frac{ki}{L}}\right\}; i = \overline{0, L-1}, b = \overline{0, p-1}, \tag{11}$$

here $C_L^{'(n)}(k,b) = C_{Nl}^{'(n)}(kp+b)$. Each vector $\overline{\zeta_L^{(n)}(b)}$ represents the set of samples on the output of b-th block L-point IDFT. b-th vector is multiplied element wise by the following function:

$$\gamma(i,b) = \exp\left(j2\pi\frac{bi}{lN}\right); i = \overline{0, L-1}.$$

The multiplication by $\gamma(i,b)$ determines the relative shift of the adjacent vectors spectra by the frequency $\Delta f_{rel} = \Delta f / l$.

Similar to $(I)FFT_N$ and $(I)FFT_L$ schemes SEFDM-signal $s(i)$ can be generated in the following way. For each SEFDM-symbol with index n the complex modulation symbols $\left\{C_N^{(n)}(k)\right\}_{k=0}^{N-1}$ are transformed according to (10). Then p L-point IDFTs are performed according to (11). Vectors $\overline{\zeta_L^{(n)}(b)}$, $b = \overline{0, p-1}$, coming from IDFT, are element wise multiplied by samples $\gamma(i,b)$ and summed up. The result is one vector with L samples.

Sequential calculation of vectors $\overline{\zeta_L^{(n)}(b)}$ instead of parallel, and transmitting symbols $\left\{C_{Nl}^{'(n)}(m)\right\}_{m=0}^{Nl-1}$ in sets of L samples, allows to use only one IDFT block, one multiplier bank and one accumulator instead of L adders with p inputs.

On the receiver side the n-th SEFDM symbol of $s(i)$ is multiplied by p discrete time exponents:

$$\gamma'(i,b) = \exp\left(-j2\pi\frac{bi}{lN}\right); i = \overline{0, L-1}, b = \overline{0, p-1}.$$

Resulting p L-point vectors go to L-point DFT block. The estimates of complex modulation symbols are made after reordering and downsampling the output from the DFT block and subcarrier interference compensator. It is done to improve BER performance of SEFDM when $\alpha < 1$.

Similar to transmitter side conversion to sequential processing of samples $s_L^{(n)}(i)$ can be done. This makes possible to use only one DFT block and one multiplier bank.

If the set of the following conditions is met

$$L = 2^q, \ q \in N; \ \alpha = 1/p, \{l, p\} \in N; \ N = Lp/l \in N. \tag{12}$$

it is possible to use fast algorithms in proposed schemes, decreasing their computation complexity. Hereinafter, the proposed method for SEFDM-signals generation and reception using p-parallel or p-sequential L-point (I)DFT is denoted as $(I)DFT_L^p$ and special case of the FFT/IFFT is denoted as $(I)FFT_L^p$.

The conditions set (12) is identical to set (8). Thus the restrictions on the possible values for α and L for schemes $(I)FFT_L^p$ and $(I)FFT_L$ are the same. Hence the curves of $M_\alpha(N)$ for these schemes are also the same. So the conclusions on applicability of $(I)FFT_L$ scheme can be used for $(I)FFT_L^p$ scheme.

The computational complexity of $(I)FFT_L^p$ scheme for one SEFDM-symbol generation expressed in a number of pairs of elementary operations of addition and multiplication is determined by the following expression (excluding subcarrier interference compensation)

$$Compl\left((I)FFT_L^p\right) = pLlog_2(L) + (p-1)L. \tag{13}$$

From (4), (8) and (12) (the conditions to use $(I)FFT_N$, $(I)DFT_L$, $(I)FFT_L^p$ schemes) it is clear to see that all three schemes can be applied only if the set of the following conditions is met:

$$N = 2^n, \ n \in N; \ L = 2^q, \ q \in N; \ \alpha = 1/p = 1/2^m, \ m \in N.$$

4   Computational Complexity and Performance Analysis

Overview of the current scientific literature on SEFDM, showed that to date a small number of SEFDM-signals generation and reception schemes has been proposed which are of practical interest in terms of their implementation in digital signal processors (DSP) and programmable logic integrated circuits (FPGAs). Some of the authors (ex. in [1], [2]) considered only analog modulation of subcarriers or IDFT based algorithm with complexity at least of $Compl\left((I)DFT\right) = NL$.

In [5], [6] two schemes for SEFDM-signals generation and reception based on FFT/IFFT are proposed. In first scheme one N/α-point FFT/IFFT block is used (hereinafter, we denote this scheme as $(I)FFT_{N/\alpha}$). The computational complexity of $(I)FFT_{N/\alpha}$ is:

$$Compl\left((I)FFT_{N/\alpha}\right) = (N/\alpha)log_2(N/\alpha). \tag{14}$$

In the second scheme p N-point FFT/IFFT blocks are used (hereinafter, we denote this scheme as $(I)FFT_N^p$). The computational complexity of $(I)FFT_N^p$ is:

$$Compl\left((I)FFT_N^p\right) = pNlog_2(N) + (p-1)N. \tag{15}$$

Considering equations (14), (15) and (5), (13) it is clear to see that proposed in this article algorithms outperform methods in [5], [6] in terms of computational complexity.

For better comparison of algorithms it is convenient to consider the computational gain obtained when using various schemes of SEFDM-signals generation and reception: $(I)FFT_N$, $(I)FFT_L$, $(I)FFT_L^p$, $(I)FFT_{N/\alpha}$ и $(I)FFT_N^p$. The computational gain is calculated regarding the scheme based on DFT/IDFT:

$$\frac{Compl\left((I)DFT\right)}{Compl(X)} = \frac{NL}{Compl(X)},$$

here X is one of the considered in this article schemes. The plots for computational gain of $(I)FFT_N$, $(I)FFT_L$, $(I)FFT_L^p$, $(I)FFT_{N/\alpha}$ and $(I)FFT_N^p$ versus number of used subcarriers are shown on fig. 1.

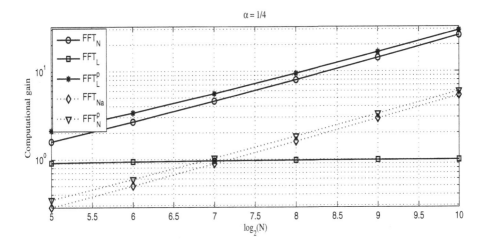

Fig. 1. Computational gain of $(I)FFT_N$, $(I)FFT_L$, $(I)FFT_L^p$, $(I)FFT_{N/\alpha}$ and $(I)FFT_N^p$ versus number of used subcarriers for $\alpha = 1/4$, $\alpha = 1/2$ and $\alpha = 2/3$

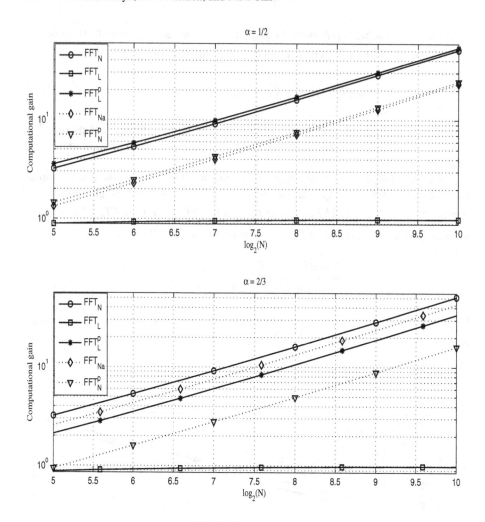

Fig. 1. (*continued*)

It can be seen from fig. 1 that efficiency of considered methods differs for various values of α. For $\alpha = 2/3$ and $N = 64...1024$ schemes $(I)FFT_N$ and $(I)FFT_L^p$ provide computational gain from 3 to 50. Moreover, the computational gain of $(I)FFT_N$ is 1,2 times more than of $(I)FFT_{N/\alpha}$. The comparison of schemes $(I)FFT_L^p$ and $(I)FFT_N^p$ shows that the first one outperforms the second one by 2,2 times. Further decrease of α leads to absolute superiority of proposed schemes $(I)FFT_N$ and $(I)FFT_L^p$ over $(I)FFT_{N/\alpha}$ and $(I)FFT_N^p$. The computational gain of $(I)FFT_N$ and $(I)FFT_L^p$ becomes 2...6 times better than of schemes in ([5], [6]). The computational gain of proposed algorithms for low values of α is 2,5...54 for $N = 64...1024$. At the same time the computational gain of proposed scheme $(I)FFT_L^p$ is quite low.

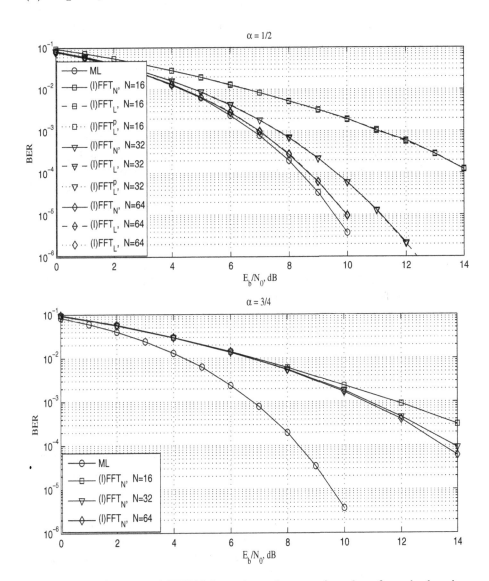

Fig. 2. BER performance of SEFDM for various schemes and number of guard subcarriers, $\alpha = 1/2, 3/4$, BPSK, AWGN

The computational complexity of scheme $(I)FFT_N$ does not depend on α. This fact comes from (5). The computational complexity of scheme $(I)FFT_L^p$ increases with α. It is important to note that computational complexity of $(I)FFT_L^p$ is lower than of $(I)FFT_N$. For $\alpha = 1$ the complexity of these schemes is equal which comes from (5) and (13). Thus when there is a choice which scheme to implement it is better to use $(I)FFT_L^p$. The most computational-intensive of all the developed methods for SEFDM-signals generation and reception is scheme $(I)FFT_L$. Its computational complexity increases linearly with α and exceeds the complexity of $(I)FFT_L^p$ 6 times for $\alpha = 0,5$ and more than 10 times for $\alpha \to 1$.

The BER performance of the proposed methods $(I)FFT_N$, $(I)FFT_L$, $(I)FFT_L^p$ in AWGN channel is shown on fig. 2. The number of used subcarriers is 10, the full number of subcarriers in SEFDM-signal including guard subcarriers is N. No interference compensation was applied. The BER performance of SEFDM-signals with BPSK modulation and corresponding α using maximum likelihood receiver is also shown on fig. 2 (marked as ML). All the proposed algorithms – $(I)FFT_N$, $(I)FFT_L$ и $(I)FFT_L^p$ – provide the same BER performance. The energy losses versus the ML scheme depend on the values of α and the number of guard subcarriers.

From fig. 2 it can be seen that the energy loss for BER = 10^{-3} lies from 0,2 dB for $\alpha = 1/2$ to 4 dB for $\alpha = 3/4$. The increase of guard part of SEFDM-signal from 19% to 42% from the used frequency range provides energy gain of up to 4 dB. Aliasing effect degrades the BER performance of SEFDM-signals significantly.

5   Conclusions

In this work three new algorithms for generation and reception of SEFDM-signals were proposed. These algorithms are based on FFT/IFFT, their complexity and conditions of applicability are analyzed.

It is shown that for the number of subcarriers $N = 64...1024$ the computational complexity of $(I)FFT_N$ и $(I)FFT_L^p$ schemes is 2,5-50 times better than of the algorithms based on DFT/IDFT. The proposed algorithms also outperform the schemes considered in [5], [6] by 1,2-6 times.

The BER performances of the proposed methods in AWGN channel are the same. For BPSK modulation the energy loss for BER = 10^{-3} lies from 0,2 dB for $\alpha = 1/2$ to 4 dB for $\alpha = 3/4$. The considerable energy loss for SEFDM-signals requires interference compensation methods to be used on the receiver side.

It is also shown that the number of the guard subcarriers significantly affects the BER performance of SEFDM. The increase of guard part of SEFDM-signal from 19% to 42% from the used frequency range provides energy gain of up to 4 dB. This is due to reducing the impact of aliasing.

References

1. Yang, X., Ai, W., Shuai, T., Li, D.: A fast decoding algorithm for non-orthogonal frequency division multiplexing signals. In: International Conference on Communications and Networking in China (CHINACOM), pp. 595–598 (August 2007)
2. Kanaras, I., Chorti, A., Rodrigues, M., Darwazeh, I.: Analysis of Sub-optimum detection techniques for a bandwidth efficient multi-carrier communication system. In: Proceedings of the Cranfield Multi-Strand Conference, pp. 505–510. Cranfield University (May 2009)
3. Kanaras, I., Chorti, A., Rodrigues, M., Darwazeh, I.: Spectrally Efficient FDM Signals: Bandwidth Gain at the Expense of Receiver Complexity. In: Proceedings of the IEEE International Conference on Communications, ICC 2009 (June 2009)
4. Kanaras, I., Chorti, A., Rodrigues, M., Darwazeh, I.: A New Quasi-Optimal Detection Algorithm for a Non Orthogonal Spectrally Efficient FDM. In: International Symposium on Communication and Information Technologies, pp. 460–465 (September 2009)
5. Ahmed, S., Darwazeh, I.: IDFT Based Transmitters for Spectrally Efficient FDM System. In: London Communication Symposium (September 2009)
6. Ahmed, S., Darwazeh, I.: Inverse discrete Fourier transform-discrete Fourier transform techniques for generating and receiving spectrally efficient frequency division multiplexing signals. American Journal of Engineering and Applied Sciences 4, 598–606 (2011)
7. Sergienko, A.B.: Digital signal processing / A. B. Sergienko – SPb, 608p. Piter (2002)

Projecting Resource Management of a Telecommunications Enterprise to Ensure Business Competitive Ability

Elena Balashova

St. Petersburg State Polytechnical University, St. Petersburg, Russia
elenabalashova@mail.ru

Abstract. Ensuring competitive ability of business under conditions of market stagnation aggravated with high competition is quite a complex managerial objective and cannot be solved by conventional means. Resource management implies simultaneous management of all available resources to achieve maximal efficiency. The goal of resource management is to project such management system that would ensure attainment of growth of quality of telecommunications services and reduction of their costs simultaneously. Due to this, organizational technologies that in-crease efficiency of management of a telecommunications company to ensure competitive ability of business acquire undisputable relevancy. Resource management is a modern methodology combining organizational technologies allowing the maximal efficient use and management of resource potential, detection and compensation of available production reserves.

Keywords: resource management, enterprise project management, lean production, reference model of business.

1   Introduction

According to research in the conditions of the Russian market for telecommunications services the growth dynamics of the key performance indicators exceed the growth of similar indicators of the global telecommunications market. Despite the slowing-down of Russia's GDP accompanied with the reduced investment and consumer demand, the telecommunications industry of Russian economy continuously grows, the most significant increment is recorded in the mobile communications and mobile internet sectors [1], the leading analysts of the telecommunications market noting that "...the market of telecommunications services of the Russian Federation is at the mature phase of development. The basic market segments (mobile communications and fixed internet access) enter the stage of stagnation..." [2]. The year 2014 challenges the basic players of the telecommunications market at least in two ways:

- joining of Rostelekom with TELE2 Russia and, as a consequence, the aggravation of competitive confrontation;

S. Balandin et al. (Eds.): NEW2AN/ruSMART 2014, LNCS 8638, pp. 502–508, 2014.

- an opportunity of a transfer of consumers from one mobile operator to another which has been already christened as the "abolition of mobile slavery".

All these factors change the picture of competitive equilibrium both for small telecommunications companies and for the "big three" - MTS, Megafon, Beeline. One of the main problems of insufficient competitive ability of domestic enterprises in the market is a problem of resource provision of not just current manufacturing activities but also innovative development, one of the key factors of low efficiency of innovative development of Russian enterprises being insufficient elaborateness of the methodology of scientific justification of balanced integrated resource provision of the basic activity.

Ensuring competitive ability of business under conditions of market stagnation aggravated with high competition is quite a complex managerial objective and cannot be solved by conventional means. The activity of any enterprise is subdivided into three types: basic (operational), financial, and investment. Management pays special attention, as a rule, to the latter two types. It raises no doubts, nevertheless, that success in the long run cannot be secured by smoothly running financing and well-thought-of investment programs only. Winning the competitive struggle requires, in the first place, an efficient mechanism of operating activities, i.e. that same basic activity due to which the business exists. The key attention of corporate management must be paid to the level and technologies of organization of basic business processes. It is here, in the sphere of organization of basic processes, that there is almost everywhere in the Russian economy the dominance of outdated conceptions and organization systems built on the principles of the bygone epoch of mass production.

The designated problem is noted at the state level. The President of the Russian Federation Vladimir Putin speaking at a meeting of the presidium of the State Council of the Russian Federation in Volgograd, said: "... one of the core reasons of today's problems is an archaic structure and organization of production..."; "... the experience of successful countries shows that a fundamentally new model of organization of production is required. A model oriented at creating innovations and relying on the competitive environment ..."

2   Lean Production in Resource Management

In the modern world the overwhelming majority of companies, leaders in their industries, acknowledge as such model the lean production model the main goal of which is to simultaneously reduce the costs of operation activities resulting in reducing the prime cost of a unit of production / service, and to better the quality of the latter. The achievement of the set goal is possible by creating and using a fundamentally new model of corporate resource management. The basic object of resource management has long been material flow at various stages of its movement. Due to this, lean production for the first time appeared on industrial enterprises (automobile building), the tools and methods of lean production were traditionally related to materials management.

The sphere of provision of services has long been excluded from the process of business reorganization according to the lean production principles. The situation changed when it became clear that a value stream rather than material flow should be the object of resource management, the presence of material flow is just an isolated occasion. Such transformation of management principles has lead to its being possible to use the lean production concept in the sphere of services.

The Savings Bank of Russia pioneered in this sphere which as long ago as in 2009 began implementation of a large-scale project for introducing the Savings Bank's Production system the basic idea of which is the orientation of the Bank's business at the customer's needs and continuous improvement of service quality including on account of improvement of in-house processes, i.e. in other words, lean production. The undisputed efficiency of such reorganization is proven by an analysis of the key performance indicators. Thus, during 2011 the Bank optimized 127 business processes in the result of which the increment of labor efficiency amounted to 11.5% [3].

Despite the objective efficiency of using the lean production concept in operational activities generally and in the sphere of resource management in particular, business is reorganized for implementing this managerial model usually as a response to unfavorable market conditions. Until nothing on the side of the market threatens a company, the management as a rule contemplate no changes in the managerial concept. Russian telecommunications companies have until presently been making stakes in issues of improving operational efficiency of business for the technological effectiveness of basic processes and the level of technological infrastructure. This is accounted for by enterprises of telecommunications industry being as a rule high-tech and science-driven. Besides, telecommunications enterprises are distinguished by:

- considerable initial costs;
- high capital and science intensity of production;
- limited production capacities;
- fast obsolescence of equipment.

The basic production processes are implemented using complex technical systems. As a consequence, the efficiency of organizational processes is often related to the quality and capacities of the equipment which is utilized. The improvement of the processes is perceived as a legitimate consequence of a company's technological infrastructure. Without belittling the significance of the latter, one would want nevertheless to note that the processes are effected by people, and improvement of their efficiency directly depends on organization of work of not only technical systems but also of a telecommunications enterprise's employees. Competition, high versatility, short lifecycle of services must all unavoidably influence corporate management system.

Considering the production system of a telecommunications company in terms of utilized resources (Fig. 1), it is expedient to single out their following basic types:

- financial,
- productive,
- labor.

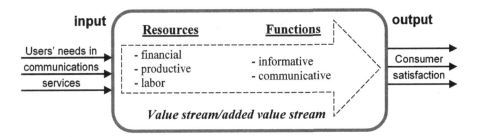

Fig. 1. Production system of a telecommunications company

The technological effectiveness of the basic production processes and level of technological infrastructure are indicators characteristic for production resources. Resource management implies simultaneous management of all available resources to achieve maximal efficiency. The goal of resource management is to project such management system that would ensure attainment of growth of quality of telecommunications services and reduction of their costs simultaneously. Such setting of an objective implies searching for and creating unconventional management models, for dependency of quality and costs was always considered direct, i.e. with the growing quality of a product or service their production costs were assumed to grow by default, which was justified by "quality requiring payment".

3   Projecting Resource Management

The projecting (from Latin. projectus, literally cast ahead) of resource management of a telecommunications enterprise to ensure business competitive ability is a process of creating a project which is a prototype, foretype of an assumed or possible object, condition [4]. One of the routes of projecting resource management of a telecommunications enterprise is to optimize business processes in accordance with international standards TeleManagement Forum (TMF). These are common processes for all companies which are embodied in a reference model of business processes.

For the first time the concept of reference model appeared in the medium of companies engaged in optimizing business processes and implementing ERP systems. A reference model (Fig. 2) is a model of an efficient business process created for an enterprise of a specific industry, implemented in practice and designed to be used when designing/reorganizing business processes at other enterprises. Essentially, reference models are templates of business organization developed for specific business processes on the basis of real experience of implementing in various companies internationally. They include field-proven procedures and methods of organization of management. Reference models allow enterprises to commence development of their own models on the basis of an already available set of functions and processes. The reference model of a business process is the aggregate of logically interrelated functions. A performer, input and output documents or information objects are specified for each function.

Many reference models of business processes are characteristic of and exist for other industries of business, but TeleManagement Forum (TMF) uniting around five hundred companies globally, has become in this occasion one of the first ones, bringing a model of eTOM[1] processes to the level of an international standard.

When designing process models, one should understand that an eTOM model is a reference one, and its uncontrollable use will yield no results, therefore for a specific organization it is just a model.

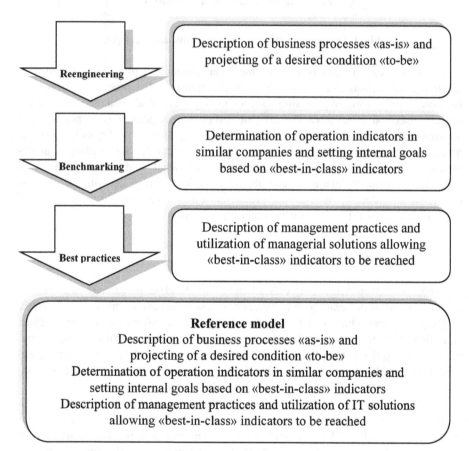

Fig. 2. Reference model of business

To build a reference model of resource management of a telecommunications company, the process of creating added value should in the first place be divided into the basic and supporting processes (Fig. 3).

---

[1]   eTOM – Enhanced Telecom Operations Map – extended structural model of business processes.

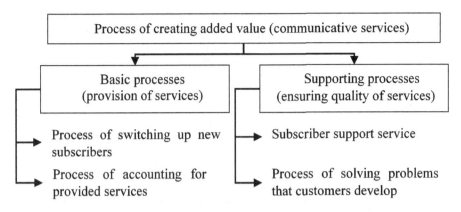

Fig. 3. Basic processes of a telecommunications company

An analysis of the current condition of business (as-is) shows that managerial attention is preferably given to the basic processes (provision of telecommunications services). The basic processes are constantly improved, the supporting ones are mainly subordinate to the logic of standard procedures, i.e. when a consumer applies to a support service, one of the possible standard algorithms is activated depending on the problem in question, the selection and management of the processes being given to the consumer. As a rule, this leads to a labor-consuming and low –efficient search for options.

All processes, basic and supporting, are communicative, therefore the traditional classification of production (resource) losses can be updated with taking into account priorities of communicative processes (Fig. 4).

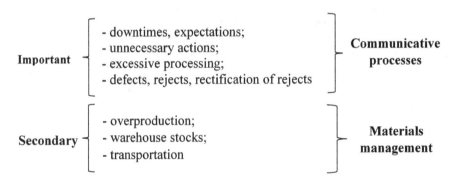

Fig. 4. Priorities of communicative processes

4 Results

Discovering the sources and reasons of occurrence of these losses and their reduction will allow the above set goals to be achieved: growth of quality of services and simultaneous reduction of losses for their creation and promotion. Thus, a reference model of resource management can in simplified form be described as follows (Fig. 5).

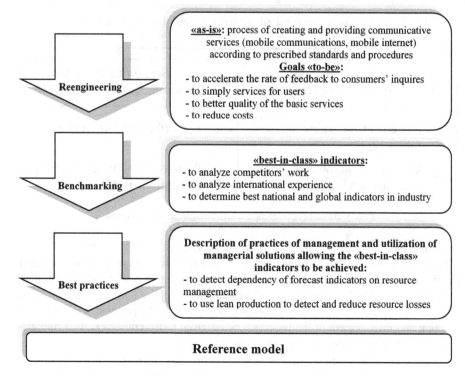

Fig. 5. Reference model of resource management

Production and economic activities of enterprises are always implemented under conditions of resource restrictions. Due to this, organizational technologies that increase efficiency of management of a telecommunications company to ensure competitive ability of business acquire undisputable relevancy. Resource management is a modern methodology combining organizational technologies allowing the maximal efficient use and management of resource potential, detection and compensation of available production reserves.

References

1. Russian market of telecommunications services grows faster than the global one, http://digit.ru/telecom/20120801/393758621.html
2. Russian market of telecommunications services: Results of 2012, forecasts of 2013-2014. NeoAnalytics. Smart marketing solutions, Moscow (2013)
3. Sberbank Production System Annual Report, http://2011.report-sberbank.ru/reports/sberbank/annual/2011/gb/English/10602010/sberbank-production-system_-version-2_0_.html
4. Contemporary vocabularies, http://slovari.yandex.ru

Project Portfolio Structure
in a Telecommunications Company

Vladimir V. Gluhov and Igor V. Ilin

Saint-Petersburg State Polytechnical University, Russia
vicerector.me@spbstu.ru, ilyin@fem.spbstu.ru

Abstract. The paper discusses the peculiarities of the composition of the project portfolio telecommunications company. Systematized set of projects aimed at development of business and solve specific business tasks. Special attention is paid to projects of simulation, implementation and optimization of business processes. The structure of the portfolio of projects in telecommunications Company is suggested.

Keywords: Project, project portfolio, business process, telecommunications company.

1 Introduction

Currently, the project management is actively developing as a field of research, as well as an organization of companies' activity of solving various levels of business problems. Enterprise activities become more project-oriented. Moreover, the project management can be viewed as an organizational business strategy. This is directly related to the telecommunications industry.

A lot of authors paid attention to different aspects of management of telecommunication companies [1]. The telecommunications industry is currently being actively developed. Development of companies focuses on business processes [2], technology, design and implementation of information technology solutions, development and promotion of new products and services [3]. It is tightly connected with project portfolio management issues.

Telecommunications companies in terms of business organization are both process- and project-oriented.

It should be noted, that the intense evolution of the environment determines the company's ability to take into account the evolving needs of market and business. When solving business problems, defined by the conditions of the environment, project management supports the companies' activities. Project approach allows creating the most effective business model, business architecture [4], herewith optimizing and maintaining internal processes' flexibility.

Business in the telecommunications industry includes two types of projects:

- Business reorganization projects,
- Projects meant to the creation and delivery of specific products and services to customers.

S. Balandin et al. (Eds.): NEW2AN/ruSMART 2014, LNCS 8638, pp. 509–518, 2014.

Project-oriented organization concurrently performs lots of projects and programs. It is necessary to comply with integration tasks. For this purpose projects and programs can be clustered into project portfolios, networks and chains. A project portfolio is a set of projects of project-oriented organization [5]. A network of projects is a set of closely-coupled projects held within the project portfolio [5]. A chain of projects is a set of sequential projects for the performance of several related business processes [5]. All these objects of management can be considered as clusters.

The determination of project company clusters is one of the most important tasks of project management of a telecommunication company.

2   Related Work

The authors distinguish between projects that are implemented for organizational change [6], process re-engineering projects [7] and projects of delivering specific products and services to customers [5]. Currently, the telecommunications industry is on the rise. The focus of the business development of telecommunications companies turns to the internal processes and technologies and the corporate project management system. The company's success in dynamic external environment determines the company's ability to adapt to the growing demands of the market and business, optimizing and maintaining flexible internal processes [2]. Project approach allows forming a business model that is the most effective in a dynamic environment: it allows you to define clear goals and manage changing business requirements [5]. Corporate project management system allows connecting project activities with the company's strategy, if necessary methodologies may consider the basic principles of eTOM/TAM (enhanced Telecom Operations Map/Telecom Applications Map) [8].

3   Business Reorganization Projects

Improvement of telecommunications companies' activities includes following projects:

- Projects of modeling, optimization and implementation of business processes
- Projects of development and implementation of information data models involved in the business processes of the company
- Projects of development of structure components (applications) of the company information environment
- Projects of architecture integration modeling and contractual interface definitions that define the interaction and integration of data and business processes in a distributed environment.

Projects of modeling, optimization and implementation of business processes extend [2], [9]:

- Projects of preparation for processes modeling
- Projects of developing a coherent processes structure

- Projects of modeling and analysis "as is" processes
- Projects of modeling "to be" processes and formation of the organizational structure
- Projects of processes implementation.

These projects can be considered as a chain of projects because they follow each other. The result of the each of first four projects is the input for the following project (Table 1). The result of the last project and the project of modeling, optimization and implementation of business processes in whole is an implemented system of business processes and a relevant organizational structure.

Table 1. An example of a project chain

Projects	Results
Project of preparation for processes modeling	<ul><li>Identification and Selection of Relevant Perspectives</li><li>Determination of communication channels</li><li>Specification of modeling technologies</li><li>Selection of modeling tools</li></ul>
Project of developing a coherent processes structure	<ul><li>Identification of core processes</li><li>Identification of support processes</li><li>Systematization of processes</li></ul>
Project of modeling and analysis "as is" processes	<ul><li>As-is processes</li><li>Identification and documentation of weaknesses and potential improvements</li></ul>
Project of modeling "to be" processes and formation of the organizational structure	<ul><li>To-be processes</li><li>Organizational structure</li></ul>
Project of processes implementation	<ul><li>Implemented system of business processes and a relevant organizational structure</li></ul>

Results of such a project chain are connected with service provision of telecommunications company. The development and optimization of communications service providers' business, as well as standardization of network management are the responsibility of the global non-profit industry association TeleManagement Forum (TM Forum). Documents developed by this association are accepted as industry standards by the International Telecommunication Union. In particular, TM Forum developed the enhanced business process map for communications service providers (eTOM) (Fig. 1), which is widely used in the area of telecommunications as a business process framework. There are also other referent models of business processes, for instance, CMMI (Capability Maturity Model Integration), set of models (methodologies) for improvement of processes in organizations of different sizes and types of activities. CMMI provides a set of recommendations in the form of practices, the implementation of which, according to developers of the model, allows realizing the goals necessary for the full realization of certain activities [10].

The standardization of eTOM in the frame of International Telecommunication Union is realized in the Recommendation Rec.ITUT M.3050.1 [8].

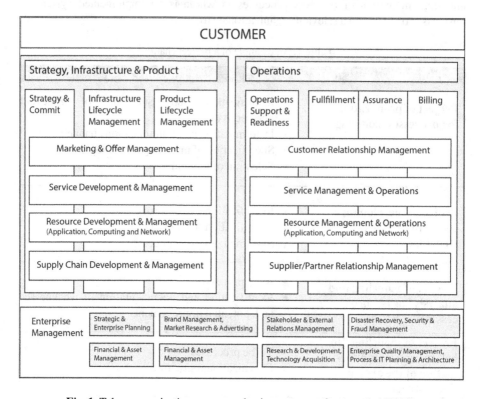

Fig. 1. Telecommunications company business process framework (eTOM)

This figure is used as a reference model. Based on this scheme, the following may be offered as an integrated business process structure of a telecommunications company:

- Main business processes (Operations)
  - Provision of telecommunications services
  - Ensuring the quality of communication services
  - Billing for telecommunications services
  - Exploitation of equipment and communication lines
- Supporting business processes
  - Buildings and structures
  - Operation of vehicles
  - Operation of non-current assets

- Management business processes
  - General company Management
  - External relations Management
  - Financial Management
  - Human Resource Management
  - Procurement Management

On the basis of the whole structure the following decomposition can be done:

- Main business processes (Operations)
  - Provision of telecommunications services
    - Interaction with potential customers
    - Informing and finding solutions according to customer needs
    - Receiving orders, applications, organization and execution of orders, customer informing
    - Market analysis of supply and demand for communication services, promotion of telecommunications services, production of promotional materials and analyzing feed-back
  - Ensuring the quality of communication services
    - Receiving messages about failures and problem situations from customers, solution of problems, informing customers of actions taken
    - Monitoring the quality of customer service
  - Billing for telecommunications services
    - Producing invoices and sending them to customers
    - Collecting payments and recovering indebtedness
    - Taking action on requests and customer complaints regarding payment of accrued amounts and receipt of invoices
  - Exploitation of equipment and communication lines
    - Provision and maintenance of facilities for equipment
    - Equipment ventilation and air conditioning
    - Daily monitoring of equipment status
    - Measuring network load
    - Inclusion of new trunks
    - Testing and calibration of equipment
- Supporting business processes
  - Buildings and structures
    - Safety and cleaning
    - Current and overhaul maintenance
    - Heating, electricity and water supply
  - Operation of vehicles
    - Current and overhaul maintenance
    - Technical support service
    - Clean-up service

- Operation of non-current assets
  - o Office equipment and personal computers service
  - o Procurement of information and automated control systems
- Management business processes
  - General Company Management
    - o Development of economic and organizational strategy
    - o Services quality management
    - o Coordination of interaction with the structural units and subsidiaries of the company
    - o Threat Assessment
  - External Relations Management
    - o Interaction with media
    - o Interaction with investors and founders
    - o Compliance dispute resolution
  - Financial Management
    - o Managing financial resources of the company
    - o Organization of accounting and report generation (financial, managerial, tax, statistical)
    - o Analysis of financial and economic activity
    - o Organizing the system of internal control and audit
  - Human Resource Management
    - o Implementing the policy of manpower allocation and promotion
    - o Development of organizational and production structure of the company, its improvement
    - o Increasing professional knowledge and skills of staff in accordance with business needs
    - o Definition of terms of employment, development of employment contracts
  - Procurement Management
    - o Conclusion and maintenance of the contracts with suppliers
    - o Monitoring of logistics and procurement
    - o Managing inventory.

This system of business processes can be used for realization of the project chain (Table 1).

The extended model of telecommunications company business processes (eTOM) is the process basis of the concept of TW Forum telecom industry organization. eTOM provides building the control systems as a set of modules – each system is fixed for different functions and business processes [8]. To ensure the integration of different systems a common information model is used. Within the NGOSS program the SID model is offered - a unified information model that describes the typical structure of an integrated information system data to support operations and business processes of the service provider. Model SID, as part of the NGOSS concept, is focused on providing data for systems of OSS/BSS class (Operation Support System/Business Support System). Its upper layer is shown in Fig. 2.

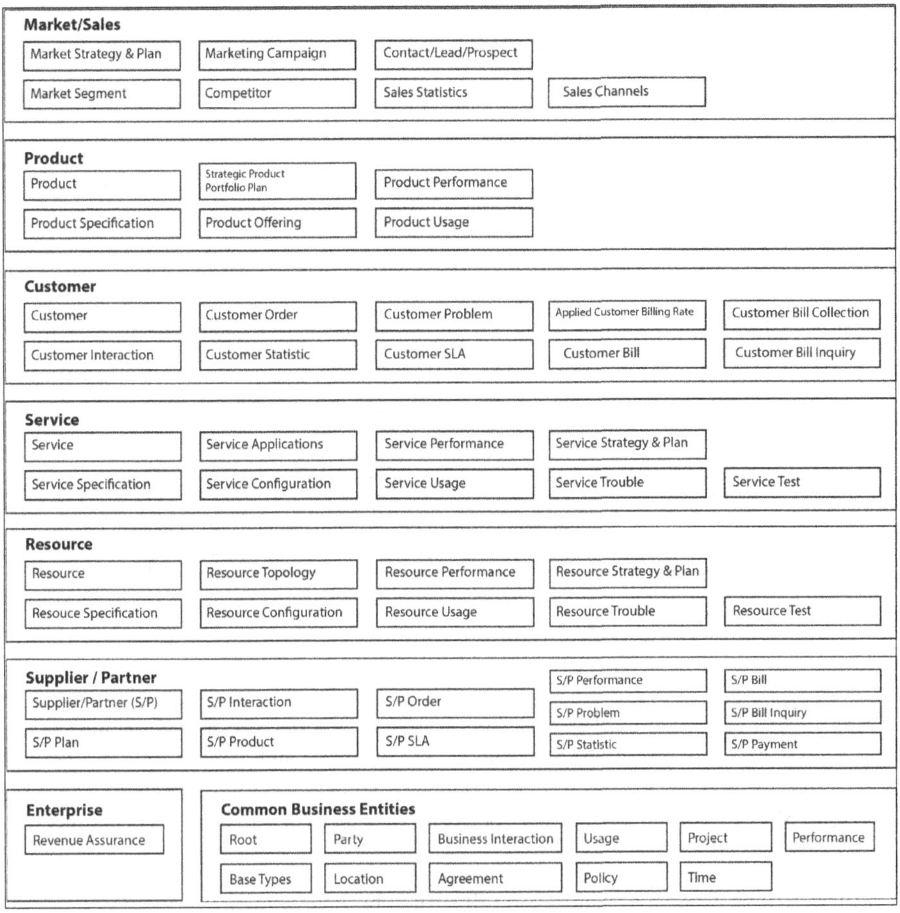

Fig. 2. Upper layer of CID model

Development of a common information model is also a project in the framework of the reorganization of the company. Together with the chain of projects (Table 1) it forms a network of projects.

Development of applications reference cards describing the structure of telecommunications company IT environment components is also one of the main projects within the reorganization of the company. TAM applications reference card can be used for this (Fig. 3).

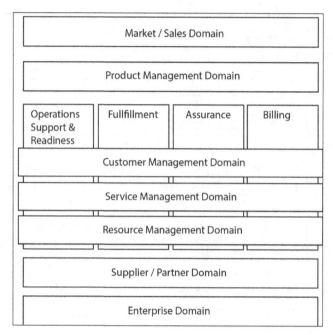

Fig. 3. TAM application reference card

Application reference card is a model of system architecture for telecommunications companies. A program of projects of forming and rolling out a system architecture follows the project of development of application reference cards.

To complete the structure of project portfolio it is necessary to add two programs:

- Program of project of hardware development
- Program of projects of locations and infrastructure development.

4   Results

A project portfolio of a telecommunications company comprises the following types of projects:

- Chain of projects of modeling, optimization and implementation of business processes
  - Projects of preparation for processes modeling
  - Projects of developing a coherent processes structure
  - Projects of modeling and analysis "as is" processes
  - Projects of modeling "to be" processes and formation of the organizational structure
  - Projects of processes implementation
- Projects of development and implementation of information data models involved in the business processes of the company

- Program of projects of development and rolling out of structure components (applications) of the company information environment
  - Quality management
  - OSS class
  - Monitoring and control of data transmission services
  - Managing faults in networks
  - Managing customer orders
  - Managing infrastructure projects
  - Managing projects of developing new products and bringing them to market
  - Managing projects of implementation and modification of billing systems
  - Formation and management of programs and portfolios of projects of telecommunications company development
- Program of projects of architecture integration modeling and contractual interface definitions that define the interaction and integration of data and business processes in a distributed environment
- Program of projects of hardware development
- Program of projects of locations and infrastructure development.

To implement the examined projects a corporate system of project management can be designed and implemented. Its formation requires:

- Development of a project management methodology. Statement of standard processes of project management, both client and internal development related
- Development of a project management organizational structure of project management for the entire company or a separate department or direction
- Organizing and forming requirements for project management process automation and implementation of these requirements using systems of any complexity.

5   Conclusions

Telecommunications companies simultaneously perform a multitude of various projects and programs. To comply with integration tasks they can be considered as a project portfolio. Synergies can be created by means of considering relationships between the projects. Understanding the structure of the portfolio of projects of the company allows to simplify considerably the formation of the portfolio of projects of a particular company and to manage effectively the development and implementation of projects and programs of business restructuring and solving specific business tasks. Moreover, it allows implementing the overall architecture model of company management more effectively [11], [2] and improving an organization's current project management maturity level [12].

References

1. Gluhov, V.V., Balashova, E.S.: Economy and management in the telecommunications. Piter, Saint-Petersburg (2012)

2. Becker, J., Kugeler, M., Rosemann, M.: Process Management. Springer, Heidelberg (2011)
3. Lankhorst, M.: Enterprise Architecture at Work. Modelling, Communication, Analysis. Springer, Berlin (2013)
4. Ilin, I.V., Antipin, A.R., Lyovina, A.I.: Business architecture modeling for process- and project-oriented companies. Economics and Management 95, 32–38 (2013)
5. Gareis, R.: Happy Projects. MANZ Verlag, Vienna (2005)
6. Gemünden, H.G., Salomo, S., Krieger, A.: The influence of project autonomy on project success. International Journal of Project Management 23, 366–373 (2005)
7. Hvam, L., Have, U.: Re-engineering the specification process. Business Process Management Journal 4(1), 25–43 (1998)
8. ITUT Recommendation M.3050.1 Enhanced Telecom Operations Map (eTOM) – The business process framework (2013)
9. Becker, J., Vilkov, L., Taratuhin, V., Kugeler, M., Rosenmann, M.: Process Management. Exmo, Moscow (2010)
10. Chrissis, M.B., Konrad, M., Shrum, S.: CMMI: Guidelines for Process Integration and Product Development. Addison-Wesley, Boston (2003)
11. Kondratiev, V.: Projecting corporate architecture. Exmo, Moscow (2007)
12. Kwak, Y.H., Ibbs, W.C.: Project Management Process Maturity (PM2) Model. Journal of Management In Engineering, 150–155 (2002)

Forecasting of Investments
into Wireless Telecommunication Systems

Tatyana Nekrasova, Valery Leventsov, and Ekaterina Axionova

St. Petersburg State Polytechnical University, Saint Petersburg, Russia
dean@fem.spbstu.ru, vleventsov@spbstu.ru, director@eei.spbstu.ru

Abstract. The method of forecasting of investments into the wireless telecom systems is suggested. The factors influencing the investments are determined.

Keywords: forecasting, investments, wireless telecom systems, base transceiver stations, subscribers, communication channels, project effectiveness.

Nowadays the investments into telecommunications including wireless telecommunications can be characterized as huge and their volume can be estimated as tens of millions dollars. The author [7] has predicted the future of telecommunications, which in turn demands for special predicting and estimating methods for investments into this sector.

To estimate the investments into the wireless telecommunication systems it is suggested that the already existing methodic approaches are to be used. But this approaches should be appropriately adjusted for the features typical to the investment process in the sphere.

To determine the overall investment volume and the investment structure is the initial stage of the investment process management. These indicators are determined by the prediction calculations.

The forecasted investment volume is determined by the following approaches:

- based on the possibilities of the company to build up the investment resources, which corresponds with conservative fundraising policy;
- based on the full satisfaction of the demand for investments.

The mechanism of forecasting of the investment volume consists of the following steps [1]:

- to determine the operational net income flow;
- to determine the participation in the total investment resources ratio;
- to determine the volume of financing with the own funds;
- to determine the volume of financing with the borrowed capital;
- to determine the volume of borrowing;
- to determine the overall investments volume during the forecasted period;
- to determine the structure of investments during the forecasted period.

S. Balandin et al. (Eds.): NEW2AN/ruSMART 2014, LNCS 8638, pp. 519–525, 2014.

The participation ratio that considered in the calculations of the forecasted volume of the investments with the own funds; can be calculated as [1]:

$$K_1 = \frac{C_1}{C_2} \qquad (1)$$

where C1 – the total amount of the investments, built up with the own funds during the period; C2 – the total amount of funds, built up with the own capital.

$$C_3 = NPV_p \times K_1 \qquad (2)$$

where C3 – the forecasted amount of the investments with the own funds; NPVp – the predicted operational net cash flow.

The overall forecasted investments volume is the following sum:

$$V_p = C_3 + C_4 + C_5 \qquad (3)$$

where C4 – predicted amount of the own investment capital, that is formed with through external sources; C5 – predicted amount of the borrowed investment capital.

The predicting method that is used for matching the investment demand completely with the investment capital works under the following assumption:

$$I_p = I_n \qquad (4)$$

where Ip – total amount of the investment resources; In – total amount of the demand for investments within a company. While forecasting the total volume and the structure of investments the following indicators are used:

- the provision index of the investment activity

$$K_2 = \frac{I_{pp}}{I_{nh}} \qquad (5)$$

where I_{pp} – forecasted amount of the investment resources; I_{np} – forecasted demand for investments.

- the self-financing index

$$K_3 = \frac{C_3 + C_4}{V_p} \qquad (6)$$

The further suggested estimation of the investment volume necessary for development of a wireless telecom company is carried out under the assumption that these investments are financed only with own funds raised through internal sources.

The investments into the wireless telecom infrastructure are influenced by the following factors: developing and implementation of new technologies; rapid change of technology; accessibility of area where wireless network is to be implemented (geographical factor); complexity of equipment; extension of the coverage area; extension of the market.

Developing and implementation of new technologies and changes of technology demand additional investments within wireless telecom services sector.

The forecasted investments needed for creating and developing of the wireless infrastructure are determined with the formula:

$$I_T^{np} = I_1 + \sum_{t=1}^{T} \frac{I_{t+1}^{np}}{(1+r)^t} \tag{7}$$

where I_1 – overall investment costs needed for organizing and construction of the wireless telecom system during the initial period; I_{t+1}^{np} – the investment costs needed for organizing and construction of the additional wireless elements during the t+1 period (at the beginning of the year); r – banking cash rate; t – number of the period.

The total investment costs for creation of the wireless telecom system in the host region that are necessary for servicing a certain number of initial subscribers can be determined as:

$$I_1 = \sum_{n=1}^{N} I_{BTS_n} + I_{MSC} + I_l \tag{8}$$

where I_{BTS_n} – initial investments in construction of cellular base transceiver station (BTS); $n=1....N$ – number of BTS; I_{MSC} – investments needed for building a mobile switching center (MSC) serving the N number of base stations; I_l – cost for licensing.

The needed investments for creating a BTS (building, purchasing and installation of the equipment) is determined by the number of BTS (N) that are connected to the MSC (the number depends on the number of subscribers (density of population), characteristics of the coverage area, technical features of base stations (number of communication channels, communication standard)).

The necessary number of BTS that depends on the coverage area can be calculated with the formula:

$$L_S \cong \frac{S}{S_i} \tag{9}$$

where S – total square of the coverage area; S_i – coverage area of the i-th base station, say, one sq.km. $L_s >> 1$.

Simultaneously, the needed number of BTS that is determined by the number of subscribers, equals:

$$L_q = \frac{Q}{n_s} \tag{10}$$

where Q – overall number of subscribers; ; n_s – average number of subscribers in the area of 1 sq .km. $L_q >> 1$. Consequently $L_s \cong L_q$.

The accomplished calculations show that the needs of the average number of subscribers (ns) are satisfied with one single BTS and the number of coverage areas is stipulated by the number of BTS with capacity sufficient for serving the total number of subscribers (Q).

The capacity of BTS varies from 1000 up to 10000 communication channels and it matches the capacity of the MSC. The needs of the subscribers can be satisfied through the following wireless organization schemes:

- setup of many BTS with low capacity (n_{cp} = 1000 pers.);
- setup of a smaller number of BTS with high capacity ($n_{cp} \geq$ 10000 pers.).

The dependence of the investment volume on the technical scheme applied is shown on the Figures 1-3. It should be pointed out that equipment (with the setup included) is not considered to be costly for wireless telecom companies.

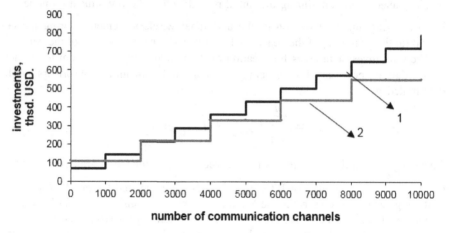

Fig. 1. Investments for creating BTS with capacities of 1000 and 2000 channels (1 – capacity of 1000 channels, 2 – capacity of 2000 channels)

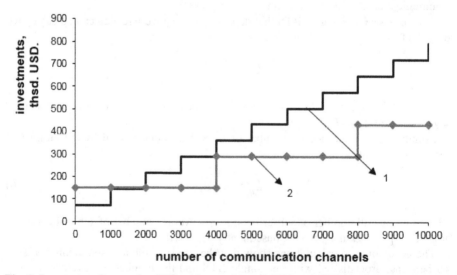

Fig. 2. Investments for creating BTS with capacities of 1000 and 4000 channels (1 – capacity of 1000 channels, 2 – capacity of 4000 channels)

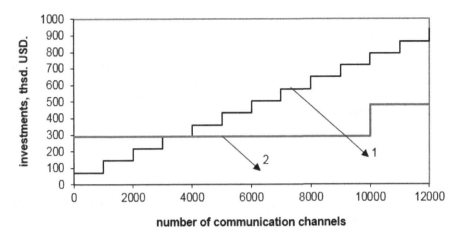

Fig. 3. Investments for creating BTS with capacities of 1000 and 10000 channels (1 – capacity of 1000 channels, 2 – capacity of 10000 channels)

The analysis of the charts shows that investing into less costly equipment with low capacity with forecasted demand higher than 10 000 subscribers is unpractical (costs ~25% higher with 4000 subscribers).

Low-capacity BTS are sensible for the areas with low population density (remote areas, coasts, along railroads and highways).

The obtained index represents one of the planning and forecasting steps in the investment activity of a telecom company. Complexity and irregularity of landscape or some types of soil requires some upgrading of BTS for maintaining of normal quality of communications according to various parameters. This leads to an increase of the needed investments.

The investments needed for creating and organizing of BTS for serving of a given area with various geological and geographical conditions can be calculated with the formula:

$$I_{BTC_{nj}} = I_{BTS_{nj}} * k_j \left(1 + \xi\right) \tag{11}$$

where $I_{BTS_{nj}}$ – the needed investments for creating a BTS in a j-th region; k_j – the investment index of the j-th region; ξ – the reserve for the costs (in decimal points of the total investments for building of a BTS in the j-th region).

The mobile switching center of the BTS depending on the technical features can coordinate and serve from N1 up to (N1)max of basic stations, which may need some technical upgrade of the equipment. The increase of the number of basic stations above the (N1)max level demands for creating and implementing of a new MSC, that will serve from N2 up to (N2)max basic stations.

The investments for building and maintenance of the MSC depend on the number of BTS within the operational radius (say, 12 km), characteristics of the coverage area, communication standard and capacity (the number of communication channels).

The changes of the investments needed for creating of mobile switching centers where each of them serves for N BTS, demonstrate the dependence shown on the Figure 4.

The investments needed for creating a MSC in the j-th region are calculated in the similar way as it was for creating a BTS in the j-th region.

Therefore, the investments for creating of the wireless telecom system in the j-th region can be determined with the formula:

$$I_{1j} = k_j * (1 + \xi) * \left(\sum_{n=1}^{N} I_{BTS_n} + I_{MSC} \right) + I_l \qquad (12)$$

The costs of licensing is viewed as a constant value that should be paid once. The law stipulates the volume of payment.

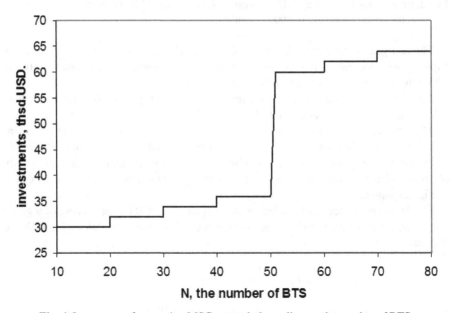

Fig. 4. Investments for creating MSC network depending on the number of BTS

Some part of the investments constitutes the fixed assets of the telecom company. They are the subject for depreciation. To calculate the amount of fixed assets one distributes the overall investments among various kinds of work:

$$I_1 = I_2 + I_3 + I_4 \qquad (13)$$

where I_2 – investments needed for financing of construction and installation work; I_3 – investments needed for purchasing of the equipment; I_4 – other costs.

The distribution of the investments among the types of work is based on the total cost statement for the building project.

The changes in the investments for building of new BTS and MSC depend on the number subscribers, number of communication channels and the coverage area. The forecasted number of subscribers can be obtained with the help of analytical method of demand estimation [5]. The method considers such factors as sales volume, the prices for the services, the income of the subscribers during the basic and the estimated time periods and share and trends of the market for a given service.

The forecasted investments build up the basis for calculations of the performance indicators of wireless telecom companies. The calculations also include calculations of cash flows, which make further estimations of the project effectiveness possible (NPV, IRR, PI, IP) [3, 6]. Also this can help to rank the project and to compare it with other possible investments.

Implementation of investment projects in the wireless telecom sector is closely connected with the emergence of some technical, technological, investment and geographical risks that influence the revenues of the companies. Despite of that, however, in the short run this investments can be viewed as perspective.

References

1. Blank, I.A.: Investment management, 469. p. Nika-tsentr Publishing House, Kyev (2003)
2. Gloukhov, V.V., Balashova, E.S.: Economics and management of telecommunications. Textbook, 272 p. Piter Publishing House, St. Petersburg (2012)
3. Kovaliov, V.V.: Budgeting techniques, 143 p. Finansy i Statistika Publishing House, Moscow (2000)
4. Makarov, S.B., Pevtsov, N.V., Popov, E.A., Sivers, M.A.: Telecommunication technologies: introduction to GSM technologies, 256 p. Academia Publishing House, Moscow (2006)
5. Smirnova, O.A.: Demand estimation and forecasting. Textbook, 52 p. Nestor Publishing House, St. Petersburg (2001)
6. Margolin, A.M., Bystriakov, A.J.: Assessment of investments, Moscow. Textbook, 240 p. (2001)
7. Berner, G.: Management in 20XX, Siemens (2004)
8. Trends in Telecommunication Reform 2009. Summary, ITU (2010)

Assessment of Possible Uses of Optimization Methods and Models in Resource Management of Telecommunications Enterprises

Elena Balashova and Evgeniy Artemenko

St. Petersburg State Polytechnical University, St. Petersburg, Russia
{elenabalashova,evg_art}@mail.ru

Abstract. An analysis of the Russian market for telecommunications evidences the relevancy of a search for new approaches to business development. Modern resource management presented by various academic schools (Lean production, Theory of constraints, Resource – based view) ensures possible multivariant solution of managerial objectives. The use of optimization methods and models in resource management allows telecommunications enterprises to considerably improve economic performance indicators. Under the conditions of diversity of possible approaches to searching for and implementing optimal solutions of resource management of telecommunications enterprises, the choice of the optimization method or model most suitable to this situation turns out to be a non-trivial problem the solution of which requires a clear understanding of advantages and drawbacks of their use when solving a specific managerial problem.

Keywords: models of resource management, lean production, theory of constraints, resource – based view, the optimum of an economic system, economic-mathematical models.

1 Introduction

The market for telecommunications in Russia has long been developing by almost the same rates as in the raw materials industries. Three years in a row (2002-2004) its turnover increased by 40% annually. The rate of the market development was accounted for by many factors, in the first place, by an evolution of new technologies and relative underdevelopment of telephone infrastructure. The actual growth rates of business of Russian mobile operators in the beginning of the 2000s exceeded the scheduled ones 2.5-3 times, which led to the connectivity of mobile communications in Russia exceeding 166% when calculating by SIM cards and 110% when calculating by active subscribers. The quantity of broadband internet subscribers exceeded 27 mln., the connectivity attained 49%, attaining 60-80% in big cities and often exceeding 50% in smaller cities with population below 500 thousand. In 2014-2018 the Russian telecommunications market will according to experts' estimate slow down the growth rates. If in 2013 the market attained 1 trillion 635 billion rubles, and its growth rates amounted to 6%, which is 1% below the similar index of the previous year, the

S. Balandin et al. (Eds.): NEW2AN/ruSMART 2014, LNCS 8638, pp. 526–534, 2014.

forecast for 2014-2018 will be 4%. Negative tendencies brought about by the rapid growth of the market in the beginning of the current decade are noted in all the spheres of the telecommunications sector. The low growth rates are accounted for by the exhaustion of sources of extensive growth in all largest facilities. The varied market realities will inevitably aggravate the competitive struggle in the sector. The companies which are used to a high level of main economic indicators of their activities have to change the management paradigm. An impossibility of further extensive development of business leads to updated options of intensive development of telecommunications companies. Of special importance are managerial models allowing multivariant solution of problems. Resource management is in this respect a unique opportunity of efficient business management.

2   The Concept and Models of Resource Management of a Telecommunications Enterprise

Resource management generally is an economic function whose implementation sphere is efficient resource support of current activities and management of resource potential of an organization in projected activities. Currently no single methodological approach to determination and formation of a management mechanism by aggregate economic resources of an enterprise has been worked out. Joint resource management is the sphere of science presented by academic schools and directions considering efficiency from various points of view and sides. The main modern models, however (Table 1) have similar goals based on the following principles:

1. minimization of aggregate costs;
2. enhancing efficiency of using available resources;
3. maximization of a company's profits.

Table 1. Modern models of resource management of an enterprise

Models	Authors, years of publication	Basic idea
Lean production	Taiti Ono, Masaaki Imai (authors of the model) [1,2], John Krafchik (author of title) [3]	Enhancing efficiency on the basis of internal reserves of an enterprise; process of continuous improvement
Theory of constraints - TOC	Eliyahu Goldratt [4,5]	Finding and management of the key constraint of the enterprise system (a weak link), focusing of the available resources of a company on a small quantity of the system's aspects
Resource – based view – RBV	K.K. Prahalad, Gary Hamel [6], David Collins, Cinty Montgomery [7]	Development of competitive advantages of a company on the basis of having unique resources and organizational abilities (key competences)

An explosive interest in resource management in the second half of the 20th century was substantiated by the market conditions which led to an almost simultaneous appearance of various approaches. The difference of models presented in (Table 1) is accounted for by, in the first place, academic schools on which the modern resource models are based (Table 2).

Table 2. Development of resource management systems

Basic academic schools	Key problems – substantiation of appearance of the school	Aspects of solutions of set problems	Modern model of resource management
Approach to management as a process (process approach)	Strengthening of managerial control over all business parameters	Projecting of processes ensuring added value	Lean production
School of quantitative approach	Assessment of solution rationality	Rational distribution of resources of an enterprise	Theory of constraints - TOC
System approach	Presentation of an organization as a single system adaptation of external environment	System building Strategic planning	Resource – based view – RBV

There are no universal models of resource management. The advantages of each model are necessarily accompanied by its drawbacks. The attainment of the stated goals (optimization of costs and maximization of profits) in each model is expected in accordance with its unique inherent algorithm. The study of unique specifics of the models of resource management, their merits and drawbacks allows the management to use the managerial patterns suitable to the respective current conditions.

Thus, for instance, the "lean production" management model is an extraordinary popular and extremely widespread system of highly efficient management. Lean production is built on simultaneous attainment of the following internal goals ensuring a company's competitive advantage:

- the constant reduction of costs of production and sales of products;
- the continuous improvement of the quality of products;
- the optimization of the speed of movement of materials flow according to the Just In Time logic.

The basic idea is to detect production losses, i.e. actions for which the consumer does not intend to pay. At that, it is necessary to note that here there is a certain rupture of the logic chain – the consumer does not pay for actions, at the moment of purchase he/she pays for the final product or provided service. An understanding of which actions result in generating this product or service allows the telecommunications activities which are carried out to be divided into conditional groups and managed in different ways (Fig. 1).

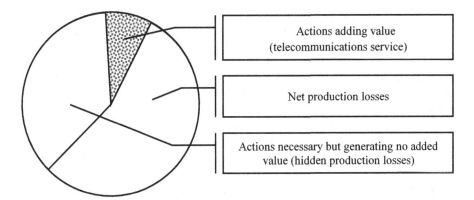

Fig. 1. Grouping of kinds of activities depending on an attitude to generating added value

It is accepted to believe that lean production is a model of a company's management generally. We believe, though, that it is expedient to refer it to the model of resource management, i.e. it is on resource management that the focus of managerial attention is. The famous classification of production losses (overproduction, expectation, excessive action, excessive processing, warehouse stocks, defects, rejects, reprocessing, transportation losses) is hinged on improvement of efficiency of using available resources of an enterprise, their optimization and attainment on this base of any stated goals.

Lean production implies simultaneous action on the entire value stream, basic tools of lean production (kaizen, statistical process control and others) are based on the idea of continuous improvement of the entire process of activities at an enterprise. At that, the basic idea can be formulated as "we improve the result by improving the process".

Another, not less famous model of resource management of a telecommunications enterprise, known as "the theory of constraints" (TOC) was developed and proposed by E. Goldratt in the 1980s of the previous century. The theory of constraints sets before management the same goals that lean production, but their attainment is intended following other principles, namely, the method of the so-called "critical chain". As a chain, an enterprise is offered for review, and the basic idea of this model consists in the focusing of managerial attention on the weakest link of the chain. The available resource base must be aimed at eliminating the "weakest link" of an enterprise's production system. Its elimination will cause the strength of the chain, the efficiency of the entire enterprise's activities, to grow, but only to the next constraint in the queue, i.e. to the nest weak link. Thus, Goldratt's theory of constraints matches the pin-point strategy of resource potential management against the wide resource management of the lean production system. In other words, lean production offers dispersal of managerial efforts and available resources for a company in general and attainment therewith of implementation of a continuous improvement process, kaizen. The theory of constraints is built in its turn on the idea of the focusing of resource management in one points of a company's production system.

The third model of modern resource management highlighted in table 1 is the newest one which appeared in the end of the 20th century.

The original specific feature of this model resource management is in the logic of organizational construction of the business. To minimize the total level of costs, those business functions must be given to outsourcing which are not "specific in respect of the firm", specific ones (it is accepted to call them "key business competences") must remain inside the business subject, i.e. the enterprise, their development must bring to the company the required level of competitive ability. Consequently, such strategy of resource management can be conditionally called a "sampling" one, for it is based in the first place on the selection of resources which are key competences.

The term "business key competences" was introduced by scientists K.K. Prahalad and G. Hamel [6]. The substantial idea of key competences is that within resource management they must be in all manners developed and safeguarded, whereas other, non-original business resources should be leased or outsourced, if a comparative analysis of costs shows that in this way the aggregate costs of production and sales of products are either reduced, or pass from the constant component to the varying one, depending in direct proportion on the production volumes and sales.

3 Methods and Models of Optimization in Resource Management of Telecommunications Enterprises

Under the existing conditions of multivariant approaches to a search for and implementation of optimal solutions in the sphere of resource management of telecommunications enterprises, substantiation and adoption of such solutions (economic, organizational, marketing, commercial, and managerial) can be effected in several ways significantly differing in efficiency:

- conventional, i.e. on the basis of accumulated experience and existing traditions which earlier did not result in any serious mistakes;
- on the basis of available qualification and common sense, which allows solutions that are far from being worst to be taken;
- on the basis of formalizing a problem to be solved using criteria and constraints allowing best solutions to be found.

The diversity of possible means of attaining a specific goal characterized with the common performance indicator is accounted for by the ambivalence of influence of various factors on the value (criterion) of efficiency, and predetermines the existence of an optimal mode of action ensuring the extreme performance indicator.

The problem of a search for an optimal option of development of an economic system which we believe to be telecommunications business generally or its separate component, resource management, is a most important problem based on the concept of an economic optimum.

An economic optimum characterizes the aggregate of the most favorable conditions for development of an economic system with the set goal and constraints to its possible states. A search for an optimum is a search for the best parameters of development of a system capable of being quantitatively described, and the parameters ensuring attainment of a goal set in an extreme form [8].

An efficient search for and finding of the optimum of an economic system is possible under the following conditions:

- the tendencies of the development of the system to the optimum;
- the multiple states of the system;
- the quantitative description of these states;
- the willful optimization, i.e. application of a certain body of methods and procedures.

The available multiple states of an economic system and its components at various levels of management are predetermined by a multiple factor and multiple resource nature of economic and managerial objectives.

One of the key ideas determining the development of the management theory over the last years is the idea of optimization. The basis of optimization methods and models is extremized economic criteria combined with constraints for resources and the required level of technical and economic indices. A doubtless advantage of the optimization approach is its resource-saving character, which allows the best option of using resources to be found both when arranging for private economic processes and when managing an economic entity in general. In combination with using computing facilities optimization methods and models allow a wide class of extreme problems to be solved at various levels of management.

It should be noted that, despite the considerable potential, the methods of optimization are not currently widely used. The basic reasons constraining their application are: insufficient theoretical development of such methods taking account of practical needs; insufficiency of initial data or the absence of their clear differentiation; insufficient assessment of efficiency of optimization methods by companies' top executives.

It is possible to choose an optimal solution from, as a rule, an unlimited number of solutions using methods allowing a problem to be formalized and in a limited number of steps to find a solution ensuring the extremum of a problem efficiency criterion. This is attained by operations research methods allowing optimal solutions to be quantitatively substantiated prior to their implementation in practice. In the context of the approach in question an operation is any measure (a system of actions) joined by a common design and aimed at attaining an earlier defined goal. An operation is always a willfully controlled measure depending on the chosen parameters characterizing it.

Applied operations research problems can use achievements of any sphere of mathematics. There is, however, a special class of methods specifically designed to solve such problems – mathematical methods of operations research [9].

Depending on the nature of factors that are taken into account and solution methods that are being used several groups of problems and economic-mathematical models of operations research are singled out.

The first group may include problems implying the finding of a solution under given conditions prescribed in the determined form, which convert the target function (an optimality criterion) to the extremum. The mathematical methods of setting and solving this type of problems have been developed in sufficient detail; their wide application, however, is restricted with a number of significant conditions. First, the number of the unknowns must be small, otherwise the system of constraints is cumbersome and inadmissibly complex for calculating procedures Second, the derivatives of the functions in question must be determined in all points, i.e. the functions being part of an economic-mathematical model must be continuous.

When these conditions are not complied with, problems of this group are solved by methods of mathematical programming – the sphere of mathematics developing the theory and numeric methods of solution of multidimensional extremal problems with constraints, i.e. the problems for the extremum of a function with a large number of variables and constraints to the these variables domain [10].

The name "mathematical programming" is related to the goal solving similar problems being the choice of a program of actions. The methods of solving linear programming problems were the first to be developed. In practice similar problems often contain the condition of integrality of variables, the obtaining of the initial nonintegral solution with subsequent substitution of integers for it leading to a shift of extremum of a target function, which is inadmissible. The methods of integer programming are designed to solve this problem.

In certain cases, when solving linear programming problems it is required to know within which limits the values of a target function and the conditions prescribing constraints can vary, taking into account that the obtained solution must remain optimal – a necessity arises to investigate the influences of these variations on the found solution. Parametric programming engages in analyses of such kind.

A wide class of nonlinear and discrete problems is solved using ideas of recurrent approach. Such methods are joined under the appellation dynamic programming. Solving optimization problems with random parameters in functions determining constraints comprises the area of application of stochastic programming.

The second group of problems is characterized by their taking, apart from the factors which problems of the first group have (the determined constants and the unknowns whose set of admissible values is the solution of a problem), into account also random factors, factors prescribed in a certain way, or unknown ones. At that, tow basic subgroups of problems are distinguished.

In problems of the first subgroup random factors have known laws of distribution, the solution takes place through extremization of mathematical expectation of the target function. In the second subgroup random factors cannot be described using statistical methods, for the parameters of their laws of distribution are unknown or there are no such laws (random parameters have no statistical stability). In this case a search for a set of locally optimal solutions is conducted. A compromise solution that, not being strictly optimal for any conditions, turns out admissible for a certain range of states, rather than a solution optimal for certain definite but improbable in a general case conditions is preferable in this case. The value of similar problems in their educational character: multiple solutions of such problems ensure attainment of required experience and an opportunity of the prompt finding of a solution in the simulated situation which occurred earlier.

The third group of problems is characterized by taking into account indefinite (random) factors related to the willful action of external forces and depending on no objective circumstances. Similar problems arising in conflict situations are solved using the game theory which is the ordered universe of mathematical models of taking optimal solutions under the conditions of contending interests of parties or uncertainty. At that, any difference between the participants of the game can be viewed as a conflict.

The contents of the mathematical game theory consist:

- in establishing the principles of optimal behavior of players;
- in proving the existence of situations that appear as a result of using such principles;
- in developing the methods of actual finding (above all quantitative) of such situations.

Most principles of optimal behavior of players have a character of stability principles: a deviation from a situation that has appeared as a result of optimal behavior must either become disadvantageous in itself, or after such deviation an opportunity must appear to change over with a benefit to a new situation attainable with the aid of optimal behavior.

As applied to coalition-free games, the stability idea is implemented as a form of stability principle leading to situations in which it is beneficial to neither player to change his/her strategy, if the remaining players retain their strategies unchanged.

4   Conclusions and Results

An analysis of modern methods and models of resource management of a telecommunications enterprise allows the following conclusions and assumptions to be made:

- having comparable, and often similar goals, various methods and models of resource management have an original algorithm of their attainment, the focus of managerial attention having to be oppositely different;
- maximization of profits is attained in one case by reduction of costs, in other models by increasing revenues;
- minimization of aggregate costs can by a consequence of implementing principally different resource strategies – pin-point, extensive, or sampling.

It is necessary to take into account that any optimization model is a symbolic representation of reality, its simplified reflection, therefore a mistake in choosing the type of utilized model can lead to a considerable reduction of practical value of the results obtained with its help. Understanding the essence of differences in using various methods and models in resource management allows those of them to be addressed in real economic practice the advantages of which can be used fully, and the influence of drawbacks minimized.

References

1. Imai, M.: Gemba Kaizen: A Commonsense, Low-Cost Approach to Management, 1st edn. McGraw-Hill (1997)
2. Ohno, T.: Toyota Production System, p. 58. Productivity Press (1988)
3. Womack, J.P., Jones, D.T.: Lean Thinking, p. 352. Free Press (2003)

4. Goldratt, E.M.: Critical Chain, p. 289. North River Press Publishing Corporation (1997)
5. Dettmer, H.W.: Goldratt's theory of constraints. A Systems Approach to Continuous Improvement. ASQ Quality Free Milwaukee, Wisconsin, 444 (1997)
6. Prahalad, C.K., Hamel, G.: The Core Competence of the Corporation. Harvard Business Review 68(3), 79–91 (1990)
7. Collis, D.J., Montgomery, C.A.: Competing on Resources: Strategy in the 1990s. Harvard Business Review 73(4), 118–128 (1995)
8. Kosorukov, O.A.: Methods of Quantitative Analysis in Business: Textbook. INFRA-M Publishing House, Moscow (2005)
9. Kremer, N.S., Putko, B.A., Trishin, I.M., Fridman, M.N.: Operations Research in Economics: Learning aid. Urite Publishing House, Moscow (2011)
10. Glukhov, V.V., Mednikov, M.D., Korobko, S.B.: Mathematical Methods and Models for Management: Textbook. Lan Publishing House, Saint-Petersburg (2000)

Distributed Packet Trace Processing Method for Information Security Analysis

Alexey Lukashin, Leonid Laboshin, Vladimir Zaborovsky,
and Vladimir Mulukha

St. Petersburg State Polytechnical University
195251, St.Petersburg, Polytechnicheskaya, 29, Russia
{lukash,laboshinl,vlad}@neva.ru, vladimir@mail.neva.ru

Abstract. Information security is an important topic today. Internet Service Providers use network traffic analysis for evaluating network performance, collecting statistics and detecting vulnerabilities. Analysing traffic traces collected from large network requires a computer system where both storage and computing resources can be scaled out to handle and process multi-Terabyte files. Cloud platforms and clustered file systems provide re-sizable compute and storage capacity. MapReduce programming model developed by Google, allows distributed processing of massive data amounts by defining map and reduce functions. In this paper, we propose a cloud computing framework based on MapReduce approach for fast internet traffic analytics.

Keywords: Network Traffic Analysis, MapReduce, Cloud Computing, Scala, Spark.

1 Introduction

As Internet continue to grow and change, increasingly important that Internet Service Providers (ISP) are aware of and have a handle on the different traffic types, traversing their networks. Traffic monitoring and analysis is essential in order to more effectively troubleshoot and resolve issues when they occur, control network security policies compliance, *etc.* Numerous tools are available to help administrators with the monitoring and analysis of network traffic. However, these tools are designed to run on a single server, and unable to efficiently handle and process huge amounts of raw packet dumps collected large network. For example, 10G Ethernet channel could generate about 100TB of network traffic in a 12 hours (100TB in each direction).

$$10 \text{ gbps} \times 3600 \times 12 \times 2 = 864000 \text{ gigabits}$$
$$864000 \div 8 = 108000 \text{ gigabytes} \tag{1}$$
$$108000 \div 1024 = 105 \text{ terabytes}$$

Jeffrey Dean *et al.* [2] has developed MapReduce programming model. It is intended to simplify the processing of vast amounts of data on large clusters in a reliable, fault-tolerant manner. Input data splits into appropriate size independent chunks, named blocks. Users specify a function that maps it's assigned

S. Balandin et al. (Eds.): NEW2AN/ruSMART 2014, LNCS 8638, pp. 535–543, 2014.

block to smaller intermediate ⟨*key*, *value*⟩ pairs. Sorted by key outputs of map functions are sent to reduce function that combine and merge all intermediate values associated with the same intermediate key as shown on Figure 1. Google has shown that search engines can be easily scaled out with MapReduce.

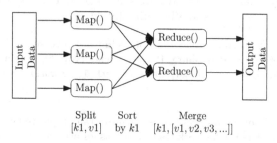

Split	Sort	Merge
$[k1, v1]$	by $k1$	$[k1, [v1, v2, v3, ...]]$

Fig. 1. Map and reduce functions

Cloud computing has recently established itself as a prominent paradigm within the realm of Distributed Systems in a short period. From cloud computing model of providing access to computational resources we could benefit such features as scalability, guaranteed Quality of Service (QoS), and cost effectiveness.

In this paper, we describe how MapReduce programming model can be extended to analyse large network traffic traces on cloud computing platform.

2   Related Work

Many tools such as Wireshark network protocol analyser or CoralReef are available for monitoring, measurement, and analysis passive Internet traffic data. However, the most of these tools designed to run on single high-performance server that is incapable of handling huge amount of traffic data captured at high-speed links. Lee *et al.* [5,4,6] proposed a Hadoop-based flow analysis system and parallel packet processor with limitation of collecting periodic statistic like total traffic amount and host/port packets/bytes count only. RIPE does not consider the parallel-processing capability of reading packet records from distributed file system, which leads to performance degradation. Vieira *et al.* [12,11] has developed a parallel packet processing system for JXTA-based applications. Qiao *et al.* [10] introduced flow record processing tool and verified that MapReduce algoritm is sutable for large traffic traces processing and could solve the problem of big storage and low efficiency of one computer computation.

A cloud system for Scientific Computing with integrated security services has been developed in Telematics department [7,14]. This system has been used for testing and evaluating algorithms as well as running simulations.

3   Scalable Data Processing Technologies

When it is required to find some piece of information in terabytes of data it becomes a challenge. We counted that 10G channel handles about 100TB in 12 hours. Our intent is to process 100TB in about one hour. We propose to use MapReduce approach for processing large amounts of network traffic dumps. Apache Hadoop is a cluster computing framework for distributed data analysis. However, the performance of each job depends on the characteristics of the underlying cluster and mapping tasks onto CPU cores and GPU devices provides significant challenges. Spark is an open source distributed computing platform for data processing on MapReduce model. Spark provides primitives for in-memory cluster computing built around distributed datasets (RDDs). RDDs support types of parallel operations: transformations, which are lazy and yield another distributed dataset (e.g., map, filter, and join), and actions, which force the computation of a dataset and return a result (e.g., count) [3]. Being compatible with Hadoop storage system (HDFS, Hbase, SequenceFiles) Spark jobs can load data into memory and query it much quicker than on disk-based systems (up to 100 times faster than Hadoop) [13]. To make programming faster, Spark integrates into the Scala language. Scala is statically typed high-level programming language designed to express common programming patterns in a concise, elegant, and type-safe way. Scala runs on JVM so it integrates features of object-oriented and functional languages.

For hardware reconfiguration we use OpenStack cloud platform and manage virtual machines by Opscode Chef scripts [7]. For massive creation of virtual infrastructure we use OpenStack Heat service, which has an interface compatible with AWS CloudFormation service.

4   Packet Trace Files Processing Model on Distributed File Systems

NetFlow is the protocol used for collecting IP traffic information by categorizing packets in flows and obtainung important flow information, such as IP address, TCP/UDP ports, byte counts. However, NetFlow summaries feature a high-level of abstraction. While this is helpful for modeling traffic across core networks, NetFlow summaries are not detailed enough to support work investigating TCP properties and other network effects as they lose a lot of the finer detail occurring at a packet level. Alternative format to save captured network data is libpcap. PCAP, short for Packet CAPture, is a protocol developed for recording and analyzing captured IP packets. PCAP-based utilities and tools are generally designed for network troubleshooting, breach and threat detection, ip traffic analysis and network forensics. A number of packet analysis programs and utilities use the PCAP file format, including Wireshark, tcpdump, Snort, ngrep, Nmap and McAfee threat detection products. Libpcap is a binary file with a global header containing some global information followed by zero or more records for each captured packet (Figure 2)

Global Header	Packet Header	Packet Data	Packet Header	Packet Data	$\cdots$

Fig. 2. Pcap file structure

MapReduce applications mostly used to process large sets of text data such as log files or web documents, so the main input file format is text format where input treats each line as a $\langle key, value \rangle$ pair: the key is the offset in the file and the value is the contents of the line. Libpcap binary file format has no mark between packet records like carriage return in text file. Since input file is chunked into fixed size blocks a packet record is usually located across two neighboring blocks. Because of the variable packet size it is difficult to parse packet records from blocks on distributed file system.

Maximum length found on any network path tend to conform to the smallest MTU permitted by any router or switch on that network path. Even on Ethernet backbone segments that are "Jumbo clean"(that is, those on which all directly connected devices are able to send and receive Jumbo frames), it is not unusual to find very few, or even no frames larger than 1518 bytes. Packets smaller than the 64 byte minimum are described as runt packets. The vast majority of Ethernet implementations in the field today will reject packets outside the 64-1518 byte range. The encapsulation of Pcap packet record and Ethernet frame is shown on Figure 3.

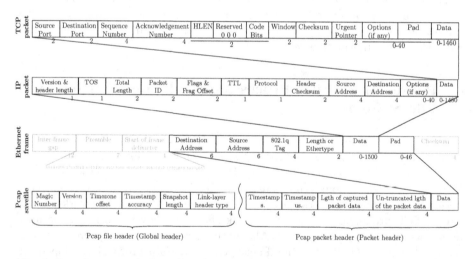

Fig. 3. Sample encapsulation of application data from TCP to a Pcap packet record

Default size of TcpDump capture record is 65535 bytes. Actual packet size is never larger than 65535 bytes, so captured and un-truncated 2-bytes long fields of each pcap record should be equal and store values in range (64-1518). We propose an algorithm for detecting packet records with searching two equal

2-byte fields from the beginning of the file system block. When this fields are found, we assume that the packet record beginning found. To ensure that our prediction is correct, we can take look at the Type/Length field of the detected packet Ethernet frame and check it is value to known EtherType values recorded in the IEEE list[1] (Figure 4).

EtherType is a two-octet field. It is used to indicate which protocol is encapsulated in the payload of an Ethernet Frame. This field was first defined by the Ethernet II framing networking standard, and later adapted for the IEEE 802.3 Ethernet networking standard. Values of 1500 (0x05DC) and below for this field indicate that the field is used as the size of the payload of the Ethernet Frame while values of 1536 and above indicate that the field is used to represent EtherType. If size of payload detected we compare it to the packet size specified in pcap record header. If it is EtherType we compare it to list of known EtherTypes. After detecting first record we start searching next packet at the place where previous record ends according to the packet size. If two equal fields found at right place we assume that all correct.

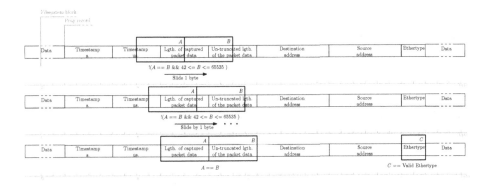

Fig. 4. Pcap packet record detection algorithm for distributed file system

5   Collecting Network Traffic Statistic Information Using MapReduce

By running simple MapReduce counting jobs it is possible to collect statistic information on packets data up to transport OSI model layer. The results of this analysis provide us with the following information:

- volume of traffic between specified subnets (bytes, packets, connections count);
- volume of local traffic;
- information on external access attempts (protocol, host);
- the identification of packets rejected due to access list;
- future traffic behavior prediction;

The problems that may be identified in this manner may include:

- The existence of a malware software in the network;
- DDoS attacks;
- access to forbidden resources identified by IP address;
- attempts to access protected network devices;

6 Application-Aware Network Content Inspection Method

Today, the most widely adopted solution for monitoring and managing network packet data is Deep Packet Inspection (DPI). It provides the ability to compare and match incoming byte streams against a database of known offending patterns called signatures. These signatures represent potential malicious content and can take the form of simple literal strings or more complex patterns such as a specific arrangement of bytes broken up by a variable amount of other possibly irrelevant data. Instead of solely checking the body or header of data packets Deep Content Inspection (DCI) reconstructs, decompresses and decodes network traffic packets into their constituting application level objects. DCI examines the entire object and detects any malicious or non-compliant intent. We develop a system which allows to perform a DPI in a large amouts of traffic data. This approach will extend existing systems by following functionality:

- Security policy a posteriori analyses;
- lookup for virus signature;
- search for text using regular expressions;
- search by applied protocol filed (e.g., HTTP request method, or URL).

In contrast to counting jobs that con be done on IP level, deep packet inspection such as analyze of HTTP transaction cannot be computed independently per file system block. TCP is a connection-oriented protocol, that means that an abstract connection is established between the two hosts. This connection is must be created with a special handshake protocol, and once active, all packets sent between the two hosts are guaranteed to be delivered reliably and in the order in which they are sent. Because the IP protocol provides no such guarantees, TCP must utilize special mechanisms in order for these properties to hold. TCP packets are given sequence numbers and every packet received by a remote host must be acknowledged by sending an ACK packet back to the original sender. If a packet is not acknowledged or it is determined that packets were not received in the correct order, then the original sender will re-transmit the packets as many times as needed to correct this. All packets sent during a single, ongoing TCP connection is known as a flow. Since it is connection-oriented, analysis requires full TCP flow extraction.

TCP port numbers used to determine which packets are relevant to which applications, certain applications have port numbers that are reserved for that application. These applications provide their own application-layer protocols on top of TCP/IP (typically) and expect packets received on their specific protocols to conform to these particular protocol specifications. A TCP connection between two hosts, A and B, consists of two half-connections: one made up of packets traveling from A to B and another consisting of packets traveling from B to A. Bidirectional TCP connection consists of two half-connections.

Packet flow is described as a set of virtual connections between users and services [8,9]. Virtual connection (VC) is a logically ordered exchange of messages between the network nodes. Virtual connections are classified as technological virtual connections (TVC) and informational virtual connections (IVC). A technological virtual connection is described by network protocols, e.g., TCP session between user and database. Information virtual connection is described by applied protocols, e.g., HTTP session with a web service. IVC might use multiple TVC, e.g., ftp session uses two TCP connections; one for data and another for control messages. And vice versa, TVC might belong to multiple IVCs, e.g., persistent connections in HTTP, as described in RFC 2616 client can reuse existing TCP connections for multiple requests, of course, resource URI might be also different. Technical virtual connection exists in parallel to and independently from other virtual connections. Virtual connections do not share any resources. It allows parallel processing of virtual connections. After reading packet record we could map packet records related to one TVC into the same reduce task. This could be done by applying commutative operation on hashes of destination and source IP addresses and ports (2).

$$F(H(IPDst:port), H(IPSrc:port)) == F(H(IPSrc:port), H(IPDst:port)) \quad (2)$$

After grouping packets related to one virtual connection we now able to perform detailed protocol analysis for traffic identified via specific ports and IP addresses. If search query requires deep packet inspection then we propose to

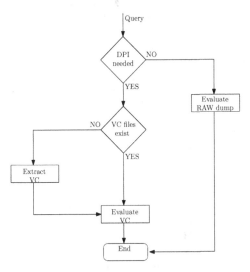

Fig. 5. Query processing algorithm

perform traffic analyses in two stages (Figure 5). First stage groups packets by virtual connections and builds context files. Second stage evaluates virtual connection context.

7 Conclusion

In our work we presented an internet traffic analysis method which is based on MapReduce paradigm. We propose to use Apache Hadoop cluster in an Open-Stack cloud and reconfigure it according to amount of data to be processed. For speeding up data processing we use Spark technology. For deep packet inspection we build session files which contain transmitted data and protocol fields. Our system can process multi-terabytes libpcap dump files and can be easily scaled out across cluster nodes in cloud infrastructure.

Future work is to develop working prototype of traffic processing cluster in a cloud and evaluate its performance and scalability. Another important work is to develop query framework similar to existing systems such as Pig or Hive which allows to perform queries to network traffic dumps which include expressions for network protocol fields and for transmitted data signatures.

References

1. IEEE 802 Numbers (2014),
   http://www.iana.org/assignments/ieee-802-numbers/ieee-802-
   numbers.xhtml
2. Dean, J., Ghemawat, S.: Mapreduce: Simplified data processing on large clusters. In: Proceedings of the 6th Conference on Symposium on Opearting Systems Design & Implementation, OSDI 2004, vol. 6, pp. 10–10 (2004)
3. Kumawat, T., Sharma, P.K., Verma, D., Joshi, K., Kumawat, V.: Implementation of spark cluster technique with scala. International Journal of Scientific and Research Publications (IJSRP) 2(11) (2012)
4. Lee, Y., Kang, W., Lee, Y.: A hadoop-based packet trace processing tool. In: Domingo-Pascual, J., Shavitt, Y., Uhlig, S. (eds.) TMA 2011. LNCS, vol. 6613, pp. 51–63. Springer, Heidelberg (2011)
5. Lee, Y., Lee, Y.: Toward scalable internet traffic measurement and analysis with hadoop. Computer Communication Review 43(1), 5–13 (2013)
6. Lee, Y., Kang, W., Son, H.: An internet traffic analysis method with mapreduce. In: 2010 IEEE/IFIP Network Operations and Management Symposium Workshops (NOMS Wksps), pp. 357–361 (April 2010)
7. Lukashin, A., Zaborovsky, V.: Secure heterogeneous cloud platform for scientific computing. In: Proceedings of The Tenth International Conference on Networking and Services, ICNS 2014, pp. 24–28 (2014)
8. Lukashin, A., Zaborovsky, V.: Dynamic access control using virtual multicore firewalls. In: Proceedings of The Fourth International Conference on Evolving Internet, INTERNET 2012, pp. 37–43 (2012)
9. Lukashin, A., Zaborovsky, V., Kupreenko, S.: Access isolation mechanism based on virtual connection management in cloud systems - how to secure cloud system using high perfomance virtual firewalls. In: ICEIS, vol. (3) (2011)

10. Qiao, Y.Y., Lei, Z.M., Yuan, L., Guo, M.J.: Offline traffic analysis system based on hadoop. The Journal of China Universities of Posts and Telecommunications 20, 97–103 (2013)
11. Vieira, T., Soares, P., Machado, M., Assad, R., Garcia, V.: Measuring distributed applications through mapreduce and traffic analysis. In: Proceedings of the 2012 IEEE 18th International Conference on Parallel and Distributed Systems, ICPADS 2012, pp. 704–705 (2012)
12. Vieira, T.P.D.B., Fernandes, S.F.D.L., Garcia, V.C.: Evaluating mapreduce for profiling application traffic. In: Proceedings of the First Edition Workshop on High Performance and Programmable Networking, HPPN 2013, pp. 45–52 (2013)
13. Xin, R.S., Rosen, J., Zaharia, M., Franklin, M.J., Shenker, S., Stoica, I.: Shark: Sql and rich analytics at scale. In: Proceedings of the 2013 ACM SIGMOD International Conference on Management of Data, SIGMOD 2013, pp. 13–24. ACM, New York (2013)
14. Zaborovskiy, V., Lukashin, A., Popov, S., Vostrov, A.: Adage mobile services for its infrastructure. In: Proceedings of The 13th International Conference on ITS Telecommunications, ITST 2013, pp. 127–132 (2013)

NetMap - Creating a Map of Application Layer QoS Metrics of Mobile Networks Using Crowd Sourcing

Lars M. Mikkelsen, Steffen R. Thomsen, Michael S. Pedersen,
and Tatiana K. Madsen

University of Aalborg, Networking and Security Section,
Fredrik Bajers vej 7, 9220 Aalborg East, Denmark
{lmm,tatiana}@es.aau.dk, {riber4,michael.soelvkaer}@gmail.com

Abstract. Based on the continuous increase in network traffic on mobile networks, the large increase in smart devices, and the ever ongoing development of Internet enabled services, we argue for the need of a network performance map. In this paper NetMap is presented, which is a measurement system based on crowd sourcing, that utilizes end user smart devices in automatically measuring and gathering network performance metrics on mobile networks. Metrics measured include throughput, round trip times, connectivity, and signal strength, and are accompanied by a wide range of context information about the device state. The potential of the NetMap approach is demonstrated by three usage examples.

Keywords: quality of service, performance, measurement, crowd sourcing.

1 Introduction

In recent years mobile networks have seen a huge increase in data transferred (and is estimated to continue to increase[4]), while other types of network usages drop (SMS and speech[6]). This trend is highly associated with the large increase in the amount of smart devices, such as tablets and smartphones, which are getting more and more powerful in terms of functionalities, processing power and connection capabilities. Along with the more powerful devices, another reason for the high increase in network traffic is based on the services being used, such as video streaming (estimated to grow to more than 69% of all consumer Internet traffic[4]), music streaming (Spotify, Grooveshark, etc.), constantly updating social networks (Facebook, Twitter, Instagram), and in general network enabled services. At the same time Internet Service Providers (ISPs) are deploying faster network technologies[2], e.g. HSPA+ and LTE/4G, enabling higher data transfer rates and QoS for the users of the networks.

Users are getting accustomed to have high speed networks and capable devices. However, when degradation in performance is experienced, it might not be easy for a user to identify a factor causing performance drop. The device he is using

S. Balandin et al. (Eds.): NEW2AN/ruSMART 2014, LNCS 8638, pp. 544–555, 2014.
© Springer International Publishing Switzerland 2014

might have poorer capabilities than expected, such as poor antennas, network chip, or processing power. The network might be poorly configured or might lack high performance capabilities in certain areas. Lastly the service might not be scaled to the amount of users using it, or it might be slow due to momentarily high load. Regarding the service, there are usually ways of estimating if the service is experiencing sub par performance at the moment[12], but generally the user has no influence on any of the performance impediments of the device or network. To make sure that he receives the best possible performance, he need to compare the performance of his device on his ISP network with other devices, or with other ISP networks in the area.

This is where a proposed system called NetMap[8] comes into play. NetMap is a mobile network performance measurement system based on crowd sourcing, which continuously gathers information about the performance of the network connection from the device to the backbone of the connected network. Based on the information gathered via NetMap a geographical Network Performance Map (NPM) of mobile networks is generated while also mapping the performance of the devices.

1.1 Advantages of NetMap

Today most ISPs have made coverage maps available to their users, where it is possible to see an estimated signal strength in an area and an estimated throughput speed[10]. These maps are based on theoretical calculations, and give a static image of the expected coverage, but true experienced performance may vary due to local conditions such as hills, trees and houses, and might be affected by dynamic factors such as network load.

Here NetMap has a clear advantage, as it offers a NPM based on actual measurements on existing networks, and by using actual end user devices in real end user scenarios. The NPM shows what network speeds to expect, based on real empirical and continuously updated data, and at the same time it provides a more realistic image of what the end user can expect as the measurements are performed with devices similar to his. Another advantage of generating a NPM using NetMap, is the way the data is generated and gathered: by using crowd sourcing. It would be very time consuming and require a large amount of resources, including equipment and man labour for an ISP to create this type of NPM. Furthermore, the client side software, i.e. an app, can be distributed by using existing infrastructure, such as Google Play or Apple App Store. Additionally, a wide range of metadata is extracted on the state of the device while the measurements are being performed, e.g. location, movement speed, battery level, connection type, etc., which can help in understanding the cause of fluctuations in the measured performance.

1.2 Existing Solutions

In [7] an approach for making a network coverage map based on crowd sourcing is proposed, which in many ways is similar to NetMap, but only measures signal strength, resulting in a network coverage map being constructed.

Open Signal[9], available as an Android app, is aimed at measuring physical connection metrics by mapping cell towers, signal strengths and WiFi access points, and the results are presented in a coverage map. Furthermore it is possible to manually perform throughput and ping measurements in the app.

Both solutions use crowd sourcing to perform and gather measurements on mobile networks, and their focus is on physical connection measurements, while the main focus of NetMap is to perform application layer network performance measurements, and creating a NPM.

1.3   Measurement Approach in NetMap

In Figure 1 the overall measurement approach used in NetMap can be seen. The concept relies on the ISP networks being connected to an Internet exchange point, and the measurement point being located on a neutral network, also connected to the Internet exchange. By neutral network is meant a non-mobile ISP network, where no mobile devices are directly connected using mobile technologies like 3G, 4G etc..

In Denmark an example of such a neutral network is the Danish Research Network[5], which does not offer any mobile connections, and is connected to the danish Internet exchange. Looking at Figure 1, by comparing measurements from Clinet #1 and #2, the network and the device performance will be compared, as the measured entity is the connection starting at the device and ending at the internet exchange point.

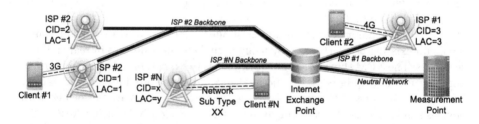

Fig. 1. Concept of the measurement approach in NetMap (measurement traffic is marked with gray). For description of CID, LAC and Network Sub Type see Section 3.2.

As the measurements are performed via an application installed on a mobile device like any other application, the measured performance is comparable with regular network usage of end users, as it is not a dedicated device, or a dedicated network interface, while other applications running will also consume the resources available on the device during measurements. Furthermore, due to the diverse market of smart devices, the measurement devices will vary in quality, and thereby vary in quality of connectivity hardware.

Both these issues are addressed by gathering metadata on the device. With the metadata it is possible to understand the situation the measurements are performed in, both in terms of context of the device and the state of the device, as it allows various filtering and sorting of the measurements.

1.4   Requirements to NetMap

In order to give NetMap a higher chance of both getting more users and keeping users longer, while also increasing the chance of generating results that will have an impact, some requirements are defined.

REQ1 Low resource usage of measurement methods (data) and application in general (power, memory, etc.).

REQ2 Unbiased measurements as there are multiple mobile networks it is important that the measurements are not biased towards any network.

REQ3 Relatable metrics as the user is not necessarily a technical person the measured metrics need to be understandable for this type of user as well to maintain interest/motivation.

REQ4 Transparency of measurement methods is important such that it can be verified that the results are obtained by using correct methodology.

REQ5 Simple to use system e.g. by making it fully automatic is important as if the user needs to go through too much hassle to perform or submit measurements he will be inclined to not use it.

REQ6 Privacy of the user is a vital point, as a lot of metadata is gathered about the context of the device, and it should not be possible to identify a single user from this.

2   General System Functionality/Architecture

In Figure 2 the main actors, components and entities of NetMap can be seen, and will in the following be described.

NetMap App: The client side software from where the measurements are initiated, scheduled, and results submitted to Collection Point. NetMap App is the front end, as seen from the end user perspective, giving the user basic control such as start/stop of measurements, selecting measurement types to run, setting maximum data usage, and other settings. NetMap App also offers immediate feed back from the measurements, i.e. latest results or aggregated results.

Measurement Point: The end point of all measurements performed in NetMap, which is responsible for performing the measurements requested from NetMap App. This is the only truly critical point in the design and implementation of NetMap, since if it is not properly scaled to accommodate all requests the measured values will be affected. In that case the measurements will not show the true state of the network, but rather the state of the resources of the server. Therefore initiatives, such as multiple servers and load balancing, and maximum allowed simultaneously active measurements, are taken to ensure scalability of the system.

Data Collection Point and Database: Measurement results are submitted to Data Collection Point, and when results are received they are stored in the database. This setup allows for version checking of result objects before adding them to the database, so if a measurement method is updated and additional

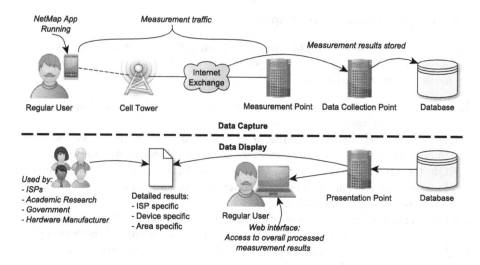

Fig. 2. Overview of the entities and data flows in NetMap

result fields are added this can be handled before adding the results to the same table as the previous version of the same measurement method.

Presentation Point: The entity responsible for processing the data in the database and presenting it according to the specific purpose. For the NPM purpose, Presentation Point extracts the data from the database and performs some preprocessing to generate a reduced dataset which is the NPM. The reduced dataset is then made available as an API, that can be used by webpages and applications that offer NPMs to end users.

2.1 End User Controls

As mentioned with NetMap App, the end users have some options for basic control. This is due to the fact that having an application constantly running and performing measurements could potentially drain the available resources on the mobile subscription. Therefore the user is able to control the data usage by setting an upper limit, selecting which measurement types to run as some use more data than others, and by setting the time between each measurement run.

2.2 End User/Crowd Motivation

The outcome of NetMap relies heavily on end users installing the app and performing measurements, why motivation of users to participate is a key aspect of the system concept. Currently users are motivated to participate only based on measurement results of NetMap, which can be used to check if they see expected or sub par performance. In an initial 4 week trial period, where end users only had access to view their own measurements, approximately 1500 users chose to participate, which shows that users are motivated as there is a general interest in evaluating network performance. However, based on the limited format of the output to users, it is likely that participants were mostly technical savvy persons, why additional motivation is needed to reach a broader user base.

3 Metrics and Measurement Methods

The measurements are scheduled according to a periodic scheme, where measurements are performed at a fixed interval of 15 minutes, which can be changed by updating the app. By changing the interval between measurements, i.e. changing the frequency of measurements, the maximum load at Measurement Point can be controlled according to number of active users of NetMap. Following are the measured metrics described along with arguments for why they are relevant.

3.1 Application Layer QoS Metrics

Bulk Transfer Capacity (BTC): BTC, or throughput, as defined by the Internet Engineering Task Force (IETF)[11], is the average amount of data that can be transmitted over a link per second, $BTC = data\_sent/elapsed\_time$, also known as bit rate or bits per second (bit/s). As the measurements are performed

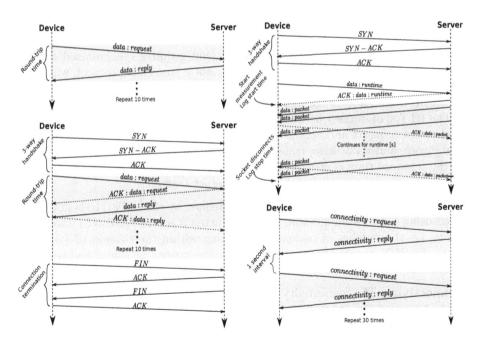

Fig. 3. Top left: UDP RTT. **Bottom left:** TCP RTT. **Top right:** Throughput Downlink. **Bottom right:** Connectivity. Measurement flows are marked with grey areas.

at the application layer, data represents the bits received excluding headers or overhead, and time is measured as the time between the first received bit and the last. The throughput is measured for both uplink and downlink using TCP, and each measurement is performed for 15 seconds. In Figure 3 the method for measuring throughput is illustrated, indicating that it is a simple procedure where as much data as possible is pushed through the connection making it rather expensive in data usage.

It is interesting to know the throughput a connection can deliver, especially in connection with content based applications such as streaming or cloud services. Furthermore, as the throughput is the most common performance metric used to describe a connection, both by users and ISPs, it is the metric giving the best perception of how well a connection or network performs. Based on these arguments throughput is considered the most important metric that is being measured in NetMap.

Round Trip Time (RTT): The RTT is, as defined by the IETF[11], the time it takes from a request is transmitted from the device to the server, and until the reply is received from the server at the device. The RTT is measured using both TCP and UDP, and only by measuring on payload packets, that is for TCP the connection is initiated before the measurements are performed. Both UDP and TCP RTT consist of 10 request/reply sequences, as illustrated in Figure 3. When measuring the RTT it should be noted that it is not necessarily the same route that is being measured, i.e. the request to the server will travel by one route, while the reply might travel via another. Alternative delay estimations include one-way latency measurements, that would make it possible to compare the uplink and downlink of the connection, but requires synchronized clocks. The ICMP protocol is not used because it is not a transport protocol, why it is difficult to compare to other use cases. This choice is supported in [13] showing that TCP has an advantage in accuracy over ICMP.

The RTT is of interest because many applications operate using remote data, where most requests and actions performed by the user will cause a connection being made and data being transfered, why the delay induced by the network on the response time is of interest. UDP and TCP were chosen based on them being the most widely used protocols in Internet traffic.

Packet Loss: or round-trip packet loss as defined by the IETF[11], is not measured using a dedicated method, but is extracted from the UDP RTT measurements, i.e. the number of packets not arriving within the timeout of 1 second set in the RTT measurements, effectively saving processing power and network traffic.

Connectivity: The connectivity is based on the IETF definition of the two-way interval temporal connectivity[11]. In the definition N request packets should be transmitted uniformly distributed between time T and time $T + \triangle T - W$, where W is the time to wait for a reply for each packet. For each sequence of N packets, if a single reply is received within the waiting time W the connectivity sequence returns true, and false otherwise.

The connectivity measurement in NetMap is only transmitting 1 UDP packet, then it waits for the reply with a timeout of 1 second, after which it waits 1 second and starts the sequence over, as illustrated in Figure 3, which is done 30 times before the results are saved. In this way the connectivity requests are not randomly distributed, but they are not completely periodic either, because each request is spaced 1 second plus the varying RTT of the connectivity request/reply. Each request/reply will be evaluated to be either true or false, which is stored as the result in a vector indicating whether the individual request/reply sequence was a success or not. Furthermore, despite the connectivity running continuously, it is a cheap metric in terms of resources only using approximately 210 kB pr hour.

The connectivity is a discrete indication of the reachability of the server from the device at any point, while giving an indication of the continuous performance experienced. With the strict timeout defined for the metric, connectivity can be interpreted as an indicator of when the network performance drops below some defined threshold, which in this case means that it takes more than a second to do a request/reply sequence. The threshold, or timeout, of 1 second, is chosen as a response slower than this would be noticeable by an average user.

3.2 Physical Layer Connection Metrics

Network Context State Switch (NCSS): A NCSS is a change in state of the Cell ID (CID), Location Area Code (LAC), Network Type (Wifi or mobile connection), or Network Sub Type (3G, 4G, etc.) (see Figure 1), and is measured, or rather registered, event based, why no data is transmitted in order to measure it. In Android the API PhoneStateListener[3] is used for this purpose.

It was chosen to register the NCSS because these are expected to impact the performance, e.g. in connection with a CID switch (in Figure 1 when client #1 moves from CID 1 to 2) it is expected that the performance of RTT, throughput and connectivity will drop. By registering the CID and LAC switches the individual cell towers, and their coverage boundaries, can be identified and described in terms of performance as seen from the connected device. The performance of the tower is also expressed via the Network Sub Type, which indicates what connection technology is supported.

Received Signal Strength Indication (RSSI): RSSI is defined in [1] with values ranges from 0 to 31, which then can be translated into dBm, and is also registered via the Phone State Listener API in Android.

Registering the RSSI value of the connection currently being used by the device, helps identify why a connection might perform poorly. Adding the empirical RSSI measurements to the NPM, enables comparison to be made with the theoretical model based maps, provided by ISPs, to see if the experienced signal strength compare to the expected.

3.3 Notes Regarding NetMap Measurements

All the application layer QoS metrics are measured with active measurements, which is why they are performed sequentially, i.e. in a measurement sequence organized as [UDP RTT; TCP RTT; throughput uplink; throughput downlink], while connectivity is running continuously and the physical connection metrics are event based. The order in the measurement sequence is irrelevant but need to be non-overlapping to avoid one method influencing another.

When measuring public mobile networks the impact of the results can potentially rank an ISP as poorly performing, which emphasizes the importance of transparency of methods to evaluate causes and influencing factors on the results. Examples of influencing factors are non-dedicated devices, device quality, mobility of users, varying network load according to user activity, local environment conditions, and other.

In creating a NPM, it is vital that a sufficient number of measurements are used. But regardless of the number of measurements the NPM is based on, a number of indicators about the measurements in that area should be presented, e.g. number of measurements, distribution of measurements according to time of day or according to device type, and other.

To obtain a map, each collected measurement is tagged with a geographical point. Location information is preferably obtained using GPS achieving accuracy down to 4-5 meters. As having GPS receiver on is expensive in terms of power, alternatively other location estimators can be used, e.g. cell tower info, but with a much worse accuracy, possibly ranging from few hundred meters to more than 1 km. Other key metadata types logged are time, battery level, and manufacturer and device model.

4   Examples of Results

NetMap has not been rolled out in full scale to the public yet, but in an initial trial approximately 1500 users have participated in covering mainly the region of Aalborg and parts of the Northern Jutland region in Denmark, as a proof of concept. Additionally, we have designed and performed a number of tests gathering extensive measurements in a limited, selected in advance, geographical area. The amount of measurements collected in this way allows for statistical analysis. Selected results of these tests are presented below.

Indication of Network Load: Figure 4 shows that the performance measured is influenced by the load on the network according to the time of day. The 2 vertical lines indicate the start and end of a typical work day, where the amount of people in the area and thereby the load on the network is higher, which is indicated by the measured throughput dropping in this timespan.

This shows that the measurements can be used to give a more precise forecast of what performance a user can expect from the network in a location at a given time. This can be utilized in a NPM but also in a prediction scenario where a network state forecast is used.

Indication of Area Type Influence: Figure 5 shows a tendency that the RTTs are lower in small town and rural areas than in urban areas. This is estimated to be due to a combination of the network resources available and the user density

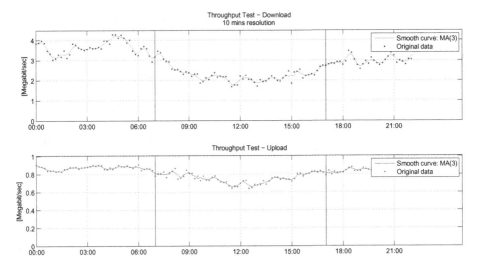

Fig. 4. The throughput measurements are performed by 4 identical devices during 24 hours placed in the same location in the middle of campus, connected to the same ISP network, with the measurements being performed continuously with a few seconds of delay between each measurement sequence. The plot is based on 6175 throughput downlink and 6191 uplink measurements averaged over 10 min intervals to create datapoints, and the curve is smoothed with a MA(3) process.

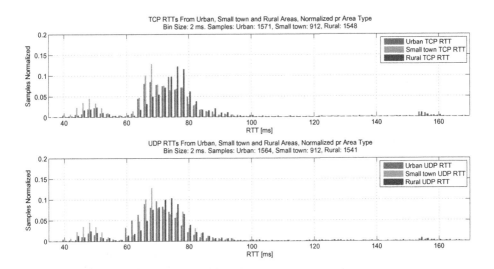

Fig. 5. The RTT measurements are performed while driving in a car, following the speed limits, through different types of areas defined as urban, small town, and rural. The plot is based on 3689 TCP RTT and 3730 UDP RTT measurement sequences, where the means of the sequences are sorted in 2 ms bins.

in the areas. This gives an indication of the balancing between network resources assigned by the ISP in the areas, and the users using the network.

Indication of Difference in Device Performance: Figure 6 shows that 3 of the 4 devices used show the same performance tendencies as in Figure 4, while the 4th device delivers such low performance that it is uninfluenced by the general network performance drop. This indicates that the device model can have a big impact on the performance measured, why it is important to include information about the device model in the metadata gathered along with the measurements.

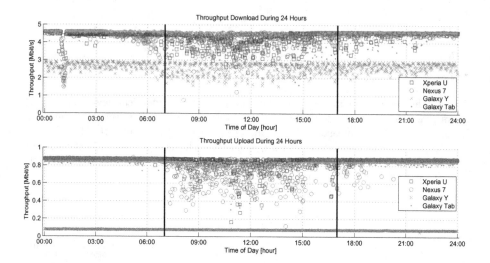

Fig. 6. Same setup as described with Figure 4, but with 4 different devices. The plot is based on approximately 1100 throughput uplink and 1100 downlink measurements pr device. Note that the sudden drop in downlink rates around 01:00 hrs is identified to be caused by backup being performed internally on the network where Measurement Point is located.

5   Conclusion

In this paper a novel system called NetMap, for measuring network performance using crowd sourcing has been described, in which measurements are performed by users who have installed the NetMap App on their devices, and results are automatically submitted to Collection Point. The system measures network QoS metrics on the application layer, while available physical connection QoS metrics also are registered. Along with the measurements a wide range of metadata including the state and context of the device while it is performing the measurements is gathered, to help in describing the measured performance and generating a NPM. The need for such a system, and such measurements, is based on the big increase in Internet traffic going through mobile networks, the increase in number of mobile devices, and that the existing network coverage maps are based on theoretical models rather than actual measurements. Finally some examples were given on the potential of the measurements obtained from NetMap, which includes indication of network load, balancing of network resources, and impact of the device on the measurements.

Outlook: Further work to be done on NetMap involves the following points:

- Ensuring scalability of Measurement Point to profit from all active users.
- Optimized measurement methods to estimate QoS metrics.
- Developing a more intelligent scheme for scheduling measurements, i.e. to accommodate if more measurements are needed in one area than another.
- Extrapolation of measurements in NPM to cover areas where few measurements are performed.
- Additional motivation of users for using NetMap, e.g. including competitions, sponsors and rewards.

References

1. 3GPP: TS 27.007 Specification Detail, chap. 8.5 Signal quality +CSQ (2014), http://www.3gpp.org/ftp/Specs/html-info/27007.htm
2. 4G Americas: 3g/4g deployment status (May 2014), http://www.4gamericas.org/index.cfm?fuseaction=page&pageid=939
3. Android Developer: Phone state listener (May 2014), http://developer.android.com/reference/android/telephony/PhoneStateListener.html
4. Cisco: Cisco visual networking index: Forecast and methodology, 20122017 (May 2014), http://www.cisco.com/c/en/us/solutions/collateral/service-provider/ip-ngn-ip-next-generation-network/white_paper_c11-481360.html
5. DeIC: Danish research network (May 2014), http://www.deic.dk/en/node/171
6. Erhvervsstyrelsen, Denmark: Mobil datatrafik fordoblet paa et aar (translate: Mobile data traffic doubled in a year) (April 2014), http://erhvervsstyrelsen.dk/pressesoeg/694899/5
7. Mankowitz, J., Paverd, A.: Mobile device-based cellular network coverage analysis using crowd sourcing. In: 2011 IEEE EUROCON - International Conference on Computer as a Tool (EUROCON), pp. 1–6 (April 2011)
8. NetMap Measurement System: Netmap measurement system wiki (May 2014), http://sourceforge.net/p/netmap-system/wiki/Home/
9. Open Signal: Open signal (May 2014), http://opensignal.com/
10. Telia Denmark: Coverage map (May 2014), http://telia.dk/mobil/kundeservice/daekning
11. The Internet Engineering Task Force: Ip performance metrics (May 2014), http://datatracker.ietf.org/wg/ippm/charter/
12. Unitz LLC: Down right now (May 2014), http://downrightnow.com/
13. Wenwei, L., Dafang, Z., Jinmin, Y., Gaogang, X.: On evaluating the differences of TCP and ICMP in network measurement. Computer Communications 30(2), 428–439 (2007), http://www.sciencedirect.com/science/article/pii/S0140366406003719

Resource Scheduler Based on Multi-agent Model and Intelligent Control System for OpenStack

Alexey Lukashin and Anton Lukashin

St. Petersburg State Polytechnical University
Polytechnicheskaya, 29, St. Petersburg, 195251, Russia
lukash@neva.ru, an.lukashin@gmail.com

Abstract. Resource consumption in virtualized computing environment is an important problem in modern cloud infrastructures. This paper introduces methods of detecting and solving problems of unbalanced load nodes in cloud cluster. We propose multi-objective optimization problem analysis as the bin packaging problem, using criteria convolution method to make it a single objective problem and neural networks with back-propagation to reduce balancing time. The considered approach is modeled in AnyLogic and the integration software for the system is implemented on OpenStack cloud platform.

Keywords: Cloud Computing, neural networks, load balancing, multi objective optimization, AnyLogic, OpenStack.

1 Introduction

Information and communications technologies are becoming an important part of the infrastructure used for the innovative development of the scientific, technical, social and educational activities. A key component of this infrastructure is data-processing systems. The complexity of creating computer systems with the desired characteristics of the operating system, hardware components and network topology strongly depends on the characteristics of problems solved by these systems. That is why the problem of developing technologies for automated reconfiguration of computer systems is important. The usage of virtualization technologies and solutions based on the paradigm of "cloud computing" allows you to enhance significantly the ability to control computing resources, telecommunication environment and heterogeneous resources. This is particularly important in organization of high-performance parallel computing. These computations require expensive hardware and complex software. For example, one of the most important practical applications of cloud technologies nowadays is to process distributed solutions for control problems. In this case "distributed" means that each computational element could be a single agent (e.g., robot) performing computational tasks that require a lot of resources. Group of computational primitives also could solve this task. Systems like this could also provide interaction between groups of agents that can be physically distant.

S. Balandin et al. (Eds.): NEW2AN/ruSMART 2014, LNCS 8638, pp. 556–566, 2014.
© Springer International Publishing Switzerland 2014

However, there is a problem of unbalanced utilization of resources. Each node (physical computing element with hypervisor) can support simultaneous work of multiple virtual machines that are processing some client's tasks. There are three basic types of resources consumed by a virtual machine:

- Computational (CPU)
- Memory (RAM)
- Input and Output operations (I/O)

Inequality of resource consumption occurs since it is impossible to determine in advance what kind of load will prevail during the virtual machine lifecycle. Hence the task can be formulated as solving a multi-objective optimization problem, where the minimized functional corresponds to the distribution of resources. On the other hand, this problem can be classified as a multi-agent system. In this case, the "agent" means hypervisor that can monitor resource utilization and send information to the control node.

The main results of this research are: a virtual machines distribution method based on neural networks, an imitation model in AnyLogic and implementation of an extension for OpenStack scheduler.

Next parts of this paper are organized as follows. Section 2 describes related works; Section 3 presents the proposed solution; Section 4 describes the performed experiments; Section 5 presents practical results; Section 6 concludes the paper and describes future work.

2   Related Works

Some techniques, methods and algorithms have been developed for virtual machine scheduling in cloud systems. Anton Beloglazov et al. [1] proposed to use virtual machine migration for distributing workload in a cloud. They developed an extension to OpenStack called OpenStack Neat. The major objective of dynamic VM consolidation is to improve the utilization of physical resources and reduce energy consumption by re-allocating VMs using live migration according to their real-time resource demands and switching idle hosts to sleep mode. But this work is related only to the migration problem.

Mike Miroliubov in his master's thesis proposed to combine live migration of virtual machines and initial optimal distribution of virtual machines between hypervisors [2]. He developed algorithms based on simulated annealing algorithm and added live migration as a possible scenario in OpenStack Compute service.

Before starting this work, a cloud system for Scientific Computing with integrated security services has been developed in Telematics department [3, 4]. This system was used for testing and evaluating algorithms as well as for running simulations.

This work extends existing methods and algorithms by applying neural algorithms to virtual machine scheduling and migration. Research is based on the optimization model proposed in [2].

3   Infrastructure as a Service Cloud Resource Model

In order to describe this model it is necessary to define the components and the interactions between them. The cloud infrastructure (cloud) is a set of compute nodes interconnected by data channels. Compute Node is a physically existing computer that controls a hypervisor. There are several basic types of nodes:

- Control nodes - nodes that run software that is responsible for managing cloud infrastructure.
- Working nodes - nodes that host one or more virtual machines.
- Resource Management Node (RMN) - this node is used to balance cloud platform resource utilization.

Hypervisor is a program or a hardware circuit that provides simultaneous execution of multiple operating systems on the same host computer. The hypervisor also provides isolation of operating systems from one another, protection and security, resource sharing between different running operating systems, and resource management. A Virtual Machine is a platform to organize user software. It emulates real computer.

Cloud is a set of nodes that are connected by network channels. The cloud has an external interface that allows a specific set of actions to manage the virtual machines (e.g., VMs creation, termination, and suspension). Nodes communicate with each other and the environment (with respect to the cloud) by sending information packets. RMN monitors resource utilization of all nodes, reacts and provokes changes in the internal structure of the system (e.g., VM migration from one node to another or shutdown of a virtual machine). Hypervisors receive information about resource utilization on current node and transmit it to the RMN.

3.1   Optimization Model of a Cloud

The task of finding an optimal allocation of virtual machines in case we want to minimize the number of nodes used by the cloud system is similar to the N- dimensional problem of packing containers (Bin Packing Problem), where N corresponds to the number of virtual machine's selected characteristics taken into account in the allocation. Backpack problem:

$$max \sum_{i=1}^{n} c_i x_i$$

Where ci is the value of the object i, xi = 1 if the object is placed into a backpack, else 0. Under the constraints:

$$\sum_{i=1}^{n} p_i x_i \leq P$$

Where pi is the weight of the object i, P - the size of a backpack. For our problem formulation of the backpack problem changes as follows. Minimize the number of physical nodes of the cloud system:

$$\min \ z = \sum_{i=1}^{n} y_i$$

Under the constraints:

$$\sum_{j=1}^{m} c_j x_{ij} \le c_{S_i} y_i, \ \ i = \{1,...,n\}, \ \ j = \{1,...,m\}$$

$$\sum_{j=1}^{m} m_j x_{ij} \le m_{S_i} y_i, \ \ i = \{1,...,n\}, \ \ j = \{1,...,m\}$$

$$\sum_{j=1}^{m} d_j x_{ij} \le d_{S_i} y_i, \ \ i = \{1,...,n\}, \ \ j = \{1,...,m\}$$

$$\sum_{i=1}^{n} x_{ij} = 1$$

Where n is the number of servers, m – the number of virtual machines, c_j, m_j and d_j - virtual machine j resources: CPU, memory and disk subsystem. Yi=1 if server is used otherwise 0. This problem is a multi-objective optimization problem. To solve it, it must be reduced to a single-objective problem by using the convolution method criteria.

Resulting single-objective problem is:

$$\min \ \frac{1}{3}\sqrt{\frac{1}{n}\sum_{i=1}^{n}\left(c_{S_i} - \sum_{j=1}^{m} c_j x_{ij}\right)^2} + \frac{1}{3}\sqrt{\frac{1}{n}\sum_{i=1}^{n}\left(m_{S_i} - \sum_{j=1}^{m} m_j x_{ij}\right)^2} + \frac{1}{3}\sqrt{\frac{1}{n}\sum_{i=1}^{n}\left(d_{S_i} - \sum_{j=1}^{m} d_j x_{ij}\right)^2}$$

Under the constraints:

$$\sum_{j=1}^{m} c_j x_{ij} \le c_{S_i}, \ \ i = \{1,...,n\}, \ \ j = \{1,...,m\}$$

$$\sum_{j=1}^{m} m_j x_{ij} \le m_{S_i}, \ \ i = \{1,...,n\}, \ \ j = \{1,...,m\}$$

$$\sum_{j=1}^{m} d_j x_{ij} \le d_{S_i}, \ \ i = \{1,...,n\}, \ \ j = \{1,...,m\}$$

$$\sum_{i=1}^{n} x_{ij} = 1$$

Solving the optimization problem described in the previous section takes a lot of time. In real production system the cloud receives hundreds of requests per minute from different agents. Therefore, it is appropriate to use intelligent control systems. In our paper we select neural network approach to solve the optimization problem described above.

Multilayer perceptron with back propagation learning algorithm was chosen as the neural network. Perceptron model of the algorithm is presented in figure 1.

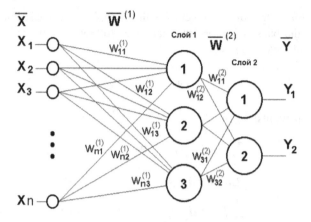

Fig. 1. Perceptron model of the algorithm

In figure 1:

- X - elements (inputs) provide information from each virtual machine
- $W^{(i)}$ – inner elements (solvers) on i-th layer.
- Y - elements (outputs) provide the results of computation.

We applied reinforcement-learning algorithm to train the neural network. Optimization problem solver was used as the reinforcement unit.

3.2   Infrastructure as a Cloud Resource Model in Terms of an Agent System for Simulation

Cloud computing system can be represented as a multi-agent system like net-centric systems. Each element is connected with other elements on the same level and can perform requests to a higher level. Model levels are presented in figure 2.

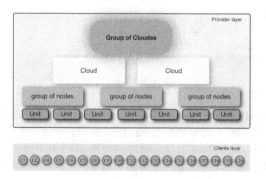

Fig. 2. Agent Model

- Virtual machine level: each virtual machine is a simple agent, whose control function is to exist within the hypervisor's resource reservation and perform calculations.
- Hypervisor level. Each hypervisor provides real resources. VMs can expand or narrow the allocated resources.
- Node group Level: nodes are combined into separate groups, which may reallocate resources among themselves. Special groups for specific calculations can be formed if necessary.
- Cloud Controllers Level: cloud controller manages the input requests and resources at all nodes. It has an ability to migrate virtual machines between nodes and other clouds.

4   Simulation Results

We used AnyLogic for performing a simulation of a cloud system. To simplify the model we assumed that all compute nodes have 8 core processors, 32 GB of memory and 1TB hard drive. With this assumption the maximum number of virtual machines created in the cloud should not exceed the number of cores.

Agent model with discrete event generation was chosen as a basic modeling type. Objects to manage - cloud cluster nodes, resources - virtual machines. We investigated two methods of resource allocation decision:

- Selecting a random node for placing a new VM
- Using a neural network algorithm.

In the first approach an uneven load of the cluster nodes of the cloud and a full system load were observed. The second approach allows to control an optimal load - the number of loaded nodes at each step does not exceed the total number of nodes. Random selection of nodes is illustrated in figure 3.

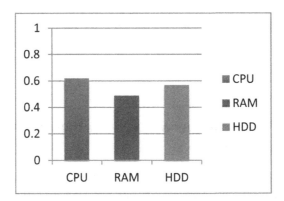

Fig. 3. Unbalanced resource provisioning results

The situation has changed after applying our algorithms (Figure 4). The illustrations show that resource reallocation allows the system to be more reactive. The system can allocate a VM with a larger amount of resources. In case of an overflow of the current group of nodes (clouds) resource allocation request can be broadcasted to the next level of control system.

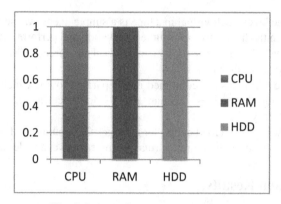

Fig. 4. Balanced resource provisioning

Simulation interface in AnyLogic is illustrated in Figure 5:

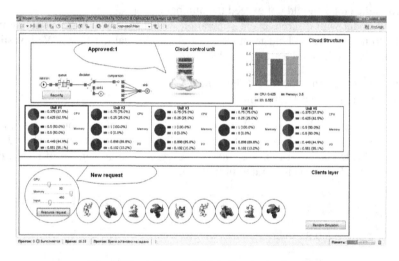

Fig. 5. Simulation model interface in AnyLogic

5   Virtual Machine Resource Scheduler Implementation for the OpenStack Cloud

Implemented algorithms were applied in OpenStack cloud for scientific computing [3]. A cloud stand based on OpenStack Havana was built for testing and evaluation. We used OpenStack for creating groups of virtual machines in a cloud environment service. This service supports description of configurations in an Amazon Cloud Formation format that ensures compatibility with public services such as Amazon AWS. This service allows creating groups of virtual machines according to various patterns, virtual networks, cloud-based routers and other components. The images of virtual machines contain a basic set of services. Any other application specific packages are installed using the automation services provided by Opscode Chef. This tool provides automated deployment of software configurations on virtual machines and bare-metal servers. When new computation segment is being created the security system spawns and configures virtual firewall, which is filtering access to the newly created network which serves for computation. Dynamic network creation is supported by OpenStack Neutron services and distributed by virtual switch Open vSwitch. After computing and receiving the results, the segment is removed, the cloud resources are released, and the results are uploaded to the cloud storage and become available to other consumers of the service. Every operation is automated: there are no steps which need to involve human operations. Reconfigurable segments of the cloud allow solving a wide range of scientific and technical tasks, among them: tasks that operate on large data sets based on the MapReduce technology; Bioinformatics tasks, including processing of genetic information in distributed systems; tasks of class CAD/CAE; calculation jobs that do not require high-speed networks.

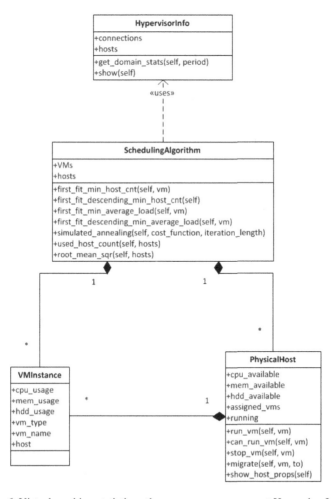

Fig. 6. Virtual machine statistic gatherer program component HypervisorInfo

For collecting information from KVM hypervisors special software was created (figure 6). This component connects to each hypervisor of the cloud and collects the following information for each virtual machine:

- Usage of CPU resources;
- Memory usage;
- Hard drive utilization (in i/o operations);

Also this module collects information about available resources in the hypervisor (cpus/memory/storage).

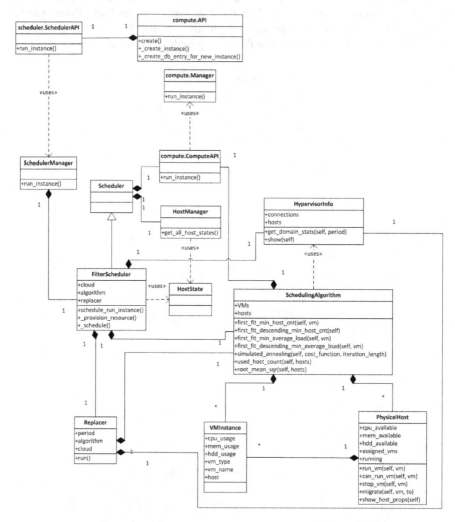

Fig. 7. Modified OpenStack nova scheduler class diagram

The implemented software was integrated with OpenStack nova-scheduler component. We changed nova scheduler and added implemented algorithms to OpenStack. Class diagram of the new scheduler is presented in figure 7.

We also evaluated the time of hot migration of virtual machines. Distributed storage based on Ceph file system was added to the OpenStack research stand. A test utility for evaluating time of virtual machine availability was implemented. This utility establishes TCP connection with the service on virtual machine and checks the timeout between ping-pong packets. The diagram of the migration procedure is shown in figure 8.

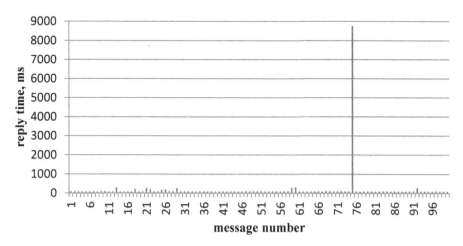

Fig. 8. Reply time histogram

The average time of virtual machine migration is about 8600 milliseconds. According to Amazon EC2 SLA (99.95%) a virtual machine can be unavailable up to 15768 seconds. This means that about 1800 migrations can be performed in a year for each virtual machine (4 migrations every day).

6   Conclusions and Future Work

In this paper a new approach for virtual machines distribution was proposed. Neural algorithm showed good results in the imitation mode. Another contribution of this paper is a new virtual machines scheduler that is able to place a virtual machine on the optimal compute node and migrate the virtual machine to another node if resource consumption state has been changed. We found that we can perform up to four migrations for each virtual machine every day and our future work is to apply this restriction to our model and algorithms. The general idea is to restrict migrations according to the given cloud SLA. Another future work is to integrate these results to our systems described in [5, 6]. An important part of our future research is comparing the scheduler based on neural algorithms and the scheduler based on simulated annealing algorithms, which were proposed in [2].

References

1. Beloglazov, A., Buyya, R.: Managing Overloaded Hosts for Dynamic Consolidation of Virtual Machines in Cloud Data Centers Under Quality of Service Constraints. In: IEEE Transactions on Parallel and Distributed Systems (TPDS), vol. 24(7), pp. 1366–1379. IEEE CS Press, USA (2013)
2. Mirolubov, M.: Optimization of load distribution in a cloud computing environment with an algorithm of dynamic allocation of virtual machines: Master's graduate work, Saint-Petersburg, vol. 115 (June 2013)
3. Lukashin, A., Zaborovsky, V.: Secure Heterogeneous Cloud Platform for Scientific Computing. In: ICNS 2014 - Proceedings of the Tenth International Conference on Networking and Services, pp. 24–28 (2014)
4. Lukashin, A., Zaborovsky, V., Kupreenko, S.: Access isolation mechanism based on virtual connection management in cloud systems: How to secure cloud system using high perfomance virtual firewalls. In: CEIS 2011 - Proceedings of the 13th International Conference on Enterprise Information Systems 3 ISAS 2011, pp. 371–375 (2011)
5. Lukashin, A., Lukashin, A., Tyutin, B., Kotlyarov, V.: Cloud service architecture for scientific and engineering computing in distributed environment. NTV Saint Petersburg State Technical University ITC #4(128) (2011)
6. Zaborovskiy, V.S., Lukashin, A.A., Popov, S.G., Vostrov, A.V.: Adage mobile services for ITS infrastructure. In: 13th International Conference on ITS Telecommunications, ITST 2013, pp. 127–132 (2013), doi:10.1109/ITST.2013.6685533

Implementation of IPv4 Reflection Scheme for Linux-Based Storage Systems

Kirill Krinkin and Mike Krinkin

St. Petersburg Academic University, Saint-Petersburg, Russia
kirill.krinkin@fruct.org, krinkin.m.u@gmail.com

Abstract. Linux program stack is very popular as a system platform in low range and middle range storage systems. On the one hand, it allows rely on powerful, well tested solution with full range of system services such as TCP/IP and Input/Output stack. Such solutions have good vertical and horizontal scalability. On the other hand, Linux is a general purpose system and requires extra efforts in and custom modification for Input-Output and Networking components in order to implement special use cases and reach maximum performance for data path. We discuss our implementation for IPv4 gateway holding mechanism (IP reflection) which allows IP storage system by fixing nearest gateway for each client bypassing regular routing scheme. Architecture and preliminary performance testing results are discussed.

Keywords: routing, linux kernel, netfilter, storage, IP reflection.

1 Introduction

Storage system (middle range storages are considering) usually contains two computers (storage processors) which are working in synchronous mode and connected by high performance data link. In normal mode operation each of storage processors (SP) are serving clients' requests. During the fail opened sessions (including sockets and I/O stack state) could be moved from broken SP to another. Each SP has several service instances of provided service (like NFS, CIFS, iSCSI, etc), each of them could serve several clients. In this paper we will use term *server* to specify this service instance.

Each service is movable (could be transferred from one SP to another) and has associated network configuration including static routing tables, DNS configuration, time configuration, and so on. This configuration should be kept after transferring server to different physical environment (another SP).

We consider a case when clients located behind the different routers. Storage system should work by following next rules:

- when client connects to the array with certain VLAN ID, storage should answer with the same VLAN ID;
- when client connects to the array by certain Ethernet socket, storage answer from the same socket.

S. Balandin et al. (Eds.): NEW2AN/ruSMART 2014, LNCS 8638, pp. 567–576, 2014.

– when client connects to us via certain gateway, storage should answer via the same gateway.

This set of rules could be implemented by rule based routing, but it requires managing settings in each subnet address and even the addresses of everything in interest which lies beyond these subnets.

At the same time, most of connections to the storage are initiated by the client. It means that the incoming packet is labeled with the client IP address and the last router MAC address, and comes through the right socket. The aim of this work is to extract all required information from incoming packets and automatically setup the first step of route for outbonding packets. In case there is no information solution should use the standard routing scheme.

2   Related Work

There are set of research about effective implementation of different schemes of routing for storages and data centers. Some works suggests particular solutions for linux network stack customization [1]; other works are considering whole review about current IP-network architecture [2].

We focus on easiest way to implement packet reflection scheme with minimum kernel modifications - *package Reflection (IPR)*. There are implementation of similar routing schemes. One of them Juniper Networks Screen OS approach to asymmetrical routing in networks [3]. Screen OS allow to route traffic to the same remote destination via more than one gateway. For this purpose Screen OS maintains MAC cache for TCP connections. Firewall creates cache entry for SYN packet, and forward the reverse packet (i. e. SYN-ACK and other) to the same gateway from where the initial SYN packet had arrived.

Another one similar technique is Packet Reflect option in EMC VNX storages [4].This option ensures that replay packets always goes through the same interface that request packet had arrived. When Packet Reflect is enabled, storage does not examine the route table when sending outbound packets within established connection, instead it uses same network interface and next-hop MAC address as the request packet. In this article we propose similar functionality for Linux.

3   Linux Implementation

IPR for Linux is implemented as kernel loadable module and based on netfilter API. It does not replace usual Linux routing, but replace route found by Linux network subsystem for packets, if it know better one.

Module maintains MAC cache as like Screen OS firewall, but it does not take in account protocol or packet type. It is why IPR works for all IPv4 protocols, but does not allow to hold route per TCP connection as Screen OS does.

3.1   Linux Network Stack Data Structures

Before dive in details of IPR implementation, get brief overview of principal data structures involved in IPR implementation.

Socket Buffer. `struct sk_buff` defined in `include/linux/skbuff.h` is the main structure in Linux network stack. This structure holds network packed data and related information. This information includes pointers to link-layer, network and transport headers, packet route, socket and all information that need to be passed between Linux kernel network subsystem layers.

To extract link-layer header from `sk_buff` Linux kernel provides function `unsigned char *skb_mac_header(const struct sk_buff *skb)`. This function returns pointer to network packet data where MAC header starts, and we have to cast this pointer to appropriate type.

Network header accessible through function `unsigned char *skb_network_header (const struct sk_buff *skb)`, as like `skb_mac_header` it returns pointer to place in a packet data where network header starts.

Route. Another one principal data structure is `struct dst_entry` defined in `include/net/dst.h`. This structure contains common information about route. There are specific structures for routes, for example, `struct rtable` for IPv4 or `struct rt6_info` for IPv6, that hold `dst_entry` as a field.

`dst_entry` holds pointer to `struct net_device`, that represents ingress or egress network interface in Linux kernel, and pointers to output (and input) function for specific network protocol, for example, `ip_output` for IPv4 or `ip6_output` for IPv6.

Neighbor Entry. Last one data structure important for IPR implementation is `struct neighbour`. This structure part of neighbor discovering subsystem, i. e. ARP implementation. This structure keeps mapping between network address and link-layer address and stores pending packets queue until link-layer address would be discovered.

3.2   Netfilter

As mentioned above IPR uses netfilter API. Netfilter is a set of hooks inside Linux kernel that allow to register callback functions with the network stack [5]. It is widely used for package handling in Linux base storage systems [6–8]

There are five hooks in IPv4 part of network stack:

- Preroute. Input packets pass this hook.
- Postroute. Output packets pass this hook.
- Forward. Transit packets pass this hook between preroute and postroute.
- Local Input. Packets send to host, i. e. one of host processes, pass this hook after preroute.
- Local Output. Packets originated from host pass this hook before postroute.

```
unsigned int netfilter_hook (const struct nf_hook_ops *ops,
                             struct sk_buff *skb,
                             const struct net_device *in,
                             const struct net_device *out,
                             int (*okfn)(struct sk_buff *));
```

(Netfilter callback signature.)

Callback signature is shown above. skb is a network packet this callback called for. in and out - ingress and egress network interfaces for the packet.

Callback returns status code, there are six common return codes:

- NF_DROP. Stop processing and drop the packet.
- NF_ACCEPT. Continue processing of the packet.
- NF_STOLEN. Stop processing like as NF_DROP, but callback takes ownership of this packet.
- NF_QUEUE. Packet queued for later processing, i. e. userspace handling.
- NF_REPEAT. Call this callback again.
- NF_STOP. Do not call other callbacks for this packet.

IPR registers two callbacks in local input and local output hooks. Input callback fills in IPR Cache, and output callback uses this cache to route packet through appropriate network interface and gateway.

3.3 IP Reflection Cache Implementation

IPR Cache is a standard Linux hashtable with 65536 buckets. IPR uses hash of client IP address to determine bucket. Cache uses pair of client and local IP address as a unique key.

Cache Entry. Cache entry also contains next-hop link-layer address field(h_ha) as well as route filed (h_route) to route packets through appropriate gateway and network interface.

```
struct hwaddr_entry
{
    struct rcu_head      h_rcu;
    struct hlist_node    h_node;
    atomic_long_t        h_stamp;
    atomic_t             h_refcnt;
    struct rtable*       h_route;
    unsigned             h_ha_len;
    u8                   h_ha [MAX_ADDR_LEN];
    __be32               h_local;
    __be32               h_remote;
};
```

(IPR Cache entry.)

Field h_rcu is used for safe memory reclamation.

Fields h_refcnt and h_stamp used by cache entries garbage collector.

Cache Entry Allocation. IPR creates Slab cache for `struct hwaddr_entry` instances. Slab cache allow to allocate entries aligned for CPU cache lines. Because cache entry used for every output packet is is important enough to access cache entry as fast as possible.

```
struct hwaddr_entry *hwaddr_alloc(const struct net_device *
    dev,
                                  __be32 remote,
                                  __be32 local,
                                  u8 const *ha,
                                  unsigned ha_len)
{
    struct hwaddr_entry *entry = NULL;

    if (ha_len > MAX_ADDR_LEN)
        return NULL;

    entry = (struct hwaddr_entry *) kmem_cache_zalloc(
        hwaddr_cache, GFP_ATOMIC);
    if (!entry)
        return NULL;

    atomic_long_set(&entry->h_stamp, (long) get_seconds());
    atomic_set(&entry->h_refcnt, 0);
    memcpy(entry->h_ha, ha, ha_len);
    entry->h_ha_len = ha_len;
    entry->h_remote = remote;
    entry->h_local = local;

    entry->h_route = hwaddr_create_route(dev, remote, local);
    if (!entry->h_route)
    {
        pr_warn("cannot create route for %pI4\n", &remote);
        kmem_cache_free(hwaddr_cache, entry);
        return NULL;
    }

    return entry;
}
```

(Cache Entry creation function.)

For every Cache Entry IPR allocate a route in `hwaddr_create_route` function. Client IP address used as a gateway in this route. If client is a direct neighbor of host it is correct to use client IP as a gateway. And it is safe to use client IP as a fake gateway for non direct neighbors because it is not disturb valid neighbor entries.

Hashtable. Hashtable is protected by one spinlock, that serializes cache modifications. Lookups require RCU protection only. We assume that after creation of cache entry modifications are rarely enough to use coarse-grained locking. So we prefer standard Linux hashtable implementation instead of custom fine-grained so far.

```
struct hwaddr_entry *hwaddr_lookup(__be32 remote,
                                   __be32 local)
{
    struct hwaddr_entry *entry = NULL;
    struct hlist_node *list = NULL;

    hwaddr_hash_for_each_possible_rcu( hwaddr_hash_table,
        entry, list, h_node, remote)
    {
        if (entry->h_remote == remote && entry->h_local ==
            local)
        {
            atomic_long_set( &entry->h_stamp, (long)
                get_seconds() );
            return entry;
        }
    }

    return NULL;
}
```

(IPR Cache lookup function.)

Cache lookup function requires external RCU or spinlock synchronization. Lookup function updates timestamp in cache entry that used by garbage collector to detect old entries.

```
void hwaddr_update(const struct net_device *dev,
                   __be32 remote,
                   __be32 local,
                   u8 const *ha,
                   unsigned ha_len)
{
    struct hwaddr_entry *entry = NULL;

    rcu_read_lock();
    entry = hwaddr_lookup(remote, local);
    if ( !entry || entry->h_ha_len != ha_len || memcmp(entry
        ->h_ha, ha, ha_len) )
        hwaddr_create_slow(dev, remote, local, ha, ha_len);
    rcu_read_unlock();
}
```

(IP Reflection Cache update function.)

Update function creates or update cache entries. IPR uses optimistic approach to synchronization. Assuming that updates are rarely enough hwaddr_update at first lookup an entry, and then if entry not found or link-layer address change detected, IPR acquires spinlock, repeats lookup and creates or updates entry if needed in hwaddr_create_slow function.

Input Netfilter callback after some sanity checks calls `hwaddr_update` function for every input IPv4 packet hereby fills in IPR Cache.

3.4 Output Netfilter Callback

In output Netfilter callback IPR reroutes packet through appropriate network interface and gateway. As mentioned above IPR uses client IP as a gateway, so ARP fails resolving link-layer address of gateway if client is not direct neighbor of host. So IPR needs to manually specify link-layer address for `neighbour` instance, for this purpose Linux kernel provides `int neigh_update(struct neighbour *neigh, const u8 *lladdr, u8 new, u32 flags)` declared in `include/net/neighbour.h`.

Route which would be used by network stack to send packet is stored in `sk_buff`, so to reroute packet IPR just replaces pointer to `dst_entry` in `sk_buff` by pointer from IPR Cache entry.

If there is no cache entry for IP addresses from a packet, IPR keeps route found by Linux network stack.

3.5 Cache Entries Reclamation Strategy

To make solution suitable for enterprise applications IPR needs way to remove outdated cache entries. For this purpose IPR uses timeout based garbage collector.

Every entry contains a timestamp, IPR updates timestamp for every successful lookup of cache entry. Garbage collector periodically removes entries with outdated timestamp. Threshold timeout value passed as module parameter.

Of course it is possible to have connection that is not active for a time longer than a threshold value. Such connection will be broken if there is no appropriate route for packets. For this reason IPR stores reference counter in a cache entries and userspace application can increase or decrease reference counter to hint that an entry still in use even if timestamp is outdated. For entries with positive reference counter value, IPR uses another timeout threshold value that should be much greater than first one. Second threshold as well as first one passed as module parameter.

To organize periodical works of garbage collector module uses Linux kernel global work queues API. IPR uses first threshold value as interval for garbage collector works.

4 Testing

IPR is designed for storage systems where performance is critical. For this reason it is needed to compare IP packet processing time with and without IPR module.

The most time consuming operations are hashtable update and lookup. In worst case update require two lookups and memory allocation. So to estimate performance we measure average update operation time and average Linux IP layer packet processing time without IPR.

4.1 Benchmark Methodology

Linux Kernel Time Measurements. For time measurements we use high resolution kernel timer API function `ktime_get` which has nanosecond resolution.

Measure Cache Updates. We measure worst case time of `hwaddr_update` function in relation to hashtable bucket filling. We do not take in account contention for hashtable spinlock. We repeat measurement a thousand times to get mean time value.

Measure IP Packet Processing Time. To measure IP layer packet processing time we made patch for Linux kernel. This patch adds timestamp field in `sk_buff`. In `ip_rcv` function kernel sets this timestamp, and then in `ip_local_deliver_finish` function, before upper layer protocol handler call, kernel calculates elapsed time. Kernel accumulates elapsed time for every packet and expose average processing time through kernel log.

4.2 Benchmark Results

IP Packet Processing Time. In Virtual Box test stand we used for measurements average IP layer packet processing time is 2585 ns. We use this as base line to estimate performance overhead of IPR.

Cache Update Time. As mentioned above we measure cache update time relatively to hashtable bucket length. Measurements results shown in table 1.

Table 1. Cache Update Time

Length	Time, ns	Ratio
0	1606	0.62
1000	26231	10.15
2000	58289	22.55
3000	92977	35.97
4000	124601	48.20
5000	158932	61.48
6000	185366	71.71
7000	215216	83.26
8000	245635	95.02
9000	276259	106.87
10000	307375	118.91

First column of table 1 contains hashtable bucket length, second column contains average update time, and last column is a ratio of update time to IP layer packet processing time.

4.3   Benchmark Summary

Cache update time as well as ration scales linearly as expected. So with bucket of length less than hundred, we expect that IP layer packet processing time would be two times slower with IPR enabled.

If average bucket length is about a hundred, then hashtable stores over six millions entries. Such number of entries is enough for many purposes.

If take in account upper layers (i. e. TCP) that has relatively large processing time, and that we measured worst case, performance of IPR is reasonable for many cases, but overhead still valuable.

5   Future Work

Measurements above do not take in account possible contention for hashtable spinlock, it is fair for single core CPU. But for modern multicore processors it is not fair, because in Linux it is possible to process network packets simultaneously on different CPU cores. So contention is possible and we need to make tests that measure performance overhead under high contention.

As mentioned earlier, IPR uses client and host IP address pair as a unique key, but Linux kernel does not allow to have several `neighbour` instances for same client IP address, so really it is not possible to have different routes for same client IP through different network interfaces with current IPR implementation, and it is significant drawback, because storage host usually has a number of network interfaces.

6   Availability

IPR implementation is open source and distributed under GPL license. All sources are available on github `https://github.com/OSLL/hwaddr-cache`.

References

1. Wang, H., Tian, L., Zhu, M., Xiao, L., Ruan, L.: A Mechanism Based on Netfilter for Live TCP Migration in Cluster. In: 2010 9th International Conference on Grid and Cooperative Computing (GCC), pp. 218–222. IEEE Computer Society, Washington, DC (2010)
2. Shanbhag, S., Wolf, T.: Implementation of End-to-end Abstractions in a Network Service Architecture. In: Proceedings of the 2008 ACM CoNEXT Conference, pp. 16:1–16:12. ACM, New York (2008)
3. Juniper Networks Knowledge Center,
   `http://kb.juniper.net/InfoCenter/index?page=content&id=KB14429`
4. Configuring and Managing Networking on VNX,
   `http://mydocs.emc.com/VNXDocs/Networking.pdf`
5. Netfilter Hacking HOWTO, `http://www.netfilter.org/documentation/HOWTO/`
   `netfilter-hacking-HOWTO-1.html`

6. Zhong, B., Huaqing, L.: Design of A New Firewall Based on Netfilter. In: Proceedings of the 2012 International Conference on Computer Science and Electronics Engineering, vol. 03, pp. 624–627. IEEE Computer Society, Washington, DC (2012)
7. Yang, Y., Yonggang, W.: A Software Implementation for a hybrid Firewall Using Linux Netfilter. In: Second WRI World Congress on Software Engineering, pp. 18–21. IEEE Computer Society, Washington, DC (2010)
8. Accardi, K., Bock, T., Hady, F., Krueger, J.: Network Processor Acceleration for a Linux Netfilter Firewall. In: Proceedings of the 2005 ACM Symposium on Architecture for Networking and Communications Systems, pp. 115–123. ACM, New York (2005)

The Algorithm for Cars License Plates Segmentation

Alexandr A. Kryachko[1], Boris S. Timofeev[2], and Alexandr A. Motyko[2]

[1] Saint-Petersburg State Polytechnical University, Saint-Petersburg, Russia
alex_k34.ru@mail.ru
[2] Saint-Petersburg State University of Aerospace Instrumentation, Saint-Petersburg, Russia
timofeev-boris36@mail.ru, motyko.alexandr@yandex.ru

Abstract. We have conducted the needs assessment that highlights the current issues requiring novel solutions in the design of the modern system of automotive license plates recognition. We propose an algorithm of segmentation the license plates on a complex background, invariant to their size, contrast and angle positioning on the image.

Keywords: ANPR system, digital image processing, road traffic monitoring.

1 Introduction

In most cases, the vehicles are identified by their license plates, which can be easy read by humans, but not by a computer system. In recent years, many companies developed systems that read license plates of vehicles and provide the probability of correct identification about 80÷85%. These systems have hard requirements for the installation of CCTV cameras, lighting conditions and purity of license plates [1]. There are following tasks in the development of new generation of these systems:

- Recognition of license plates of all vehicles involved in the field of view of the camera;
- Segmentation of license plates on a complex background under different road conditions;
- Provide invariance to the size and the angle and position of the license plates;
- Improve the recognition of contaminated license plates with low contrast of their images.

2 Physical Aspects of the License Plate Recognition Systems

For the technical system the license plate is a gray box with dark text that defined the two-dimensional function $L(x,y)$, where x and y are spatial coordinates, and L – brightness at this point. We need to develop a reliable mathematical algorithm that can extract the semantics of the spatial region of the captured image on a complex background. These functions are implemented in the so-called "ANPR systems"

S. Balandin et al. (Eds.): NEW2AN/ruSMART 2014, LNCS 8638, pp. 577–590, 2014.

where ANPR abbreviation stands for "Automatic Number Plate Recognition". ANPR system performs data conversion between the real environment and information system. ANPR systems algorithms belong to the research field of the artificial intelligence, computer vision, pattern recognition and neural networks [2].

One of the main problems is the development of algorithm for detecting the rectangular areas of license plates in a digital image of the transport situation. People identify the license plate as "a small rectangular plate with numbers and letters attached to the vehicle for purposes of identification", but the computer does not understand the definition.

We need to find an alternative definition of the license plate based on descriptors that are understandable to computers. We can define the following signs of the license plates:

- White (gray) plate, the brightness differs from the neighboring fragments of the image;
- Rectangular plate with an aspect ratio of 5:1 , with black frame (for Russian license plates);
- Rectangular area with a predominance of vertical strokes, because dark numbers and letters numbers are elongated in the vertical direction.

These signs are extracted via the preprocessing operations. Implementation of these principles may fail if the plates are too dirty, deformed or distorted. Then we should use more sophisticated segmentation algorithms with different methods of normalization and detection of symbols.

Automatic number plate recognition system ANPR is a set of hardware and software for creating and processing digital images or video. The hardware of the system usually consists of cameras, image processing block, camera control block, module for record and store information. Whenever the sensor detects a vehicle at a predetermined distance from the camera, he activates the detection mechanism. An alternative solution is based on the detection of entering the detection zone of the vehicle and continuous processing of the received video signal when the vehicle passes through the zone. Direct processing of video data consumes more system resources, but allows the accumulation of results and increase the probability of correct recognition, for example, through the use of the majority principle. ANPR system can have its own processor (all-in-one solution), or send the data to the CPU for further processing (generic ANPR). Image processor receives static images taken with the camera, and returns a textual representation of the detected license plate.

Basis for further digital processing efficiency is to obtain high quality images of vehicles in the camera view. Capturing must be done with fast shutter speed. Otherwise, the picture quality will deteriorate due to the undesirable effect of blur caused by movement of the vehicle. For example, during the 1/100 sec. a car that is moving at 80 km/h travels a distance of 0.22 m This will cause a significant degradation of the system's ability to recognize numbers ANPR. It is also necessary to ensure the invariance of the camcorder relative to the light conditions. Ordinary camera should not be used to capture images in dark or night, as well as working in the visible spectrum of light. Using the infrared camera in conjunction with the infrared illumination significantly improves the recognition results. Under infrared lighting plates, which are made of reflective material, will well stand out against the rest of the image. In addition, it is possible to identify highly contaminated plates.

3 Mathematical Aspects of License Plates Segmentation Algorithms

The first step in the process of automatic license plate recognition is the detection of the plate. Implementation involves the luminance feature quantization of the original image by a predetermined number of levels n and subsequent image analyses on each of them (Fig. 1).

Black letters and numbers make the brightness of license plate inhomogeneous. Therefore, the image should be processed with Gaussian smoothing filter (Fig. 1b). At the final stage plates-candidates are subjected to checking on the basis of shape: a true plate should be the rectangle with aspect ratio 5:1.

The disadvantage of the proposed method is the dependence of the results on the license plate size on the original image. Smoothing of the filter on a large image may not be sufficient and the segmentation of license plate fails (bottom row of images in Fig. 1). Another disadvantage is the substantial computational cost, as it is necessary to analyze the candidates for each of the n levels.

a)   b)   c)   d)

Fig. 1. Image processing stages: a) input images; b) after Gaussian smoothing; c) quantized images; d) segmentation of plate-candidate

More promising [3] is the algorithms with extraction of vertical lines in the image (Fig. 2) using the Sobel FS or Prewitt FP filter masks, or the filter FV for vertical lines segmentation

$$F_S = \begin{bmatrix} -1 & 0 & 1 \\ -2 & 0 & 2 \\ -1 & 0 & 1 \end{bmatrix}; \quad F_P = \begin{bmatrix} -1 & 0 & 1 \\ -1 & 0 & 1 \\ -1 & 0 & 1 \end{bmatrix}; \quad F_V = \begin{bmatrix} -1 & 2 & -1 \\ -1 & 2 & -1 \\ -1 & 2 & -1 \end{bmatrix}.$$

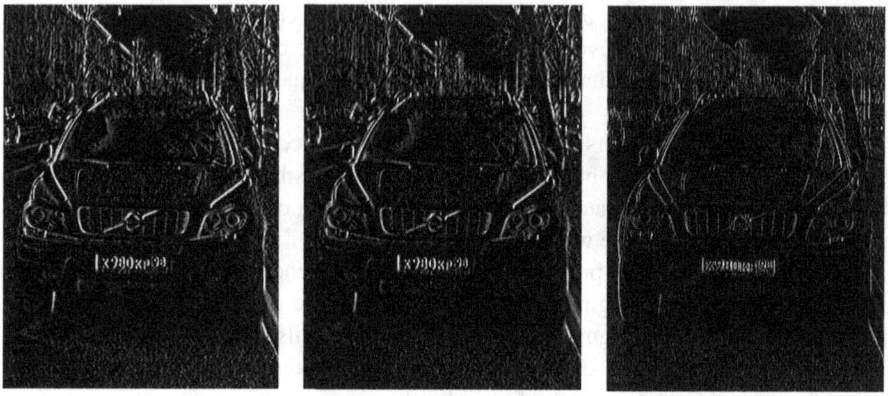

a)   b)   c)

Fig. 2. Images of the edges, obtained via different filters: F_S – a), F_P – b), и F_V – c)

The third filter is more preferable (Fig. 2 – c), because the use of the filter FV allows better preserve the shape of the plate and the license plate gives less intense responses from lines of other destinations.

To increase the contrast of the contour the threshold is used. The standard way is for example, the global threshold computed via algorithm k-means or more perfect Otsu method, which allows dividing the image pixels into two classes with a threshold k, which maximizes the interclass variance

$$\sigma_B^2(k) = \frac{\left[m_G P_1(k) - m(k) \right]}{P_1(k)\left[1 - P_1(k) \right]}, \tag{1}$$

where m_G – global average brightness across the image, $P_1(k)$ – the probability that a certain pixel will be assigned to the first class, $m(k)$ – the cumulative sum of pixel brightness to a level k [4]. To find the optimal threshold value, we should vary k and evaluate the right side of the expression (1). Then we choose a value of k, at which $\sigma_B^2(k)$ is maximal.

It is possible that there are two or more cars on the image with license plates of varying degrees of pollution, which leads to different contrast of their images. The use of the global threshold in that case does not give satisfactory results (Fig. 3, center).

Fig. 3. Results of vertical edges extraction: top) input image; center) filtering with F_v using Otsu method; bottom) filtering with F_v using double local threshold

In order to determine a local threshold for each pixel of the original image is performed statistical analysis of neighboring pixels. Let us assume that $L(x,y)$ is the brightness of the pixel, for which we try to compute the local threshold k. For simplicity, examine the square neighborhood with the width $2r + 1$, where

$$\left[x-r, y-r\right], \ \left[x-r, y+r\right], \ \left[x+r, y-r\right], \ \left[x+r, y+r\right]$$

are coordinates of the corners. There are several approaches for local threshold calculation [5]: The average brightness of pixels neighborhood:

$$k(x, y) = average\{L(i, j)\}$$
$$x - r \leq i \leq x + r; y - r \leq j \leq y + r$$

The median of the neighborhood

$$k(x,y) = median\{L(i,j)\}$$
$$x-r \leq i \leq x+r; y-r \leq j \leq y+r$$

The average between a minimum and maximum brightness values in the neighborhood

$$k(x,y) = 1/2\left(\min\{L(i,j)\} + \max\{L(i,j)\}\right)$$
$$x-r \leq i \leq x+r; y-r \leq j \leq y+r$$

It is desirable that the neighborhood boundaries were not sharp. To estimate the average over the neighborhood is advisable to apply a Gaussian filter: calculate the convolution of the original image with the impulse response of the digital filter **G**, which is a very rough approximation of the Gaussian function:

$$k(x,y) = \sum_{i=-r}^{r} \sum_{j=-r}^{r} L_{x-j,y-i} G_{i,j} \, .$$

To reduce the number of multiplications of real numbers, the impulse response of the filter is composed of binary numbers, for example:

$$\mathbf{G} = \begin{bmatrix} 0 & 1 & 1 & 1 & 0 \\ 1 & 1 & 2 & 1 & 1 \\ 1 & 2 & 8 & 2 & 1 \\ 1 & 1 & 2 & 1 & 1 \\ 0 & 1 & 1 & 1 & 0 \end{bmatrix} / 32 \, .$$

Next, the original image is quantized into two levels according to the rule:

$$L_n(x,y) = \begin{cases} 0, \text{if } L(x,y) < k(x,y) \\ 1, \text{if } L(x,y) \geq k(x,y) \end{cases} \, .$$

Because of the local threshold managed to confidently extract vertical strokes belonging to the numbers of both cars. However, in case of the complex background the simple local threshold technique can give poor results. Therefore we propose a double local threshold (Fig. 3, bottom). The algorithm is described below.

- To process the image of vertical edges with Gaussian filter with different standard deviation and to obtain two new images with relative small and large blurring.
- To calculate for each pixel of vertical edges image (with the value of brightness B) the cutoff threshold T, i.e. find the maximum brightness B_{max} in small location around the pixel on the image with large blurring and compute the value of threshold via the exponential logistic function:

$$T = (1 - \exp(-B_{max}^2)) \, .$$

- To compare B with the brightness of correspondent pixel B' on the image with small blurring and set the binarized value according to the rule:

$$L_n(x, y) = \begin{cases} 1, if\ B \geq B' \geq T \\ 0, else \end{cases}.$$

In the next step of processing the image is partitioned into clusters [3]. In order to separate each cluster with a rectangular border, the image projection on the vertical axis is calculated (Fig. 4b):

$$\Pr[y] = \sum_{x=1}^{b} L(x, y),$$

where b – is the width of the image. Then find horizontal bands corresponding to the local maxima in the vertical projection (Fig. 4b), and calculate the projections of these areas of the image on the horizontal axis (Fig. 4c, d):

$$\Pr_Z[y] = \sum_{y=y_{z\min}}^{y_{z\max}} L(x, y),$$

where z – number of slice, $y_{z\min}$, $y_{z\max}$ – coordinates of the beginning and the end of the slice.

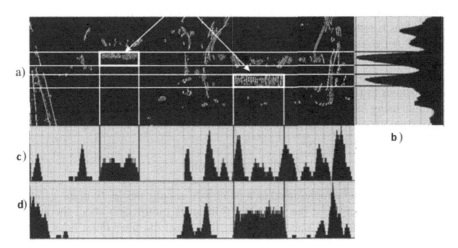

Fig. 4. Projections: a) input image; b) projection on vertical axis; c), d) projections on horizontal axis

Implementation of this method can give poor results if the original image is full of details with vertical edges that mask objects of interest. Another method involves the clustering of homogeneous brightness areas on the basis of connectivity. For this purpose, the original image is pre-treated with morphological filter "dilatation" with the structural element, elongated in the horizontal direction (Fig. 5). For each cluster a corresponded model in form of ellipse or rectangle is given. The model in the form of ellipse or rectangle is set for each cluster.

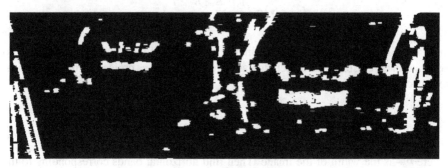

Fig. 5. Image after dilatation

The resulting models are sorted by geometric characteristics: aspect ratio should at least roughly follow the format of the license plate – (5:1).

The captured image of a rectangular plate can be rotated and twisted by positioning the vehicle with respect to the camera. Since this fact makes it difficult to recognize, it is important to implement additional mechanisms that can detect and correct skew of the plate. The main challenge is to defining the angle of the skewed plate. Plate number is a three-dimensional object which is projected onto the two-dimensional plane, which can lead to distortion of the angles and proportions. If the optical axis of the camera lens is orthogonal to the plane of the license plate, but the vertical lines on the plate did not coincide with the same lines in the image, the image is only rotated. If the orthogonality condition is not met, additional perspective distortions arise.

For the evaluation of the angle of rotation is advisable to use Hough transform, which allows to find straight lines on the monochrome image. Straight-line equation is written using the parameters ρ and θ. The parameter ρ is the Euclidean norm vector drawn from the origin to the nearest point on the line and θ – is the angle between the vector and the x-axis:

$$\rho(\theta) = x\cos\theta + y\sin\theta . \tag{1}$$

Straight line can be represented as a point with coordinates (ρ, θ) in the parameter space (Hough space), and the parameter θ includes information about the angle of inclination of the straight line (Fig. 6). By the use of the filter FG a lines, that are close to horizontal, can be extracted (Fig. 7).

$$F_G = \begin{bmatrix} -1 & -1 & -1 \\ 2 & 2 & 2 \\ -1 & -1 & -1 \end{bmatrix} .$$

For the points, that belong to that lines (Fig. 6a and Fig. 7b) we can build their representation in a Hough space (Fig. 6b). If we take several points of one straight line and build in a (ρ, θ) space a sinusoidal curve (according to equation 1), then the point of the intersection will determine the line (Fig. 6b). Thus, the problem of detection of straight lines can be reduced to finding the intersection points of the sinusoidal curves and counting the number of their common points. Parallel straight lines are characterized by the same parameter θ (Fig. 6b), which allows us to determine the angle of inclination.

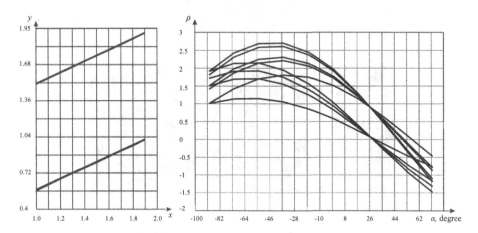

Fig. 6. a) Two parallel lines; b) their representation in the Hough space

In the image of the car, there are enough horizontal lines to implement the procedure of voting in the Hough space to determine their angle and then the angle of rotation of the license plate (Fig. 7).

Straight lines of other areas may distort the voting results; in addition, it may be several cars on the image, with different angles of the plates. Therefore, the angle correction procedure is advantageously carried out individually for each of the license plate.

On the second stage of correction a geometric transformation of the image of the plate is applied. Algorithm for determining the angle based on Hough transform does not allow split image rotation, and skew due to perspective distortion, which is often encountered in practice. To correct the shape of the license plate based on the data of the angle α the affine transformation matrix should be used.

$$\mathbf{A} = \begin{bmatrix} 1 & S_y & 0 \\ S_x & 1 & 0 \\ 0 & 0 & 1 \end{bmatrix} = \begin{bmatrix} 1 & -\tan \alpha & 0 \\ 0 & 1 & 0 \\ 0 & 0 & 1 \end{bmatrix},$$

where S_x и S_y – skew factors, and S_x is zero, because only S_y should be corrected.

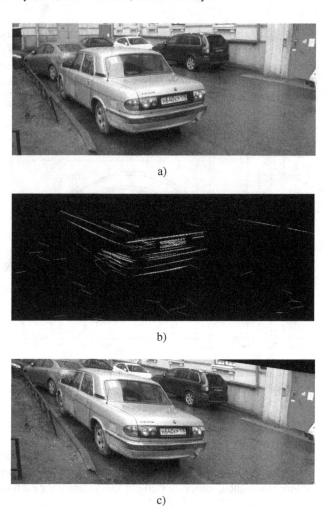

a)

b)

c)

Fig. 7. Angle correction a) initial image; b) Hough lines; c) corrected image

The new coordinate vector of the point of image $L(x, y)$ is:

$$\begin{bmatrix} x_s \\ y_s \\ 1 \end{bmatrix} = \begin{bmatrix} 1 & -\tan\alpha & 0 \\ 0 & 1 & 0 \\ 0 & 0 & 1 \end{bmatrix} * \begin{bmatrix} x \\ y \\ 1 \end{bmatrix}.$$

Applying this procedure to the extracted license plates of two vehicles, it is possible to eliminate distortion for each of them (Fig. 8).

Most of the algorithms for the letters and numbers recognition is performed for the normalized size of the binary image of license plate. In ihe initial image the sizes of license plates may be different.

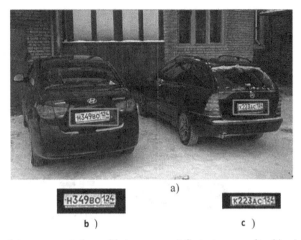

Fig. 8. License plates segmentation with two cars: a) first stage results; b), c) post processing results

It is worthwhile to convert the sizes of the plates to a standard value (for example 768x160 pixels) with an effective method of interpolation (for example bicubic) (Fig. 9).

Fig. 9. Three cars with various sizes of the license plates

Very often the top or another part of the license plate is in shadow. Therefore the use of global threshold for binarization leads to a significant distortion of the license plate, that may obstruct or eliminate its recognition (Fig. 10a). Using the proposed double threshold can significantly improve the image of the license plate (Fig. 10b).

Due to local irregularities of the brightness on the image of license plate after binarization it may be difficult to recognize letters and digits. To eliminate unwanted artifacts the contour analysis is implemented. In most cases, the forms of artifacts are different from the forms of characters, that allows to successfully filtering them (Fig. 10c).

Fig. 10. License plate post processing: a) with global threshold b) with local threshold, c) after contour filtering

Complex use of all the above methods of segmentation and image processing allows us to prepare the images of license plates for the subsequent numbers recognition (Fig. 9), for all vehicles caught in the field of view video camera (Fig. 11).

$$B\,894cc\,\underset{RUS}{77}$$

$$K\,583_{HH}\,96$$

$$C\,093\,vA\,98$$

Fig. 11. Images prepared for recognition

The last step before the recognition of individual characters of license plate is the finding of the positions of symbols. We calculate the projection of the image of license plate on the horizontal axis (Fig. 12). Positions of the local maxima indicate the intervals between characters. The symbols on the license plate are standardized and the font is known. This allows using relatively simple algorithms for character recognition [6], [7].

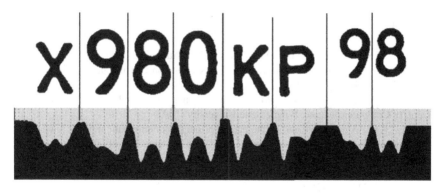

Fig. 12. Finding the individual positions of symbols

4   Conclusion

We proposed an efficient algorithm for licensed plate's segmentation.

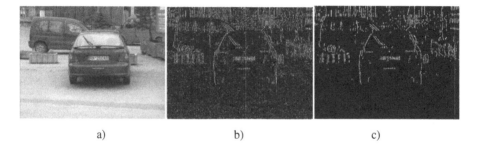

a)   b)   c)

Fig. 13. Initial image (a), vertical edges with local threshold (b), and with double threshold (c)

Thus the main advantages of the proposed algorithm are:

- the ability to detect the license plate on the complex background;
- the invariant to the irregularity of the contrast on the image;
- the invariant to the angle of rotation and size of the license plate.
- the ability to detect the plates of all vehicles on the image with different scales

The carried out experiments gave the following result. The success rate of segmentation was higher than 90%.

The feature of the algorithm is the double local threshold of binarization. This is the core of stability of the algorithm in terms of different negative factors. The comparison of the results of ordinary local threshold (based on the convolution with Gaussian filter) and proposed double threshold is shown below (Fig 13). The noise on the obtained image is significantly decreased.

References

1. Shan, D., Ibrahim, M., Shehata, M., Badawy, W.: Automatic License Plate Recognition (ALPR): A State-of-the-Art Review. IEEE Transactions on Circuits and Systems for Video Technology (February 1, 2013)
2. Sorin, D.: A neural network based artificial vision system for license plate recognition, Dept. of Computer Science. Wayne State University (1997)
3. Martinsky, O.: Algorithmic and mathematical principles of automatic number plate recognition systems, B.Sc. thesis, Department of Intelligent Systems, Faculty of Information Technology, Brno University of Technology (2007)
4. Liao, P.-S., Chen, T.-S., Chung, P.-C.: A Fast Algorithm for Multilevel Thresholding. J. Inf. Sci. Eng. 17, 713–727 (2001)
5. Gonzalez, R.C., Woods, R.E., Eddins, S.L.: Digital Image Processing using MATLAB. Pearson Education (2004) ISBN 978-81-7758-898-9
6. OCR Introduction. Dataid.com (June 16, 2013),
   http://www.dataid.com/aboutocr.htm (retrieved)
7. Draghici, S.: A neural network based artificial vision system for license plate recognition. International Journal of Neural Systems 8(1), 113–126 (1997)

Improvement of Finite Difference Method Convergence for Increasing the Efficiency of Modeling in Communications

Elena N. Velichko[1,2], Aleksey Grishentsev[2], Konstantin Korikov[1], and Anatoliy Korobeynikov[3]

[1] Saint Petersburg State Polytechnical University, St. Petersburg, Russia
velichko-spbstu@yandex.ru, korikov.constantine@spbstu.ru
[2] Saint Petersburg National Research University of Information Technologies, Mechanics and Optics, St. Petersburg, Russia
tigerpost@yandex.ru
[3] St. Petersburg Institute of Terrestrial Magnetism, Ionosphere and Radio Wave Propagation of the RAS, St. Petersburg, Russia
Korobeynikov_A_G@mail.ru

Abstract. The finite difference method for differential equation solution is considered. To increase the efficiency of the finite difference method, a new method of intermediate solution computation was developed and tested. The principles, software implementation of the method, and the results are considered. Our method can be used to significantly increase the efficiency of the finite difference method in solving applied problems of modeling for communication lines. It can also prove useful for the efficient design of key electronic components, such as antennas for wireless communication devices, high-speed digital and microwave circuits, and integrated optics.

Keywords: channel modeling, finite difference method, differential equation.

1 Introduction

Finite difference methods continue to provide an important approach to many numerical simulation problems. In particular, this method is used for solving applied problems of modeling for communication lines and for the efficient design of key electronic components, such as antennas for wireless communication devices, high-speed digital and microwave circuits, and integrated optics [1,2]. A number of modeling tasks in the communication field are related to solution of differential equations which, as a rule, cannot be solved analytically. So, development of new approaches with the aim of increasing the computation efficiency is a topical problem.

The finite difference method (FDM) is used to solve differential equations by replacing them with finite differences [1,3-5]. But if the number of iterated elements is high the convergence rate becomes low, and this is the main limitation of the FDM. Quantity reduction of these elements by discretization often causes decrease of exact

S. Balandin et al. (Eds.): NEW2AN/ruSMART 2014, LNCS 8638, pp. 591–597, 2014.

solution convergence. Sometimes, instead of FDM a more rapid finite element method (FEM) can be used. It also allows solving differential equations and has a number of advantages. For example, improving the solution accuracy and elimination the effect of aliasing boundaries may be achieved by high discretization. But in the FEM the separation of space to finite elements is required. Currently there are a number of algorithms that are effective for triangulation in n-dimensional spaces, and sometimes the FEM method may be more effective in comparison with FDM

Despite of all the advantages of the FEM, there are a number of limitations of the widespread practical application of FEM instead of FDM. These include the fact that the majority of algorithms and programs are designed for triangulation in 2-dimensional space. For the 3-dimensional space available triangulation tools are significantly poorer and for the more-than-three-dimensional spaces, for example, four-dimensional space (the space-time continuum) commonly used in problems of mathematical physics, such tools are not available in open sources at all. Developing of algorithms libraries for plotting Voronoi diagrams in n-dimensional space is rather time- and source-consuming and not always justified.

The goal of our work was to develop the method of FDM optimization for reducing the computation time via simple algorithms and accessible implementation of the method for a random n-dimensional space. We suggested a new method based on acceleration of the FDM by increasing of the convergence rate of the iterations. The method of optimization we developed is easy for software implementation for differential equations in a random n-dimensional space.

2   The Method

Let us consider the method of reducing the computation time in FDM for soft differential equations [5] in a domain **W** with boundary conditions **G** (bounding for physical areas and initial for time). The essence of the method is an enlargement of step h in the grid **U** in the domain **W+G**. In this case the solution accuracy decreases and the solution becomes wrong. For example, electrical charges in the fundamental problem of electrostatics can be described as single points. But with an enlargement of the grid spacing such boundary conditions can be lost, which fundamentally changes the essence of the intermediate solution.

To avoid such mistakes a solution with several sequentially applied grids is usually used. At first a grid U_1 with largest spacing is used to obtain an approximation of the solution. Then the grids U_k, $k > 1$ with a smaller spacing are used to refine the solution. The number of the grids is usually 2-3. We should note that the grid spacing h can vary in different directions and areas of the grid **U**. Thus the acceleration of the procedure of convergence to the problem's solution with a given precision p happens due to a more rapid formation of an intermediate solution in the domain **W**.

There is an equation

$$Au(x) = f(x) \tag{1}$$

where A – is differential operator, $x \in$ **W+G** (**W** – is solution domain, **G** – are boundary conditions), $f(x)$ – is prescribed function.

According to the theorem on the convergence [6] of a difference solution $y(x)$ to
the exact solution $u(x)$ with a grid spacing h for the equation (1), and in case the
condition

$$\|y(x) - u(x)\| \to 0, h \to 0$$

is satisfied, and for the specified accuracy p the assumption

$$\|y(x) - u(x)\| \to O(h^p),$$

is satisfied, it follows that this method of reducing the grid spacing h, is a direct consequence of this theorem.

Each solution refinement is iterative and has the computational complexity equal to

$$O(k \prod_{i-1}^{n} N_i), \tag{2}$$

where n – is dimension of the problem, N_i – is number of grid nodes in each direction,
k – is the number of iterations providing the specified accuracy at a step. The value
(2) can be very high in multidimensional problems even with rarefied grid.

The proposed method of the initial solution computation is based on the approximation of an intermediate solution in the domain **W**, but not on a consequent iteration
approximation. The approximation is based on the data on the boundary conditions **G**
taking into consideration the equation (1). In this case the computational complexity
can be estimated as

$$O(n \prod_{i-1}^{n} N_i), \tag{3}$$

where usually $n \ll k$ and n, k have the same meaning as in (2).

The specified problem of finding intermediate solution in the domain **W** can be
solved by approximation the values of $f(x)$ from the right side of equation (1). These
values successively form a partitioning uniform grid U along the lines with boundary
conditions **G**. Approximation can be produced with different functions depending on
the type of (1) and the physical nature of the problem, for example, with a polynomial

$$a_0 + a_1 z + a_2 z^2 + \ldots + a_p z^p, \tag{4}$$

where $a_0, a_1, a_2, \ldots, a_p$ – are polynomial coefficients, z – is the integration variable.

In general, if the grid lines are not parallel to the coordinate axes, forming the domain **W**, z can differ from the set of variables of the initial problem. In this case it is
necessary to take into consideration the rotation of the coordinate system where the
argument z is considered in relation to the initiative coordinate system. For n-
dimensional $(n > 1)$ problem the approximation is performed consequently for all
lines forming a grid in each direction. Approximation is followed by evaluation of the
mean values for each grid node. The mean value of grid node is calculated as the
mean value of approximation at the points of the grid lines belonging to the chosen
node. It can be said that to find an intermediate solution the FDM problem is divided
into a set of one-dimensional FEM problems equal to a number of grid lines in FDM.

We should note that it is also possible to use the proposed method and the method of grid spacing enlargement together because even for the grid enlarged spacing acceleration of convergence process remains urgent. Application of the proposed method is based on the physical nature of processes and conditions of each specific problem.

After the grid spacing enlargement not all the boundary conditions are used, but only part of them or averaged boundary conditions. Generally in case of perturbation of boundary conditions, these conditions can be lost after grid enlargement, which fundamentally changes the essence of the intermediate solution. For example, electrical charges in the fundamental problem of electrostatics can be described as single points. If the boundary conditions, for example, talking about an isolated point charge, are reflected in the nodes of the grid with enlarged spacings by approximating and getting new boundary conditions, then in fact the proposed method is applied for this task but in a local area. Actually the proposed method uses all the boundary conditions with the default frequency sufficient to obtain the required accuracy of the solution.

An intermediate solution obtained by the proposed method in some cases (depending on the features of the boundary conditions), can be "not smooth". Practically there is a possibility to use additional steps increasing smoothness. For example, during the approximation it is possible to take into an account the values from the neighboring nodes that weren't included as a value or an argument in an approximating function applied to the grip lines.

3   Testing of the Method

The method was tested on various engineering and communication modeling tasks which may be described by the Laplace's and Poisson's equations in spaces with a number of dimensions $n = 2$ and $n = 3$, including:

— modeling of electromagnetic field in natural (for example, distance between ionosphere and earth's surface) and human-induced waveguides;
— solving the fundamental problem of electrostatics (including the case with the charges described by the one boundary element);
— calculating the field of a complex shape capacitor;
— modeling of periodic electromagnetic interferences in communication lines;
— modeling of GPGPU (General-purpose graphics processing units);
— modeling of antennas for wireless communication devices;
— design of key electronic components for wireless communications devices and high-speed digital circuits.

In addition the method has been tested on several other types of problems described by partial differential equations.

In all the cases the obtained intermediate solution allowed considerable reducing of a number of iterations in FDM required for reaching the solution of the specified accuracy. It should be noted that the special benefit from reducing the number of iterations is found for smooth boundary conditions. For a certain sufficient value of the smooth the solution can be obtained only by using this method without FDM iteration process.

For partitioning domain **W** a rectangular grid formed by lines parallel to the coordinate axes and with a step h equal in all directions was chosen. The equation of the line $a_0 + a_1x$ was used as a polynomial (4). Its coefficients were calculated based on two elements of the boundary conditions **G**.

The results (Fig. 1) show a significant acceleration of convergence for the iterative process in FDM.

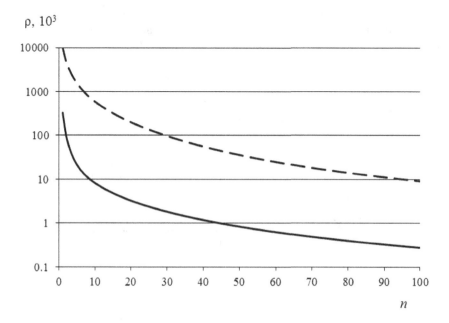

Fig. 1. The total residuals over all grid nodes depending on the number of iterations without (dashed line) and with (solid line) searching for prior decision in two-dimensional space (in logarithmic scale)

Application of the considered method for achieving a specified accuracy of the solution reduced the number of iterations in 10-100 times in a number of two-dimensional problems. This indicates a significant increase in the efficiency of FDM.

FDM stability for soft equations and the Laplace's equation in particular is shown in [4-6].

As a result of finding an intermediate solution $y'(x)$ FDM the convergence improves. Acceleration of the iterative process depends on the value of

$$\|y'(x) - u(x)\| \tag{5}$$

in the solution domain **W**.

The lower the difference (5) the closer the result to the exact solution and the less iteration are required to refine the solution (to reduce the residual to the specified value). The intermediate solution $y'(x)$ depends on the method of approximation (the choice of approximation function and its dimension). Practically the additional effect is obtained from taking into an account the values belonging to the grid lines neighboring with the one for which the iteration is performed.

In Fig. 2 the points $A,B,C \in G$ – are the elements of the boundary conditions, $u(x)$ – is the exact solution, $y'(x)$ – is an intermediate solution and $y^0(x)$ – are the initial values. The areas between the elements of G belong to the domain W in which we are looking for the solution. The initial values $y^0(x)$ are prescribe arbitrarily. Usually $y^0(x) = const$ as it shown in Fig. 2. The value $\|y'(x) - u(x)\| = d'$ is the difference between the exact and the intermediate solutions, and $\|y^0(x) - u(x)\| = d^0$ is the difference between the exact solution and the initial value.

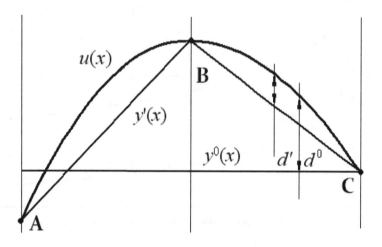

Fig. 2. The calculation of an intermediate solution

The strong mathematical evidence of the possibility of approximation of the function $u(x)$ with the function $y'(x)$ (in particular, using an algebraic polynomial (4)) is given in [7].

4   The Method Implementation in Two-Dimensional Space

We consider a two-dimensional problem with boundary conditions described by the Laplace's equation, in the case of a uniform grid with a step h formed by lines parallel to the axes of the coordinate as an example of method's implementation. Searching of the intermediate solution is realized consistently – at first in the direction of the axis X and then in the direction of the axis Y or vice-versa. We make the pass along the first line and if we find an element that is a part of the boundary conditions we remember its value. Then we move on until we meet the next element of the boundary conditions.

When the values of two boundary elements within a vector are known we use an approximation to calculate the values of the elements between them. In the same way we approximate all the values within a selected column. Sequentially we pass through all the columns in the X direction and then make the similar pass over all the lines in the Y direction. The actual value is calculated as a mean value between the passes in X and Y directions. To make an intermediate solution closer to the exact the similar passes along the diagonals can be added.

5   Conclusion

To achieve an improvement in convergence of the finite difference method, the intermediate solution in n-dimensional modeling tasks with boundary conditions has been obtained. The application of the approach to two-dimensional computations has demonstrated that our method is faster than the commonly employed finite difference method and the number of iterations for achieving a specified residual is 10-100 times less.

Thus, the results show that the suggested approach can appreciably reduce the execution time if large-scale computations. This method significantly increases the efficiency of the finite difference method in applied problems of modeling for communication lines.

References

1. Grishentsev, A.U., Korobeynikov, A.G.: Solution model of inverse problem of ionosphere vertical sounding. Scientific and Technical Journal of Information Technologies, Mechanics and Optics 2(72), 109–113 (2011)
2. Taflove, A., Hagness, S.C.: Computational Electrodynamics: The Finite-Difference Time-Domain Method. Arthech House (2000)
3. Demidovich, B.P., Maron, I.A., Shuvalova, I.Z.: Numerical analysis computation. Function approximation, differential and integral equations, 400 p. Saint-Petersburg, Lan (2010)
4. Isakov, V.: Inverse Problems for Partial Differential Equations. Applied Mathematical Sciences, vol. 127. Springer (1997)
5. Bahvalov, N.S., Voevodin, V.V.: Modern problems of computational mathematics and mathematical modeling, vol. 1, 343 p. Nauka, Moskow (2005)
6. Formalev, V.F., Reveznikov, D.L.: Numerical computation, 400 p. Phismatlit, Moskow (2004)
7. Shen, Z., Li, Z., Yew, P.C.: An empirical study of Fortran programs for paralleizing compilers. IEEE Trans. on Parallel and Distributed Systems, 350–364 (1990)

Hardware and Software Equipment for Modeling of Telematics Components in Intelligent Transportation Systems

Serge Popov, Mikhail Kurochkin, Leonid M. Kurochkin, and Vadim Glazunov

Telematics Department, St. Petersburg State Polytechnical University,
29, Polytechnicheskaya, St. Petersburg, 195251, Russia
popovserge@spbstu.ru,
{kurochkin.m,kurochkinl,neweagle}@gmail.com

Abstract. The paper deals with hardware and software solutions for development of the testbed of simulation of wireless network data transfer of vehicles by intelligent transportation system (ITS) on basis of multiprotocol unit. The testbed's hardware meets requirements of scaling, openness and compatibility of all components. Software structure consists of five modules. It allows to prepare and carry out the research data traffic dynamics of ITS, time of message delivery, reliability of message transfer, probability of message delivery via a dedicated channel and some others. The testbed allows working out technologies of dynamic multi-protocol routing in heterogeneous mobile networks of ITS vehicles.

Keywords: Wireless network, routing protocols, intelligent transportation system, real-world simulation, hardware and software solution, heterogeneous mobile networks, cloud services.

1 Introduction

An increased interest of developers and users in upgrading of functions of ITS creates new requirements for quality and reliability of message transfer via wireless networks. Presentation of new control and safety services tht provide traffic and incident messaging, "vehicle-cloud service" data transfer introduce high requirements for data transfer reliability among all traffic subjects. These services are used in in "car 2 car" [1], "ecall" [2], "ERAGLONASS" [3]. In conditions of no access available to global data transfer network (LTE, Wi-Fi, GSM) following issues of the research are of great importance:

— construction variants of message reception and transfer multiprotocol units;
— characteristics of dynamic routing algorithms;
— methods of ITS services application;
— effective coordination of vehicles and road infrastructure via DSRC.

S. Balandin et al. (Eds.): NEW2AN/ruSMART 2014, LNCS 8638, pp. 598–608, 2014.
© Springer International Publishing Switzerland 2014

Absence of total coverage of transport networks by global data transfer networks makes it impossible to provide fixed access to ITS cloud services. Use of multi-protocol units increases reliability of data transfer. Performance assessment of multi-protocol peer-to-peer (MESH) mobile network with dynamically changing structure is of special interest in researching ITS characteristics. Figure 1. shows the organization chart of data transfer via MESH-network.

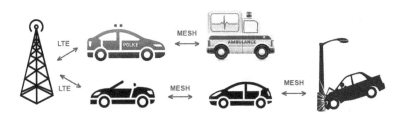

Fig. 1. Multiprotocol units

Today there are a lot of research focused on traffic characteristics studies [4], routing protocols and data transfer [5],[7], data transfer channels [6] used in ITS. For example, the paper [8] describes methods of imitation simulation and key characteristics of dynamic routing protocols. Simulation was made with ns-2 simulator. The paper [9] performs a real-word experiment to define characteristics of OLSR, B.A.T.M.A.N. and BABEL protocols. 802.11a wireless standard was used in the experiment; iperf and fping utilities were used for performance assessment. PWAN (Portable Wireless Ad-Hoc Node) software is also assessed. At the same time hardware construction and software research complex of mobile peer-to-peer networks are of special interest. Experimental conditions in this complex are maximum close to those in a reality. The given work considers structure and functional provision of hardware and software complex providing real-world experiments with components of mobile peer-to-peer data transfer networks in ITS.

2   Functional Requirements

Hardware-program implementation of multiprotocol data transmission network model is intended for research of mobile wireless networks descriptions of data transmission in the conditions of dynamically changing network structures, volumes of transferrable data, rates of data, message loss probabilities. The basic requirements in development of hardware-program implementation are scalability and run-time reconfiguration of hardware and software components of the testbed. A dynamically changing network structure results in the decline of data transmission reliability, complicates protocols of routing, strengthens the unevenness of network load. Therefore it is necessary to develop methods of assured delivery of urgent reports from the mobile network participants to a service center. The increase of reliability of reports delivery is due to the use of multiprotocol units, which support usage of MESH-networks and co-operating with the objects of road infrastructure.

Basic Solution. There are some base structures of data, describing characteristics of ITS objects in architecture of the testbed for raising and implementation of experiments:

- composition of telematic tools aboard a vehicle;
- infrastructure of telematic tools on a roadway network;
- the composition of access interfaces to cloud services.

Basic structures are described parametrically and allow to implement the different scenarios of the use of ITS.

Developed hardware-program implementation allows us to carry out the following classes of experiments:

- reseach of ITS data traffic dynamics;
- assessment of the real data rate;
- assessment the size of the transmitted service and useful information;
- reliability research messaging to determine the probability of message delivery through a dedicated channel;
- research of delivery time of reports (messages);
- research of the dynamic routing algorithms;
- development of the methods for efficient interaction of network layer protocols, testing of control and protocol's algorithms and etc.

3   Hardware Implementation of the Testbed

Hardware implementation of the testbed includes two multiprotocol units and load simulator with the connected wireless routers. Multiprotocol units have the same configuration and model behavior of intelligent transport network using various protocols for wireless communication: 802.11a/b/g/n (Wi-Fi), 802.11s (MESH) and LTE. Configuration of multiprotocol units includes two wireless Wi-Fi/MESH adapters and one LTE-modem. This allows using all of them for transmitting data simultaneously and independently. A load simulator is used to simulate the cloud, as well as an additional multiprotocol unit which is also equipped with a wireless adapter Wi-Fi/MESH and two wireless routers. A load simulator is used to control the test bench by the means of additional Ethernet channel. Compatibility between hardware and software components have been examined. Wireless network adapters TP-Link TL-WDN 4800 and provide data transmission in Wi-Fi and MESH networks. These adapters support two frequency bands: 2.4 GHz and 5 GHz, and provide operate at peak bandwidth of up to 300 Mbit/s in Wi-Fi networks. TP-Link TL-WDN 4800 network adapters are implemented on Atheros chipset family AR9300. Adapters based on this chipset are provided with stable drivers for the Linux operating system with support for the MESH - mode. Wireless routers Netgear WNDR 3800 were used as Wi-Fi access points, supporting two frequency bands and providing a peak bandwidth of up to 300 Mbit/s. These routers are based on Atheros chipset family AR9200 which is also compatible with the variety of wireless equipment. These devices provide a flexible way to customize a variety of parameters to ensure stable operation in Wi-Fi networks at high loads and noisy channel. Huawei E398 modem for which stable drivers for the Linux OS exist provides connection to the LTE network. The modem switches automatically between the modes (LTE/3G/EDGE/GPRS) depending on the capabilities of the network and channel quality. Multiprotocol units and load simulator equipped with USB 2.0/3.0 ports, which can be used to connect additional devices such as receivers navigation systems GPS/GLONASS and Bluetooth adapters. The structure of the hardware implementation of the testbed is shown on Figure 2.

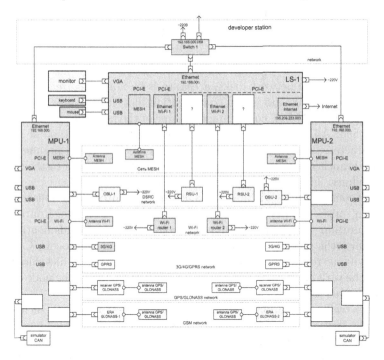

Fig. 2. Structure of a hardware implementation of the testbed

The solution shown in Figure 2 includes two MPU blocks supporting 4 types of data transmission technologies. The given solution meets the requirements of scalability, openness and hardware compatibility of all components.

4   Software Implementation of the Testbed

Equipment of testbed running on Debian GNU/Linux operating system version 7.4/Kernel 3.2.0-4. For testing in various modes a specialized software is used:

- Mgen - an application that creates network traffic transport layer with the given parameters;
- WEB and FTP-servers - which shares user files with specified sizes. WEB Server: Lighttpd and FTP: Vsftpd;
- Netflow - an application for analysis and visualization of test results;
- Iptables - software package is used for management and customization channel characteristics;
- Wget - the client side application which supported FTP and HTTP; Configuration parameters for test scenarios are located in pre-prepared files. They allow to simulate various daily interaction models of moving objects between themselves and with the cloud environment.

Experiment parameters:

- broadcast/multicast/unicast network traffic types;
- transport layer protocols (TCP/UDP), packet size, application layer protocol (HTTP, FTP, NTP);
- loss percentage in data channel, channel rate.

Software is stored on network disks, access to it by nodes realized true through NFS. Software settings are in the configuration files, files loading based on the variable ${HOSTNAME}, containing the name of the node on which the module was launched. All participants of experiment used the same type of software. Node characteristic settings vary due to changes in the model parameters and execution branches (defined by the host name). Software part of testbed consists of five modules. The modules are shown in Figure 3.

Module configure network interfaces - assigns IP addresses, establishes a VPN connection, raises the internal and external interfaces of nodes. Routing module configuration - configures the dynamic routing protocol, adds floating static routes, starts the services (daemons) of dynamic routing protocols. Configurations of dynamic routing protocols are in separate configuration files, files loading based on the variable ${HOSTNAME} containing the name of the node on which the module was launched.

Network traffic generation module - runs file services (daemons), such as FTP, HTTP. The server side parts of these services are launched on an loading imitator (simulator). FTP and HTTP clients run on the client (MPU) to upload the files of the selected size, the files must be generated first on the server side. Additional support traffic generation transport layer provided through mgen: TCP, UDP. In the case with using UDP traffic it is possible to select the broadcast or multicast traffic types.

Module of the experiment completion and statistics collection - Records file download time, given the recovery time after disconnection. All traffic which is passing through the nodes is counted using the module net flow, which consists of several components: a kernel part (ipt-netflow) and a userspace daemon that collects network flow data.

A firewall module - causes interference at the link layer, which prevents the routes and the data from spreading through the selected communication channel.

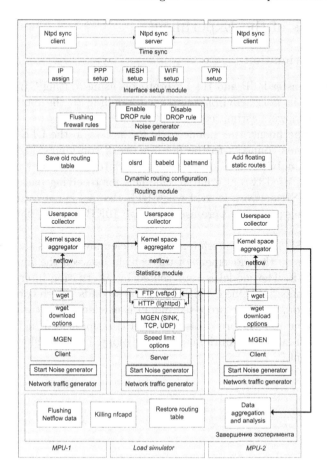

Fig. 3. Testbed software modules

This module is used by the noise generator to create a mobile traffic patterns (movement scheme) of the node. The module allows you to specify statistical or probabilistic models in traffic noise, i.e. discarding each n-th packet, or the probability of a packet passing through the filter. The firewall algorithm is based on Module of the experiment completion and statistics.

Furthermore, the module includes means for calculating total numerical data characteristics for visualization. Upon experiment complete, the module executes remote completion on other nodes of network. A service data channel is used to prevent user co,,ands from influencing measurement values. Once this is complected, net flow fixes all measurements and waits a minute to complete the maximum data collection cycle through netflow.

The proposed structure of the software allows to preparing and conducting experiments of all classes listed above, as well as to storing and processing the collected data. A hardware-program testbed implemented in a stationary and mobile version/static and dynamic modes.

5   Examples of the Experiments

5.1   Delay Estimation

Delays of file transfer FTP and HTTP protocols using routing OLSR and BAT-MAN protocols in noisy environments corresponding to the movement of the vehicle in the city were estimated.

Under the experiment, the mobile node has access to the LTE channel, and at regular intervals connects to the mesh-network containing high gateway to the Internet. A mobile node downloads the update system and user software. The experiments used two alternative channels for receiving packets, LTE and MESH. The MESH network is used when LTE channel is not available. OLSR, BATMAN routing protocols were used, file size 10KB-100MB, data rate server 128Kbit/s.

Model of MESH network interference at vehicle movement is shown in Figure 4. It corresponds to the movement of the vehicle between the junctions at the speed of 60 km/h. The model is represented by alternating intervals of stable and unstable connection. One cycle equals to 60 seconds, which corresponds to a region of one kilometer. The proposed model is efficient for networks which are based on dynamic routing protocols and data transmission in network charac-terized with high instability of the network topology. Figure 4. shows the results of this experiment.

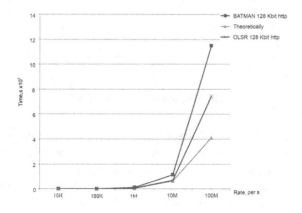

Fig. 4. Time of file transfer dependence of the file size, protocols BATMAN and OLSR to rate 128kbits/s

The obtained results reflect dynamics of the data at slow speeds. This speed is typical for large data includes information about the road network and other network nodes. An OLSR dynamic routing protocol is preferred over a BATMAN protocol under these conditions.

5.2 Analysis of Stability of Clock Synchronization over the Networks of MESH and LTE Technologies

Custom applications smooth operating require high precision clock multiprotocol unit vehicle. High accuracy is achieved through the network synchronization of the system clock. Synchronization can be executed at random time points using different local and global technologies. In order to confirm the accuracy of network synchronization technology application for purposes of estimation of the clock adjustments in unit, four batches of tests were performed using the test bench. Tests consisted of synchronization of multiprotocol unit's clock with the cloud service over following networks:

- wired local network with 100 Mbit/s bandwidth, no other traffic;
- wireless local MESH network, some other traffic present;
- wireless LTE network under real world conditions;
- the combination of wireless local MESH network and LTE network with multiprotocol unit number 2 acting as a router between them.

Paths for transmission of synchronization messages are marked with numbers 1–4 at Figure 5.

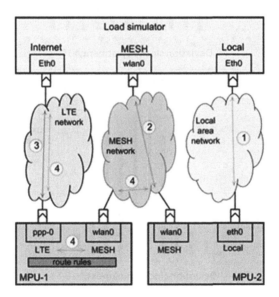

Fig. 5. Routes of synchronization

Input:

- wireless networking technology: Ethernet, 802.11s, LTE;
- static routing;

– route's length : 1,2 hops;
– user-level protocol: NTP;
– network-level protocol: UDP;
– frequency synchronization: 0,0125 Hz.

Series of clock adjustment values was obtained during the experiment for every formed path. Figure 6 shows a histograms of distribution of clock adjustment values depending on various network channels. Correction values are presented in 19 intervals.

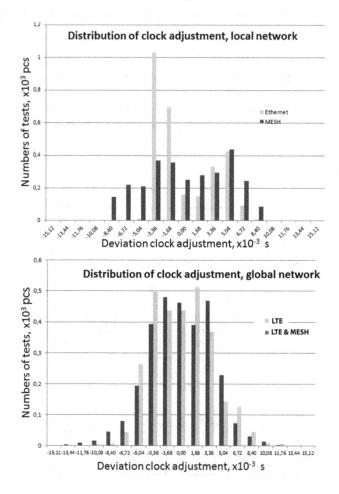

Fig. 6. Distribution of clock adjustmenet, global and local network technology

These results shows that the best accuracy and stability of adjustment values are achieved through such local network technologies like Ethernet and MESH, providing synchronization accuracy of $10^{-4}s$. Use of global communication technology reduces the accuracy 10 times, down to 10^{-3}. The combination of MESH and LTE provides comparable to LTE synchronization one. Thus, we can conclude that synchronization of multiprotocol unit's clock over multi-hop networks provides the accuracy of vehicle's clock at $10^{-3}s$ level with 75% probability. These data agree well with the results of the authors listed in the work [10].

Examples of experiments conducted on the testbed, confirm the correctness of proposed hardware and software solutions and prospects for its use in the study of various characteristics of telematic components of intelligent transportation systems.

6 Conclusions

In this paper, we discussed the concept and implementation of hardware and software for wireless network simulation in intelligent transportation systems. The developed concept operates with wireless technologies such as: Wi-Fi, MESH and LTE used in intelligent transportation systems. By using this solution we can vary number of data transfer techniques and scalability on multiprotocol units. Efficiency is achieved through: alternate network interfaces and routes, additional modules of network technology, combining dynamic routing protocols and support flexible network topologies. Experiments can be carried out using proposed hardware and software solutions to determine the properties of traffic in wireless intelligent transportation systems, as well as a comparison of the usu dynamic routing protocols. The proposed solution can serve as a prototype to create a real hardware.

Acknowledgments. This research was supported by a grant from the Ford Motor Company. This paper funded by RFBR grant 13-07-12106.

References

1. CAR 2 CAR communication consortium,
   http://www.car-2-car.org
2. eCall European Comission,
   http://ec.europa.eu/digital-agenda/en/ecall-time-saved-lives-saved
3. NIS GLONASS, http://www.nis-glonass.ru/projects/era_glonass/
4. Glazunov, V., Kurochkin, L., Popov, S.: Instrumental environment of multi-protocol cloud-oriented vehicular MESH network. In: ICINCO 1, pp. 568–574. SciTePress (2013)
5. Zaborovsky, V., Chuvatov, M., Gusikhin, O., Makkiya, A., Hatton, D.: Heterogeneous multiprotocol vehicle controls systems in cloud computing environment. In: ICINCO 1, pp. 555–561. SciTePress (2013)
6. Boeglen, H., Hilt, B., Lorenz, P., Ledy, J., Poussard, A., Vauzelle, R.: A survey of V2V channel modeling for VANET simulations. In: Wireless on Demand Network Systems and Service - WONS (2011)
7. Murray, D., Dixon, M., Koziniec, T.: An Experimental Comparison of Routing Protocols in Multi Hop Ad Hoc Networks. In: Australasian Telecommunication Networks and Applications Conference (2010)

8. Naski, S.: Performance of Ad Hoc Routing Protocols Characteristics and Comparison. In: IEEE Seminar on Internetworking. Helsinki University of Technology (2004)
9. Abolhasan, M., Hagelstein, B., Chun-Ping Wang, J.: Real-world performance of current proactive multihop mesh protocols (University of Wollongong Research Online) (2009), http://ro.uow.edu.au/cgi/viewcontent.cgi?article=1747&context=infopapers
10. NTP: The Network Time Protocol. Basic Concepts,
    http://www.ntp.org/ntpfaq/NTP-s-algo.htm

Comparison of Stepwise and Piecewise Linear Models of Congestion Avoidance Algorithm

Olga Bogoiavlenskaia

Petrozavodsk State University,
Lenin st., 33, 185910, Petrozavodsk, Russia
olbgvl@cs.karelia.ru

Abstract. This paper describes probabilistic analysis of New Reno version of TCP protocol AIMD algorithm. The model is formulated as a stepwise semimarkovian stochastic process. Theorems on ergodic properties of the process and two embedded Markov chains are proved and functional equation for generating function of sliding window size distribution is obtained. On its base the estimation of the average window size is constructed. The estimation obtained is compared with those provided by piecewise linear AIMD models.

Keywords: Stochastic analysis, Semimarkovian process, Data communication, AIMD.

1 Introduction

Methods of distributed control provide one of the fundamental Internet concepts. Essential part of the methods responsible for congestion control of the Internet infrastructure (links and routers) are performed at transport layer where data sources follow congestion control algorithms. Thus a source increases throughput if there is end-to-end-route capacity available and decreases throughput if it receives a congestion signal. Most of these algorithms are implemented in Transmission Control Protocol (TCP). Since distributed control and congestion control methods in particular play key role for data communication networks integrity and performance they are the object of intensive research. This research is focused on congestion avoidance Additive Increase Multiplicative Decease (AIMD) algorithm described by [1], [2], i.e. New Reno TCP version. Recently the algorithm's disadvantages on the end-to-end paths that include high-speed, high BDP value or wireless links are widely discussed. Nevertheless the algorithm provided exponential growth of the Internet during more than twenty years. Also its performance is used as a measure of fairness and for tuning parameters of the experimental TCP versions, see e. g. [3]. Thus better understanding of New Reno behavior provides foundation for further research and a tool for an administrative solutions for the different networking environments as well. At present NewReno version of TCP is implemented in a wide variety of modern OS kernels' networking modules.

S. Balandin et al. (Eds.): NEW2AN/ruSMART 2014, LNCS 8638, pp. 609–618, 2014.
© Springer International Publishing Switzerland 2014

At the moment most publications evaluate the estimations of TCP sliding window size or throughput expectation for one of the following forms $\lim\limits_{t \to \infty} \mathsf{E}[w(t)]$ and $\dfrac{1}{T} \int\limits_0^T w(t)dt$. Among them the most oftenly cited works are [4], [5] and [6]. An example of further development of the approach proposed by [5] one can find in [7]. Studies of TCP behavior in wireless networks form, in fact, particular wide area of the research. The area provides special class of analytical models where matrix exponential methods could extend the models to more complex and realistic assumptions about networking environment properties, see e.g. [8] — [13]. Significant amount of the research is also devoted to simulations and/or experimental evaluation of TCP behavior in different networking environments [14] — [20]. Generally the area of TCP behavior research and modeling is rather big, therefore for further information about state-of-art in the area one can address to the survey [21].

There has formed two different approaches in the practice of modeling for the description of TCP data loss process. The first one considers the losses as a random flow. So if τ_k and τ_{k+1} are the moments of two consequent data loss, then the form of distribution functions of $[\tau_{k+1} - \tau_k]$ intervals is one of the most essential assumptions of the model, e.g. [6]. The second approach treats the sequence of data sent and demands the definition of the distribution function of r. v. s_n which is amount of data sent during $[\tau_{k+1} - \tau_k]$ interval, e.g. [5].

Most recent researches consider TCP congestion window size as a piecewise linear process, where growth periods are defined by linear function with derivative proportional to $1/d$ and d is average Round Trip Time (RTT). Then at the congestion signal the process decreases proportionally to multiplicative coefficient. The piecewise linear models allow avoiding many analytical problems, make models simpler and tractable, but they ignore discrete nature of TCP sliding window, since the real values provided by AIMD variables are rounded before further processing of the data to send. Also these models cannot take into account RTT distribution since RTT is deterministic value for all of them. At the moment there is no research published that study analytically the precision of piecewise linear models and characterize explicitly the difference between their results and the real behavior of congestion window stepwise process.

This work considers stepwise process of TCP New Reno AIMD algorithm with the assumption that $[\tau_{k+1} - \tau_k]$ follows the exponential distribution. For this model we formulate semimarkovian process and two Markov chains embedded in this process. The theorems on the process ergodic distribution are proved, the connection between embedded chains is derived and the equation which defines generating function of stationary distribution for one of the Markov chains is obtained. Further analysis yields interval estimation for the congestion window size expectation of the kind $\lim\limits_{t \to \infty} \mathsf{E}[w(t)]$. The interval estimation provides the estimation error and hence could be used to compare the results of stepwise and piecewise linear analysis for different RTT distributions.

The work is organized as follows. Section 2 describes semimarkovian process, defines the Markov chains embedded in it and provides ergodic theorems, Section 3 presents the generating function, the interval estimation and the example. Section 4 contains conclusions.

2   Semimarkovian Model of Congestion Window Size

Let us define the random process which describes a behavior of AIMD congestion window size when TCP segment losses form a Poisson flow.

Let t_n are the moments of AIMD-rounds end-points. Hence $[t_{n-1}, t_n]$ are round trip intervals, and RTT length is $\xi_n = t_n - t_{n-1}$. Let us denote $w(t)$ a congestion windows size under AIMD control at the time moment t. Then $\{w(t)\}_{t>0}$ is a stepwise process such that

$$w(t_n + 0) = \begin{cases} \left\lfloor \dfrac{w(t_n)}{2} \right\rfloor, & \text{if during the interval (TCP round) } [t_{n-1}, t_n] \\ & \text{one or more TCP segment losses has happened,} \\ w(t_n) + 1, & \text{if all data in the round were successfully delivered.} \end{cases}$$

Between the moments t_n the process $\{w(t)\}_{t>0}$ stays constant.

Let us suppose that ξ_n are independent identically distributed (i.i.d.) random variables with distribution $R(x)$, which is absolutely continuous on the set $\mathbb{R}^+$ excepted possibly finite number of points and $\mathsf{E}[\xi_n] < \infty$. If TCP-sender under consideration does not contribute significantly into the end-to-end route workload then the losses it observes are caused due to the events generated by external sources, i. e. by other data flows which share the same end-to-end paths and/or its fragments or the routers queues. We assume that segment losses generated by each external source form a weak loss flow visible by the sender considered and according to the central limit theorem for flows the summary loss flow is close to Poisson flow with the parameter $\lambda > 0$. Then the sequence $\{w_n = w(t_n)\}$ form the Markov chain and $\{w(t)\}_{t>0}$ is semimarkovian random process (see Fig. 1). Let us denote

$$q = \int_0^\infty e^{-\lambda\tau} dR(\tau) \tag{1}$$

the probability that there were no segment losses on the interval $[t_{n-1}, t_n]$.

Theorem 1. *Markov chain $\{w_n\}$ has steady state distribution.*

P r o o f. Let's define the function $f(i) = i$, then

$$\mathsf{E}[f(w_{n+1})|w_n = i] = q(i+1) + (1-q)\left\lfloor \frac{i}{2} \right\rfloor \leq \tag{2}$$

$$\leq \frac{i}{2}(q+1) + q.$$

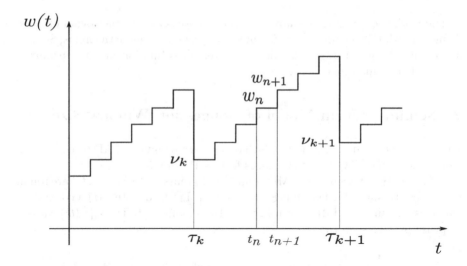

Fig. 1. Evolution of TCP congestion window size

For some $\epsilon > 0$ consider the condition

$$\frac{i}{2}(q+1) + q \leq i - \epsilon. \tag{3}$$

Then

$$i \geq (\epsilon + q)\frac{2}{1-q} \tag{4}$$

By definition $q < 1$ and, hence, $\forall \epsilon > 0 \ \exists$ finite k, defined by formula (4), such as $\forall i > k$

$$\mathsf{E}[f(w_{n+1})|w_n = i] \leq f(i) - \epsilon. \tag{5}$$

At the same time $\forall i \leq k$ expectation defined by (2) is limited on the finite set $[0, \ldots, k]$.

So irreducible aperiodic Markov chain $\{w_n\}$ satisfies Foster criterion in the form formulated by [22, § 3.1, p.2], and, hence, is ergodic. □

Theorem 2. *If* $\exists \ \epsilon > 0 : \ R(\epsilon) \leq 1 - \epsilon$ *then the random process* $\{w(t)\}_{t>0}$ *is ergodic.*

P r o o f. According to all assumptions made and the theorem condition the process $\{w(t)\}_{t>0}$ is regular and the Markov chain $\{w_n\}$ is ergodic according to the theorem 1. Hence the semimarkovian process is ergodic too [22, § 3.1 p.5]. □

We define that data *loss event* has happened if at the time interval $[t_{n-1}, t_n]$ m TCP segments were lost, where $m \geq 1$. Let τ_k be equal to the first moment t_n, arrived after kth data loss event, i.e.

$$\tau_k = t_n : \ w(t_n + 0) = \left\lfloor \frac{w(t_n)}{2} \right\rfloor, \ k = 1, 2, \ldots.$$

and $\tau_{k_1} < \tau_{k_2}$, if $k_1 < k_2$. Then the sequence $\{\nu_k = w(\tau_k + 0)\}_{k \geq 0}$ is the Markov chain embedded in both the Markov chain $\{w_n\}$, and in the semimarkovian process $\{w(t)\}_{t > 0}$ as well. Lets denote

$$f_j = \mathbf{P}\left\{\nu_{k+1} = \left\lfloor \frac{1}{2}\left(\left\lfloor \frac{\nu_k}{2}\right\rfloor + j\right)\right\rfloor\right\}, \ j = 0, 1\ldots, \tag{6}$$

i.e. f_j is the distribution of the number of the consequentive rounds at the end of which $w(t)$ size increased. Hence the probabilities $\{f_j\}$ completely define the Markov chain $\{\nu_k\}$. This Markov chain and its properties are important for practice and further analysis as well. Therefore we consider ergodic properties of $\{\nu_k\}$ independently from the chain $\{w_n\}$.

Let's define the expectation determined by the sequence f_j as $B = \sum_{j=1}^{\infty} j f_j$.

Theorem 3. *If B is finite then the Markov chain $\{\nu_k\}$ has steady state distribution.*

P r o o f. Let us compute $g(i) = \mathsf{E}[\nu_{k+1} | \nu_k = i]$ as follows

$$g(i) = \sum_{k=0}^{\infty}\left(\left\lfloor \frac{i}{2}\right\rfloor + k\right) f_k = \left\lfloor \frac{i}{2}\right\rfloor + B. \tag{7}$$

Henceforth if the condition of the theorem holds then

$$\mathsf{E}\{\nu_{k+1} | \nu_k = i\} = \left\lfloor \frac{i}{2}\right\rfloor + B, \tag{8}$$

where B is finite. Now we consider following condition

$$\left\lfloor \frac{i}{2}\right\rfloor + B \leq i - \epsilon, \tag{9}$$

or

$$i > 2(B + \epsilon). \tag{10}$$

Henceforth, $\forall \epsilon > 0 \ \exists$ finite $n: \ \forall i > n \ \mathsf{E}[\nu_{k+1} | \nu_k = i] \leq i - \epsilon$. Since n is defined by the formula (10) then $\forall i$ from the finite set $\{0, \ldots, n\}$ the function $g(i)$ is limited as well. Hence according to Foster criterion used above irreducible aperiodic Markov chain $\{\nu_k\}$ is ergodic. $\qquad\square$

3   Steady State Distributions of the Embedded Chains

For the random process $\{w(t)\}_{t > 0}$ the sequence of the loss events that do or do not happen on the time intervals $[t_n, t_{n-1}]$ form the sequence of Bernoulli tries with parameter q, and henceforth

$$f_j = (1 - q)q^j.$$

Let us denote p_m for steady state distribution of the Markov chain $\{w_n\}$ and π_r denote steady state distribution of the Markov chain $\{v_k.\}$ Then correspondent Kolmogorov equations have the following form

$$p_0 = (1-q)p_0 + (1-q)p_1 \tag{11}$$
$$p_m = qp_{m-1} + (1-q)p_{2m} + (1-q)p_{2m+1}, \qquad m > 1$$

for the Markov chain $\{w_n\}$ and

$$\pi_0 = \pi_0(f_0 + f_1) + \pi_1 f_0 \tag{12}$$
$$\pi_r = \sum_{j=0}^{2r} \pi_j f_{2r-j} + \sum_{j=0}^{2r+1} \pi_j f_{2r+1-j}$$

for the Markov chain $\{v_k\}$. The following theorem formulates the connection between equation systems (11) and (12).

Theorem 4. *Let the sequence $\{\pi_r\}$ is the solution of the equations (12), then*

$$p_m = \sum_{j=0}^{m} \pi_j q^{m-j}. \tag{13}$$

P r o o f. Let us substitute expression (13) in the equations (11)

$$p_m = q \sum_{j=0}^{m-1} \pi_j q^{m-1-j} + (1-q) \sum_{j=0}^{2m} \pi_j q^{2m-j} + \tag{14}$$
$$+ (1-q) \sum_{j=0}^{2m+1} \pi_j q^{2m+1-j}.$$

Notice that the sum of the last two summands in the right part of (14) is equal to π_n, according to equations (12). Henceforth

$$p_m = \sum_{j=1}^{m-1} \pi_j q^{m-j} + \pi_m = \sum_{j=0}^{m} \pi_j q^{m-j}. \tag{15}$$

$\square$

Lets denote $W = \lim_{t\to\infty} E[w(t)]$.

Corollary 1

$$W = \lim_{k\to\infty} E[v_k] + B. \tag{16}$$

P r o o f. Let $P_w = \lim_{t\to\infty} \mathbf{P}\{w(t) = w\}$. Then according to [22, § 3.1]

$$P_w = \frac{p_w \alpha_w}{\sum\limits_{i=0}^{\infty} p_i \alpha_i},$$

where α_w is the expectation of the sojourn time of the process $\{w(t)\}_{t>0}$ in the state w. Notice that $\forall w \ \alpha_w = \mathsf{E}\xi_n$, since $\{\xi_n\}$ is i.i.d. and hence $P_w = p_w$. According to the theorem 4 the distribution p_w is the convolution of the distributions π_r and f_j, that implies the statement of the corollary.   □

Now further analysis of the system of equations (12) follows. The system can be reformulated as

$$\pi_r = \sum_{j=0}^{2r} \pi_j(f_{2r-j-1} + f_{2r-j}), \ f_{-1} = 0. \tag{17}$$

Notice that right part of (17) is the convolution of the sequences $\{\pi_r\}$ and $\{f_i + f_{i+1}\}_{i\geq-1}$ in the points of $2r$. The necessity to consider the equation in the discrete space of state $\mathbb{Z}^+$ of the semimarkovian process $\{w(t)\}_{t>0}$, where the floor function does not have the reverse one, does not let one using standard transform methods.

Nevertheless let's denote $P(z)$ generating function of the sequence $\{\pi_r\}$. Let us multiply each equation of (17) with z^{2k} and sum all of them. This yields

$$\sum_{i=0}^{\infty} \pi_i z^{2i} = \left(\sum_{j=0}^{\infty} \pi_{2j} z^{2j}\right) \left(\sum_{k=0}^{\infty}(f_{2k} + f_{2k+1})z^{2k}\right) + \tag{18}$$

$$+ \left(\sum_{j=0}^{\infty} \pi_{2j+1} z^{2j+1}\right) \left(\sum_{k=0}^{\infty}(f_{2k-1} + f_{2k})z^{2k-1}\right).$$

After series of the algebraic transformations one obtains

$$P(z^2) = \frac{1+z}{2z}F(z)P(z) + \frac{z-1}{2z}F(-z)P(-z), \tag{19}$$

where $F(z)$ is the generating function of the sequence $\{f_j\}$. The derivatives of both parts of the equation (19) yield in the point $z = 1$

$$2P'(1) = -\frac{1}{2} + F'(1) + P'(1) + \frac{1}{2}F(-1)P(-1).$$

Since $|P(-1)| \leq 1$, then using the corollary of the theorem 4 one obtains

$$2F'(1) - \frac{1}{2}(|F(-1)|+1) \leq W \leq 2F'(1) + \frac{1}{2}(|F(-1)| - 1). \tag{20}$$

The last formula defines interval estimation (upper and lower bounds) for steady state expectation of the semimarkovian process $\{w(t)\}_{t>0}$. The estimation derived holds under condition $F'(1) > 1$. If the condition does not hold then it means that expected number of the rounds between two consequent loss events is smaller or equal to one. In the case AIMD performance is unacceptable for the applications anyway. In the following example the interval estimation is compared with the piecewise linear model described in [6].

The Example

Let assume that ξ_n is deterministic variable which value is d and $\lambda d < 1$. Then $q = e^{-\lambda d}$ and

$$F(z) = \frac{1 - e^{-\lambda d}}{1 - ze^{-\lambda d}}.$$

Hence

$$F'(1) = \frac{e^{-\lambda d}}{1 - e^{-\lambda d}} \tag{21}$$

and

$$F(-1) = \frac{1 - e^{-\lambda d}}{1 + e^{-\lambda d}}.$$

To compare the estimation (20) with the results of the piecewise linear models we use the result provided by [6] which states

$$\mathsf{E}[X_n] = \frac{\alpha}{\lambda(1 - \nu)}, \tag{22}$$

where $X_n \in \mathbb{R}^+$ is the congestion window size just prior the multiplicative decrease moment, α is a rate of the window growth in the absence of random losses, ν is multiplicative decrease factor and λ is the loss intensity. Following stepwise model described above one sets $\alpha = 1/d$ and $\nu = 1/2$. Then for the first summand of upper and lower bounds of (20) considering first order of series expansion for both nominator and denominator yields

$$2F'(1) \approx \frac{2}{\lambda d}, \tag{23}$$

which is equal to (22) and $-0,5 < \frac{1}{2}(|F(-1)| - 1) < 0$. Thus if RTT variability is low and segment losses form Poisson flow, then (22) is first order approximation of (20) and both estimations are close enough for practical purposes.

4   Conclusion

Piecewise linear random processes are widely used for the modeling of TCP congestion avoidance algorithms. Meanwhile real TCP implementations support discrete arithmetics for congestion window size control. Therefore piecewise linear models provide approximate results and do not take into account variability of RTT which they treat as a deterministic value. In this paper the stepwise model of AIMD New Reno congestion avoidance is analyzed. With the aim the semi-markovian random process is formulated, theorems on its ergodic properties are proved. There is obtained functional equation which defines generating function of Markov chain embedded in the semimarkovian process. Further processing of the equation yields upper and lower bounds of the steady state expectation of the congestion window size. The estimation obtained treats RTT as i.i.d. variables and may evaluate bounds of the error for other average congestion window size estimates.

References

1. Allman, M., Paxson, V., Stevens, W.: TCP Congestion Control. RFC 2581 (1999)
2. Floyd, S., Hendersin, T.: The NewReno Modification to TCP's Fast Recovery Algorithm. RFC 2582 (1999)
3. Ha, S., Rhee, I., Xu, L.: CUBIC: A New TCP-Friendly High-Speed TCP Variant. ACM SIGOPS Operating System Review (2008)
4. Floyd, S., Fall, F.: Promoting the use of end-to-end congestion control in the Internet. IEEE/ACM Transactions on Networking (August 1999)
5. Padhey, J., Firoiu, V., Towsley, D., Kurose, J.: Modeling TCP Throughput: A Simple Model and its Empirical Validation. IEEE/ACM Transactions on Networking 8(N2), 133–145 (2000)
6. Altman, E., Avrachenkov, K., Barakat, C.: A Stochastic model of TCP/IP with Stationary Random Losses. Proceedings of ACM SIGCOMM 2000, Stockholm, 231–242 (2000)
7. Parvez, N., Mahanti, A., Wiliamson, G.: An Analytic Throughput Model for TCP NewReno. IEEE/ACM Transactions on Networking 18(N2), 447–461 (2010)
8. Shi, Z., Beard, C., Mitchell, K.: Analytical models for understanding misbehavior and mac friendliness in CSMA networks. Performance Evaluation Archive 66(9-10), 469–487 (2009)
9. Shi, Z., Beard, C., Mitchell, K.: Analytical Models for Understanding Space, Backoff and Flow Correlation in CSMA Wireless Networks. Wireless Networks (WINET) Journal (2012), doi:10.1007/s11276-012-0474-8
10. Shi, Z., Beard, C., Mitchell, K.: Competition, cooperation, and optimization in multihop CSMA networks with correlated traffic. International Journal of Next-Generation Computing (IJNGC) 3(3) (2012)
11. Shi, Z., Beard, C., Mitchell, K.: Competition, Cooperation, and Optimization in Multi-Hop CSMA Networks(Link). In: Proceedings of the 8th ACM Symposium on Performance evaluation of wireless ad hoc, sensor, and ubiquitous networks, PE-WASUN 2011 (November 2011)
12. Shi, Z., Beard, C., Mitchell, K.: Misbehavior and MAC Friendliness in CSMA Networks. In: IEEE Wireless Communications and Networking Conference, WCNC 2007, IEEE, Hong Kong (March 2007)
13. Shi, Z., Beard, C., Mitchell, K.: Tunable traffic control for multihop csma networks. In: MILCOM 2008, pp. 1–7 (November 2008)
14. Eun, D.Y., Wang, X.: Achieving 100% Throughput in TCP/AQM Under Aggressive Packet Marking With Small Buffer. IEEE/ACM Transactions on Networking 16(4), 945–956 (2008)
15. Khan, N.I., Ahmed, R., Aziz, T.: A Survey of TCP Reno, New Reno and SACK over mobile ad-hoc Network. IJDPS 3(1) (January 2012)
16. Carofiglio, M., Muscariello, L.: On the impact of TCP per-flow scheduling on internet performance. IEEE/ACM Transactions on Networking 20(2), 620–633 (2012)
17. Dumas, V., Guillemin, F., Robert, P.: A Markovian analysis of AIMD algorithms. Advances in Applied Probability 34(1), 85–111 (2002)
18. Lopker, A.H., Leeuwaarden, J.S.H.: Transient Moments of the TCP window size process. J. Appl. Prob. 45(1), 163–175 (2008)
19. Lestas, M., et al.: A new estimation sheme for the effective number of users in internet congestion control. IEEE/ACM Transactions on Networking 19(5), 447–461 (2011)

20. Barbera, M., et al.: Queue stability analysis and performance evaluation of a TCP-compliant window. IEEE/ACM Transactions on Networking 18(4), 1275–1288 (2010)
21. Afanasyev, A., Tilley, N., Reiher, P., Kleinrock, L.: Host-to-Host Congestion Control for TCP. IEEE Communications Surveys & Tutorials 12(3) (2010)
22. Gnedenko, B.V., Kovalenko, I.N.: Introduction to Queuing Theory, Moscow, 'Radio i Sviaz' (1987)

Cyber-Physical Approach
to the Network-Centric Robot Control Problems

Vladimir Zaborovsky, Mikhail Guk, Vladimir Muliukha, and Alexander Ilyashenko

Saint-Petersburg State Polytechnical University, Russia
vlad@neva.ru, mgook@stu.neva.ru, vladimir@mail.neva.ru,
ilyashenko.alex@gmail.com

Abstract. The paper analyzes features of the cybernetic methods' application to the distributed physical objects control task. Some parts of such cyber-physical object interact with each other by transmitting information via computer networks. In the framework of this cyber-physical approach proposed a structure for interactive control system for on-surface robot from International Space Station. This system implements the circuit-torque sensitization algorithms for network delays while transferring data over the computer telecommunication network.

Keywords: cyber-physics, open systems, robots, force-torque feedback, dual-contour control system.

1   Introduction

In the near future new generation of artificial physical devices will be created. They will be characterized by the flexibility, elasticity and sensitivity that are common to living organisms, but will have greater strength and durability because of used materials.

Such devices would be able to receive, store and transmit information about their surrounding environment, which will be used during their operations. Information is transmitted between physical objects and also between objects and the human operator.

Complex engineering tasks concerning control for groups of mobile robots are developed poorly. In our work for their formalization we use cyber-physical (CPh) approach, which extends the range of engineering and physical methods for a design of complex technical objects by researching the informational aspects of communication and interaction between objects and with an external environment.

The selection of CPh systems as a special class of designed objects is due to the necessity of integrating various components responsible for computing, communications and control processes («3C» – computation, communication, control). Although in modern science there are different approaches to the use of information aspects of the physical objects, but only within cybernetics, such approaches have had structural engineering applications. The conceptual distinction between closed and open

S. Balandin et al. (Eds.): NEW2AN/ruSMART 2014, LNCS 8638, pp. 619–629, 2014.

systems in terms of information and computational aspects requires the use of new models, which take into account the characteristics of information processes that are generated during the operation of the physical objects and are available for monitoring, processing and transmission via computer network.

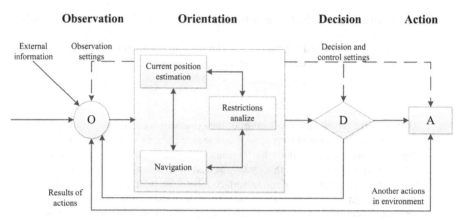

Fig. 1. Cyber-physical model of proposed control system

According to Fig.1, CF model of control system can be represented as a set of components, including following units: information about the characteristics of the environment (Observation), analysis of the parameters of the current state for the controlled object (Orientation), decision-making according to the formal purpose of functioning (Decision), organization and implementation of the actions that are required to achieve the goal (Action). The interaction of these blocks using information exchange channels allows us to consider this network structure as a universal platform, which allows using various approaches, including: the use of algorithms and feedback mechanisms or reconfiguration of the object's structure for the goal's restrictions entropy reduction or the reduction of the internal processes' dissipation.

Centric solutions allow using universal means for the organization of information exchange to integrate different technologies for the both observed and observable components of the control system. The main differences between which are the property "part – whole" and the ratio "system – environment". The parameters and the structure of such control system can quickly be adjusted according to the current information about the internal state of the object and the characteristics of the environment, which are in a form of digital data. Reported features open up the new prospects for the development of intelligent cyber-physical systems that will become in the near future an integral part of the human environment in the information space "Internet of Things." According to the estimates [1,2], network-centric cyber-objects in the global information space of the Internet will fundamentally change the social and productive components of people's lives. That will accelerate of the knowledge accumulation and the intellectualization for all aspects of the human activity. However, this process requires not only innovative engineering ideas, but also the development of scientific concepts uniting universal scientific paradigm [3,4]. Within this paradigm, the information should be considered as a fundamental concept of objective reality, in which physical reality has "digital" basis and therefore is computable. The idea of integrating the physical concepts with the theory of computation has led to the new conceptual schema for nature descriptions, known as «it from bit» [5]. In this scheme, all physical objects, processes and phenomena of nature, which are available to be read and understood by a person, are inherently informational and therefore they are isomorphic to some digital computing devices. Within this paradigm information acts as an objective attribute of matter that characterizes the fundamental distinctiveness of the potential states of the real object. The distinctiveness, according to the Landauer's principle [4], is an energy factor of the object's states and that is why it gives an explanation of what are the states and how they are perceived by other objects. This distinctiveness appears while creating the systems that are capable to ensure the autonomy of the existence during the interaction with the external environment by the self-reproduction of its characteristics. It should be noted that on the way to the widespread use of "digital reality" for the control problems there are some limitations that reflect the requirements for the existence of the special state of physical objects reflecting its changes as a result of the information exchange processes. So cyber-physical approach now often used to describe the properties of the so-called non-Hamiltonian systems in which the processes of self-organization are described by dissipative evolution of the density states matrix. However, the cyber-physical methodology may be successfully used to create complex robotic systems, the components which are capable for reconfiguration as the result of transmitting and processing digital data or metadata. The control tasks that are considered in this paper cover the actual scope of the cyber-physical approach, which is the basis of cloud computing technology and develop the methodology of cybernetics in the direction of the metadata control.

2   Cyber-Physics in Control Systems

Modern technical systems have clear boundaries separating them from the environment or other objects and systems (see Fig. 2).

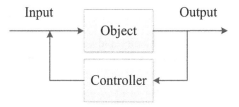

Fig. 2. Classical automatic control scheme

Therefore, the description of the processes in such systems is local and the change of its state can be described by the laws of physics, which are, in its most general form, the deterministic form of the laws of conservation, for example, energy, mass, momentum, etc. The mathematical formalization of these laws allows computationally determine the motion parameters of the physical systems, using position data on the initial condition, the forces in the system and the properties of the external environment. Although the classical methodology of modern physics, based on abstraction of "closed system" is significantly modified by studying the mechanisms of dissipation in the so-called "open systems", but such aspect of reality as the information is still not used to build the control models and to describe the properties of complex physical objects. In the modern world, where the influence of the Internet, supercomputers and global information systems on all aspects of the human activity becomes dominant, accounting an impact of information on physical objects cannot be ignored, for example, while realizing sustainability due to the information exchange processes. The use of cyber-physical methods becomes especially important while studying the properties of systems, known as the "Internet of Things" [6,7], in which robots, network cyber-objects and people interact with each other by sharing data in the single information space for the characterization of which are used such concepts as "integrity", "structure", "purposeful behavior", "feedback", "balance", "adaptability", etc.

The scientific bases for the control of such systems have become called Data Science. The term "Big Data" describes the process of integration technologies for digital data processing from the external physical or virtual environment, which are used to extract useful information for control purposes. However, the realization of the Data Science potential in robotics requires the creation of new methods for use information in control processes based on sending data in real time at the localization point of moving objects (the concept of "Data in motion").

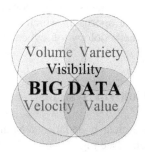

Fig. 3. Big Data structure

In general, the "Big Data" are characterized by a combination of four main components (four "V"): volume, variety, velocity and value (see Fig. 3). The general «V» is visibility of data and it is also a key defining characteristic of Big Data.

As a result, "Big Data" (Fig. 3) in modern science has become a synonymous for the complexity of the system control tasks, combining such factors of the physical processes that characterize the volume, velocity and variety and value of data generated by them.

3   Robot Control from ISS

We consider the application of the proposed above principles to control physical objects: on-surface robot, the motion of which is set and controlled from an orbital space station (Fig. 4).

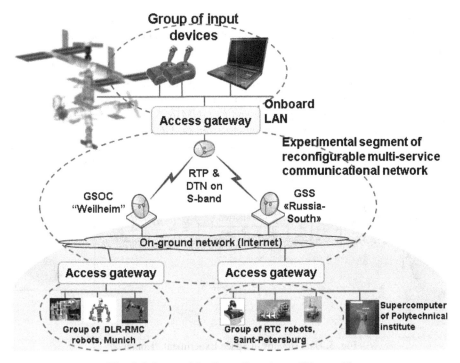

Fig. 4. Scheme of the Space Experiment "Kontur-2"

The feature of this Space Experiment (SE) "Kontur-2", which allows to carry it to the class of cyber-physical experiments, is that it uses the telepresence technology for the operator to simulate the robotic movement while the parameters of the environment (delays in communication channels and obstacles on the planet's surface) may vary in sporadic way. The designed control system allows in real-time to make palpable the results of robotic operations by analyzing the information about the values of the current axes of movement or moments in the joints of the manipulator, which are transmitted via computer communication network with a frequency of 500 packets per second.

The use of the circuit-torque delays sensitization effect allows the operator to adjust the speed and movement direction of the robot and by using force feedback effects on the joystick feel the impact of the network environment and generate an assessment of the environment state in which operates on-surface robot. In the considered control system the processes of the information exchange between joystick and robot can be decomposed into two processes of local commands realization in hard real-time and the commands' delivery process via network infrastructure using the TCP/IP stack.

Physical structure of the data streams in such control system is shown in Fig. 5 and includes:

1. The local loop, in which the software module "Joystick controller" (JC) provides cyclic polling of the current joystick coordinates, calculating and sending in joystick

the force vector depending on the current position and velocity of movement of the joystick's handle, as well as feedback information (T'), obtained from the controlled object (CO) ;

2. Network loop, in which the software components are used to organize the transfer of the control vector (C) and the telemetry (T) between the JC and the CO.

The basis of the network control loop is a software module "Transporter" consisting of the network modules of joystick and CO (NMJo, NMCO, see Fig. 5) which are connected by the virtual transport channel, based on UDP [6,7,8]. With the end systems (JC and CO), the network modules are connected through the adaptation modules (AM) to the properties of the communication media (AMJo and AMCO).

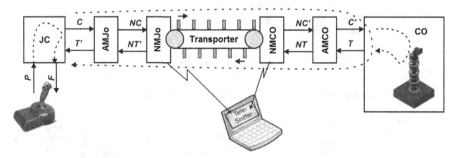

Fig. 5. Data streams in Space Experiment "Kontur-2"

Software module "Transporter" delivers data using UDP, providing isochronous communication for local controllers: vector control samples, that are uniformly received from the JC should also be uniformly (but, of course, delayed) transferred to the CO. Similarly, in the opposite direction is transferred the vector data obtained from the telemetry system of the robot. Thus, the "digital" reality of the physical processes is updated with a sampling frequency equal to the frequency of sending packets. Taking into account that the data delivery delay in digital communication channels is not constant, some packets may be lost, and delivery order may be disturbed, adaptation module provides recovery of the missing packets using automata model of data transfer processes, the adequacy of which is verified using the probabilistic model of packet delivery.

Module "Transporter" also ensures the delivery of asynchronous messages about events that are relevant to the remote control. The example of such events may be the pressing the button on the joystick's handle. These buttons may be used to control the operating mode of the CO or for some other action. Asynchronous event occurs sporadically at any given time. However, it should be guaranteed to reach the operator, saving the time reference to the transmitted isochronous packets' stream.

The computing resources of adaptation module allow to implement the methods of predictive modeling and, in the case of insufficient data, to predict the behavior of the control object or operator without any delay in the transmission of control signals, and thereby ensure the smooth control for the robot. Control circuit scheme is presented in Fig. 6.

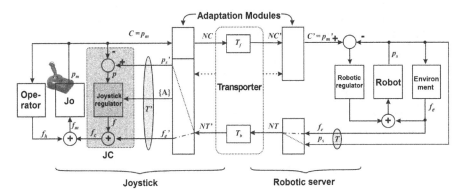

Fig. 6. "Kontur-2" control circuit scheme

In Fig. 6, there is the separation of the control circuit into three main parts:

1. The loop of Joystick (Jo) (the left part of the scheme), whose output is the control vector c that is showing the current position of the Jo handle p_m. Movement of the handle is determined by the force f, which is summable from the operator's impact forces f_h, force f_c, which is formed by Joystick, and force f_e', displaying the impact of the environment f_e on robot. P input of JC is the mismatch between the current Jo position (p_m) and the display of the current position of the robot (p_s). The controller parameters vector $\{A\}$ can be changed dynamically.
2. The means of communication (Transporter), which implements delays in delivery, and adaptation modules that are counteract the effect of these delays. In general, the delays in the direct (T_f) and the reverse channels (T_b) can be different.
3. The loop of the controlled device (the right part of the scheme), generating the impact on the robot using a mismatch between its current position p_s and displayed position of the Jo p_m'. A robot may also be affected by the environment with force f_e. As the feedback, the vector of current robot position p_s can be used, and if there are appropriate sensors, the force vector of environmental impact f_e also can be used.

For such network control problem, queueing theory is a powerful tool [9-11]. The Transporter was realized using preemptive queueing system [7]. The remote robot control with force-torque sensitization requires that during the control process, operator has to be able to "feel" the current state of the robot, and communication network. Therefore, to describe the configuration space of the controlled system is proposed to use a model of the virtual spring that is fixed at the one end to the base of the joystick, and the initial position of which coincides with the current position of the controlled robot. In the initial position, while not moving, the robot does not affect the joystick and the operator. However, if the operator starts the control process and move the joystick's handle from the initial position, the virtual string will try to return it to the initial position. Farther the joystick will be deflected from the initial position, the greater force will the operation of stretching virtual springs require from the operator. Virtual elastic force of the spring is calculated based on analysis of the current situation, taking into account current positions of joystick and information from the robot. It is calculated using the formula (1):

$$p = p'_s - p_m,$$ (1)

where p'_s is a vector of robot's coordinates, which is currently available for the adaptation module of joystick , p_m is a current position of the joystick. The impact force felt by the operator is calculated according to the formula (2):

$$f(t) = A_0 \int_0^t p(\tau)d\tau + A_1 p + A_2 \dot{p} + A_3 \ddot{p},$$ (2)

where A_0 is an astatism in system , A_1 is virtual spring stiffness, A_2 is viscosity of the control medium, A_3 is a virtual mass of the handle.

During the control process, the initial position of the virtual spring will change to coincide with the current position of the robot. The proposed model allows taking into account the information about the state of the communication channel by making adjustments during the process of sensitization. In other words, the virtual stiffness of the spring must be increased not only according to the divergence between positions of the robot and joystick, but also using the value of the data delay in the communication channel, thus increasing the inertia of the network control loop.

To implement such interaction mode we will need to handle the additional information about the state of the communication channel: sequence of RTT samples that is ordered by time and the percentage of the lost network packets (PLP) [7,8]. Thus, the stiffness of the virtual spring and a moving speed of the robot will vary proportionally average values of RTT and PLP, which are calculated for the time period according to the formula (3). This period is commensurate with the time constant for the closed-loop control system.

$$f(t) = (A_0 \int_0^t p(\tau)d\tau + A_1 p + A_2 \dot{p} + A_3 \ddot{p}) + A_4 \cdot RTT \cdot PLP \cdot \frac{p}{\|p\|},$$ (3)

where A_4 is a coefficient of influence of the environment on the force-torque feedback. As a result, while increasing the latency, the virtual spring will not allow the operator to change robot's position quickly. This circumstance reduces speed of the robot while moving, but allow the operator to adjust the results of mining operations, analyzing data of control actions, despite the fact that this data will be available to the operator is delayed.

The proposed organization of the remote control system was designed for ground testing of algorithms and software debugging for scientific equipment in the Space Experiment "Kontur-2". On the experimental stand the parameters of the communication system and software module "Transporter" are modified according to the constraints imposed by S-band communication channel's bandwidth. This constraints are estimated according to the necessity of the telepresence mode, which requires every 2 ms to transmit IP packets containing 22 bytes of application data. These requirements helped to modify the format of the messages and to consider properties that reflect the specifics of network processes [7,12]. To reduce the bit length of the packet's number field and timestamps we propose to divide numbers of transmitted packets on the even and odd. The even packets are used for the transmission of isochronous traffic (5 parameters in 32-bit floating-point format), the odd ones are for transmission of asynchronous messages and proprietary information, which provides the definition of the channel state with the implementation of network protocols and the use of "lightweight" LwIP stack.

According to the requirements of the bilateral control, the software for a two loop telecontrol system is symmetrical, namely, the structure of software modules on the side of the robot and joystick coincide. The local controllers in joystick and robot are implemented as PID controllers and differ from each other only by settings. In Fig. 8 are shown the structure of the developed software for joystick, which is implemented by the 32-bit ARM microcontroller.

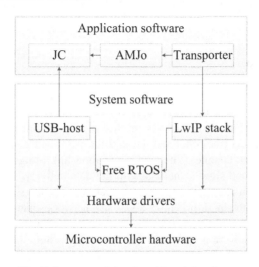

Fig. 7. Structure of developed joystick software

JC software module reads current joystick position and state of its buttons, as well as forms and transmits the force feedback control commands. The module implements the PID controller function, for which the input value and the coefficients in the control law are formed by the software AMJo.

Transporter module provides the regular transmission for network packets, requesting the adaptation module to convert control vector and telemetry, as well as to process the current information about communication delays in the channel.

Adaptation module (AMJo) performs the conversion, ensure stability of the closed system and the quality of control. The system software in ARM microcontroller generates runtime and includes the following components: USB-host, LwIP, FreeRTOS, and a set of drivers for configuring and using network interfaces.

Free RTOS module is used for the sharing of different modules' tasks listed below:

- Robot control server which receives commands from personal computer and forms counter-flow data to control 3D-model of the robot and to register control processes parameters;
- USB-host is providing regular request for current joystick's coordinates and transmitting force control commands;
- LwIP stack, performs primary processing of IP packets (checksums, statistics, etc.) and setting them into the input queue of the program module Transporter;
- Transporter, providing two-way exchange with isochronous packets and asynchronous messages.

Transporter module is activated at the double frequency of data exchange that allows to call the adapting module and to transmit it alternatively the telemetry vector, received from the robot, and the control vector, obtained from JC. The result from AMJo, are transmitted to the JC or sent via the network to robot. If there is no input message at the proper time, in this case, the AMJo generates a control action, predicted using equation (3).

4   Conclusion

The paper describes the structure of a cyber-physical system for remote force-torque control of a robot located at the Earth's surface, from the manned orbital space station. The feature of the proposed solution is the use of dual-circuit system in which the "almost analog" objects (the robot and the joystick with their control loops) are connected by the network channel. This communication link brings in the control loop the variable discrete and significant delays during data delivery. Force-torque sensitization implementation allows the operator to feel not only the state of the controlled device, but also to obtain information about the network channel, which indicates about possible unsafe further control. The control method and software architecture for its implementation were proposed. As the operating system for network nodes was chosen freely distributed real-time system FreeRTOS, which provides a high frequency cycles in local control loops and network interaction between the operator and the robot. Software systems for the interactive remote control are used in bilateral mode for both joystick and robot that allows developing effectively algorithms for circuit-torque feedback to the various options for network infrastructure.

References

1. Fradkov, A.L.: Cybernetics physics: Principles and Examples. SPb. Nauka, 208 p. (2003) (in Russian)
2. Zaborovsky, V.S., Kondratiev, A.S., Silinenko, A.V., Muliukha, V.A., Ilyashenko, A.S., Filippov, M.S.: Remote Control for Robotic Objects in Space Experiments of "Kontur" series Scientific and Technical Statements of SPbSPU. Computer. Telecommunications. Control 6(162), 23–32 (2012) (in Russian)

3. Artigas, J., Jee-Hwan, R., Preusche, C., Hirzinger, G.: Network Representation and Passivity of Delayed Teleoperation Systems. In: 2011 IEEE/RSJ International Conference on Intelligent Robots and Systems (IROS), pp. 177–183 (September 2011)
4. Niemeyer, G., Slotine, J.-J.E.: Telemanipulation with Time Delays. Int. Journal of Robotics Research 23(9), 873–890
5. Wheeler, J.A.: Information, Physics, Quantum: The Search for Links. In: Zurek, W. (ed.) Complexity, Entropy, and the Physics of Information, Addison-Wesley, Redwood City (1990)
6. Zaborovsky, V., Lukashin, A., Kupreenko, S., Mulukha, V.: Dynamic Access Control in Cloud Services. International Transactions on Systems Science and Applications 7(3/4), 264–277 (2011)
7. Zaborovsky, V., Zayats, O., Mulukha, V.: Priority Queueing With Finite Buffer Size and Randomized Push-out Mechanism. In: Proceedings of The Ninth International Conference on Networks, ICN 2010, Menuires, The Three Valleys, French Alps, April 11-16, pp. 316–320. Published by IEEE Computer Society (2010)
8. Zaborovsky, V., Gorodetsky, A., Muljukha, V.: Internet performance: TCP in stochastic network environment. In: 1st International Conference on Evolving Internet, INTERNET 2009, pp. 21–26 (2009), doi:10.1109/INTERNET.2009.36
9. Shi, Z., Cory, B.: QoS in the Mobile Cloud Computing Environment. In: Book: Mobile Networks and Cloud Computing Convergence for Progressive Services and Applications, Ch. 11, pp. 200–217 (2013), doi:10.4018/978-1-4666-4781-7.ch011]
10. Shi, Z., Beard, C., Mitchell, K.: Analytical models for understanding misbehavior and MAC friendliness in CSMA networks. Performance Evaluation 66(9-10), 469–487 (2009), doi:10.1016/j.peva.2009.02.002
11. Shi, Z., Beard, C., Mitchell, K.: Analytical models for understanding space, backoff, and flow correlation in CSMA wireless networks. 2013 19(3), 393–409 (2013), doi:10.1007/s11276-012-0474-8
12. Zaborovsky, V., Muliukha, V., Popov, S., Lukashin, A.: Heterogeneous Virtual Intelligent Transport Systems and Services in Cloud Environments. In: Proceedings of The Thirteenth International Conference on Networks, ICN 2014, Nice, France, February 23 - 27, pp. 236–241 (2014)

On the Comparison between the Bit-by-Bit Technique and the Viterbi Technique for Demodulation of Single-Carrier Signals Affected by Clipping

Alexey M. Markov

St. Petersburg State Polytechnical University, St. Petersburg, Russia
markov@cee.spbstu.ru

Abstract. To decrease out-of-band radiation of signals, based on single-carrier modulation, root-raised cosine (RRC) filters are usually used. To decrease a peak-to average-power ratio (PAPR) we have clipped signals, passed RRC filters. To retain the constant peak power of a power amplifier for different clipping levels we have increased the values of clipped signals before the power amplifier. In this paper, we have studied the bit-by-bit coherent reception technique and the Viterbi technique and also calculated bit-error-rate (BER) for BPSK, QMSK, 8-PSK and 16-QAM. We have found the optimal clipping probability and the maximum gain in the peak signal-noise ratio (PSNR) of the Viterbi technique in comparison with the bit-by-bit technique.

Keywords: Viterbi technique, power amplifier, coefficient of performance, peak signal-noise ratio, RRC filter.

1   Introduction

Single-carrier modulation is used in wireless networks together with OFDM [1]. An increase of capabilities of mobile stations causes an increase of the power consumption of systems. Therefore it is relevant problem to increase of the coefficient of performance (COP) of a transmitter power amplifier. On the other hand it is desirable to decrease the peak power of the amplifier, as a less powerful amplifier costs less. In this paper we propose to modernize the existent spectrally efficient systems by clipping signals and a digital gain of clipped signals before a power amplifier to obtain greater COP and improve the BER performance.

Classic single-carrier signals have a large spectral bandwidth. To decrease the bandwidth it is applied filtering of signal quadrature components by RRC filters [1]. If we use the same RRC filter at the receiver, the intersymbols interference will be eliminated at symbol points. The disadvantage of such filtering is an increase in the PAPR. It is resulted in a small COP of the power amplifier. For the non-distortion amplification of such signals it is required the linear static modulation characteristic of a power amplifier.

S. Balandin et al. (Eds.): NEW2AN/ruSMART 2014, LNCS 8638, pp. 630–639, 2014.

To increase the COP of the chosen power amplifier and improve the BER performance we have clipped signal quadrature components, passed RRC filters, and then we have digitally increase those clipped components. It resulted in an increase in the average power of the power amplifier input signal, but the peak power remains fixed. As it is usually chosen the power amplifier on the basis of the peak power, we can use the same amplifier for any clipping levels. On the other hand in case of clipping it can be used the less powerful amplifier to obtain the same BER performance as in the case of no clipping.

We consider that the power amplifier is working at the understressed mode under conditions of class B and its static modulation characteristic is linear, so that distortions caused by the power amplifier can be neglected. The BER performance depends on the compromise between the increase of the signal average power and the increase of the intersymbols interference due to clipping. For coherent reception we have used the bit-by-bit technique and the Viterbi technique [2], based on «in whole» reception.

The first goal of the paper is the improvement in the BER performance by clipping and following digital gain of clipped signals. The second goal is the comparison between the bit-by-bit reception technique and the Viterbi reception technique in terms of the BER performance and the techniques speed on a channel with additive white Gaussian noise (AWGN).

2 The Signals Forming

Now we consider the signals forming. Fig.1 shows the block diagram of the signals forming.

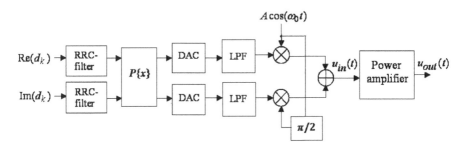

Fig. 1. The block diagram of the signals forming

The real and imaginary parties of complex channel symbols d_k are inputted in RRC-filters. The index k is the sequence number of symbol, transmitted on $[kT, (k+1)T]$, $k = 0, 1, 2,..., N-1$, where T is the symbol duration. The N value is the count of symbols to transmit. Those symbols are selected from the set of m constellation points, $d_k \in S_m = \{s_0, s_1,..., s_{m-1}\}$.

The impulse response of the RRC filter, defined over time interval $t \in [0, 2T_{shift}]$, is $h(t) = h_0(t - T_{shift})$, where $h_0(t)$ expressed as [3]

$$h_0(t) = \frac{\sin(\frac{\pi t}{T}(1-\alpha)) + 4\alpha \frac{t}{T}\cos(\frac{\pi t}{T}(1+\alpha))}{\frac{\pi t}{T}(1-(\frac{4\alpha t}{T})^2)}, \tag{1}$$

where α is the roll factor of the RRC-filter, T_{shift} is the chosen time shift. Fig. 2 shows $h(t)$ for $\alpha = 0.22$, $T_{shift} = 6T$.

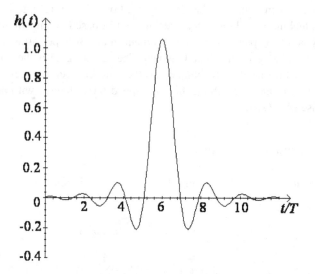

Fig. 2. The impulse response of the RRC filter

The operator $P\{x\}$ realizes clipping and gain:

$$P\{x\} = G \cdot \begin{cases} L\dfrac{x}{|x|}, & |x| \ge L \\ x, & |x| < L \end{cases} \tag{2}$$

where L is the clipping level and G is the gain factor. The digital-analog converters (DAC) are followed by the low-pass filters. These filters eliminate the signal spectrum copies, caused by sampling.

The input voltage of the amplifier is $u_{in}(t) = \mathrm{Re}(U_{in}(t)e^{j\omega_0 t})$, where j is the imaginary unit, ω_0 is a carrier frequency, $U_{in}(t)$ is the complex envelope of $u_{in}(t)$, expressed as

$$U_{in}(t) = k_T P\{\sum_{k=0}^{N-1} d_k h(t-kT)\},\tag{3}$$

where k_T is the transformation coefficient of DACs. The output transmitter voltage can be written as $u_{out}(t) = \mathrm{Re}(U_{out}(t)e^{j\omega_0 t})$. The output voltage complex envelope is $U_{out}(t) = K_U U_{in}(t)$, the K_U value is the amplifier voltage gain.

Now we choose G. The G value is chosen so that the maximum value of $|U_{in}(t)|$, denoted as $U_{in\,mg}$, will be equal to the input voltage amplitude of the power amplifier, appropriated to the marginal conditions. Hence the G value is equal to $U_{in\,mg}/L$. Fig. 3 shows the static modulation characteristic, i.e. the dependence of the output voltage amplitude $|U_{out}(t)|$ on the input voltage amplitude $|U_{in}(t)|$ of the power amplifier and its approximation.

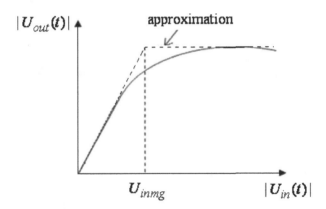

Fig. 3. The static modulation characteristic of the power amplifier

We can accept that if $|U_{in}(t)| \le U_{in\,mg}$, the power amplifier hardly introduce distortions. Hence if the peak power of the amplifier is fixed we can use the same power amplifier for any L.

Fig. 4 shows the $|U_{in}(t)|$ for arbitrary d_k for the case without clipping (a) and clipped by level L and gained in $G = U_{in\,mg}/L$ times (b).

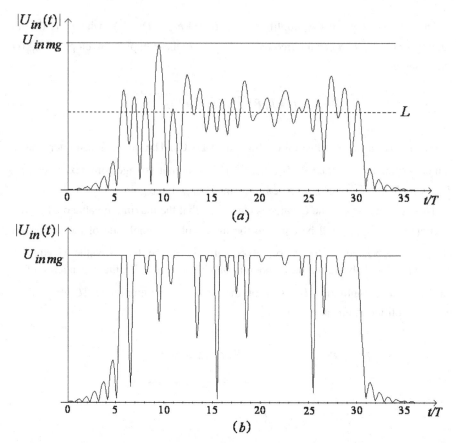

Fig. 4. The input amplitude voltage of the power amplifier: (*a*) without clipping, (*b*) clipping by level L and $G = U_{in\,mg}/L$

3   The Signals Reception

The input receiver signal is $x(t) = \mu u_{out}(t - t_s) + n(t)$, where μ is the transfer channel coefficient, t_s is the time of signals spreading from the transmitter to the receiver, $n(t)$ is an additive white Gaussian noise. We suppose that μ and t_s are precise known.

Fig. 5 shows the block diagram of the signal reception.

The cosine and sine quadrature signals components, extracted from $x(t)$ by LPFs, are sampled at the frequency $16/T$. Those discrete signal components are inputted in analog-digital converters (ADC) and then in RRC – filters, similar to those at the transmitters. The components $A_{xc}(t)$ and $A_{xs}(t)$ are sampled at frequency $1/T$ and inputted in the detector. If there are no clipping at the transmitter, intersymbols interference at the points of time $t_k = t_s + kT + 2T_{shift}/T$, $k=0,1,2...$ is absent.

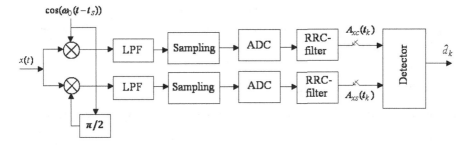

Fig. 5. The block diagram of the signals reception

However if there are clipping at the transmitter, intersymbols interference at those points is presence. We have used two different reception techniques. The first technique is the bit-by-bit technique. Its decision technique rule can be written as follows:

$$\hat{d}_k = \arg \min_{d_k \in S_m} \{ \, | F_x(t_k) - D d_k |^2 \}, \tag{4}$$

where $F_x(t) = A_{xc}(t) + jA_{xs}(t)$. The D coefficient is estimated before reception to obtain minimum BER. To improve the BER performance, we propose the near Maximum Likelihood Detection (MLD) technique, based on the Viterbi technique, explained as follows:

1. Form each possible symbols sequences of length M: $(d_0^{(i)}, ..., d_M^{(i)})$, where i is the sequence index number ($i = 1, ..., m^{M+1}$). The M value is much less than the length of sequence N and is chosen to obtain the necessary technique speed.   Set iteration count $k=0$.

2. For each sequence $(d_0^{(i)}, ..., d_{k+M}^{(i)})$ calculate the values

$$Q^{(i)}(t_k) = Q^{(i)}(t_{k-1}) + | F_x(t_k) - F^{(i)}(t_k) |^2, \tag{5}$$

where $F^{(i)}(t_k) = A_c^{(i)}(t_k) + jA_s^{(i)}(t_k)$ is formed at the receiver. The $A_c^{(i)}(t_k)$ and $A_s^{(i)}(t_k)$ values are signals quadrature components, which would be at the detector input in suggestion that transmitted symbols are $(0, ..., 0, d_{k-M}^{(i)}, ..., d_k^{(i)}, ..., d_{k+M}^{(i)}, 0, ..., 0)$ and noise is absence. Consider $d_r^{(i)} = 0$, if $r<0$ and $Q^{(i)}(t_{-1}) = 0$.

3. Divide all sequences $(d_0^{(i)}, ..., d_k^{(i)}, ..., d_{k+M}^{(i)})$ into subsets so that sequences, consisted of $2M$ symbols $(d_{k-M+1}^{(i)}, ..., d_k^{(i)}, ..., d_{k+M}^{(i)})$, will be equal in each subset.

4. Remain only one sequence in each subset. This sequence must minimize $Q^{(i)}(t_k)$ in its subset. Other sequences are eliminated from the subsets. If $k \geq M$ the number of sequences reduces by m.

5. Add each of constellation point symbols $\{s_0, s_1, ..., s_{m-1}\}$ to each survivor sequence. The number of sequences multiplies by m. We have the sequences: $(d_0^{(i)}, ..., d_k^{(i)}, ..., d_{k+M}^{(i)}, s_0), ..., (d_0^{(i)}, ..., d_k^{(i)}, ..., d_{k+M}^{(i)}, s_{m-1})$, where i is the index number of a sequence.

6. If $k \geq k_0$ (for example $k_0 = 5$), find the sequence which minimizes $Q^{(i)}(t_k)$ among all survivor sequences. Pass the symbol $d_{k-k_0}^{(i_{min})}$ to the detector output, where i_{min} is the index number of the best sequence.

7. Set iteration count $k=k+1$.

8. If $k < N$, go to step 2 else go to step 9.

9. Find the sequence, which minimizes $Q^{(i)}(t_k)$ among all m^{2M+1} sequences. Pass the symbols $d_{N-k_0}^{(i_{min})}, ..., d_{N-1}^{(i_{min})}$ to the detector output, where i_{min} is the index number of the best sequence.

The state number of the Viterbi technique is equal to m^{2M}. The increase in M causes an improvement of the BER performance, but the technique speed decreases. We set $M=3$ for $m=2$ (BPSK) and $M=1$ for $m>2$ (QPSK, 8-PSK, 16-QAM). This technique is named near Maximum Likelihood Detection as we set the value of M less than the half length of the RRC-filter impulse response in terms of the symbol duration.

4   The BER Performance and Discussion

We have evaluated the BER performance on an AWGN channel with one-sided power spectral density N_0 by computer simulation. The noise samples at the reception RRC-filter outputs are independent with zero mean and variance $N_0/(4T)$. Fig. 6 shows the BER performance as a function of the received peak signal-to-noise ratio (PSNR) for clipping probability $p_{clip} = 0.4$ and 16-QAM. The clipping probability is the ratio of the number of clipping signal samples to the total number of samples. The PSNR is $h = \mu\sqrt{P_{max}T/N_0}$, where $P_{max} = K_U^2 |U_{inmg}|^2/2R$ is the peak power of the transmitter amplifier, R is the equivalent resistance of the power amplifier active device at the carrier frequency ω_0.

Fig. 6 shows that for BER$>10^{-2}$ the difference in PSNR is negligible. At BER=10^{-3} the Viterbi technique gains in PSNR about 3.5 dB in comparison with the bit-by-bit technique. Fig. 7 shows the curves of PSNR, required to obtain the BER at 10^{-3}, versus the clipping probability p_{clip} for BPSK, QPSK, 8-PSK and 16-QAM.

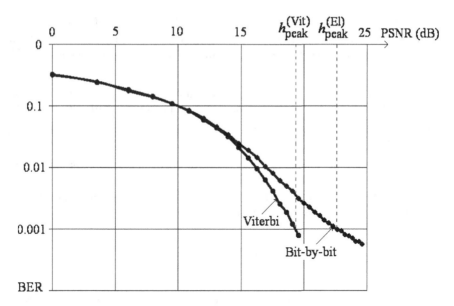

Fig. 6. The BER versus PSNR for 16-QAM, p_{clip}=0.4

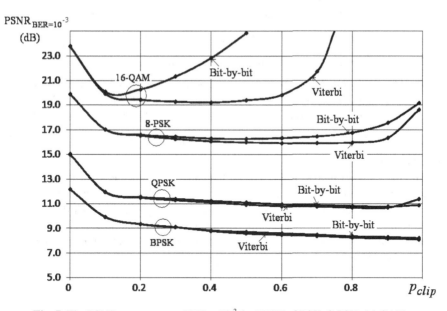

Fig. 7. The PSNR versus p_{clip} at BER = 10^{-3} for BPSK, QPSK, 8-PSK, 16-QAM

Fig. 7 shows that the bit-by-bit technique of BPSK and QPSK signals provides almost the same PSNR as the Viterbi technique at any clipping level. For BPSK and QPSK the optimal clipping levels are close zero ($p_{clip} \approx 1$) and the gain in PSNR is 4 dB comparison with $p_{clip} = 0$.

For 8-PSK modulation the optimal clipping probability is 0.5 for the bit-by-bit technique and 0.8 for the Viterbi technique. The gain in PSNR at BER=10^{-3} is 3.6 dB for the bit-by-bit technique and 3.9 dB for the Viterbi technique comparison with $p_{clip} = 0$.

For 16-QAM modulation the optimal clipping probability is 0.15 for the bit-by-bit technique and 0.4 for the Viterbi technique. The gain in PSNR at BER=10^{-3} is 3.9 dB for the bit-by-bit technique and 4.6 dB for the Viterbi technique comparison with $p_{clip} = 0$.

Hence for 8-PSK the Viterbi technique gains in PSNR of 0.3 dB comparison with the bit-by-bit technique and 0.7 dB for 16-QAM.

Now we will examine the coefficient of efficiency of the power amplifier. The coefficient of efficiency is expressed as [4]

$$
\eta = \frac{\eta_{max}}{\max\{|U_{in}(t)|\}} \frac{< |U_{in}(t)|^2 >}{< |U_{in}(t)| >},
\tag{6}
$$

where η_{max} is the coefficient of efficiency at $|U_{in}(t)| = \max\{|U_{in}(t)|\}$, $<x>$ denotes the mean value of x.

We have found that the change $p_{clip} = 0$ to $p_{clip} = 1$ provides the increase in η value of 30% for BPSK and QPSK. Also we have found that $\eta_{8PSK}^{(0.8)} - \eta_{8PSK}^{(0.5)} = 3\%$ and $\eta_{16QAM}^{(0.4)} - \eta_{16QAM}^{(0.15)} = 7\%$, where the upper index denotes p_{clip}.

We have calculated by computer simulation the spectral bandwidth, contained 99% of the total signals power. We have found that for 8-PSK the change $p_{clip} = 0.5$ to $p_{clip} = 0.8$ provides the increase of bandwidth by 1.5 times, but for 16-QAM the change $p_{clip} = 0.15$ to $p_{clip} = 0.4$ provides the increase of bandwidth only by 1.05 times.

The main disadvantage of the Viterbi technique is the decrease of its speed comparison with the bit-by-bit technique. We have studied that the Viterbi technique is 5 and 50 times slower than the bit-by-bit technique for 8-PSK and 16-QAM, respectively. To reduce the computational complexity of the Viterbi technique it can be used the Reduced-state sequence estimation (RSSE) and the Sphere Detection techniques [5], instead of the Viterbi technique without a significant loss in BER.

5 Conclusion

In this paper we have analyzed the BER performance of clipping signals, based on single-carrier modulation at the fixed peak signal-to-noise ratio. We proposed bit-by-bit and Viterbi techniques for coherent reception. We have found that for optimal clipping levels the gain in PSNR is about 4 dB comparison with no clipping. Therefore we can decrease the peak power of the amplifier by 2.5 times without a loss in BER.

References

1. IEEE Standard for Local and metropolitan area networks. Part 16: Air Interface for Broadband Wireless Access Systems. IEEE Std 802.16 (2009)
2. Makarov, S.B., Markov, A.M.: On the whole. reception algorithm of arbitrary sequences of frequency-shift keyed signals with intersymbol interference. In: Radio Engineering, vol. 3, pp. 46–51 (2011)
3. 3GPP TS 25.104 V6.8.0 (2004-12). Technical Specification. 3rd Generation Partnership Project; Technical Specification Group Radio Access Network; Base Station (BS) radio transmission and reception (FDD) (Release 6)
4. Shakhgil'djan V.V: Designing of radio transmitters: Moscow, Radio and Communication (2000)
5. Wu, X.: Softbit Detector / Equalizer for GSM release 7: Technical. University of Denmark (2008)

Informational Properties of a DWDM Electrically-Controllable Integrated Optical Filters with an Additional Polarizer

Alexey N. Petrov[1,2,*], Alexander V. Shamray[2], and Viktor M. Petrov[1]

[1] Saint Petersburg State Polytechnical University, Polytechnicheskaya Str, 29,
St.-Petersburg, 195251, Russia
alexey-np@yandex.ru, vikpetroff@mail.ru
[2] Ioffe Institute, Polytechnicheskaya Str, 28, St.-Petersburg, 194021, Russia
achamrai@mail.ioffe.ru

Abstract. The paper discusses the design and implementation of the integrated electrically controllable filter based on Bragg gratings with high wavelength and polarization selectivity. The filter we suggest combine a high spectral selectivity (to 0.1 nm) of the Bragg gratings, a high polarization extinction ratio (up to 40 dB) of plasmon polariton polarizer, and wide possibilities to electrically control of the spectral transfer function. It has a bandwidth of about 0.25 nm and time of switching from one spectral channel to the other below 1 μs. The proposed filters might be used both in the optical communication lines (including those relying on the DWDM principles) and in the optical coders/decoders and narrow-band rapidly tunable optical sources.

Keywords: DWDM networks, optical filters, informational throughput.

1 Introduction

Integrated optical filters based on Bragg gratings in lithium niobate attract attention of designers of different devices used in optical communication lines, including those relying on the DWDM principles, because these filters can be used to develop fast and "fully transparent" controllable optical multiplexers and demultiplexers.

The discussed in the paper filters combine a high spectral selectivity (to 0.1 nm) of the Bragg gratings, and wide possibilities to electrically control diffraction conditions in these gratings because a grating is recorded in an electrooptic material [1,2]. For example, it was shown in [3,4] that tuning of a selected wavelength can be carried out, and transfer functions with several spectral bandwidths and also transfer functions with different shapes can be synthesized. Since such a control is accomplished via the electrooptic effect, the time needed for changes of the transfer function shape of the filter (i.e., the switching time) is determined by the electric capacity of the interelectrode spacing and can be as short as fractions of a microsecond for open circuit electrodes. If the travelling wave electrodes are used a very high modulation frequencies (several tens of gigahertz) will be possible. [5]

---

* Corresponding author.

S. Balandin et al. (Eds.): NEW2AN/ruSMART 2014, LNCS 8638, pp. 640–646, 2014.

Among significant disadvantages of these filters is their sensitivity to the input polarization. It is caused by a considerable birefringence of LiNbO$_3$, i.e., different refractive indices for the ordinary and extraordinary waves and as consequence different diffraction conditions for different incident light polarization. This sensitivity gives significant polarization dependent crosstalk if the filter is used for DWDM channel selection and high polarization dependent noise for electro optical switching and modulation.

Different methods for reducing the influence of input polarization on operation of the filters based on birefringent materials have been suggested in literature. E.g., a filter containing two gratings with different periods was suggested in [5]. The period of one grating corresponds to the Bragg conditions for the ordinary polarization, and the period of the other grating corresponds to the Bragg conditions for the extraordinary polarization. In spite of an undoubtful efficiency of this method of reducing the polarization influence, it is rather difficult to implement it especially for tunable and switchable filters.

This paper reports about simple damping TM polarization which allows the polarization dependent crosstalk and noise effect can be appreciably weakened and its influence on the filter throughput.

2   Experimental Results

2.1   Design of Filters and Experimental Setup

We fabricated and studied the samples of filters shown in Fig.1. As a substrate (1), a LiNbO$_3$ crystal approximately 10 x 30 x 1 mm^3 in size was used. The optical waveguide (2) was formed on the crystal surface along the long side.

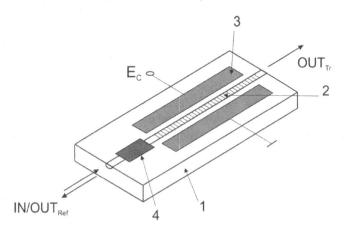

Fig. 1. Optical DWDM filter with an additional polarizer. 1- LiNbO$_3$ substrate, 2 – optical waveguide with a reflective Bragg grating, 3 – control electrodes, 4 – polarizer.

The single mode channel waveguide was produced by the Ti-ion diffusion technique. A reflective Bragg grating was recorded in such a way that the grating wave vector was parallel to the waveguide. The photorefractive effect and thermal fixing of the grating were used for recording. The grating length in the light propagation direction was 10 – 12 mm. The Bragg grating period Λ_B was chosen so that in the reflection geometry the grating reflected light with wavelength λ_{ref} in the 1555–1556 nm region, i.e., in the C-band. This can be achieved under the condition

$$\lambda_{ref} = 2n\, \Lambda_B \tag{1}$$

where n is the effective refractive index of the waveguide mode. To provide light launching and exit, optical fibers were glued to the waveguide from the substrate ends. We fabricated filters of two types, i.e., (i) modified filters with an additional metallic layer (4) above the waveguide that played the role of a polarizer and (ii) conventional filters without an additional metallic layer (Fig. 1). The polarizer length was about 7 mm. The polarizer was placed close to the input of the filter.

A structure of single mode channel waveguide and thin metal film over it supports hybrid eigen modes. The modes are TE and TM polarization modes of channel waveguide and two plasmon-polariton modes at the metal-substrate and metal-air interfaces. If the conditions of the phase matching are fulfilled the efficient energy transfer from the TM waveguide mode to the plasmon-polariton mode occurs, where intensity of the light wave decays due to ohmic absorption in metal film. The base materials for the plasmon-polariton polarizer were a metal (aluminum) film and a dielectric (Al_2O_3) buffer layer which was used for phase matching. This combination of materials allows the polarizer to be manufactured in a single technological cycle in a vacuum magnetron sputtering setup. At the first stage, an Al_2O_3 dielectric buffer layer was formed on the wavegude surface by reactive sputtering of an aluminum target in argon–oxygen atmosphere. At the second stage, a metallic aluminum film was deposited onto the dielectric buffer layer by sputtering the target in a pure argon atmosphere. The use of a single two stage deposition cycle ensures high purity of the deposited materials and small (±5 %) uncertainty in the thicknesses of deposited layers. The relatively low cost of aluminum and simplicity of the process make the proposed technology highly attractive for the large-scale commercial production. Optimum technological conditions have been established for manufacturing polarizers on singlemode channel waveguide for a working wavelength of 1550 nm, which are matched to the standard SMF28 type optical fiber [6]. For a metallic aluminum films thickness above 100 nm and a dielectric buffer layer thickness of 15 nm, the polarization extinction ratio amounts to 19 dB/mm and the excess losses are on a level of 0.1 dB/mm. These values are comparable with those reported for the world best analogs.

We studied the transmission and reflection geometries of the filter operation. To this end, the powers of the reflected OUT_{Ref} or transmitted OUT_{Tr} light were measured as functions of wavelength. The Stokes parameters S_1, S_2 and S_3 were measured by a state of polarization analyzer with a tunable wavelength. The Stokes parameters were measured for the filters with a polarizer, and for the filters without a polarizer. The experimental setup is shown schematically in Fig. 2.

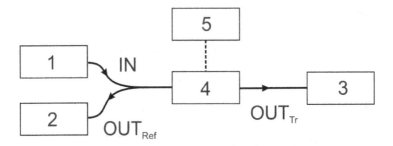

Fig. 2. Experimental setup. 1 – SOP analyzer with tunable wavelength, 2 and 3 – photodetectors, 4 – filter, 5 – control unit

2.2 Transfer Functions of the Filters and Polarization Properties

Figs. 3-a,b show the transfer functions of a single filter in the transmission geometry (left panel) and dependence of the Stokes parameters on wavelength for the filter with an additional polarizer (a) and without polarizer (b) (right panel).

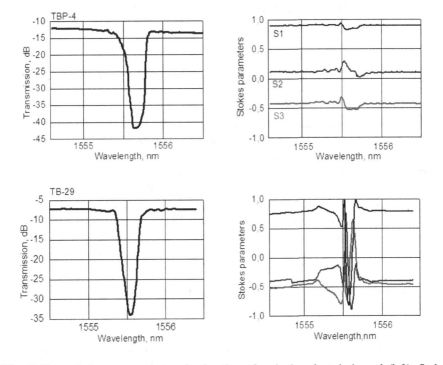

Fig. 3. Transmission geometry: transfer function of a single selected channel (left), Stokes parameters (right), A: with polarizer, B: without polarizer

Fig. 4, a,b shows similar dependences for the same samples of filters in the reflective geometry.

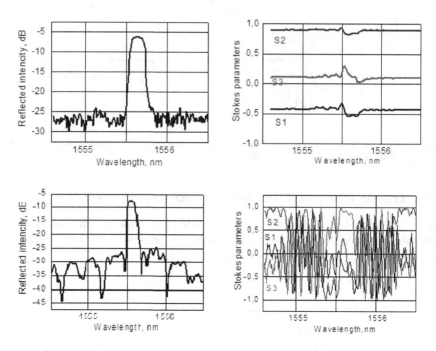

Fig. 4. Reflective geometry: transfer function of a single selected channel (left), Stokes parameters (right), A: with polarizer, B: without polarizer

The experimental dependences obtained in the study show that the using of the polarizer in the filter construction allowed a nearly complete suppression of the polarization dependent noise: variations in the Stokes parameters decreased in approximately by an order of magnitude as compared with the filter without an additional metallic layer. An expected increase in insertion losses of the device were observed. In both cases (reflection and transmission geometries), loss increased from ~-7 to 12 dB. The filter isolation was about 30 dB in transmission configuration, and approximately 20 dB in reflection configuration. The bandwidth of the filters by the FWHM level was about 0.25 nm, which made them applicable to the DWDM standard.

2.3   Demonstration of Switching Time of the Filter

Fig. 5 presents experimental data that demonstrate the switching time of the filter. The left-hand figure shows transfer functions of the filter for channel A and channel B. The wavelength shift is approximately 0.25 nm, which corresponds to the shift per one spectral channel in the standard frequency grid of DWDM. At the next step, the operating laser was tuned to the wavelength corresponding to the transfer function maximum of channel A. Then the control voltage was applied to the electrodes, and the transfer function of the filter was switched to channel B. From the oscilloscope traces shown in the right-hand figure, the switching time of the filter can be estimated to be not more than 1 µs and was limited by available electronic filter driver.

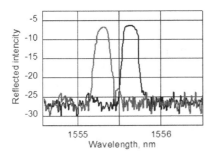 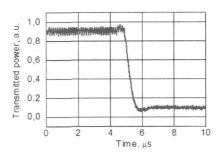

Fig. 5. Demonstration of switching time. The transfer function has been switched from channel A to channel B. The operating laser has been adjusted to the central wavelength of channel A.

2.4   Demonstration of Throughput

We also investigated the informational properties of the filter. Fig. 6 presents an "eye diagram" for the throughput 10.7 Gbit/s.

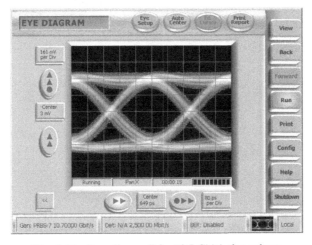

Fig. 6. The "eye diagram" for 10.7 Gbit/s throughput

3   Discussion

We demonstrated an electrically controllable filter with appreciably reduced polarization dependent noise as compared with similar filters based on electrooptic materials. The filter we suggest has a bandwidth of about 0.25 nm and time of switching from one spectral channel to the other of not more than 1 µs. These filters and devices based on them can be of interest for not only optical communication lines but also for development of optical coders/decoders and narrow-band rapidly tunable optical sources

References

1. Petrov, V.M., Denz, C., Chamray, A.V., Petrov, M.P., Tschudi, T.: Electrically controlled volume LiNbO3 holograms for wavelength demultiplexing systems. Optical Materials 18, 191–194 (2001)
2. Petrov, V.M., Karaboue, C., Petter, J., Tschudi, T., Bryksin, V.V., Petrov, M.P.: A dynamic narrow-band tunable optical filter. Appl. Phys. B 76, 41–44 (2003)
3. Petrov, V.M., Lichtenberg, S., Petter, J., Tschudi, T., Chamrai, A.V., Bryksin, V.V., Petrov, M.P.: Optical on-line controllable filters based on photorefractive crystals. J. Opt. A. Pure and Appl. Opt. 5, 471–476 (2003)
4. Arora, P., Petrov, V.M., Petter, J., Tschudi, T.: Integrated optical Bragg filter with fast electrically controllable transfer function. Opt. Comm. 281(N8), 2067–2072 (2008)
5. Greshnov, A.A., Lebedev, V.V., Shamrai, A.V.: High-frequency modulation of light in diffraction by the Bragg grating with a refractive index traveling wave. Tech. Phys. 57, 1219–1224 (2012)
6. Breer, S., Buse, K.: Wavelenght demultiplexing with volume phase holograms in photorefractive lithium niobate. Appl. Phys. B 66, 339–344 (1998)
7. Il'ichev, I.V., Toguzov, P.V., Shamray, A.V.: Plasmon-polariton polarizers on the surface of single-mode channel optical waveguides in lithium niobate. Tech. Phys. Lett. 35, 831–833 (2009)

Investigation of Analog Photonics Based Broadband Beamforming System for Receiving Antenna Array

Alexander P. Lavrov, Sergey I. Ivanov, and Igor I. Saenko

St-Petersburg State Polytechnical University, St-Petersburg, Russia
{lavrov,s.ivanov,saenko}@cef.spbstu.ru

Abstract. An approach to broadband beamforming for receiving microwave antenna arrays based on photonic true time delays (TTD) is considered. Many architectures have been proposed earlier for the implementation of TTD with fiber optic components usage to provide control of one- or multibeam antenna array pattern. Some beamforming arrangements can currently provide the required TTD capabilities by using the units and elements available at the market of modern components of fiber-optical telecommunication systems. The essential parameters of accessible analog fiber optic link main components are considered. The technique is developed and calculations of the main performance data of the chosen beamforming scheme – the signal/noise ratio and dynamic range are carried out. Influence of the scheme components' parameters and their operation modes on the estimates of the beamforming scheme main performance characteristics is analysed.

Keywords: phased array antenna, beamforming, photonic true time delay, fiber optic link, dynamic range, signal to noise ratio.

1 Introduction

It is well known that beamformers for the conventional phased array antennas (PAA) based on electrical phase shifters have the limited performance bandwidth due to the beam squint problem [1,2]. Meanwhile for many applications (e.g. in modern radars) it is highly desirable that the phase array antennas can operate in a broad band (some octaves or even decades). An effective solution for this problem is to use true-time delay (TTD) beamforming. During the last years TTD beamforming based on photonic technologies has been extensively investigated due to its advanced features: extremely wide bandwidths, fast switching, light weight and immunity to electromagnetic interference [3,4,5]. Many architectures have been proposed for the implementation of TTD with fiber optic components usage to provide control of one- or multibeam antenna array pattern. Most of them require the specially developed components, such as chirped fiber Bragg gratings, fast tunable lasers with narrow spectral linewidths, reflection fiber segments, filter arrays and so on. However, some beamforming arrangements can currently provide the required TTD capabilities by using the units and elements available at the market of modern components of fiber-optical telecommunication systems.

S. Balandin et al. (Eds.): NEW2AN/ruSMART 2014, LNCS 8638, pp. 647–655, 2014.

For receiving antenna array a key problem is whether photonic beamformer can satisfy the noise figure and dynamic range demands. So the fiber link components that comprise the specific beamforming scheme need to be analysed from this point of view.

The goal of this paper is to analyse the essential parameters of accessible analog fiber optic link main components and to estimate the main performance characteristics of a chosen beamforming scheme (BFS). Section 2 briefly discusses the specifics of beamforming scheme to be analysed. Section 3 summarizes technique developed and calculations of the main performance data of the chosen beamforming scheme – the signal-to-noise ratio and dynamic range. Section 4 briefly discusses some results of fiber-optic link investigation. Finally, conclusions are presented in Section 5.

2   TTD Photonic Beamformer Based on Fiber Chromatic Dispersion

Among various configurations for photonic BFS with fiber optic components that we have examined, e.g. [6,7,8] the approach based on the optoelectronic switching of fiber segment's lengths to provide requisite time delays [9,10] is the most applicable and simple. This switching along with optical fiber's chromatic dispersion and WDM technique permits a stepped true time delay control of a phased array antenna pattern.

Configuration of the BFS under consideration is shown in Fig. 1. This scheme is designed for N elements line phased array antenna and uses N lasers with different (but constant), equidistant (step $\Delta\lambda$) wavelengths, the total wavelength band is $(N-1)$ $\Delta\lambda$. After independent modulation by the received microwave signals (circuit elements M) intensity-modulated optical carriers are combined into a single fiber by a multiplexer (MUX N x 1 element) and fed into a time delay unit (TDU).

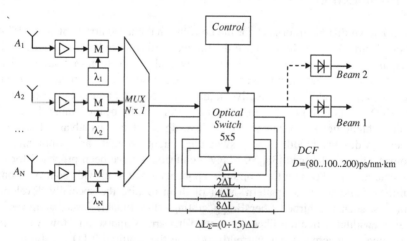

Fig. 1. Configuration of optical switched TTD beamformer with fiber chromatic dispersion based delays

TDU includes one switch n x n (Optical Switch) and (n - 1) fiber segments whose lengths vary exponentially: ΔL, 2 ΔL, 4 ΔL, etc. (The figure shows a 5x5 switch and four segments with lengths from ΔL to 8 ΔL.) Switch control allows to modify the total length of the fibers from ΔL to $(2^{n-1} - 1) \Delta L$ in increments of ΔL. As all the optical carriers share with the same light paths, the time delay differences between adjacent channels are produced by the fiber chromatic dispersion D measured in ps/(nm*km). The time delay of the optical carrier in the k-th channel relative to the 1st "reference" channel is linearly dependent on the difference between the wavelengths $(k-1) \Delta\lambda$ and the fiber length L established by the switching control, $T_k = (k-1) \Delta\lambda L D$

After passing the TDU the aggregate radiation of all wavelengths (with aligned microwave envelopes delays) is detected by the following high-speed photodetector. Photodetector output replicates the microwave signal received by a phased array antenna from the corresponding direction. Switching to another set of fibers by Optical Switch produce the new mutual delays and, accordingly, the new beam direction. Switching beam angles may be arbitrary with invariable beamsteering speed. Chromatic dispersion of standard single mode fiber SMF - 28 D is about 17 ps /(nm*km) at a wavelength of 1550 nm. By using special dispersion compensating fibers (DCF) having a large negative dispersion (D at a wavelength of 1550 nm is equal to - (100 .. 200) ps/(nm*km) or more), the length of the TDU segments can be reduced by an order.

3   Photonic BFS Modeling Results

For PAA in a receiving mode their statistic characteristics such as the output signal to noise ratio and dynamic range are of most importance. That's why the calculations have to be made to get the estimations of those parameters for chosen photonic BFS. We have developed the modeling technique and performed calculations of signal to noise ratio (SNR) at the output of some embodiments of the BFS for the linear phased array operating in the receiving mode [11]. The approach used in our calculations is based on a mathematical model considering the specific BFS, reflecting all linear and nonlinear signal conversions in different modules of radiofrequency and optical parts of the scheme in accordance with the general approach presented in [12]. In the BFS model the laser power coupled into the fiber is assumed from 1 to 20 mW and using of external integrated optical Mach-Zehnder modulators is considered. The noise sources that have been taken into account are: thermal noise of the input antenna and receiver, the instability of the laser radiation intensity RIN, intrinsic noise of microwave amplifiers and different noise components of the photodiode current [12]. In calculations we also determined BFS elements that make the main contribution to the output noise of the whole system.

Table 1 provides a summary of the BFS elements specifications used in calculation.

Fig. 2 shows the BFS transfer characteristic $P_{OUT}(P_{IN})$, where P_{IN} is an input RF harmonic signal power in each BFS channel (at modulator input). Compression point of the transfer characteristic for its 1 dB deviation from linearity is $P_{IN\,1\,dB} = 16$ dBm. Transfer coefficient for BFS (BFS transfer gain) is $G_{BF} = P_{OUT}/P_{IN}$, and it is equal to - 22.4 dB for $N = 8$, and is increased by 11 dB when the number of channels increases by 4 times (from 8 to 128).

Table 1. Specifications of BFS elements

Parameter	Value
Input RF amplifier of PAA receiver and BFS output RF amplifier	
Input RF amplifier gain G_{IN}	30 dB
Output RF amplifier gain G_{PD}	10 dB
Physical temperature of all elements T_{EL}	300 K
Instantaneous radio bandwidth Δf_{RF}	1..12 GHz
RF amplifier noise figure NF_{AMP}	3 dB
Number of channels in BFS (number of PAA elements) N	8, 32, 128
Laser	
Relative intensity noise RIN	-165 dB/Hz@4 GHz
Wavelength range λ_n	$\lambda_n \sim 1550$ nm
Optical module forming controlled time delays	
Insertion optical loss L_n (including an optical switch loss)	7 dB
Optical combiner of microwave modulated optical carriers	
WDM Nx1 multiplexer loss (per 1 channel) L_{EX_WDM}	≤8 dB

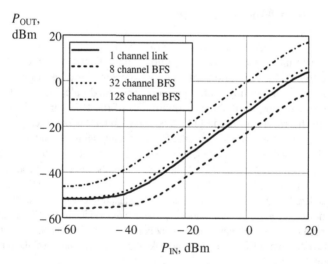

Fig. 2. BFS transfer curve for different numbers N of PAA receiving elements

Fig. 3 shows the output SNR versus input RF signal power – $SNR_{OUT}(P_{IN})$. Output noise power is calculated across total RF bandwidth. The threshold sensitivity corresponds to input power $P_{IN MIN}$ giving $SNR_{OUT} = 0$ dB, and calculations are resulted in $P_{IN MIN} = -46.5$ dBm for $N = 128$; $P_{IN MIN} = -41$ dBm for $N = 32$; $P_{IN MIN} = -33.5$ dBm for $N = 8$. Knowing the compression point and threshold sensitivity one can calculate BFS dynamic range $DR = P_{IN\,1\,dB}/P_{IN\,MIN}$: 49.5 dB for $N = 8$, 57 dB for N = 32, and 62.5 dB for $N = 128$.

When increasing number N of receiving elements in PAA the BFS output SNR and dynamic range increase, owing to incoherent addition of noise in the optical scheme. Characteristics in Fig. 2 and 3 are calculated for lasers with 10 mW power and modulators with 5 V half-wave voltage.

Note that Fig. 2 and 3 also show calculated transfer characteristic and SNR for one channel link comprising elements, which specifications corresponding to that measured for the link model (see Section 4).

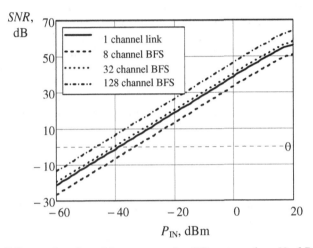

Fig. 3. Output SNR as a function of input power for different numbers N of PAA receiving elements

The calculations also show that with increasing laser power I_0 the transfer gain G_{BF}, SNR_{OUT}, threshold sensitivity and, therefore, the BFS dynamic range DR increase. However, the optical power at the photodetector after summation all channels RF modulated signals must not exceed the allowed maximum for the used photodiode type.

Results of calculation of the dynamic range, the main BFS noise characteristics and optical power budget allow to estimate the possibilities and limitations of beam forming schemes for broadband – up to 3 or more octaves – receiving PAA using modern analog microwave photonics (radiophotonics) components.

4   Measurements of Model Fiber-Optic Link Characteristics

4.1   Model Fiber-Optic Link

On the market there are many companies producing components for radio frequency over fiber links (RFoFL), but we note that some manufacturers, for example, [13,14], use direct current modulated lasers, limiting their implementation in selected BFS.

Therefore, we investigated the model of fiber-optic transmission link with laser transmitter having a wavelength corresponding ITU grid and an external intensity modulator, and measured its main characteristics - transfer gain, dynamic range and noise performance.

Investigations were carried out on the model link consisting of a transmitter module based on the DFB-laser from Lasercom [15], external integrated-optical Mach-Zehnder interferometer modulator from Optilab [16] and photodetector module (PDM) from Miteq [14]. These modules are characterized by the following parameters. Laser diode module LDI-DFB-1550-20: wavelength 1550 nm, power coupled into the fiber in continuous mode $I = 20$ mW at a current 95..110 mA, there is a photodiode for monitoring and stabilizing the laser power, temperature sensor and a Peltier element for laser temperature stabilization. The laser has a single-mode polarization-holding fiber output. Intensity modulator IM-1550-20-a: material $LiNbO_3$, the operating wavelength range 1530..1600 nm, insertion loss $\leq$ 5.0 dB, operating frequencies for RF input (S21, $\pm$ 3 dB) 0..20 GHz, $U\pi$ on RF input $\leq$ 5.7 V typ., $U\pi$ on DC input $\leq$ 10 V. PDM – from the fiber optic link MITEQ-SCML-100M18G (data given by the company are only for the whole link) frequency range 0.1..18 GHz, transimpedance amplifier in the transmitter and the receiver, dynamic range free from intermodulation products, for the frequency band 15 GHz, $SFDR = 101$ dB/Hz$^{2/3}$.

4.2   Measurements Results

Performance of a modulator combined with a FO link is largely determined by its modulation characteristic (on DC) and operating point position on it. The measured modulation characteristic is shown in Fig. 4 (note that the manufacturer does not provide this one). Two important parameters were calculated by using the approximation of measured modulation characteristic: the half-wave voltage $U\pi = 5.3$ V and phase angle $\varphi_0 = -1.11\pi/2$, showing the operating point position on the modulation characteristic for bias voltage $U = 0$. The characteristic is described by $I(U) = I_{0\,MAX} \cdot [1 + \cos(\pi U / U\pi + \varphi_0)]$ and is determined by the internal structure of the modulator – two-arm interferometer.

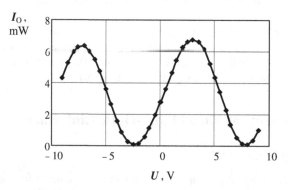

Fig. 4. Modulation characteristic of IM-1550-20-a

Fig. 4 shows the modulator output optical power with the input provided by the laser module The maximal power at the modulator input was $I_{O\,MAX}$ = 18.5 mW.

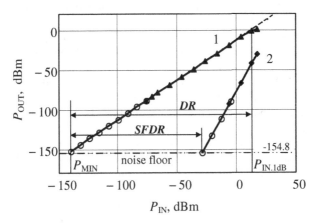

Fig. 5. Optical link output RF power versus input RF power

Fig. 5 shows the measured transfer characteristic of model link for harmonic signal with a frequency f = 2.5 GHz (line 1). Here the level of the third harmonic f_3 = 7.5 GHz (line 2), as well as noise N_{FLOOR} = -154.8 dB (noise floor) are shown also; noise floor is the PDM output reduced to a bandwidth of 1 Hz. The measured link transfer coefficient G = - 13 dB. Compression point $P_{IN.1dB}$ = 13 dBm (for the transfer characteristic 1 dB deviation from the linear dependence), and $P_{MIN} = P_{IN\,MIN}$ = - 142 dBm when line 1 intersects noise floor. So the dynamic range is estimated as

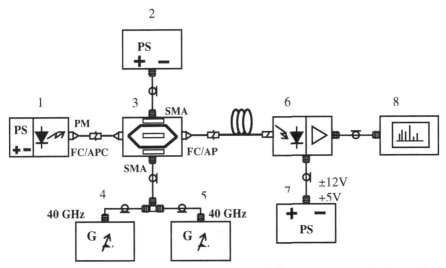

Fig. 6. Scheme for the link transfer and noise characteristics measurements. 1 - laser diode module LDI-DFB-1550-20/80; 2 and 7 - DC power supply sources; 3 - modulator IM-1550-20-a; 4 and 5 - RF signal generators Rohde & Schwarz SMB 100A; 6 - photodetector module Miteq; 8 - spectrum analyzer Antritsu MS 2726S.

$DR = P_{IN.1dB}/P_{MIN} = 13 - (- 142) = 155$ dB. The dynamic range free of input signal harmonics, $SFDR = (- 29$ dBm$) - P_{MIN} = 113$ dB.

Measurements were carried out at the test installation shown in Fig. 6, and modulator bias voltage was $U = 0$ V.

Noise figure NF for FO link is calculated by the formula [12]: $NF = N_{FLOOR} - 10\log (kT) - 30 - G$, where N_{FLOOR} – noise level of link output, in dBm/Hz; k – Boltzmann constant; T – temperature of 300 K; G – RF power transmission gain from input to output of FO link. Above measured data give $NF = 32.5$ dB.

Fig. 7 shows the measured PSD at the PDM output as a function of input unmodulated optical power I_O: overall level (line 1) and the level caused by photodetector loading optical power(line 2): shot-noise component and relative intensity noise (laser RIN) component. Measurements were made near the frequency 4 GHz. Here position of the working point on the IM-1550-20-a modulation characteristics is displayed (line 3) and the spectrum analyzer noise floor also.

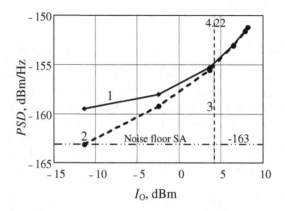

Fig. 7. Output RF noise power spectral density as a function of optical power at PDM input

Made calculations of the basic performances of our model fiber-optic link showed satisfactory agreement with the experimental values.

The results of measuring the link transfer and noise characteristics are used to calculate the basic performance of the BFS for receiving PAA in a manner we have considered in Section 3 above. For further studies of the optical BFS and engineering of its operating sample one can use components of radio frequency over fiber links using DWDM technology.

5   Conclusions

We have investigated the essential parameters of accessible analog fiber optic link main components with the view of their application in true time delay based broadband beamforming system for receiving microwave phased array antenna. For future development the BFS based on fiber optic dispersion and DWDM technology is proposed. The technique is developed and calculations of the main performance data of the chosen beamforming scheme – the signal-to-noise ratio and dynamic range are carried out. The results obtained permits to determine the parameters and operation modes of microwave fiber link's components limiting the photonic BFS performance characteristics.

References

1. Yao, J.P.: A Tutorial on Microwave Photonics – Part II. IEEE Photonics Society Newsletter 26(3), 5–12 (2012)
2. Mitchella, M., Howarda, R., Tarran, C.: Adaptive digital beamforming (ADBF) architecture for wideband phased array radars. Proc. SPIE 3704, 36–47 (1999)
3. Chazelas, J., Dolfi, D., Tonda, S.: Optical Beamforming Networks for Radars & Electronic Warfare Applications. RTO-EN-028 9, 1–14 (2003)
4. Levine, A.M.: Fiber Optics For Radar And Data Systems. Laser and Fiber Optics Communications. Proc. SPIE 0150, 185 (1978)
5. Raz, O., Barzilay, S., Rotman, R., Tur, M.: Submicrosecond Scan-Angle Switching Photonic Beamformer With Flat RF Response in the C and X Bands. Journal of Lightwave Technology 26(15), 2774–2881 (2008)
6. Soref, R.: Optical dispersion technique for time-delay beam steering. Appl. Opt. 31, 7395–7397 (1992)
7. Goutzoulis, A.P., Zomp, J.M.: Development and field demonstration of an eight-element receive wavelength-multiplexed true-time-delay steering system. Appl. Opt. 36, 7315–7326 (1997)
8. Kaman, V., Zheng, X., Helkey, R.J., Pusarla, C., Bowers, J.E.: A 32-Element 8-Bit Photonic True-Time-Delay System Based on a 288x288 3-D MEMS Optical Switch. IEEE Photonics Technology Letters 15, 6 (2003)
9. Fathpour, S., Riza, N.A.: Silicon-photonics-based wideband radar beamforming: basic design. Optical Engineering 49(1), 018201 (2010)
10. Yang, Y., Dong, Y., Liu, D., He, H., Jin, Y., Hu, W.: A 7-bit photonic true-time-delay system based on an 8×8 MOEMS optical switch. Chinese Optics Letters, 7 7(2), 118–120 (2009)
11. Ivanov, S.I., Lavrov, A.P., Saenko, I.I.: Dynamic range and S/N ratio in microwave PAA beamforming systems based on analog photonics elements. In: Proceedings. Vserossiiskaya Konferentsiya po Fotonike i Informatsionnoi Optike. Sbornik Nauchnykh Trudov, pp. 195–196. Moscow, MEPhI (2014); (Иванов С.И., Лавров А.П., Саенко И.И. Динамический диапазон и отношение сигнал/шум диаграммообразующих систем микроволновых ФАР на базе элементов аналоговой фотоники // Сб. научных трудов III Всероссийской конференции по фотонике и информационной оптике. 2014, Москва.- М.: Изд-во МИФИ, 2014.- С.195-196.)
12. Froberg, N.M., Ackerman, E.I., Cox, C.: Analysis of Signal to Noise Ratio in Photonic Beamformers. IEEE Aerospace Conference, Paper 1067 (2006)
13. Ivanov, A.V., Isaev, D.S., Kurnosov, V.D., Kurnosov, K.V., Simakov, V.A., Chernov, R.V.: Investigation of the noise characteristics and dynamic range of the POM-27 and PROM-15 set and also POM-27 and PROM-15 with preamplifier set. In: Ivanov, A.V., Isaev, D.S., Kurnosov, V.D., Kurnosov, K.V., Simakov, V.A., Chernov, R.V. (eds.) 21-st Int. Crimean Conf. Microwave & Telecommunication Technology, CriMiCo 2011, Sevastopol, pp. 357–358 (2011)
14. 18 GHz SCM Fiber Optic Link (Miteq), http://www.miteq.com/docs/MITEQ-SCM_-18G.PDF
15. Hybrid laser module LDI-DFB-1550-20/80 (LasersCom), http://www.laserscom.com/#gibridnye-moduli/cjep
16. Intensity Modulator IM-1550-20-a (Optilab), http://www.optilab.com/images/datasheets/IM-1550.pdf

Ultimate Information Capacity
of a Volume Photosensitive Media

Yuriy I. Kuzmin and Viktor M. Petrov

St.-Petersburg State Polytechnical University, Polytechnicheskaya Str, 29,
St.-Petersburg, 195251, Russia
iourk@yandex.ru, vikpetroff@mail.ru

Abstract. The ultimate information capacity of a volume photosensitive media for the case of an optimal use of the dynamic range of, number of pages, the readout conditions is considered. The volume hologram is regarded as an object of the information theory. For the first time the formalism of the reciprocal lattice has been introduced in order to estimate the informational properties of the hologram. The diffraction-limited holographic recording is analyzed in the framework of the reciprocal lattice formalism. Calculations of the information capacity of a three-dimensional hologram involve analysis of a set of multiplexed holograms, each of which has a finite signal-to-noise ratio determined by the dynamic range of the holographic medium and the geometry of recording and readout. An optimal number of pages that provides a maximum information capacity at angular multiplexing is estimated.

Keywords: information capacity, three-dimensional holograms.

1   Theory

Information capacity and throughput of different optical objects, including holograms, is of vital importance for optical information processing and storage systems [1- 5]. Here we analyze the ultimate information capacity of a three-dimensional (volume) hologram for the case of an optimal use of the dynamic range of the storage medium.

Information capacity is directly related to the physical properties of the optical media. It can be defined as the maximum amount of information that can be recorded and then read out with an arbitrarily small probability of error. According to the Kotelnikov-Shannon sampling theorem, the information recorded in a hologram is fully determined by $4A\Delta^2W$ pixels, where A is the hologram cross section, and Δ^2W is the two-dimensional width of the spectrum of recorded spatial frequencies. The factor $4A\Delta^2W$ is a two-dimensional analog of the Nyquist frequency. The upper limit of information capacity of a three-dimensional and two-dimensional hologram can be found by the analogy with the Shannon formula for the channel capacity of a communication link in the presence of white noise [6-9]

$$C_{3D} = 4A\Delta^2WN \log_2\left(1 + R_{3D}\left(\Delta^2W, N\right)\right), \text{ [bit]} \tag{1}$$

S. Balandin et al. (Eds.): NEW2AN/ruSMART 2014, LNCS 8638, pp. 656–661, 2014.

$$C_{2D} = 4A\Delta^2 W \log_2\left(1 + R_{2D}\left(\Delta^2 W\right)\right), \text{ [bit]} \tag{2}$$

where N is the number of multiplexed holograms (pages), $R = P_s/P_n$ is the signal-to-noise ratio at readout of one pixel, P_s is the upper boundary of the average power of the image-forming signal, and P_n is the average power of optical noises.

Let us find the maximum number of pixels that can be recorded in a photosensitive media as a volume hologram in the case only diffraction limitations exist. We assume that the elementary holographic grating is a spatial distribution of the recorded physical parameter invariant with respect to the translation of the type

$$T_{3D} = n_1 e_1 + v_2 e_2 + v_3 e_3, \quad \forall n_1 \in \mathfrak{S}, \quad \forall v_2, v_3 \in \mathfrak{R} \tag{3}$$

$$T_{2D} = n_1 e_1 + v_2 e_2 + v_0 e_3, \quad \forall n_1 \in \mathfrak{S}, \quad \forall v_2 \in \mathfrak{R} \tag{4}$$

for the three- and two-dimensional holograms, respectively; where e_i are the basis vectors of translation, v_0 is the grating plane coordinate, $\mathfrak{S}$ and $\mathfrak{R}$ are the sets of integer and real numbers. In the three-dimensional space the vector of translation (3) describes the set of parallel planes $T_{3D} \in \mathfrak{S} \otimes \mathfrak{R}^2$ (Fig. 1,a), the vector in (4) describes the set of collinear and complanar lines $T_{2D} \in \mathfrak{S} \otimes \mathfrak{R}$ (Fig. 1, c)

The reciprocal lattices corresponding to translations (3) and (4) in the k space are

$$Q_{3D} = m^1 q^1, \quad \forall m^1 \in \mathfrak{S} \tag{5}$$

$$Q_{2D} = m^1 q^1 + \mu^0 q^2 + \mu^3 q^3, \quad \forall m^1 \in \mathfrak{S}, \quad \forall \mu^3 \in \mathfrak{R} \tag{6}$$

where μ^0 is the coordinate of the reciprocal grating plane, q^j are the basis vectors of the reciprocal grating satisfying the orthogonality relation $e_i \cdot q^j = 2\pi\delta_i^j$, where δ_i^j is the Kronecker symbol.

In the k space the reciprocal grating (5) is a set of equidistant points $Q_{3D} \in \mathfrak{S}$ (Fig.1,b) and the reciprocal grating (6) is a set of collinear and complanar lines $Q_{2D} \in \mathfrak{S} \otimes \mathfrak{R}$ the orientation of which is defined by the orthogonality relation (Fig. 1, d).

When information is readout from a hologram, the orientation of the reconstructed beam is determined by the points of intersection of the Ewald sphere (the radius of which is equal to the readout light wave vector k_0) and the reciprocal grating. The scattering vector coincides in this case with the vector of reciprocal grating (5) or (6), as shown in Fig. 1. Independent states in the k space at the Ewald sphere correspond to diffraction-resolvable Fourier components of the reconstructed image. Therefore, the maximum number of pixels that can be recorded in the hologram of any dimension is equal to the number of states at 1/2 of the Ewald sphere surface

$$\sup\left(4A\Delta^2 W\right) = \frac{1}{2}\left(\frac{4\pi k_0^2}{\Delta^2 k}\right) = 8\pi\frac{A}{\lambda^2} \tag{7}$$

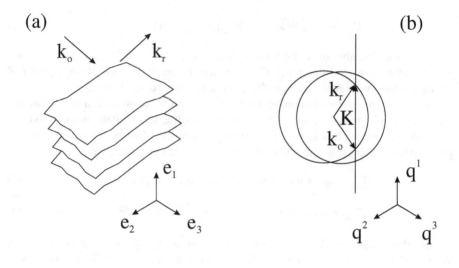

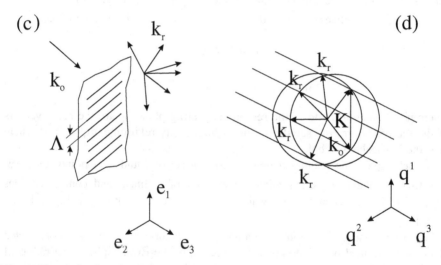

Fig. 1. Translation-invarient sets and reciprocal gratings for optical diffraction from three-dimensional (a,b) and two-dimensional (c,d) holographic gratings: (a, c) - the coordinate space, (b,d) - the k space; K is the scattering vector (shown only for scattering "forward"), k_0 and k_r are the wave vectors of the readout and reconstructed beams, respectively, $\Lambda = 2\pi/K$ – the grating period

where $\Delta^2 k = \pi/A$ is the square of the minimum uncertainty of the wave vector in the diffraction limit of resolution, and λ is the readout light wavelength. The calculations of the states at 1/2 of the Ewald sphere surface corresponds to summation over spatial frequencies in the limits of the entire Fourier plane. Expression (7) is written in for the case when there is no polarization multiplexing. If it is taken into account, the results should be doubled.

The maximum number of pages recorded in a three-dimensional hologram at angular multiplexing can be calculated by summation over all vectors of the reciprocal grating sitting on the Ewald sphere (i.e., over all the wave vectors of the recorded holographic gratings): $\max(N) = 2k/\Delta k = 4L\lambda$, where $\Delta k = \pi/L$ is the minimum uncertainty of the wave vector, and L is the hologram thickness. Taking into account (7), it is easy to find the maximum number of pixels that potentially could be recorded at all pages of a three-dimensional hologram in the case of an unlimited dynamic range

$$\sup\!\left(4A\Delta^2 W\right)\max(N) = 32\pi A\, L/\lambda^3 \tag{8}$$

Estimates of the type of "volume"$/\lambda^3$ are often given for the ultimate information capacity of a hologram [1,2,5]; but an unjustified assumption of information storage in the form of stored elements of volume ("voxels" [8]) is frequently made in this case, and the two-dimensionality of the spectrum of spatial frequencies of the recorded image is ignored. The number of multiplexed holograms is determined by the finite dynamic range of the holographic medium on which the signal-to-noise ratio depends. This is the reason, while the estimation (8) practically is unachievable. At multiplexing, the information capacity does not grow in N times, as it could be concluded from a shallow analysis of expressions (1) and (8) that did not take into account the $R_3D = R_3D\,(N)$ dependence.

Let us consider now, how the number of pages affects the information capacity. In the case of a two-dimensional hologram the entire dynamic range is used to code each pixel with the maximum word length. For a three-dimensional hologram an exchange of the word length on the number of pages in the limits of the same dynamic range is possible. An increase in the number of pages is achieved by decreasing R_{3D} up to the word length of one bit per pixel. Let us show that there is an optimal number of pages at which the information capacity is the highest. The number of multiplexed holograms can be presented in the form $N = \sqrt{P_s(1)}/\sqrt{P_s(N)}$, where $P_s(1)$ is the maximum signal power at recording of only one page for which the entire dynamic range is used, $P_s(N)$ is the signal power at recording of one page from N multiplexed pages. Now we can relate $R_{3D}(N)$ and $R_{3D}(1) = \max R_{3D}(N)$

$$R_{3D}(N) \equiv \frac{P_s(N)}{P_n} = \frac{P_s(1)}{N^2 P_n} = \frac{P_{3D}(1)}{N^2}.$$

Under the condition $R_{3D}(1) \gg N^2$ expression (1) acquires the form

$$C_{3D}(N) = N C_{3D}(1) - 8A\Delta^2 W N \log_2 N \tag{9}$$

This function has a maximum a

$$N_0 = 2\!\left(\frac{C_{3D}(1)}{8A\Delta^2 W} - \frac{1}{\ln 2}\right) \tag{10}$$

Therefore, there is an optimal number of pages $N_{opt} = entier(N_0)$ above which the information capacity will decrease because of a reduction in the signal-to-noise ratio $R_{3D}(N)$.

It is interesting to note that in the case of a sufficiently high $R_{3D}(1)$ the information capacity $C_{3D}(N)$ of the three-dimensional hologram in which N pages are recorded is lower than the information capacity $NC_{3D}(1)$ of the set consisting of N holograms in each of which only one page is recorded, all other conditions being equal. As follows from Eq. (9), the difference in the information capacity per one pixel is described by the function

$$L(N) \equiv \frac{NC_{3D}(1) - C_{3D}(N)}{4A\Delta^2 W} = 2N \log_2 N, \quad \text{[bit]} \quad (11)$$

The numerical calculations are shown in Fig.2.

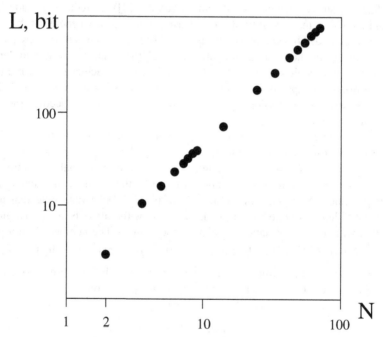

Fig. 2. Function L(N) versus number of multiplexed holograms

2   Conclusion

To summarize, the hologram was regarded in our study as an object of the information theory. Calculations of the information capacity of a three-dimensional hologram involved analysis of the set of multiplexed holograms, each of which had a finite signal-to-noise ratio determined by the dynamic range of the storage medium. The problem of an optimal use of the dynamic range at angular multiplexing was solved. Analysis of diffraction-limited holographic information recording was carried out in the framework of the reciprocal grating formalism, which allowed us to use such a basic property of an optical image as two-dimensionality of the spectrum of its spatial frequencies to a full extent.

References

1. Van Heerden, P.J.: Theory of Optical Information Storage in Solids. Appl.Opt. 2(4), 393–400 (1963)
2. Ramberg, R.G.: Holographic Information Storage. RCA Rev. 33, 5–52 (1972)
3. Tiemann, M., Schmidt, M., Petrov, V.M., Petter, J., Tschudi, T.: Information throughput of photorefractive spatial solitons in the telecommunication range. Applied Optics 46(N14), 2683–2687 (2007)
4. Petrov, M.P., Stepanov, S.I., Khomenko, A.V.: Photorefractive crystals in coherent optical systems, 275 p. Springer (1991)
5. Wullert, J., Lu, Y.: Limits of the Capacity and Density of Holographie Storage. Appl. Opt. 33(11), 2192–2196 (1994)
6. Shannon, C.E.: A Mathematical Theory of Communication. Bell Syst. Tech. J. 27, 379-423, 623–656 (1948)
7. Shannon, C.E.: Communication in the Presence of Noise. Proc. IRE 37(1), 10–21 (1949)
8. Fellgett, P.B., Linfoot, E.H.: On the Assessment of Optical Images. Phil. Trans. Roy. Soc. A.247, 369–407 (1955)
9. Brady, D., Psaltis, D.: Control of Volume Holograms. J. Opt. Soc. Amer. A. 9(7), 1169–1182 (1992)

Transmitting of a Full-TV Image
by a Pulsed One-Dimensional Serial Optical
Frames: A Non-linear Approach

Constantine Korikov[1], Yuri Mokrushin[1], and Roman Kiyan[2]

[1] St. Petersburg State Polytechnical University, St. Petersburg, Russia
{korikov.constntine,yuri.mokrushin}@spbstu.ru
[2] Laser Zentrum Hannover, Hannover, Germany
r.kiyan@lzh.de

Abstract. A non-linear case of the transmitting of the TV-image by a pulsed laser light has been considered. An analytical solution for the diffracted light field for the practically important geometry of scattering corresponding to the wideband anisotropic light diffraction from a slow shear elastic wave propagating in the [110] direction in the TeO$_2$ crystal near its optical axis has been obtained. The third-order of nonlinear interaction of the first order diffracted light has been demonstrated. The possibility to control the amplitude distortions produced by nonlinearities were discussed and experimentally demonstrated. The optimal geometry for the image transmission has been presented and experimentally investigated.

Keywords: optical informational systems, acousto-optics.

1 Introduction

Image transmission using laser acousto-optic systems are well-known for many informational applications [1-4]. Owing to a fast and exact control of the direction of laser radiation propagation, acousto-optic systems attract more and more attention of designers of precision optical devices for such applications as, for example, development systems for two-photon fabrication in nano-scale [5] and control displacement of macroobjects under a light pressure and Casimir force [6]. Of particular interest for the solution of the problems indicated above are acousto-optic systems that transmit two-dimensional images. It can be obtained by two methods (Fig.1). In the first case (Fig.1,a) the image is built in the form of a two-dimensional array of $N \times M$ independent pixels. To this end, two acousto-optic cells each of which scans the beam along a proper axis are used. The readout is carried out by a single light pulse. Therefore, one light pulse forms one pixel in the image. In this method, $N \times M$ light pulses are needed to obtain a complete two-dimensional image.

In the second case (Fig.1,b) the image is built in the form of lines. Each of the lines contains N elements of the image which, strictly speaking, are not

S. Balandin et al. (Eds.): NEW2AN/ruSMART 2014, LNCS 8638, pp. 662–675, 2014.

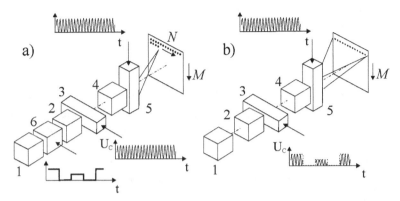

Fig. 1. Transmitting of a two-dimensional array of MxN pixels. a): the array of MxN independent pixels, b) the array of M lines (1-D frames), each line contains N resolved pixels. 1- pulse laser, 2, 4 beam forming systems, 3 acousto-optical modulator: provides the scanning of the beam along N (case "a"), and provides the modulation of the line (case "b"), 5 acousto-optic modulator, provides the scanning along M (cases "a" and "b" both), 6 amplitude modulator (Sometimes use the galvanometric scanner instead of acousto-optical modulator).

independent. One of the acousto-optic cells moves the beam along a proper coordinate, thereby playing the role of frame scan in a TV receiver. The second cell plays the role of horizontal scan. The radio-frequency signal corresponding to the light distribution along the line is launched into this cell. In this method of image formation one light pulse forms a line in the image. To obtain an entirely two-dimensional image, light pulses are needed.

In order to take full advantage of both methods and achieve the images that will comply with the most strict requirements, it is necessary to take into consideration the nonlinearities caused by interaction of the readout light with the sound wave and a pulsed mode of readout.

2   Geometry of Efficient Transmission

Let us consider a homogeneous infinite anisotropic nonmagnetic nonconducting medium with a weak spatial dispersion. The material equation for such a medium is

$$D_i = \left(\epsilon_{ij}^0 + i\gamma_{ijk}k_i\right)E_i \tag{1}$$

where $\mathbf{D}$ is the displacement vector; $\mathbf{E}$ is the electric field vector; $\hat{\epsilon}^0$ is permeability tensor of the medium; $\hat{\gamma}$ is the third-rank tensor (antisymmetric with respect to the first two indexes) that characterizes the gyrotropy of the medium; and $\mathbf{k}$ is the wave vector of the electromagnetic field.

Let us determine in this medium the coordinate system $x_1 = X$, $x_2 = Y$, $x_3 = Z$ related to the crystallophysical system x_1^0, x_2^0, x_3^0 by the transfer matrix $\alpha_{ij} : x_i = \alpha_{ij} \times x_j^0$. In the general case, two elliptically polarized plane light waves can propagate in the $X0Z$ plane of this medium (it will be referred to as the scattering plane) in direction m

$$\mathbf{E}_{k_d} = \mathbf{U}_{k_d} e^{i(\mathbf{k}_d \cdot \mathbf{r} - \omega t)} \tag{2}$$

where $\mathbf{k}_d \cdot \mathbf{r} = \frac{2\pi}{\lambda_0} n_d (z \cos\theta - x \sin\theta)$, $d = 1,2$; $\mathbf{r}$ is the radius-vector, λ_0 is the light wavelength in vacuum; ω is the light wave frequency; $\mathbf{U}_{k_d}$ is the vector of the light wave amplitude; n_d are the refractive indexes of the medium for two elliptically polarized waves propagating in direction $\mathbf{m}$, θ is the angle between $\mathbf{m}$ and Z

We consider in this medium volume V limited by plane surfaces $S_1(z = -\frac{L}{2})$ and $S_2(z = \frac{L}{2})$ having infinite sizes along X and Y (Fig. 2).

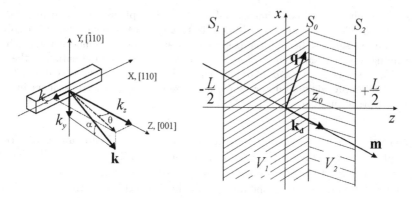

Fig. 2. The geometry of acousto-optical interaction

Let as consider an ultrasonic wave

$$u(\mathbf{r}, t) = A(\mathbf{r}' - \nu t) \mathbf{u}^0 e^{i(\mathbf{q} \cdot \mathbf{r} - \Omega t)} \tag{3}$$

which propagates at $\left| x \le \frac{L}{2} \right|$ in the direction $\mathbf{r}'$. Here, $u(\mathbf{r}, t)$ is the displacement vector; $A(\mathbf{r}' - \nu t)$ is the amplitude distribution that moves in the crystal with velocity ν and the law of variation of which is controlled externally; $\mathbf{u}^0$ is the polarization vector that characterizes the direction of local displacements of particles in the medium; $\mathbf{q}$ is the wave vector of the elastic wave; and $\Omega = 2\pi f$ is the circular frequency.

This wave causes deformation, which in its turn leads to a change in the components of the dielectric impermeability tensor of the medium given by

$$\Delta \eta_{fp} = p_{ijfp} \xi_{fp} = \xi p_{ijfp} \xi_{fp}^0 \tag{4}$$

where

$$\xi = A(\mathbf{r}' - \nu t) e^{i(\mathbf{q} \cdot \mathbf{r} - \Omega t)} \tag{5}$$

$$\xi_{fp}^0 = \frac{1}{2} \left[\frac{\partial u_f^0}{\partial x_p} + \frac{\partial u_p^0}{\partial x_f} \right] = i q_f u_p^0 \tag{6}$$

p_{ijfp} and ξ_{fp}^0 are the components of tensors of photoelasticity $\hat{p}$ and deformation $\hat{\xi}$ in the coordinate system chosen.

If we neglect the acoustic nonlinearity of interaction, the perturbed tensor of dielectric permeability of the medium $\hat{\epsilon}'$ under the conditions $\epsilon_{ik}\Delta\eta_{kj} << 1$ and $\gamma_{ijl} << 1$ is

$$\hat{\epsilon}' = \epsilon' - \hat{\epsilon} \cdot \Delta\hat{\eta} \cdot \hat{\eta} \approx \hat{\epsilon} - \hat{\epsilon}^0 \cdot \Delta\hat{\eta} \cdot \hat{\epsilon}^0 \tag{7}$$

where $\hat{\epsilon} = \hat{\epsilon}^0 + i\hat{\gamma}k$ is the nonperturbed dielectric permeability tensor of the medium in the absence of gyrotropy.

Let us write Maxwells equations for the electromagnetic field in the anisotropic nonmagnetic nonconducting medium that exhibits gyrotropy in the presence of perturbation $\Delta\hat{\eta}$ under under the following assumptions: $\Omega << \omega, \epsilon_{ik}\Delta\eta_{kj} << 1$ and $\gamma_{ijl} << 1$. Then we have

$$\begin{cases} \text{rot}\mathbf{E} = \frac{i\omega}{c}\mathbf{H} \\ \text{rot}\mathbf{H} = -\frac{i\omega}{c}\hat{\epsilon} \cdot \mathbf{E} + \frac{i\omega}{c}\hat{\epsilon}^0 \cdot \Delta\hat{\eta} \cdot \hat{\epsilon}^0 \cdot \mathbf{E} \end{cases} \tag{8}$$

We introduce equivalent currents

$$\frac{4\pi}{c}j_{\text{eq}} = \begin{cases} \frac{i\omega}{c}\hat{\epsilon}^0 \cdot \Delta\hat{\eta} \cdot \hat{\epsilon}^0 \cdot \mathbf{E}, |z| \leq \frac{L}{2} \\ 0, |z| > \frac{L}{2} \end{cases} \tag{9}$$

into the consideration and reduce the solution of the Maxwells equations to the solution of the integral equation of the form

$$\mathbf{E} = \mathbf{E}_i + Q(j_{\text{eq}}) = E_i + Q\left(\frac{i\omega}{c}\hat{\epsilon}^0 \cdot \Delta\hat{\eta} \cdot \hat{\epsilon}^0 \cdot \mathbf{E}\right) \tag{10}$$

where $\mathbf{E}_i$ is the incident light wave, $i = 1, 2$, and Q is an integral operator.

The derivation of the integral equation for the diffracted field is based on the application of the Lorentz lemma to volumes V_1 and V_2 (see Fig. 1). For a gyrotropic medium this lemma can be written as

$$\oint_S (E_1 \times H_2^* + E_2^* \times H_1)\, ndS = -\frac{4\pi}{c}\oint_V (j_2^* E_1 + j_1 E_2^*)\, dV \tag{11}$$

where V is the volume occupied by equivalent currents; S is the surface that limits this volume; n is the external normal to the surface S; $E_{1,2}$, $H_{1,2}$ are the fields satisfying the Maxwells equations with extrinsic currents $j_{1,2}$; and $(*)$ is the complex conjugation sign. The final form of the integral equation for the diffracted field has the form:

$$E = E_i - \sum_{d=1}^{2}\frac{i\omega}{4\pi}\int_{k_{d_z}>0}\frac{U_{k_d}e^{i(k_d r - \omega t)}}{N_{k_d}}\left[\lim_{x_0 \to \infty}\int_{-\frac{L}{2}}^{z_0}\int_{-x_0}^{x_0}\left(\hat{\epsilon}^0\hat{\rho}\hat{\xi}\hat{\epsilon}^0 E\right)U_{k_d}^* e^{-i(k_d - \omega t)}\, dx\, dz\right]dk_{d_x} \tag{12}$$

The particular form of integral equation (12) depends on the choice of the geometry of acoustooptic interaction, i.e., on the transfer matrix α_{ij} interacting light and sound waves. To find the solution of integral equation (12), we use the method of successive approximations

$$E = \sum_{n=0}^{\infty} E^n \tag{13}$$

We assume that the incident wave in the scattering plane is given by

$$E^{(0)} = E_i = U_{d_0} e^{i(k_{d_0} r - \omega t)} = U_{d_0} e^{4i\left[\frac{2\pi}{\lambda_0} n_d^0 (z \cos\theta_0 - x \sin\theta_0) - \omega t\right]}, d_0 = 1, 2 \tag{14}$$

This wave is taken to be a zero approximation. For the n-th approximation, corresponding transformations yield

$$E^{(n)} = \sum_{d_0} \sum_{d_1} \cdots \sum_{d_n} \left(-\frac{i}{2}\right)^2 \int_{k_{d_n}} U_{d_n} e^{i(k_{d_n} r - \omega t)} \int_{k_{d_{n-1}}} \cdots \int_{k_1} \prod_{m=0}^{n-1} \chi_{d_{m+1} d_m}^{(m)} \times \tag{15}$$

$$\times e^{-i\left(\Delta k_{d_x}^{m+1} \cdot q_x\right)\nu t} F\left(\Delta k_{d_x}^{m+1} \cdot q_x\right) \int_{-1}^{z'_{m+1}} e^{-i\left(\Delta k_{d_x}^{m+1} \cdot q_x\right)z'_m} dz_m \, dk_{d_{m+1}x} \tag{16}$$

Here, the following designations are used

$$\chi_{d_{m+1} d_m}^{(m)} = \frac{U_{k_{d_{m+1}}}^* \epsilon_{k_{d_{m+1}}} i p_{ijfp} \xi_{fp}^0 \epsilon_{jk_{d_m}} U_{k_{d_m}} \omega L}{4\pi N_{k_{d_{m+1}}}} \tag{17}$$

$U_{k_{d_{m+1}}}^*, U_{k_{d_m}}$ are the complex electric field amplitudes for the waves with the wave vectors $k_{d_{m+1}}$ and k_{d_m}; ϵ_{mm} are the components of the dielectric permeability tensor

$$F(\Delta k_{d_x}^{m+1} \cdot q_x) = \lim_{x_{m+1} \to 1} \int_{-x_{m+1}}^{x_{m+1}} A(r'_m) e^{-i\left(\Delta k_{d_x}^{m+1} \cdot q_x\right)k_m} \tag{18}$$

is the instantaneous Fourier spectrum of the ultrasonic perturbation $A(r'_m)$;

$$\Delta k_{d_x}^{m+1} = \frac{2\pi}{\lambda_0} \left(n_d^{m+1} \sin\theta_{m+1} - n_d^m \sin\theta_m\right) \tag{19}$$

$$\Delta k_{d_z}^{m+1} = \frac{2\pi}{\lambda_0} \left(n_d^{m+1} \cos\theta_{m+1} - n_d^m \cos\theta_m\right) \tag{20}$$

q_x, q_z are the projections of the wave vector of the elastic wave on the X and Z axes; z'_m is the current coordinate satisfying condition $-1 \leq z'_m \leq 1$.

The choice of the sign (plus) or (minus) in Eqs. (15) and (17) at a definite stage of the scattering process is dictated by fulfillment of the law of conservation of momentum for the incident and diffracted light waves. This choice is governed by the fact that real displacements in local regions of the crystal are described by the real part of Eq. (3).

According to the essence of this method, light diffraction can be thought to result from successive acts of scattering into different diffraction orders that occur in accordance with the types of the light and sound waves. Summation in Eq.(15) is carried out over all possible types of light modes with indexes dn arising at each stage of scattering.

For each m, $\chi_{d_{m+1}d_m}^{(m)}$ is the matrix of relative scattering coefficients

$$\begin{pmatrix} \chi_{11}^m & \chi_{12}^m \\ \chi_{21}^m & \chi_{22}^m \end{pmatrix} \tag{21}$$

the elements of which characterize their types of acoustooptic diffraction at frequency ω_n defined as

$$\omega_n = \omega + \kappa\Omega, \tag{22}$$

where

$$\kappa = \begin{cases} \pm 1, \pm 3, \ldots, \pm(2m+1) \text{at } n = 2m+1 \\ 0, \pm 2, \ldots, \pm 2m, \text{ at } n = 2m \end{cases}$$

$$m = 0, 1, 2, \ldots$$

It follows from expression (15) that the major contribution into the diffracted field comes from the spectral components of ultrasonic perturbation for which conditions of synchronism are fulfilled

$$\begin{cases} \Delta k_{d_x}^{m+1} \cdot q_x = 0 \\ \Delta k_{d_z}^{m+1} \cdot q_z = 0 \end{cases} \tag{23}$$

If we take into account conditions (12), the solution of (13), (15) can be described by a scattering diagram. The term $\prod_{m=0}^{n-1} \chi_{d_{m+1}d_m}^m$ defines the branch of this diagram that makes a contribution into the diffraction order considered. Estimation of the terms of series (25) for each of the branches of the scattering diagram shows that $|E_m^{(n)}| \leq M_m \frac{(\chi_m)^n}{n!}$ where M_m, χ_m are constant values.

This provides convergence of series (13). Thus, the field in the k-th diffraction order can be presented as a sum of fields that take into account contributions from each successive interaction order. Both the amplitude and polarization of these additional terms will be determined by the coefficients $\chi_{d_{m+1}d_m}^{(m)}$ depending on the choice of the scattering geometry and acoustic wave type.

3   Image Transmission by the Use Anisotropic Diffraction and Amplitude Modulation of Ultrasonic Signal

Let us assume that the ultrasonic perturbation amplitude in the crystal varies as

$$A(r) = A_0 \left[1 + m_0 \cos\left(2\pi f_0 \frac{\mathbf{r}}{\nu}\right)\right] = A_0 \left[1 + m_0 \cos\left(\Phi_0 x\right)\right] \tag{24}$$

where m_0 and f_0 are the modulation depth and frequency; ν is the velocity vector; and $\mathbf{r}$ is the direction of the ultrasonic perturbation propagation in the medium.

We also assume that at small β, i.e., the angle of deviation of the propagation direction from the X axis, the ultrasonic wave velocity changes only slightly.

One can show, that:

$$F\left(\Delta k_{1x}^1 \cdot q_x\right) = 2\pi A_0 Y\left(\beta\right)\left[\delta\left(\Delta k_{1x}^1 \cdot \frac{2\pi f}{\nu}\cos\beta\right) + \right.$$

$$\left. +\frac{m_0}{2}\delta\left(\Delta k_{1x}^1 \cdot \frac{2\pi(f-f_0)}{\nu}\cos\beta\right) + \frac{m_0}{2}\delta\left(\Delta k_{1x}^1 \cdot \frac{2\pi(f+f_0)}{\nu}\cos\beta\right)\right] \quad (25)$$

By assuming that the central spectral component of the amplitude-modulated signal obeys condition at $\alpha = 0, \beta = 0, \theta_0 = \theta_0^{(0)}, \theta_1 = \theta_0^{(1)}$ and also by taking into account only the interaction between the $(+1)$st and (0)th diffraction orders, we calculate the field in the $(+1)$st diffraction order at the ultrasonic column boundary. The calculations are carried out according to the scattering diagram shown in Fig. 3.

The horizontal arrows at the diagram characterize partial plane waves that propagate at angles θ_p^m for $m = 0, 1, 2$ and 3 (zero, first, second, and third interaction orders). The inclined arrows show possible directions of scattering of these waves under the action of ultrasonic perturbation. It follows from the diagram that in the approximation of the 3rd interaction order there are 27 possible combinations of scattering giving contribution into the diffracted field near the $(+1)$st order for a signal with three spectral components. The spectrum of this field consists of 7 components differing in angles of diffraction θ_p^3 and frequencies by

$$\Delta\Omega_p = p2\pi f_0 \quad (26)$$

where $p = -3, -2, -1, 0, +1, +2, +3$.

The diffracted field is calculated by using the general solution (12) by successively finding angles $\theta_p^{(m+1)}$ and $\beta_{p,q}^{(m+1)}$ satisfying system of equations, in which the following substitution is to be made $\theta_{m+1} = \theta_p^{(m+1)}, \theta_m = \theta_p^{(m)}, \beta_{m+1,m} = \beta_{p,q}^{(m+1)}$, where $p, q = -3, -2, -1, 0, +1, +2, +3$. The angles θ_p^{m+1} obtained at the preceding stage are used as initial angles in the calculations of angles in the next interaction order. Simultaneously with angles θ_p^{m+1}, angles $\beta_{p,q}^{(m+1)}$ and weighting factor $Y(\beta_{p,q}^{(m+1)})$ are found from Eq.(12). By summing up all the combinations of scattering shown in Fig. 3, we obtain the field in the $(+1)$st diffraction order at the output AOM aperture taking into account the approximation of the third interaction order

$$E_{+1}^3 = -i\chi_0 e^{i\left[(k_0 r)|_{\alpha=0} -\omega t+\frac{2\pi f}{\nu}(x-\nu t)+2L_0\alpha^2\right]} \times \sum_{p=-3}^{3} C_p\left(\theta_p^3 e^{i\left[p\frac{2\pi f_0}{\nu}(x-\nu t)-2L_0\gamma_p^{(3)}\alpha^2\right]}\right)$$

$$(27)$$

Where

$$C_0\left(\theta_0^{(3)}\right) = U_{k_1}\left(\theta_0^{(3)}\right)\left[K_{12}\left(\theta_0^{(1)},\theta_0^{(0)}\right) \times \sin\left(c\left[\frac{L_0}{\pi}\alpha^2\left(\gamma_0^{(1)}+\gamma_0^{(0)}\right)\right]\right) - \frac{\chi_0^2}{8}\left(S(0,0,0)+\right.\right.$$

$$\left.\left.+\frac{m_0^2}{4}\{S(0,0,+1)+S(0,0,-1)+S(0,+1,+1)+S(0,+1,0)+S(0,-1,0)+S(0,-1,-1)\}\right)\right]$$

$$(28)$$

$$C_{\pm 1}\left(\theta_{\pm 1}^{(3)}\right) = U_{k_1}\left(\theta_{\pm 1}^{(3)}\right)\frac{m_0}{2}\left[K_{12}\left(\theta_{\pm 1}^{(1)},\theta_0^{(0)}\right)\times\sin\left(c\left[\frac{L_0}{\pi}\alpha^2\left(\gamma_{\pm 1}^{(1)}+\gamma_0^{(0)}\right)\right]\right) -\right.$$

$$(29)$$

$$-\frac{\chi_0^2}{8}\left(S(\pm 1,\pm 1,\pm 1,\pm 1)+S(\pm 1,\pm 1,0)+S(\pm 1,0,0)+\right.$$

$$\left.\left.+\frac{m_0^2}{4}\{S(\pm 1,0,\pm 1)+S(\pm 1,0,*1)+S(\pm 1,\pm 2,\pm 1)\}\right)\right]\qquad(30)$$

$$C_{\pm 2}\left(\theta_{\pm 2}^{(3)}\right) = -U_{k_1}\left(\theta_{\pm 2}^{(3)}\right)\frac{\chi_0^2}{8}\frac{m_0}{2}\left[S(\pm 2,\pm 2,\pm 1)+S(\pm 2,\pm 1,\pm 1)+S(\pm 2,\pm 1,0)\right]$$

$$(31)$$

$$C_{\pm 3}\left(\theta_{\pm 3}^{(3)}\right) = -U_{k_1}\left(\theta_{\pm 2}^{(3)}\right)\frac{\chi_0^2}{8}\frac{m_0}{2}S(\pm 3,\pm 2,\pm 1)\qquad(32)$$

$$L_0 = \frac{\pi L n_0 c}{\lambda_0} = \frac{\pi L n_0\left(n_e^2-n_o^2\right)}{4\lambda_0 n_e^2}\qquad(33)$$

It is seen from Eqs. (27)-(31) that in the general case the amplitudes of the spectral components with positive and negative indexes p become unequal. The symmetry of the spectrum with respect to the direction $\theta_1 = \theta_0^{(1)} = \theta_0^{(3)}$ is violated. The shape of the spectrum envelope depends on the choice of the carrier frequency f and tuning of the central spectral component of the spectrum to the conditions of Bragg diffraction. The amplitudes of spectral components with "parasitic" modulation frequencies ($p = \pm 2, \pm 3$) grow with increasing modulation index χ^0 and exert a strong dependence on the modulation depth m0. If the modulation depth is small, the contribution of these components into the total field can be neglected. Nevertheless, in order to get a good contrast in the case of image visualization, it is necessary to employ $m_0 = 1$, and thus these components cannot be neglected at high modulation indexes. Let us to note, that the considered here case when many different orders of diffraction are propagate in one direction quite a similar to the presented in [7].

4   The Influence of Nonlinearity on the Image Transmission: Numerical and Experimental Results

Let us estimate the influence of the nonlinearity of the light diffraction on the ultrasonic wave. For this we will take into account the third order of interaction.

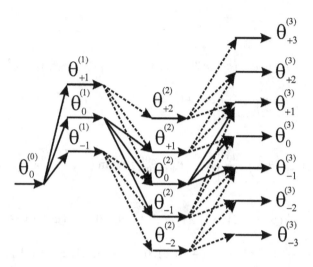

Fig. 3. Scattering diagram for light diffraction into the (+1)st order in the approximation of the third interaction order

Let us assume that the amplitude of the ultrasonic input signal can be written in the following form:

$$A(r) = A_0 \left[1 + m_0 \cos \left(2\,pi f_0 \frac{r}{\nu} \right) \right] \tag{34}$$

Now we can find the average distribution of the light intensity of the line on the screen:

$$I_1(x_1) = \tilde{\Gamma}_0 \chi^2 \left\{ C_0 + 2 \sum_{n=1}^{6} e^{-\left(\frac{\pi n f_0 \tau_0}{2\sqrt{\ln(2)}} \right)^2} \left[C_{n1} \cos \left(\frac{2\pi n f_0 x_1}{\nu} \right) + C_{n2} \sin \left(\frac{2\pi n f_0 x_1}{\nu} \right) \right] \right\} \tag{35}$$

where:

$$C_0 = \sum_{k=1}^{2} \sum_{l=-3}^{3} \left(A_{kl}^2 + B_{kl}^2 \right) \tag{36}$$

$$C_{n1} = \sum_{k=1}^{2} \sum_{l=-3}^{3} \sum_{m=-3}^{3} \left(A_{kl} A_{mk} + B_{lk} B_{mk} \delta(l - m - n) \right) \tag{37}$$

$$C_{n2} = \sum_{k=1}^{2} \sum_{l=-3}^{3} \sum_{m=-3}^{3} \left(A_{kl} A_{mk} - B_{lk} B_{mk} \delta(l - m - n) \right) \tag{38}$$

$$A_{11} = \cos\phi(\theta_1)D_1 - \rho(\theta_1)\sin\phi(\theta_1)H_1(\theta_1)$$
$$A_{11} = \cos\phi(\theta_1)H_1 + \rho(\theta_1)\sin\phi(\theta_1)D_1(\theta_1)$$
$$A_{11} = \sin\phi(\theta_1)H_1 + \rho(\theta_1)\cos\phi(\theta_1)H_1(\theta_1)$$
$$A_{11} = \sin\phi(\theta_1)H_1 - \rho(\theta_1)\cos\phi(\theta_1)D_1(\theta_1) \tag{39}$$

Here $\delta(k)$ is the delta-function, $D_1 = \mathrm{Re}\left(\bar{S}_1\right)$ and $H_1 = \mathrm{Im}\left(\bar{S}_1\right)$.

$$\tilde{\Gamma}_0 = \frac{cE_0^2\tau_0 a_0^2}{16\sqrt{\pi\ln 2}\lambda_0 M_1' M_2' TF} \tag{40}$$

Where $\chi_0 = \frac{\pi A_0 \xi^0 n_0^3 L(p_{11}-p_{12})}{2\lambda_0}$ is the modulation index, $n = 1-6; l,m = -3, -2, \ldots, +3$; ξ^0 is the normalized deformation of the media, $p_{11}p_{12}$ is the photoelastic constants for TeO_2; L is the width of a piezotransducer, E_0 is the electric field of the wave, τ_0 is the duration of the light pulse, a_0 is the input beam diameter, F is the focus length of the lens.

For the numerical calculations of nonlinearities let us introduce the coefficient η. It characterizes the standard deviation of the intensity

$$\eta = \frac{\sqrt{\frac{1}{x_0}\int_0^{x_0}\left[I_1(x_1) - I_1(x_1)|_{n=1}\right]dx_1}}{|I_1(x_1)|_{n=1}} = \frac{\sqrt{\sum_{n=2}^6 e^{-(\phi_0 n)^2}\left(C_{n1}^2 + C_{n2}^2\right)}}{\sqrt{2}e^{-\phi_0^2}\sqrt{C_{11}^2 + C_{12}^2}} \tag{41}$$

where $\phi_0 = \frac{\pi f_0 \tau_0}{2\sqrt{\ln 2}}$, $x_0 = \frac{v}{f_0}$

The calculations were performed for the case of $m_0 = 1, \lambda = 510.6$ nm, $f = 80$ MHz, $L = 4$ mm.

Fig.4 (left) shows the dependencies of η as a function of the index modulation χ_0 for different frequency modulation f_0. One can see that the nonlinear distortions sharply increase with the increasing of index modulation beginning with a certain value of ξ_0. The distortions are especially strong at a low frequency modulation. Addition analysis shows that the increasing the duration of read-out light pulse yields to some reduction of nonlinearities simultaneously with reduction of the contrast of the image. Fig.4 (right) shows the dependencies of the contrast for M and N as a function of the frequency modulation f_0. The parameters M and N are defined as:

$$M(f_0) = \frac{I_{\max(f_0)} - I_{\min(f_0)}}{I_{\max(f_0)} + I_{\min(f_0)}}, N(f_0) = \ln\frac{I_{\max(f_0)}}{I_{\min(f_0)}} \tag{42}$$

where $I_{\max}$, $I_{\min}$ are the absolute values of the light intensity. From the presented pictures one can see that the contrast M weakly depends on f_0. At the same time the number of gray levels significantly reduces. One can estimate the number gray levels N_{GL} as much as $400-500$ at the conditions: $f_0 = 2-5$ MHz, $\chi_0 = 0.6-0.8$.

(left) demonstrates variations of the shape of a sinusoidal signal for $f_0 = 1$ MHz, $f = 80$ MHz, $\tau_0 = 10$ ns , $L = 4$ mm. Here: along x-axis is the distance in mm, along the y-axis is the function $I(x) = \frac{I_1(x)}{\Gamma_0}$. Fig.5 (right) shows the

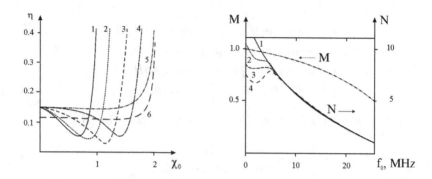

Fig. 4. Left: The dependencies of nonlinear distortions η as a function of 0. The calculations χ_0. The calculations were performed for the $\tau_0 = 10$ ns, and for $f_0 == 0.5MHz(1), 3MHz(2), 5MHz(3), 7MHz(4), 10MHz(5)$ and for $\tau_0 = 20ns, 10MHz(6)$, Right: The dependencies of the contrast for M and N i as as a function of the frequency modulation f_0. The calculations were performed for the $\chi_0 = 0.1(1), 0.2(2), 0.4(3), 0.8(4)$.

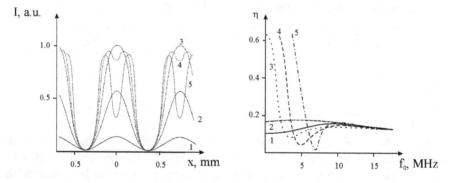

Fig. 5. Left: The variations of the shape of a sinusoidal signal for $f_0 = 1$ MHz, $f = 80$ MHz, $\tau_0 = 10$ ns, $L = 4$ mm. The calculations were performed for the $\chi_0 = 0.2(1), 0.3(2), 0.6(4), 1.0(5)$. Right: The dependencies η on f_0 calculated for different χ_0 for the case of $f = 80MHz, \tau_0 = 10$ ns, $L = 4$ mm. The calculations were performed for the $\chi_0 = 0.2(1), 0.6(2), 1.0(4), 1.2(5)$.

dependencies η on f_0 calculated for different χ_0 for the case of $f = 80$ MHz, $\tau_0 = 10$ ns, $L = 4$ mm. One can see, the increase of the χ_0 results to the distortions cover all larger frequency range.

One of the possible ways to reduce distortions is an increasing of the length of acousto-optical interaction L. Fig. 5 (right) shows the calculated dependencies η on frequency f_0 for the fixed parameter $\chi_0 = 1$ and different values of L. From one side the Fig. 5 (left) demonstrates variations of the shape of a sinusoidal signal for $f_0 = 1$ MHz, $f = 80$ MHz, $\tau_0 = 10$ ns , $L = 4$ mm. Here: along x-axis is the distance in mm, along the y-axis is the function $I(x) = \frac{I_1(x)}{I_0}$. Fig.5 (right) shows the dependencies η on f_0 calculated for different χ_0 for the case of $f = 80$ MHz, $\tau_0 = 10$ ns, $L = 4$ mm. One can see, the increase of the χ_0 results to the distortions cover all larger frequency range.

One of the possible ways to reduce distortions is an increasing of the length of acousto-optical interaction L. Fig. 5 (right) shows the calculated dependencies η on frequency f_0 for the fixed parameter $\chi_0 = 1$ and different values of L. From one side the length L should be selected as long as possible, from another side, the increasing of the length should not significantly reduce the contrast.

The experimental investigations were performed with a Cu-vapor laser with $\lambda = 510.6$ nm, the duration of light pulse $\tau_0 \approx 7$ ns and repetition frequency $f_{\text{rep}} = 15.625$ kHz, and with a Ya:KGW laser with $\lambda = 518.0$ nm, the duration of light pulse $\tau_0 \approx 75.6 fs$ and repetition frequency $f_{\text{rep}} = 75.6 MHz$. The obtained experimental results quite a similar, here we present the results, obtained with the first one. Fig.6 shows the experimentally measured distribution of diffracted light intensity of the first order of diffraction along x-axis for different UC. The current frequency of UC here and below was 82.5 MHz. The manifestation of nonlinearity as an appearance of the local minima with the increasing of UC is obvious.

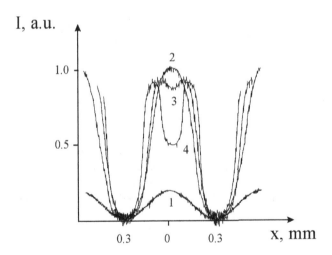

Fig. 6. The diffracted light distribution along the x-axis

Fig. 7 (left) shows the oscilloscopic traces and Fig. 7 (right) shows their spectrums for different amplitudes UC. The distortions are increase with the increasing of UC, but the increasing of the signal is slowing down.

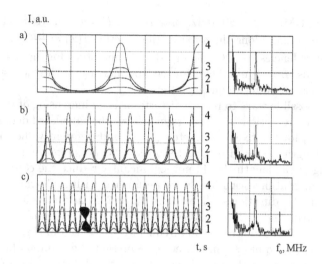

Fig. 7. Left: The oscilloscopic traces along the x-axis for $f_0 = 1$ MHz (a), $f_0 = 5$ MHz (b), $f_0 = 10.8$ MHz (c), $U_c = 0.05$ V(1), 0.1 V(2), 0.4 V(3), 0.7 V(4). Right: the spectra of the same traces.

5   Conclusion

As a result of solution of the integral equation for electromagnetic field obtained by introducing equivalent currents and expanding the field sought into plane waves, the general solution for the light field diffracted by ultrasound which is valid for the anisotropic dielectric medium exhibiting gyrotropic properties has been found. Based on the general solution of the integral equation, the procedure for calculating the diffracted light field for the practically important geometry of scattering corresponding to the wideband anisotropic light diffraction from a slow shear elastic wave propagating in the [110] direction in the TeO_2 crystal near its optical axis has been suggested. As an example, the expression for the diffracted light field in the (+1)st order at the ultrasonic column boundary for the case of a constant amplitude of ultrasonic perturbation and also its amplitude modulation according to the harmonic law has been obtained for this scattering geometry in the approximation to the third interaction order inclusive. The incidence angles of light waves on the AOM in the plane orthogonal to the diffraction plane have been taken into consideration in the expressions for the diffracted fields.

The procedure of calculation of the light field at the output AOM aperture by increasing interaction orders allows one to calculate the intensity of the amplitude- modulated light at its diffraction in the AOM made from TeO_2 (with due account of gyrotropy of this crystal) and also to solve the problem of nonlinear distortions in the video signal for a laser projection system that employs the AOM made from TeO_2.

The main reasons of the nonlinear distortions are revealed and practical ways to reduce it were demonstrated. The good agreement between the theory and the experiment has been demonstrated.

References

[1] Petrov, V.: Digital Holography for Bulk Image Acoustooptical Reconstruction. In: Poon, T.-C. (ed.) Digital Holography and Three-Dimensional Displays: Principles and Applications, vol. 430, pp. 73–99. Springer (2006)

[2] Koechlin, M., Poberaj, G., Gunter, P.: High-resolution laser lithography system based on two-dimensional acousto-optic deflection. The Review of Scientific Instruments 80(8), 85105 (2009)

[3] Salzenstein, P., Voloshinov, V.B., Trushin, A.S.: Investigation in acousto-optical Laser stabilization for crystal resonator-based optoelectronic oscillator. Optical Engineering 52(N2), 24603 (2013)

[4] Sapirel, J., Charissoux, D., Voloshinov, V., Molchanov, V.: Tunable acousto-optic filter and equalizer for WDM applications. Journal of Lightwave Technology 20, 864 (2002)

[5] Farsari, M., Chichkov, B.: Two-photon fabrication. Nature Photonics 3, 450–452 (2009)

[6] Petrov, V.M., Petrov, M.P., Bryksin, V.V., Petter, J., Tschudi, T.: Optical detection of the Casimir Force between the macroscopic objects. Optics Letters N21, 3167–3169 (2006)

[7] Petrov, M.P., Petrov, V.M., Zouboulis, I.S., Xu, L.P.: Two-wave and induced three-wave mixing on a thin $Bi_{12}TiO_{20}$ hologram. Optics Communications 134, 569–579 (1997)

Application of Optimal Spectrally Efficient Signals in Systems with Frequency Division Multiplexing

Sergey V. Zavjalov, Sergey B. Makarov, and Sergey V. Volvenko

St. Petersburg State Polytechnical University, St. Petersburg, Russia
zavyalov_sv@spbstu.ru, {makarov,volk}@cee.spbstu.ru

Abstract. A study of optimal form of envelope for Spectrally Efficient Orthogonal Frequency Division Multiplexing (SEOFDM) signals and corresponding energy spectra of signals has been conducted. Increasing of additional energy losses of BER performance of SEOFDM signals in the case of application optimal coherent bit-by-bit processing algorithm for signals with OFDM forces to explore more effective processing technique. An algorithm based on decision diagram can significantly improve BER performance of SEOFDM signals with smoothed optimal form of envelope signals on subcarriers.

Keywords: SEOFDM, spectrally efficient signals, out-of-band emissions reduction, interchannel interference, bit-by-bit coherent algorithm, algorithm based on decision diagram, BER performance.

1 Introduction

One of the disadvantages of OFDM signals is high level out-of-band emissions (OOBE) at the edges of the occupied bandwidth. This is a serious obstacle to increase the data rate by increasing the number of subcarriers in allocated bandwidth of telecommunications systems. It is suggested in [1] usage of optimal function, obtained as the solution of the functional equation in the form of spheroidal functions, as envelope on each subcarrier frequency of signals with a quadrature phase shift keying. Obtained envelop of signals, as follows from the results of [1], provides evenly distributed low level OOBE. However, the reduction of OOBE of such signals is small. This fact complicates the use of such signals for increase in number of transmission channels by their frequency multiplexing. In [2] it is suggested to use narrow-band filtering to generate signals of duration T; resulting envelope has the form of a Nyquist pulse. Significant intersymbol interference, which leads to a degradation of BER performance is manifested in higher orders narrowband filter [2]. Reduction of OOBE of OFDM signals is considered in [3], where we can see a proposition to use a window Gabor. The signals have smoothed shape of the envelope in each quadrature channel of subcarrier. However, the decreasing rate of the energy spectrum for different parameters of OFDM signals is unchanged.

Characteristics of the spectrally efficient PSK (SEPSK) signals which are optimal by the criterion of maximum OOBE reduction, located on each subcarrier of OFDM

S. Balandin et al. (Eds.): NEW2AN/ruSMART 2014, LNCS 8638, pp. 676–685, 2014.

signals will be considered below. We will call such signals of Spectrally Efficient Orthogonal Frequency Division Multiplexing (SEOFDM).

2   Criterion of a Minimum of OOBE Level

Optimal SEPSK signals can be obtained by solving the optimization problem, which is based on minimization procedure of functional, which determines the spectral characteristics of the energy spectrum

$$G(\omega) = \lim_{N \to \infty} \frac{1}{NT} m_1 \left\{ |S(\omega)|^2 \right\},$$

where $S(\omega)$ is the spectrum of an arbitrary finite realization (consisting of N members) of signal's sequence and mathematical expectation is determined by averaging over all possible finite sequences of transmitted symbols. Boundary conditions and various restrictions, such as restriction on the energy, duration of the signal, the correlation coefficient, and the value of peak-to-average power ratio (PAPR) of signals are used in the formulation of the optimization problem.

The solution of a problem of synthesis of optimum signals with the envelope $a(t)$ providing the maximum OOBE reduction is connected with the numerical solution of a minimization problem of functional:

$$J = \frac{1}{2\pi} \int_{-\infty}^{+\infty} g(\omega) |S(\omega)|^2 \, d\omega, \tag{1}$$

where $g(\omega)$ – weighing function that determines the reduction of OOBE.

Energy spectrum of SEPSK signal with envelope $a(t)$ can be written as follows:

$$|S_a(\omega)|^2 = S_a(\omega) S_a *(\omega) = \int_{-T/2}^{T/2} \int_{-T/2}^{T/2} a(t) a(s) e^{-j(\omega - \omega_0)(t-s)} \, dt \, ds. \tag{2}$$

We will use the quadratic function as a weighting function:

$$g(\omega) = \omega^{2n}. \tag{3}$$

It can be shown that we must find numerical solution of the functional of the following form:

$$J = (-1)^n \int_{-\infty}^{+\infty} a(t) a^{(2n)}(t) \, dt, \tag{4}$$

where $a^{2n}(t)$ – $2n$-th derivative of $a(t)$. Thus, the optimization problem is reduced to the problem of finding the function $a(t)$ in the time interval $[-T/2; T/2]$, which provides the minimum of functional (4).

Restrictions on the energy of the signal and the following boundary conditions are used in solving this optimization problem:

$$a^{(k)}\Big|_{t=\pm T/2} = 0 \, , \, k = 1, \, ..., \, (n-1) \tag{5}$$

(all derivatives of envelope $a(t)$ until the derivative of $(n-1)$-th order $a^{(n-1)}(t)$ have no discontinuities and $a^{(n)}(t)$ is finite on the whole time interval).

Energy restriction may be defined as follows:

$$\int_{-T/2}^{T/2} a^2(t)dt = E = 1 \, . \tag{6}$$

The required envelope $a(t)$ is even function in the interval $[-T/2; T/2]$ and can be represented by the following expression:

$$a(t) = \frac{a_0}{2} + \sum_{k=1}^{m} a_k \cos\left(\frac{2\pi}{T}kt\right), \tag{7}$$

where $a_0 = \dfrac{1}{T}\displaystyle\int_{-T/2}^{T/2} a(t)dt$, $a_k = \dfrac{1}{T}\displaystyle\int_{-T/2}^{T/2} a(t)\cos\left(\dfrac{2\pi}{T}kt\right)dt$, $b_k = \dfrac{1}{T}\displaystyle\int_{-T/2}^{T/2} a(t)\sin\left(\dfrac{2\pi}{T}kt\right)dt$.

Restriction on energy of the signal (6) can be rewritten in the following form:

$$\int_{-T/2}^{T/2}\left[\frac{a_0}{2} + \sum_{k=1}^{m} a_k \cos\left(\frac{2\pi}{T}kt\right)\right]^2 dt = \frac{T}{2}\left(\frac{a_0^2}{2} + \sum_{k=1}^{m} a_k^2\right) = E = 1 \tag{8}$$

or

$$\frac{T}{2}\left(\frac{a_0^2}{2} + \sum_{k=1}^{m} a_k^2\right) = E = 1 \, . \tag{9}$$

It can be shown that the expression (4) can be rewritten as follows:

$$J\left(\{a_k\}_{k=1}^{m}\right) = \frac{T}{2}\sum_{k=1}^{m}\left(\frac{2\pi}{T}k\right)^{2n} a_k^2 \, . \tag{10}$$

Thus, transition from the functional (4) to the function of many variables $J\left(\{a_k\}_{k=1}^{m}\right)$ (10) is made. Note that boundary conditions (5) are converted into additional constraint equations (with this transition):

$$a(t)\Big|_{t=\pm T/2} = \frac{a_0}{2} + \sum_{k=1}^{m}(-1)^k a_k = 0 \, , \tag{11}$$

$$a^{(2k)}(t)\Big|_{t=\pm T/2} = (-1)^k \sum_{k=1}^{m} a_k \left(\frac{2\pi}{T}k\right)^{2k} = 0 \, , \, k = 1...n,$$

$$a^{(2k-1)}(t)\Big|_{t=\pm T/2} \equiv 0 \, , \, k = 1...n.$$

Fig. 1 represents view of the envelopes and the corresponding energy spectra, which is obtained as the solution of the optimization problem (1) for $n = 1, 2, 3$. Reduction of OOBE of energy spectrum $G(\omega)$ is equal to $1/f^6$ for $n = 3$.

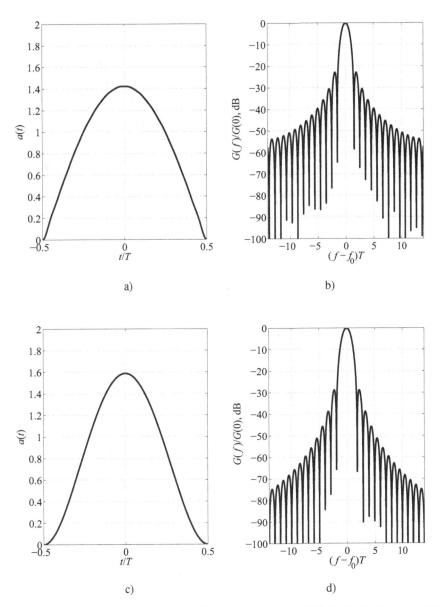

a)   b)

c)   d)

Fig. 1. The envelopes and the corresponding energy spectra, which is obtained as the solution of the optimization problem (1) for $n = 1$ (a, b), 2 (c, d), 3 (e, f)

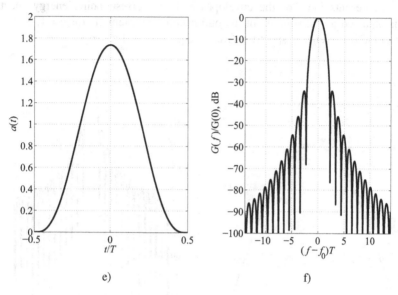

e)   f)

Fig. 1. (*continued*)

3   Time and Spectral Characteristics of the Signals

The instantaneous power of OFDM and SEOFDM signals (with usage of envelope $a(t)$ for $n = 1$ on each subcarrier) for the same information sequence is shown in fig. 2 (number of subcarriers is equal to 64).

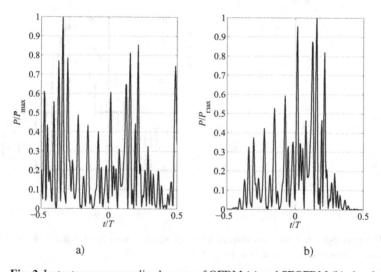

a)   b)

Fig. 2. Instantaneous normalized power of OFDM (a) and SEOFDM (b) signals

It can be seen that the SEOFDM signals have smooth reduction of the emission power at the edges of the time interval of signals existence.

Normalized energy spectra of optimal SEOFDM signals (number of subcarriers is equal to 64) for the case of fixed average power of emitted signals are shown in fig. 3.

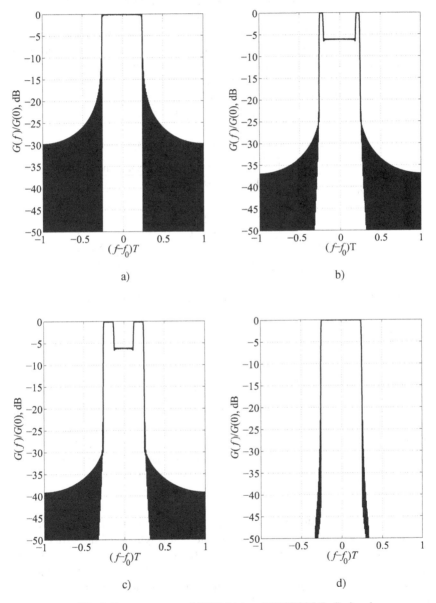

Fig. 3. Energy spectra of OFDM (a) and SEOFDM (b-d) signals

The form of the energy spectra of random sequences of SEOFDM signals with different ratio of subcarriers, on which we use signals SEPSK with the envelope shown in fig. 1, and subcarriers, on which we use rectangular form of envelop, is shown in fig. 3. It used 20% (fig. 3, b), 50% (fig 3, c) and 100% (fig. 4, d) of SEPSK signals. Thus subcarriers, where SEPSK signals are used, are located symmetrically along the edges of the bandwidth of SEOFDM signal. Energy spectrum of OFDM signals is shown in the same fig. 3, a, for comparison. As it can be seen, only usage of 100% SEPSK signals on all subcarriers allows obtaining the highest reduction of OOBE. This reduction is equal to the value of $1/f^6$. In other cases, the reduction of OOBE levels will be equal to the $1/f^2$, because the spectral components of PSK signals with a rectangular form of the envelope has greater influence.

The occupied bandwidth ΔF for normalized energy spectrum $G(f)/G(0)$ for the case of fixed average power of emitted signal is shown in table 1.

Table 1. Values of ΔF for normalized energy spectrum

$G(f)/G(0)$, dB	Value of portion of SEPSK signals			
	0%	20%	50%	100%
	ΔFT			
−20	0.630	0.506	0.506	0.506
−30	1.910	0.690	0.540	0.526
−40	−	−	−	0.584
−50	−	−	−	0.680

As seen from table 1, the bandwidth ΔF of SEOFDM signals with 100% of the smoothed envelopes on subcarriers is less than $0.68/T$ on the level of −50 dB that is significantly less than optimum signal shown in [1].

4   BER Performance

Let`s consider optimal coherent bit-by-bit processing algorithm for signals with OFDM as demodulation algorithm of SEOFDM signals. The dependences of the average error probability on the value E_b/N_0 (E_b – signal energy per bit of information, N_0 – average power spectral density of the additive normal random process) are shown in fig. 4. Increasing of portion of SEPSK signals on subcarriers of SEOFDM signal (this parameter is equal to 0; 0.2, 0.5, 1.0 in fig. 4) leads to the error probability increase and additional energy losses at fixed error probability (table 2). It is caused by the level increase of interchannel interference generated by the manifestation of nonorthogonality between SEPSK signals, located on neighboring subcarriers. Level of nonorthogonality is determined by function $a(t)$ type, obtained from (10) with restriction on energy (9) and the boundary conditions (11). SEOFDM signals with 100% smoothed SEPSK signals on subcarriers have the worst BER performance: additional energy losses are equal to 5.3 dB in area of BER performance $p=10^{-2}$ and 10.6 dB in area of BER performance $p=10^{-3}$ (fig. 4 and table 2).

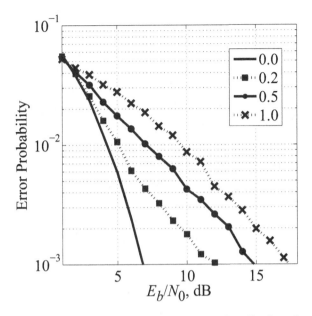

Fig. 4. BER performance of SEOFDM signals in the case of application of optimal coherent bit-by-bit processing algorithm for signals with OFDM

Table 2. BER performance of SEOFDM signals in the case of application of optimal coherent bit-by-bit processing algorithm for signals with OFDM

Additional energy losses, dB			
BER performance	Portion of SEPSK signals		
	0	50%	100%
10^{-2}	0	2,8	5,3
10^{-3}	0	8,0	10.6

Let`s consider coherent processing algorithm on the basis of the decision diagram for improving BER performance of SEOFDM signals. This algorithm is based on the Viterbi algorithm. Significant interchannel interference is manifested in case of application of smoothed envelope on subcarriers of SEOFDM signals. The algorithm uses information about already received SEPSK signals, located on neighboring subcarrier, at each step. At each step weighting correlation processing is performed and decisions estimation on received signals is done by selecting minimum of Euclid distance between a received signal and all possible realizations of neighboring signals at this step of the algorithm.

BER performance of SEOFDM signals for the case of fixed average power of emitted signals with application of coherent processing algorithm on the basis of the decision diagram is shown in fig. 5. Application of the coherent processing algorithm on the basis of the decision diagram provides a significant increase in BER performance (from comparing dependences in fig. 4 and fig. 5). Additional energy losses are equal to 2.8 dB in area of BER performance p=10^{-2} and 3.1 dB in area of BER performance p=10^{-3} (table 3) for SEOFDM signals with 100% smoothed SEPSK signals on subcarriers.

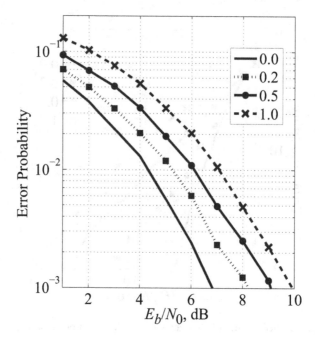

Fig. 5. BER performance of SEOFDM signals in the case of application of the algorithm on the basis of decision diagram

Table 3. BER performance of SEOFDM signals in the case of application of the algorithm on the basis of decision diagram

Additional energy losses, dB			
BER performance	Portion of SEPSK signals		
	0	50%	100%
10^{-2}	0	1,8	2,8
10^{-3}	0	2,3	3,1

Usage of coherent processing algorithm on the basis of decision diagram provides energy gain of more than 7 dB in area of BER performance 10^{-3} in comparison with the case of usage the optimal coherent bit by bit processing algorithm for signals with OFDM.

5   Conclusions

Application of optimal spectrally efficient signals in systems with orthogonal frequency division multiplexing provides significant reduction of the occupied frequency bandwidth. The bandwidth ΔF of SEOFDM signals with 100% of the smoothed envelopes on subcarriers is less than 0.68/T on the level of −50 dB that is significantly less than optimum signal in [1]. Reduction of additional energy losses, which are associated with the application of SEOFDM signal with smoothed envelope, is possible due to partial (e.g., 50%) usage of smoothed form of envelope with rectangular form of envelope in frequency bandwidth. Note, that at this situation additional energy losses are equal to 8 dB in area of BER performance $p = 10^{-3}$ for the case of application of known processing algorithm for OFDM signals.

A more significant reduction in additional energy losses is possible through the usage of coherent processing algorithm on the basis of decision diagram. For SEOFDM signals with 100% smoothed SEPSK signals on subcarriers usage of coherent processing algorithm on the basis of decision diagram provides energy gain of more than 7 dB in area of BER performance 10^{-3} in comparison with the case of usage the optimal coherent bit-by-bit processing algorithm for signals with OFDM.

References

1. Vahlin, N., Holte, N.: Optimal finite duration pulses for OFDM. IEEE Transaction on Communications 44(11), 10–14 (1996)
2. Nicolas, J., Baas, D.P.: Taylor: Pulse Shaping for Wireless Communication Over Time-or Frequency - Selective Channels. IEEE Transaction on Communications 52(9), 1477–1479 (2004)
3. Sriram, S., Vijajakumar, N., Adityakumar, P., Shetty, A.S.: Spectrally Efficient Multy-Carrier Modulation Using Gabor Transform. Wireless Engineering and Technology 4, 112–116 (2013), doi:10.4236/wet.2013.420017

Signal Processing for Mobile Mass-Spectrometry Data Transfer via Wireless Networks

Alexey N. Petrov, Elena N. Velichko, and Oleg Yu. Tsybin

Saint Petersburg State Polytechnical University, Polytechnicheskaya Str, 29,
St. Petersburg, 195251, Russia
{alexey-np,velichko-spbstu}@yandex.ru, otsybin@rphf.spbstu.ru

Abstract. Non-Fourier transformation algorithm and computer program for efficient analysis of short broadband signals for the mass-spectrometry data transmission via wireless networks to remote data users have been developed and tested. It has been shown that even in the case no preliminary data on the signal phase parameters are available, short broad-band multicomponent transients can be transformed into high-resolution signals with high signal-to-noise ratios. Such signals are suitable for high-speed transmission via wireless networks.

Keywords: signal processing, mobile mass-spectrometry, wireless data transfer.

1 Introduction

Modern mass spectrometry (MS) employs high-efficiency instruments for basic research in physics, chemistry, biology, medicine, proteomics, ecology, and many others and solution of applied problems [1-4]. The emerging applications require development of transportable or/and mobile mass spectrometry stations capable of operation out of stationary laboratories. Such mobile/transportable MS stations produce in-situ sampling, fast mass-analysis of samples and data acquisition, and time-domain high-dynamic-range signal (transient) generation. In addition, owing to a simplified sampling procedure, shortened acquisition time, and specific hardware of analyzers, electronic devices, and vacuum system, transients have a low signal-to-noise ratio (SNR) and excess amplitude/phase/frequency modulation. Unfortunately, such transients are not appropriate for the classic Fourier transformation (FT) into the frequency domain. Being transformed into the frequency domain, the spectra may contain time-dependent frequencies, wide bands, harmonics, and sidebands [5-7]. Finally, the mass-spectrometric parameters presented by such frequency-domain signals must be transmitted, sometimes immediately after acquisition, via Wireless network to the remote data users.

The goal of our investigations was to develop and test more effective non-FT algorithms and computer programs appropriate for the MS data transmission via WiFi or Internet.

S. Balandin et al. (Eds.): NEW2AN/ruSMART 2014, LNCS 8638, pp. 686–693, 2014.
© Springer International Publishing Switzerland 2014

The simplest way to improve SNR is to accumulate scans, but because SNR increases only as the square root of the number of scans, this option is often not appropriate for time scales of mobile stations.

Post-acquisition algorithmic processing has been shown to fairly improve the resolution of FT-MS spectra [8], particularly when the ion cloud phase is known [9].

2   The Method

We have recently suggested a non-FT algorithm for wide-band spectral function analysis that does not need a priori information on phase parameters. The main idea of our investigation was to develop a method based on correlation definition of signal periodicities by using instrument function in real time (transient signal in the time domain) and in the specific "shift-time" domain [10]. The algorithms and program for computer analysis of mass-spectrometric signals that allow one to reduce the time needed for a high-resolution data acquisition have been developed on this basis and tested.

At first signal periodicities were sought for by expanding the initial function in terms of basis of reference pulse sequence in the time window. The reference signal was simulated by the $F_a(t)$ function as a convolution of the comb Sha-function and instrument function. The Sha-function was defined as

$$H(t) = \sum_{i=1}^{I} \sum_{p=0}^{N} (-1)^p \delta(t - pT_i) , \tag{1}$$

where T_i is a half of the cyclotron period of the i-th ion in the transient, t is the time, N is the number of pulses in the signal, and δ is the finite height function with the width equal to the discretization interval.

If we take into account a typical exponential decay of the mass spectrometric signal amplitude [11,12] with a decay constant γ defined by the destruction of the ion cluster coherence, Eq.(1) used for calculation of the Sha-function becomes

$$L(t) = H(t)e^{-\gamma t}. \tag{2}$$

To specify the instrument function, the ion signal peaks were assumed to be described by the Gauss function (which corresponds to typical mass-spectrometric data) with the dispersion σ determined by the characteristic time (e. q. the time period) of ion cluster flight along the detector

$$D(x) = \frac{1}{\sqrt{2\pi}\sigma} e^{-\frac{(t-T_0)^2}{(2\sigma)^2}} , \tag{3}$$

where t is the time, T_0 is the position of the peak maximum $T_0 = {}^{T_c}/_4$, and T_c is the cyclotron period.

The instrument function was given by

$$F_a(t) = \int_0^\infty \big(D(\tau)L(t-\tau)\big)d\tau. \tag{4}$$

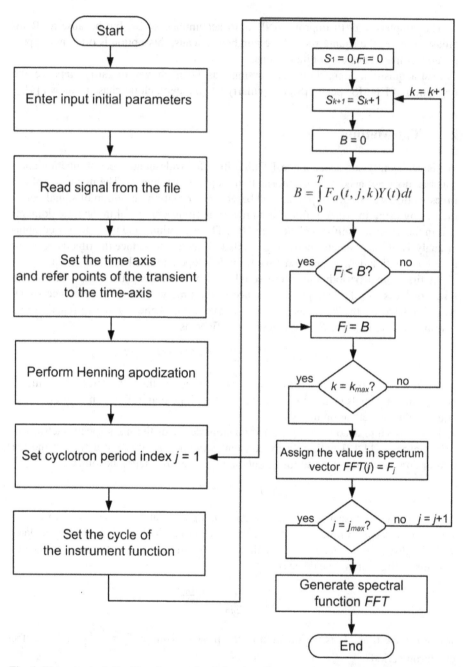

Fig. 1. Flow chart of algorithm for search of signal cyclotron periods by finding approximation coefficients: S_k sets the phase shift between instrument function and signal, F_j is the approximation coefficient optimized with respect to phase shift, $F1$ is the value of the integral for the specified phase shift; FFT is the final spectral function

When transforming the function into the τ-shift time domain, the coefficients of expansion in basis functions F were calculated. The initial signal phase was chosen by maximizing the integral values

$$F = \int_0^T F_a(t)Y(t)dt. \tag{5}$$

Then sequential scanning over repetition periods of basis functions was performed.

The scanning range was preset as an initial parameter on the basis of a priori estimates of the sample composition, owing to which the signal processing time was shorter. In the case such a priori information was absent, the scanning was carried out over all possible periods.

The flow chart of the computer program algorithm for revealing periodicities in the signal we have developed is presented in Fig.1. The main designations and brief description are given in the figure and in the caption. The input parameters were

— the sampling rate (which was set by the regime of measurements for each mass analyzer);
— the time range of search periods (which was determined on the basis of preliminary estimates of the sample mass composition).

3   Results and Discussion

During the test period the program was applied to the model signal which was formed as a transient with exponential amplitude decay. To form the model signal, superposition of ion components, magnetic fields and trap radius, and also sizes of detector plates of a particular mass analyzer were specified. In contrast to typical transients, the signal comprised of short pulses (0.01 of the cyclotron period) with long repetition periods had a much broader bandwidth.

An example of the distribution of signal amplitudes over periods which was calculated for the superposition of three ion components in the transient is presented in Fig. 2. Such periodogram can readily be converted into a frequency spectrum of the signal

$$f_{sign} = {}^1\!/_{T_{sign}}$$

or into the spectrum of specific weights M:

$$M\!/_q = \frac{B}{2\pi f_{sign}}$$

where B is the magnetic field and q is the ion charge.

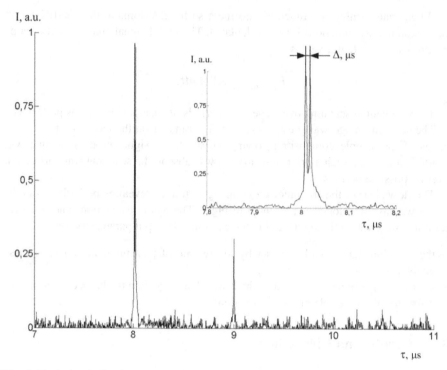

Fig. 2. Typical calculated spectrum of model signal consisting of three ion components as a distribution of correlation coefficients over periods: I is the amplitude of the signal, τ is the period, $\Delta = 0.005$ µs

The abscissa axis in Fig. 2 shows the calculated signal periods in the spectrum in milliseconds and the ordinate axis gives amplitudes of correlation coefficients in relative units. The calculations demonstrated a high resolution (10^3-10^4) of the single and double peaks and high signal-to-noise ratios (in our case they reached approximately 70). The data were obtained for the transient duration of 1 ms, which can be regarded as a good example of processing of short signals. These results look attractive because they show that a possibility of using shorter signals as compared with ordinary signals having duration of 100 ms and more arises. The signal frequency band was 4 MHz. In the case of an ordinary Fourier transformation, a useful signal could not be separated out of this transient because the noise level in the spectral function obtained was too high.

The signal periods determined by the method we suggest allowed us to calculate characteristic frequencies and, hence, ion masses in the mixture. In addition to the main peaks corresponding to ion components, noise and cross-talk peak signals with low amplitudes were observed. Cross-talk signals with low amplitudes were related to the pulse shapes and reflect some inter-correlation frequencies. The amplitudes and positions of such peaks depended on the duration and on the shapes of the reference signal peaks. Calculations for different peak shapes, including rectangular, triangular, delta function and other shapes, were carried out. From the point of view of attainable parameters (resolution and signal-to-noise ratio) the peaks with the Gaussian envelope were found to be optimal and were used in simulation.

Fig.3 shows resolution and signal-to-noise ratio as functions of parameter K of the reference signal shape.

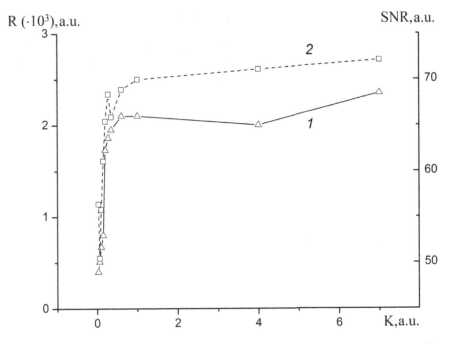

Fig. 3. Frequency resolution R (solid curve 1) and signal-to-noise ratio SNR (dashed curve 2) as functions of parameter K that determines the width of the reference signal peak

The peak width was set by a standard deviation σ from the Gaussian envelope defined by

$$\sigma = {T_c}/{K}$$

where T_c is the cyclotron period of the reference function and K is the coefficient that determines the peak width.

As one can see from the curves, as the width of the reference function peaks decreases, the signal-to-noise ratio and resolution increase almost simultaneously, and as the width becomes ≤ 0.5 of the peak width in the signal being analyzed attain saturation. From the point of view of a reasonable ratio between the required computational effort and attainable parameters, the optimal width of the reference peak was found to be ≈ 0.5-0.7 of the peak width in the signal being analyzed.

Let us consider specific features of the structure of a single signal peak. Fig. 4 shows results of transformation of a single-frequency model transient into the shift-time domain. A resolution of more than 10^4 was obtained for the duration of a wide-band transient of 1 ms, which greatly exceeded the possibilities of Fourier transform for such a signal. The signal in the shift-time domain was represented by not only the main peak with a broadened base but also by a definite set of sidebands (Fig. 4). The nature of such sidebands is explained by a typical *sinc* expansion of the input pulsed signal. The Henning apodization given by

$$\omega(n) = 0.5 \left(1 - \cos\left(\frac{2\pi n}{N-1}\right)\right), \tag{6}$$

where N is the number of points in the signal and n is the integer from 0 to (N-1), allowed us to decrease the width of the peak base and amplitudes of sidebands and also to increase the signal-to-noise ratio in the region near the main peak (in the range from 7.8 to 8.8 μs) in 2 times.

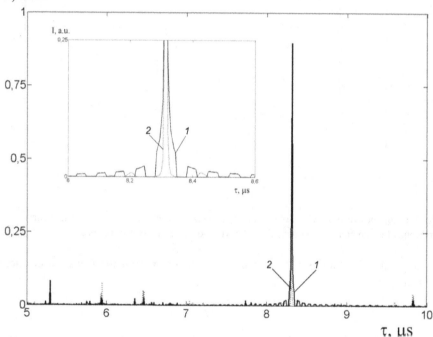

Fig. 4. Signal being analyzed in the shift-time domain before (solid curve) and after (dashed curve) application of apodization (the inset shows the region of the spectrum at an enlarged scale): I is the amplitude of the signal, τ is the period

4   Conclusion

To summarize, we have developed and tested the non-FT algorithm and computer program for efficient analysis of short broadband signals for the MS data transmission via wireless network to remote data users. It has been shown that even in the case no preliminary data on the signal phase parameters are available, short broad-band multi-component transients can be transformed into the high resolution and signals with high signal-to-noise ratio suitable for high-speed data transmission.

References

1. Senko, M.W., Cantenbury, J.D., Guan, S., Marshall, A.G.: A high-performance modular data system for Fourier transform ion cyclotron resonance mass spectrometry. Rapid Commun. Mass Spectrom. 10, 1839–1844 (1996)
2. Marshall, A.G., Hendrickson, C.L., Jackson, G.S.: Fourier transform ion cyclotron resonance mass spectrometry: a primer. Mass Spec. Rev. 17, 1–35 (1998)
3. Amster, I.J.: Fourier transform mass spectrometry. J. Mass Spectrom. 31, 1325–1337 (1996)
4. Marshall, A.G., Verdun, F.R.: Fourier Transforms in NMR, Optical and Mass Spectrometry. A User's Handbook. Elsevier Sci. Pub (1990)
5. Miladinović, S.M., Kozhinov, A.N., Tsybin, O.Y., Tsybin, Y.O.: Sidebands in Fourier transform ion cyclotron resonance mass spectra. Int. J. of Mass Spectrom. 325, 10–18 (2012)
6. Tsybin, O.Y., Tsybin, Y.O.: Non-linear time-dependent phase phenomena in FT-ICR MS. In: Innovations in Mass Spectrometry Instrumentation Conference. Book of Abstracts, pp. 34–35. Saint Petersburg, Russia (2013)
7. Tsybin, O.Y., Kozhinov, A.N., Nagornov, K.O., Tsybin, Y.O.: Instantaneous frequency theory and applications in FTMS. In: 11th European FTMS Workshop, p. 74. Institute Pasteur, Paris (2014)
8. Kozhinov, A.N., Tsybin, Y.O.: Filter Diagonalization Method-Based Mass Spectrometry for Molecular and Macromolecular Structure Analysis. Anal. Chem. 84, 2850–2856 (2012)
9. Greenwood, J.B., et al.: A comb-sampling method for enhanced mass analysis in linear electrostatic ion traps. Rev. of Scientific Instr. 82, 043103-1 – 043103-12 (2011)
10. Tsybin, O.Y., Petrov, A., Tsybin, Y.O.: Analysis of transient signal amplitude and frequency modulation in FT-ICR MS. In: Proceedings of 9th European FTMS Conference, Lausanne, Switzerland, p. 102 (2010)

Fiber-Optic Super-High-Frequency Signal Transmission System for Sea-Based Radar Station

Vadim V. Davydov, Sergey V. Ermak, Anton U. Karseev,
Elina K. Nepomnyashchaya, Alexander A. Petrov, and Elena N. Velichko

St.Petersburg State Polytechnical University, St.Petersburg, Russia
{davydov_vadim66,serge_ermak}@mail.ru,
{antonkarseev,elina.nep}@gmail.com,
{alexandrpetrov.spb,velichko-spbstu}@yandex.ru

Abstract. A fiber-optic system for transmission of super-high-frequency signals in the 8-12 GHz range from a multi-component active phased antenna array of a sea-based radar station to an information processing system located in a ship compartment is suggested and investigated. Power loss and total attenuation of optical signals are calculated. The power loss margin and signal transmission time of the system are analyzed. Dependences of the characteristics of the fiber-optic data transmission system on the input signal frequency and temperature are experimentally measured. The fiber-optic data transmission system we have developed can prove an efficient tool for improving interaction between the wireless radar and wired signal transmission systems.

Keywords: super-high-frequency signal, radar station, fiber-optic data transmission system, optical insulator.

1 Introduction

Fiber-optic data transmission (FODT) systems are widely used on ships and submarines [1-3]. One of topical problems in the area of modern fiber-optic data transmission systems is development of devices of the super-high-frequency (SHF) range for radar systems. Such systems can be efficiently used in different channels of multi-component phased antenna arrays [4,5].

An optical signal modulated by a SHF signal can be transmitted by an optical fiber at distances of tens of kilometers almost without attenuation (less than 0.5 dB per km) and without the necessity to retransmit the signal. An analog fiber-optic communication line can be used to transmit ultra-broadband signals because its frequency range is from units of megahertz to 12–14 GHz (in future – to 30 GHz). Neither of the SHF signal transmission lines (coaxial, waveguide, open-wire) is capable of accomplishing this. This necessitates development of fiber-optic systems for spaced radar systems, antennas which are remote from the control and information processing points, and systems for routing heterodyne, control, and synchronizing signals in phased antenna arrays. Delay lines based on optical fibers exhibit unique properties because the delay of the analog signal in a fiber is frequency -independent, which allows creation of

S. Balandin et al. (Eds.): NEW2AN/ruSMART 2014, LNCS 8638, pp. 694–702, 2014.

false targets even in the case ultra-broadband sounding signals are used. These advantages are not offered by the delay lines relying on other physical principles.

The use of fiber-optic systems on air and sea vehicle reduces energy consumption, radically decreases the sensitivity to any electromagnetic interference, appreciably increases the reliability of information transmission, and considerably reduces weight. For example, a replacement of even most modern vehicle-borne data transmission cables based on copper by vehicle-borne fiber-optic cables reduces the weight of cable interconnections in more than 10 times.

All modern military ships and submarines encounter the problem of overloading of the upper part of the mast where receiving-transmitting antennas are located with equipment and auxiliary mechanisms and devices (especially screening systems). In addition, masts are overloaded with a great number of cable lines with different systems for their screening from interferences of various types. Since copper cables are not flexible enough, and additional connectors induce extra cross talks and interferences, the cables are often laid in a bypass manner, which increases their length by more than 50 - 70% as compared with the shortest distance and, hence, the weight and energy consumption increase.

One of the most promising ways to solve this problem is to develop new lines for transmission of SHF signals in the ranges 0.1-2 GHz; 2-4 GHz; 4-8 GHz; 8-12 GHz; 18-26 GHz; and 26-40 GHz on the basis of FODT systems.

The use of the FODT systems has many advantages as compared with cables. They include a smaller thickness of the line, the number of channels being the same, a higher flexibility and easy wiring, immunity to electromagnetic interferences, a broad bandwidth and a more uniform amplitude-frequency response.

The use of a fiber-optic system comprising optoelectronic transducers and cladded optical waveguide for transmission of SHF signals not only solves the problem indicated above but also provides an increase in the reliability of the radar station complex. In the case the FODT system is employed, the upper part of the mast contains only antennas, preamplifiers, optoelectronic transducers, and auxiliary devices. Owing to a less dense packing, an easier access to the equipment is provided if repair works are needed, which greatly facilitates the technical maintenance and adjustment of the radar station. The remaining equipment can be located in a ship compartment, which reduces the load on the mast and increases the space for antenna motion.

The goal of our work was to develop and investigate a fiber-optic system for transmission of SHF signals in the 8-12-GHz frequency range from a sea-based radar station to a processing system located in a ship compartment.

2 Fiber-Optic Communication Line for Transmission of SHF Radar Signals

The choice of basic components of the FODT system, i.e., transmitting and receiving optical modules, fiber-optic line, and connectors depends on the distance of information transmission and the frequency range of transmitted SHF signals. Calculations of the FODT system parameters were carried out for the distance of information transmission of no more than 130 m and the frequency range from 8 to 12 GHz. For this reason, after preliminary analysis, transmitters with a direct modulation at a wavelength of 1310 nm were chosen. Transmitters and receivers are, as a rule, presented in pairs in the market. As analysis showed, the optimal choice was a high-frequency Dilas DMPO131-23 laser module and DFDMSh40-16 receiving optical module. They have the following parameters: the frequency range is from 0.1 to 16.2 GHz; the transmission coefficient is -32 dB; the nonuniformity of amplification within the entire transmission bandwidth is ±2 dB; the compression point is 12 dBm; the maximum input signal level is 10 dBm; the noise factor is 25 dB; the input and output resistance is 50 Ohm; the laser power is 4-9 mW; the working wavelength of the photoreceiver is 980-1650 nm; the working frequency bandwidth of the photoreceiver is $\Delta F = 15.99$ GHz; and the range of working temperatures is from -40 to +60 $^{\circ}$C.

The power supply of the transmitting optical module was controlled by a power supply driver. A standard regime of the driver operation is maintenance of constant temperature and average power of the laser diode in the transmitting optical module by maintaining constant resistance of the thermoresistor and photocurrent of the photomonitor of feedback communication built in the transmitting optical module. An additional protective structure with the regime of thermal stabilization that chambered the receiving and transmitting modules was developed.

As preliminary investigations have shown, a single-mode fiber should be used for transmission of analog SHF signals. A multimode fiber is characterized by much higher loss and dispersion. The major disadvantage is a low fiber modal bandwidth of 1200 MHz·km. To provide transmission in the SHF window of 1310 nm, the fiber conforming to the ITU-T G.652 standard should be used. We chose the Corning® SMF-28e+® fiber conforming to this standard and readily available in the Russian market.

The main factors in choosing connectors for the FODT system we developed were the return loss and strength of the connector. The FC/APC connectors were found to be the most reliable and convenient.

3   Estimation of Operating Capability of the FODT System

The main factors used in analysis of the FODT system are the optical power and signal transmission time of the system. For this reason, calculation of the system includes two stages: energy calculations and estimation of signal transmission time [6,7].

3.1   Energy Calculations for the FODT System

The required frequency range ΔF is 500 MHz. The required signal/noise ratio (SNR) is 50 dB. The attenuation at the working wavelength $\lambda = 1300$ nm is $\alpha_2 = 0.33$ dB/km. The fiber length L is 300 m. The transmitting optical module is DMPO131-23 (laser power $P_{ld} = 4$ mW). So, the power in dBm is $P_{ld} = 10 \lg(4) = 6.02$ dBm.

The average optical power in the case of analog intensity modulation is one half of the working laser power, $\alpha_{3b} = $ dB, so the average power at the line input is

$$P_{in} = P_d - \alpha_{2b} = 3.02 \ dBm \tag{1}$$

The required optical power in the photoreceiver is estimated from the specified parameters as

$$P_{min,W} = NER_W\sqrt{\Delta F * SNR} = 0.013 \text{ mW} = -18.73 \text{ dBm} \tag{2}$$

The allowed loss is

$$\alpha_4 = P_{in} - P_{min} = 21.75 \text{ } dB \tag{3}$$

The total power loss is

$$\alpha_5 = NL\alpha_2 = 0.105 \text{ } dB \tag{4}$$

The loss at laser radiation launching into the fiber is $\alpha_6 = 0$ dB. The loss at the fiber- detector connection is $\alpha_7 = 7$ dB. The tolerance for temperature variations under the specified conditions is $\alpha_{8a} = 1$ dB. The tolerance for deterioration of the parameters of the assembly consisting of the laser and pin-photodiode is $\alpha_{8b} = 4$ dB. The total additional tolerance for the loss is

$$\alpha_8 = \alpha_{8a} + \alpha_{8b} = 5 \text{ } dB \tag{5}$$

No couplers are used in the scheme, so $\alpha_9 = 0$. To connect the laser, receiver, and optical insulator with the fiber, six connectors with the loss typical of a single-mode fiber $\alpha_{pc} = 0.2$ dB are used. The loss in the optical insulator is 0.41 dB. The total loss at the connectors is

$$\alpha_{10} = 6 \cdot 0.2 + 0.41 = 1.61 \text{ } dB \tag{6}$$

Therefore, the total attenuation in the system is

$$\alpha_{11} = \alpha_5 + \alpha_6 + \alpha_7 + \alpha_8 + \alpha_9 + \alpha_{10} = 6.716 \text{ } dB \tag{7}$$

The power loss margin is

$$\alpha_{12} = \alpha_4 - \alpha_{11} = 15.034 \text{ } dB \tag{8}$$

Thus, the calculations show that the system is capable of efficient operation.

3.2   Calculation of Signal Transmission Time

The total signal transmission time of the system for the transmitted signal bandwidth ΔF = 500 MHz is given by

$$\tau_1 = \frac{0.35}{\Delta F} = 0.7 \text{ } ns \tag{9}$$

The signal transmission times of the light source τ_2 and photoreceiver τ_3 can be found from the working modulation frequency range ΔF_M equal to 15.99 GHz

$$\tau_2 = \tau_3 = \frac{0.35}{\Delta F_M} = 0.0219 \text{ } ns \tag{10}$$

The spectrum width of the transmitting optical DMPO131-23 module is $\Delta\lambda$ = 3 nm.

According to the technical information provided by the manufacturer of the SMF-28 fiber, the chromatic dispersion coefficient can be estimated as

$$M(\lambda) = \frac{M_0}{4}\left(\lambda - \frac{\lambda_0^4}{\lambda^3}\right)$$ (11)

where $M_0 = 0.092$ ps·nm^{-1}·km^{-1}, and $\lambda_0 = 1317$ nm. Then the chromatic dispersion coefficient for the wave with $\lambda = 1310$ nm is $M = 0.65$ ps·nm^{-1}·km^{-1}

Since the FODT system employs a single-mode fiber, $\tau_6 = 0$.

The signal transmission time of the system caused by the chromatic dispersion is given by

$$\tau_7 = M \cdot \Delta\lambda \cdot L \cdot N = 0.000585 \; ns$$ (12)

The signal transmission time of the system is

$$\tau_8 = \sqrt{\tau_2^2 + \tau_3^2 + \tau_6^2 + \tau_7^2} = 0.0309 \; ns$$ (13)

Thus, it can be concluded that the FODT system conforms to the requirements to the signal transmission time.

4   Experimental Investigations of the FODT System

A block diagram of the experimental setup for SHF signal transmission is presented in Fig. 1. The FODT system comprises a transmitting module in the form of a laser with a direct modulation, a receiving module in the form of a photodetector, a fiber-optic cable 250 m in length, two 25 m-long patch-cords, and also a set of optical connectors.

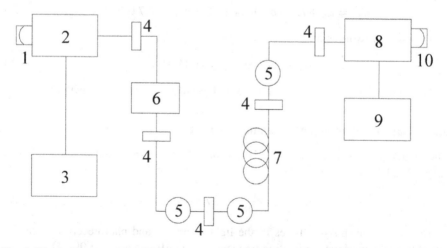

Fig. 1. Block diagram of the FODT system for transmission of analog SHF signals: 1 – SHF input; 2 – transmitting optical module; 3 – power supply driver; 4 – optical FC/APC connector; 5 – optical patch-cord 25 m in length; 6 – optical insulator; 7 – 250-m long optical fiber; 8 – receiving optical module; 9 – 12V power supply; 10 – SHF output

To determine the loss in the fiber, it is necessary to measure the level of optical power at the output and then subtract it from the level of the power launched into the fiber. The value obtained will give the true loss in the optical fiber [1,2,8]. Fig. 2 shows, as examples, experimental dependences of the power at the output P_{out} on the power at the input P_{in}.

As measurements showed, the power loss was 0.119 dB. The discrepancy with the theoretical calculations can be attributed to inaccuracy of measurements.

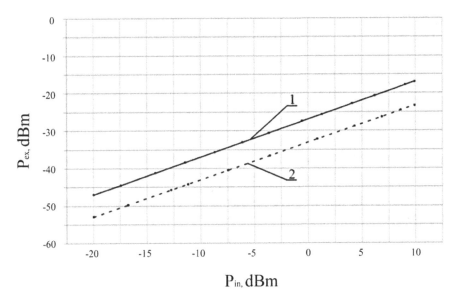

Fig. 2. Output power of optical signal as a function of input power. Curve 1 is for input signal frequency F = 1 GHz. Curve 2 - F = 12 GHz

Fig. 3 shows, as an example, the experimental dependence of the loss factor K_l in the FODT system on frequency F of the input signal.

The data shown in Fig. 3 lead to the following conclusions. The loss in the line is nearly independent of the optical fiber length. It is mainly determined by the conversion loss (in the transmitting and receiving optical modules) and losses at optical connectors (0.2 - 0.4 dB at each connection). It is also clear that the transmission coefficient of the FODT system has a small nonuniformity, about 4 dB, in the range 10 – 12 GHz.

The noise factor (K_n) of the FODT system was determined by the Y-factor method in the frequency range from 8 to 12 GHz with the help of an Agilent N9030A spectrum analyzer and an Agilent N4002A SNS Series noise source. Fig. 4 presents the experimental dependence of the noise factor K_n on input signal frequency F.

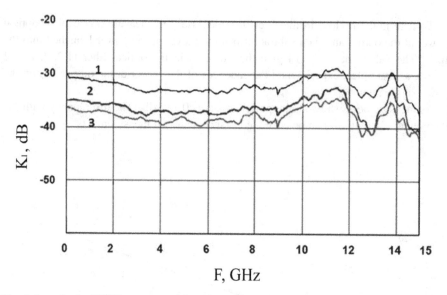

Fig. 3. Loss in the FODT system as a function of input SHF signal frequency. Curve 1 corresponds to the case when only optical insulator and connectors are placed between transmitting module and receiver. Curve 2 corresponds to the case when 250-m long optical fiber is also present. Curve 3 is the case when all components given in Fig. 2 are present between transmitting module and receiver.

Fig. 4. Noise factor as a function of input signal frequency

The experimental data shown in Fig.4 indicate that the noise factor varies from 23 to 27 dB. The result obtained is typical of the systems of such a type.

Since the fiber-optic communication line is designed for operation at sea-based stations, possibility of operation in different temperature ranges must be envisaged. Changes in temperature of the environment cause variations in the index of refraction of the fiber and an increase in the fiber length due to thermal expansion. This leads to a change in the light phase and, hence, a change in the modulation phase of the radiation transmitted through the fiber. For this reason, the experimental estimation of the temperature-induced shift of the modulation phase in the fiber was carried out.

Fig. 5 presents a block diagram of the experimental setup used for these investigations.

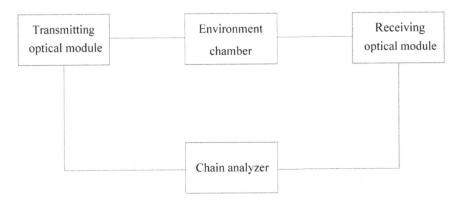

Fig. 5. Block diagram of experimental setup for direct measurements of phase stability of modulation frequency

The laser radiation modulated by a SHF signal at frequency 10 GHz was transmitted through the fiber, the 25-m long section of which was placed into a "Mini Sub-Zero" MS-71 environment chamber. The temperature in the chamber was varied from room temperature to 60 °C during approximately 3.5 hours on the first day and from room temperature to -50 °C on the second day. Simultaneously with the temperature measurements, the modulation phase of the output signal was measured and compared with the control signal phase.

Fig. 6 shows the experimental dependence of the shift in the light modulation phase $\Delta\varphi_m$ on environmental temperature T.

The investigations showed that the modulation phase in the Corning SMF - 28e fiber varied by about 6 degrees in the temperature interval from -50 to + 60 °C.

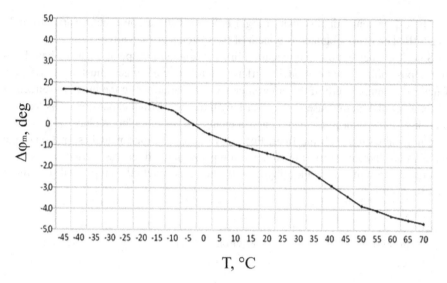

Fig. 6. Modulation phase shift $\Delta\varphi_m$ as a function of temperature

5   Conclusion

Our experimental and theoretical results have shown that the fiber-optic system for transmission of SHF radar signals in the 8-12 GHz range we have developed satisfies all the requirements to parameters of fiberoptic systems for radar stations. The fiber-optic data transmission system can prove an efficient tool for improving interaction between the wireless radar and wired signal transmission systems.

References

1. Agrawal, G.P.: Light wave technology: telecommunication systems, 480 p. Wiley-Inter science, NJ (2005)
2. Friman, R.K.: Fiber-optic communication systems, 496 p. Tekhnosfera, Moscow (2012)
3. Davydov, V.V., Dudkin, V.I., Karseev, A.U.: Nuclear Magnetic Flowmeter-Spectrometer with Fiber-Optical Communication Line in Cooling Systems of Atomic Energy Plants. Optical Memory & Neural Networks (Information Optics) 22(2), 112–117 (2013)
4. O'Mahony, M.J.: Future optical networks. IEEE OSA Journal of Light wave Technology 24(12), 4684–4696 (2006)
5. Ushakov, V.: Optical devices in radio engineering. Radiotechnika, Moskow (2009)
6. Zak, E.A.: Fiber-optic transducers with external modulation. Tekhnosfera, Moscow (2012)
7. Velichko, M.A., Naniy, O.E., Susyan, A.A.: New modulation formats in optical communication systems. Light wave Russian Edition (4), 21–30 (2005)
8. Dudkin, V.I., Pakhomov, L.N.: Kvantovaya Elektronika. Polytechnical Institute Publ., Saint-Petersburg (2012)

Nonlinear Coherent Detection Algorithms of Nonorthogonal Multifrequency Signals

Sergey V. Zavjalov, Sergey B. Makarov, and Sergey V. Volvenko

St. Petersburg State Polytechnical University, St. Petersburg, Russia
zavyalov_sv@spbstu.ru, {makarov,volk}@cee.spbstu.ru

Abstract. A study of detection algorithms of random sequences of nonorthogonal Spectrally Efficient Frequency Division Multiplexing (SEFDM) signals has been conducted. Realization complexity of optimal coherent algorithms forces to explore more simple processing techniques. A nonlinear bit-by-bit coherent algorithm with interchannel interference compensation and an algorithm based on decision diagram have been developed and studied. Application of the proposed nonlinear algorithms can significantly improve BER performance of multifrequency signals with nonorthogonal frequency spacing between subcarriers.

Keywords: Nonorthogonal frequency spacing, SEFDM, interchannel interference, nonlinear bit-by-bit coherent algorithm, nonlinear algorithm based on decision diagram.

1   Introduction

Use of algorithms of coherent bit-by-bit detection in communication systems with signals of SEFDM leads to decrease in information processing reliability [1]. Results of modeling show, that even at a small frequency spacing Δf between subcarriers smaller $1/T$, where T is the duration of the signaling interval, power losses are very essential.

In paper [2] it is shown that at a frequency spacing $0.5/T$ the BER performance of SEFDM signals coincides with a BER performance at an orthogonal frequency spacing subcarriers, but only for BPSK modulation on subcarriers.

In paper [3] efficiency of various algorithms of coherent bit-by-bit detection for special cases of $\Delta f = 5/6/T$ and $\Delta f = 4/5/T$ is considered while using QPSK. Results of modeling show that power losses when using the considered signals grow at increase SNR.

The objective of this paper is to design nonlinear algorithms of coherent detection of SEFDM signals when using QPSK manipulation on each subcarrier at various Δf values (from $0.5/T$ to $1/T$), providing a better BER performance. The algorithm optimum by criterion of a maximum likelihood is taken as an initial algorithm.

S. Balandin et al. (Eds.): NEW2AN/ruSMART 2014, LNCS 8638, pp. 703–713, 2014.

2   Algorithm for Optimal Coherent Detection

J-th realization of N QPSK signals shifted on frequency with amplitude A_0, an arbitrary frequency spacing between subcarriers Δf, medium subcarrier frequency ω_0 can be expressed in general as follows (if quadrature components are formed independently):

$$s_j(t) = A_0 \sum_{n=-(N-1)/2}^{(N-1)/2} a(t)\Big[d_{in} \cos\big((\omega_0 + n\Delta\omega)t\big) + \tag{1}$$

$$+ d_{qn} \sin\big((\omega_0 + n\Delta\omega)t\big)\Big]; (-T/2 \le t \le T/2),$$

where $j = 1, 2, 3, \dots 4^N$, values of symbols d_{in} depend on value of index $i = 1, 2$ and index $n = -(N-1)/2, \dots, (N-1)/2$. In particular, $d_{1n} = 1$; $d_{2n} = -1$.

Form of signal (1) on each subcarrier depends on the transmitted information symbol and the information symbols from all neighboring subcarriers. Each value d_{rn} symbol on subcarrier will meet one of $4N$ possible waveforms of signal (1) in the period in question.

Signal (1) on the interval $[-T/2; T/2]$ may be represented as follows:

$$s_j(t) = \sum_{n=-(N-1)/2}^{(N-1)/2} s\big(t, d_{in}, d_{qn}\big) = s\Big(t; d_{i(-(N-1)/2)}, d_{q(-(N-1)/2)}; \dots; d_{i((N-1)/2)}, d_{q((N-1)/2)}\Big) \tag{2}$$

where $s\big(t, d_{in}, d_{qn}\big) = s\big(t, d_{in}\big) + s\big(t, d_{qn}\big) =$

$$= A_0 a(t)\Big[d_{in} \cos\big((\omega_0 + n\Delta\omega)t\big) + d_{qn} \sin\big((\omega_0 + n\Delta\omega)t\big)\Big].$$

The form of the signal at the receiver`s input:

$$x(t) = A_0 a(t) \sum_{n=-(N-1)/2}^{(N-1)/2} \Big[d_{in} \cos\big((\omega_0 + n\Delta\omega)t\big) + d_{qn} \sin\big((\omega_0 + n\Delta\omega)t\big)\Big] + n(t) \tag{3}$$

Generally, at synchronous message transmission, receiver`s task is to make a decision about which set of code symbols $d_{i(-(N-1)/2)}, d_{q(-(N-1)/2)}, \dots, d_{i((N-1)/2)}, d_{q((N-1)/2)}$ in (1) corresponds to the input realization of the process $x(t)$ in (3).

Optimal detection algorithm for a maximum likelihood ratio criterion can be presented by the following expression (on the condition of a priori equal probabilities of transmitted symbols): *signal $s(t; d_{i(-(N-1)/2)}, d_{q(-(N-1)/2)}, \dots, d_{i((N-1)/2)}, d_{q((N-1)/2)})$ will be registered if the following system of inequalities is fulfilled*:

$$\int_{-T/2}^{T/2} x(t) s_j(t)\,dt - \frac{1}{2}E_j > \int_{-T/2}^{T/2} x(t) s_r(t)\,dt - \frac{1}{2}E_r \tag{4}$$

where E_j – energy of the signal $s(t; d_{i(-(N-1)/2)}, d_{q(-(N-1)/2)}, \dots, d_{i((N-1)/2)}, d_{q((N-1)/2)})$.

Realization of the algorithm (4) involves usage of 4^N correlators, each of them using signal $s(t; d_{i(-(N-1)/2)}, d_{q(-(N-1)/2)}, ..., d_{i((N-1)/2)}, d_{q((N-1)/2)})$ as a reference voltage of the multiplier. The output voltages are weighed by using coefficients E_j and transmitted to the comparator`s inputs.

The algorithm of this device working determines the most reliable sequence of symbols $d_{i(-(N-1)/2)}, d_{q(-(N-1)/2)}, ..., d_{i((N-1)/2)}, d_{q((N-1)/2)}$ (in accordance with (4)). Algorithm (4) realizes the optimal coherent detection "as a whole" of signal sequence at a predetermined time interval. Complexity of realization (4) forces to research more simple detection algorithms of signals $s(t; d_{i(-(N-1)/2)}, d_{q(-(N-1)/2)}, ..., d_{i((N-1)/2)}, d_{q((N-1)/2)})$.

3   Nonlinear Bit-by-Bit Coherent Detection Algorithm

Let us consider the bit-by-bit detection of QPSK signal, located on frequency ω_0. The shape of this signal is influenced by the signals on subcarrier frequencies $\omega_0 + n\Delta\omega$ ($n = \pm1, ..., \pm(N-1)/2$). Then, a functional of likelihood ratio can be written as follows (according to (4)):

$$\Delta_{pr} = \frac{\exp\left\{\frac{2}{N_0}\int_{-\frac{T}{2}}^{\frac{T}{2}} x(t)s_{yp}(t)dt - \frac{1}{N_0}\int_{\frac{T}{2}}^{\frac{T}{2}} s_{yp}^2(t)dt\right\}}{\exp\left\{\frac{2}{N_0}\int_{-\frac{T}{2}}^{\frac{T}{2}} x(t)s_{yr}(t)dt - \frac{1}{N_0}\int_{\frac{T}{2}}^{\frac{T}{2}} s_{yr}^2(t)dt\right\}} \tag{5}$$

where

$$s_{yp}(t) = y_l^{(-)}(t) + s_p(t) + y_l^{(+)}(t)$$

$$s_{yr}(t) = y_l^{(-)}(t) + s_r(t) + y_l^{(+)}(t)$$

$$s_p(t) = A_0 a(t)\left[d_{i0}\cos(\omega_0 t) + d_{q0}\sin(\omega_0 t)\right]; p = 1,2,3,4,$$

$$y_l^{(-)}(t) = A_0 \sum_{n=-(N-1)/2}^{-1}\left[d_{in}z_{in}(t)\cos(\omega_0 t) + d_{qn}z_{qn}(t)\sin(\omega_0 t)\right]; l = 1,2,...,4^{N/2},$$

$$y_l^{(+)}(t) = A_0 \sum_{n=+1}^{(N-1)/2}\left[d_{in}z_{in}(t)\cos(\omega_0 t) + d_{qn}z_{qn}(t)\sin(\omega_0 t)\right]; l = 1,2,...,4^{N/2},$$

$z_{in}(t)$ and $z_{qn}(t)$ – real envelop of QPSK signals in (2) $s(t, d_{in}, d_{qn})$ ($n \neq 0$)) on subcarrier ω_0. So we obtain the optimal bit-by-bit algorithm by averaging the numerator and the denominator over all combinations $y_l^{(-)}(t)$ and $y_l^{(+)}(t)$ in (5):

$$\sum_{\substack{\{d_{in},d_{qn}\}\\n\neq 0}} \exp\left\{B_{pl}(t) - \frac{1}{N_0}E_{pl}\right\} > \sum_{\substack{\{d_{in},d_{qn}\}\\n\neq 0}} \exp\left\{B_{rl}(t) - \frac{1}{N_0}E_{rl}\right\} \tag{6}$$

where $B_{pl}(t) = \dfrac{2}{N_0}\displaystyle\int_{-\frac{T}{2}}^{\frac{T}{2}} x(t)\left[s_p(t) + y_l(t)\right]dt$; $E_{pl} = \displaystyle\int_{-\frac{T}{2}}^{\frac{T}{2}}\left[s_p(t) + y_l(t)\right]^2 dt$

and $y_l(t) = y_l^{(-)}(t) + y_l^{(+)}(t)$.

Now let us examine the bit-by-bit coherent detection algorithm of signals on each subcarrier, where the signals on neighbouring subcarriers are interferences. This algorithm for single frequency ω_0 and for one quadrature can be written as follows (provided that $d_j = -d_r$):

$$A_0 \int_{-T/2}^{T/2} x(t)a(t)\cos(\omega_0 t)\,dt \underset{d_{20}}{\overset{d_{10}}{\underset{<}{\overset{>}{=}}}} 0\,;\ A_0 \int_{-T/2}^{T/2} x(t)a(t)\sin(\omega_0 t)\,dt \underset{d_{20}}{\overset{d_{10}}{\underset{<}{\overset{>}{=}}}} 0, \tag{7}$$

where $x(t)$ is determined by (3).

Then let us look at the nonlinear bit-by-bit coherent detection algorithm with interchannel interference compensation. We assume all the symbols of quadrature component $s(t, d_{i0})$ (or $s(t, d_{q0})$) of signal $s(t, d_{i0}, d_{q0})$ in (2) (without considering the frequency ω_0) to be detected correctly. So we can get the following algorithm (according to (5)):

$$\int_{-\frac{T}{2}}^{\frac{T}{2}} x(t)s_{i0}(t)\,dt - E_{ip}^* \underset{d_{20}}{\overset{d_{10}}{\underset{<}{\overset{>}{=}}}} 0, \tag{8}$$

where d_{in}^* – evaluations   of   received   information   symbols,

$$E_{il}^* = \int_{-\frac{T}{2}}^{\frac{T}{2}} \left(s(t,d_{i0}) + y_{il}^*(t)\right)^2 dt\,,\ i = 1, 2;\ y_{il}^*(t) = A_0 \sum_{\substack{n=-(N-1)/2\\n\neq 0}}^{(N-1)/2} d_{in}^* z_{in}(t)\cos(\omega_0 t)\,;$$

$s_{i0}(t) = A_0 a(t)\cos(\omega_0 t)$; $E_{ip}^* = \left(E_{1l}^* - E_{2l}^*\right)/4$. Algorithm for the other quadrature component of signal $s(t, d_{i0}, d_{q0})$ can be obtained similarly.

Let us consider multistep modification of nonlinear bit-by-bit coherent detection algorithm with interchannel interference compensation. Algorithm (7) is used on first step and algorithm (8) is used on next steps. Symbols decisions, which were obtained in the previous step, are used as estimates of the received symbols on next steps.

Decisions on received information symbols are formed on the last step of algorithm. Evaluation of the received symbol on r-th step is denoted as $d_{in}^*(r)$. So:

on the first step

$$\int_{-\frac{T}{2}}^{\frac{T}{2}} x(t) s_{i0}(t) dt \underset{\underset{d_{20}^*(1)}{<}}{\overset{\overset{d_{10}^*(1)}{>}}{}} 0;$$

on r-th step

$$\int_{-\frac{T}{2}}^{\frac{T}{2}} x(t) s_{i0}(t) dt - E_{ip}^*(r-1) \underset{\underset{d_{20}^*(r)}{<}}{\overset{\overset{d_{10}^*(r)}{>}}{}} 0, \, r = 2...L-1; \qquad (9)$$

on the last L-th step

$$\int_{-\frac{T}{2}}^{\frac{T}{2}} x(t) s_{i0}(t) dt - E_{ip}^*(L-1) \underset{\underset{d_{20}}{<}}{\overset{\overset{d_{10}}{>}}{}} 0,$$

where $\quad s_{i0}(t) = A_0 a(t) \cos(\omega_0 t); \qquad E_{ip}^*(r-1) = \left(E_{1i}^*(r-1) - E_{2i}^*(r-1) \right)/4,$

$$E_{il}^*(r-1) = \int_{-\frac{T}{2}}^{\frac{T}{2}} \left(s(t, d_{i0}) + A_0 \sum_{\substack{n=-(N-1)/2 \\ n \neq 0}}^{(N-1)/2} d_{in}^*(r-1) z_{in}(t) \cos(\omega_0 t) \right)^2 dt, \, i = 1, 2.$$

Realization of algorithm (8) would consist of the N channels for QPSK signal detection for the N subcarriers (fig. 1, a). Correlators are used in each channel (fig. 1, b), signals $\cos((\omega_0 + k\Delta\omega)t)$ and $\sin((\omega_0 + k\Delta\omega)t)$ are used as reference voltage of the multipliers. Correlator output voltages are transmitted through delay line $r\Delta T_0$ $(r = 0,1,...,L-1)$ to the inputs of the decision unit (DU), where the output voltage from the correlator is compared with zero threshold. Evaluations of received information symbols are generated at the output of the decision unit $d_{in}^*(r)$.

Let us examine the nonlinear bit-by-bit coherent detection algorithm on the decision diagram. It is necessary to find a sequence of characters $d_{i(-(N-1)/2)}, d_{q(-(N-1)/2)}, ..., d_{i((N-1)/2)}, d_{q((N-1)/2)}$, which minimizes the target function:

$$Q = \int_{-T/2}^{T/2} \left| x(t) - \sum_{n=-(N-1)/2}^{(N-1)/2} s(t, d_{in}, d_{qn}) \right| dt, \qquad (10)$$

where $x(t)$ is determined by (3).

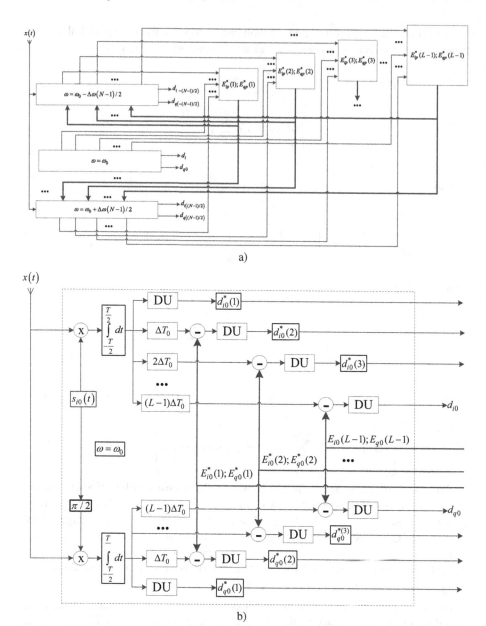

Fig. 1. Block diagram of multistep modification of nonlinear bit-by-bit coherent detection algorithm with interchannel interference compensation

Decision diagram is presented in fig. 2, a. All possible combinations of symbols d_{in} and d_{qn} (+1+1; −1+1; +1−1; −1−1), which are transmitted on subcarrier $\omega_0 + n\Delta\omega$, where $(n = -(N-1)/2, ..., (N-1)/2)$, are located at the decision diagram's nodes. Total number of nodes in this diagram is equal to $4N$. There are only four transitions of each selected node to nodes corresponding to the combinations of symbols transmitted on neighbouring subcarriers. Each path on the decision diagram is formed from node $d_{i(-(N-1)/2)}$, $d_{q(-(N-1)/2)}$ to $d_{i((N-1)/2)}$, $d_{q((N-1)/2)}$ and corresponds to a sequence of characters which is determined by the target function (10). The number of possible transitions between the states of nodes on the diagram grows exponentially with N.

We now consider the procedure of sequential evaluation of the target function (10). One can represent (10) in the form of calculations sequential procedure which is similar to the Viterbi algorithm.

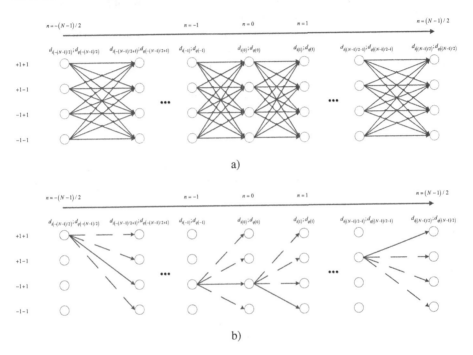

Fig. 2. a) Decision diagram, b) decision diagram for usage of recurrent form of expression for calculation of target function (11)

Intermediate values calculation of the target function is performed at each step of processing (which is performed from the subcarrier with index $-(N-1)/2$ to the subcarrier with index $(N-1)/2$).

$$Q(r) = \int_{-T/2}^{T/2} \left| x(t) - \sum_{n=-(N-1)/2}^{r} s(t, d_{in}, d_{qn}) \right| dt, \; r = -(N-1)/2, \ldots, (N-1)/2. \quad (11)$$

Recurrent form of expression (11) can be written as follows:

$$Q(r) = Q(r-1) + V(r) \; ; \; V(r) = \int_{-T/2}^{T/2} \left| x(t) - s\left(t, d_{ir} d_{qr}\right) \right| dt \; ; \; r = -(N-1)/2, \dots, (N-1)/2 ;$$

$$Q(-(N-1)/2) = V(-(N-1)/2) = \int_{-T/2}^{T/2} \left| x(t) - s\left(t, d_{i(-(N-1)/2)} d_{q(-(N-1)/2)}\right) \right| dt \; .$$

Thus, the signals from already received symbols and all the possible symbols from the next subcarrier must be generated on each transition between neighbouring subcarriers for calculation values of target function.

In fig. 2, b one can see a decision diagram for usage of recurrent form of expression for calculation of target function (11). Target function intermediate value (11) is calculated at each step for each of the four possible transitions. A transition, realizing the minimum value of the target function, is selected. Four paths, one of which is the most probable, always exist in the processing unit memory.

4   BER Performance

Let us examine algorithm (7). If $a(t)$ has a rectangular form, BER performance of algorithm (7) for one quadrature component can be presented by the following expression:

$$p_0(d_{in}|_{n \neq 0}) = \frac{1}{2} - \frac{1}{4} \Phi \left\{ -\sqrt{2} h_0 \left[-1 + \beta \left(d_{in}|_{n \neq 0} \right) \right] \right\} + \frac{1}{4} \Phi \left\{ -\sqrt{2} h_0 \left[1 + \beta \left(d_{in}|_{n \neq 0} \right) \right] \right\}, \; (12)$$

where $h_0 = \sqrt{E_b / N_0}$;

$$\beta(d_{in}|_{n \neq 0}) = \sum_{\substack{n = -(N-1)/2 \\ n \neq 0}}^{(N-1)/2} \frac{d_{in}}{n \Delta \omega \frac{T}{2}} \sin\left(n \Delta \omega \frac{T}{2} \right).$$

The total BER performance for a single subcarrier frequency ω_0 can be obtained by averaging (11) over all possible realizations of symbols transmitted on neighbouring subcarriers.

Dependences of the total error probability p on the h_0^2 for $N = 5$ for different values of Δf applicable to algorithm (7) are shown in fig. 3, a.

Frequency spacing reduction between subcarriers leads to additional energy loss for h_0^2 at a fixed BER performance. For example, an additional energy loss is equal to 5 dB at the change of value Δf for $0.9/T$ to $0.825/T$ in area of BER performance $p = 10^{-3}$-10^{-4}.

BER performance for $N = 5$ for algorithm (8) or for algorithm (9) for $L = 1$ usage can be estimated by simulation (fig. 3, b). As a result, additional energy losses are equal to 1 dB for the algorithm (8) usage ($\Delta f = 0.875/T$ in area of BER performance $p = 10^{-3}$-10^{-4}). Additional energy losses reach 13 dB for value $\Delta f = 0.75/T$ for $p = 10^{-3}$.

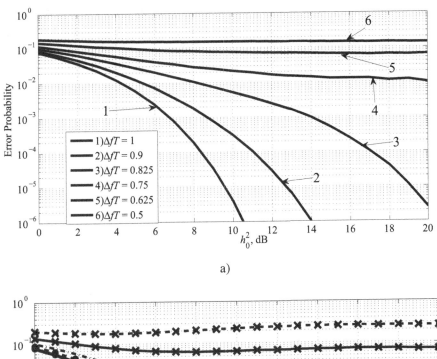

a)

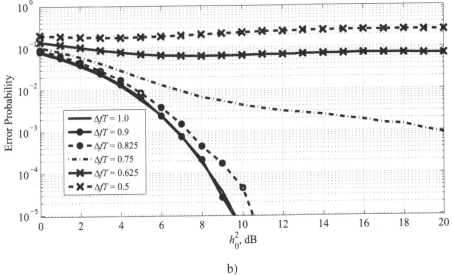

b)

Fig. 3. a) BER performance for algorithm (7) and $N = 5$; b) Simulation results for the algorithm (8) for $N = 5$

These results are consistent with [1], [3], where the limit value of Δf, at which BER performance is comparable to BER performance of orthogonal signals, is found. This value is equal to $\Delta f = 0.802/T$ (we can see a significant increase in additional energy loss with further reduction of frequency spacing as opposed to the orthogonal one).

BER performance for different number of iteration for multistep modification of nonlinear bit-by-bit coherent detection algorithm with interchannel interference compensation can be seen in fig. 4, a ($N = 5$, $h_0^2 = 14$ dB, $\Delta f = 0.75/T$). As we can see, error probability can be reduced by increasing number of iteration to 2 or more. But we can see threshold effect of BER performance, which is related with the fact of limited ability of improvement symbol decisions on each step of algorithm (9).

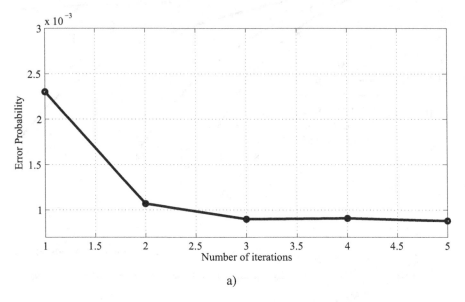

a)

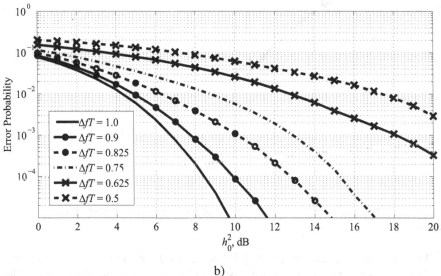

b)

Fig. 4. a) BER performance for different number of iteration for multistep modification of nonlinear bit-by-bit coherent detection algorithm with interchannel interference compensation for $N = 5$, $h_0^2 = 14$ dB, $\Delta f = 0.75/T$; b) BER performance for algorithm (11) for $N = 5$

We proceed to estimation of BER performance for usage nonlinear detection algorithm based on decision diagram. Simulation results for $N = 5$ are shown in fig. 4, b.

This algorithm is effective in reducing Δf to $0.5/T$ in contrast to algorithm with interchannel interference compensation. Additional energy loss is equal to 15 dB in area of BER performance $p = 10^{-3}$ for $\Delta f = 0.5/T$. Additionally, algorithm based on decision diagram (11) is less effective than algorithm with interchannel interference compensation at value $\Delta f = 0.9/T$ (see fig. 3 and fig. 4, b). It can be explained by the fact that algorithm's (11) application cannot fully compensate the interchannel interference as opposed to one of the algorithm (8).

5   Conclusions

Application of nonlinear algorithm with interchannel interference compensation allows to increase significantly BER performance of SEFDM signals when using at each subcarrier of QPSK modulation.

For nonlinear algorithm with interchannel interference compensation the BER makes $2 \cdot 10^{-2}$ for value of the $E_b/N_0 = 5$ dB that will be corresponded with BER performance results of SEFDM signals for algorithm 04MMSE [1]. At the same time practical realization of nonlinear algorithm appears simpler. The optimum number of iterations for algorithm with interchannel interference compensation is equal to two. For $\Delta f = 0.825/T$ probability BER made $4 \cdot 10^{-5}$ at E_b/N_0 value 10 dB that is more than twice better than the results received in [3] for close Δf values.

The nonlinear algorithm with interchannel interference compensation becomes inefficient at Δf values less than $0.75/T$. In this case it is offered to use nonlinear algorithm on the basis of the decision diagram which allows to carry out signal processing up to value $\Delta f = 0.5/T$. Probability of BER 10^{-2} for $\Delta f = 0.5/T$ it is reached at value $E_b/N_0 \approx 18$ dB. Results are received in lack of forward error correction.

References

1. Kanaras, Y., Chorti, A., Rodrigues, M., Darwazeh, I.: An overview of optimal and suboptimal detection techniques for a non orthogonal spectrally efficient FDM. In: LCS/NEMS 2009, London, UK, September 3-4 (2009)
2. Li, K., Darwazeh, I.: System performance comparison of Fast-OFDM system and overlapping Multi-carrier DS-CDMA scheme. In: London Communications Symposium (2006), http://www.ee.ucl.ac.uk/lcs/previ-ous/LCS2006/54.pdf
3. Clegg, R.G., Isam, S., Kanaris, I., Darwazeh, I.: A practical system for improved efficiency in frequency division multiplexed wireless networks. IET Communications 6(4), 449–457 (2012)

Author Index

.